W9-BIN-544

Periodic Table of the Elements

1 Group IA	2 Group IIA	3 Group IIIB	4 Group IVB	5 Group VB	6 Group VIB	7 Group VIIB	8 Group VIIIB	9 Group VIIIB	10 Group	11 Group IB	12 Group IIB	13 Group IIIA	14 Group IVA	15 Group VA	16 Group VIA	17 Group VIIA	18 Group VIIIA
1 **H** 1.01																	2 **He** 4.00
3 **Li** 6.94	4 **Be** 9.01											5 **B** 10.81	6 **C** 12.01	7 **N** 14.01	8 **O** 16.00	9 **F** 19.00	10 **Ne** 20.18
11 **Na** 22.99	12 **Mg** 24.30											13 **Al** 26.98	14 **Si** 28.09	15 **P** 30.97	16 **S** 32.06	17 **Cl** 35.45	18 **Ar** 39.95
19 **K** 39.10	20 **Ca** 40.08	21 **Sc** 44.96	22 **Ti** 47.87	23 **V** 50.94	24 **Cr** 52.00	25 **Mn** 54.94	26 **Fe** 55.84	27 **Co** 58.93	28 **Ni** 58.69	29 **Cu** 63.55	30 **Zn** 65.41	31 **Ga** 69.72	32 **Ge** 72.64	33 **As** 74.92	34 **Se** 78.96	35 **Br** 79.90	36 **Kr** 83.80
37 **Rb** 85.47	38 **Sr** 87.62	39 **Y** 88.91	40 **Zr** 91.22	41 **Nb** 92.91	42 **Mo** 95.94	43 **Tc** (98)	44 **Ru** 101.07	45 **Rh** 102.91	46 **Pd** 106.42	47 **Ag** 107.87	48 **Cd** 112.41	49 **In** 114.82	50 **Sn** 118.71	51 **Sb** 121.76	52 **Te** 127.60	53 **I** 126.90	54 **Xe** 131.29
55 **Cs** 132.91	56 **Ba** 137.33	57 **La** 138.91	72 **Hf** 178.49	73 **Ta** 180.95	74 **W** 183.84	75 **Re** 186.21	76 **Os** 190.23	77 **Ir** 192.22	78 **Pt** 195.08	79 **Au** 196.97	80 **Hg** 200.59	81 **Tl** 204.38	82 **Pb** 207.2	83 **Bi** 208.98	84 **Po** (209)	85 **At** (210)	86 **Rn** (222)
87 **Fr** (223)	88 **Ra** (226)	89 **Ac** (227)	104 **Rf** (261)	105 **Db** (262)	106 **Sg** (266)	107 **Bh** (264)	108 **Hs** (269)	109 **Mt** (268)	110 **Ds** (271)	111 – (272)	112 – (277)		114 – (289)				

58 **Ce** 140.12	59 **Pr** 140.91	60 **Nd** 144.24	61 **Pm** (145)	62 **Sm** 150.36	63 **Eu** 151.96	64 **Gd** 157.25	65 **Tb** 158.93	66 **Dy** 162.50	67 **Ho** 164.93	68 **Er** 167.26	69 **Tm** 168.93	70 **Yb** 173.04	71 **Lu** 174.97
90 **Th** 232.04	91 **Pa** 231.04	92 **U** 238.03	93 **Np** (237)	94 **Pu** (242)	95 **Am** (243)	96 **Cm** (248)	97 **Bk** (247)	98 **Cf** (251)	99 **Es** (252)	100 **Fm** (257)	101 **Md** (260)	102 **No** (259)	103 **Lr** (262)

Nonmetals

Metals

Atomic Numbers and Atomic Masses of the Elements

Elememt	Atomic Symbol	Atomic Number	Mass	Elememt	Atomic Symbol	Atomic Number	Mass
Actinium	Ac	89	(227)	Manganese	Mn	25	54.938049
Aluminum	Al	13	26.981538	Meitnerium	Mt	109	(268)
Americium	Am	95	(243)	Mendelevium	Md	101	(260)
Antimony	Sb	51	121.760	Mercury	Hg	80	200.59
Argon	Ar	18	39.948	Molybdenum	Mo	42	95.94
Arsenic	As	33	74.92160	Neodymium	Nd	60	144.24
Astatine	At	85	(210)	Neon	Ne	10	20.1797
Barium	Ba	56	137.327	Neptunium	Np	93	(237)
Berkelium	Bk	97	(247)	Nickel	Ni	28	58.6934
Beryllium	Be	4	9.012182	Niobium	Nb	41	92.90638
Bismuth	Bi	83	208.98038	Nitrogen	N	7	14.0067
Bohrium	Bh	107	(264)	Nobelium	No	102	(259)
Boron	B	5	10.811	Osmium	Os	76	190.23
Bromine	Br	35	79.904	Oxygen	O	8	15.9994
Cadmium	Cd	48	112.411	Palladium	Pd	46	106.42
Calcium	Ca	20	40.078	Phosphorus	P	15	30.973761
Californium	Cf	98	(251)	Platinum	Pt	78	195.078
Carbon	C	6	12.0107	Plutonium	Pu	94	(244)
Cerium	Ce	58	140.116	Polonium	Po	84	(209)
Cesium	Cs	55	132.90545	Potassium	K	19	39.0983
Chlorine	Cl	17	35.453	Praseodymium	Pr	59	140.90765
Chromium	Cr	24	51.9961	Promethium	Pm	61	(145)
Cobalt	Co	27	58.933200	Protactinium	Pa	91	231.03588
Copper	Cu	29	63.546	Radium	Ra	88	(226)
Curium	Cm	96	(247)	Radon	Rn	86	(222)
Darmstadtium	Ds	110	(271)	Rhenium	Re	75	186.207
Dubnium	Db	105	(262)	Rhodium	Rh	45	102.90550
Dysprosium	Dy	66	162.500	Rubidium	Rb	37	85.4678
Einsteinium	Es	99	(252)	Ruthenium	Ru	44	101.07
Element 111	—	111	(272)	Rutherfordium	Rf	104	(261)
Element 112	—	112	(277)	Samarium	Sm	62	150.36
Element 114	—	114	(289)	Scandium	Sc	21	44.955910
Erbium	Er	68	167.259	Seaborgium	Sg	106	(266)
Europium	Eu	63	151.964	Selenium	Se	34	78.96
Fermium	Fm	100	(257)	Silicon	Si	14	28.0855
Fluorine	F	9	18.9984032	Silver	Ag	47	107.8682
Francium	Fr	87	(223)	Sodium	Na	11	22.989770
Gadolinium	Gd	64	157.25	Strontium	Sr	38	87.62
Gallium	Ga	31	69.723	Sulfur	S	16	32.065
Germanium	Ge	32	72.64	Tantalum	Ta	73	180.9479
Gold	Au	79	196.96655	Technetium	Tc	43	(98)
Hafnium	Hf	72	178.49	Tellurium	Te	52	127.60
Hassium	Hs	108	(269)	Terbium	Tb	65	158.92534
Helium	He	2	4.002602	Thallium	Tl	81	204.3833
Holmium	Ho	67	164.93032	Thorium	Th	90	232.0381
Hydrogen	H	1	1.00794	Thulium	Tm	69	168.93421
Indium	In	49	114.818	Tin	Sn	50	118.710
Iodine	I	53	126.90447	Titanium	Ti	22	47.867
Iridium	Ir	77	192.217	Tungsten	W	74	183.84
Iron	Fe	26	55.845	Uranium	U	92	238.02891
Krypton	Kr	36	83.798	Vanadium	V	23	50.9415
Lanthanum	La	57	138.9055	Xenon	Xe	54	131.293
Lawrencium	Lr	103	(262)	Ytterbium	Yb	70	173.04
Lead	Pb	82	207.2	Yttrium	Y	39	88.90585
Lithium	Li	3	6.941	Zinc	Zn	30	65.409
Lutetium	Lu	71	174.967	Zirconium	Zr	40	91.224
Magnesium	Mg	12	24.3050				

Introduction to
Chemical Principles

Introduction to Chemical Principles

EIGHTH EDITION

H. Stephen Stoker

Weber State University

PEARSON

Prentice Hall

Upper Saddle River, New Jersey 07458

Library of Congress Cataloging-in-Publication Data

Stoker, H. Stephen (Howard Stephen).
  Introduction to chemical principles / H. Stephen Stoker. — 8th ed.
    p.   cm.
  Includes index.
  ISBN 0-13-185006-7
  1. Chemistry.   I. Title
QD33.2.S76   2005
540—dc22   2003070722

Project Manager: Kristen Kaiser
Editor-in-Chief, Science: John Challice
Vice President of Production and Manufacturing: David W. Riccardi
Executive Managing Editor: Kathleen Schiaparelli
Assistant Managing Editor: Beth Sweeten
Assistant Managing Editor, Media: Nicole Bush
Assistant Managing Editor, Supplements: Becca Richter
Media Editor: Paul Draper
Senior Marketing Manager: Steve Sartori
Manufacturing Buyer: Alan Fischer
Manufacturing Manager: Trudy Pisciotti
Creative Director: Carole Anson
Director of Creative Services: Paul Belfanti
Art Director: Kenny Beck
Interior and Cover Designer: Susan Anderson-Smith
Cover Photo: Warren Bolster/Getty Images, Inc.
Managing Editor of AV Management and Production: Patty Burns
AV Editor: J.C. Morgan
Illustrator: Precision Graphics
Editorial Assistants: Nancy Bauer/Jacquelyn Howard
Text Composition: Laserwords

© 2005, 2002, 1999, 1996 by Pearson Education, Inc.
Pearson Prentice Hall
Pearson Education, Inc.
Upper Saddle River, New Jersey 07458

Previous editions copyright 1993, 1990, 1986, and 1983 by Macmillan Publishing Company.

Pearson Prentice Hall® is a trademark of Pearson Eduction, Inc.

Printed in the United States of America

10 9 8 7 6 5 4 3 2 1

ISBN 0-13-185006-7

Pearson Education Ltd., *London*
Pearson Education Australia PTY, Limited, *Sydney*
Pearson Education Singapore, Pte. Ltd.
Pearson Education North Asia Ltd., *Hong Kong*
Pearson Education Canada, Ltd., *Toronto*
Pearson Educacíon de Mexico, S.A. de C.V.
Pearson Education—Japan, *Tokyo*
Pearson Education Malaysia, Pte. Ltd.

Contents

7   Chemical Bonds   197

8   Chemical Nomenclature   254

11   States of Matter   391

12   Gas Laws   435

13 Solutions   501

14 Acids, Bases, and Salts   547

15 Oxidation and Reduction   598

the input numbers used for the calculation, for example, in molecular mass values. To minimize such frustration, operational rules have been introduced for "standardizing" uncertainty in input numbers. The standard mode of operation is always (1) to round all atomic masses to hundredths before using them in molecular mass calculations, and (2) to specify frequently used numbers, such as Avogadro's number, molar volume, and the ideal gas constant to four significant figures. Using these operational rules for input numbers, student answers will match the back-of-the-book answers *to the last significant digit.*

6. **Defined terms always appear in self-standing complete sentences.** All definitions are highlighted in the text when they are first presented, using boldface and italic type. Each defined term appears as a complete sentence; students are never required to deduce a definition from context. In addition, the definitions of all terms appear in a separate glossary found at the end of the text. A major emphasis in this new edition has been "refinements" in the defined-terms arena. All defined terms have been reexamined to see if they could be stated with greater clarity. The result is a rewording of many defined terms. In addition, the number of defined terms has been increased.

7. **All end-of-chapter exercises occur in matched pairs.** In essence, each chapter has two independent, but similar, problem sets. Counting subparts to problems, there are over 5000 questions and problems available for students to use in their journey to proficient problem solving. Answers to all of the odd-numbered problems are found at the end of the text. Thus, two problem sets exist, one with answers and one without.

8. **Each end-of-chapter problem set,** except for Chapters 1 and 2, **is divided into three sections:** (1) Practice Problems, (2) Additional Problems, and (3) Cumulative Problems. The Practice Problems are categorized by topic and are arranged in the same sequence as the chapter's textual material. These problems, which are always single-concept, are "drill" problems that most students will find routine. The Additional Problem section contains problems that involve more than one concept from the chapter and are usually more difficult than the Practice Problems. The cumulative-skills section draws not only on materials from the current chapter, but also on concepts discussed in previous chapters. The working of problems in this third group allows students to continue to use, rather than forget, problem-solving techniques presented earlier.

9. **Historical vignettes are used to address some of the "people aspects" of chemistry.** These vignettes, entitled "The Human Side of Chemistry," are brief biographies of scientists who helped develop the foundations of modern chemistry. In courses such as the one for which this text is written, it is very easy for students to completely lose any feeling for the people involved in the development of the subject matter they are considering. If it were not for the contributions of these people, many of whom worked under adverse conditions, chemistry would not be the central science that it is today.

10. **"Chemical Extensions" are used to bridge the gap between mathematics and chemistry.** These "extensions," which are appended to most of the worked-out examples in the text, focus on the chemical compound that is the subject of the calculation. They give information on the compound's occurrence, its properties and uses, its relationship to the environment, its relationship to living systems (biochemistry), and so on. It is easy for students to become so involved in the mathematics of problem solving that they completely forget about the "realness" of the compound or compounds that are the subject of their calculation.

11. **Marginal notes are used extensively.** The two main functions of the marginal notes are (1) to summarize key concepts and often give help for remembering concepts or distinguishing between similar concepts, and (2) to provide additional details, links between concepts, or historical information about the concepts under discussion.

Preface

Introduction to Chemical Principles is a text for students who have had little or no previous instruction in chemistry or who had such instruction long enough ago that a thorough review is needed. The text's purpose is to give students the background (and confidence) needed for a subsequent successful encounter with a main sequence, college-level, general chemistry course.

Many texts written for preparatory chemistry courses are simply "watered down" versions of general chemistry texts: they treat almost all topics found in the general chemistry course, but at a superficial level. *Introduction to Chemical Principles* does not fit this mold. My philosophy is that it is better to treat fewer topics extensively and have the student understand those topics in greater depth. I resisted the very real temptation to include lots of additional concepts in this new edition. Instead, my focus for this edition was on rewriting selected portions to improve the clarity of presentation.

Important Features of the Eighth Edition

1. **Development of each topic starts out at "ground level."** Because of the varied degrees of understanding of chemical principles possessed by students taking a preparatory chemistry course, each topic is developed step by step from "ground level" until the level of sophistication required for a further chemistry course is attained.

2. **Problem-solving pedagogy is based on dimensional analysis.** Thirty-five years of teaching experience suggest to me that student "troubles" in general chemistry courses are almost always centered on the inability to set up and solve problems. Whenever possible, I use dimensional analysis in problem solving. This method, which requires no mathematics beyond arithmetic and elementary algebra, is a powerful and widely applicable problem-solving tool. Most important, it is a method that an average student can master with an average amount of diligence. Mastering dimensional analysis also helps build the confidence that is so valuable for future chemistry courses.

3. **Detailed commentary accompanies all worked-out example problems.** In all chapters, one or more worked-out example problems follow the presentation of key concepts. These examples "walk" students through the thought processes involved in solving the particular type of problem. Detailed commentary accompanies all of the steps involved in solving a problem. In addition, an unworked practice exercise is coupled to each worked-out example. It is intended that students work this exercise immediately after examining the worked-out example. A section at the end of each chapter gives the answers to these unworked practice exercises. In total, the number of worked-out examples is significantly greater than that found in most texts and has increased from that in the previous edition of this work.

4. **Significant-figure concepts are emphasized in all problem-solving situations.** Routinely, electronic calculators display answers that contain more digits than are needed or acceptable. In all worked-out examples, students are reminded about these "unneeded digits" by the appearance of two answers to the example: the calculator answer (which does not take into account significant figures) and, in color, the correct answer (which is the calculator answer adjusted to the correct number of significant figures).

5. **Operation rules for "standardizing" uncertainty in numbers are used.** Students often experience a relatively high degree of frustration when they correctly solve a problem and yet obtain an answer that differs *slightly* from the one given in the answer section at the back of the book. They want to get the "exact" number shown in the answer section. Most often the discrepancy is due to differing degrees of uncertainty in

Supplements

For the Instructor

Instructor's Solutions Manual (**ISBN 0-13-144942-7**) by Nancy J. Gardner, California State University-Long Beach. Contains full solutions to all of the end-of-chapter problems in the text.

Test Item File (**ISBN 0-13-144992-3**) by Bobby J. Stanton, University of Georgia. Contains approximately 1000 multiple-choice and short-answer questions, all referenced to the text.

TestGen (**ISBN 0-13-144939-7**). This powerful testing and grade management software creates exams from an electronic database version of the Test Item File. Instructors can generate alternate versions of the same test, add their own material, and edit existing tests effortlessly with this program.

PowerPoint™ Images Hundreds of illustrations from the textbook have been digitized and pre-inserted in PowerPoint files for instructor use. Instructors may reformat these slides and add their lecture presentation notes to suit classroom needs.

Annotated Instructor's Manual to Prentice Hall Laboratory Manual for Introductory Chemistry 3e (**ISBN 0-13-096883-8**) by Charles H. Corwin, American River College. Contains a complete listing of chemicals and reagent preparation directions for each experiment. It also provides suggested unknowns and answers to the post-laboratory assignments and a quiz item file with more than 500 class-tested multiple-choice questions.

WebAssign Online Homework. Our partnership with WebAssign gives instructors the opportunity to assign algorithmically generated homework, based on end-of-chapter problems, have it automatically graded, and have the grades entered into a central gradebook. Students get instant feedback and the opportunity for repeated practice.

For the Student

Student Solutions Manual (**ISBN 0-13-144941-9**) by Nancy J. Gardner, California State University-Long Beach. Includes full solutions to all odd-numbered end-of-chapter problems in the text.

Math Review Toolkit (**ISBN 0-13-144993-1**) by Gary Long, Virginia Tech. Designed to provide assistance to students with weaker math skills. This supplement includes a chapter-by-chapter math review keyed to problems in the text as well as a brief self-assessment test.

Introductory ChemIST CD (**ISBN 0-13-033729-3**). This *Interactive Student Tutorial* gives students a media-rich, engaging overview of all of the topics and fundamental skills in an introductory chemistry course, including animations, simulations, video clips, and self-assessment questions.

Prentice Hall Laboratory Manual for Introductory Chemistry 3e (**ISBN 0-13-062333-4**) by Charles H. Corwin, American River College. This comprehensive collection of 25 laboratory experiments with detailed safety information offers an exciting, hands-on introduction to the world of chemistry.

Acknowledgments

I'd like to gratefully acknowledge the valuable contributions of my accuracy reviewers: Boyd Beck of Snow College, David B. Shaw of Madison Area Technical College, and Andreas Lippert of Weber State University.

Every effort has been made to rid this text of any typographical errors. I encourage my readers who notice anything suspicious, or who have other questions or comments, to e-mail me at the address below.

H. Stephen Stoker
e-mail: hstoker1@weber.edu

Reviewers of the Seventh Edition of *Introduction to Chemical Principles*, Stoker

Ralph Benedetto, Jr., *Wayne Community College*
Nelson De Leon, *Indiana University*
Roberta Eddy, *Indiana University of Pennsylvania*
Stanley Grenda, *University of Nevada*
M. Elizabeth Gurnack, *Georgia Southwestern State University*
Pamela K. Kerrigan, *College of Mount Saint Vincent, Manhattan College*
Laurie LeBlanc, *Cuyamaca College*
Andreas Lippert, *Weber State University*
Panayiotis Meleties, *Bronx Community College, City University of New York*
Eric L. Trump, *Emporia State University*
William Vanderbout, *Sierra College*
Catherine Woytowicz, *George Washington University*

1

The Science of Chemistry

1.1 Chemistry—A Scientific Discipline

Chemistry is part of a larger body of knowledge called *science*. **Science** *is the study in which humans attempt to organize and explain, in a systematic and logical manner, knowledge about themselves and their surroundings.*

Because of the enormous scope of science, the sheer amount of accumulated knowledge, and the limitations of human mental capacity to master such a large and diverse body of knowledge, science is divided into smaller subdivisions called *scientific disciplines*. A **scientific discipline** *is a branch of science limited in size and scope to make it more manageable.* Examples of scientific disciplines are *chemistry*, astronomy, botany, geology, physics, and zoology.

Figure 1.1 shows an organizational chart, with emphasis on chemistry, for the various scientific disciplines. These disciplines can be grouped into *physical sciences* (the study of matter and energy) and *biological sciences* (the study of living organisms). Chemistry is a physical science.

Rigid boundaries between scientific disciplines *do not exist*. All scientific disciplines borrow information and methods from each other. No scientific discipline is totally independent. Environmental problems that scientists have encountered in the last two decades particularly show the interdependence of the various scientific disciplines. For example, chemists attempting to solve the problems of chemical contamination of the environment find that they need some knowledge of geology, zoology, and botany. It is now common to talk not only of chemists, but also of geochemists, biochemists, chemical physicists, and so on. The middle portion of Figure 1.1 shows the overlap of the other scientific disciplines with chemistry.

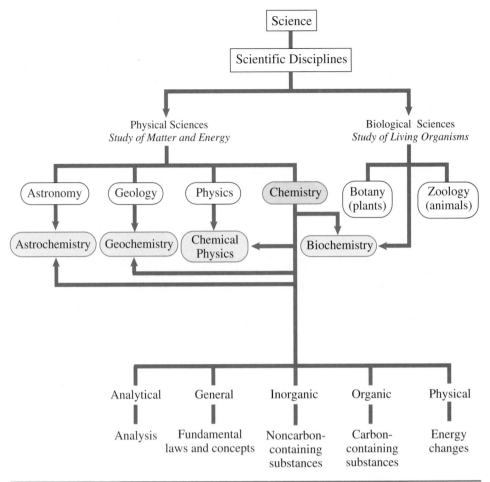

Figure 1.1

An organizational chart showing the relationship of the scientific discipline called chemistry to other scientific disciplines and also the substructuring that occurs within the discipline of chemistry.

Discipline overlap requires that scientists, in addition to having in-depth knowledge of a selected discipline, also have limited knowledge of other disciplines. Discipline overlap also explains why a great many college students are required to study chemistry. One or more chemistry courses are required because of their applicability to the disciplines in which the student has more specific interest.

The body of knowledge found within the scientific discipline of chemistry is itself vast. No one can hope to master completely all aspects of chemical knowledge. However, the fundamental concepts of chemistry can be learned in a relatively short period of time.

The vastness of chemistry is sufficiently large that it, like most scientific disciplines, is partitioned into *subdisciplines*. The lower portion of Figure 1.1 shows the five fundamental branches of chemistry: analytical, general, inorganic, organic, and physical. Most of the subject matter of this textbook falls within the realm of *general chemistry*, the fundamental laws and concepts of chemistry.

1.2 Scientific Disciplines and Technology

> **Science is the lifeline for technology, continually supplying it with new ideas to be worked on.**

Science is the pursuit of knowledge for its own sake. This pursuit results naturally from the curiosity and fascination of human beings for their surroundings. Science itself, however, does not cause a change in "our conditions" unless something is done with the body of knowledge that accumulates from scientific endeavors. Change is brought about by *technology*. **Technology** *is the application of scientific knowledge to the production of new products to improve human survival, comfort, and quality of life.* Technology manipulates nature for advantage. Technological advances began affecting our society about 200 years ago, and new advances still continue, at an accelerating pace, to have a major impact on human society. Section 1.3 considers numerous contributions of chemical technology to human well-being.

Both benefits and detriments can be obtained from the same "piece" of scientific knowledge depending on the technology used to put it to work. For example, knowledge concerning the structure of atoms—the fundamental building blocks for all substances (Chapter 5)—can be applied using one form of technology to build nuclear weapons and another form of technology to detect and cure disease in the human body (Chapter 17).

1.3 The Scope of Chemistry and Chemical Technology

Although chemistry is concerned with only a part of the scientific knowledge that has been accumulated, it is in itself an enormous and broad field. Chemistry touches all parts of our lives.

Many of the clothes we wear are made from synthetic fibers produced by chemical processes. Even natural fibers, such as cotton or wool, are the products of naturally occurring chemical reactions within living systems. Our transportation usually involves vehicles powered with energy obtained by burning chemical mixtures such as gasoline, diesel, and jet fuel. The drugs used to cure many of our illnesses are the result of chemical research. The paper on which this textbook is printed was produced through a chemical process, and the ink used in printing the words and illustrations is a mixture of many chemicals. The movies we watch are possible because of synthetic materials called "film." The images on film are produced through the interaction of selected chemicals. Almost all of our recreational pursuits involve objects made of materials produced by chemical industries. Skis, boats, basketballs, bowling balls, musical instruments, and television sets all contain materials that do not occur naturally, but are products of human technological expertise.

Our bodies are a complex mixture of chemicals. The principles of chemistry are fundamental to an understanding of all processes of the living state. Chemical secretions (hormones) produced within our bodies help determine our outward physical characteristics such as height, weight, and appearance. Digestion of food involves a complex series of chemical reactions. Food itself is an extremely complicated array of chemical substances. Chemical reactions govern our thought processes and how knowledge is stored in and retrieved from our brains. In short, chemistry runs our lives.

A formal course in chemistry can be a fascinating experience because it helps us understand ourselves and our surroundings. We cannot truly understand or even know very much about the world we live in or about our own bodies without being conversant with the fundamental ideas of chemistry.

1.4 How Chemists Discover Things— The Scientific Method

The word *chemistry* conjures up images of people in white lab coats shaking test tubes or peering into a complicated apparatus. Why is this generally valid image associated with chemists? The reason is simple. Chemists, as well as all other scientists, discover the general principles that govern the physical world (both its seen and unseen parts) through experimentation and observation (see Fig. 1.2).

Whether or not a given piece of scientific knowledge is technologically used for beneficial or detrimental purposes depends on the motives of those men and women, whether in industry or government, who have the decision-making authority. In democratic societies, citizens (the voters) can influence many technological decisions. It is important for citizens to become informed about scientific and technological issues.

Figure 1.2

Chemistry is an experimental science. Most discoveries in chemistry are made through analysis of data obtained from experiments carried out in laboratories. (Jeff Greenberg/Visuals Unlimited)

A majority of the scientific and technological advances of the twentieth century are the result of systematic experimentation using a method of problem solving known as the *scientific method*. The **scientific method** *is a set of procedures used to acquire knowledge and explain phenomena*. The procedural steps in the scientific method are

1. Identify the problem, break it into small parts, and carefully plan procedures to obtain information about all aspects of this problem.
2. Collect data concerning the problem through observation and experimentation.
3. Analyze and organize the data in terms of general statements (generalizations) that summarize the experimental observations.
4. Suggest probable explanations for the generalizations.
5. Experiment further to prove or disprove the proposed explanations.

On occasion, a great discovery is made by accident, but the majority of scientific discoveries are the result of the application of these five steps over long periods of time. There are no instantaneous steps in the scientific method: applying them requires considerable amounts of time. Even in those situations where luck is involved, it must be remembered that "chance favors the prepared mind." To take full advantage of an accidental discovery, a person must be well trained in the procedures of the scientific method.

The imagination, creativity, and mental attitude of a scientist using the scientific method are always major factors in scientific success. The procedures of the scientific method must always be enhanced with the abilities of a thinking scientist.

There are special vocabulary terms associated with the scientific method and its use. This vocabulary includes the terms *experiment, fact, natural law, hypothesis*, and *theory*. An understanding of the relationships among these terms is the key to a real understanding of how to obtain chemical knowledge.

> **Although two different scientists rarely approach the same problem in exactly the same way, there are always similarities in their approaches. These similarities are the procedures associated with the scientific method.**

The beginning step in the search for chemical knowledge is the identification of a problem concerning some chemical system that needs study. After determining what other chemists have already learned about the selected problem, a chemist sets up *experiments* for obtaining more information. An **experiment** *is a well-defined, controlled procedure for obtaining information about a system under study.*

Performing an experiment involves making careful observations about a system and recording the information, that is, data, obtained. Such data may be *qualitative* or *quantitative*, with the latter being preferred. **Qualitative data** *is non-numerical data consisting of general observations about a system under study.* The observation that ice is less dense than liquid water is an example of qualitative information about a system. **Quantitative data** *is numerical data obtained by various measurements on a system under study.* The information that ice has a density of 0.9170 grams per cubic centimeter at 0°C whereas liquid water has a density of 0.9999 grams per cubic centimeter at the same temperature represents quantitative data. Quantitative observations are more useful than qualitative ones because they can be compared with each other and trends or patterns in information can be seen.

An experiment typically involves study of at least two quantities, that is, variables, that have changing values. Usually, the effect of change in one variable on another variable, with all other variables held constant, is measured. For example, the effect that temperature change has on the density of a fixed quantity of a gas, with pressure held constant, can be measured.

A well-designed experiment is always performed under controlled conditions, that is, the values of all variables are always noted, not just those that are changing. When such is the case, the experimental data can be reproduced, if needed, by repeating the experiment.

The individual pieces of new information (data) about a system under study, obtained by carrying out experimental procedures, are called *facts*. A **fact** *is a valid observation about some natural phenomenon.* Facts are reproducible pieces of information. If a given experiment is repeated, under exactly the same conditions, the same results (facts) should be obtained. To be acceptable, all facts must be verifiable by anyone who has the time, means, and knowledge needed to repeat the experiments that led to their discovery. It is important that scientific data be published so that other scientists have the opportunity to critique and double-check both the data and experimental design.

As a next step, the scientist makes an effort to determine ways in which the facts about a given system relate both to each other and to facts known about similar systems. Repeating patterns often emerge among the collected facts. These patterns that describe the behavior of chemical systems under specific conditions are called *natural laws*. A **natural law** *is a generalization that summarizes facts about natural phenomena.*

Do not assume that natural laws are easy to discover. Often, many years of work and thousands of facts are needed before the true relationships among variables in the area under study emerge.

A natural law is a description of what happens in a given type of experiment. No new understanding of nature results from simply stating a natural law. A natural law merely summarizes already known observations (facts).

A natural law can be expressed either as a verbal statement or as a mathematical equation. An example of a verbally stated natural law is "If hot and cold pieces of metal are placed in contact with each other, the temperature of the hot piece always decreases and the temperature of the cold piece always increases."

It is important to distinguish between the use of the word *law* in science and its use in a societal context. Natural laws are *discovered* by research (see Figure 1.2), and researchers have *no control* over what the laws turn out to be. Societal laws, which are designed to control aspects of human behavior, are *arbitrary conventions* agreed upon (in a democracy) by the majority of those to whom the laws apply. These laws *can be* and *are changed* when necessary. For example, the speed limit for a particular highway (a societal law) can be decreased or increased for various safety or political reasons.

A contrast exists between the ways in which scientific facts and the results of technology (Section 1.2) are shared. Scientists publish their observations (facts) as widely, openly, and quickly as possible. Technological breakthroughs, on the other hand, are usually kept secret by an individual or company until patent rights for the new process or product are obtained. Even then, only limited information is released.

There is no mention in a natural law about why the occurrence described happens. The natural law simply summarizes experimental observations without attempting to clarify the reasons for the occurrence. Chemists, and other scientists, are not content with such a situation. They want to know *why* a certain type of observation is always made. Thus, after a natural law is discovered, scientists work out *plausible, tentative* explanations of the behavior encompassed by the natural law. These explanations are called *hypotheses*. A **hypothesis** *is a tentative model or statement that offers an explanation for a natural law*.

Once a hypothesis has been proposed, experimentation begins again. Scientists run more experiments under varied, but controlled, conditions to test the reliability of the proposed explanation. The hypothesis must be able to predict the outcome of as-yet-untried experiments. The validity of the hypothesis depends upon its predictions being true.

In practice, scientists usually start with a number of alternative hypotheses for a given law. Evaluation proceeds by demonstrating that certain proposals are *not* valid. A successful experiment is one in which one or more of the alternative hypotheses are demonstrated to be inconsistent with experimental observation and are thus rejected. Scientific progress is made in the same way a marble statue is: unwanted bits of marble are chipped away. Example 1.1 contains a simple illustration of this "chipping away" principle in a scientific context.

> It is much easier to disprove a false hypothesis than to prove a true one. A negative result from an experiment indicates that the hypothesis is not valid as formulated and must be modified. Obtaining positive results supports the hypothesis, but it does not definitely prove it. There is always the chance that someone will carry out a previously unthought of experiment that disproves the hypothesis.

EXAMPLE 1.1

Relating Hypotheses to Experimental Information

Suppose you encounter a situation involving two unopened books with no identification on their covers and four alternative hypotheses about these books, which are (1) the thinner book is a chemistry textbook, (2) the thicker book is a chemistry textbook, (3) both books are chemistry textbooks, and (4) neither book is a chemistry textbook. What evaluative information about these hypotheses can be obtained by opening the thicker book and determining that it is a chemistry textbook?

SOLUTION

This experiment (opening the thicker book) proves hypothesis 2 and disproves hypothesis 4; it does not, however, prove that *only one* of the hypotheses is true. The fact that the thicker book is a chemistry textbook does not rule out the possibility the thinner book is also a chemistry textbook.

Practice Exercise 1.1

Based on the same "two-book, four-hypothesis" situation stated in Example 1.1, what evaluative information about the hypotheses is obtained from the single observation that the thinner book is *not* a chemistry textbook?

> Answers to practice exercises are located at the end of the chapter.

As further experimentation continues to validate a particular hypothesis, its acceptance in scientific circles increases. If, after extensive testing, the reliability of a hypothesis is still very high, confidence in it increases to the extent that it is accepted by the scientific community at large. After more time has elapsed and more positive support has accumulated, the hypothesis assumes the status of a *theory*. A **theory** *is a hypothesis that has been tested and validated over a long period of time*. The dividing line between a hypothesis and a theory is arbitrary and cannot be precisely defined. There is no set number of supporting experiments that must be performed to give theory status to a hypothesis.

Theories serve two important purposes: (1) they allow scientists to predict what will happen in experiments that have not yet been run, and (2) they simplify the very real problem of being able to remember all the scientific facts that have already been discovered.

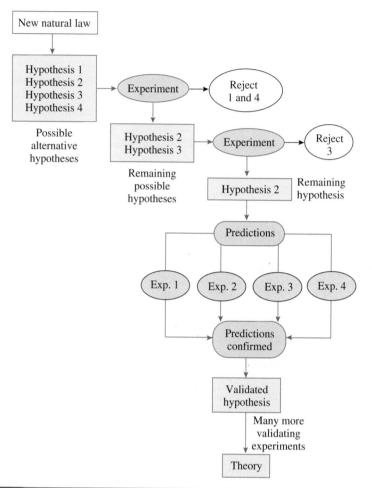

As a problem-solving approach, the scientific method is used by many people who do not call themselves scientists. An automobile mechanic uses the scientific method approach when working on a car. First, based on tests (experimental observations), the automobile mechanic deduces a probable cause of the problem (a hypothesis). Then parts are adjusted or replaced, and the car is checked (more experimental observations) to see if the problem has been corrected (testing of the hypothesis). An experienced automobile mechanic learns that certain observations almost always indicate a specific problem (a validated hypothesis).

Figure 1.3

A possible sequence of events experienced by scientists doing experimental research based on a newly discovered natural law. (1) A number of alternative hypotheses are proposed to explain the new natural law; (2) experiments are carried out to eliminate invalid hypotheses; (3) predictions are made based on the surviving hypothesis and further experiments are carried out to test these predictions; (4) confirmed predictions produce a validated hypothesis; and (5) many more validating experiments give the hypothesis the status of a theory.

Figure 1.3 shows the interplay that must occur between hypotheses and experimentation before an acceptable theory is obtained.

Theories must often undergo modification. As scientific tools, particularly instrumentation, become more accurate, there is an increasing probability that some experimental observations will not be consistent with all aspects of a given theory. A theory inconsistent with new observations must either be modified to accommodate the new results or be restated in such a way that scientists know where it is useful and where it is not. Most theories in use have known limitations. These "imperfect" theories are simply the best ideas anyone has found *so far* to describe, explain, and predict what happens in the world in which we live. Theories with limitations are generally not abandoned until a better theory is developed.

Facts that have been verified by repeated experiments will never be changed, but the theories that were invented to explain these facts are subject to change. In this sense, facts are more important than the theories devised to explain them. It is a mistake to believe that, by

The term *theory* is often misused by nonscientists in everyday contexts. "I have a theory that such and such is the case" is a frequently heard comment. In this case, "theory" means a "speculative guess," which is not what a theory is. The terminology *unvalidated hypothesis* would be closer to what is meant.

knowing all the natural laws and theories that are derived from experimental observations, the experimental facts are not needed. New theories can only be developed by people who have a wide knowledge of the facts relating to a particular field, especially those facts that have not been satisfactorily accounted for by existing theories.

Scientists do not view scientific theories as "absolute truth." All theories in science are considered provisional—subject to change in the light of new experimental observations. A science is like a living organism; it continues to grow and change. Figure 1.4 highlights the central role that experimentation plays in the scientific method as well as summarizes the general steps, procedures, and terminology associated with this most important "pattern of action" for acquiring scientific knowledge.

General Steps in the Scientific Method

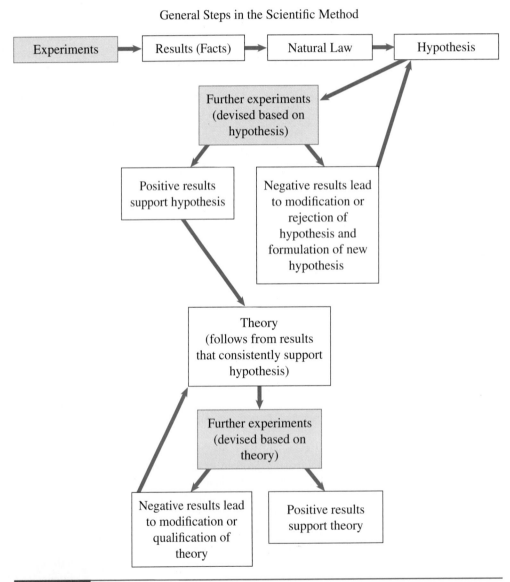

Figure 1.4

A diagram showing the central role that experimentation plays in the scientific method for obtaining new scientific information.

Summary

1. **Science and Scientific Disciplines** Science is the study in which humans attempt to organize and explain, in a systematic manner, knowledge about themselves and their surroundings. Scientific disciplines, of which chemistry is one, are branches of science limited in size and scope to make them more manageable.

2. **Technology** Scientific disciplines represent abstract bodies of knowledge. Technology is the physical application of scientific knowledge to the production of new products to improve human survival, comfort, and quality of life.

3. **Scope of Chemistry** The scope of chemistry is extremely broad, and it touches every aspect of our lives. The principles of chemistry are fundamental to an understanding of all processes of the living state. Chemical processes produce the products needed for our clothes, housing, transportation, medications, and recreational pursuits.

4. **Scientific Method** The scientific method is a set of systematic procedures used to acquire knowledge and explain phenomena. The terminology associated with the scientific method and its use includes the terms *experiment, fact, natural law, hypothesis*, and *theory*. An experiment is a well-defined, controlled procedure for obtaining information about a system under study. A fact is a valid observation about some natural phenomenon obtained by carrying out experiments. A natural law is a generalization that summarizes facts about natural phenomena. A hypothesis is a tentative model or statement that offers an explanation for a natural law. A theory is a hypothesis that has been tested and validated over a long period of time. A natural law addresses how matter behaves and a theory addresses why it behaves that way.

Key Terms

The new terms defined in this chapter are

experiment   *Sec. 1.4*
fact   *Sec. 1.4*
hypothesis   *Sec. 1.4*
natural law   *Sec. 1.4*
qualitative data   *Sec. 1.4*

quantitative data   *Sec. 1.4*
science   *Sec. 1.1*
scientific discipline   *Sec. 1.1*
scientific method   *Sec. 1.4*
technology   *Sec. 1.2*
theory   *Sec. 1.4*

Practice Problems

Scientific Disciplines (Sec. 1.1)

1.1 Indicate whether each of the following statements is true or false.

(a) Science is the study in which humans attempt to organize and explain, in a systematic and logical manner, knowledge about themselves and their surroundings.

(b) Scientific disciplines are limited in size and scope to the extent that subdividing of subject matter within a discipline is not necessary.

(c) Boundaries between scientific disciplines are very rigid.

(d) Collectively, the knowledge in scientific disciplines constitutes the whole of scientific knowledge currently known.

1.2 Indicate whether each of the following statements is true or false.

(a) Scientific disciplines are branches of scientific knowledge limited in size and scope to make them more manageable.

(b) Scientific disciplines are defined in such a manner that each discipline is totally independent of other disciplines.

(c) Complete mastery of all concepts within a scientific discipline is difficult (but possible) because of limited size and scope for the discipline.

(d) The scientific discipline of chemistry has some overlap with other physical sciences but no overlap with biological sciences.

The Scientific Method (Sec. 1.4)

1.3 Arrange the following steps in the scientific method in the sequence in which they normally occur.

(a) Suggest probable explanations for generalizations obtained from data.

(b) Collect data concerning a problem through observation and experimentation.

(c) Identify a problem and carefully plan procedures to obtain information about all aspects of this problem.

(d) Experiment further to prove or disprove proposed explanations.

(e) Analyze and organize data in terms of general statements that summarize experimental observations.

1.4 Arrange the following terms associated with the scientific method in the order in which they are normally encountered as the scientific method is applied to a problem.

(a) natural law

(b) fact

(c) theory

(d) experiment

(e) hypothesis

1.5 Classify each of the following statements as a *fact*, a *natural law*, or a *hypothesis*.

(a) In northern climates, cars rust faster during the winter months than during the summer months.

(b) The author of this chemistry textbook is bald because he chewed his food too fast as a child.

(c) A sample of oxygen gas expanded when it was heated.

(d) The boiling point of water is 100°C at sea level.

1.6 Classify each of the following statements as a *fact*, a *natural law*, or a *hypothesis*.

(a) A man's hair turned gray because of the driving habits of his teenage children.

(b) All samples of gaseous substances expand when heated.

(c) The force of gravity upon an object depends on the color of the object.

(d) The diameter of the moon is 3476 kilometers.

1.7 Indicate whether each of the following statements is true or false.

(a) A theory is a summary of experimental observations.

(b) A hypothesis is a summary of experimental facts.

(c) A theory is subject to modification in light of new experimental observations.

(d) An experiment is a well-defined, controlled procedure for obtaining facts.

1.8 Indicate whether each of the following statements is true or false.

(a) A theory is a hypothesis that has not yet been subjected to experimental testing.

(b) It is much easier to disprove a false hypothesis than it is to prove a valid one.

(c) Established theories eventually become natural laws.

(d) A natural law is an explanation of why a particular phenomenon occurs.

1.9 Constructively criticize the statement "You needn't take it too seriously; after all, it's only a theory."

1.10 Constructively criticize the statement "The results of the experiment do not agree with the theory. Something must be wrong with the experiment."

1.11 Assume that you have four pennies with unknown mint dates and four hypotheses concerning these dates: (1) all dates are the same, (2) two different dates are present, (3) three different dates are present, and (4) all dates are different. Which of the listed hypotheses could be eliminated by determining that

(a) two pennies have the same date?

(b) two pennies have different dates?

(c) two of three pennies have the same date?

(d) three pennies have different dates?

1.12 Assume that you have four red balls of equal size and four hypotheses concerning the masses of the balls: (1) each ball has a different mass, (2) there are balls of two masses, (3) balls of three different masses are present, and (4) all balls have the same mass. Which of the listed hypotheses could be eliminated by determining that

(a) two balls have the same mass?

(b) three balls have the same mass?

(c) there are two masses among three balls?

(d) there are two masses among four balls?

1.13 Indicate whether each of the following statements represents *qualitative data* or *quantitative data*.

(a) The sun sets in the west.

(b) The automobile tire pressure is 32 pounds per square inch.

(c) The length of the copper rod is 6.37 meters.

(d) The colorless liquid has an alcohol-like odor.

1.14 Indicate whether each of the following statements represents *qualitative data* or *quantitative data*.

(a) The empty vial weighs 54.2 grams.

(b) The density of oxygen gas increases as its temperature increases.

(c) The density of gold at 20°C is 19.3 grams per cubic centimeter.

(d) The odorless liquid has a reddish-brown color.

1.15 A researcher studies the behavior of a fixed amount of a gas under constant temperature conditions with the following results:

(a) At a pressure of 4.0 atmospheres the gas occupies a volume of 2.0 liters.

(b) At a pressure of 1.0 atmosphere the gas occupies a volume of 8.0 liters.

(d) At a pressure of 2.0 atmospheres the gas occupies a volume of 4.0 liters.

(d) At a pressure of 8.0 atmospheres the gas occupies a volume of 1.0 liter.

What generalization (natural law) concerning the relationship between volume and pressure, under the conditions of the experiments, can be obtained from these data?

1.16 A researcher studies the behavior of a gas under constant temperature and constant volume conditions with the following results:

(a) 10.0 grams of gas exerted a pressure of 4.0 atmospheres.

(b) 40.0 grams of gas exerted a pressure of 16.0 atmospheres.

(c) 5.0 grams of gas exerted a pressure of 2.0 atmospheres.

(d) 20.0 grams of gas exerted a pressure of 8.0 atmospheres.

What generalization (natural law) concerning the relationship between amount of gas and pressure, under the conditions of the experiments, can be obtained from these data?

1.17 What are the differences between a natural law and a societal law?

1.18 What is the reason for repeating experiments several times before developing a natural law based on the experiments?

1.19 Why is it important that scientific data be published?

1.20 The phrase "it has been proved scientifically" is rarely used by scientists. Explain why.

1.21 Why is it useless to conduct an experiment under uncontrolled conditions?

1.22 What are the two important purposes that theories serve?

1.23 What is the difference between a qualitative observation and a quantitative observation?

1.24 If a theory is false, how will the scientific method, applied over time, reveal that such is the case?

Answers to Practice Exercise

1.1 Hypotheses 1 and 3 are disproved.

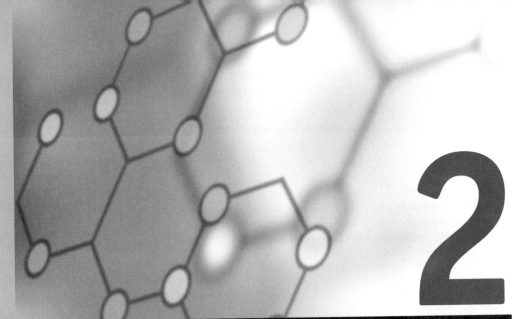

2

Numbers from Measurements

2.1 | The Importance of Measurement

It would be extremely difficult for a carpenter to build cabinets without being able to use tools such as hammers, saws, and drills. They are a carpenter's "tools of the trade." Chemists also have "tools of the trade." Their most used tool is the one called *measurement*. **Measurement** *is the determination of the dimensions, capacity, quantity, or extent of something.* In chemical laboratories, the most common types of measurements are those of mass, volume, length, time, temperature, pressure, and concentration.

Understanding measurement is indispensable in the study of chemistry. Questions such as "how much . . . ?", "how long . . . ?", and "how many . . . ?" simply cannot be answered without resorting to measurements. It is the purpose of this chapter and the next to help students acquire the necessary background to deal properly with measurement. Almost all of the material of these two chapters is mathematical. An understanding of this mathematics is a necessity for students of chemistry who want their encounters with the subject to be successful. The following analogy is appropriate for the situation. Physical exertion in sports can be fun, relaxing, and challenging for those in good physical shape. But for those not in good physical condition, such exertion is not satisfying and may even be downright painful (especially the day after). Being "in good shape" mathematically has the same effect on the study of chemistry. It can cause that study to be a very satisfying and enjoyable experience. On the other hand, a lack of the necessary mathematical skills can cause "chemical exercise" to be somewhat painful. The message should be clear. The contents of this chapter (and Chapter 3) must be taken very seriously. Skimming over this material is a sure invitation to frustration and struggle with the chemical topics that follow.

2.2 Exact and Inexact Numbers

In scientific work, numbers are grouped in two categories: *exact numbers* and *inexact numbers*. An **exact number** *is a number that has a value with no uncertainty in it; that is, it is known exactly.* Exact numbers occur in definitions (for example, there are exactly 12 objects in a dozen, not 12.01 or 12.02); in counting (for example, there can be 7 people in a room, but never 6.99 or 7.02); and in simple fractions (for example, 1/3, 3/5, and 5/9).

An **inexact number** *is a number that has a value with a degree of uncertainty in it.* Inexact numbers result anytime a measurement is made. It is impossible to make an *exact* measurement; some uncertainty will always be present. Flaws in measuring-device construction, improper calibration of an instrument, and the skills (or lack of skills) possessed by a person using a measuring device all contribute to error (uncertainty).

2.3 Accuracy, Precision, and Error

Two important terms relating to the uncertainties associated with measurement values are *precision* and *accuracy.* Although these terms are used somewhat interchangeably in nonscientific discussions, they have distinctly different meanings in science.

Precision *refers to how close a series of measurements on the same object are to each other.* Note that precision determination involves a series of measurements; it is not proper to speak of the precision of a single measurement made on an object. **Accuracy** *refers to how close a measurement (or the average of multiple measurements) comes to a true or accepted value.* The activity of throwing darts at a target illustrates nicely the difference between these two terms (see Fig. 2.1). Accuracy refers to how close the darts are to the center (bull's-eye) of the target. Precision refers to how close the darts are to each other.

Both the precision and accuracy of a series of measurements usually relate directly to the actual physical measuring device used. You would expect, and it is most often the case, that the precision and accuracy of temperature readings obtained from a thermometer with a scale marked in tenths of a degree would be better than readings obtained from a thermometer whose scale has only degree marks. A series of readings obtained from a stopwatch whose dial shows tenths of a second will usually be more precise and accurate than readings from a stopwatch that shows only seconds.

Both the precision and accuracy depend not only on the measuring device used but also on the technical skill of the person making the measurement. How well can that person read the numerical scale of the instrument? How well can that person calibrate the instrument before its use?

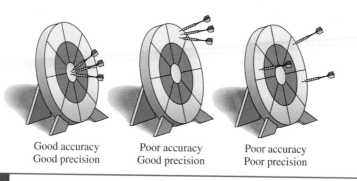

Good accuracy   Poor accuracy   Poor accuracy
Good precision   Good precision   Poor precision

Figure 2.1

The difference between precision and accuracy.

Normally, high accuracy accompanies high precision. However, high precision and low accuracy are also possible. For example, results obtained using a poorly calibrated instrument could give high precision but low accuracy. All measurements would be off by a constant amount as a result of the improper calibration.

Errors in measurement can be classified as either *random errors* or *systematic errors*. **Random errors** *are errors originating from uncontrolled variables in an experiment.* Such errors result in experimental values that fluctuate about the true value. A variation in the angle from which a measurement scale is viewed will cause random error. Momentary changes in air currents, atmospheric pressure, or temperature near a sensitive balance for weighing would cause random errors. The net result of random errors, which can never be completely eliminated, is a decrease in the precision of a series of measurements.

Systematic errors *are errors originating from controllable variables in an experiment.* They are "constant" errors that occur again and again. A flaw in a piece of equipment, such as a chipped weight in a balance, would cause systematic error. All readings would be off by a specific amount because of that flaw. Systematic errors affect the accuracy of measurements. Results are consistently either too high or too low compared to the true value.

EXAMPLE 2.1

Determining the Accuracy and Precision of a Series of Measurements

Two teams of three students each count the number of people entering the main gate of a football stadium in the 5-minute period just prior to kickoff. Their individual and team counts are

Team A	Team B
Miranda: 577 people	Spencer: 577 people
Brinley: 579 people	Taylor: 585 people
Breiana: 581 people	Brandi: 593 people

An "electronic counter" indicates that 581 people passed through the gate in the designated time period.

(a) Which person's count is the most accurate?

(b) Which team's count is the most accurate?

(c) Which person's count is the most precise?

(d) Which team's count is the most precise?

SOLUTION

(a) Breiana's count is the most accurate because his count is the same as the electronic count.

(b) The average count of 579 for team A is closer to 581 than the average count of 585 for team B. Thus, team A made the more accurate count.

(c) Because each person made only one count, the term *precision* does not apply.

(d) Team A's counts range from a high of 581 to a low of 577, which gives a spread of 4. Team B's counts range from a high of 593 to a low of 577, which gives a spread of 16. The smaller spread makes Team A's counts more precise.

Practice Exercise 2.1

Two teams of three students each count the number of cars passing a certain point on a highway during rush hour. Their individual and team counts are

Team A	Team B
Jessica: 473 cars	Cassidy: 469 cars
Timothy: 483 cars	Matthew: 475 cars
Benjamin: 484 cars	Brayden: 481 cars

An "electronic counter" indicates that 485 cars passed this location during the time interval of the count.

(a) Which person's count is the most accurate?
(b) Which team's count is the most accurate?
(c) Which person's count is the most precise?
(d) Which team's count is the most precise?

> Answers to practice exercises are located at the end of the chapter.

2.4 Uncertainty in Measurements

As noted in Section 2.2, every measurement carries a degree of uncertainty or error. Even when very elaborate and expensive measuring devices are used, some degree of uncertainty will be present in the measurement.

To illustrate how measurement uncertainty arises, let us consider how two different thermometer scales, illustrated in Figure 2.2, are used to measure a given temperature. Determining the temperature involves determining the height of the mercury column in the thermometer. The scale on the left in Figure 2.2 is marked off in one-degree intervals. Using this scale, we can say with certainty that the temperature is between 29 and 30 degrees. We can further say that the actual temperature is closer to 29 degrees than to 30 and estimate it to be 29.2 degrees. The scale on the right has more subdivisions, being marked off in tenths of a degree rather than in degrees. Using this scale, we can definitely say that the temperature is

> Every measurement has some degree of uncertainty associated with it.

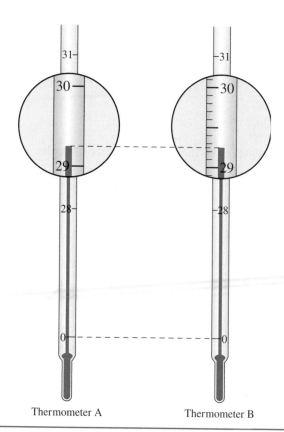

Thermometer A   Thermometer B

Figure 2.2

Measuring a temperature. A portion of the degree scale on each of the two differently scaled thermometers has been magnified.

between 29.2 and 29.3 degrees and can estimate it to be 29.25 degrees. Note how both temperature readings contain some digits (all those except the last one) that are exactly known and one digit (the last one) that is estimated. It is this last digit, the estimated one, that produces uncertainty in the measurement. Note also that the uncertainty in the second temperature reading is less than that in the first reading—an uncertainty in the hundredths place compared to an uncertainty in the tenths place.

Only one estimated digit is ever recorded as part of a measurement. It would be incorrect for a scientist to report that the height of the mercury column in Figure 2.2, as read on the scale on the right, corresponds to a temperature of 29.247 degrees. The value 29.247 contains two estimated digits (the 4 and the 7) and would indicate a measurement with smaller uncertainty than is actually obtainable with that particular measuring device.

> **In reading a measurement scale, all digits known for certain are recorded plus one estimated digit. It is wrong to record more than one estimated digit.**

The magnitude of the uncertainty in the last recorded digit in a measurement (the estimated digit) may be indicated using a "plus–minus" notation. The following three time measurements illustrate this notation.

$$15 \pm 1 \text{ second}$$

$$15.3 \pm 0.1 \text{ second}$$

$$15.34 \pm 0.03 \text{ second}$$

Most often the uncertainty in the last recorded digit is one unit (as in the first two time measurements), but it may be larger (as in the third time measurement). In this text we will follow the almost universal practice of dropping the "plus–minus" notation if the magnitude of the uncertainty is one unit. Thus, in the absence of "plus–minus" notation you will be expected to assume that there is an uncertainty of one unit in the last recorded digit. A measurement reported simply as 27.3 inches means 27.3 ± 0.1 inch. Only in the situation where the uncertainty is greater than one unit in the last recorded digit will the amount of the uncertainty be explicitly shown.

Example 2.2 relates measurement uncertainty to actual measuring-device scales.

EXAMPLE 2.2

Recording Measurements to the Proper Number of Digits

Determine the numerical value of the volume to be recorded in each of the following volume measurements.

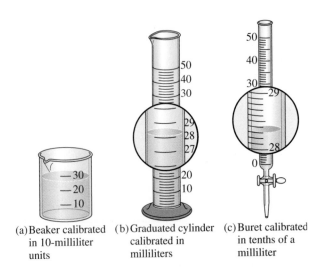

(a) Beaker calibrated in 10-milliliter units

(b) Graduated cylinder calibrated in milliliters

(c) Buret calibrated in tenths of a milliliter

SOLUTION

(a) We know definitely that the volume of liquid is between 20 and 30 milliliters. We estimate the final digit (to the closest milliliter) to be 8, giving a reading of 28 milliliters.

(b) The level of the liquid is between 28 and 29 milliliters. We estimate the level to be at 28.2 milliliters.

(c) The buret is calibrated in tenths of a milliliter. We know for certain that the liquid level is between 28.3 and 28.4 milliliters. Adding one estimated digit (hundredths of a milliliter) gives a reading of 28.31 milliliters.

Practice Exercise 2.2

Determine the numerical value of the volume to be recorded in each of the following volume measurements.

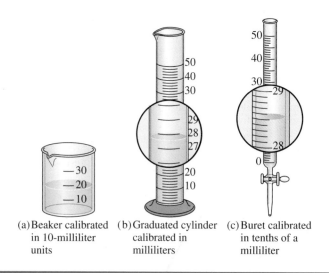

(a) Beaker calibrated in 10-milliliter units (b) Graduated cylinder calibrated in milliliters (c) Buret calibrated in tenths of a milliliter

2.5 Significant Figures

Because measurements are never exact (Sec. 2.4), two types of information must be conveyed whenever a numerical value for a measurement is recorded: (1) the magnitude of the measurement and (2) the uncertainty of the measurement. The magnitude is indicated by the digit values. Uncertainty is indicated by the number of *significant figures* recorded. **Significant figures** *are the digits in any measurement that are known with certainty plus one digit that is uncertain.* To summarize, in equation form,

> The term *significant figures* is often verbalized in shortened form as "sig figs."

Number of significant figures = All certain digits + One uncertain digit

Determining the number of significant figures in a measurement is not always as straightforward as Example 2.2 infers. In this example, you knew the type of instrument used for each measurement and its limitations because you made the measurement. Quite often when someone else makes a measurement, such information is not available. All that is known is the reported final result—the numerical value of the measured quantity. In this situation questions often arise about the "significance" of various digits in the measurement. For example, consider the published value of the distance from the Earth to the sun, which is 93,000,000 miles. Intuition tells you that it is highly improbable that this distance is known to the closest mile. You suspect that this is an estimated distance. To what digit has this number been estimated? Is it to the nearest million miles, the closest hundred thousand miles, the nearest ten thousand miles, or what?

A set of guidelines has been developed to aid scientists in interpreting the significance of reported measurements or results calculated from measurements. Four rules constitute the guidelines, one rule for the digits 1 through 9 and three rules for the digit 0. A zero in a measurement may or may not be significant depending on its location in the sequence of digits forming the numerical value for the measurement. There is a rule for each of three classes of zeros—leading zeros, confined zeros, and trailing zeros.

Nonzero digits are always significant.

RULE 1 The digits 1 through 9 inclusive (all of the nonzero digits) always count as significant figures.

14.232	five significant figures
3.11	three significant figures
244.6	four significant figures

Leading zeros are never significant.

RULE 2 *Leading zeros* are zeros that occur at the start of a number, that is, zeros that precede all nonzero digits. Such zeros do not count as significant figures. Their function is simply to indicate the position of the decimal point.

0.00045	two significant figures
0.0113	three significant figures
0.000000072	two significant figures

Leading zeros are always to the left of the first nonzero digit.

Confined zeros are always significant.

RULE 3 *Confined zeros* are zeros between nonzero digits. Such zeros always count as significant figures.

2.075	four significant figures
6007	four significant figures
0.03007	four significant figures

Trailing zeros are sometimes significant.

RULE 4 *Trailing zeros* are zeros at the end of a number. They are significant if (a) there is a decimal point present in the number or (b) they carry overbars. Otherwise trailing zeros are not significant.

The following numbers, all containing decimal points, illustrate condition (a) of rule 4.

62.00	four significant figures
24.70	four significant figures
0.02000	four significant figures
4300.00	six significant figures

By condition (b), trailing zeros in numbers lacking an explicitly shown decimal point become significant when marked with a bar above the zero(s).

$$36,\overline{000} \quad \text{five significant figures}$$

$$36,0\overline{00} \quad \text{four significant figures}$$

$$36,00\overline{0} \quad \text{three significant figures}$$

$$10,02\overline{0} \quad \text{five significant figures}$$

In cases involving trailing zeros where neither a decimal point nor overbar(s) is present, "confusion exists" because these zeros may or may not be significant. For example, it is not possible to definitely know how many significant figures are present in the measurement 5600 grams (no decimal place explicitly shown). There are three possible interpretations—two, three, or four significant figures—depending on the uncertainty associated with the measurement (5600 ± 100, 5600 ± 10, and 5600 ± 1). Standard operating procedure in such cases, where no other information about the measurement is available, is to assume the largest of the uncertainties possible for the measurement. Thus, the measurement 5600 grams contains two significant figures. Generally, then, in cases involving trailing zeros where neither a decimal point nor overbar(s) is present, the trailing zeros are assumed to be nonsignificant.

> Another method, more convenient than rule 4, for dealing with the significance of trailing zeros involves expressing the number in scientific notation. In this notation, to be presented in Section 2.7, only significant digits are shown.

93,000,000   two significant figures

360,000   two significant figures

330,300   four significant figures

6310   three significant figures

EXAMPLE 2.3

Determining the Number of Significant Figures in a Numerical Value

Determine the number of significant figures in the numerical value in each of the following statements.

(a) A wire has a diameter of 0.05082 inch.
(b) The mass of the Earth is 6,600,000,000,000,000,000,000 tons.
(c) A hospital patient's blood glucose level was determined to be 4850 micrograms per milliliter of blood.
(d) Normal body temperature for a chickadee is 41.0°C.

SOLUTION

(a) There are four significant figures. The leading zeros are not significant (rule 2) and the confined zero is significant (rule 3).
(b) There are two significant figures. The trailing zeros are not significant because no decimal point or overbar notation is present (rule 4).
(c) There are three significant figures. The trailing zero is not significant (rule 4).
(d) There are three significant figures. The trailing zero is significant because a decimal point is present (rule 4).

Practice Exercise 2.3

Determine the number of significant figures in the numerical value in each of the following statements.

> **(a)** A regular-issue U.S. postage stamp has a width of 0.021 meter.
> **(b)** The melting point of the metal gold is 1064°C.
> **(c)** The Earth's oceans and seas contain 330,000,000 cubic meters of seawater.
> **(d)** The volume of a drop of water is 0.000050 liter.

It is important to remember what is "significant" about significant figures. The number of significant figures in a measurement conveys information about the uncertainty associated with the measurement. The "location" of the last significant digit in the numerical value for a measurement specifies the measurement's uncertainty: Is the last significant digit located in the hundredths, tenths, ones, or tens position, etc.? Consider the following measurement values (with the last significant digit in color for emphasis).

4620.0 has five significant figures and an uncertainty of tenths.

4620 has three significant figures and an uncertainty in the tens place.

462,000 has three significant figures and an uncertainty in the thousands place.

All numbers obtained by measurement have uncertainty. By convention, this uncertainty is assumed to be in the number's final significant digit.

EXAMPLE 2.4

Determining the Number of Significant Figures and Magnitude of Uncertainty in a Numerical Value

The number of carbon monoxide molecules in a sample of automobile exhaust is verbally reported as two hundred and five thousand. What meaning, in terms of significant figures and magnitude of uncertainty, is conveyed by each of the following written notations for this number?

(a) 205,000   **(b)** 205,$\overline{0}$00   **(c)** 205,000.   **(d)** 205,$\overline{000}$

SOLUTION

(a) Three significant figures are present in this number; the confined zero is significant but the trailing zeros are not. Since the last significant digit, the 5, is located in the fourth place to the left of the understood decimal point (the thousands place), the uncertainty is ±1000.

(b) This number has four significant figures. The overbar above the first of the three trailing zeros makes this zero significant. The last significant digit, the zero with the overbar, occupies the hundreds place in the number. Thus, the uncertainty is ±100.

(c) There are six significant figures present. Explicitly placing a decimal point at the end of the number makes all the trailing zeros significant. The uncertainty is ±1 since the last of the trailing zeros is in the ones position.

(d) With the overbar notation present on all trailing zeros, all six digits present are significant. The uncertainty is ±1 since the last of the trailing zeros is in the ones position.

Practice Exercise 2.4

The population of a town in southern England is verbally reported to be one hundred and thirty thousand. What meaning, in terms of significant figures and magnitude of uncertainty, is conveyed by each of the following written notations for this number?

(a) 130,000   **(b)** 130,$\overline{0}$00   **(c)** 13$\overline{0,0}$00   **(d)** 130,000.

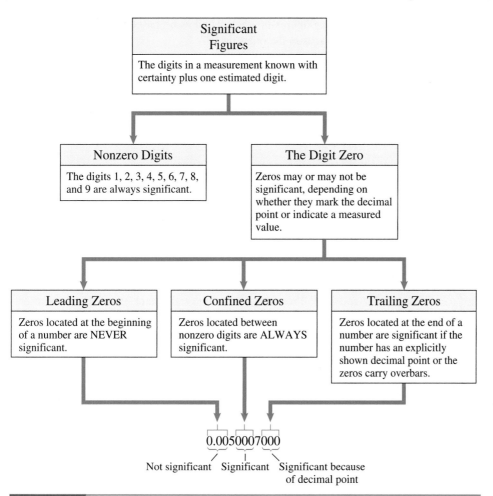

Figure 2.3

The rules for determining how many significant figures are present in a measured number.

Figure 2.3 summarizes, in diagram form, the four rules used in determining how many significant figures a measured number possesses.

2.6 Significant Figures and Mathematical Operations

When measurements are added, subtracted, multiplied, or divided, consideration must be given to the number of significant figures in the computed result. *Mathematical operations should not increase (or decrease) the uncertainty of experimental measurements.*

Concern about the number of significant figures in a calculated number is particularly critical when an electronic calculator is used to do the arithmetic of the calculation. Hand calculators now in common use are not programmed to take significant figures into account. Consequently, the digital readouts on them more often than not display more digits than are needed. It is a mistake to record these extra digits, since they have no significance; that is, they are not significant figures (see Fig. 2.4).

To record correctly the numbers obtained through calculations, students must be able to (1) round off numbers to a specified number of significant figures and (2) determine the

Figure 2.4

The digital readout on an electronic calculator usually shows more digits than are needed or justified. Electronic calculators are not programmed to take significant figures into account. (Texas Instruments, Inc. Data Systems Group)

allowable number of significant figures in a calculated result based on a set of operational rules. We will consider these skills in the order listed.

Rounding Off Numbers

When calculator answers are obtained that contain too many digits, it is necessary to drop the unneeded (nonsignificant) digits, a process that is called *rounding off*. **Rounding off** *is the process of deleting unwanted (nonsignificant) digits from a calculated number.* Three simple rules govern the process.

RULE **1**   If the first digit to be dropped is less than 5, that digit and all digits that follow it are simply dropped.

Thus, 62.312 rounded off to three significant figures becomes 62.3.

RULE **2**   If the first digit to be dropped is a digit greater than 5, or a 5 followed by digits other than all zeros, the excess digits are all dropped and the last retained digit is increased in value by one unit.

Thus, 62.782 and 62.558 rounded off to three significant figures become, respectively, 62.8 and 62.6.

RULE **3**   If the first digit to be dropped is a 5 not followed by any other digit or a 5 followed only by zeros, an odd–even rule applies. Drop the 5 and any zeros that follow it and then

(**a**) increase the last retained digit by one unit if it is *odd*, or

(**b**) leave the last retained digit the same if it is *even*.

Thus, 62.650 and 62.350 rounded to three significant figures become, respectively, 62.6 (even rule) and 62.4 (odd rule). The number zero as a last retained digit is always considered an even number; thus, 62.050 rounded to three significant figures becomes 62.0.

These rounding rules must be modified slightly when digits to the left of the decimal point are to be dropped. To maintain the inferred position of the decimal point in such situations, nonsignificant zeros must replace all of the dropped digits that are to the left of the inferred decimal point. Parts (c) and (f) of Example 2.5 illustrate this point.

> An additional point about rounding off is that numbers are not rounded *sequentially*. In rounding the number 4.548 to two significant figures we do not round it first to 4.55 and then to 4.6. When rounding we look at the first number to the right of the last retained digit and base the rounding operation on its value. The number 4.548 rounds to 4.5 because the first number to the right of the last retained digit, 4, has a value of less than 5 (rule 1).

EXAMPLE 2.5

Rounding Numbers to a Specified Number of Significant Figures

Round off each of the following numbers to two significant figures.

(**a**) 25.7   (**b**) 25.37   (**c**) 432,117

(**d**) 0.435   (**e**) 62.50   (**f**) 13,500

SOLUTION

(**a**) Rule 2 applies. The last retained digit (the 5) is increased in value by one unit.

$$25.7 \quad \text{becomes} \quad 26$$

(**b**) Rule 1 applies. The last retained digit (the 5) remains the same, and all digits that follow it are simply dropped.

$$25.37 \quad \text{becomes} \quad 25$$

(c) Since the first digit to be dropped is a 2, rule 1 applies.

432,117   becomes   430,000

Note that to maintain the position of the inferred decimal point, nonsignificant zeros must replace all of the "dropped" digits. This will always be the case when digits to the left of the inferred decimal place are "dropped."

(d) Rule 3 applies. The first and only digit to be dropped is a 5. The last retained digit (the 3) is an odd number, so, using the odd–even rule, its value is increased by one unit.

0.435   becomes   0.44

(e) Rule 3 applies again. This time the last digit retained is even (the 2), so its value is not changed.

62.50   becomes   62

(f) This is a rule 3 situation again. Since an odd digit (the 3) occupies the second significant figure place, its value is increased by one.

13,500   becomes   14,000

Note again that nonsignificant zeros must take the place of all digits that are "dropped" to the left of the inferred decimal place.

Practice Exercise 2.5

Round off each of the following numbers to three significant figures.

(a) 432.87   **(b)** 432.17   **(c)** 655,234

(d) 0.03315   **(e)** 352.50   **(f)** 162,500

Operational Rules for Mathematical Operations

Calculations should not increase or decrease the uncertainty in measurements. Two operational rules exist to help ensure that this is the case. One rule covers the operations of multiplication and division and the other the operations of addition and subtraction.

RULE **1** *Multiplication and Division.* In multiplication and division, the number of significant figures in the product or quotient is the same as in the number in the calculation that contains the fewest significant figures.

$$6.038 \times \mathbf{2.57} = 15.51766 \quad \text{(calculator answer)}$$
$$= 15.5 \quad \text{(correct answer)}$$

This number limits the answer
to three significant figures.

RULE **2** *Addition and Subtraction.* In addition and subtraction of a series of measurements, the uncertainty in the answer should be the same as that of the measurement in the series that has the greatest uncertainty.

Position of uncertainty

$$
\begin{array}{rl}
347 & \longleftarrow \text{Ones position} \\
+\quad 2.03 & \longleftarrow \text{Hundredths position} \\
+\quad 23.6 & \longleftarrow \text{Tenths position} \\
\hline
372.63 & \text{(calculator answer)} \\
373 & \text{(correct answer)}
\end{array}
$$

The measurement with the greatest uncertainty in this series of numbers is 347, which is uncertain to the ones position. Thus, the calculator answer must be rounded to the ones position, the position of greatest uncertainty.

Note that for multiplication and division (rule 1), significant figures are counted, and that for addition and subtraction (rule 2), uncertainties are considered. The answers from either addition or subtraction can have more or fewer significant figures than any of the numbers that have been added or subtracted, as is shown in Example 2.8.

EXAMPLE 2.6

Predicting the Number of Significant Figures That Should Be Present in the Answer to a Multiplication Problem

Without actually doing any multiplications, indicate the number of significant figures that should be present in the answer to each of the multiplications. Assume that all numbers are measured quantities.

(a) $6.00 \times 6.00 \times 6.00$ (b) $6.00 \times 0.600 \times 60.60$

(c) $0.006 \times 0.060 \times 0.600$ (d) $60{,}600 \times 6060 \times 606$

SOLUTION

(a) Each number to be multiplied contains three significant figures. Thus, the answer should also contain three significant figures.

(b) The first two input numbers contain three significant figures and the third number contains four significant figures. Only three significant figures should be present in the answer.

(c) The input numbers contain, respectively, one, two, and three significant figures. The one significant figure input number limits the answer to one significant figure.

(d) All three input numbers contain three significant figures. Thus, the answer should contain three significant figures.

Practice Exercise 2.6

Without actually doing any multiplications, indicate the number of significant figures that should be present in the answer to each of the multiplications. Assume that all numbers are measured quantities.

(a) $2.0 \times 2.0 \times 3.0$ (b) $2.00 \times 2.00 \times 3.0$

(c) $2.000 \times 2.00 \times 3.00$ (d) $2.0000 \times 2.00 \times 3.0000$

EXAMPLE 2.7

Expressing Multiplication/Division Answers to the Proper Number of Significant Figures

Perform the following computations, all of which involve multiplication and/or division. Express your answers to the proper number of significant figures. Assume that all numbers are measured quantities.

(a) 3.751×0.42 　　(b) $\dfrac{1,810,000}{3.1453}$

(c) $\dfrac{1800.0}{6.0000}$ 　　(d) $\dfrac{3.130 \times 3.140}{3.15}$

SOLUTION

(a) The calculator answer to this problem is

$$3.751 \times 0.42 = 1.57542$$

The input number with the least number of significant figures is 0.42, which has two significant figures. Thus the calculator answer must be rounded off to two significant figures.

$$1.57542 \quad\quad \text{becomes} \quad\quad 1.6$$
　　(calculator answer) 　　　　　(correct answer)

(b) The calculator answer to this problem is

$$\frac{1,810,000}{3.1453} = 575,461.8$$

The input number 1,810,000 contains three significant figures and the input number 3.1453 has five significant figures. Thus the correct answer is limited to three significant figures and is obtained by rounding the calculator answer to three significant figures.

$$575,461.8 \quad\quad \text{becomes} \quad\quad 575,000$$
　　(calculator answer) 　　　　　(correct answer)

A decimal point is not explicitly shown in the number 575,000 since doing so would make the trailing zeros significant.

(c) The calculator answer to this problem is

$$\frac{1800.0}{6.0000} = 300$$

Both input numbers contain five significant figures. Thus the correct answer must also contain five significant figures.

$$300 \quad\quad \text{becomes} \quad\quad 300.00$$
　　(calculator answer) 　　　　　(correct answer)

Note here how the calculator answer had too few significant figures. Most calculators cut off zeros after the decimal point even if they are significant. *Using too few significant figures in an answer is just as wrong as using too many.*

(d) This problem involves both multiplication and division. The calculator answer is

$$\frac{3.130 \times 3.140}{3.15} = 3.1200634$$

The input number with the least number of significant figures is 3.15, which contains three significant figures. Thus, the calculator answer must be rounded off to three significant figures.

3.1200634   becomes   3.12

(calculator answer)   (correct answer)

> In a calculation that involves more than one multiplication/division operation, such as part (d) of this example, carry all of the digits that show on your calculator until you arrive at the final answer and then round off using the rounding rules. Do not round off at each step in the calculation.

Practice Exercise 2.7

Perform the following computations, all of which involve multiplication and/or division. Express your answers to the proper number of significant figures. Assume that all numbers are measured quantities.

(a) 6.7321×0.0021

(b) $\dfrac{16{,}240}{23.42}$

(c) $\dfrac{120.0}{4.000}$

(d) $\dfrac{5.444 \times 8.670}{2.321 \times 3.27}$

EXAMPLE 2.8

Expressing Addition/Subtraction Answers to the Proper Number of Significant Figures

Perform the following computations, all of which involve addition or subtraction. Express your answers to the proper number of significant figures. Assume that all numbers are measured quantities.

(a) $13.01 + 13.001 + 13.010$ **(b)** $10.2 + 3.4 + 6.01$

(c) $0.6700 - 0.6644$ **(d)** $34.7 + 0.0007$

SOLUTION

(a) The calculator answer to this problem is

$$13.01 + 13.001 + 13.010 = 39.021$$

Since this is an addition problem, in going from the calculator answer to the correct answer we must consider uncertainties rather than significant figures. (The multiplication–division rule is based on significant figures and the addition–subtraction rule is based on uncertainties.)

The uncertainties in the input numbers are

13.01   hundredths

13.001   thousandths

13.010   thousandths

The number 13.01 has the greatest uncertainty (hundredths) and so the last retained digit in the correct answer should reflect this uncertainty. Hence, the calculator answer is rounded off to hundredths.

$$39.021 \qquad \text{becomes} \qquad 39.02$$

(calculator answer) (correct answer)

(b) The calculator answer to this problem is

$$10.2 + 3.4 + 6.01 = 19.61$$

The uncertainty in the first two input numbers is tenths, and the third input number involves an uncertainty of hundredths. Thus, the last retained digit in the correct answer will be in the tenths place, the largest uncertainty among the input numbers.

$$19.61 \qquad \text{becomes} \qquad 19.6$$

(calculator answer) (correct answer)

Note that the input number 3.4 possesses two significant figures and yet the correct answer contains three significant figures. Why? The number of significant figures is not the determining factor in addition and subtraction (rule 2) as it is in multiplication and division (rule 1).

(c) The calculator answer to this problem is

$$0.6700 - 0.6644 = 0.0056$$

Both input numbers are known to the ten-thousandths place. Thus, the answer should also have an uncertainty involving the ten-thousandths place.

In this particular problem the calculator answer and the correct answer are the same, a situation which does not occur very often. The correct answer is 0.0056. Note that two significant figures were "lost" in the subtraction. The answer has two significant figures. The two input numbers each have four significant figures. For addition and subtraction this is allowable; for multiplication and division it would not be allowable.

(d) The calculator answer to this problem is

$$34.7 + 0.0007 = 34.7007$$

The uncertainty in the input number 34.7 is tenths and the uncertainty in the input number 0.0007 is ten-thousandths. Thus, the calculator answer must be rounded off to the tenths place.

$$34.7007 \qquad \text{becomes} \qquad 34.7$$

(calculator answer) (correct answer)

The correct answer, 34.7, is the same as one of the input numbers. The message of this situation is that the number 0.0007 is negligible when added to the number 34.7.

Practice Exercise 2.8

Perform the following computations, all of which involve addition or subtraction. Express your answers to the proper number of significant figures. Assume that all numbers are measured quantities.

(a) $28.7 + 7.01 + 22$ **(b)** $8.3 + 1.2 + 1.7$

(c) $0.4378 - 0.4367$ **(d)** $4200 + 14.7$

Occasionally, a mathematical problem is encountered that involves both multiplication/division and addition/subtraction, such as

$$23.77 \times (1.3 + 2.58 + 6.671)$$

For such a problem, both calculational significant-figure rules must be applied by using a two-step approach. First we add the three numbers in parentheses, obtaining an intermediate answer that is based on the addition/subtraction rule. This intermediate answer is then multiplied by 23.77 to generate a final answer that is then adjusted for significant figures using the multiplication/division rule. Following this plan, we solve the preceding problem in this manner:

$$\text{Addition: } 1.3 + 2.58 + 6.671 = 10.551 \text{ (calculator answer)}$$

$$= 10.6 \text{ (intermediate answer, uncertainty of tenths based on the addition/subtraction rule)}$$

$$\text{Multiplication: } 23.77 \times 10.6 = 251.962 \text{ (calculator answer)}$$

$$= 252 \text{ (correct answer, three significant figures based on the multiplication/division rule)}$$

Significant Figures and Exact Numbers

In Section 2.2 we noted that some numbers are exact. Counted numbers, defined numbers, and simple fractions all fall into this category. The conventions of significant figures do not apply to exact numbers because there is no uncertainty associated with them. Therefore, such numbers, when present in a calculation, will never limit the number of significant figures in the computational answer; that is, they will have no effect on the number of significant figures in a calculated result.

EXAMPLE 2.9

Significant Figures in Calculations Where Exact Numbers Are Present

What would be the total combined length, in centimeters, of 143 new pencils, each of which has a length of 19.13 centimeters (cm)?

SOLUTION

Since a pencil has a length of 19.13 cm, the length of 143 such pencils will be

$$143 \times 19.13 \text{ cm} = 2735.59 \text{ cm} \quad \text{(calculator answer)}$$

The number 143 is an exact number (a counted number) and can be considered to have an infinite number of significant figures. Thus, the correct answer will have four significant figures, the number of significant figures in the measurement 19.13.

$$2735.59 \text{ cm} \quad \text{becomes} \quad 2736 \text{ cm (correct answer)}$$

Practice Exercise 2.9

A nickel is found to weigh 5.0715 grams (g). What would be the mass in grams of 97 such coins?

> An exact number has no effect on the number of significant figures in an answer obtained by calculation.

An alternative approach to the topic of significant figures and exact numbers is to view exact numbers as containing an infinite number of significant figures. From this viewpoint, such numbers will never be the limiting factors in significant-figure considerations. Thus, when we say that there are 3 feet in 1 yard (a definition), both the numbers 3 and 1 are considered to have an infinite number of significant figures.

2.7 Scientific Notation

Up to this point in the chapter, we have expressed all numbers in decimal notation, the everyday method for expressing numbers. Such notation becomes cumbersome for very large and very small numbers (which occur frequently in scientific work). For example, in one drop of blood, which is 92% water by mass, there are approximately

$$1,600,000,000,000,000,000,000$$

molecules (Sec 5.2) of water, each of which has a mass of

$$0.000000000000000000000030 \text{ gram}$$

Recording such large and small numbers is not only time consuming but also open to error; often, too many or too few zeros are recorded. Also, it is impossible to enter such numbers into most calculators because they cannot accept that many digits. (Most calculators accept eight or ten digits.)

A method called *scientific notation* exists for expressing in compact form multidigit numbers that involve many zeros. **Scientific notation** *is a numerical system in which an ordinary decimal number is expressed as a product of a number between 1 and 10 and 10 raised to a power.* The number between 1 and 10 is called a *coefficient* and is written first. The number 10 raised to a power (exponent) is called an *exponential term.* The coefficient is always multiplied by the exponential term. The scientific notation form of the number 703 is

> Scientific notation is also called *exponential notation.*

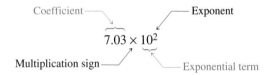

The two previously cited numbers that deal with molecules of water are expressed in scientific notation as

$$1.6 \times 10^{21} \text{ molecules} \quad \text{and} \quad 3.0 \times 10^{-23} \text{ grams}$$

Such scientific notation is compatible with most calculators.

Exponents

A brief review of exponents and their use is in order before we consider the rules for converting numbers from ordinary decimal notation to scientific notation and vice versa. An **exponent** *is a number written as a superscript following another number and indicates how many times the first number, the base, is to be multiplied by itself.* The following examples illustrate the use of exponents.

$$6^2 = 6 \times 6 = 36$$
$$3^5 = 3 \times 3 \times 3 \times 3 \times 3 = 243$$
$$10^3 = 10 \times 10 \times 10 = 1000$$

Exponents are also frequently referred to as *powers* of numbers. Thus, 6^2 may be verbally read as "six to the second power" and 3^5 as "three to the fifth power." Raising a number to the second power is often called "squaring," and raising it to the third power "cubing."

Scientific notation exclusively uses powers of ten. *When 10 is raised to a positive power, its decimal equivalent is the number 1 followed by as many zeros as the power.* This one-to-one correlation between power magnitude and number of zeros is shown in color in the following examples.

$$10^2 = 100 \qquad \text{(two zeros and a power of 2)}$$

$$10^4 = 10,000 \qquad \text{(four zeros and a power of 4)}$$

$$10^6 = 1,000,000 \qquad \text{(six zeros and a power of 6)}$$

The notation 10^0 is a defined quantity.

$$10^0 = 1$$

The preceding generalization easily explains why. Ten to the zero power is the number 1 followed by no zeros, which is simply 1.

All of the examples of exponential notation presented so far have had positive exponents. This is because each example represented a number of magnitude greater than one. Negative exponents are also possible. They are associated with numbers of magnitude less than one.

A negative sign in front of an exponent is interpreted to mean that the base and the power to which it is raised are in the denominator of a fraction in which 1 is the numerator. The following examples illustrate this interpretation.

$$10^{-1} = \frac{1}{10^1} = \frac{1}{10} = 0.1$$

$$10^{-2} = \frac{1}{10^2} = \frac{1}{10 \times 10} = \frac{1}{100} = 0.01$$

$$10^{-3} = \frac{1}{10^3} = \frac{1}{10 \times 10 \times 10} = \frac{1}{1000} = 0.001$$

When the number 10 is raised to a negative power, the absolute value of the power (the value ignoring the minus sign) is always one more than the number of zeros between the decimal point and the one. This correlation between power magnitude and number of zeros is shown in color in the following examples.

$$10^{-2} = 0.01 \qquad \text{(one zero and a power of } -2)$$

$$10^{-4} = 0.0001 \qquad \text{(three zeros and power of } -4)$$

$$10^{-6} = 0.000001 \qquad \text{(five zeros and power of } -6)$$

Differences in magnitude between numbers that are powers of 10 are often described with the phrase *orders of magnitude*. An **order of magnitude** *is a single exponential value of the number 10.* Thus, 10^6 is four orders of magnitude larger than 10^2, and 10^7 is three orders of magnitude larger than 10^4.

Converting from Decimal to Scientific Notation

The procedure for converting a number from decimal notation to scientific notation involves two operational rules:

RULE 1 The coefficient must be a number between 1 and 10 that contains the same number of significant figures as are present in the original decimal number. The coefficient is obtained by rewriting the decimal number with a decimal point after the first nonzero digit and deleting all *nonsignificant* zeros.

> The decimal and scientific notation forms of a number *always* contain the same number of significant figures.

> For 233,000, the coefficient is 2.33
>
> For 0.00557, the coefficient is 5.57
>
> For 0.35500, the coefficient is 3.5500

RULE 2 The value of the exponent for the power of ten is obtained by counting the number of places the decimal point in the coefficient must be moved to give back the original decimal number. If the decimal point movement is to the *right*, the exponent has a *positive* value and if the decimal point movement is to the *left*, the exponent has a *negative* value.

Numerous applications of these two rules are found in Example 2.10.

EXAMPLE 2.10

Expressing Decimal Numbers in Scientific Notation

Without use of a calculator, express in scientific notation the number in each of the following statements.

(a) Light travels at a speed of 186,000 miles per second.

(b) A person exhales approximately 320,000,000,000,000,000,000 molecules of carbon dioxide in one breath.

(c) The diameter of a human hair is 0.0016 inch.

(d) The maximum allowable amount of chromium in drinking water (EPA standard) is 0.00000010 gram per milliliter of water.

SOLUTION

(a) For the number 186,000 the scientific notation coefficient is 1.86. This coefficient meets the requirement that it contain the same number of significant figures as the original decimal number.

The value of the exponent in the exponential term of the scientific notation is $+5$ since the decimal point in the coefficient must be moved 5 places to the right to generate the original decimal number.

$$1.86000$$
5-place movement

Note that zeros are added to the coefficient as the decimal point is moved in order to obtain the original decimal point position.

Multiplying the coefficient by the exponential term 10^5 gives the scientific notation form of the number, which is

$$1.86 \times 10^5$$

(b) The coefficient for the number 320,000,000,000,000,000,000 is 3.2. It, like the original number, contains two significant figures.

The value of the exponent for the exponential term is +20 since the decimal point in the coefficient must be moved 20 places to the right to generate the original decimal number.

$$3.\underset{\text{20-place movement}}{\underline{20000000000000000000}}$$

Movement of the decimal point in the coefficient to the *right* always results in a *positive* exponent.

The scientific notation form of the number, obtained by multiplying the coefficient and exponential term, is

$$3.2 \times 10^{20}$$

(c) The scientific notation coefficient for the number 0.0016 is 1.6. The exponent for the power of ten is −3, since moving the decimal point in the coefficient three places to the left generates the original decimal number.

$$0.\underset{\text{3-place movement}}{\underline{001}}.6$$

Movement of the decimal point in the coefficient to the *left* always means that the exponent will be *negative*. The scientific notation form of the number is, thus, 1.6×10^{-3}.

(d) The scientific notation coefficient for the number 0.00000010 is 1.0. The coefficient is 1.0 rather than 1 because the original number has two significant figures. The exponent for the power of ten is −7.

$$0.\underset{\text{7-place movement}}{\underline{0000001}}.0$$

The number in scientific notation is, thus, 1.0×10^{-7}.

Practice Exercise 2.10

Without use of a calculator, express in scientific notation the number in each of the following statements.

(a) The Yangtze River, which flows through China, is 3915 miles long.
(b) The distance from the Earth to the sun is 93,000,000 miles.
(c) The maximum allowable amount of mercury in drinking water (EPA standard) is 0.0000015 gram per milliliter of water.
(d) The mass of a single carbon monoxide molecule is 0.0000000000000000000000465 gram.

> Some numbers, such as 2.4, 0.911, and 57, are simpler in their original form than in scientific notation. Such numbers are usually left in decimal notation. Although there are no fixed rules as to when scientific notation should be used, it is generally not used for numbers between 0.1 and 1000.

When a number is expressed in scientific notation, *only significant digits become part of the coefficient*. Because of this there is never any confusion (ambiguity) in determining the number of significant figures in a number expressed in scientific notation. There are five possible precision interpretations for the number 10,000 (one, two, three, four, or five significant figures). In scientific notation each of these interpretations assumes a different form.

1×10^4 (10,000 with one significant figure)

1.0×10^4 ($1\bar{0}$,000 with two significant figures)

1.00×10^4 ($1\overline{0,0}00$ with three significant figures)

1.000×10^4 ($1\overline{0,00}0$ with four significant figures)

1.0000×10^4 ($1\overline{0,000}$ with five significant figures)

Most scientists use the preceding scientific notation method for designating significant figures in an "ambiguous" number instead of the overbar notation discussed in Section 2.5. The overbar notation is used only in those situations where there is reason for not expressing the number in scientific notation.

Converting from Scientific to Decimal Notation

To convert a number in scientific notation, such as 6.02×10^{23}, into a regular decimal number, we start by examining the exponent. The value of the exponent tells how many places the decimal point must be moved. If the exponent is positive, movement is to the right to give a number greater than one; if it is negative, movement is to the left to give a number less than one. Zeros may have to be added to the number as the decimal point is moved.

EXAMPLE 2.11

Expressing Scientific Notation Numbers in Decimal Notation

Without use of a calculator, convert the scientific notation number in each of the following statements to a decimal number.

(a) The announced attendance at a football game was 5.3127×10^4.
(b) The circumference of the Earth is 2.5×10^4 miles.
(c) The concentration of gold in seawater is 1.1×10^{-8} gram per liter.
(d) The mass of a hydrogen atom is 1.67×10^{-24} gram.

SOLUTION

(a) The exponent $+4$ tells us the decimal is to be located four places to the right of where it is in 5.3127.

$$5.3127$$
Decimal point shift

The decimal number is 53,127.

(b) The exponent $+4$ tells us the decimal is to be located four places to the right of where it is in 2.5. Trailing zeros will have to be added to accommodate the decimal point change.

         Added zeros
$$2.5000$$
Decimal point shift

These added "trailing zeros" are not significant zeros. Thus, the number of significant digits remains at two. The decimal form of the number is

$$25,000$$

(c) The exponent -8 tells us the decimal is to be located eight places to the left of where it is in 1.1. Leading zeros will have to be added to accommodate the decimal point change.

Added zeros ⟶

$$0.00000001.1$$

Decimal point shift

These added "leading zeros" are not significant zeros. Thus, the number of significant digits remains at two. The decimal form of the number is

$$0.000000011$$

(d) The exponent -24 tells us the decimal is to be located 24 places to the left of where it is in 1.67. This will produce an extremely small number.

$$0.0000000000000000000000001.67$$

Decimal point shift

Twenty-three leading zeros were needed to mark the new decimal place. (In numbers with negative exponents, the number of added leading zeros will always be one less than the value of the exponent.) The decimal form of this number is, thus,

$$0.00000000000000000000000167$$

The number of significant figures present is three, the same number as in the original scientific notation form of the number.

Practice Exercise 2.11

Without the use of a calculator, convert the scientific notation number in each of the following statements to a decimal number.

(a) When the supersonic transport (SST) airplane flew it consumed about 1.8×10^4 liters of fuel per hour of flight.

(b) Approximately 4.5×10^{10} aspirin tablets are consumed annually in the United States.

(c) The naked eyes can detect an object that is 1×10^{-4} meter in diameter.

(d) The diameter of an influenza virus is 1×10^{-8} meter.

2.8 Mathematical Operations in Scientific Notation

A major advantage of writing numbers in scientific notation is that it greatly simplifies the mathematical operations of multiplication and division.

Multiplication in Scientific Notation

Multiplication of two or more numbers expressed in scientific notation involves two separate operations or steps.

STEP 1   Multiply the coefficients (the decimal numbers between 1 and 10) together in the usual manner.

STEP 2   *Add* algebraically the exponents of the powers of ten to obtain a new exponent.

> An electronic calculator combines steps 1 and 2 into one operation.

In general terms, we can represent the multiplication of two scientific notation numbers as follows.

$$(a \times 10^x) \times (b \times 10^y) = ab \times 10^{x+y}$$

EXAMPLE 2.12

Multiplication of Scientific Notation Numbers

Carry out the following multiplications in scientific notation. Be sure to take into account significant figures in obtaining your final answer.

(a) $(1.113 \times 10^3) \times (7.200 \times 10^5)$ **(b)** $(2.05 \times 10^{-3}) \times (1.19 \times 10^{-7})$

(c) $(4.21 \times 10^{-9}) \times (2.107 \times 10^6)$ **(d)** $(7.92 \times 10^{10}) \times (2.3 \times 10^{-4})$

SOLUTION

(a) Multiplying the two coefficients together gives

$$1.113 \times 7.200 = 8.0136$$

Since each of the input numbers for the multiplication has four significant figures, the answer should also have four significant figures, not five as given by the calculator. With rounding,

8.0136	becomes	8.014
(calculator answer)		(correct answer)

> An electronic calculator is not programmed to take into account significant figures. It cannot completely substitute for your brain.

Multiplication of the two powers of ten to give the exponential part of the answer requires adding the exponents to give a new exponent.

$$10^3 \times 10^5 = 10^{3+5} = 10^8$$

Combining the new coefficient with the new exponential term gives the answer

$$8.014 \times 10^8$$

(b) Multiplying the two coefficients together gives

$$2.05 \times 1.19 = 2.4395$$

Since both input numbers for the multiplication contain three significant figures, the calculator answer must be rounded to three significant figures.

2.4395	becomes	2.44
(calculator answer)		(correct answer)

Next, the exponents of the powers of ten are added to generate the new power of ten.

$$10^{-3} \times 10^{-7} = 10^{(-3)+(-7)} = 10^{-10}$$

(To add two numbers of the same sign, either positive or negative, just add the numbers and place the common sign in front of the sum.) Combining the coefficient and the exponential term gives the answer of

$$2.44 \times 10^{-10}$$

(c) Multiplying the two coefficients together gives

$$4.21 \times 2.107 = 8.87047$$

Because the number 4.21 contains only three significant figures, the answer is also limited to three significant figures. Thus,

8.87047	becomes	8.87
(calculator answer)		(correct answer)

In combining the exponential terms, we will have to add exponents with different signs, -9 and $+6$. To do this, we first determine the larger number (9 is larger than 6) and then subtract the smaller number from it ($9 - 6 = 3$). The sign is always the sign of the larger number (minus in this case). Thus,

$$(-9) + (+6) = (-3)$$

and

$$10^{-9} \times 10^6 = 10^{(-9)+(+6)} = 10^{-3}$$

Combining the coefficient and the exponential term gives

$$8.87 \times 10^{-3}$$

(d) Multiplying the coefficients gives

$$7.92 \times 2.3 = 18.216$$

The answer must be rounded to two significant figures because the input number 2.3 has only two significant figures.

18.216	becomes	18
(calculator answer)		(correct answer)

Again, in combining the exponential terms, we have exponents of different signs. The smaller exponent (4) is subtracted from the larger exponent (10), and the sign of the larger exponent ($+$) is used. Thus,

$$(+10) + (-4) = (+6)$$

and

$$10^{10} \times 10^{-4} = 10^6$$

Combining the coefficient and the exponential term gives

$$18 \times 10^6$$

This answer has something wrong with it; it is not in correct scientific notation form. The coefficient should be a number between 1 and 10. This problem is corrected by recognizing that 18 is equal to 1.8×10^1, making this a substitution for 18, and then combining exponential terms.

An electronic calculator automatically makes this coefficient–exponent adjustment for you.

$$18 \times 10^6 = 1.8 \times 10^1 \times 10^6 = 1.8 \times 10^7$$

The correct answer is 1.8×10^7.

Practice Exercise 2.12

Carry out the following multiplications in scientific notation. Be sure to take into account significant figures in obtaining your final answer.

(a) $(2.543 \times 10^3) \times (2.003 \times 10^6)$

(b) $(1.15 \times 10^{-2}) \times (4.52 \times 10^{-3})$

(c) $(4.210 \times 10^{-5}) \times (2.0 \times 10^2)$

(d) $(5.329 \times 10^{-1}) \times (3.11 \times 10^9)$

Division in Scientific Notation

Division of two numbers expressed in scientific notation involves two separate operations or steps.

STEP 1 Divide the coefficients (the decimal numbers between 1 and 10) in the usual manner.

STEP 2 *Subtract* algebraically the exponent in the denominator (bottom) from the exponent in the numerator (top) to give the exponent of the new power of ten.

Note that in multiplication we add exponents, and in division we subtract exponents.

EXAMPLE 2.13

Division of Scientific Notation Numbers

Carry out the following divisions in scientific notation. Be sure to take significant figures into account in obtaining your final answer.

(a) $\dfrac{2.05 \times 10^5}{1.19 \times 10^3}$ **(b)** $\dfrac{7.200 \times 10^{-3}}{1.113 \times 10^{-7}}$

(c) $\dfrac{4.21 \times 10^{-9}}{2.107 \times 10^6}$ **(d)** $\dfrac{3.92 \times 10^{10}}{9.1 \times 10^{-4}}$

SOLUTION

(a) Performing the indicated division involving the coefficients gives

$$\frac{2.05}{1.19} = 1.722689$$

Since both input numbers for the division have three significant figures, the calculator answer must be rounded off to three significant figures.

$$1.722689 \quad \text{becomes} \quad 1.72$$
$$\text{(calculator answer)} \qquad \text{(correct answer)}$$

Dividing exponential terms involves the algebraic subtraction of exponents.

$$\frac{10^5}{10^3} = 10^{(5)-(+3)} = 10^2$$

Algebraic subtraction involves changing the sign of the number to be subtracted and then following the rules for addition (as outlined in Example 2.12). In this problem the number to be subtracted ($+3$) becomes, upon changing the sign, (-3). Then we add $(+5)$ and (-3). The answer is $(+2)$, as shown in the preceding equation. Combining the coefficient and the exponential term gives

$$1.72 \times 10^2$$

(b) The new coefficient is obtained by dividing 1.113 into 7.200:

$$\frac{7.200}{1.113} = 6.4690026$$

The correct answer will contain four significant figures, the same number as in both input numbers.

6.4690026   becomes   6.469
(calculator answer)   (correct answer)

The exponential part of the answer is obtained by subtracting -7 from -3. Changing the sign of the number to be subtracted gives $+7$. Adding $+7$ and -3 gives $+4$. Therefore,

$$\frac{10^{-3}}{10^{-7}} = 10^{(-3)-(-7)} = 10^4$$

Combining the coefficient and the exponential term gives

$$6.469 \times 10^4$$

(c) One input number for the coefficient division has three significant figures and the other has four significant figures. Therefore, the answer should contain three significant figures.

$$\frac{4.21}{2.107} = 1.9981015$$

1.9981015   becomes   2.00
(calculator answer)   (correct answer)

The exponential term division involves subtracting $+6$ from -9. Changing the sign of the number to be subtracted gives a -6. Adding -6 and -9 gives -15.

$$\frac{10^{-9}}{10^6} = 10^{(-9)-(+6)} = 10^{-15}$$

Combining the two parts of the problem gives the number

$$2.00 \times 10^{-15}$$

(d) The new coefficient, obtained by dividing 3.92 by 9.1, should contain two significant figures, the same number as in the input number 9.1.

$$\frac{3.92}{9.1} = 0.43076923$$

0.43076923   becomes   0.43
(calculator answer)   (correct answer)

Performing the exponential term division by subtracting the powers of the exponential terms gives

$$\frac{10^{10}}{10^{-4}} = 10^{(+10)-(-4)} = 10^{14}$$

Combining the coefficient and the exponential term gives

$$0.43 \times 10^{14}$$

which is not in correct scientific notation form because the coefficient is a number less than one. This problem is remedied by recognizing that 0.43 is equal to 4.3×10^{-1}, making this substitution for 0.43, and then combining the two exponential terms.

$$0.43 \times 10^{14} = (4.3 \times 10^{-1}) \times 10^{14} = 4.3 \times 10^{13}$$

An electronic calculator automatically makes this coefficient–exponent adjustment for you.

The correct answer is, thus, 4.3×10^{13}.

Practice Exercise 2.13

Carry out the following divisions in scientific notation. Be sure to take significant figures into account in obtaining your final answer.

(a) $\dfrac{3.76 \times 10^9}{1.23 \times 10^6}$ (b) $\dfrac{9.98 \times 10^{-3}}{2.341 \times 10^{-7}}$

(c) $\dfrac{5.1 \times 10^{-10}}{8.76 \times 10^7}$ (d) $\dfrac{3.43 \times 10^5}{3.93 \times 10^{-4}}$

Addition and Subtraction in Scientific Notation

To add or subtract numbers written in scientific notation, *the power of ten for all numbers must be the same*. More often than not, one or more exponents must be adjusted. Adjusting the exponent requires rewriting the number in a form where the coefficient is a number greater than 10 or less than 1. With exponents all the same, *the coefficients are then added or subtracted and the exponent is maintained at its now common value*. Although any of the exponents may be changed, changing the smaller exponent to a larger one will usually produce a coefficient in the answer that is a number between 1 and 10.

EXAMPLE 2.14

Addition and Subtraction of Scientific Notation Numbers

Perform the following additions or subtractions with all numbers expressed in scientific notation. Answers will need to be checked for the correct number of significant figures and for the correct scientific notation form (coefficient is a number between 1 and 10).

(a) $(2.661 \times 10^3) + (3.011 \times 10^3)$ (b) $(2.66 \times 10^4) - (1.03 \times 10^3)$

(c) $(9.98 \times 10^{-3}) + (8.04 \times 10^{-5})$

SOLUTION

(a) The exponents are the same to begin with. Therefore, we can proceed with the addition immediately.

$$\begin{aligned} 2.661 &\times 10^3 \\ \underline{3.011} &\times 10^3 \\ 5.672 &\times 10^3 \ \text{(calculator and correct answer)} \end{aligned}$$

Both input numbers have uncertainties in the thousandths place. The calculator answer has the same uncertainty. Thus, the calculator answer and correct answer are the same. Note that the exponent values are not added. They are maintained at their common value.

(b) The exponents are different, so before subtracting we must change one of the exponents. Let us change 10^3 to 10^4:

$$10^3 \quad \text{can be written as} \quad 10^{-1} \times 10^4$$

Then, by substitution, we have

$$1.03 \times 10^3 = 1.03 \times 10^{-1} \times 10^4$$

The coefficient and the first exponent are then combined to give a new coefficient.

$$1.03 \times 10^{-1} \times 10^4 = 0.103 \times 10^4$$

We are now ready to make the subtraction called for in the original statement of the problem.

$$
\begin{aligned}
2.66 \times 10^4 = \quad & 2.66 \ \times 10^4 \\
-1.03 \times 10^3 = & -0.103 \times 10^4 \\
\hline
& 2.557 \times 10^4 \quad \text{(calculator answer)}
\end{aligned}
$$

common exponent

The calculator answer must be adjusted for significant figures. On the common exponent basis of 10^4, the uncertainty in 2.66 lies in the hundredths place and that in 0.103 lies in the thousandths place. The correct answer, therefore, is limited to an uncertainty of hundredths. (Recall, from Example 2.8, the rules on addition and significant figures.) Thus,

$$2.557 \times 10^4 \quad \text{becomes} \quad 2.56 \times 10^4 \quad \text{(correct answer)}$$

(c) The exponents are 10^{-3} and 10^{-5}. Since -5 is smaller than -3, let us have 10^{-3} as the common exponent. (Always use the larger of the two exponents as the common exponent.)

$$10^{-5} \quad \text{can be rewritten as} \quad 10^{-2} \times 10^{-3}$$

Then by substitution we have

$$8.04 \times 10^{-5} = 8.04 \times 10^{-2} \times 10^{-3}$$

The coefficient and the first exponent are then combined to give a new coefficient.

$$8.04 \times 10^{-2} \times 10^{-3} = 0.0804 \times 10^{-3}$$

We are now ready to make the addition called for in the original statement of the problem.

$$9.98 \times 10^{-3} = 9.98 \quad \times 10^{-3}$$

common exponent

$$8.04 \times 10^{-5} = 0.0804 \times 10^{-3}$$
$$10.0604 \times 10^{-3} \quad \text{(calculator answer)}$$

The calculator answer must be adjusted for significant figures and also changed into correct scientific notation since the coefficient has a value greater than 10. The significant figure adjustment rounds the answer to hundredths, giving

$$10.06 \times 10^{-3}$$

The correct scientific notation adjustment involves rewriting 10.06 as a power of ten and then simplifying the resulting expression.

$$10.06 \times 10^{-3} = 1.006 \times 10^{1} \times 10^{-3} = 1.006 \times 10^{-2} \text{ (correct answer)}$$

Practice Exercise 2.14

Perform the following additions and subtractions with all numbers expressed in scientific notation. Answers will need to be checked for the correct number of significant figures and for the correct scientific notation form (coefficient is a number between 1 and 10).

(a) $(2.723 \times 10^{4}) + (6.045 \times 10^{4})$ **(b)** $(3.73 \times 10^{3}) - (3.73 \times 10^{2})$

(c) $(9.97 \times 10^{-2}) + (5.89 \times 10^{-4})$

Summary

1. **Exact and Inexact Numbers** Numbers are of two kinds: exact and inexact. An exact number has a value that has no uncertainty associated with it. Exact numbers occur in definitions, in counting, and in simple fractions. An inexact number has a value that has a degree of uncertainty associated with it. Inexact numbers result anytime a measurement is made.

2. **Precision and Accuracy of Measurements** Precision refers to how close a series of measurements on the same object are to each other. Accuracy refers to how close a measurement (or the average of multiple measurements) comes to a true or accepted value. The precision and accuracy of a measurement depend not only on the measuring device used but also on the technical skill of the person making the measurement.

3. **Uncertainty in Measurements** Every measurement has a degree of uncertainty associated with it. In reading a measurement scale, all digits known for certain are recorded plus one estimated digit (uncertainty). It is wrong to record more than one estimated digit in a measurement.

4. **Significant Figures** Significant figures in a measurement are those digits that are certain, plus a last digit that has been estimated. The maximum number of significant figures possible in a measurement is determined by the design of the measuring device. All nonzero digits in a measurement value are always significant. Zeros in a measurement value may or may not be significant depending on their placement within the measurement value.

5. **Significant Figures and Mathematical Operations** Calculations should never increase (or decrease) the degree of uncertainty in measurements. In multiplication and division, the number of significant figures in the answer is the same as that in the input number containing the fewest significant figures. In addition and subtraction of a series of measurements, the result can be no more certain than the least-certain measurement in the series.

6. **Scientific Notation** Scientific notation is a system for writing decimal numbers in a more compact form that greatly simplifies the mathematical operations of multiplication and division. In this system, numbers are expressed as the product of a number between 1 and 10 and 10 raised to a power.

Key Terms

The new terms defined in this chapter are

accuracy   *Sec. 2.3*
exact number   *Sec. 2.2*
exponent   *Sec. 2.7*
inexact number   *Sec. 2.2*
measurement   *Sec. 2.1*

order of magnitude   *Sec. 2.7*
precision   *Sec. 2.3*
random error   *Sec. 2.3*
rounding off   *Sec. 2.6*
scientific notation   *Sec. 2.7*
significant figures   *Sec. 2.5*
systematic error   *Sec. 2.3*

Practice Problems

Exact and Inexact Numbers (Sec. 2.2)

2.1 Indicate whether the number in each of the following statements is an *exact* or an *inexact* number.

(a) A classroom contains 24 chairs.

(b) There are 60 seconds in a minute.

(c) A bag of cherries weighs 4.1 pounds.

(d) A newspaper article contains 421 words.

2.2 Indicate whether the number in each of the following statements is an *exact* or an *inexact* number.

(a) A classroom contains 44 students.

(b) The car is traveling at a speed of 63 miles per hour.

(c) The temperature on the beach is 93°F.

(d) The child is 6 years old.

2.3 Classify each of the following as an *exact* or an *inexact* number.

(a) 7 railroad cars   **(b)** 14 gallons of gasoline

(c) 547 marbles   **(d)** $23.54

2.4 Classify each of the following as an *exact* or an *inexact* number.

(a) 25 pounds of sugar   **(b)** 12 dozen apples

(c) $23.00   **(d)** 25 watermelon seeds

Accuracy and Precision (Sec. 2.3)

2.5 With a high-grade volumetric measuring device, the volume of a liquid sample is determined to be 6.321 L (liters). Three students are asked to determine the volume of the same liquid sample using a lower-grade measuring device. How do you evaluate the following work of the three students with regard to precision and accuracy?

	Students		
Trials	A	B	C
1	6.35 L	6.31 L	6.36 L
2	6.31 L	6.32 L	6.36 L
3	6.38 L	6.33 L	6.35 L
4	6.32 L	6.32 L	6.36 L

2.6 With a high-grade measuring device, the length of an object is determined to be 13.452 mm (millimeters). Three students are asked to determine the length of the same object using a lower-grade measuring device. How do you evaluate the following work of the three students with regard to precision and accuracy?

	Students		
Trials	A	B	C
1	13.6 mm	13.4 mm	13.9 mm
2	13.9 mm	13.5 mm	13.9 mm
3	13.3 mm	13.5 mm	13.3 mm
4	13.6 mm	13.4 mm	14.3 mm

Uncertainty in Measurements (Sec. 2.4)

2.7 Indicate to what uncertainty readings should be recorded (nearest 0.1, nearest 0.01, etc.) for measurements made with the following devices.

(a) a thermometer with smallest scale marking of 1 degree

(b) a cup for measuring volume with smallest scale marking of 1 fluid ounce

(c) a volumetric device with smallest scale marking of 10 milliliters

(d) a ruler with smallest scale marking of 1 millimeter

2.8 Indicate to what uncertainty readings should be recorded (nearest 0.1, nearest 0.01, etc.) for measurements made with the following devices.

(a) a ruler with smallest scale marking of 1 centimeter

(b) a protractor with smallest scale marking of 1 degree

(c) a thermometer with smallest scale marking of 10 degrees

(d) a graduated cylinder with smallest scale marking of 0.1 milliliter

2.9 What is the difference in meaning between the times 3.3 seconds and 3.30 seconds?

2.10 What is the difference in meaning between the lengths 0.54 inch and 0.540 inch?

2.11 Which of the following measurements are consistent with the uncertainty of the thermometers in Figure 2.2?

(a) thermometer A: 36.72°, 42.1°, 39°, 61.5°

(b) thermometer B: 35.03°, 45.1°, 62°, 47.98°

2.12 Which of the following measurements are consistent with the uncertainty of the thermometers in Figure 2.2?

(a) thermometer A: 16.7°, 16.83°, 42°, 35.4°

(b) thermometer B: 75.3°, 65.30°, 65.03°, 57°7

Significant Figures (Sec. 2.5)

2.13 Determine the number of significant figures in each of the following measured values.

(a) 0.00043   **(b)** 0.0220022
(c) 0.30303030   **(d)** 0.03030303

2.14 Determine the number of significant figures in each of the following measured values.

(a) 0.111101   **(b)** 0.0000007
(c) 0.013013013   **(d)** 0.130130130

2.15 Determine the number of significant figures in each of the following measured values.

(a) 4700   **(b)** 37,540
(c) 67.010   **(d)** 3000.00

2.16 Determine the number of significant figures in each of the following measured values.

(a) 4000   **(b)** 4.000
(c) 67,000,100   **(d)** 43,200

2.17 Determine the number of significant figures in each of the following measured values.

(a) 3031.02   **(b)** 3.0030
(c) 0.706050   **(d)** 46,000,300

2.18 Determine the number of significant figures in each of the following measured values.

(a) 113.00   **(b)** 2002
(c) 0.00500500   **(d)** 1,350,000

2.19 For each of the numbers in Problem 2.17, tell how many of the zeros present are

(a) confined zeros.
(b) leading zeros.
(c) trailing zeros that are significant.
(d) trailing zeros that are not significant.

2.20 For each of the numbers in Problem 2.18, tell how many of the zeros present are

(a) confined zeros.
(b) leading zeros.
(c) trailing zeros that are significant.
(d) trailing zeros that are not significant.

2.21 Identify the *estimated digit* in each of the measured values in Problem 2.17.

2.22 Identify the *estimated digit* in each of the measured values in Problem 2.18.

2.23 What is the magnitude of the uncertainty (±10, ±0.1, etc.) associated with each of the measured values in Problem 2.17?

2.24 What is the magnitude of the uncertainty (±10, ±0.1, etc.) associated with each of the measured values in Problem 2.18?

2.25 In the following pairs of numbers, tell whether both members of the pair contain the same number of significant figures.

(a) 11.01 and 11.00   **(b)** 2002 and 2020
(c) 0.05700 and 0.05070   **(d)** 0.000066 and 660,000

2.26 In the following pairs of numbers, tell whether both members of the pair contain the same number of significant figures.

(a) 2305 and 2350   **(b)** 0.6600 and 0.0066
(c) 23,000 and 23,001   **(d)** 936,000 and 0.000936

2.27 In the pairs of numbers of Problem 2.25, tell whether both members of the pair have the same uncertainty.

2.28 In the pairs of numbers of Problem 2.26, tell whether both members of the pair have the same uncertainty.

2.29 Using standard arithmetic notation and overbars (if needed), write the number twenty-three thousand in a manner such that it has the following numbers of significant figures.

(a) 2   **(b)** 4   **(c)** 6   **(d)** 8

2.30 Using standard arithmetic notation and overbars (if needed), write the number six hundred thousand in a manner such that it has the following numbers of significant figures.

(a) 1   **(b)** 3   **(c)** 5   **(d)** 7

2.31 Determine the number of significant figures in each of the following measured values.

(a) 2600 ± 10 **(b)** 1.375 ± 0.001
(c) 42 ± 1 **(d)** $73,000 \pm 1$

2.32 Determine the number of significant figures in each of the following measured values.

(a) 700 ± 100 **(b)** 700 ± 10
(c) 43.57 ± 0.01 **(d)** $64,000 \pm 1$

Rounding Off (Sec. 2.6)

2.33 Round off each of the following numbers to the number of significant figures indicated in parentheses.

(a) 0.350763 (three)   **(b)** 653.899 (four)
(c) 22.55555 (five)   **(d)** 0.277654 (four)

2.34 Round off each of the following numbers to the number of significant figures indicated in parentheses.

(a) 3883 (two)   **(b)** 0.00003011 (two)
(c) 4.4050 (three)   **(d)** 2.1000 (three)

2.35 Round off the number 3.6305023 to the indicated number of significant figures.

(a) seven   **(b)** five   **(c)** four   **(d)** three

2.36 Round off the number 4.7205059 to the indicated number of significant figures.

(a) seven   **(b)** five   **(c)** four   **(d)** three

2.37 Round off the number 30,427.29 to the indicated number of significant figures.

(a) six   **(b)** five   **(c)** four   **(d)** two

2.38 Round off the number 50,125.09 to the indicated number of significant figures.

(a) six   (b) five   (c) four   (d) two

2.39 Using proper rounding techniques, decrease by two the number of significant figures in each of the following numbers.

(a) 0.03455   (b) 2.5003
(c) 1,456,000   (d) 100.0

2.40 Using proper rounding techniques, decrease by two the number of significant figures in each of the following numbers.

(a) 0.50505   (b) 2,000,567
(c) 2.335   (d) 1234.5

2.41 Rewrite each of the following numbers so that it contains two significant figures.

(a) 0.123   (b) 123,000   (c) 12.3   (d) 0.000123

2.42 Rewrite each of the following numbers so that it contains two significant figures.

(a) 21.000   (b) $21\overline{0,0}00$ (c) 0.0210   (d) 2.100

Significant Figures in Multiplication and Division (Sec. 2.6)

2.43 Without actually solving the problems, indicate the number of significant figures that should be present in the answers to the following multiplications and divisions. Assume that all numbers are measured quantities.

(a) $4.5 \times 4.05 \times 4.50$ (b) $0.100 \times 0.001 \times 0.010$
(c) $\dfrac{655,000}{6.5500}$ (d) $\dfrac{6.00}{33.000}$

2.44 Without actually solving the problems, indicate the number of significant figures that should be present in the answers to the following multiplications and divisions. Assume that all numbers are measured quantities.

(a) $3.33 \times 3.03 \times 0.0333$
(b) $300,003 \times 20,200 \times 1.33333$
(c) $\dfrac{333,000}{3.33000}$ (d) $\dfrac{0.0666}{1.3457}$

2.45 How many significant figures must the number Q possess, in each case, to make the following mathematical equations valid from a significant-figure standpoint?

(a) $7.312 \times Q = 4.13$ (b) $7.312 \times Q = 0.0022$
(c) $7.312 \times Q = 20.44$ (d) $7.312 \times Q = 0.1100$

2.46 How many significant figures must the number Q possess, in each case, to make the following mathematical equations valid from a significant-figure standpoint?

(a) $94,461 \times Q = 33,003$ (b) $94,461 \times Q = 1.03$
(c) $94,461 \times Q = 0.6200$ (d) $94,461 \times Q = 233,620,000$

2.47 Carry out the following multiplications and divisions, expressing your answers to the correct number of significant figures. Assume that all numbers are measured quantities.

(a) 4.2337×0.00706 (b) 3700×37.00
(c) $\dfrac{5671}{4.44}$ (d) $\dfrac{5.01}{5.07}$

2.48 Carry out the following multiplications and divisions, expressing your answers to the correct number of significant figures. Assume that all numbers are measured quantities.

(a) 350.00×0.00072 (b) $620,000 \times 620.000$
(c) $\dfrac{3554}{2.22}$ (d) $\dfrac{0.000623}{0.000632}$

2.49 Carry out the following mathematical operations, expressing your answers to the correct number of significant figures. Assume that all numbers are measured quantities.

(a) $\dfrac{4.5 \times 6.3}{7.22}$ (b) $\dfrac{5.567 \times 3.0001}{3.45}$
(c) $\dfrac{37 \times 43}{4.2 \times 6.0}$ (d) $\dfrac{112 \times 20}{30 \times 63}$

2.50 Carry out the following mathematical operations, expressing your answers to the correct number of significant figures. Assume that all numbers are measured quantities.

(a) $\dfrac{2.322 \times 4.00}{3.200 \times 6.73}$ (b) $\dfrac{7.403}{3.220 \times 5.000}$
(c) $\dfrac{11.2 \times 11.2}{3.3 \times 6.5}$ (d) $\dfrac{5600 \times 300}{22 \times 97.1}$

Significant Figures in Addition and Subtraction (Sec. 2.6)

2.51 Without actually solving the problems, indicate the uncertainty (tenths, hundredths, etc.) that should be present in the answers to the following additions and subtractions. Assume that all numbers are measured quantities.

(a) $12.1 + 23.1 + 127.01$ (b) $43.65 - 23.7$
(c) $1237.6 + 23 + 0.12$ (d) $4650 + 25 + 200$

2.52 Without actually solving the problems, indicate the uncertainty (tenths, hundredths, etc.) that should be present in the answers to the following additions and subtractions. Assume that all numbers are measured quantities.

(a) $0.06 + 1.32 + 7.901$ (b) $4.72 - 3.908$
(c) $23.6 + 33 + 17.21$ (d) $46,230 + 325 + 45$

2.53 Perform the following additions or subtractions. Report your results to the proper number of significant figures. Assume that all numbers are measured quantities.

(a) $12 + 23 + 127$ (b) $3.111 + 3.11 + 3.1$
(c) $1237.6 + 23 + 0.12$ (d) $43.65 - 23.7$

2.54 Perform the following additions or subtractions. Report your results to the proper number of significant figures. Assume that all numbers are measured quantities.

(a) $237 + 37 + 7$ (b) $4.000 + 4.002 + 4.20$
(c) $235.45 + 37 + 36.4$ (d) $4.111 - 3.07$

2.55 Perform the following additions or subtractions. Report your results to the proper number of significant figures. Assume that all numbers are measured quantities.

(a) 999.0 + 1.7 − 43.7   **(b)** 345 − 6.7 + 4.33

(c) 1200 + 43 + 7   **(d)** 132 − 0.0073

2.56 Perform the following additions or subtractions. Report your results to the proper number of significant figures. Assume that all numbers are measured quantities.

(a) 1237.6 + 1237.4   **(b)** 1237.6 − 1237.4

(c) 23,000 + 457 + 23   **(d)** 3.12 − 0.00007

Significant Figures and Exact Numbers (Sec. 2.6)

2.57 A rubber heel for a man's shoe is found to have a thickness of 1.12 centimeters. What would be the height, in centimeters, of a stack of 13 such identical rubber heels?

2.58 A thumbtack is found to weigh 0.482 gram. What would be the total mass, in grams, of 125 such identical thumbtacks?

2.59 Each of the following calculations involves the numbers 4.3, 230, 20, and 13.00. The numbers 4.3 and 13.00 are measurements; 230 and 20 are exact numbers. Express each answer to the proper number of significant figures.

(a) 4.3 + 230 + 20 + 13.00   **(b)** 4.3 × 230 × 20 × 13.00

(c) 4.3 + 230 − 20 − 13.00   **(d)** $\dfrac{4.3 \times 230}{20 \times 13.00}$

2.60 Each of the following calculations involves the numbers 200, 17, 24, and 40. The numbers 200 and 17 are exact; 24 and 40 are measurements. Express each answer to the proper number of significant figures.

(a) 200 + 17 + 24 + 40   **(b)** 200 × 17 × 24 × 40

(c) 200 − 17 − 24 − 40   **(d)** $\dfrac{200 \times 17}{24 \times 40}$

Exponents and Orders of Magnitude (Sec. 2.7)

2.61 Write each of the following expressions as an exponential term.

(a) 5 × 5 × 5   **(b)** 10 × 10 × 10 × 10

(c) $\dfrac{1}{3 \times 3 \times 3 \times 3}$ **(d)** $\dfrac{1}{10 \times 10 \times 10}$

2.62 Write each of the following expressions as an exponential term.

(a) 7 × 7 × 7 × 7 × 7   **(b)** 10 × 10

(c) $\dfrac{1}{2 \times 2 \times 2 \times 2 \times 2}$ **(d)** $\dfrac{1}{10 \times 10 \times 10 \times 10}$

2.63 Perform the indicated changes on the following exponential terms.

(a) 10^4; increase by three orders of magnitude.

(b) 10^6; decrease by seven orders of magnitude.

(c) 10^{-5}; increase by two orders of magnitude.

(d) 10^{-3}; decrease by four orders of magnitude.

2.64 Perform the indicated changes on the following exponential terms.

(a) 10^2; increase by two orders of magnitude.

(b) 10^3; decrease by four orders of magnitude.

(c) 10^{-4}; increase by three orders of magnitude.

(d) 10^{-1}; decrease by five orders of magnitude.

Scientific Notation (Sec. 2.7)

2.65 For each of the following numbers, will the exponent be positive, negative, or zero when the number is expressed in scientific notation?

(a) 0.0320   **(b)** 321.7

(c) 6.87   **(d)** 63,002

2.66 For each of the following numbers, will the exponent be positive, negative, or zero when the number is expressed in scientific notation?

(a) 0.323   **(b)** 10.23

(c) 623,000   **(d)** 9.003

2.67 For each of the following numbers, by how many places does the decimal point have to be moved in order to express the number in scientific notation?

(a) 0.000300   **(b)** 30.300

(c) 333,000   **(d)** 333

2.68 For each of the following numbers, by how many places does the decimal point have to be moved in order to express the number in scientific notation?

(a) 402.2   **(b)** 6300

(c) 0.11101   **(d)** 0.00000000030

2.69 For each of the following numbers, how many significant figures should be present in the scientific notation form of the number?

(a) 34,120   **(b)** 34.120

(c) 0.01010   **(d)** 1.0012

2.70 For each of the following numbers, how many significant figures should be present in the scientific notation form of the number?

(a) 34,000,000   **(b)** 45.00

(c) 0.00012   **(d)** 0.100010

2.71 How many digits will there be in the coefficient when each of the following numbers is expressed in scientific notation?

(a) 55.00   **(b)** 55,000

(c) 0.0001000   **(d)** 0.10010

2.72 How many digits will there be in the coefficient when each of the following numbers is expressed in scientific notation?

(a) 6.5000   **(b)** 672,000,000

(c) 0.10003   **(d)** 200.01

2.73 Express the following numbers in scientific notation.

(a) 473.2   **(b)** 0.001234

(c) 231.00   **(d)** 231,000,000

2.74 Express the following numbers in scientific notation.

(a) 787.6   **(b)** 0.01798   **(c)** 40.0   **(d)** 675,000

2.75 Using scientific notation, express the number sixty-seven thousand to the following number of significant figures.

(a) 1   (b) 3   (c) 5   (d) 7

2.76 Using scientific notation, express the number six-hundred-and-seventy-four thousand to the following number of significant figures.

(a) 2   (b) 4   (c) 6   (d) 8

2.77 Express the following numbers in decimal notation.

(a) 1.70×10^{-4} (b) 5.73×10^2
(c) 5.550×10^{-1} (d) 1.110×10^{10}

2.78 Express the following numbers in decimal notation.

(a) 3.57×10^{-8} (b) 3.500×10^{-3}
(c) 6.2134×10^3 (d) 2.0200×10^9

2.79 Each of the following numbers is expressed in nonstandard (incorrect) scientific notation. Convert each number to standard (correct) scientific notation.

(a) 342×10^4 (b) 23.6×10^{-4}
(c) 0.0032×10^5 (d) 0.12×10^{-3}

2.80 Each of the following numbers is expressed in nonstandard (incorrect) scientific notation. Convert each number to standard (correct) scientific notation.

(a) 47.23×10^2 (b) 23.60×10^{-2}
(c) 0.100×10^5 (d) 0.023×10^{-3}

Multiplication and Division in Scientific Notation (Sec. 2.8)

2.81 Without using a calculator, carry out the following multiplications of exponential terms.

(a) $10^5 \times 10^3$ (b) $10^{-5} \times 10^{-3}$
(c) $10^5 \times 10^{-3}$ (d) $10^{-5} \times 10^3$

2.82 Without using a calculator, carry out the following multiplications of exponential terms.

(a) $10^7 \times 10^4$ (b) $10^{-7} \times 10^{-4}$
(c) $10^7 \times 10^{-4}$ (d) $10^{-7} \times 10^4$

2.83 Carry out the following multiplications, making sure that your answer is expressed in correct scientific notation form and to the correct number of significant figures.

(a) $(1.171 \times 10^6) \times (2.555 \times 10^2)$
(b) $(5.37 \times 10^{-3}) \times (1.7 \times 10^5)$
(c) $(9.0 \times 10^{-5}) \times (3.000 \times 10^{-5})$
(d) $(3.0 \times 10^5) \times (9.000 \times 10^5)$

2.84 Carry out the following multiplications, making sure that your answer is expressed in correct scientific notation form and to the correct number of significant figures.

(a) $(2.340 \times 10^{-3}) \times (2.60 \times 10^6)$
(b) $(1.110 \times 10^5) \times (3.333 \times 10^{-7})$

(c) $(9.8 \times 10^2) \times (7.00 \times 10^2)$
(d) $(8.77 \times 10^{-6}) \times (5.030 \times 10^2)$

2.85 Without using a calculator, carry out the following divisions of exponential terms.

(a) $\dfrac{10^5}{10^3}$ (b) $\dfrac{10^5}{10^{-3}}$ (c) $\dfrac{10^{-5}}{10^3}$ (d) $\dfrac{10^{-5}}{10^{-3}}$

2.86 Without using a calculator, carry out the following divisions of exponential terms.

(a) $\dfrac{10^2}{10^3}$ (b) $\dfrac{10^2}{10^{-3}}$ (c) $\dfrac{10^{-2}}{10^3}$ (d) $\dfrac{10^{-2}}{10^{-3}}$

2.87 Carry out the following divisions, making sure that your answer is expressed in correct scientific notation form and to the correct number of significant figures.

(a) $\dfrac{9.51167 \times 10^{-2}}{3.32 \times 10^{-3}}$ (b) $\dfrac{4.500 \times 10^{10}}{5.0005 \times 10^{-8}}$

(c) $\dfrac{3.32 \times 10^{-3}}{9.51167 \times 10^{-2}}$ (d) $\dfrac{5.0005 \times 10^{-8}}{4.500 \times 10^{10}}$

2.88 Carry out the following divisions, making sure that your answer is expressed in correct scientific notation form and to the correct number of significant figures.

(a) $\dfrac{3.5608 \times 10^3}{5.71 \times 10^5}$ (b) $\dfrac{3.300 \times 10^{-5}}{4.0003 \times 10^2}$

(d) $\dfrac{5.71 \times 10^5}{3.5608 \times 10^3}$ (d) $\dfrac{4.0003 \times 10^2}{3.300 \times 10^{-5}}$

2.89 Without using a calculator, carry out the following mathematical operations involving exponential terms.

(a) $\dfrac{10^2 \times 10^3}{10^4}$ (b) $\dfrac{10^{-2} \times 10^{-3}}{10^{-4}}$

(c) $\dfrac{10^6}{10^{-5} \times 10^{-9}}$ (d) $\dfrac{10^{-3} \times 10^2 \times 10^5}{10^{-6} \times 10^8}$

2.90 Without using a calculator, carry out the following mathematical operations involving exponential terms.

(a) $\dfrac{10^4 \times 10^5}{10^6 \times 10^3}$ (b) $\dfrac{10^{-3} \times 10^{-3} \times 10^{-3}}{10^{-6}}$

(c) $\dfrac{10^2 \times 10^3 \times 10^4}{10^{-2} \times 10^{-3} \times 10^{-4}}$ (d) $\dfrac{10^{-6} \times 10^4}{10^3 \times 10^{-5}}$

2.91 Perform the following mathematical operations. Be sure your answer contains the correct number of significant figures and that it is in correct scientific notation form.

(a) $\dfrac{(6.0 \times 10^3) \times (5.0 \times 10^3)}{2.0 \times 10^7}$

(b) $\dfrac{2.0 \times 10^7}{(6.0 \times 10^3) \times (5.0 \times 10^3)}$

(c) $\dfrac{(3.571 \times 10^{-5}) \times (4.5113 \times 10^{-9})}{(5.10 \times 10^{-6}) \times (3.71300 \times 10^{10})}$

(d) $\dfrac{(5 \times 10^{10}) \times (6.0 \times 10^{7}) \times (3.111 \times 10^{-5})}{(3 \times 10^{3}) \times (4.00 \times 10^{-6})}$

2.92 Perform the following mathematical operations. Be sure your answer contains the correct number of significant figures and that it is in correct scientific notation form.

(a) $\dfrac{(3.00 \times 10^{5}) \times (6.00 \times 10^{3}) \times (5.00 \times 10^{6})}{2.00 \times 10^{7}}$

(b) $\dfrac{4.1111 \times 10^{-3}}{(3.003 \times 10^{-6}) \times (9.8760 \times 10^{-5})}$

(c) $\dfrac{(6 \times 10^{5}) \times (6 \times 10^{-5})}{(3 \times 10^{2}) \times (1 \times 10^{-10})}$

(d) $\dfrac{(3.00 \times 10^{6}) \times (2.7 \times 10^{3}) \times (8.50 \times 10^{3})}{(2.22 \times 10^{2}) \times (8.504 \times 10^{6})}$

Addition and Subtraction in Scientific Notation (Sec. 2.8)

2.93 Carry out the following additions and subtractions, expressing each answer in correct scientific notation form to the correct number of significant figures.

(a) $(3.245 \times 10^{3}) + (1.17 \times 10^{3})$

(b) $(9.870 \times 10^{-2}) - (5.7 \times 10^{-3})$

(c) $(9.356 \times 10^{5}) + (3.27 \times 10^{4})$

(d) $(2.030 \times 10^{4}) - (1.111 \times 10^{3})$

2.94 Carry out the following additions and subtractions, expressing each answer in correct scientific notation form to the correct number of significant figures.

(a) $(5.405 \times 10^{6}) + (3.09 \times 10^{5})$

(b) $(7.777 \times 10^{-1}) - (5.3 \times 10^{-1})$

(c) $(8.219 \times 10^{2}) - (1.901 \times 10^{1})$

(d) $(3.45 \times 10^{3}) + (3.45 \times 10^{2})$

2.95 Carry out the following subtractions, expressing each answer in correct scientific notation form to the correct number of significant figures.

(a) $(8.313 \times 10^{7}) - (6.00 \times 10^{6})$

(b) $(8.313 \times 10^{7}) - (6.00 \times 10^{5})$

(c) $(8.313 \times 10^{7}) - (6.00 \times 10^{4})$

(d) $(8.313 \times 10^{7}) - (6.00 \times 10^{2})$

2.96 Carry out the following subtractions, expressing each answer in correct scientific notation form to the correct number of significant figures.

(a) $(7.431 \times 10^{8}) - (4.00 \times 10^{7})$

(b) $(7.431 \times 10^{8}) - (4.00 \times 10^{6})$

(c) $(7.431 \times 10^{8}) - (4.00 \times 10^{5})$

(d) $(7.431 \times 10^{8}) - (4.00 \times 10^{3})$

Additional Problems

2.97 A person is told that 12 pizzas were ordered for a banquet and that a particular piece of rope is 12 inches long. What is the fundamental difference between the values of 12 in these two pieces of information?

2.98 A person is told that there are 24 apples in a bag and also that a particular circle has a diameter of 24 inches. What is the fundamental difference between the values of 24 in these two pieces of information?

2.99 How many significant figures does each of the following numbers have?

(a) 7770   (b) 7.770

(c) 7.770×10^{-3} (d) 7.770×10^{3}

2.100 How many significant figures does each of the following numbers have?

(a) 0.4500   (b) 4.500

(c) 4.500×10^{-3} (d) 4.500×10^{3}

2.101 In the following pairs of numbers, tell whether both members of the pair contain the same number of significant figures.

(a) 11.0 and 11.00

(b) 600.0 and 6.000×10^{3}

(d) 6300 and 6.3×10^{3}

(d) 0.300045 and 0.345000

2.102 In the following pairs of numbers, tell whether both members of the pair contain the same number of significant figures.

(a) 0.000066 and 660,000

(b) 1.500 and 1.500×10^{2}

(c) 54,000 and 5400.0

(d) 0.05700 and 0.0570

2.103 Using scientific notation, write the measurement 600 pounds with an uncertainty of

(a) ±1 pound   (b) ±0.1 pound

(c) ±10 pounds   (d) ±0.001 pound

2.104 Using scientific notation, write the measurement 1300 miles with an uncertainty of

(a) ±1 mile   (b) ±0.1 mile

(c) ±100 miles   (d) ±0.01 mile

2.105 In the following pairs of numbers, tell whether both members of the pair have an uncertainty of less than ±0.01.

(a) 0.006 and 0.016

(b) 2.700 and 2.700×10^{2}

(c) 3300.00 and $33\overline{00}$

(d) 5.750×10^{1} and 5.750×10^{-1}

2.106 In the following pairs of numbers, tell whether both members of the pair have an uncertainty of less than ±0.01.

(a) 3.71 and 3.50

(b) 4.500 and 4.500 × 10^{-2}

(c) 270.0 and 0.27000

(d) 4.31 × 10^{-2} and 4.31 × 10^{2}

2.107 Write each of the following numbers in scientific notation to the number of significant figures indicated in parentheses.

(a) 632,567 (4) (b) 0.312546 (3)

(c) 63,000,023 (4) (d) 0.500000 (4)

2.108 Write each of the following numbers in scientific notation to the number of significant figures indicated in parentheses.

(a) 0.00300300 (3) (b) 936,000 (2)

(c) 23.5003 (3) (d) 450,000,001 (6)

2.109 How many significant figures must the number Q possess, in each case, to make the following mathematical equations valid from a significant-figure standpoint?

(a) 6.000 × Q = 4.0 (b) $\dfrac{5.000}{Q} = 3.175$

(c) 5.250 + Q = 7.03 (d) 0.7777 − Q = 0.011

2.110 How many significant figures must the number Q possess, in each case, to make the following mathematical equations valid from a significant-figure standpoint?

(a) 450.0 × Q = 3.00 (b) $\dfrac{Q}{5.1256} = 1.703$

(c) 9.13 + Q = 10.2 (d) Q − 0.111 = 9.25

2.111 Perform the following mathematical operations, expressing your answers to the correct number of significant figures. Assume all numbers are measured quantities.

(a) 4.0 × (2.3 + 4.5) (b) 3.0 × (3.7 − 3.4)

(c) 6.0 × (34 − 4.23) (d) 7.02 × (0.0001 + 0.01)

2.112 Perform the following mathematical operations, expressing your answers to the correct number of significant figures. Assume all numbers are measured quantities.

(a) 5.5 × (12.3 − 3.2) (b) 6.0 × (3.7 + 0.001)

(c) 2.0 × (3.43 − 2.415) (d) 3.0 × (37 − 3.7)

2.113 What is wrong with the statement "The number of objects is 12.00 exactly"?

2.114 What is wrong with the statement "Through counting, it was determined that the basket contained 6.70 × 10^{1} peaches"?

2.115 Each of the following calculations contains the numbers 4.2, 5.30, 11, and 28. The numbers 4.2 and 5.30 are measured quantities, and 11 and 28 are exact numbers. Do each calculation and express each answer to the proper number of significant figures.

(a) (4.2 + 5.30) × (28 + 11)

(b) 4.2 × 5.30 × (28 − 11)

(c) $\dfrac{28 - 4.2}{5.30 \times 11}$ (d) $\dfrac{28 - 4.2}{11 - 5.30}$

2.116 Each of the following calculations contains the numbers 3.111, 5.03, 100, and 33. The numbers 3.111 and 5.03 are measured quantities, and 100 and 33 are exact numbers. Do each calculation and express each answer to the proper number of significant figures.

(a) (3.111 + 5.03) × (100 + 33)

(b) 3.111 × 5.03 × (100 + 33)

(c) $\dfrac{3.111 + 5.03}{100 \times 33}$ (d) $\dfrac{5.03 - 3.111}{100 + 33}$

2.117 Arrange the following sets of numbers in ascending order (from smallest to largest).

(a) 2.07 × 10^{2}, 243, 1.03 × 10^{3}

(b) 0.0023, 3.04 × 10^{-2}, 2.11 × 10^{-3}

(c) 23,000, 2.30 × 10^{5}, 9.67 × 10^{4}

(d) 0.00013, 0.000014, 1.5 × 10^{-4}

2.118 Arrange the following sets of numbers in ascending order (from smallest to largest).

(a) 350, 3.51 × 10^{2}, 3.522 × 10^{1}

(b) 0.000234, 2.341 × 10^{-3}, 2.3401 × 10^{-4}

(c) 965,000,000, 9.76 × 10^{8}, 2.03 × 10^{8}

(d) 0.00010, 0.00023, 3.4 × 10^{-2}

Answers to Practice Exercises

2.1 (a) Benjamin; (b) Team A; (c) The term *precision* does not apply to a single count; (d) Team A

2.2 (a) 18; (b) 27.7; (c) 28.55

2.3 (a) 2; (b) 4; (c) 2; (d) 2

2.4 (a) 2, ±10,000; (b) 4, ±100; (c) 5, ±10; (d) 6, ±1

2.5 (a) 433; (b) 432; (c) 655,000; (d) 0.0332; (e) 352; (f) 162,000

2.6 (a) 2; (b) 2; (c) 3; (d) 3

2.7 (a) 0.014; (b) 693.4; (c) 30.00; (d) 6.22

2.8 (a) 58; (b) 11.2; (c) 0.0011; (d) 4200

2.9 491.94 g

2.10 (a) 3.915 × 10^{3}; (b) 9.3 × 10^{7}; (c) 1.5 × 10^{-6}; (d) 4.65 × 10^{-23}

2.11 (a) 18,000; (b) 45,000,000,000; (c); 0.0001; (d) 0.00000001

2.12 (a) 5.094 × 10^{9}; (b) 5.20 × 10^{-5}; (c) 8.4 × 10^{-3}; (d) 1.66 × 10^{9}

2.13 (a) 3.06 × 10^{3}; (b) 4.26 × 10^{4}; (c) 5.8 × 10^{-18}; (d) 8.73 × 10^{8}

2.14 (a) 8.768 × 10^{4}; (b) 3.36 × 10^{3}; (c) 1.003 × 10^{-1}

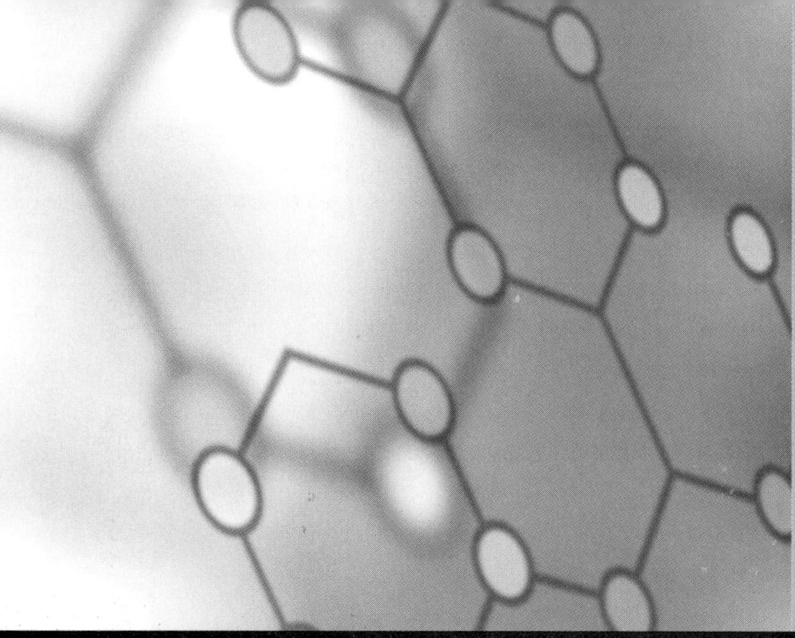

3

Unit Systems and Dimensional Analysis

3.1 The Metric System of Units

All measurements consist of three parts: (1) a number that tells the amount of the quantity measured, (2) an error that produces an amount of uncertainty, and (3) a unit that tells the nature of the quantity being measured. Chapter 2 dealt with the interpretation and manipulation of the number and error parts of a measurement. We now turn our attention to units.

A unit is a "label" that describes (or identifies) what is being measured (or counted). It can be almost anything: quarts, dimes, dozen frogs, bushels, inches, or pages, for example. Having units for a measurement is an absolute necessity. If you were to ask a neighbor to lend you six sugar, the immediate response would be "How much sugar?" You would then have to indicate that you wished to borrow six pounds, six ounces, six cups, six teaspoons, or whatever amount of sugar you needed.

Two formal systems of units of measurement are used in the United States today. Common measurements in commerce—in supermarkets, lumberyards, gas stations, and so on—are made in the *English system*. The units of this system include the familiar inch, foot, pound, quart, and gallon. A second system, the *metric system*, is used in scientific work. Units in this system include the gram, meter, and liter. The United States is one of only a very few countries that use different unit systems in commerce and scientific work. On a worldwide basis, almost universally, the metric system is used in both areas.

Metric system use has become more common in the United States in recent years. Metric units now appear on many consumer products. Road signs in some states display distances in both miles and kilometers, and soft drinks are now sold in 1-, 2-, and 3-liter containers (see Fig. 3.1). Automobile engine sizes are often given in liters. Canned and

Metric measurements are becoming increasingly common in the United States as exemplified by highway mileage signs and consumer products. (Tony Freeman/PhotoEdit; Ken Lax/Simon & Schuster/PH College)

> **What do the United States, Liberia, and Myanmar (formerly Burma) have in common? They are the only three countries left where the English system of units is used in preference to the metric system.**

packaged goods (cereals, mixes, fruits, etc.) on grocery store shelves now have the content masses listed in grams as well as ounces or pounds.

The metric system is superior to the English system. Its superiority lies in the area of interrelationships between units of the same type (volume, length, etc.). Metric unit interrelationships are less complicated than English unit interrelationships because the metric system is a decimal unit system. In the metric system, conversion from one unit size to another can be accomplished simply by moving the decimal point to the right or left an appropriate number of places.

The most recent modification of the metric system is called the International System of Units and is abbreviated SI (after the French name Système International). Since SI units are still not in universal use, we shall use the more traditional metric system in this text.

Because the SI and metric systems are very similar, switching to the SI system sometime in the future, when its use is more extensive, will present no major difficulties to the student who properly understands the traditional metric system.

Metric System Prefixes

In the metric system there is one basic unit for each type of measurement—length, volume, mass, and so on. These basic units are then multiplied by appropriate powers of ten to form smaller or larger units. The names of the larger and smaller units are constructed from the basic unit name by attaching to it a prefix that tells which power of ten is involved. These prefixes are given in Table 3.1, along with their symbols or abbreviations and mathematical meanings. The prefixes in color are those most frequently used.

The use of numerical prefixes should not be new to you. Consider the use of the prefix *tri-* in the following words: triangle, tricycle, trio, trinity, triple. Every one of these words conveys the idea of three of something. We will use the metric system prefixes in the same way.

The meaning of a prefix always remains constant; it is independent of the base unit it modifies. For example, a kilosecond is a thousand seconds; a kilowatt, a thousand watts; and a kilocalorie, a thousand calories. The prefix kilo- will always mean a thousand.

Table 3.1	Metric System Prefixes and Their Mathematical Meanings				
Prefix	**Symbol**	**Mathematical Meaning**	**Pronunciation**	**Word Meaning**
Tera-	T	$1,000,000,000,000 = 10^{12}$	TER-uh	trillion
Giga-	G	$1,000,000,000 = 10^{9}$	GIG-uh	billion
Mega-	M	$1,000,000 = 10^{6}$	MEG-uh	million
Kilo-	k	$1,000 = 10^{3}$	KIL-oh	thousand
Hecto-	h	$100 = 10^{2}$	HEK-toe	hundred
Deca-	da	$10 = 10^{1}$	DEK-uh	ten
		$1 = 10^{0}$		one
Deci-	d	$0.1 = 10^{-1}$	DES-ee	tenth
Centi-	c	$0.01 = 10^{-2}$	SEN-tee	hundredth
Milli-	m	$0.001 = 10^{-3}$	MIL-ee	thousandth
Micro-	μ*	$0.000\,001 = 10^{-6}$	MY-kro	millionth
Nano-	n	$0.000\,000\,001 = 10^{-9}$	NAN-oh	billionth
Pico-	p	$0.000\,000\,000\,001 = 10^{-12}$	PEE-koh	trillionth

This is the Greek letter mu (pronounced "mew," rhymes with "you").

Students often ask if other metric system prefixes exist besides those in Table 3.1. The answer is yes. The other approved prefixes are

10^{15}	peta (P)
10^{18}	exa (E)
10^{21}	zetta (Z)
10^{24}	yotta (Y)
10^{-15}	femto (f)
10^{-18}	atto (a)
10^{-21}	zepto (z)
10^{-24}	yocto (y)

EXAMPLE 3.1

Recognizing the mathematical meanings of metric system prefixes

Write the name of the power of ten associated with the listed metric system prefix or the metric system prefix associated with the listed power of 10.

(a) nano-   **(b)** micro-   **(c)** deci-   **(d)** 10^{-2} **(e)** 10^{6} **(f)** 10^{9}

SOLUTION

(a) *nano-* denotes 10^{-9} (one-billionth)
(b) *micro-* denotes 10^{-6} (one-millionth)
(c) *deci-* denotes 10^{-1} (one-tenth)
(d) 10^{-2} (one-hundredth) is denoted by the prefix *centi-*
(e) 10^{6} (one million) is denoted by the prefix *mega-*
(f) 10^{9} (one billion) is denoted by the prefix *giga-*

Practice Exercise 3.1

Write the name of the power of ten associated with the listed metric system prefix or the metric system prefix associated with the listed power of 10.

(a) milli-   **(b)** pico-   **(c)** mega-   **(d)** 10^{-6} **(e)** 10^{3} **(f)** 10^{-1}

Answers to practice exercises are located at the end of the chapter.

3.2 Metric Units of Length

The **meter** *(m) is the base unit of length in the metric system.* (*Metre* has been adopted as the preferred international spelling for the unit, but *meter* is the spelling used in the United States and in this book.) A meter is about the same size as the English yard unit; 1 meter equals 1.09 yards (Fig. 3.2a). The prefixes listed in Table 3.1 enable us to derive other units of length from the meter. The kilometer (km) is 1000 times larger than the meter; the centimeter (cm) and millimeter (mm) are, respectively, one-hundredth and one-thousandth of a meter. Most laboratory measurements are made in centimeters or millimeters rather than meters because of the meter's relatively large size.

Length is measured by determining the distance between two points.

(a) Length

A meter is slightly longer than a yard.
1 meter = 1.09 yards
A baseball bat is about 1 meter long.

(b) Mass

A gram is a small unit compared to a pound.
1 gram = 1/454 pound
Two pennies have a mass of about 5 grams.

(c) Volume

A liter is slightly larger than a quart.
1 liter = 1.06 quart
Most beverages are now sold by the liter rather than by the quart.

Figure 3.2

Comparisons of the metric base units of length (meter), mass (gram), and volume (liter) with common objects. (Rim light/PhotoDisc, Inc.; PhotoDisc, Inc.; Frank Labua/Simon & Schuster/PH College)

3.3 | Metric Units of Mass

The **gram** *(g) is the base unit of mass in the metric system.* It is a very small unit compared with the English pound and ounce (Fig. 3.2b). It takes approximately 28 grams to equal 1 ounce and nearly 454 grams to equal 1 pound. Both grams and milligrams (mg) are commonly used in the laboratory, where the kilogram (kg) is generally too large.

The terms *mass* and *weight* are frequently used interchangeably in measurement discussions. Although in most cases this practice does no harm, technically it is incorrect to interchange the terms. Mass and weight refer to different properties of matter, and their difference in meaning should be understood.

Mass **is measured by determining the amount of matter in an object. The determination is made using a balance.**

Mass *is a measure of the total quantity of matter in an object.* **Weight** *is a measure of the force exerted on an object by gravitational forces.* The mass of a substance is a constant; the weight of an object is a variable dependent upon the geographical location of that object.

Matter at the equator weighs less than it would at the North Pole because the Earth is not a perfect sphere but bulges at the equator. As a result, an object at the equator is farther from the center of the Earth. It therefore weighs less because the magnitude of gravitational attraction (the measure of weight) is inversely proportional to the distance between the centers of the attracting objects; that is, the gravitational attraction is larger when the objects' centers are closer together and smaller when the objects' centers are farther apart. Gravitational attraction also depends on the masses of the attracting bodies; the greater the masses, the greater the attraction. For this reason, an object would weigh much less on the moon than on Earth because of the smaller size of the moon and the correspondingly lower gravitational attraction. Quantitatively, a 22.0-pound object weighing 22.0 pounds at the Earth's North Pole would weigh 21.9 pounds at the Earth's equator and only 3.7 pounds on the moon. In outer space an astronaut may be weightless but never massless. In fact, he or she has the same mass in space as on Earth (see Fig. 3.3).

3.4 Metric Units of Volume

Before specific metric units of volume are considered, a brief review of how the quantities *area* and *volume* are calculated is in order.

Area *is a measure of the extent of a surface.* The units for area are squared units of length. Common area units include square inches (in.^{2}), square feet (ft^2), square meters (m^2), and square centimeters (cm^2). Note that a squared unit is just that unit multiplied by itself. The unit cm^2 means centimeter $\times$ centimeter in the same way that 3^2 means 3×3. Figure 3.4 lists the formulas needed for determining the areas of commonly encountered geometrical objects. Notice that in each case the formula involves taking the product of two lengths. Note also that the constant pi (π) is needed in calculating the area of a circle. Its value to three, four, and five significant figures, respectively, is 3.14, 3.142, and 3.1416.

Volume *is a measure of the amount of space occupied by an object.* It is a three-dimensional measure and thus involves units that have been cubed: in.^{3}, ft^3, m^3, cm^3, and so on.

Volume **is measured by determining the amount of space occupied by a three-dimensional object.**

Square	Area = side $\times$ side $A = s \times s$ $\quad = s^2$	
Rectangle	Area = length $\times$ width $A = l \times w$	
Circle	Area = $\pi \times$ (radius)^2 $A = \pi \times r^2$	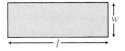$\pi = 3.1416$
Triangle	Area = $\frac{1}{2} \times$ base $\times$ height $A = \frac{1}{2} \times b \times h$	

Figure 3.4

Formulas for calculating the areas of various two-dimensional objects.

Cube
Volume = side × side × side
$V = s \times s \times s$
$= s^3$

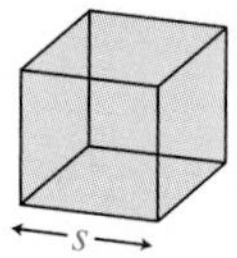

Rectangular solid
Volume = length × width × height
$V = l \times w \times h$
$= lwh$

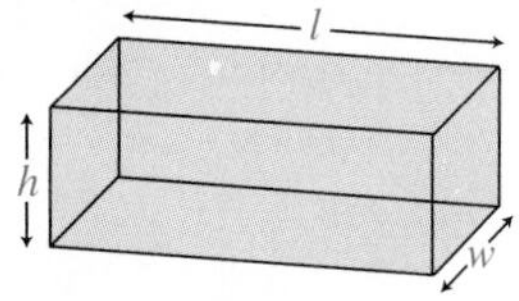

Cylinder
Volume = π × (radius)^2 × height
$V = \pi \times r^2 \times h$
$= \pi r^2 h$
$\pi = 3.1416$ 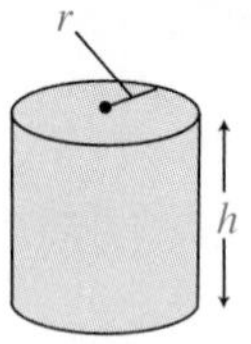

Sphere
Volume = $\frac{4}{3}$ × π × (radius)^3
$V = \frac{4}{3} \times \pi \times r^3$
$= \frac{4}{3} \pi r^3$ 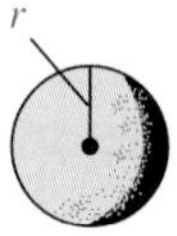
$\pi = 3.1416$

Figure 3.5

Formulas for calculating the volumes of various three-dimensional objects.

Again, a cubed unit is just that unit multiplied by itself three times in the same manner that 3^3 is $3 \times 3 \times 3$. Figure 3.5 lists volume formulas for commonly encountered three-dimensional objects. Note particularly the manner in which the volume of a cube is calculated—side × side × side. This is the key to understanding metric units of volume whose definitions are cube related.

A **liter** *(L) is the base unit of volume in the metric system.* (As with meter, we will use the U.S. spelling rather than the international spelling, which is litre.) The abbreviation for liter is a capital L rather than a lowercase l because a lowercase l is easily confused with the number 1.

A liter is a volume equal to that occupied by a cube that is 10 centimeters on each side. Because the volume of a cube is calculated by multiplying length times width times height (which are all the same for a cube), we have

$$1 \text{ liter } = \text{ volume of a cube with sides of 10 cm}$$
$$= 10 \text{ cm} \times 10 \text{ cm} \times 10 \text{ cm}$$
$$= 1000 \text{ cm}^3$$

A liter is also equal to 1000 milliliters; the prefix *milli-* means one-thousandth. Therefore,

$$1000 \text{ mL } = 1000 \text{ cm}^3$$

Dividing both sides of this equation by 1000 shows that

$$1 \text{ mL } = 1 \text{ cm}^3$$

Consequently, the units milliliter and cubic centimeter are one and the same. In practice, mL is used for volumes of liquids and gases, cm^3 for volumes of solids. Figure 3.6 shows the relationship between 1 mL (1 cm^3) and its parent unit, the liter, in terms of cubic measurements.

A liter and a quart have approximately the same volume; 1 liter equals 1.06 quarts (Fig. 3.2c). The milliliter and deciliter (dL) are commonly used in the laboratory. Deciliter

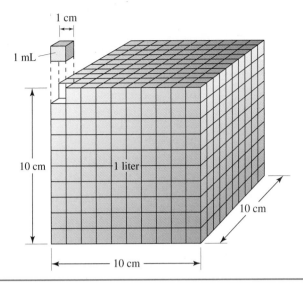

Figure 3.6

The relationship between the units liter and milliliter in terms of cubic centimeters.

units are routinely encountered in clinical laboratory reports detailing the composition of body fluids. A deciliter is equal to 100 mL.

3.5 Units in Mathematical Operations

Just as numbers can have exponents associated with them (3^2, 6^3, 10^2, 10^4, etc.), so also can units have exponents associated with them. In Section 3.4 in the discussion of volume, the unit cubic centimeters (cm^3) was introduced. As noted, cm^3 denotes the unit centimeter multiplied by itself three times ($cm \times cm \times cm$). Other commonly encountered units with exponents include

$in.^2$	square inches
ft^3	cubic feet
cm^2	square centimeters
m^2	square meters

In mathematical problems the powers on units are manipulated in the same manner as powers of ten (Sec. 2.8). Exponents are added during multiplication and subtracted during division. Just as

$$10^2 \times 10 = 10^3$$

so likewise

$$km^2 \times km = km^3$$

Similarly,

$$\frac{10^3}{10^2} = 10$$

and

$$\frac{km^3}{km^2} = km$$

> To avoid confusion with the word *in,* the abbreviation for inches, in., includes a period. This is the only unit abbreviation in which a period appears.

EXAMPLE 3.2

Manipulation of Units in Multiplication and Division

Carry out the following multiplications and divisions of measurements and units:

(a) $2.4 \text{ cm} \times 3.6 \text{ cm}$ **(b)** $3.6 \text{ km}^2 \times 6.0 \text{ km}$

(c) $\dfrac{6.3 \text{ ft}^3}{1.0 \text{ ft}}$ **(d)** $\dfrac{6.0 \text{ in.}^2 \times 3.0 \text{ in.}}{2.0 \text{ in.}}$

SOLUTION

Each of these problems can be considered in two stages: the numerical parts of the measurements are multiplied or divided as indicated, and the units are then similarly combined.

(a) $2.4 \text{ cm} \times 3.6 \text{ cm} = (2.4 \times 3.6) \times (\text{cm} \times \text{cm})$

$$= 8.64 \times \text{cm}^2$$

$$= 8.64 \text{ cm}^2 \text{ (calculator answer)}$$

$$= 8.6 \text{ cm}^2 \text{ (correct answer)}$$

The *input* numbers dictate that the answer contain only two significant figures. For the units, multiplying a unit by itself gives the unit squared.

(b) $3.6 \text{ km}^2 \times 6.0 \text{ km} = (3.6 \times 6.0) \times (\text{km}^2 \times \text{km})$

$$= 21.6 \times \text{km}^3$$

$$= 21.6 \text{ km}^3 \text{ (calculator answer)}$$

$$= 22 \text{ km}^3 \text{ (correct answer)}$$

Note that the exponents on the units km^2 and km^1 were added to give km^3. Only two significant figures are allowed in the numerical answer.

(c) $\dfrac{6.3 \text{ ft}^3}{1.0 \text{ ft}} = \dfrac{6.3}{1.0} \times \dfrac{\text{ft}^3}{\text{ft}}$

$$= 6.3 \times \text{ft}^2$$

$$= 6.3 \text{ ft}^2 \text{ (calculator and correct answer)}$$

In this case, according to the rule for division of exponential terms (Sec. 2.8), the unit exponents were subtracted. No adjustment for significant figures needs to be made.

(d) $\dfrac{6.0 \text{ in.}^2 \times 3.0 \text{ in.}}{2.0 \text{ in.}} = \dfrac{6.0 \times 3.0}{2.0} \times \dfrac{\text{in.}^2 \times \text{in.}}{\text{in.}}$

$$= 9 \times \text{in.}^2$$

$$= 9 \text{ in.}^2 \text{ (calculator answer)}$$

$$= 9.0 \text{ in.}^2 \text{ (correct answer)}$$

In this case the calculator gives too few significant figures.

Practice Exercise 3.2

Carry out the following multiplications and divisions of measurements and units. Remember that significant figures are a consideration for all measurements.

(a) $6.2 \text{ ft} \times 3.7 \text{ ft}$ (b) $3.01 \text{ mm} \times 6.007 \text{ mm}^2$

(c) $\dfrac{5.000 \text{ yd}^3}{2.00 \text{ yd}}$ (d) $\dfrac{7.0 \text{ cm}^2 \times 6.0 \text{ cm}^3}{3.0 \text{ cm}^2}$

3.6 Conversion Factors

With both the English unit and metric unit systems in common use in the United States, we often must change measurements from one system to their equivalent in the other system. The mathematical tool we use to accomplish this task is a general method of problem solving called *dimensional analysis* (Sec. 3.7). Central to the use of dimensional analysis is the concept of *conversion factors*. A **conversion factor** *is a ratio that specifies how one unit of measurement is related to another unit of measurement*.

Conversion factors are derived from equations (equalities) that relate units. Consider the quantities "1 minute" and "60 seconds," both of which describe the same amount of time. We may write an equation describing this fact:

$$1 \text{ min} = 60 \text{ sec}$$

This fixed relationship is the basis for the construction of a pair of conversion factors that relate seconds and minutes:

$$\frac{1 \text{ min}}{60 \text{ sec}} \quad \text{and} \quad \frac{60 \text{ sec}}{1 \text{ min}}$$

Note that conversion factors always come in pairs, one member of the pair being the reciprocal of the other. Also note that the numerator and the denominator of a conversion factor always describe the same amount of whatever we are considering. One minute and 60 seconds denote the same amount of time.

Conversion factors have a value of unity, that is, one. That this is the case can be seen by taking the equation

$$1 \text{ min} = 60 \text{ sec}$$

and dividing each side of it by the quantity "60 seconds," which gives

$$\frac{1 \text{ min}}{60 \text{ sec}} = \frac{60 \text{ sec}}{60 \text{ sec}}$$

Since the numerator and denominator of the fraction on the right are identical, we have

$$\frac{1 \text{ min}}{60 \text{ sec}} = 1$$

In general, we will always be able to construct a set of two conversion factors, each with a value of unity, from any two terms that describe the same "amount" of whatever we are considering. The two conversion factors will always be reciprocals of each other. (Conversion factors can also be constructed from two quantities that are equivalent rather than equal. We will consider this situation in Sec. 3.8.)

A valid conversion factor requires a correct relationship between units. For example, even though the statement

1 minute = 45 seconds

is false, you can foolishly make a conversion factor from it. However, any answers you obtain using this incorrect conversion factor will be incorrect. Correct answers require correct conversion factors, which require correct relationships between units.

Three important categories of conversion factors are (1) English to English, (2) metric to metric, and (3) metric to English or English to metric. Further details concerning these types of conversion factors follow.

English-to-English Conversion Factors

English-to-English conversion factors are used to change a measurement in one English system unit into another English unit. Both the numerator and the denominator of such conversion factors involve English units.

Most students are familiar with and have memorized numerous conversion factors within the English system of measurement. Some of these factors, with only one member of a conversion factor pair listed, are

$$\frac{12 \text{ in.}}{1 \text{ ft}} \qquad \frac{1 \text{ yd}}{3 \text{ ft}} \qquad \frac{1 \text{ gal}}{4 \text{ qt}} \qquad \frac{16 \text{ oz}}{1 \text{ lb}}$$

Such conversion factors are exact; that is, they contain an unlimited number of significant figures because the numbers within them arise from definitions (Sec. 2.2).

Metric-to-Metric Conversion Factors

Metric-to-metric conversion factors are used to change a measurement in one metric unit into another metric unit. Such conversion factors are similar to English-to-English conversion factors in that they arise from definitions and are thus exact.

Individual metric-to-metric conversion factors are derived from the meanings of the metric system prefixes (Table 3.1). For example, the set of conversion factors involving kilometer and meter come from the equality

$$1 \text{ kilometer} = 10^3 \text{ meters}$$

and those relating centigram and gram come from the equality

$$1 \text{ centigram} = 10^{-2} \text{ gram}$$

> **To write metric-to-metric conversion factors, you need to know the meaning of the metric system prefixes in terms of powers of ten (see Table 3.1).**

These two pairs of conversion factors are

$$\frac{10^3 \text{ m}}{1 \text{ km}} \quad \text{and} \quad \frac{1 \text{ km}}{10^3 \text{ m}}$$

$$\frac{10^{-2} \text{ g}}{1 \text{ cg}} \quad \text{and} \quad \frac{1 \text{ cg}}{10^{-2} \text{ g}}$$

A recommended general rule to follow in writing metric-to-metric conversion factors is that the numerical equivalent of the prefix (the power of ten) is always associated with the base (unprefixed) unit and the number one is always associated with the prefixed unit.

The number 1 always goes with the *prefixed* unit.

$$\frac{1 \text{ km}}{10^3 \text{ m}}$$

The power of 10 always goes with the *unprefixed* unit.

Metric-to-English and English-to-Metric Conversion Factors

Conversion factors that relate metric units to English units must be established by measurement since they involve two different unit systems. Such conversion factors are thus *not exact*. Table 3.2 lists commonly used "metric–English" conversion factors. These few conversion factors are sufficient to solve most of the problems that we will encounter.

Because they are not exact, metric-to-English conversion factors can be specified to differing numbers of significant figures. For example,

$$1.00 \text{ lb} = 454 \text{ g}$$

$$1.000 \text{ lb} = 453.6 \text{ g}$$

$$1.0000 \text{ lb} = 453.59 \text{ g}$$

In a problem-solving context, which "version" of a conversion factor is used depends on how many significant figures there are in the other numbers of the problem. *Conversion factors should never limit the number of significant figures in the answer to a problem.* The conversion factors in Table 3.2 are given to four significant figures, which generally is sufficient for the applications we will make of them.

Figure 3.7 summarizes the major concepts about conversion factors considered in this section of the text.

To help avoid confusion about the accuracy of metric–English conversion factors, scientists have agreed on common "experiment-based" *definitions* for English units in terms of metric units. They have also agreed that these definitions should be taken to be *exact*. The three *exact* definitions are

1 in. = 2.540005 cm
1 lb = 453.59237 g
1 qt = 0.94633343 L

Conversion factors obtained from these *exact* definitions are almost always rounded to fewer digits. These rounded conversion factors are not exact.

Table 3.2	Conversion Factors That Relate the English and Metric Systems of Measurement to Each Other

	Factor to Convert from Metric to English Unit
Length	
Inch and centimeter	$\dfrac{1 \text{ in.}}{2.540 \text{ cm}}$
Inch and meter	$\dfrac{1 \text{ in.}}{0.02540 \text{ m}}$
Mile and kilometer	$\dfrac{1 \text{ mi}}{1.609 \text{ km}}$
Mass	
Pound and gram	$\dfrac{1 \text{ lb}}{453.6 \text{ g}}$
Pound and kilogram	$\dfrac{1 \text{ lb}}{0.4536 \text{ kg}}$
Ounce and gram	$\dfrac{1 \text{ oz}}{28.35 \text{ g}}$
Volume	
Quart and liter	$\dfrac{1 \text{ qt}}{0.9463 \text{ L}}$
Pint and liter	$\dfrac{1 \text{ pt}}{0.4732 \text{ L}}$
Fluid ounce and milliliter	$\dfrac{1 \text{ fl oz}}{29.57 \text{ mL}}$

Characteristics of Conversion Factors
• Ratios that specify how units are related to each other
• Derived from equations that relate units 1 minute = 60 seconds
• Come in pairs, one member of the pair being the reciprocal of the other $$\frac{1 \text{ min}}{60 \text{ sec}} \quad \text{and} \quad \frac{60 \text{ sec}}{1 \text{ min}}$$
• Conversion factors originate from two types of relationships: (1) defined relationships   and   (2) measured relationships

Conversion Factors from DEFINED Relationships	Conversion Factors from MEASURED Relationships
• All English-to-English and metric-to-metric conversion factors • Have an unlimited number of significant figures 12 inches = 1 foot (exactly) 4 quarts = 1 gallon (exactly) 1 kilogram = 10^3 grams (exactly) • Metric-to-metric conversion factors are derived using the meanings of the metric system prefixes	• All English-to-metric and metric-to-English conversion factors • Have a specific number of significant figures, depending on the uncertainty in the defining relationship 1.00 lb = 454 g (three sig figs) 1.000 lb = 453.6 g (four sig figs) 1.0000 lb = 453.59 g (five sig figs)

Figure 3.7

English-to-English and metric-to-metric conversion factor are based on defined relationships. Metric-to-English and English-to-metric conversion factors are based on measured relationships.

3.7 Dimensional Analysis

Dimensional analysis *is a general problem-solving method in which the units associated with numbers are used as a guide in setting up the calculations.* In this method, units are treated in the same way as numbers; that is, they can be multiplied, divided, or canceled. For example, just as

$$3 \times 3 = 3^2 \quad (3 \text{ squared})$$

we have

$$\text{km} \times \text{km} = \text{km}^2 \quad (\text{km squared})$$

Also, just as the twos cancel in the expression

$$\frac{\cancel{2} \times 3 \times 6}{\cancel{2} \times 5}$$

the inches cancel in the expression

$$\frac{\cancel{\text{in.}} \times \text{cm}}{\cancel{\text{in.}}}$$

"Like units" found in the numerator and denominator of a fraction will always cancel, just as "like numbers" do.

The following steps indicate how to set up a problem using dimensional analysis.

STEP 1 *Identify the known or given quantity (both a numerical value and units) and the units of the new quantity to be determined.*

This information will always be found in the statement of the problem. Write an equation with the given quantity on the left and the units of the desired quantity on the right.

STEP 2 *Multiply the given quantity by one or more conversion factors in a manner such that the unwanted (original) units are canceled out, leaving only the new desired unit.*

The general format for the multiplication is

information given $\times$ conversion factor(s) = information sought

The number of conversion factors used depends on the individual problem. Except in the simplest of problems, it is a good idea to predetermine formally the sequence of unit changes to be used. This sequence will be called the unit "pathway."

STEP 3 *Perform the mathematical operations indicated by the conversion factor setup.*

In performing the calculation, you need to double check that all units except the desired set have canceled out. You also need to check the numerical answer to see that it contains the *proper number of significant figures.*

> Significant figures are not something we learn about in Chapter 2 and then forget about for the remainder of the course. Significant figures will be part of every calculation done in this course.

Now let us work a number of sample problems using dimensional analysis and the steps just outlined. Our first two examples involve only metric system units and thus will involve only metric-to-metric conversion factors.

EXAMPLE 3.3

One-Step "Metric-to-Metric" Conversion Problem

A vitamin C tablet is found to contain 0.500 g of vitamin C. How many milligrams of vitamin C does this tablet contain?

SOLUTION

STEP 1 The given quantity is 0.500 g, the mass of vitamin C in one tablet. The unit of the desired quantity is milligrams.

$$0.500 \text{ g} = ? \text{ mg}$$

STEP 2 Only one conversion factor will be needed to convert from grams to milligrams— one that relates grams to milligrams. Two forms of this factor exist.

$$\frac{1 \text{ mg}}{10^{-3} \text{ g}} \quad \text{and} \quad \frac{10^{-3} \text{ g}}{1 \text{ mg}}$$

The first factor is used because it allows for the cancellation of the gram units, leaving us with milligrams as the new units.

$$0.500 \, \cancel{g} \times \frac{1 \, mg}{10^{-3} \, \cancel{g}}$$

For cancellation, a unit must appear in both the numerator and the denominator. Because the given quantity (0.500 g) has grams in the numerator, the conversion factor used must be the one with grams in the denominator.

If the other conversion factor had been used, we would have

$$0.500 \, g \times \frac{10^{-3} \, g}{1 \, mg}$$

No unit cancellation is possible in this setup. Multiplication gives g^2/mg as the final units, which is certainly not what we want. In all cases, only one of the two conversion factors of a reciprocal pair will correctly fit into a dimensional analysis setup.

STEP 3 Step 2 takes care of the units. All that is left is to combine numerical terms to get the final answer; we still have to do the arithmetic. Collecting the numerical terms gives

$$\frac{0.500 \times 1}{10^{-3}} \, mg = 500 \, mg \qquad \text{(calculator answer)}$$

$$= 5.00 \times 10^2 \, mg \quad \text{(correct answer)}$$

Since the conversion factor used in this problem is derived from a definition, it contains an unlimited number of significant figures and will not limit in any way the allowable number of significant figures in the answer. Therefore, the answer should have three significant figures, the same number as in the given quantity.

CHEMICAL EXTENSION

Vitamin C is necessary for the proper formation of skin, ligaments, and tendons as well as for bone and teeth. Specifically, it is necessary for the formation of the protein *collagen*, a substance present in all these substances.

Although many people think citrus fruits are the best source of vitamin C (50 mg per 100 g), peppers (128 mg per 100 g), cauliflower (70 mg per 100 g), strawberries (60 mg per 100 g), and spinach or cabbage (60 mg per 100 g) are all richer in vitamin C.

Practice Exercise 3.3

Analysis shows the presence of 203 μg of cholesterol in a sample of blood. How many grams of cholesterol are present in this blood sample?

EXAMPLE 3.4

Two-Step "Metric-to-Metric" Conversion Problem

The *ozone layer* is a region in the upper atmosphere, at altitudes between 25 and 35 km, where the concentration of ozone is several times higher than at ground level. Express the altitude 35 km (the upper limit of the ozone layer) in centimeter units.

SOLUTION

STEP 1 The given quantity is 35 km, and the units of the desired quantity are centimeters.

$$35 \text{ km} = ? \text{ cm}$$

STEP 2 In dealing with metric–metric unit changes where both the original and desired units carry prefixes (which is the case in this problem), it is recommended that you always channel units through the base unit (unprefixed unit). If you do that, you will not need to deal with any conversion factors other than those resulting from prefix definitions. Following this recommendation, the unit pathway for this problem is

$$\text{km} \rightarrow \text{m} \rightarrow \text{cm}$$
prefixed   base   prefixed
unit   unit   unit

In the setup for this problem we will need two conversion factors, one for the kilometer-to-meter change and one for the meter-to-centimeter change.

$$35 \text{ km} \times \frac{10^3 \text{ m}}{1 \text{ km}} \times \frac{1 \text{ cm}}{10^{-2} \text{ m}}$$

This conversion factor converts km to m.

This conversion factor converts m to cm.

The units cancel except for the desired centimeters.

STEP 3 Carrying out the indicated numerical calculation gives

$$\frac{35 \times 10^3 \times 1}{1 \times 10^{-2}} \text{ cm} = 3.5 \times 10^6 \text{ cm} \quad \text{(calculator and correct answer)}$$

Numbers from first conversion factor

Numbers from second conversion factor

The correct answer is the same as the calculator answer for this problem. The given quantity has two significant figures, and both conversion factors are exact. Thus, the correct answer should contain two significant figures.

CHEMICAL EXTENSION

The substance oxygen exists in two gaseous forms: (1) dioxygen, which is the oxygen we breathe, and (2) trioxygen, which is also known as *ozone*. In contrast to the color-less and odorless dioxygen we breathe, ozone has a light-blue color and its pungent odor is that associated with thunderstorms; lightning discharges produce ozone.

Naturally occurring ozone present in the upper atmosphere (the "ozone layer") screens out 95–98% of the ultraviolet light that comes from the sun. Excessive exposure to ultraviolet light causes sunburn and can contribute to the development of certain types of skin cancer. The presence of ozone in the upper atmosphere is a very desirable situation.

Ozone, under certain conditions, can also be found in the lower atmosphere. Here, its presence is considered undesirable. In the lower atmosphere, ozone is produced during photochemical smog formation. As the "active ingredient" in such smog, it is damaging to plants and to the human lungs.

Practice Exercise 3.4

The average diameter of a coronary artery is 0.32 cm. What is this diameter in nanometers?

As shown in the detailed solutions for Examples 3.3 and 3.4, when using dimensional analysis to set up and solve problems, three questions are always asked:

1. What data is given in the problem?
2. What quantity do we wish to obtain in the problem?
3. What conversion factors are needed to facilitate the transition from the given quantity to the desired one?

Our next two worked-out example problems involve the use of English-to-English conversion factors. The conversion factors themselves should pose no problems for you. The examples are intended to give you further insights into dimensional analysis, particularly in the area of setting up the pathway for unit change. In addition, Example 3.6 exposes you to the complication of having to deal with "compound" units.

EXAMPLE 3.5

Multistep "English-to-English" Conversion Problem

If a person's stomach produces 87 fl oz of gastric juice in a day, what is the equivalent volume of the gastric juice in gallons?

SOLUTION

STEP 1 The given quantity is 87 fl oz, and the unit of the desired quantity is gallons.

$$87 \text{ fl oz} = ? \text{ gal}$$

STEP 2 The logical pathway to follow to accomplish the desired change is

$$\text{fl oz} \rightarrow \text{qt} \rightarrow \text{gal}$$

Always use logical steps in setting up the pathway for a unit change. It does not have to be done in one big jump. Use smaller steps for which you know the conversion factors. Big steps usually get you involved with unfamiliar conversion factors. Most people do not carry around in their head the number of fluid ounces in a gallon, but they do know that there are 32 fl oz in a quart.

The setup for this problem will require two conversion factors: fluid ounces to quarts and quarts to gallons.

$$87 \text{ fl oz} \times \frac{1 \text{ qt}}{32 \text{ fl oz}} \times \frac{1 \text{ gal}}{4 \text{ qt}}$$
$$\text{fl oz} \rightarrow \text{qt} \rightarrow \text{gal}$$

The units all cancel except for the desired gallons.

STEP 3 Performing the indicated multiplications, we get

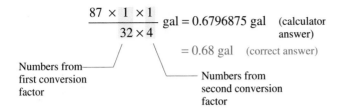

$$\frac{87 \times 1 \times 1}{32 \times 4} \text{ gal} = 0.6796875 \text{ gal} \quad \text{(calculator answer)}$$

$$= 0.68 \text{ gal} \quad \text{(correct answer)}$$

Numbers from first conversion factor

Numbers from second conversion factor

The calculator answer contains too many significant figures. The correct answer should contain only two significant figures, the number in the given quantity. Again, both conversion factors involve exact definitions and will not enter into significant figure considerations.

CHEMICAL EXTENSION

Secreted by the gastric glands of the stomach, gastric juice is a mixture of water, enzymes, and hydrochloric acid. The presence of hydrochloric acid in gastric juice causes the sensation of heartburn if it is refluxed into the esophagus.

The acidity of gastric juice (hydrochloric acid) prevents bacterial growth in the stomach, kills most bacteria that enter the body with food, and effects the partial breakdown of protein by activating protein-digesting enzymes. (Digestion of carbohydrates and fats occur primarily in the small intestine rather than in the stomach.)

The concentration of hydrochloric acid in gastric juice is approximately 5% by mass, creating an acidity effect about the same as that of lemon juice. Hydrochloric acid is known commercially as *muriatic acid* and can be purchased under this name in hardware stores.

Practice Exercise 3.5

The estimated amount of recoverable oil from the field at Prudhoe Bay in Alaska is 4.0×10^{11} gal. What is the amount of oil in fluid ounces (fl oz)?

EXAMPLE 3.6

Unit Conversion Problem Involving "Compound" Units

Worldwide emissions of carbon dioxide into the atmosphere are estimated at 2×10^9 tons per year. What is this emission rate in pounds per hour?

SOLUTION

STEP 1 The given quantity is 2×10^9 ton/year, and the units of this quantity are to be changed to lb/hr.

$$2 \times 10^9 \, \frac{\text{ton}}{\text{yr}} = ? \frac{\text{lb}}{\text{hr}}$$

STEP 2 This problem is more complex than previous examples because two different types of units are involved: mass and time. However, the approach we use to solve it is similar, other than that we must change two units instead of one. The tons must be converted to pounds and the years to hours.

The logical pathway for the mass change is the direct one-step path

$$\text{ton} \rightarrow \text{lb}$$

For time, it is logical to make the change in two steps.

$$\text{yr} \rightarrow \text{day} \rightarrow \text{hr}$$

It does not matter whether time or mass is handled first in the conversion-factor setup. We will arbitrarily choose to handle time first. The setup becomes

$$2 \times 10^9 \; \frac{\text{ton}}{\cancel{\text{yr}}} \times \frac{1 \; \cancel{\text{yr}}}{365 \; \cancel{\text{day}}} \times \frac{1 \; \cancel{\text{day}}}{24 \; \text{hr}}$$

The units at this ———— point are ton/day. ———The units at this point are ton/hr.

Note that in the first conversion factor, years had to be in the numerator to cancel the years in the denominator of the given quantity.

We are not done yet. The time conversion from years to hours has been accomplished, but nothing has been done with mass. To take care of mass we do not start a new conversion factor setup. Rather, an additional conversion factor is tacked onto those we already have in place.

$$2 \times 10^9 \; \frac{\cancel{\text{ton}}}{\cancel{\text{yr}}} \times \frac{1 \; \cancel{\text{yr}}}{365 \; \cancel{\text{day}}} \times \frac{1 \; \cancel{\text{day}}}{24 \; \text{hr}} \times \frac{2000 \; \text{lb}}{1 \; \cancel{\text{ton}}}$$

The tons in the denominator of the last factor cancel the tons in the numerator of the given quantity. With this cancellation the units now become lb/hr.

STEP 3 Collecting the numerical factors and performing the indicated math gives

$$\frac{2 \times 10^9 \times 1 \times 1 \times 2000}{365 \times 24 \times 1} \; \frac{\text{lb}}{\text{hr}} = 4.56621 \times 10^8 \frac{\text{lb}}{\text{hr}} \qquad \text{(calculator answer)}$$

$$= 5 \times 10^8 \; \frac{\text{lb}}{\text{hr}} \qquad \text{(correct answer)}$$

The given quantity possesses only one significant figure. Rounding the calculator answer to one significant figure produces the correct answer of 5×10^8 lb/hr. Again, none of the conversion factors plays a role in significant-figure considerations since they all originated from definitions.

CHEMICAL EXTENSION

There are three main sources for carbon dioxide discharged into the atmosphere: respiration of plants and animals (93%), forest fires and other burning of plant materials (2%), and the burning of fossil fuels such as oil, coal, and natural gas (5%). Most (95%) of this released carbon dioxide is removed by the natural processes of photosynthesis and uptake of carbon dioxide by the oceans. There is, however, a net gain in atmospheric carbon dioxide each year.

The environmental impact of increased carbon dioxide levels is often called the *greenhouse effect*, an effect that could produce global warming (an increase in average air temperatures). Carbon dioxide traps some of the energy radiated by the Earth as it

cools at night, thus preventing the radiation from escaping to outer space. The trapped
energy causes an increased warming of the atmosphere.

Practice Exercise 3.6

The average human heart pumps blood at the rate of 6.8 fl oz/sec. What is this pump rate
in gallons per hour?

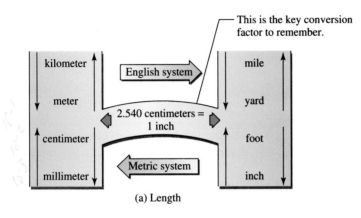

(a) Length

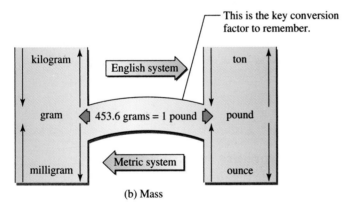

(b) Mass

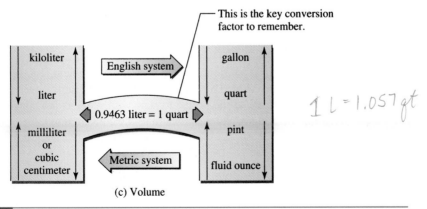

(c) Volume

Figure 3.8

Measurement bridges connecting the metric and English systems of measurement for (a) length,
(b) mass, and (c) volume.

We will now consider some sample problems that involve both the English and metric systems of units. As mentioned previously (Sec. 3.6), conversion factors between these two unit systems do not arise from definitions, but rather are determined experimentally. Hence they are not exact. Some of these experimentally determined conversion factors were given in Table 3.2.

Instead of trying to remember all of the conversion factors listed in Table 3.2, memorize only one factor for each type of measurement (mass, volume, length). Knowing only one factor of each type is sufficient information to work metric-to-English or English-to-metric conversion problems. The relationships that are the most useful for you to memorize are the following.

$$\text{Length:} \quad 1 \text{ in.} = 2.540 \text{ cm} \quad \text{(four significant figures)}$$

$$\text{Mass:} \quad 1 \text{ lb} = 453.6 \text{ g} \quad \text{(four significant figures)}$$

$$\text{Volume:} \quad 1 \text{ qt} = 0.9463 \text{ L} \quad \text{(four significant figures)}$$

These three equalities can be considered "bridge relationships" connecting English and metric system measurement units of various types. These bridge relationships are depicted in Figure 3.8.

These bridge relationships are always applicable in problem solving. For example, no matter what mass units are given or asked for in a problem, we can convert to pounds or grams (bridge units), cross the bridge with our memorized conversion factor, and then convert to the desired final unit. The only advantage that would be gained by memorizing all the factors in Table 3.2 would be that you could work some problems with fewer conversion factors, usually only one factor fewer. The reduction in the number of conversion factors used is usually not worth the added complication of keeping track of the additional conversion factors.

EXAMPLE 3.7

"English-to-Metric" Conversion Problem

Your feisty neighbor Handy Uppert weighs 244 lb. What is his mass in kilograms?

SOLUTION

STEP 1 The given quantity is 244 lb, and the units of the desired quantity are kilograms.

$$244 \text{ lb} = ? \text{ kg}$$

STEP 2 This is an English–metric mass-unit conversion problem. The mass-unit "measurement bridge" (Fig. 3.8b) involves pounds and grams. Since pounds are the given units, we are at the bridge to start with. Pounds are converted to grams (crossing the bridge), and then the grams are converted to kilograms.

$$\text{lb} \rightarrow \text{g} \rightarrow \text{kg}$$

The conversion factor setup for this problem is

$$244 \text{ lb} \times \frac{453.6 \text{ g}}{1 \text{ lb}} \times \frac{1 \text{ kg}}{10^3 \text{ g}}$$

The units all cancel except for kilograms.

STEP 3 Performing the indicated arithmetic gives

$$\frac{244 \times 453.6 \times 1}{1 \times 10^3} \text{ kg} = 110.6784 \text{ kg} \quad \text{(calculator answer)}$$

$$= 111 \text{ kg} \quad \text{(correct answer)}$$

The calculator answer must be rounded off to three significant figures, since the given quantity (244 lb) has only three significant figures. The first conversion factor has four significant figures, and the second one is exact.

An alternative pathway for working this problem makes use of the conversion factor in Table 3.2 involving pounds and kilograms. If we use this conversion factor, we can go directly from the given to the desired unit in one step. The setup is

$$244 \text{ lb} \times \frac{0.4536 \text{ kg}}{1 \text{ lb}} = 110.6784 \text{ kg} \quad \text{(calculator answer)}$$

$$= 111 \text{ kg} \quad \text{(correct answer)}$$

As is the case here, for most problems there is more than one pathway that can be used to get the answer. When alternative pathways exist, it cannot be said that one way is more correct than another. The important concept is that you select a pathway, choose a correct set of conversion factors consistent with that pathway, and get the answer.

Although both pathways used in this problem are "correct," we prefer the first solution because it uses our mass measurement bridge. Use of the measurement bridge system cuts down on the number of conversion factors you are required to know (or look up in a table).

Practice Exercise 3.7

A piece of copper metal has a mass of 17.62 lb. What is the copper mass in kilograms?

> Conversion factors based on measurement, such as English–metric conversion factors, should never be the limiting quantity in determining the number of significant figures in a final answer. For this reason, and also to minimize rounding error, we will always use four-significant-figure English–metric conversion factors (Fig. 3.8 and Table 3.2) in problem solving unless the given quantity dictates the use of conversion factors with even more significant figures.

EXAMPLE 3.8

Multistep "Metric-to-English" Conversion Problem

The average direct daily use of water in the United States per person is 3.0×10^5 mL. This includes water for drinking, cooking, bathing, dishwashing, laundry, flushing toilets, watering lawns, and so on. What is the volume, in gallons, of this amount of water?

SOLUTION

STEP 1 The given quantity is 3.0×10^5 mL, and the units of the desired quantity are gallons.

$$3.0 \times 10^5 \text{ mL} = ? \text{ gal}$$

STEP 2 The "bridge relationship" for volume (Fig. 3.8c) involves liters and quarts. Thus, we need to convert the milliliters to liters, cross the bridge to quarts, and then convert quarts to gallons.

$$\text{mL} \rightarrow \text{L} \rightarrow \text{qt} \rightarrow \text{gal}$$

Following the pathway, the setup becomes

$$3.0 \times 10^5 \text{ mL} \times \frac{10^{-3} \text{ L}}{1 \text{ mL}} \times \frac{1 \text{ qt}}{0.9463 \text{ L}} \times \frac{1 \text{ gal}}{4 \text{ qt}}$$

STEP 3 The numerical calculation involves the following collection of numbers.

$$\frac{3.0 \times 10^5 \times 10^{-3} \times 1 \times 1}{1 \times 0.9463 \times 4} \text{ gal} = 79.256049 \text{ gal} \quad \text{(calculator answer)}$$

$$= 79 \text{ gal} \quad \text{(correct answer)}$$

The calculator answer must be rounded to two significant figures because the input number 3.0×10^5 has only two significant figures.

Again, as in Example 3.7, to illustrate that there is more than one way to set up almost any unit conversion problem, let us work this problem two alternative ways using the other two volume conversion factors from Table 3.2.

To use the conversion factor based on 1pt = 0.4732 L requires a pathway of

$$mL \rightarrow L \rightarrow pt \rightarrow qt \rightarrow gal$$

The conversion factor setup is

$$3.0 \times 10^5 \text{ mL} \times \frac{10^{-3} \text{ L}}{1 \text{ mL}} \times \frac{1 \text{ pt}}{0.4732 \text{ L}} \times \frac{1 \text{ qt}}{2 \text{ pt}} \times \frac{1 \text{ gal}}{4 \text{ qt}} = 79.247675 \text{ gal}$$
$$\text{(calculator answer)}$$

$$= 79 \text{ gal}$$
$$\text{(correct answer)}$$

Note that the calculator answers obtained from this setup and the previous one for this problem are different—79.256049 and 79.247675—but each gives the same answer after rounding to two significant figures. The slight difference arises from the fact that the English-to-metric conversion factors used are not exact definitions. The two conversion factors used have different rounding-off errors in them.

The conversion factor based on 29.57 mL = 1 fl oz requires a pathway of

$$mL \rightarrow fl\ oz \rightarrow qt \rightarrow gal$$

The conversion factor setup for this pathway is

$$3.0 \times 10^5 \text{ mL} \times \frac{1 \text{ fl oz}}{29.57 \text{ mL}} \times \frac{1 \text{ qt}}{32 \text{ fl oz}} \times \frac{1 \text{ gal}}{4 \text{ qt}} = 79.261075 \text{ gal}$$
$$\text{(calculator answer)}$$

$$= 79 \text{ gal}$$
$$\text{(correct answer)}$$

Note that all three methods give the same correct answer after rounding to the correct number of significant figures. Although each of the three setups is correct, we still prefer the method of having a specific bridge relationship for volume, mass, and length to use in crossing the metric-to-English bridge.

CHEMICAL EXTENSION

On average, per day, approximately 2400 mL of water enters and leaves the body.

Water enters the body in three different ways: (1) the liquids we drink (1500 mL), (2) water in foods (700 mL), and (3) water produced in cells as a by-product of the reactions by which food is broken down (200 mL). Water normally leaves the body by four mechanisms: (1) urine produced in the kidneys (1400 mL), (2) water in expired air from the lungs (350 mL), (3) perspiration from the skin (450 mL), and (4) feces formed in the intestines (200 mL).

For a body to function normally, the total volume of water entering the body must equal the total volume of water leaving the body. Serious dehydration can occur in an adult if there is a 10% net loss in total body fluid, and a 20% loss of fluid can be fatal. An infant suffers dehydration with a 5–10% loss in body fluid.

Practice Exercise 3.8

Typical water usage by a household washing machine, per load, is 140,000 mL. What is the water usage in gallons?

EXAMPLE 3.9

"Metric-to-English" Conversion Problem Involving Cubic Units

A large flask contains 375 mL of strained (no seeds) watermelon juice. (Tests are to be run on the juice to see how much vitamin A it contains.) Specify the volume of this amount of juice in

(a) cubic centimeters   **(b)** cubic feet

SOLUTION

(a) From Section 3.4, we note that

$$1 \text{ mL} = 1 \text{ cm}^3$$

Therefore,

$$375 \text{ mL} \times \frac{1 \text{ cm}^3}{1 \text{ mL}} = 375 \text{ cm}^3 \quad \text{(calculator and correct answer)}$$

(b) We will use the answer from part (a), 375 cm^3, as the starting point for this calculation.

STEP 1 The given quantity is 375 cm^3, and the units for the desired quantity are ft^3:

$$375 \text{ cm}^3 = ? \text{ ft}^3$$

STEP 2 If this problem were a problem involving just length rather than (length)^3, the pathway would be

$$\text{cm} \rightarrow \text{in.} \rightarrow \text{ft}$$

The pathway for (length)^3 is just an adaptation of this pathway.

$$\text{cm}^3 \rightarrow \text{in.}^3 \rightarrow \text{ft}^3$$

The conversion factors needed for this pathway are simply those for length raised to the third power; that is,

$$375 \text{ cm}^3 \times \left(\frac{1 \text{ in.}}{2.540 \text{ cm}} \right)^3 \times \left(\frac{1 \text{ ft}}{12 \text{ in.}} \right)^3$$

Note that the *entire* conversion factor, in each case, must be cubed, not just the units on the conversion factor. The notation $\left(\frac{1 \text{ ft}}{12 \text{ in.}}\right)^3$ means that the conversion factor within the parentheses is multiplied by itself three times. Thus,

$$\left(\frac{1 \text{ ft}}{12 \text{ in.}}\right)^3 = \frac{1 \text{ ft}}{12 \text{ in.}} \times \frac{1 \text{ ft}}{12 \text{ in.}} \times \frac{1 \text{ ft}}{12 \text{ in.}}$$

$$= \frac{1 \text{ ft}^3}{1728 \text{ in.}^3}$$

The complete setup for this problem, showing the removal of parentheses and cancellation of units, is

$$375 \text{ cm}^3 \times \left(\frac{1 \text{ in.}}{2.540 \text{ cm}}\right)^3 \times \left(\frac{1 \text{ ft}}{12 \text{ in.}}\right)^3$$

$$= 375 \text{ cm}^3 \times \frac{1^3 \text{ in.}^3}{(2.540)^3 \text{ cm}^3} \times \frac{1^3 \text{ ft}^3}{(12)^3 \text{ in.}^3}$$

Note that the numbers as well as the units must be cubed with removal of parentheses.

STEP 3 The numerical setup is

$$\frac{375 \times 1^3 \times 1^3}{(2.540)^3 \times (12)^3} \text{ ft}^3 = 0.013243 \text{ ft}^3 \quad \text{(calculator answer)}$$

$$= 0.0132 \text{ ft}^3 \quad \text{(correct answer)}$$

Practice Exercise 3.9

A graduated cylinder contains 37.5 mL of water. Express this water volume in

(a) cubic centimeters   **(b)** cubic inches

Although the emphasis in this section has been on using conversion factors to change units within the English or metric systems or from one to the other, the applications of conversion factors go far beyond this type of activity. We will resort to using conversion factors time and time again throughout this textbook in solving problems. What has been covered in this section is only the "tip of the iceberg" relative to dimensional analysis and conversion factors.

3.8 Density

Density *is the ratio of the mass of an object to the volume occupied by that object; that is,*

$$\text{density } (d) = \frac{\text{mass}}{\text{volume}}$$

The most frequently encountered density units in chemistry are grams per cubic centimeter (g/cm^3) for solids, grams per milliliter (g/mL) for liquids, and grams per liter (g/L) for gases. Use of these units avoids the problem of having density values that are extremely small or extremely large numbers. Table 3.3 gives density values for a number of substances.

Table 3.3	Densities of Selected Solids, Liquids, and Gases

Solids	Density (g/cm³ at 25°C)*	Liquids	Density (g/mL at 25°C)*	Gases	Density (g/L at 25°C, 1 atm)*
Gold	19.3	Mercury	13.55	Chlorine	3.17
Lead	11.3	Milk	1.028–1.035	Carbon dioxide	1.96
Copper	8.93	Blood plasma	1.027	Oxygen	1.42
Aluminum	2.70	Urine	1.003–1.030	Air (dry)	1.29
Table salt	2.16	Water	0.997	Nitrogen	1.25
Bone	1.7–2.0	Olive oil	0.92	Methane	0.66
Table sugar	1.59	Ethyl alcohol	0.79	Hydrogen	0.08
Wood, pine	0.30–0.50	Gasoline	0.56		

Density changes with temperature. (In most cases it decreases with increasing temperature, since almost all substances expand when heated.) Consequently, the temperature must be recorded along with a density value. In addition, the pressure of gases must be specified.

People often speak of one substance being "heavier" or "lighter" than another. For example, it is said that "lead is a heavier metal than aluminum." What is actually meant by this statement is that lead has a higher density than aluminum; that is, there is more mass in a specific volume of lead than there is in the same volume of aluminum. The density of an object is a measure of how tightly the object's mass is packed into a given volume. Even though the density of lead (11.3 g/cm^3) is greater than that of aluminum (2.70 g/cm^3), 1 g of lead weighs exactly the same as 1 g of aluminum—1 g is 1 g. Because the aluminum is less dense than the lead, the mass in the 1 g of aluminum will occupy a larger volume than the mass in the 1 g of lead. Said another way, if equal-volume samples of lead and aluminum are weighed, the lead will have the greater mass. When we say lead is "heavier" than aluminum, we actually mean that lead is more dense than aluminum.

The densities of solids and liquids are often compared to the density of water. Anything less dense ("lighter") than water floats in it (see Fig. 3.9), and anything more dense ("heavier")

Metals are classified as "light" or "heavy" based on density. Light metals are metals whose densities are less than 4.00 g/cm³. Included among the light metals are

calcium	1.55 g/cm³
magnesium	1.74 g/cm³
aluminum	2.70 g/cm³

The "lightest" of the light metals is lithium, with a density of 0.53 g/cm³.

Figure 3.9

Icebergs float in water because ice (solid water) is less dense than liquid water. (Delphine Star/Tony Stone Images)

sinks. In a similar vein, densities of gases are compared to that of air. Any gas less dense ("lighter") will rise in air, and anything more dense ("heavier") will sink in air.

To obtain an object's density, we must make two measurements: one involves determining the object's mass, and the other its volume.

EXAMPLE 3.10

Using Mass and Volume to Calculate Density

A student determines that the mass of a 20.0-mL sample of olive oil (to be used in oil-and-vinegar salad dressing) is 18.4 g.

(a) What is the density of the olive oil in grams per milliliter?

(b) Predict whether the olive oil layer will be on top or bottom in unshaken oil-and-vinegar salad dressing.

SOLUTION

(a) Substituting the given mass and volume values into the formula

$$\text{density} = \frac{\text{mass}}{\text{volume}}$$

we have

$$\text{density} = \frac{18.4 \text{ g}}{20.0 \text{ mL}} = 0.92 \frac{\text{g}}{\text{mL}} \quad \text{(calculator answer)}$$

$$= 0.920 \frac{\text{g}}{\text{mL}} \quad \text{(correct answer)}$$

Because both input numbers contain three significant figures, the density is specified to three significant figures.

(b) Since vinegar is a water-based solution, its density will be slightly greater than that of water (1.00 g/mL). The olive oil will be the top layer because its density is less than that of water or vinegar.

CHEMICAL EXTENSION

Medical research shows a correlation exists between dietary fat intake and heart and circulatory system problems. Additionally, some dietary fats have greater adverse effects on the human body than others. In general, the greater the degree of saturation of a fat, the greater its undesirable effects.

How do liquid-state dietary fats (plant oils) compare to solid-state dietary fats (animal fats) in terms of degree of saturation? Generally, plant oils are less saturated than animal fats, although coconut oil is an exception to the rule.

A measure of the degree of saturation of a fat (solid or liquid) is given by its *iodine number*, that is, by the number of grams of iodine that react with 100 g of fat. The *higher* the iodine number, the *lower* the degree of saturation (a desirable situation in terms of diet). Iodine numbers of selected solid and liquid fats are as follows:

coconut oil	8–10	cottonseed oil	100–117
butter	25–40	corn oil	115–130
beef tallow	30–45	canola oil	125–135
lard	45–70	soybean oil	125–140
olive oil	75–95	safflower oil	130–140
peanut oil	85–100	sunflower oil	130–145

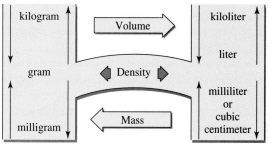

$1\,mL = 1\,cm^3$

Figure 3.10

Density, a measurement bridge connecting mass and volume.

Practice Exercise 3.10

Osmium is the densest of all metals. What is its density, in grams per cubic centimeter, if 50.00 g of the metal occupies a volume of 2.22 cm³?

In a mathematical sense, density can be thought of as a conversion factor that relates the volume and mass of an object. Interpreting density in this manner enables us to calculate a substance's volume from its mass and density or its mass from its volume and density. Density is thus a "bridge" connecting mass and volume (see Fig. 3.10).

Density may be used as a conversion factor to convert from mass to volume or vice versa

EXAMPLE 3.11

"Volume-to-Mass" Conversion Using Density

Methane, a gas that is naturally present in the Earth's atmosphere in small amounts, has a density of 0.714 g/L at a particular temperature and pressure. What is the mass, in grams, of 10.0 L of methane?

SOLUTION

We will use density as a conversion factor in solving this problem by dimensional analysis.

STEP 1   The given quantity is 10.0 L of methane. The unit of the desired quantity (mass) is grams. Thus,

$$10.0 \text{ L} = ? \text{ g}$$

STEP 2   The pathway going from liters to grams involves a single step, since density, used as a conversion factor, directly relates grams and liters.

$$10.0 \text{ L} \times \frac{0.714 \text{ g}}{1 \text{ L}}$$

STEP 3   Doing the indicated math gives the following answer.

$$\frac{10.0 \times 0.714}{1} \text{ g} = 7.14 \text{ g} \quad \text{(calculator and correct answer)}$$

The calculator answer and correct answer turn out to be the same.

CHEMICAL EXTENSION

The small amount of methane present in Earth's atmosphere comes from terrestrial sources. The decomposition of animal and plant matter in an oxygen-deficient environment—swamps, marshes, bogs—produces methane. A common name for methane, marsh gas, refers to the production of methane in this matter.

Bacteria that live in termites and in the digestive tracts of plant-eating animals have the ability to produce methane from plant materials. The methane output of a large cow (belching and flatulence) can reach 20 liters per day.

Methane gas is also found associated with coal and petroleum deposits. That associated with petroleum is most often recovered, processed, and marketed as *natural gas*. The processed natural gas is 85–95% methane by volume. Because methane is odorless, an odorant (smelly substance) must be added to the processed natural gas used in home heating. Otherwise, natural gas leaks could not be detected.

Practice Exercise 3.11

The density of octane, a component of gasoline, is 0.702 g/mL. What is the mass, in grams, of 875 mL of octane?

EXAMPLE 3.12

"Mass-to-Volume" Conversion Using Density

Common table sugar has a density of 1.587 g/cm^3. What would be the volume, in cubic centimeters, of 2.500 g of table sugar?

SOLUTION

STEP 1 The given quantity is 2.500 g of table sugar. The unit of the desired quantity (volume) is cubic centimeters. Thus,

$$2.500 \text{ g} = ? \text{ cm}^3$$

STEP 2 The pathway going from grams to cubic centimeters involves a single step, since density, used as a conversion factor, directly relates grams and cubic centimeters.

$$2.500 \text{ g} \times \frac{1 \text{ cm}^3}{1.587 \text{ g}}$$

Note that the conversion factor used is actually the reciprocal of the density. Use of the inverted form is necessary in order for the gram units to cancel.

STEP 3 Doing the indicated math gives the following answer.

$$\frac{2.500 \times 1}{1.587} \text{ cm}^3 = 1.5752993 \text{ cm}^3 \quad \text{(calculator answer)}$$

$$= 1.575 \text{ cm}^3 \quad \text{(correct answer)}$$

The correct answer is the calculator answer rounded to four significant figures, the number in each input number.

Practice Exercise 3.12

The density of ethyl alcohol is 0.789 g/mL. What volume of ethyl alcohol, in milliliters, would have a mass of 25.0 g?

Density is a conversion factor of a different type than those previously used in this chapter. It involves a ratio whose numerator and denominator are *equivalent* rather than *equal*; that is, it is an *equivalence* conversion factor rather than the previously used *equality* conversion factors. An **equivalence conversion factor** *is a ratio that converts one type of* *measure to a different type of measure.* Density involves mass and volume, two different types of measure. This contrasts with equality conversion factors that have ratios in which the numerator and denominator involve the same measure. An **equality conversion factor** *is a ratio that converts one unit of a given measure to another unit of the same measure.* Twelve inches and one foot, which are both measures of length, generate an equality conversion factor.

A major difference between *equivalence* conversion factors and *equality* conversion factors is that the former have applicability only in the particular problem setting for which they were derived, whereas the latter are applicable in all problem-solving situations. Many different gram-to-cubic centimeter (mass-to-volume) relationships (densities) exist, but only one foot-to-inch relationship exists. Mathematically, equivalence conversion factors can be used the same way equality conversion factors are, and such conversion factors will be used often in later chapters.

> **Density is a property peculiar to a particular substance. The density of water applies only to water and to no other substance. Different substances have different densities at a given set of conditions.**

3.9 Equivalence Conversion Factors Other than Density

In the previous section we noted that density was an *equivalence* conversion factor rather than an *equality* conversion factor. Equivalence conversion factors have application only in a limited setting, the setting for which they were derived. Besides density, numerous other equivalence-type conversion factors exist.

An example from the financial world involves the relationships between various currencies (which vary from day to day). On August 22, 2003, it was true that

$$1 \text{ dollar} = 0.92 \text{ European Euro}$$

$$1 \text{ dollar} = 117.47 \text{ Japanese yen}$$

$$1 \text{ dollar} = 0.64 \text{ British pound}$$

Conversion factors obtainable from these relationships include

$$\frac{0.92 \text{ euro}}{1 \text{ dollar}} \qquad \frac{1 \text{ dollar}}{117.47 \text{ yen}} \qquad \frac{0.64 \text{ pound}}{1 \text{ dollar}}$$

Equivalence conversion factor examples from the world of chemistry involve concentration and dosage relationships. If the concentration of salt in a salt water solution is 4.5 mg/mL, then we have

$$4.5 \text{ mg salt} = 1 \text{ mL of solution}$$

and the two conversion factors are

$$\frac{4.5 \text{ mg salt}}{1 \text{ mL solution}} \quad \text{and} \quad \frac{1 \text{ mL solution}}{4.5 \text{ mg salt}}$$

If the dosage for an antibiotic is 125 mg/kg of body weight, then we have

$$125 \text{ mg antibiotic} = 1 \text{ kg body weight}$$

and the two conversion factors are

$$\frac{125 \text{ mg antibiotic}}{1 \text{ kg body weight}} \quad \text{and} \quad \frac{1 \text{ kg body weight}}{125 \text{ mg antibiotic}}$$

Examples 3.13 and 3.14 illustrate how conversion factors such as the preceding ones are used in problem-solving contexts.

EXAMPLE 3.13

Using Currency Conversion Factors

If one U.S. dollar is equivalent to 117.47 Japanese yen, what is the value in dollars of 100 yen?

SOLUTION

STEP 1 The given quantity is 100 yen. The unit of the desired quantity is dollars.

$$100 \text{ yen} = ? \text{ dollars}$$

STEP 2 The pathway between yen and dollars involves a single step since the relationships between yen and dollars is directly known (given in the problem statement).

$$100 \text{ yen} \times \frac{1 \text{ dollar}}{117.47 \text{ yen}}$$

STEP 3 Doing the indicated mathematics gives the following answer.

$$\frac{100 \times 1}{117.47} \text{ dollars} = 0.85128117 \text{ dollars} \quad (\text{calculator answer})$$

$$= 0.85128 \text{ dollars} \quad (\text{correct answer})$$

Note that the answer can be specified to five significant figures because 117.47 has five significant figures. The numbers 100 and 1 are considered to be defined numbers.

Practice Exercise 3.13

If one British pound is equivalent to 1.56 dollars, what is the value, in pounds, of a 50-dollar bill?

EXAMPLE 3.14

Using Concentration Conversion Factors

A blood analysis shows a vitamin C concentration of 0.2 mg/100 mL of blood. How many grams of vitamin C are there in 5.0 L of this blood?

SOLUTION

STEP 1 The given quantity is 5.0 L of blood. The unit of the desired quantity is grams of vitamin C.

$$5.0 \text{ L blood} = ? \text{ g vitamin C}$$

STEP 2 The pathway in going from liters of blood to grams of vitamin C is

$$\text{L blood} \rightarrow \text{mL blood} \rightarrow \text{mg vitamin C} \rightarrow \text{g vitamin C}$$

The conversion factors needed for this pathway are

$$1 \text{ L blood} \times \frac{1 \text{ mL blood}}{10^{-3} \text{ L blood}} \times \frac{0.2 \text{ mg vitamin C}}{100 \text{ mL blood}} \times \frac{10^{-3} \text{ g vitamin C}}{1 \text{ mg vitamin C}}$$

S TEP 3 The numerical calculation involves the following collection of numbers.

$$\frac{5.0 \times 1 \times 0.2 \times 10^{-3}}{10^{-3} \times 100 \times 1} \text{ g vitamin C} = 0.01 \text{ g vitamin C}$$

(calculator and correct answer)

The given number of 0.2 mg vitamin C limits the answer to one significant figure.

Practice Exercise 3.14

A blood analysis indicates that there are 33 ng of testosterone in 75 mL of blood. How many grams of testosterone are there in 2.0 L of this blood?

3.10 Percentage and Percent Error

Percent *is the number of items of a specified type in a group of 100 total items.* The quantity 45% means 45 items per 100 total items. In general, percent is parts per 100 total parts.

A mathematical statement of the percent concept is

$$\text{percent} = \frac{\text{number of items of interest}}{\text{total number of items}} \times 100$$

Example 3.15 shows a simple calculation of a percent.

EXAMPLE 3.15

Percentage Calculation

A professor proctoring an examination notices that 8 students out of a class of 80 students write with their left hand. What is the percentage of *right-handed* students in the class?

SOLUTION

We first calculate the number of right-handed students in the class.

right-handed students $= 80 - 8 = 72$ (calculator and correct answer)

The percentage of right-handed students is equal to the number of right-handed students divided by the total number of students times the factor 100.

percentage right-handed students $= \dfrac{72}{80} \times 100$

$$= 90\% \quad \text{(calculator answer and correct answer)}$$

The answer to this problem is exactly 90% since all of the numbers used in the calculation are exact numbers.

Practice Exercise 3.15

The composition of a 14-karat gold ring is found to be 10.68 g gold and 7.62 g copper. What is the percentage by mass of copper in the ring?

$$\frac{62.1 \text{ left-handed, bald-headed adult males}}{100 \text{ bald-headed adult males}}$$

$$\frac{91.3 \text{ handsome, left-handed, bald-headed adult males}}{100 \text{ left-handed, bald-headed adult males}}$$

The unit-conversion pathway for this problem will be

$$\text{adult males} \rightarrow \begin{array}{c}\text{bald-headed}\\\text{adult males}\end{array} \rightarrow \begin{array}{c}\text{left-handed,}\\\text{bald-headed}\\\text{adult males}\end{array} \rightarrow \begin{array}{c}\text{handsome,}\\\text{left-handed,}\\\text{bald-headed}\\\text{adult males}\end{array}$$

The dimensional analysis setup for this problem is

$$184 \text{ adult males} \times \frac{20.1 \text{ bald-headed}}{100 \text{ adult males}} \times \frac{62.1 \text{ left-handed}}{100 \text{ bald-headed}} \times \frac{91.3 \text{ handsome}}{100 \text{ left-handed}}$$

All units cancel except those in the numerator of the last conversion factor.

STEP 3 Performing the indicated arithmetic gives

$$\frac{184 \times 20.1 \times 62.1 \times 91.3}{100 \times 100 \times 100} \text{ handsome, left-handed, bald-headed adult males}$$

$$= 20.968929 \text{ handsome, left-handed, bald-headed adult males} \qquad \text{(calculator answer)}$$

$$= 21.0 \text{ handsome, left-handed, bald-headed adult males} \quad \text{(correct answer)}$$

The calculator answer must be rounded off to three significant figures since all three of the given percentages contain only three significant figures.

Practice Exercise 3.17

A 5-lb box of chocolates contains 112 chocolates. Dark chocolates are more prevalent than light chocolates—75% versus 25%. Exactly 25% of the dark chocolates are cream filled. How many cream-filled dark chocolates are there in the 5-lb box of chocolates?

Percent Error

Percent error *is the ratio of the difference between a measured value and the accepted value for the measurement and the accepted value itself all multiplied by 100.* A mathematical statement of the percent error concept is

$$\text{percent error} = \frac{\text{measured value} - \text{accepted value}}{\text{accepted value}} \times 100$$

A percent error can be either positive or negative. If the measured value is greater than the accepted value, the difference (measured value − accepted value) will be positive, and the percent error will be positive. If the measured value is less than the accepted value, the difference will be negative and the percent error will be negative.

EXAMPLE 3.18

Calculation of a Percent Error

The accepted value for the density of copper is 8.93 g/cm^3. In a laboratory setting, three students are asked to experimentally determine the density of copper. Their results are

$$\text{student 1:} \quad 8.91 \text{ g/cm}^3$$
$$\text{student 2:} \quad 8.23 \text{ g/cm}^3$$
$$\text{student 3:} \quad 8.99 \text{ g/cm}^3$$

Calculate the percent error associated with each student's reported density.

SOLUTION

We will use the equation

$$\text{percent error} = \frac{\text{measured value} - \text{accepted value}}{\text{accepted value}} \times 100$$

to calculate the percent error in each student's result.

STUDENT 1

$$\text{percent error} = \frac{(8.91 - 8.93) \text{ g/cm}^3}{8.93 \text{ g/cm}^3} \times 100$$
$$= \frac{-0.02}{8.93} \times 100$$
$$= -0.22396416 \quad \text{(calculator answer)}$$
$$= -0.2 \quad \text{(correct answer)}$$

The negative sign associated with the result indicates that the measured value was less than the accepted value.

STUDENT 2

$$\text{percent error} = \frac{(8.23 - 8.93) \text{ g/cm}^3}{8.93 \text{ g/cm}^3} \times 100$$
$$= \frac{-0.70}{8.93} \times 100$$
$$= -7.8387458 \quad \text{(calculator answer)}$$
$$= -7.8 \quad \text{(correct answer)}$$

The percent error is again negative, reflecting that the measured value was less than the accepted value.

STUDENT 3

This time we will have a positive percent error because the measured value is greater than the accepted value.

$$\text{percent error} = \frac{(8.99 - 8.93) \text{ g/cm}^3}{8.93 \text{ g/cm}^3} \times 100$$
$$= \frac{0.06}{8.93} \times 100$$
$$= 0.67189249 \quad \text{(calculator answer)}$$
$$= 0.7 \quad \text{(correct answer)}$$

Note that there are no units associated with percent error; g/cm^3, present in both numerator and denominator, cancel.

Practice Exercise 3.18

The accepted length of an aluminum rod is 34.7 cm. In a laboratory setting, three students are asked to experimentally determine the rod's length. Their results are

student 1: 34.4 cm

student 2: 35.0 cm

student 3: 34.8 cm

Calculate the percent error associated with each student's reported length.

3.11 Temperature Scales

Temperature *is a measure of the hotness or coldness of an object*. The most common instrument for measuring temperature is the mercury-in-glass thermometer, which consists of a glass bulb containing mercury sealed to a slender glass capillary tube. The higher the temperature, the farther the mercury will rise in the capillary tube. Graduations on the capillary tube indicate the height of the mercury column in terms of defined units, usually called *degrees*. A tiny superscript circle is used as the symbol for a degree.

Three different temperature scales are in common use—Celsius, Kelvin, and Fahrenheit. Both the Celsius and Kelvin scales are part of the metric measurement system, and the Fahrenheit scale belongs to the English measurement system. Different degree sizes and different reference points are what produce the various temperature scales.

The Celsius scale, named after Anders Celsius (1701–1744), a Swedish astronomer, is the scale most commonly encountered in scientific work. On this scale the boiling and freezing points of water serve as reference points, with the former having a value of 100°C (degrees Celsius) and the latter, 0°C. Thus, there are 100 degree intervals between the two reference points. The Celsius scale was formerly called the centigrade scale.

The Kelvin scale is a close relative of the Celsius scale. The size of the temperature unit is the same on both scales, as is the interval between the reference points. The two scales differ only in the numerical values assigned to the reference points and the names of the units. On the Kelvin scale the boiling point of water is at 373.15 K (kelvins), and the freezing point of water is at 273.15 K. The scale, proposed by the British mathematician and physicist William Kelvin (1824–1907) in 1848, is particularly useful when working with relationships between temperature and pressure–volume behavior of gases (see Sec. 12.3). The degree sign is not used in specifying Kelvin temperatures (321 K instead of 321°K).

A unique feature of the Kelvin scale is that negative temperature readings never occur. The lowest possible temperature thought to be obtainable occurs at 0 on the Kelvin scale. This temperature, known as *absolute zero*, has never been produced experimentally, although scientists have come within a fraction of a degree of reaching it.

The Fahrenheit scale was designed by the German physicist Gabriel Fahrenheit (1686–1736) in the early 1700s. After proposing several scales he finally adopted a system that used a salt–ice mixture and boiling mercury as the reference points. These were the two extremes in temperature available to him at that time. A reading of 0 was assigned to the salt–ice mixture and 600 to the boiling mercury. The distance between these two points was divided into 600 equal parts or degrees. On this scale, water freezes at 32°F (degrees Fahrenheit) and boils at 212°F. Thus, there are 180 degrees between the freezing and boiling points of water on this scale as contrasted to 100 degrees on the Celsius scale and 100 kelvins on the Kelvin scale. Figure 3.11 shows a comparison of the three temperature scales.

The Celsius Scale

Thirty is hot,
Twenty is pleasing;
Ten is quite cool,
And zero is freezing.

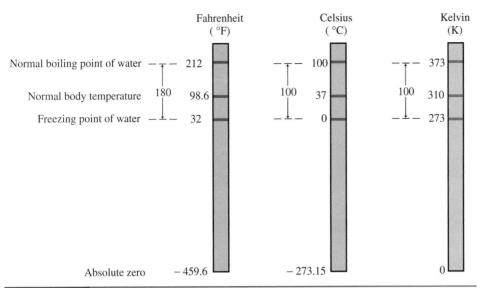

Figure 3.11

The relationships among the Fahrenheit, Celsius, and Kelvin temperature scales.

When changing a temperature reading on one scale to its equivalent on another scale, we must take two factors into consideration: (1) the size of the unit on the two scales may differ, and (2) the zero points on the two scales do not coincide.

Difference in unit size will be a factor any time the Fahrenheit scale is involved in a conversion process. The conversion factors necessary to relate the size of the Fahrenheit degree to the size of the Celsius degree or the kelvin are obtainable from the information in Figure 3.11. From that figure we see that 180 Fahrenheit degrees are equivalent to 100 Celsius degrees or kelvins. Using this relationship and the fact that $\frac{180}{100} = \frac{9}{5}$, we obtain the following equalities.

$$5 \text{ Celsius degrees} = 9 \text{ Fahrenheit degrees}$$

$$5 \text{ kelvins} = 9 \text{ Fahrenheit degrees}$$

Conversion factors derived from these equalities will contain an infinite number of significant figures; that is, they are exact conversion factors.

Adjustment for differing zero-point locations is carried out by considering how many degrees above or below the freezing point of water (the ice point) the original temperature is. Examples 3.19 and 3.20 show how this zero-point adjustment is carried out in addition to illustrating the use of temperature scale conversion factors.

EXAMPLE 3.19

"Fahrenheit-to-Celsius" Temperature Conversion

An oven for baking pizza operates at approximately 525°F. What is the equivalent temperature on the Celsius scale?

SOLUTION

First, we determine the number of degrees between the ice point (freezing point of water) and the given temperature on the original scale.

$$525°F - 32°F = 493°F \text{ above the ice point}$$

How hot is hot?

100°C	Boiling point of water
800°C	Campfire
875°C	Cigarette ember
1,600°C	Gas stove fire
2,300°C	Filament of light bulb
7,500°C	Surface of the sun
30,000°C	Typical lightning bolt

Second, we convert from Fahrenheit units to Celsius units.

$$493 \text{ Fahrenheit degrees} \times \frac{5 \text{ Celsius degrees}}{9 \text{ Fahrenheit degrees}} = 273.88888 \text{ Celsius degrees}$$
$$\text{(calculator answer)}$$

$$= 274 \text{ Celsius degrees}$$
$$\text{(correct answer)}$$

Third, taking into account the ice point on the new scale, we determine the new temperature. On the Celsius scale, the temperature will be 274 degrees above the ice point. Since the ice point is 0°C, the new temperature will be 274°C.

Practice Exercise 3.19

A comfortable temperature for bath water is 95°F. What is the equivalent temperature on the Celsius scale?

EXAMPLE 3.20

"Celsius-to-Fahrenheit" Temperature Conversion

The ozone layer over Antarctica thins dramatically every year during spring. Antarctica's unusual winter weather conditions of extreme cold (-85°C) and total darkness are necessary prerequisites for the occurrence of the chemical reactions that lead to ozone depletion. What is the equivalent temperature on the Fahrenheit scale of the Antarctica's winter temperature of -85°C?

SOLUTION

First, we determine how many degrees there are between the original temperature and the ice point. On the Celsius scale this will always be equal numerically to the original temperature, since the ice point is 0°C.

$$-85°C = 85 \text{ Celsius degrees below the ice point}$$

Second, we change this number of degrees from Celsius units to Fahrenheit units.

$$85 \text{ Celsius degrees} \times \frac{9 \text{ Fahrenheit degrees}}{5 \text{ Celsius degrees}} = 153 \text{ Fahrenheit degrees}$$

Significant-figure considerations for changes in size of degree (Celsius to Fahrenheit or vice versa) represent an exception to the general rules for handling significant figures. Although size of degree changes involve a multiplication, position of uncertainty (addition–subtraction rules) is used to determine how the calculator answer should be modified to give the correct answer. If the original number of degrees is known to tenths, then the new number of degrees is also specified to tenths. Or, as in our specific problem here, if the original number of degrees is known to the closest degree (± 1), then the new number of degrees is specified to the closest degree.

    Third, taking into account the ice point on the new scale, we determine the new temperature. The new temperature will be 153 Fahrenheit degrees below the ice point on the Fahrenheit scale.

$$32°F - 153°F = -121°F \quad \text{(calculator and correct answer)}$$

CHEMICAL EXTENSION

Since the mid-1970s, scientists have observed a seasonal thinning (depletion) of ozone in the upper atmosphere (stratosphere) above Antarctica. This phenomenon, which is commonly called the *ozone hole*, occurs in September and October of each year, the beginning of the Antarctic spring. Up to 70% of the ozone above Antarctica is lost during these two months.

Winter conditions in Antarctica include extreme cold (coldest location on Earth) and total darkness. When sunlight appears in the spring, it triggers the chemical reactions that lead to ozone depletion. By the end of November, weather conditions are such that the ozone-depletion reactions stop. Then the ozone hole disappears as air from nonpolar areas flows into the polar region, replenishing the depleted ozone levels.

Practice Exercise 3.20

The temperature at the bottom of a blast furnace used in the production of iron (steel-making) is measured at 1935°C. What is the equivalent temperature on the Fahrenheit scale?

Examples 3.19 and 3.20 point out that Fahrenheit–Celsius temperature-scale conversions are more complicated than the unit conversions of the last three sections. Not only is multiplication by a conversion factor required, but also addition and subtraction.

The relationship between the Kelvin and Celsius scales is very simple because the sizes of the degree unit is the same. No conversion factors are needed. All that is required is an adjustment for the differing zero points. This adjustment involves the number 273, the number of units by which the two scales are offset from each other. The adjustment factor is specifically 273, 273.2, or 273.15 depending on the uncertainty in the temperature measurement. Since temperatures are most often stated in terms of a whole number of units (31°C, 43°C, etc.), 273 is the most used adjustment factor. However, the other two factors are needed when dealing with temperatures involving tenths or hundredths of a degree (31.5°C, 43.12°C, etc.).

To change a Celsius temperature to the Kelvin scale we add the adjustment factor 273.

$$K = °C + 273$$

To change a Kelvin temperature to Celsius we subtract this same adjustment factor.

$$°C = K - 273$$

Note that the symbol for the kelvin is K, not °K.

The relationship between the Fahrenheit scale and the Celsius scale can also be stated in an equation format.

$$°F = \frac{9}{5}(°C) + 32$$

$$°C = \frac{5}{9}(°F - 32)$$

Some students prefer to use these equations rather than the dimensional analysis approach used in Examples 3.19 and 3.20. The use of these equations is illustrated in Example 3.21.

EXAMPLE 3.21

Temperature Scale Conversions

A person suffering from heat stroke is found to have a body temperature of 41.1°C. What is this temperature on (a) the Fahrenheit scale? (b) the Kelvin scale?

SOLUTION

(a) Substituting into the Celsius-to-Fahrenheit equation, we get

$$°F = \frac{9}{5}(°C) + 32 = \frac{9}{5}(41.1) + 32$$

$$= 74.0 + 32$$

$$= 106°F \quad \text{(calculator answer)}$$

$$= 106.0°F \quad \text{(correct answer)}$$

The multiplication $\frac{9}{5}(41.1)$ gives the calculator answer 73.98, which is rounded to 74.0, the closest tenth of a degree (see the significant-figure discussion in Example 3.20). Adding 32 (an exact number by definition) to 74.0 gives 106.0 as the correct answer.

(b) Substituting into the Celsius-to-Kelvin equation, we get

$$K = °C + 273.2 = 41.1 + 273.2$$

$$= 314.3 \text{ K} \quad \text{(calculator and correct answer)}$$

Note that we used the adjustment factor 273.2 rather than simply 273 because the given temperature involved tenths of a degree.

Practice Exercise 3.21

A hospital patient with a fever has a temperature of 39.4°C. What is this temperature on **(a)** the Fahrenheit scale? **(b)** the Kelvin scale?

Temperature Readings and Significant Figures

Temperatures encountered in everyday situations are generally determined using a thermometer (or equivalent electronic device). Standard procedure in reading a thermometer is to estimate the temperature to the closest degree. Thus, thermometer-obtained temperature readings have an uncertainty in the "ones places," that is, to the closest degree.

When considering significant figures for temperature readings, the preceding operational procedure must be taken into account. Celsius or Fahrenheit temperatures of 10°, 20°, 30°, etc., are considered to have two significant figures even though no decimal point is explicitly shown after the zero (Sec. 2.5). A temperature reading of 100°C or 100°F is considered to possess three significant figures.

Note that the previous paragraph applies only to "ordinary" temperatures obtained using a thermometer (or equivalent electronic device). The temperature at the bottom of a blast furnace (used in steelmaking) often reaches 3600°F. Such a temperature is not a four-significant-figure temperature but rather would be assumed, without further information, to be a two-significant-figure temperature. Obviously, this temperature was not determined by inserting a standard thermometer into the bottom of the blast furnace.

Summary

1. **Metric System of Units** The metric system, the unit system preferred by scientists, is a decimal system in which larger and smaller units of a quantity are related by factors of 10. Prefixes are used to designate relationships between the base unit and larger or smaller units. Base units in the metric system include the gram (mass), liter (volume), and meter (length).

2. **Conversion Factors** A conversion factor is a ratio that converts a measure expressed in one unit to a measure expressed in another unit. English-to-English and metric-to-metric conversion factors are defined quantities. Metric-to-English conversion factors are obtained by measurement and therefore have an uncertainty factor associated with them.

3. **Dimensional Analysis** Dimensional analysis is a general problem-solving method in which the units associated with numbers are used as a guide in setting up calculations. A given quantity is multiplied by one or more conversion factors in such a manner that the

unwanted (original) units are canceled, leaving only the desired units.

4. **Density** Density is the ratio of the mass of an object to the volume occupied by that object. A correct density expression includes a numerical value, a mass unit, and a volume unit. Density can be used as a conversion factor to relate mass to volume or vice versa.

5. **Percentage and Percent Error** Percent means parts per hundred; that is, the number of items of a specified type in a group of 100 items. Percentage values find use as conversion factors in problem-solving situations where dimensional analysis is employed. Percent error compares a measured value with its accepted value.

6. **Temperature Scales** The three major temperature scales are the Celsius, Kelvin, and Fahrenheit scales. The size of the degree for the Celsius and Kelvin scales is the same. They differ only in the numerical values assigned to the reference points. The Fahrenheit scale has a smaller degree size than the other two temperature scales.

Key Terms

The new terms or concepts defined in this chapter are

area   *Sec. 3.4*
conversion factor   *Sec. 3.6*
density   *Sec. 3.8*
dimensional analysis   *Sec. 3.7*
equality conversion factor   *Sec. 3.8*
equivalence conversion factor   *Sec. 3.8*
gram   *Sec. 3.3*

liter   *Sec. 3.4*
mass   *Sec. 3.3*
meter   *Sec. 3.2*
percent   *Sec. 3.10*
percent error   *Sec. 3.10*
temperature   *Sec. 3.11*
volume   *Sec. 3.4*
weight   *Sec. 3.3*

Practice Problems

Metric System Units (Secs. 3.1–3.4)

3.1 Indicate whether each of the following quantities is expressed in metric units.

(a) area of a field: 2.4 acres
(b) thickness of a sheet of paper: 0.0106 centimeter
(c) speed of light: 186,000 miles/second
(d) amount of anti-inflammatory in a capsule: 0.500 gram

3.2 Indicate whether each of the following quantities is expressed in metric units.

(a) recommended daily intake of thiamine: 1.4 milligrams
(b) highway speed limit: 80 kilometers/hour
(c) volume of a copper rod: 67.3 cubic inches
(d) amount of milk in a container: 2.33 deciliters

3.3 Identify the metric prefixes corresponding to each of the following powers of ten or vice versa.

(a) 10^3 (b) 10^{-2} (c) 10^{-6}
(d) nano   (e) mega   (f) milli

3.4 Identify the metric prefixes corresponding to each of the following powers of ten or vice versa.

(a) 10^{-3} (b) 10^{-1} (c) 10^6
(d) pico   (e) micro   (f) centi

3.5 Write the symbol (abbreviation) for each of the following metric system units or vice versa.

(a) microgram   (b) kilometer   (c) centiliter
(d) dm   (e) mL   (f) pg

3.6 Write the symbol (abbreviation) for each of the following metric system units or vice versa.

(a) megagram   (b) microliter   (c) millimeter

(d) cL   (e) nm   (f) kg

3.7 Use the appropriate metric prefix abbreviation to replace the power of ten in each of the following values.

(a) 6.8×10^{-9} m   (b) 3.2×10^{-6} L

(c) 7.23×10^{3} L   (d) 6.5×10^{9} g

3.8 Use the appropriate metric prefix abbreviation to replace the power of ten in each of the following values.

(a) 4.1×10^{-3} L   (b) 9.9×10^{-12} g

(c) 8.721×10^{-2} g   (d) 4.4×10^{6} m

3.9 For each of the pairs of units listed, indicate whether the first unit is larger or smaller than the second unit, and then indicate how many times larger or smaller it is.

(a) centigram, gram

(b) nanogram, microgram

(c) kilogram, decigram

(d) milligram, megagram

3.10 For each of the pairs of units listed, indicate whether the first unit is larger or smaller than the second unit, and then indicate how many times larger or smaller it is.

(a) milliliter, liter

(b) kiloliter, microliter

(c) nanoliter, deciliter

(d) centiliter, megaliter

3.11 What type of quantity (length, mass, area, or volume) do each of the following units represent?

(a) cm^{3} (b) mm   (c) ML   (d) km^{2}

3.12 What type of quantity (length, mass, area, or volume) do each of the following units represent?

(a) L   (b) cm^{2} (c) kg   (d) km^{3}

3.13 For each of the pairs of units listed, indicate which quantity is larger.

(a) 1 centimeter, 1 inch

(b) 1 meter, 1 yard

(c) 1 gram, 1 pound

(d) 1 liter, 1 gallon

3.14 For each of the pairs of units listed, indicate which quantity is larger.

(a) 1 kilometer, 1 mile

(b) 1 milliliter, 1 fluid ounce

(c) 1 kilogram, 1 pound

(d) 1 liter, 1 quart

Units and Mathematical Operations (Secs. 3.4 and 3.5)

3.15 Carry out the following mathematical manipulations of units.

(a) mm × mm × mm   (b) $\dfrac{nm^{3}}{nm}$

(c) $\dfrac{km^{3} \times km}{km^{2} \times km}$ (d) $\dfrac{cm}{sec} \times sec$

3.16 Carry out the following mathematical manipulations of units.

(a) cm × cm^{2} (b) $\dfrac{mm^{2}}{mm}$

(c) $\dfrac{m^{2} \times m}{m^{3}}$ (d) $\dfrac{km}{min} \times min$

3.17 Calculate the area of the following surfaces.

(a) a square surface whose side is 4.52 cm

(b) a rectangular surface whose width is 3.5 m and whose length is 9.2 m

(c) a circle whose radius is 4.579 mm

(d) a triangle whose height is 3.0 mm and whose base is 5.5 mm

3.18 Calculate the area of the following surfaces.

(a) a rectangular surface whose dimensions are 24.3 m and 32.1 m

(b) a circle of radius 2.7213 cm

(c) a triangular object whose base is 12.0 mm and whose height is 8.00 mm

(d) a square surface with sides of 6.7 cm

3.19 Calculate the volume of each of the following objects, each of which has a regular geometrical shape.

(a) a copper block 5.4 cm long, 0.52 cm high, and 3.4 cm wide

(b) a cylindrical piece of cheese that has a height of 7.5 cm and a radius of 2.4 cm

(c) a spherical piece of styrofoam with a radius of 87 mm

(d) a piece of gold in the shape of a cube whose edge is 7.2 cm

3.20 Calculate the volume of each of the following objects, each of which has a regular geometrical shape.

(a) a cube of steel whose edge is 3.5175 mm

(b) a spherical marble with a radius of 1.212 cm

(c) a bar of iron 6.0 m long, 0.10 m wide, and 0.20 m high

(d) a cylindrical rod of copper whose length is 62 mm and whose radius is 3.2 mm

Conversion Factors (Sec. 3.6)

3.21 Write an equation that relates the members of each of the following pairs of time units and also write the two conversion factors associated with the equation.

(a) days and hours   (b) minutes and seconds

(c) decades and centuries   (d) days and years

3.22 Write an equation that relates the members of each of the following pairs of time units and also write the two conversion factors associated with the equation.

(a) days and weeks   **(b)** hours and minutes

(c) months and years   **(d)** years and centuries

3.23 Give the two forms of the conversion factor that relate each of the following pairs of units.

(a) kL and L   **(b)** mg and g

(c) m and cm   **(d)** μsec and sec

3.24 Give the two forms of the conversion factor that relate each of the following pairs of units.

(a) ng and g   **(b)** dL and L

(c) m and Mm   **(d)** psec and sec

3.25 Indicate how each of the following conversion factors should be interpreted in terms of significant figures present.

(a) $\dfrac{1.609 \text{ km}}{1 \text{ mi}}$ **(b)** $\dfrac{10^{-2} \text{ m}}{1 \text{ cm}}$

(c) $\dfrac{28.35 \text{ g}}{1 \text{ oz}}$ **(d)** $\dfrac{12 \text{ in.}}{1 \text{ ft}}$

3.26 Indicate how each of the following conversion factors should be interpreted in terms of significant figures present.

(a) $\dfrac{2.540 \text{ cm}}{1 \text{ in.}}$ **(b)** $\dfrac{453.6 \text{ g}}{1 \text{ lb}}$

(c) $\dfrac{2.113 \text{ pt}}{1 \text{ L}}$ **(d)** $\dfrac{10^{-9} \text{ m}}{1 \text{ nm}}$

3.27 Indicate whether each of the following equations relating units would generate an *exact* set of conversion factors or an *inexact* set of conversion factors relative to significant figures.

(a) 1 dozen = 12 objects

(b) 1 kilogram = 2.20 pounds

(c) 1 minute = 60 seconds

(d) 1 millimeter = 10^{-3} meter

3.28 Indicate whether each of the following equations relating units would generate an *exact* set of conversion factors or an *inexact* set of conversion factors relative to significant figures.

(a) 1 gallon = 16 cups

(b) 1 week = 7 days

(c) 1 pint = 0.4732 liter

(d) 1 mile = 5280 feet

Dimensional Analysis—Metric–Metric Unit Conversions (Sec. 3.7)

3.29 Perform the following metric system conversions using dimensional analysis and one conversion factor.

(a) 25 mg to g   **(b)** 323 km to m

(c) 25.0 L to dL   **(d)** 0.010 g to pg

3.30 Perform the following metric system conversions using dimensional analysis and one conversion factor.

(a) 3.50 nm to m   **(b)** 20,000 μg to g

(c) 250 L to cL   **(d)** 0.225 g to Mg

3.31 Perform the following metric system conversions using dimensional analysis and two conversion factors.

(a) 23 dL to cL   **(b)** 6.00 kg to mg

(c) 6×10^{-3} μL to nL   **(d)** 25 Mm to nm

3.32 Perform the following metric system conversions using dimensional analysis and two conversion factors.

(a) 3.00 km to μm   **(b)** 35.7 cL to nL

(c) 4×10^{4} pm to dm   **(d)** 5×10^{-8} Mg to mg

3.33 A certain chemical process consumes water at a rate of 55 L/sec. Express this water consumption rate in the following units.

(a) L/hr   **(b)** kL/sec   **(c)** dL/min   **(d)** mL/day

3.34 A certain petroleum refinery operation uses hydrogen gas at the rate of 5×10^{4} g/min. Express this hydrogen consumption rate in the following units.

(a) dg/min   **(b)** g/sec   **(c)** kg/hr   **(d)** ng/day

3.35 A surface has an area of 365 m^2. What is this area in each of the following units?

(a) km^2 **(b)** cm^2 **(c)** dm^2 **(d)** Mm^2

3.36 An object has a volume of 365 m^3. What is this volume in each of the following units?

(a) km^3 **(b)** cm^3 **(c)** dm^3 **(d)** Mm^3

3.37 Perform the following metric system conversions using dimensional analysis.

(a) 6.0 cm^2 to m^2 **(b)** 7.2 mm^3 to m^3

(c) 25 $μm^2$ to dm^2 **(d)** 0.023 km^3 to nm^3

3.38 Perform the following metric system conversions using dimensional analysis.

(a) 3.25 km^2 to m^2 **(b)** 0.30 pm^3 to m^3

(c) 9.552 dm^2 to mm^2 **(d)** 5.6 cm^3 to $μm^3$

Dimensional Analysis—Metric–English Unit Conversions (Sec. 3.7)

3.39 The length of a football field, between goal lines, is 100.0 yd. Express this length in the following units.

(a) meters   **(b)** centimeters

(c) kilometers   **(d)** inches

3.40 The length of a football field, between goal posts, is 120.0 yd. Express this length in the following units.

(a) meters   **(b)** millimeters

(c) megameters   **(d)** miles

3.41 A spray steam iron has a capacity of 75 mL of water. Express this water capacity in the following units.

(a) qt   (b) gal   (d) fl oz   (c) cm^3

3.42 An automobile's gasoline tank has a capacity of 64 L. Express this capacity in the following units.

(a) qt   (b) gal   (c) fl oz   (d) cm^3

3.43 The mass of the Earth is estimated to be 6.6×10^{21} tons. Express the mass of the Earth in the following units.

(a) g   (b) kg   (c) ng   (d) oz

3.44 A defensive lineman on a professional football team has a mass of 295 lb. Express the mass of this football player in the following units.

(a) kg   (b) Mg   (c) mg   (d) ton

3.45 A regular-issue U.S. postage stamp is 2.1 cm wide and 2.5 cm long. Express the surface area of this postage stamp in

(a) square centimeters   (b) square inches

3.46 A rectangular piece of concrete has dimensions of 3.6 m and 1.2 m. Express its surface area in

(a) square meters   (b) square yards

3.47 The luggage compartment of an automobile has the dimensions 95 cm $\times$ 105 cm $\times$ 145 cm. What is the volume of this compartment in cubic feet?

3.48 A copper block is 65 cm long, 3.0 cm high, and 4.0 cm wide. What is the volume of this block in cubic inches?

Density (Sec. 3.8)

3.49 Calculate the density, in grams per milliliter, for each of the following.

(a) 25.0 g of ethyl alcohol having a volume of 31.7 mL
(b) 25.0 g of chromium metal having a volume of 3.48 cm^3
(c) 25.0 mL of olive oil having a mass of 22.9 g
(d) 25.0 L of chloroform having a mass of 37,200 g

3.50 Calculate the density, in grams per milliliter, for each of the following.

(a) 15.0 g of sea water having a volume of 14.6 mL
(b) 15.0 g of cork having a volume of 60.0 cm^3
(c) 15.0 mL of kerosene having a mass of 12.3 g
(d) 15.0 L of helium gas having a mass of 2.67 g

3.51 Calculate the mass, in grams, for each of the following.

(a) 22.2 mL of blood plasma ($d = 1.027$ g/mL)
(b) 22.2 cm^3 of gold metal ($d = 19.3$ g/cm^3)
(c) 22.2 L of dry air ($d = 1.29$ g/L)
(d) 22.2 L of urine ($d = 1.027$ g/mL)

3.52 Calculate the mass, in grams, for each of the following.

(a) 33.3 mL of milk ($d = 1.03$ g/mL)
(b) 33.3 cm^3 of bone ($d = 1.8$ g/cm^3)
(c) 33.3 L of hydrogen gas ($d = 0.087$ g/L)
(d) 33.3 L of lead metal ($d = 11.3$ g/cm^3)

3.53 Calculate the volume, in milliliters, for each of the following.

(a) 50.0 g of acetone ($d = 0.791$ g/mL)
(b) 50.0 g of silver metal ($d = 10.40$ g/cm^3)
(c) 50.0 g of carbon monoxide gas ($d = 1.25$ g/L)
(d) 50.0 g of rock salt ($d = 2.18$ g/cm^3)

3.54 Calculate the volume, in milliliters, for each of the following.

(a) 75.0 g of gasoline ($d = 0.56$ g/mL)
(b) 75.0 g of sodium metal ($d = 0.93$ g/cm^3)
(c) 75.0 g of ammonia gas ($d = 0.759$ g/L)
(d) 75.0 g of mercury ($d = 13.6$ g/mL)

3.55 A small bottle contains 2.171 mL of a red liquid. The total mass of the bottle and liquid is 5.261 g. The empty bottle weighs 3.006 g. What is the density, in grams per milliliter, of the liquid?

3.56 A piece of metal weighing 187.6 g is placed in a graduated cylinder containing 225.2 mL of water. The combined volume of solid and liquid is 250.3 mL. What is the density, in grams per milliliter, of the metal?

3.57 An automobile gasoline tank holds 13.0 gal when full. How many pounds of gasoline will it hold, if the gasoline has a density of 0.56 g/mL?

3.58 Liquid sodium metal has a density of 0.93 g/cm^3. How many pounds of liquid sodium are needed to fill a container whose capacity is 15.0 L?

3.59 What mass of the metal chromium (density $= 7.18$ g/cm^3) occupies the same volume as 100.0 g of aluminum (density $= 2.70$ g/cm^3)?

3.60 What volume of the metal nickel (density $= 8.90$ g/cm^3) has the same mass as 100.0 cm^3 of lead (density $= 11.3$ g/cm^3)?

3.61 Water has a density of 1.0 g/mL at room temperature. State whether each of the following will sink or float when dropped in water.

(a) paraffin wax ($d = 0.90$ g/cm^3)
(b) limestone ($d = 2.8$ g/cm^3)

3.62 Air has a density of 1.29 g/L at room temperature. State whether each of the following will rise or sink in air.

(a) helium gas ($d = 0.18$ g/L)
(b) argon gas ($d = 1.78$ g/L)

Equivalence Conversion Factors (Sec. 3.9)

3.63 Classify each of the following as an *equality* conversion factor or an *equivalence* conversion factor.

(a) $\dfrac{1.28 \text{ g salt solution}}{1 \text{ mL salt solution}}$ (b) $\dfrac{10^{-3} \text{ L salt solution}}{1 \text{ mL salt solution}}$

(c) $\dfrac{4 \text{ qt salt solution}}{1 \text{ gal salt solution}}$ (d) $\dfrac{4.5 \text{ mg salt}}{1 \text{ g salt solution}}$

3.64 Classify each of the following as an *equality* conversion factor or an *equivalence* conversion factor.

(a) $\dfrac{1 \text{ mg sugar}}{10^{-3} \text{ g sugar}}$

(b) $\dfrac{1.20 \text{ mg sugar}}{10.0 \text{ kg sugar solution}}$

(c) $\dfrac{2 \text{ pt sugar solution}}{1 \text{ qt sugar solution}}$

(d) $\dfrac{1.20 \text{ mg sugar solution}}{10.0 \text{ mL sugar solution}}$

3.65 If one U.S. dollar is equal to 7.65 Norwegian kroner, what is the value in dollars of 225 kroner?

3.66 If one U.S. dollar is equal to 8.50 Swedish kronor, what is the value in dollars of 25 kronor?

3.67 A pediatric dosage of a certain antibiotic is 32 mg/kg of body weight per day. How much antibiotic, in milligrams per day, should be administered to a child who weighs 15.9 kg?

3.68 A pediatric dosage of a certain analgesic is 225 mg/kg of body weight per day. How much analgesic, in milligrams per day, should be administered to a child who weighs 12.3 kg?

3.69 An arthritis medication dosage is 6.00 mg/kg of body weight. If you give 375 mg of medication to a patient, what is the patient's weight in pounds?

3.70 A narcotic painkiller dosage is 3.00 mg/kg of body weight. If you give 245 mg of medication to a patient, what is the patient's weight in pounds?

3.71 The concentration of sugar in a sugar solution is found to be 2.30 μg/L. What is the sugar concentration in the following units?

(a) mg/L   (b) μg/mL   (c) cg/cL   (d) kg/m^3

3.72 The concentration of salt in a salt solution is found to be 4.5 mg/mL. What is the salt concentration in the following units?

(a) mg/L   (b) pg/mL   (c) g/L   (d) kg/m^3

Percentage and Percent Error (Sec. 3.10)

3.73 An assortment of coins contains 17 pennies, 5 nickels, 2 dimes, 15 quarters, and 1 half dollar. Calculate, to three significant figures, the percentage of the coins that

(a) are nickels

(b) are quarters

(c) have a face value of 10¢ or less

(d) are smaller in diameter than a nickel

3.74 An assortment of coins contains 6 pennies, 14 nickels, 9 dimes, 16 quarters, and 5 half dollars. Calculate, to three significant figures, the percentage of the coins that

(a) are dimes

(b) are pennies

(c) have a face value of 10¢ or more

(d) are larger in diameter than a dime

3.75 A 1980 U.S. penny (a zinc–copper alloy) with a mass of 3.053 g contains 2.902 g of copper. What is the mass percentage in the penny of

(a) copper   (b) zinc

3.76 A 2004 U.S. penny (zinc plated with a thin layer of copper) with a mass of 2.552 g contains 2.488 g of zinc. What is the mass percentage in the penny of

(a) copper   (b) zinc

3.77 How many grams of water are contained in 65.3 g of a mixture of alcohol and water that is 34.2% water by mass?

3.78 How many grams of alcohol are contained in 467 g of a mixture of alcohol and water that is 23.0% alcohol by mass?

3.79 A solution of table salt in water contains 15.3% by mass of table salt. If 437 g of solution are evaporated to dryness, how many grams of table salt will remain?

3.80 A solution of table salt in water contains 15.3% by mass of table salt. In a 542-g sample of this solution, how many grams of water are present?

3.81 Consider the following facts about a candy mixture containing "Gummi bears" and "Gummi worms": (1) 30.9% of the 661 items present are Gummi bears, (2) 23.0% of the Gummi bears are orange, and (3) 6.4% of the orange Gummi bears have only one ear. How many one-eared, orange Gummi bears are present in the candy mixture?

3.82 An analysis of the makeup of a beginning chemistry class gives the following facts: (1) 47.1% of the 87 students are female; (2) 43.9% of the female students are married; and (3) 33.3% of the married female students are sophomores. How many students in the class are sophomore female students who are married?

3.83 The accepted value for the normal boiling point of ethyl alcohol is 78.5°C. In a laboratory setting, three students are asked to experimentally determine the normal boiling point of ethyl alcohol. Their results are

> student 1: 78.0°C
>
> student 2: 77.9°C
>
> student 3: 79.7°C

Calculate the percent error associated with each student's reported boiling point.

3.84 The accepted value for the normal boiling point of benzaldehyde, a substance used as an almond flavoring, is 178°C. In a laboratory setting, three students are asked to experimentally determine the normal boiling point of benzaldehyde. Their results are

> student 1: 175°C
>
> student 2: 190°C
>
> student 3: 181°C

Calculate the percent error associated with each student's reported boiling point.

3.85 The life expectancy for men in the United States rose from 53.6 years in 1920 to 72.6 years in 1995. What is the percentage increase in life expectancy for men during this time period?

3.86 The life expectancy for women in the United States rose from 54.6 years in 1920 to 79.0 years in 1995. What is the percentage increase in life expectancy for women during this time period?

Temperature Scales (Sec. 3.11)

3.87 State the freezing point of water on each of the following temperature scales.

(a) Fahrenheit   **(b)** Celsius   **(c)** Kelvin

3.88 State the boiling point of water on each of the following temperature scales.

(a) Fahrenheit   **(b)** Celsius   **(c)** Kelvin

3.89 Convert each of the following Celsius temperatures to the Fahrenheit scale.

(a) 1251°C   **(b)** 23.2°C   **(c)** −2°C   **(d)** −87°C

3.90 Convert each of the following Celsius temperatures to the Fahrenheit scale.

(a) 950°C   **(b)** 37.3°C   **(c)** −9°C   **(d)** −53°C

3.91 Convert each of the following Fahrenheit temperatures to the Celsius scale.

(a) 2450°F   **(b)** 337°F   **(c)** 11°F   **(d)** −37°F

3.92 Convert each of the following Fahrenheit temperatures to the Celsius scale.

(a) 1530°F   **(b)** 117°F   **(c)** 2°F   **(d)** −133°F

3.93 Convert each of the following temperature readings to the Kelvin scale.

(a) 231°C   **(b)** 231.7°C   **(c)** 231.74°C   **(d)** 37.3°F

3.94 Convert each of the following temperature readings to the Kelvin scale.

(a) 137°C   **(b)** 137.2°C   **(c)** 137.23°C   **(d)** 79.0°F

3.95 Carry out the following temperature scale conversions.

(a) The temperature on a hot summer day is 101°F. What is this temperature in degrees Celsius?

(b) Oxygen, the gas necessary to sustain life, freezes to a solid at −218.4°C. What is this temperature in degrees Fahrenheit?

(c) The melting point of sodium chloride (table salt) is 804°C. What is this temperature in kelvins?

(d) Liquefied nitrogen boils at 77 K. What is this temperature in degrees Fahrenheit?

3.96 Carry out the following temperature scale conversions.

(a) Mercury freezes at 234.3 K. What is this temperature in degrees Celsius?

(b) Normal body temperature for a chickadee is 41.0°C. What is this temperature in degrees Fahrenheit?

(c) A recommended temperature setting for household hot water heaters is 140°F. What is this temperature in degrees Celsius?

(d) The metal aluminum melts at 934 K. What is this temperature in degrees Fahrenheit?

3.97 Which is the higher temperature, −10°C or 10°F?

3.98 Which is the higher temperature, −15°C or 4°F?

3.99 Which is the lower temperature, 223 K or −60°F?

3.100 Which is the lower temperature, 381 K or 98°C?

Additional Problems

3.101 When each of the following measurements of length is converted to miles using the conversion factor (1 mile/5280 feet), how many significant figures should the answer have?

(a) 3.2 ft   **(b)** 3.02 ft   **(c)** 0.33031 ft   **(d)** 2.311 ft

3.102 When each of the following measurements of mass is converted to pounds using the conversion factor (1 pound/16 ounces), how many significant figures should the answer have?

(a) 2 oz   **(b)** 3.3333 oz   **(c)** 23.5 oz   **(d)** 24 oz

3.103 By unit change within the metric system, write each of the following quantities in a form in which the numerical value is a number between 1 and 10.

(a) 6301 m   **(b)** 1442 msec

(c) 1327 μg   **(d)** 0.021 L

3.104 By unit change within the metric system, write each of the following quantities in a form in which the numerical value is a number between 1 and 10.

(a) 3333.0 km   **(b)** 0.003 mg

(c) 1373.00 ML   **(d)** 234 cL

3.105 What power of ten should replace the question mark in each of the following equalities?

(a) 1 km = ? m   **(b)** 1 nm = ? m

(c) 1 m = ? pm   **(d)** 1 m = ? Mm

3.106 What power of ten should replace the question mark in each of the following equalities?

(a) 1 Mg = ? g   **(b)** 1 pg = ? g

(c) 1 g = ? cg   **(d)** 1 g = ? Gg

3.107 The heights of the starting five players on a basketball team are 20.9 dm, 2030 mm, 1.90 m, 0.00183 km, and 203 cm. What is the average height, in centimeters, of these five basketball players? (The average is the sum of the individual values divided by the number of values.)

3.108 The masses for the five heaviest defensive linemen on a football team are 141,000 g, 0.133 Mg, 1.28×10^{8} mg, 126 kg, and 1.22×10^{11} μg. What is the average mass, in grams, of these defensive linemen? (The average is the sum of the individual values divided by the number of values.)

3.109 Using dimensional analysis, convert the following measurements to gallons.

(a) 4.67 L   **(b)** 4.670 L

(c) 4.6700 L   **(d)** 4.67000 L

3.110 Using dimensional analysis, convert the following measurements to pounds.

(a) 4.67 g   **(b)** 4.670 g

(c) 4.6700 g   **(d)** 4.67000 g

3.111 Using the dimensional analysis method of problem solving, set up and solve the following problem: "If your heart beats at a rate of 69 times per minute, how many times will your heart have beat by the time you reach your 8th birthday?" (Ignore the fact that two of the years were leap years.)

3.112 Using the dimensional analysis method of problem solving, set up and solve the following problem: "A chemistry course meets for 50-minute sessions four times a week for 15 weeks (a semester). How many seconds will a student with perfect attendance spend in class during the semester?"

3.113 If your blood has a density of 1.05 g/mL at 20°C, how many grams of blood would you lose if you donated 1.00 pint of blood?

3.114 If your urine has a density of 1.030 g/mL at 20°C, how many pounds of urine would you lose if you eliminated 0.500 pint of urine?

3.115 The density of a solution of sulfuric acid is 1.29 g/cm^3, and it is 38.1% acid by mass. What volume, in milliliters, of the sulfuric acid solution is needed to supply 325 g of sulfuric acid?

3.116 The density of a solution of sulfuric acid is 1.29 g/cm^3, and it is 38.1% acid by mass. What is the mass, in grams, of 25.0 mL of sulfuric acid?

3.117 Levels of blood glucose higher than 400 mg/dL are life threatening. Is either of the following laboratory-measured glucose levels life threatening?

(a) 5000 μg/mL   **(b)** 0.5 g/L

3.118 Levels of blood glucose lower than 40 mg/dL are life threatening. Is either of the following laboratory-measured glucose levels life threatening?

(a) 2000 μg/mL   **(b)** 20,000,000 ng/cL

3.119 The concentration of carbon monoxide, a common air pollutant, is measured at 5.7×10^{-3} μg/cm^3 inside a room. How many grams of carbon monoxide are present in the room if the room's dimensions are 3.5 m × 3.0 m × 3.2 m?

3.120 If the concentration of mercury in a polluted lake is 0.39 μg/mL, what is the total mass of mercury, in kilograms, present in the lake if the lake has a surface area of 125 mi^2 and an average depth of 35 ft?

3.121 A square piece of aluminum foil, 4.0 in. on a side, is found to weigh 0.466 g. What is the thickness of the foil, in millimeters, if the density of the foil is 269 cg/cm^3?

3.122 The density of osmium (the densest metal) is 2260 cg/cm^3. What is the mass, in grams, of a block of osmium with dimensions 5.00 in. × 4.00 in. × 0.25 ft?

3.123 A scientist invented a new temperature scale called the Howard scale (°H) and assigned the boiling and freezing points of water the values 200°H, and −200°H, respectively. What is a temperature of 50°F equivalent to on the Howard scale?

3.124 A scientist invented a new temperature scale called the Scott scale (°S) and assigned the boiling and freezing points of water the values 150°S and −50°S, respectively. What is a temperature of 50°C equivalent to on the Scott scale?

3.125 Given the following information about a bag of Gummi bears, calculate the number of Gummi bears in the bag: (1) 30.3% of the Gummi bears are orange; (2) 8.10% of the orange Gummi bears have only one ear; (3) there are 3 one-eared orange Gummi bears in the bag.

3.126 Given the following information about a class of students, calculate the number of students in the class: (1) 37.3% of the students have blue eyes; (2) 20.0% of the blue-eyed students are left handed; (3) there are 5 blue-eyed, left-handed students in the class.

Cumulative Problems

3.127 A sample of a colorless liquid has a mass of two grams and a volume of four milliliters. Calculate the liquid's density using the following uncertainty specifications and express your answers in scientific notation.

(a) 2.000 g and 4.000 mL

(b) 2.00 g and 4.0 mL

(c) 2.0000 g and 4.0000 mL

(d) 2.000 g and 4.0000 mL

3.128 A one-gram sample of a powdery white solid is found to have a volume of two cubic centimeters. Calculate the solid's density using the following uncertainty specifications and express your answers in scientific notation.

(a) 1.0 g and 2.0 cm^3

(b) 1.000 g and 2.00 cm^3

(c) 1.0000 g and 2.0000 cm^3

(d) 1.000 g and 2.0000 cm^3

3.129 A rectangular box measures 10 cm wide, 200 cm long, and 4 cm high. Calculate the volume of the box, in cubic centimeters, given that all the dimensions are known to

(a) the closest centimeter

(b) the closest tenth of a centimeter

(c) the closest hundredth of a centimeter

(d) two significant figures

3.130 A rectangular room has a width of 9 m and a length of 21 m. Calculate the area of this room, in square meters, given that both dimensions are known to

(a) the closest meter

(b) the closest tenth of a meter

(c) the closest hundredth of a meter

(d) three significant figures

3.131 Indicate which measurement in each of the following sets of measurements has the least uncertainty.

(a) 3.256×10^3 g, 3.256×10^4 g, 3.256×10^5 g

(b) 3.34 g, 3.34 kg, 3.34 mg

(c) 4.31 g, 4.31×10^{-3} kg, 4.31×10^3 mg

(d) 325.0 cg, 3.2500 g, 0.00325 kg

3.132 Indicate which measurement in each of the following sets of measurements has the least uncertainty.

(a) 2.53×10^{-3} m, 2.53×10^{-4} m, 2.53×10^{-5} m

(b) 7.612 m, 7.612 km, 7.612 cm

(c) 6.73 m, 6.73×10^2 cm, 6.73×10^6 μm

(d) 35.300 mm, 3.530 cm, 0.0353 m

3.133 Indicate whether each of the following measurements is equivalent in all aspects to the measurement 1.2120 g.

(a) 0.00121 kg   **(b)** 121.20 cg

(c) 12120 mg   **(d)** 1212.0 μg

3.134 Indicate whether each of the following measurements is equivalent in all aspects to the measurement 3.4020 L.

(a) 0.034020 kL   **(b)** 340.20 cL

(c) 3402 mL   **(d)** 0.0000034020 ML

Answers to Practice Exercises

3.1 **(a)** 10^{-3} **(b)** 10^{-12} **(c)** 10^6
     (d) micro-   **(e)** kilo-   **(f)** deci-

3.2 **(a)** 23 ft^2 **(b)** 18.1 mm^3 **(c)** 2.50 yd^2 **(d)** 14 cm^3

3.3 2.03×10^{-4} g

3.4 3.2×10^6 nm

3.5 5.1×10^{13} fl oz

3.6 190 gal/hr

3.7 7.992 kg

3.8 37 gal

3.9 **(a)** 37.5 cm^3 **(b)** 2.29 in.^{3}

3.10 22.5 g/cm^3

3.11 614 g

3.12 31.7 mL

3.13 32.1 British pounds

3.14 8.8×10^{-7} g

3.15 41.6% copper

3.16 5.2 g

3.17 21 cream-filled dark chocolates

3.18 student 1, -0.9%; student 2, 0.9%; student 3, 0.3%

3.19 35°C

3.20 3515°F

3.21 **(a)** 102.9°F   **(b)** 312.6 K

4

Basic Concepts about Matter

4.1 Chemistry—The Study of Matter

Chemistry *is the scientific discipline concerned with the characteristics, composition, and transformations of matter.* What is matter? What is it that chemists study? Intuitively, most people have a general feeling for the meaning of the word *matter*. They consider matter to be the materials of the physical universe—that is, the "stuff" from which the universe is made. Such an interpretation is a correct one.

Matter *is anything that has mass and occupies space.* Matter includes all things—both living and nonliving—that can be seen (such as plants, soil, and rocks), as well as things that cannot be seen (such as air and bacteria). Not considered to be matter are the various forms of energy, such as heat, light, and electricity. However, chemists must be concerned with energy as well as with matter, because nearly all changes that matter undergoes involve the release or absorption of energy.

The scope of chemistry is extremely broad, and it touches every aspect of our lives. An iron gate rusting, a chocolate cake baking, the diagnosis and treatment of a heart attack, the propulsion of a jet airliner, and the digesting of food all fall within the realm of chemistry. The key to understanding such diverse processes is an understanding of how matter can be classified into a surprisingly small number of categories. The naturally occurring materials of the universe and the synthetic materials humans have fashioned from them are, indeed, much simpler in makeup than they outwardly appear.

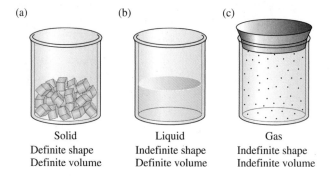

Solid
Definite shape
Definite volume

Liquid
Indefinite shape
Definite volume

Gas
Indefinite shape
Indefinite volume

Figure 4.1

(a) A solid has a definite shape and a definite volume. (b) A liquid has an indefinite shape—it takes the shape of its container—and a definite volume. (c) A gas has an indefinite shape and an indefinite volume—it assumes the shape and volume of its container.

4.2 Physical States of Matter

Three physical states exist for matter: *solid*, *liquid*, and *gas*. The classification of a given matter sample in terms of physical state is based on whether its shape and volume are definite or indefinite.

Solid *is the physical state characterized by a definite shape and a definite volume.* Sugar cubes have the same shape and volume whether they are placed in a large container or on a table top (Fig. 4.1a). For solids in powdered or granulated forms, such as sugar or salt, a quantity of the solid takes the shape of the portion of the container it occupies, but each individual particle has a definite shape and volume. **Liquid** *is the physical state characterized by an indefinite shape and a definite volume.* A liquid always takes the shape of its container to the extent that it fills the container (Fig. 4.1b). **Gas** *is the physical state characterized by an indefinite shape and an indefinite volume.* A gas always completely fills its container, adopting both the container's volume and its shape (Fig. 4.1c).

The state of matter observed for a particular substance is always dependent on the temperature and pressure under which the observation is made. Because we live on a planet characterized by relatively narrow temperature extremes, we tend to fall into the error of believing that the commonly observed states of substances are the only states in which they occur. Under laboratory conditions, states other than the "natural" ones can be obtained for almost all substances. Oxygen, which is nearly always thought of as a gas, can be obtained in the liquid and solid states at very low temperatures. People seldom think of the metal iron as being a gas, its state at extremely high temperatures (above 3000°C). At intermediate temperatures (1535–3000°C), iron is a liquid. Water is one of the very few substances familiar to everyone in all three of its physical states: solid ice, liquid water, and gaseous steam (see Fig. 4.2).

Chapter 11 will consider in detail further properties of the different physical states of matter, changes from one state to another, and the question of why some substances decompose. Suffice it to say at present that physical state is one of the qualities by which matter can be classified.

> **Most people think of rocks as solid, hence the sayings "solid as a rock" and "it is written in stone." But at the high temperatures and pressures within the Earth, rock can turn to molten magma.**

4.3 Properties of Matter

How are the various kinds of matter differentiated from each other? The answer is simple—by their *properties*. **Properties** *are the distinguishing characteristics of a substance that are used in its identification and description.* Just as we recognize a friend by characteristics

such as hair color, walk, tone of voice, or shape of nose, we recognize various chemical substances by how they look and behave. Each chemical substance has a unique set of properties that distinguishes it from all other substances. If two samples of matter have every property identical, they must necessarily be the same substance.

Knowledge of the properties of substances is useful in

1. *Identifying an unknown substance.* Identifying a confiscated drug such as cocaine involves comparing the properties of the drug to those of known cocaine samples.
2. *Distinguishing between different substances.* A dentist can quickly tell the difference between a real tooth and a false tooth because of property differences.
3. *Characterizing a newly discovered substance.* Any new substance must have a unique set of properties different from those of any previously characterized substance.
4. *Predicting the usefulness of a substance for specific applications.* Water-soluble substances obviously should not be used in the manufacture of bathing suits.

There are two general categories of properties of matter: *physical* and *chemical.* A **physical property** *is a characteristic of a substance that can be observed without changing the substance into another substance.* Color, odor, taste, size, physical state, boiling point, melting point, and density are all examples of physical properties.

The physical appearance of a substance may change while a physical property is being determined, but the substance's identity will not. For example, the melting point of a solid cannot be measured without melting the solid, changing it to a liquid. Although the liquid's appearance is much different from that of the solid, the substance is still the same. Its chemical identity has not changed. Hence, melting point is a physical property.

A **chemical property** *is a characteristic of a substance that describes the way the substance undergoes or resists change to form a new substance.* When copper objects are exposed to moist air for long periods of time, they turn green; this is a chemical property of copper. The green coating formed on the copper is a new substance; it results from the reaction of copper metal with the oxygen, carbon dioxide, and water in air. The properties of this green coating are very different from those of metallic copper. On the other hand, gold objects resist change when exposed to air for long periods of time. The lack of reactivity of gold with air is a chemical property of gold.

Most often the changes associated with chemical properties result from the interaction (reaction) of a substance with one or more other substances. However, the presence of a second

> **Chemical properties describe the ability of a substance to form new substances, either by reaction with other substances or by decomposition. Physical properties are associated with a substance's physical existence. They can be determined without reference to any other substance, and determining them causes no change in the identity of the substance.**

Table 4.1	Selected Physical and Chemical Properties of Water

Physical Properties	Chemical Properties
1. Colorless	1. Reacts with bromine to form a mixture of two acids.
2. Odorless	2. Can be decomposed by means of electricity to form hydrogen and oxygen.
3. Boiling point $= 100°C$	3. Reacts vigorously with the metal sodium to produce hydrogen.
4. Freezing point $= 0°C$	4. Does not react with gold even at high temperatures.
5. Density $= 1.000$ g/mL at $4°C$	5. Reacts with carbon monoxide at elevated temperatures to produce carbon dioxide and hydrogen.

substance is not an absolute requirement. Sometimes the presence of energy (usually heat or light) can trigger the change called *decomposition*. The fact that hydrogen peroxide, in the presence of either heat or light, decomposes into the substances water and oxygen is a chemical property of hydrogen peroxide.

When we specify chemical properties, we usually give conditions such as temperature and pressure because they influence the interactions between substances. For example, the gases oxygen and hydrogen are unreactive toward each other at room temperature, but they interact explosively at a temperature of several hundred degrees Celsius.

Selected physical and chemical properties of water are contrasted in Table 4.1. Note how the chemical properties of water cannot be described without reference to other substances. It does not make sense to say simply that a substance reacts. The substance that it interacts with must be specified because it might interact with many different substances.

The properties of a substance may be classified in a second manner—as *intensive* or *extensive* properties—based on whether they depend on the *amount* of substance present. An **intensive property** *is a property that is independent of the amount of substance present.* Temperature, color, melting point, and density are all intensive properties; they are the same for a small sample and for a large one of the same substance. All chemical properties of a substance are intensive properties. In general, intensive properties are useful in the identification of a pure substance. An **extensive property** *is a property that depends on the amount of substance present.* The mass, length, and volume of a substance are examples of extensive properties. Values of the same extensive property are additive. For example, two pieces of aluminum have a combined mass that is the sum of the individual masses. Intensive properties, such as melting point and density, are not additive.

> Color is an intensive property that cannot always be used for identification. This property can "fool you" if particle size for a substance is extremely small. For example, silver is a silvery-white metal with a high luster; however, in the finely divided state, metallic silver appears black.

4.4   Changes in Matter

Changes in matter are common and familiar occurrences. Changes take place when food is digested, paper is burned, and iron rusts (see Fig. 4.3). Like properties of matter, changes in matter are classified into two categories: *physical* and *chemical*.

A **physical change** *is a process in which a substance changes its physical appearance but not its chemical composition.* A new substance is never formed as a result of a physical change.

A change in physical state is the most common type of physical change. The melting of ice, the freezing of liquid water, the conversion of liquid water into steam (evaporation), the condensation of steam to water, the sublimation of ice in cold weather, and the formation of snow crystals in clouds in the winter (deposition) all represent changes of state. The terminology used in describing changes of state, with the exception of the terms *sublimation* and *deposition*, should be familiar to almost everyone. Although the processes of sublimation

> When roasted coffee is ground, the rich aroma is due to the sublimation of coffee components.

> Frost on a windowpane is an example of deposition.

Figure 4.3

The rusting of an iron pipe is an example of a naturally occurring chemical change. (Jack Denmid/Photo Researchers, Inc.)

and deposition—going from a solid directly to the gaseous state or vice versa—are not common, they are encountered in everyday life. Dry ice sublimes, as do mothballs placed in a clothing storage area. As mentioned previously, ice or snow forming in clouds is an example of deposition. Figure 4.4 summarizes the terminology used in describing changes of state.

In any change of state, the composition of the substance undergoing change remains the same even though its physical state and outward appearance have changed. The melting of ice does not produce a new substance. The substance is water before and after the change. Similarly, the steam produced from boiling water is still water. Changes such as these illustrate that matter can change in appearance without undergoing a change in chemical composition.

A **chemical change** *is a process in which a substance undergoes a change in chemical composition.* The creation of one or more new substances is always a characteristic of a chemical change. Carbon dioxide and water are two new substances produced when the chemical change associated with the burning of gasoline occurs. Ashes, carbon dioxide, and water are among the new substances produced when wood is burned. Chemical changes are often called *chemical reactions*. A **chemical reaction** *is a process in which at least one new substance is produced as a result of chemical change.*

> **Not all physical changes involve a change of state. Pulverizing an aspirin tablet into a fine powder and cutting a piece of adhesive tape into small pieces are examples of physical changes that involve only the solid state.**

Figure 4.4

Terminology associated with physical changes of state.

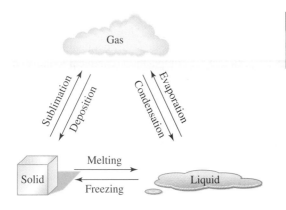

Table 4.2	Classification of Changes as Physical or Chemical
Change	**Classification**
Rusting of iron	chemical
Melting of snow	physical
Sharpening a pencil	physical
Digesting food	chemical
Taking a bite of food	physical
Burning gasoline	chemical
Slicing an onion	physical
Detonation of dynamite	chemical
Souring of milk	chemical
Breaking of glass	physical

Table 4.2 classifies a number of changes for matter as being either physical or chemical.

Most changes for matter can easily be classified as either physical or chemical. However, not all changes are "black" or "white." There are some "gray" areas. For example, the formation of certain solutions falls in the "gray" area. Common salt dissolves easily in water to form a solution of saltwater. The salt can easily be recovered by the physical process of evaporating the water. When gaseous hydrogen chloride is dissolved in water, again a solution results; but in this case the starting materials cannot be easily recovered by evaporation. The formation of saltwater is considered a physical change because the original components can be recovered in an unchanged form using physical methods. The second solution presents classification problems because of the possibility that a chemical reaction took place.

The changes involved in the cooking of an egg also present classification problems. The cooked egg contains the same structural units as the uncooked egg. However, some changes in structural arrangement have taken place, so is the change physical or chemical? Despite the existence of "gray" areas, we shall continue to use the concepts of physical and chemical change because their usefulness far outweighs the problems created by a few exceptions.

Use of the Terms "Physical" and "Chemical"

Generalizations concerning the use of the terms "physical" and "chemical" are in order.

The message of the "modifiers" *physical* and *chemical* is constant: *physical* denotes no change in composition, and *chemical* denotes change in composition or resistance to such change.

1. Whenever the term *physical* is used to modify another term, as in physical property or physical change, it always conveys the idea that the composition (chemical identity) of the substance involved *did not change*.

2. Whenever the term *chemical* is used to modify another term, as in chemical property or chemical change, it always conveys the idea that the composition (chemical identity) of the substance(s) involved *did change* or *successfully resisted change* as the result of an external challenge to identity.

Based on these generalizations, we note that techniques used to accomplish physical change are called physical methods or physical means. Chemical methods and chemical means are used to bring about chemical change. A physical separation would be a separation process in which none of the components experienced composition changes. Composition changes would be part of a chemical separation process.

EXAMPLE 4.1

Correct Use of the Terms *Physical* and *Chemical*

Correctly complete each of the following sentences by placing the word *physical* or *chemical* in the blank.

(a) The fact that pure aspirin melts at 143°C is a _____ property of aspirin.

(b) The fact that sodium metal explosively interacts with water to produce hydrogen gas is a _____ property of sodium.

(c) Straightening a bent nail with a hammer is an example of a _____ change.

(d) Draining the water off from a water-and-hard-boiled-egg mixture is an example of a _____ separation.

SOLUTION

(a) Physical. Changing solid aspirin to liquid aspirin (melting) does not produce any new substances. We still have aspirin.

(b) Chemical. A new substance, hydrogen, is produced.

(c) Physical. The nail is still a nail.

(d) Physical. We started out with water and eggs and after the separation we still have water and eggs.

Practice Exercise 4.1

Correctly complete each of the following sentences by placing the word *physical* or *chemical* in the blank.

(a) The fact that the metal gold is yellow in color is a _____ property of gold.

(b) The process of water evaporating from a lake represents a _____ change.

(c) The stirring of orange juice using a wooden spoon is an example of a _____ technique.

(d) Lighting a match is an example of a _____ change.

> **Answers to practice exercises are located at the end of the chapter.**

4.5 Pure Substances and Mixtures

In addition to its classification by physical state (Sec. 4.2), matter can also be classified in terms of its chemical composition as a *pure substance* or a *mixture*. A **pure substance** *is a single kind of matter that cannot be separated into other kinds of matter using physical means*. All samples of a pure substance contain only that substance and nothing else. Pure water is water and nothing else. Pure sucrose (table sugar) contains only that substance and nothing else.

A pure substance always has a definite and constant composition. This invariant composition dictates that the properties of a pure substance are always the same under a given set of conditions. Collectively, these definite and constant physical and chemical properties constitute the means for identification of the pure substance.

A **mixture** *is a physical combination of two or more pure substances in which each substance retains its own chemical identity*. Components of a mixture retain their identity because they are physically mixed rather than chemically combined. Consider, for example, a mixture of salt and pepper (see Fig. 4.5). Close examination of such a mixture will show distinct particles of salt and pepper with no obvious interaction between them. The salt particles in the mixture are identical in properties and composition to the salt particles in the

> *Substance* is a general term used to denote any variety of matter. *Pure substance* is a specific term that applies only to matter that contains a single substance.

> All samples of a pure substance, no matter what their source, have the same properties under the same conditions.

Commercial table salt is not just the pure substance sodium chloride. It is a mixture containing small amounts of potassium iodide as a nutritional supplement, dextrose as a stabilizer, and calcium silicate as an anticaking agent (to keep it pouring when it rains) in addition to the sodium chloride.

Most naturally occurring samples of matter are mixtures. Gold and diamond are two of the few naturally occurring pure substances. Despite their scarcity in nature, numerous pure substances are known. They are obtained from natural mixtures by using various types of separation techniques or are synthesized in the laboratory from naturally occurring materials.

salt container, and the pepper particles in the mixture are no different from those in the pepper container.

Once a particular mixture is made up, its composition is constant. However, mixtures of the same components with different compositions can also be made up; thus, mixtures are considered to have variable compositions. Consider the large number of salt and pepper mixtures that could be produced by varying the amounts of the two substances present.

An additional characteristic of any mixture is that its components can often be retrieved intact from the mixture by physical means, that is, without a chemical change. In many cases, the differences in properties of the various components make the separation relatively easy. For example, in our salt–pepper mixture, if the pepper grains were large enough, the two components could be separated manually by picking out all the pepper grains. Alternatively, the separation could be carried out by dissolving the salt particles in water, removing the insoluble pepper particles, and then evaporating the water to recover the salt.

Most mixture separations are not as easy as that for a salt–pepper mixture; expensive instrumentation and numerous separation steps are required. Imagine the logistics involved in the separation of the components of blood, a water-based mixture of varying amounts of proteins, sugar (glucose), salt (sodium chloride), oxygen, carbon dioxide, and other components.

Figure 4.6 summarizes the differences between mixtures and pure substances. Further considerations about mixtures are found in Section 4.6, and Section 4.7 contains further details about pure substances.

Hetero is a prefix that means "different." Its use in the classification term *heterogeneous mixture* focuses on the different properties associated with different parts of the mixture.

4.6 Types of Mixtures: Heterogeneous and Homogeneous

Mixtures are subclassified as *heterogeneous* or *homogeneous*. This subclassification is based on visual recognition of the mixture's components. A **heterogeneous mixture** *is a mixture that contains two or more visually distinguishable phases (parts), each of which has different properties*. A heterogeneous mixture of sand and sugar is said to be

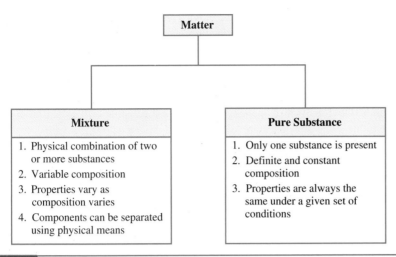

Matter

Mixture	**Pure Substance**
1. Physical combination of two or more substances 2. Variable composition 3. Properties vary as composition varies 4. Components can be separated using physical means	1. Only one substance is present 2. Definite and constant composition 3. Properties are always the same under a given set of conditions

Figure 4.6

A comparison chart contrasting the differences between mixtures and pure substances.

composed of two phases: a sand phase and a sugar phase. Because two or more phases are always present, a nonuniform appearance is characteristic of all heterogeneous mixtures. Naturally occurring heterogeneous mixtures include rocks, soils, and wood (see Fig. 4.7).

The phases in a heterogeneous mixture may or may not be in the same physical state. Set concrete contains a number of phases, all of which are in the solid state. A mixture of sand and water contains two phases, and each is in a different state (solid and liquid). It is possible to have heterogeneous mixtures in which all components are liquids. In order for these mixtures to occur, the mixed liquids must have limited solubility in each other. When this is the case, the mixed liquids form separate layers with the least dense liquid on top. An oil-and-vinegar salad dressing is an example of such a liquid–liquid mixture. Oil-and-vinegar dressing consists of two phases (oil and vinegar) regardless of whether the mixture consists of two separate layers or of oil droplets dispersed throughout the vinegar, a condition caused by shaking the mixture. All the oil droplets together are considered to be a single phase.

American dimes and quarters in current use are heterogeneous mixtures of metals. A copper phase is sandwiched between a copper–nickel phase. Older "silver" dimes and quarters are a homogeneous mixture of silver (90%) and copper (10%).

Figure 4.7

Common materials such as rocks and wood are heterogeneous mixtures. (John Schultz, PAR/NYC, Grant Heilman Photography, Inc.)

Homo is a prefix that means "the same." Its use in the classification term *homogeneous mixture* focuses on the same properties throughout the mixture.

Americans are accustomed to buying products like chocolate bars and peanut butter that look homogeneous. Manufacturers add substances called emulsifying agents to keep these products homogeneous. Without emulsifiers, the ingredients would slowly separate into phases and look unpalatable.

A **homogeneous mixture** *is a mixture that contains only one visually distinguishable phase (part), which has uniform properties throughout.* A sugar–water mixture in which all the sugar has dissolved is a one-phase (homogeneous) system with an appearance that cannot be distinguished from that of pure water. Air is a homogeneous mixture of gases, motor oil and gasoline are multicomponent homogeneous mixtures of liquids, and metal alloys such as 14-karat gold (a mixture of copper and gold) are examples of homogeneous mixtures of solids. Obviously, homogeneous mixtures are possible only when all components present are in the same physical state.

A thorough intermingling of the components in a homogeneous mixture is required in order for a single phase to exist. Sometimes this occurs almost instantaneously during the preparation of the mixture, as in the addition of alcohol to water. At other times, an extended period of mixing or stirring is required. For example, when a hard sugar cube is added to a container of water, it does not instantaneously dissolve to give a homogeneous solution. Only after much stirring does the sugar completely dissolve. Prior to that point, the mixture is heterogeneous, containing a solid phase (the undissolved sugar cube) and a liquid phase (sugar dissolved in water).

Figure 4.8 contrasts the properties common to all mixtures with those specific for heterogeneous and homogeneous mixtures.

Use of the Terms "Homogeneous" and "Heterogeneous"

A summary of the major concepts developed in both this section and Section 4.5 is presented in Figure 4.9. This summary is based on the interplay between the terms *heterogeneous* and *homogeneous* and the terms *chemical* and *physical*. From this interplay come the new expressions *chemically homogeneous, chemically heterogeneous, physically homogeneous,* and *physically heterogeneous*.

All pure substances are *chemically homogeneous*. Only one substance can be present in a chemically homogeneous material. Mixtures, which by definition must contain two or

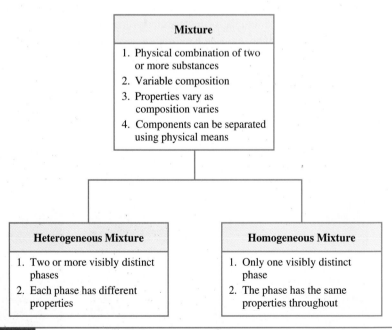

Mixture

1. Physical combination of two or more substances
2. Variable composition
3. Properties vary as composition varies
4. Components can be separated using physical means

Heterogeneous Mixture

1. Two or more visibly distinct phases
2. Each phase has different properties

Homogeneous Mixture

1. Only one visibly distinct phase
2. The phase has the same properties throughout

Figure 4.8

A comparison chart contrasting the properties common to all mixtures with those specific for heterogeneous and homogeneous mixtures.

Chemically homogeneous
Physically homogeneous

One substance and one phase

(a) Pure water

Chemically heterogeneous
Physically homogeneous

Two substances and one phase

(b) Sugar water

Chemically heterogeneous
Physically heterogeneous

Two substances and two phases

(c) Oil and water

Chemically homogeneous
Physically heterogeneous

One substance and two phases

(d) Ice and water

Figure 4.9

Describing a sample of matter using the terms *chemically heterogeneous, physically heterogeneous, chemically homogeneous*, and *physically homogeneous*.

more substances, are always *chemically heterogeneous*. The term *physically homogeneous* describes materials consisting of only one phase. If two or more phases are present, then the term *physically heterogeneous* applies.

Pure water (Fig. 4.9a) is both chemically homogeneous and physically homogeneous because only one substance and one phase are present. Water with some sugar dissolved in it (Fig. 4.9b) is physically homogeneous with only one phase present, but is chemically heterogeneous because two substances, sugar and water, are present. A mixture of oil and water (Fig. 4.9c) is both chemically and physically heterogeneous because it contains two substances and two phases. An ice cube in liquid water (Fig. 4.9d) represents the somewhat unusual situation of chemical homogeneity and physical heterogeneity—one substance and two phases. This combination occurs only when a single substance is present in two or more physical states (solid, liquid, gas).

> **Blood appears homogeneous to the naked eye. But when looked at under a microscope, it can be seen to be a heterogeneous mixture of red and white blood cells and liquid called plasma.**

4.7 Types of Pure Substances: Elements and Compounds

Chemists have isolated and characterized an estimated 9 million pure substances. A very small number of these pure substances, 113 to be exact, are different from all the others. They are *elements*. All the rest, the remaining millions, are *compounds*. What distinguishes an element from a compound?

An **element** *is a pure substance that cannot be broken down into simpler pure substances using ordinary chemical means such as a chemical reaction, an electric current, heat, or a beam of light.* The metals gold, silver, and copper are elements as are the gases hydrogen, oxygen, and nitrogen.

A **compound** *is a pure substance that can be broken down into two or more simpler pure substances using chemical means.* Water is a compound. By means of an electric current, water can be broken down into the gases hydrogen and oxygen, both of which are

> **The definition for the term *element* given here will do for now. After considering the concept of atomic number (Sec. 5.6), we will present a more precise definition.**

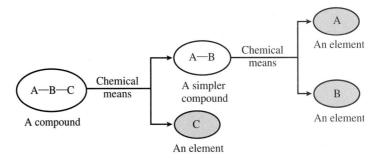

Figure 4.10

Figure 4.10

Stepwise breakdown of a compound containing three elements (A, B, and C) to yield its constituent elements.

> Every compound that exists is made up of some combination of two or more of the 113 known elements. Usually, four or fewer elements are present in a given compound.

elements. Hydrogen peroxide is a compound. Light can be used to decompose it into water and gaseous oxygen.

Ultimately the products from the breakdown of any compound are elements. In practice the breakdown often occurs in steps, with simpler compounds resulting from the intermediate steps, as illustrated in Figure 4.10.

Before a substance can be classified as an element, all possible attempts must be made chemically to subdivide it into simpler substances. If a sample of pure substance, S, is subjected to a decomposition process and two new substances, X and Y, are produced, S would be classified as a compound. If, on the other hand, a number of attempts made chemically to

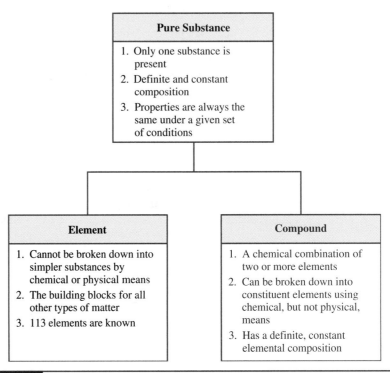

Figure 4.11

A comparison chart contrasting the properties common to all pure substances with those specific for elements and compounds.

subdivide S proved unsuccessful, we might correctly call it an element, but until all possible reactions have proved unsuccessful, such a classification could be in error.

Figure 4.11 contrasts the properties general to all pure substances with those specific for the pure substance subclassifications of elements and compounds.

Even though two or more elements are obtained from decomposition of compounds, compounds are not mixtures. Why is this so? Remember, substances can be combined either physically or chemically. Physical combination of substances produces a mixture. Chemical combination of substances produces a compound, a substance in which combining entities are *bound together*. No such binding occurs during physical combination. Three important distinctions between compounds and mixtures are:

1. Compounds have properties distinctly different from those of the substances that combined to form the compound. The components of mixtures retain their individual properties.
2. Compounds have a definite composition. Mixtures have a variable composition.
3. Physical methods are sufficient to separate the components of a mixture. The components of a compound cannot be separated by physical methods; chemical methods are required.

Example 4.2, which involves two comparisons involving ballpoint pens and their caps, further illustrates the difference between compounds and mixtures.

EXAMPLE 4.2

"Composition" Difference Between a Mixture and a Compound

Consider two boxes with the following contents: the first contains 25 ballpoint pens each with its cap on; the second contains 25 ballpoint pens without caps and 25 ballpoint pen caps. Which box has contents that would be an analogy for a mixture and which box has contents that would be an analogy for a compound?

SOLUTION

The box containing the ballpoint pens with their caps on represents the compound. Two samples withdrawn from this box will always be the same; each will be a ballpoint pen with its cap on. Each item in the box has the same "composition."

The box containing separated ballpoint pens and caps represents the mixture. Two samples withdrawn from this box need not be the same; results could be two ballpoint pens, two caps, or a cap and a ballpoint pen. All items in the box do not have the same "composition."

Practice Exercise 4.2

Consider two boxes with the following contents: the first contains 30 bolts and 30 nuts that fit the bolts; the second contains the same number of bolts and nuts with the difference that each bolt has a nut screwed on it. Which box has contents that would be an analogy for a mixture and which box has contents that would be an analogy for a compound?

Figure 4.12 presents the "thought processes" that a chemist goes through in classifying a sample of matter as a heterogeneous mixture, a homogeneous mixture, a compound, or an element. This figure, which serves as a final summary of matter classification, is based on the following three questions about a sample of matter:

1. Does the sample of matter have the same properties throughout?
2. Are two or more different substances present?
3. Can the pure substance be broken down into simpler substances?

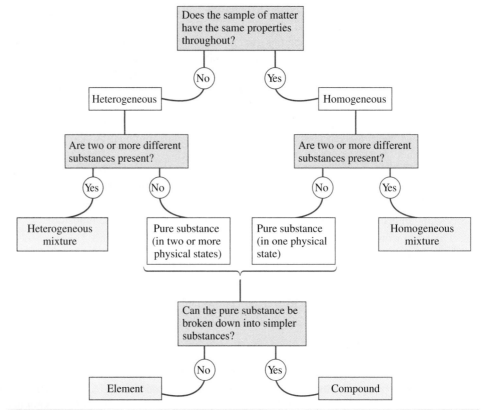

Figure 4.12

The "thought processes" used in classifying matter into various categories.

A student who attended a "university" in the year 1700 would have been taught that 11 elements existed. In 1750, he or she would have learned about 17 elements; in 1800 about 43; in 1850 about 58; in 1900 about 83; in 1950 about 96. Today's total of 113 elements was reached in 1999.

Any increase in the number of known elements from 113 will result from the production of additional synthetic elements. Current chemical theory strongly suggests that all naturally occurring elements have been identified. The isolation of the last of the known naturally occurring elements, rhenium, occurred in 1925.

4.8 Discovery and Abundance of the Elements

The discovery and isolation of the 113 known elements, the building blocks for all matter, have taken place over a period of several centuries. Most of the discoveries have occurred since 1700, with the 1790s and 1880s being the most active decades.

Not all the elements are naturally occurring. Eighty-eight of the 113 elements occur naturally, and the remaining 25 have been synthesized in the laboratory by bombarding samples of naturally occurring elements with small particles. The synthetic (laboratory-produced) elements are all unstable (radioactive) and usually quickly revert back to naturally occurring elements (see Sec. 17.6).

The naturally occurring elements are not evenly distributed on Earth or in the universe as a whole. What is startling is the degree of inequality in the distribution. A very few elements account for the majority of elemental particles (atoms). (An atom is the smallest particle of an element that can exist; see Sec. 5.1.)

Studies of the radiation emitted by stars enable scientists to estimate the elemental composition of the universe. Results indicate that two elements, hydrogen and helium, are absolutely dominant (Fig. 4.13a). All other elements are mere "impurities" when their abundances are compared with those of these two dominant elements. In this big picture, in which Earth is but a tiny microdot, 91% of all elemental particles (atoms) are hydrogen, and nearly all of the remaining 9% are helium.

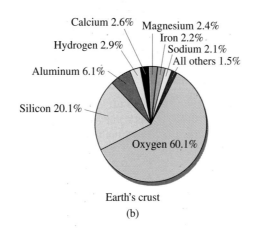

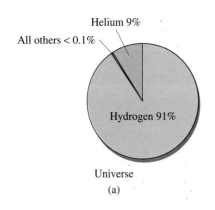

Figure 4.13

Abundance of elements in the universe and in Earth's crust (in atom percent).

The Human Side of Chemistry 1

Joseph Priestley (1733–1804)

Oxygen, the most abundant element within Earth's crust, was isolated in gaseous form for the first time on Monday, August 7, 1774, by the English theologian and "part-time chemist" Joseph Priestley. The experiment leading to oxygen's isolation involved heating an oxygen-containing compound of mercury until it decomposed.

Born in Yorkshire, England, in 1733, Priestley was the eldest son of a nonconformist minister. As a youth he studied languages, logic, and philosophy but never formally studied science. He developed strong religious beliefs of his own and became a Unitarian minister.

As a result of meeting Benjamin Franklin in London in 1766, when Franklin was attempting to settle the taxation dispute between the British government and the American colonists, Priestley became interested in science. At first his studies involved electricity (one of Franklin's interests), and then his interest turned to research on gases.

Coincident with his focus on gases was a move to Leeds, England, to take over a pastorate there. Here he lived next door to a brewery from which he could readily obtain carbon dioxide, the gas produced by fermenting grain products. His carbon dioxide studies led him to the idea of dissolving this gas under pressure in water. The resulting "soda water" became famous all over Europe.

Only three gases were known when Priestley began his gas studies: air, carbon dioxide, and the then recently discovered hydrogen. Numerous new gases were discovered by him, including ammonia, hydrogen chloride, sulfur dioxide, carbon monoxide, hydrogen sulfide, and his most notable discovery, oxygen.

During the period of his scientific studies, Priestley remained an outspoken man on religion and also openly supported the American colonists in their revolt against the British king. In 1791, while living in Birmingham, England, his church, home, and laboratory were burned by an angry mob upset because of his sympathetic attitude toward the American and French revolutions. After fleeing the area in disguise, Priestley eventually emigrated to the United States, where he lived the remaining ten years of his life in relative seclusion in Northumberland, Pennsylvania.

Priestley was the author of more than 150 books, mostly on theological subjects. He always considered theology more important than science. Concerning his scientific accomplishments, a contemporary wrote that "no single person ever discovered so many new and curious substances."

In 1874, on the hundredth anniversary of the discovery of oxygen, the American Chemical Society, now the largest chemical organization in the world, was organized at the residence that Priestley had occupied in Northumberland, Pennsylvania.

If we narrow our view of elemental abundances to the chemical world of humans—the Earth's crust (its waters, atmosphere, and outer solid surfaces)—a different perspective emerges. Again, two elements dominate, but this time they are oxygen and silicon (Fig. 4.13b). The numbers in Figure 4.13 are atom percents, that is, the percentage of total atoms that are of a given type. Note that eight elements (the only elements with atom percents greater than 1%) account for over 98% of total atoms in the Earth's crust. Furthermore, two elements (oxygen and silicon) account for 80% of the atoms that make up the chemical world of humans.

Oxygen, the most abundant element in Earth's crust, was isolated in pure form for the first time in 1774 by the English chemist and theologian Joseph Priestley (see "The Human Side of Chemistry 1"). Discovery years for the other "top five" elements of Earth's crust are 1824 (silicon), 1827 (aluminum), 1766 (hydrogen), and 1808 (calcium).

> **In the materials that humans routinely come in contact with, the Earth's crust, there are more oxygen atoms than all other atoms combined. The vast majority of these oxygen atoms are found in the solid part of the Earth's crust, within compounds present in rocks and minerals, rather than in the atmosphere.**

4.9   Names and Chemical Symbols of the Elements

Each element has a unique name, which in most cases was selected by its discoverer. A wide variety of rationales for choosing a name has been applied. Some elements bear geographical names. Germanium was named after the native country of its German discoverer. The elements francium and polonium acquired names in a similar manner. The elements mercury, uranium, neptunium, and plutonium are all named for planets. Helium gets its name from the Greek word *helios* for "sun," since it was first observed spectroscopically in the Sun's corona during an eclipse. Some elements carry names that relate to specific properties of the element or compounds containing it. Chlorine's name is derived from the Greek *chloros* denoting "greenish yellow," the color of chlorine gas. Iridium gets its name from the Greek *iris* meaning "rainbow" because of the various colors of the compounds from which it was isolated.

The Human Side of Chemistry 2

Jöns Jakob Berzelius (1779–1848)

"These symbols are horrifying. A young student in chemistry might as well learn Hebrew as make himself acquainted with them." Such was the response of a contemporary when, in 1814, Jöns Jakob Berzelius first proposed the system of elemental symbols that forms the basis for the symbol system we use today. This initial opposition turned later to acceptance as the advantages of his symbol system became apparent, although acceptance was slow during his lifetime.

Born in Vaversunda, Sweden, in 1779, Berzelius was the son of a clergyman-schoolmaster. The death of both his parents, before he was 9, caused his youth to be a constant shuffling between relatives.

In 1796, at age 17, partly to get away from his unhappy home "situa-tion," he began the study of medicine in Uppsala, obtaining a degree six years later. His first position was physician to the poor in several neighborhoods in Stockholm. A university appointment in Stockholm as professor of Medicine and Pharmacy came in 1807. Berzelius's true interests were, however, in the field of chemistry, and he left medicine shortly thereafter. In 1815 he obtained an appointment as professor of chemistry.

Berzelius loved experimental (laboratory) work and was probably the best experimental chemist of his generation. His experimental contributions to chemistry are considered even more important than his "chemical symbols." Working in a laboratory with facilities no more elaborate than those of a kitchen, he performed more than 2000 experiments over a ten-year period to determine accurate atomic masses (Sec. 5.8) for 43 of the known 48 elements. The values he obtained are remarkably accurate, as measured by today's stan-dards, an amazing accomplishment considering the crudeness of his laboratory equipment. He reported atomic masses of 35.41 and 63.00 for chlorine and copper, respectively. Today's accepted values for these two elements are, respectively, 35.45 and 63.55. Many laboratory innovations, among them the wash bottle, filter paper, and rubber tubing, came from the work of Berzelius.

In addition to atomic symbols and atomic masses, Berzelius also discovered the elements cerium, thorium, selenium, and silicon. Silicon is the second most abundant element in Earth's crust, exceeded only by oxygen (Sec. 4.8). Silicon–oxygen compounds (silicates) make up most of the Earth's rock, soil, and sand. In recent times, "silicon chips" have become the basis for the computer industry. Indeed, a region of California, near San Francisco, where many different computer-related industries are located, is called Silicon Valley.

In the early 1800s chemists adopted the practice of assigning *chemical symbols* to the elements. A **chemical symbol** *is a one- or two-letter designation for an element derived from the element's name*. In written communications, chemical symbols are used more frequently than names in referring to the elements. The system of chemical symbols now in use originated in 1814 with the Swedish chemist Jöns Jakob Berzelius (see "The Human Side of Chemistry 2").

A complete list of the known elements and their chemical symbols is given in Table 4.3. The chemical symbols and names of the more frequently encountered elements are shown in color in this table.

Fourteen elements have one-letter symbols, and the rest have two-letter symbols. If a symbol consists of a single letter, it is capitalized. In all two-letter symbols the first letter is always capitalized, and the second letter is always lowercase. Double-letter symbols usually, but not always, start with the first letter of the element's English name. The second letter of the symbol is frequently, but not always, the second letter of the name. Consider the elements terbium, technetium, and tellurium, whose symbols are, respectively, Tb, Tc, and Te. Obviously, a variety of choices of second letters is necessary because the first two letters are the same in all three elements' names.

Eleven elements have chemical symbols that bear no relationship to the element's English-language name. In ten of these cases, the symbol is derived from the Latin name of the element; in the case of tungsten, a German name is the symbol source. Most of these elements have been known for hundreds of years and date back to the time when Latin was the language of scientists. Table 4.4 shows the relationship between the chemical symbol and the non-English name of these 11 elements.

The chemical symbols of the elements are also found on the inside front cover of this book. The chart of elements on the left side of the inside front cover is called a *periodic table*. More will be said about it in later chapters. Both cover listings also give other information about the elements. This additional information will be discussed in Chapter 5.

Learning the chemical symbols of the more common elements is an important key to having a successful experience in studying chemistry. Knowledge of chemical symbols is essential for writing chemical formulas (Sec. 5.4), naming compounds (Chapter 8), and writing chemical equations (Sec. 10.2).

Table 4.4	Elements Whose Chemical Symbols Are Derived from a Non-English Name of the Element	
English Name of Element	**Non-English Name of Element**	**Chemical Symbol**
Chemical Symbols from Latin		
Antimony	stibium	Sb
Copper	cuprum	Cu
Gold	aurum	Au
Iron	ferrum	Fe
Lead	plumbum	Pb
Mercury	hydragyrum	Hg
Potassium	kalium	K
Silver	argentum	Ag
Sodium	natrium	Na
Tin	stannum	Sn
Chemical Symbol from German		
Tungsten	wolfram	W

Gold gets its symbol from the Latin *aurum*, which means "shiny." The liquid metal mercury gets its symbol from the Latin *hydragyrum*, which means "runs like water." Lead gets its symbol from the Latin *plumbum*, from which we get the name "plumber," because pipes used to be made out of lead.

The distribution of elements in the human body and other living systems is very different from that found in the Earth's crust. This results from living systems *selectively* taking up matter from their external environment rather than simply accumulating matter representative of their surroundings.

The four most abundant elements in the human body, in terms of atom percent, are hydrogen (63%), oxygen (25.5%), carbon (9.5%), and nitrogen (1.4%). The high abundances of hydrogen and oxygen in the body reflect the high water content of the body. Hydrogen is over twice as abundant as oxygen, largely because water contains hydrogen and oxygen in a 2-to-1 atom ratio.

Table 4.3	**The Chemical Symbols for the Elements***						
Ac	actinium	Ge	germanium	Pt	platinum		
Ag	silver†	H	hydrogen	Pu	plutonium		
Al	aluminum	He	helium	Ra	radium		
Am	americium	Hf	hafnium	Rb	rubidium		
Ar	argon	Hg	mercury†	Re	rhenium		
As	arsenic	Ho	holmium	Rf	rutherfordium		
At	astatine	Hs	hassium	Rh	rhodium		
Au	gold†	I	iodine	Rn	radon		
B	boron	In	indium	Ru	ruthenium		
Ba	barium	Ir	iridium	S	sulfur		
Be	beryllium	K	potassium†	Sb	antimony†		
Bh	bohrium	Kr	krypton	Sc	scandium		
Bi	bismuth	La	lanthanum	Se	selenium		
Bk	berkelium	Li	lithium	Sg	seaborgium		
Br	bromine	Lu	lutetium	Si	silicon		
C	carbon	Lr	lawrencium	Sm	samarium		
Ca	calcium	Md	mendelevium	Sn	tin†		
Cd	cadmium	Mg	magnesium	Sr	strontium		
Ce	cerium	Mn	manganese	Ta	tantalum		
Cf	californium	Mo	molybdenum	Tb	terbium		
Cl	chlorine	Mt	meitnerium	Tc	technetium		
Cm	curium	N	nitrogen	Te	tellurium		
Co	cobalt	Na	sodium†	Th	thorium		
Cr	chromium	Nb	niobium	Ti	titanium		
Cs	cesium	Nd	neodymium	Tl	thallium		
Cu	copper†	Ne	neon	Tm	thulium		
Db	dubnium	Ni	nickel	U	uranium		
Ds	darmstadtium	No	nobelium	V	vanadium		
Dy	dysprosium	Np	neptunium	W	tungsten†		
Er	erbium	O	oxygen	Xe	xenon		
Es	einsteinium	Os	osmium	Y	yttrium		
Eu	europium	P	phosphorus	Yb	ytterbium		
F	fluorine	Pa	protactinium	Zn	zinc		
Fe	iron†	Pb	lead†	Zr	zirconium		
Fm	fermium	Pd	palladium				
Fr	francium	Pm	promethium				
Ga	gallium	Po	polonium				
Gd	gadolinium	Pr	praseodymium				

*Only 110 elements are listed in this table. Yet to be named are element 111—discovered (synthesized) in 1994, element 112—synthesized in 1996, and element 114—synthesized in 1999.

†These elements have symbols that were derived from non-English sources.

Summary

1. **Chemistry and Matter** Chemistry is the branch of science concerned with the characterization, composition, and transformations of matter. Matter, the substances of the physical universe, is anything that has mass and occupies space.

2. **Physical States of Matter** Matter exists in three physical states: solid, liquid, and gas. Physical-state classification is based on whether a substance's shape and volume are definite or indefinite.

3. **Properties of Matter** Properties, the distinguishing characteristics of a substance used in its identification and description, are of two types: physical and chemical. Physical properties can be observed without changing a substance into another substance. Chemical properties are properties that matter exhibits as it undergoes or resists changes in chemical composition. The failure of a substance to undergo change in the presence of another substance is considered a chemical property.

4. **Changes in Matter** Changes that can occur in matter are of two types: physical and chemical. A physical change is a process that does not alter the chemical composition of the substance. No new substances are ever formed as a result of a physical change. A chemical change is a process that involves a change in the chemical composition of the substance. Such changes always involve conversion of the material or materials under consideration into one or more new substances that have properties and composition distinctly different from those of the original materials. Chemical changes are also called chemical reactions.

5. **Pure Substances and Mixtures** All specimens of matter are either pure substances or mixtures. A pure substance is a form of matter that always has a definite and constant composition. A mixture is a physical combination of two or more pure substances in which the pure substances retain their identity.

6. **Types of Mixtures** Mixtures can be classified as heterogeneous or homogeneous on the basis of the visual recognizability of the components present. A heterogeneous mixture contains visibly different parts or phases, each of which has different properties. A homogeneous mixture contains only one phase, which has uniform properties throughout.

7. **Types of Pure Substances** A pure substance can be classified as an element or a compound on the basis of whether or not it can be broken down into two or more simpler substances by ordinary chemical means. Elements cannot be broken down into simpler substances. Compounds yield two or more simpler substances when broken down. There are 113 pure substances that qualify as elements. Not all of the elements are naturally occurring. There are millions of compounds.

8. **Chemical Symbols** Chemical symbols are a shorthand notation for the names of the elements. Most chemical symbols consist of two letters; a few involve a single letter. The first letter of a chemical symbol is always capitalized, and the second letter is always lowercase.

Key Terms

The new terms or concepts defined in this chapter are

chemical change   *Sec. 4.4*

chemical property   *Sec. 4.3*

chemical reaction   *Sec. 4.4*

chemical symbol   *Sec. 4.9*

chemistry   *Sec. 4.1*

compound   *Sec. 4.7*

element   *Sec. 4.7*

extensive property   *Sec. 4.3*

gas   *Sec. 4.2*

heterogeneous mixture   *Sec. 4.6*

homogeneous mixture   *Sec. 4.6*

intensive property   *Sec. 4.3*

liquid   *Sec. 4.2*

matter   *Sec. 4.1*

mixture   *Sec. 4.5*

physical change   *Sec. 4.4*

physical property   *Sec. 4.3*

properties   *Sec. 4.3*

pure substance   *Sec. 4.5*

solid   *Sec. 4.2*

Practice Problems

Physical States of Matter (Sec. 4.2)

4.1 What physical characteristic

(a) distinguishes liquids from solids?
(b) is common to the gaseous and liquid states?

4.2 What physical characteristic

(a) distinguishes gases from liquids?
(b) is common to the liquid and solid states?

4.3 Indicate whether each of the following substances does or does not take the shape of its container and also has a definite volume.

(a) copper wire
(b) oxygen gas
(c) a hard sugar cube
(d) bulk granulated sugar

4.4 Indicate whether each of the following substances does or does not take the shape of its container and also has an indefinite volume.

(a) aluminum powder
(b) carbon dioxide gas
(c) a single grain of table salt
(d) bulk-granulated table salt

4.5 Gold has a melting point of 1063°C and a boiling point of 2966°C. Specify the physical state of gold at each of the following temperatures.

(a) 500°C   **(b)** 1000°C
(c) 2000°C   **(d)** 3000°C

4.6 Oxygen has a melting point of −218°C and a boiling point of −183°C. Specify the physical state of oxygen at each of the following temperatures.

(a) −250°C   **(b)** −200°C
(c) −100°C   **(d)** 100°C

Properties of Matter (Sec. 4.3)

4.7 The following are properties of the metal beryllium. Classify them as physical or chemical.

(a) In powdered form, it burns brilliantly on ignition.
(b) Bulk metal does not react with steam even when red hot.
(c) It has a density of 1.85 g/cm^3 at 20°C.
(d) It is a relatively soft silvery-white metal.

4.8 The following are properties of the metal aluminum. Classify them as physical or chemical.

(a) It generates a colorless, odorless gas when added to sulfuric acid.
(b) It can easily be formed into thin foils.

(c) It is a solid at room temperature.
(d) It is a good conductor of heat.

4.9 Indicate whether each of the following statements describes a physical or chemical property.

(a) Silver compounds discolor the skin by reacting with skin protein.
(b) Hemoglobin gives blood its red color.
(c) Lithium metal is light enough to float on water.
(d) Mercury is a liquid at room temperature.

4.10 Indicate whether each of the following statements describes a physical or chemical property.

(a) Charcoal lighter fluid can be ignited with a match.
(b) Magnesium metal does not react with cold water.
(c) Carbon monoxide is a colorless gas.
(d) Sodium metal is so soft that it can be cut with a sharp knife.

4.11 Classify each of the following as an intensive property or an extensive property.

(a) length   **(b)** density
(c) color   **(d)** boiling point

4.12 Classify each of the following as an intensive property or an extensive property.

(a) mass   **(b)** temperature
(c) volume   **(d)** melting point

4.13 Classify each of the following pairs of characterizations as (1) differing in extensive properties, (2) differing in intensive properties, or (3) differing in both extensive and intensive properties.

(a) 20 g phosphorus at 25°C
    65 g phosphorus at 25°C
(b) 65 g liquid bromine at 25°C
    65 g liquid water at 25°C
(c) 35 g copper metal at 25°C
    35 g copper metal at 100°C
(d) 42 g gold metal at 25°C
    35 g gold metal at 15°C

4.14 Classify each of the following pairs of characterizations as (1) differing in extensive properties, (2) differing in intensive properties, or (3) differing in both extensive and intensive properties.

(a) 20 g sulfur at 25°C
    65 g phosphorus at 25°C
(b) 65 g liquid water at 35°C
    65 g liquid water at 25°C
(c) 65 g silver metal at 25°C
    65 g copper metal at 100°C
(d) 42 g gold metal at 25°C
    35 g gold metal at 25°C

Changes in Matter (Sec. 4.4)

4.15 Classify each of the following changes as physical or chemical.

(a) crushing a dry leaf

(b) hammering a metal into a thin sheet

(c) burning your chemistry textbook

(d) slicing a ham

4.16 Classify each of the following changes as physical or chemical.

(a) evaporation of water from a lake

(b) "scabbing over" of a skin cut

(c) reflection of light by a shiny metal object

(d) melting of candle wax

4.17 Indicate whether each of the following methods for obtaining various substances involves physical or chemical change.

(a) Sodium chloride (salt) is obtained from saltwater by evaporation of the water.

(b) Nitrogen gas is obtained from air by letting the nitrogen boil off from liquid air.

(c) Oxygen gas is obtained by decomposition of the oxygen-containing compound potassium chlorate.

(d) Water is obtained by the high-temperature reaction of gaseous hydrogen with gaseous oxygen.

4.18 Indicate whether each of the following methods for obtaining various substances involves physical or chemical change.

(a) Mercury is obtained by decomposing a mercury–oxygen compound, liberating the oxygen and leaving the mercury behind.

(b) Sand is obtained from a sand–sugar mixture by adding water to the mixture and pouring off the resulting sugar–water solution.

(c) Ammonia is obtained by the high-temperature, high-pressure reaction between hydrogen and nitrogen.

(d) Water is obtained from a sugar–water solution by evaporating off and then collecting the water.

4.19 Complete each of the following sentences by placing the word *chemical* or *physical* in the blank.

(a) The freezing over of a pond's surface is a _____ process.

(b) The crushing of some ice to make ice chips is a _____ procedure.

(c) The destruction of a newspaper through burning it is a _____ process.

(d) Pulverizing a hard sugar cube using a wooden mallet is a _____ procedure.

4.20 Complete each of the following sentences by placing the word *chemical* or *physical* in the blank.

(a) The reflection of light by a shiny metallic object is a _____ process.

(b) The decomposing of a blue powdered material to produce a white glassy-like substance and a gas is a _____ procedure.

(c) A burning candle produces light by _____ means.

(d) The grating of a piece of cheese is a _____ technique.

4.21 Give the name of the change of state associated with each of the following processes.

(a) Water is made into ice cubes.

(b) The inside of your car window fogs up.

(c) Mothballs in the clothes closet disappear with time.

(d) Perspiration dries.

4.22 Give the name of the change of state associated with each of the following processes.

(a) Dry ice disappears without melting.

(b) Snowflakes form.

(c) Dew on the lawn disappears when the sun comes out.

(d) Ice cubes in a soft drink disappear with time.

Pure Substances and Mixtures (Secs. 4.5 and 4.6)

4.23 Consider the following classifications of matter: heterogeneous mixture, homogeneous mixture, and pure substance.

(a) In which of these classifications must two or more substances be present?

(b) In which of these classifications must the composition be uniform throughout?

4.24 Consider the following classifications of matter: heterogeneous mixture, homogeneous mixture, and pure substance.

(a) In which of these classifications must the composition be constant?

(b) In which of these classifications is separation into simpler substances using physical means possible?

4.25 Classify each of the following statements as true or false.

(a) Heterogeneous mixtures must contain three or more substances.

(b) Pure substances cannot have a variable composition.

(c) Substances maintain many of their properties in a heterogeneous mixture but not in a homogeneous mixture.

(d) Pure substances are seldom encountered in the "everyday" world.

4.26 Classify each of the following statements as true or false.

(a) Homogeneous mixtures must contain at least two substances.

(b) Heterogeneous mixtures but not homogeneous mixtures can have a variable composition.

(c) Pure substances cannot be separated into other kinds of matter using physical means.

(d) The number of known pure substances is less than one hundred thousand.

4.27 Assign each of the following descriptions of matter to one of the following categories: heterogeneous mixture, homogeneous mixture, pure substance.

(a) two substances present, two phases present

(b) two substances present, one phase present

(c) three substances present, one phase present

(d) three substances present, three phases present

4.28 Assign each of the following descriptions of matter to one of the following categories: heterogeneous mixture, homogeneous mixture, pure substance.

(a) one substance present, one phase present

(b) one substance present, two phases present

(c) one substance present, three phases present

(d) three substances present, two phases present

4.29 Classify each of the following as a heterogeneous mixture, a homogeneous mixture, or a pure substance. Also indicate how many phases are present. (In each case the substances are present in the same container.)

(a) water and dissolved salt

(b) water and dissolved sugar

(c) water and sand

(d) water and oil

4.30 Classify each of the following as a heterogeneous mixture, a homogeneous mixture, or a pure substance. Also indicate how many phases are present. (In each case the substances are present in the same container.)

(a) liquid water and ice

(b) liquid water, oil, and ice

(c) carbonated water (soda water) and ice

(d) oil, ice, saltwater solution, sugar–water solution, and pieces of copper metal

4.31 In each of the following situations two of the four phrases *chemically homogeneous, chemically heterogeneous, physically homogeneous,* and *physically heterogeneous* apply. Select the two correct phrases for each situation.

(a) pure water

(b) tap water

(c) oil-and-vinegar salad dressing with spices

(d) oil-and-vinegar salad dressing without spices

4.32 In each of the following situations two of the four phrases *chemically homogeneous, chemically heterogeneous, physically homogeneous,* and *physically heterogeneous* apply. Select the two correct phrases for each situation.

(a) water and dissolved table salt

(b) water and white sand

(c) carbonated beverage right after opening container

(d) carbonated beverage that has gone flat

Elements and Compounds (Sec. 4.7)

4.33 Based on the information given, classify each of the pure substances A through D as elements or compounds, or indicate that no such classification is possible because of insufficient information.

(a) Analysis with an elaborate instrument indicates that substance A contains two elements.

(b) Substance B decomposes upon heating.

(c) Heating substance C to 1000°C causes no change in it.

(d) Heating substance D to 500°C causes it to change from a solid to a liquid.

4.34 Based on the information given, classify each of the pure substances A through D as elements or compounds, or indicate that no such classification is possible because of insufficient information.

(a) Substance A cannot be broken down into simpler substances by chemical means.

(b) Substance B cannot be broken down into simpler substances by physical means.

(c) Substance C readily dissolves in water.

(d) Substance D readily reacts with the element chlorine.

4.35 Indicate whether each of the following statements is true or false.

(a) Both elements and compounds are pure substances.

(b) A compound results from the physical combination of two or more elements.

(c) For matter to be heterogeneous, at least two compounds must be present.

(d) Compounds, but not elements, can have a variable composition.

4.36 Indicate whether each of the following statements is true or false.

(a) Compounds can be separated into their constituent elements using chemical means.

(b) Elements can be separated into their constituent compounds using physical means.

(c) A compound must contain at least two elements.

(d) A compound is a physical mixture of different elements.

4.37 Based on the information given in the following equations, classify each of the pure substances A through G as elements or compounds, or indicate that no such classification is possible because of insufficient information.

(a) $A + B \rightarrow C$ **(b)** $D \rightarrow E + F + G$

4.38 Based on the information given in the following equations, classify each of the pure substances A through G as elements or compounds, or indicate that no such classification is possible because of insufficient information.

(a) $A \rightarrow B + C$ **(b)** $D + E \rightarrow F + G$

4.39 Consider two boxes with the following contents: the first contains 50 individual paper clips and 50 individual rubber bands; the second contains the same number of paper clips and rubber bands with the difference that each paper clip is interlocked with a rubber band. Which box has contents that would be an analogy for a mixture and which has contents that would be an analogy for a compound?

4.40 Consider the characteristics of the two breakfast cereals "Crispy Wheat 'N Raisins" and "Crispix." The first cereal contains wheat flakes and raisins. The second cereal contains a fused two-layered flake, one side of which is rice and the other side corn. Characterize the properties of these two cereals that make one an analogy for a mixture and the other an analogy for a compound.

4.41 Based on the notation used in Figure 4.10, classify each of the following pairs of substances as (1) two elements, (2) two compounds, (3) an element and a compound, or (4) a single pure substance.

(a) (Q—X) and (Q) **(b)** (Q—X) and (X)

(c) (Q) and (X) **(d)** (Q—X) and (Q—X)

4.42 Based on the notation used in Figure 4.10, classify each of the following pairs of substances as (1) two or more elements, (2) two compounds, (3) an element and a compound, or (4) a single pure substance.

(a) (Q—X) and (C) **(b)** (Q—X) and (Q—C)

(c) (Q) and (X) and (C) **(d)** (Q—X—C) and (Q)

Discovery and Abundance of the Elements (Sec. 4.8)

4.43 Indicate whether each of the following statements about elements is true or false.

(a) All except three of the elements are naturally occurring.

(b) New elements have been identified within the last ten years.

(c) Oxygen is the most abundant element in the universe as a whole.

(d) Two elements account for over 75% of the elemental particles (atoms) in Earth's crust.

4.44 Indicate whether each of the following statements about elements is true or false.

(a) The two most active decades for discovery of new elements were the 1790s and the 1840s.

(b) The majority of the known elements have been discovered since 1900.

(c) Silicon is the second most abundant element in Earth's crust.

(d) At present, 116 elements are known.

4.45 For each of the pairs of elements listed, indicate whether the first listed element is more abundant or less abundant in the Earth's crust, in terms of atom percent, than the second listed element.

(a) silicon and aluminum

(b) calcium and hydrogen

(c) iron and oxygen

(d) sodium and potassium

4.46 For each of the pairs of elements listed, indicate whether the first listed element is more abundant or less abundant in the Earth's crust, in terms of atom percent, than the second listed element.

(a) oxygen and hydrogen

(b) iron and aluminum

(c) calcium and magnesium

(d) copper and sodium

Names and Chemical Symbols of the Elements (Sec. 4.9)

4.47 Give the name of the element associated with each of the following chemical symbols or vice versa.

(a) N **(b)** Ni **(c)** Pb **(d)** Sn
(e) aluminum **(f)** neon **(g)** hydrogen **(h)** uranium

4.48 Give the name of the element associated with each of the following chemical symbols or vice versa.

(a) Li **(b)** He **(c)** F **(d)** Zn
(e) mercury **(f)** chlorine **(g)** gold **(h)** selenium

4.49 Write the chemical symbol for each member of the following pairs of elements.

(a) sodium and sulfur

(b) magnesium and manganese

(c) calcium and cadmium

(d) arsenic and argon

4.50 Write the chemical symbol for each member of the following pairs of elements.

(a) copper and cobalt

(b) potassium and phosphorus

(c) iron and iodine

(d) silicon and silver

4.51 Several elements have chemical symbols that begin with the letter B. For each of the following chemical symbols, give the name of the corresponding element.

(a) B **(b)** Ba **(c)** Be **(d)** Bi **(e)** Bk **(f)** Br

4.52 Several elements have chemical symbols that begin with the letter T. For each of the following chemical symbols, give the name of the corresponding element.

(a) Ta **(b)** Tb **(c)** Tc **(d)** Te **(e)** Th **(f)** Tl

4.53 Each of the following names of elements is spelled incorrectly. Correct the misspellings.

(a) flourine **(b)** zink
(c) potasium **(d)** sulfer

4.54 Each of the following names of elements is spelled incorrectly. Correct the misspellings.

(a) phosphorous **(b)** murcury
(c) clorine **(d)** argone

4.55 Give the English name and symbol for each of the following elements, whose Latin name is

(a) ferrum   **(b)** stannum

(c) natrium   **(d)** aurum

4.56 Give the English name and symbol for each of the following elements, whose Latin name is

(a) kalium   **(b)** argentum

(c) plumbum   **(d)** stibium

4.57 Certain words can be viewed whimsically as sequential combinations of symbols of the elements. For example, the given name Stephen is made up of the following chemical symbol sequence: S-Te-P-He-N. Analyze each of the following given names in a similar manner.

(a) Rebecca   **(b)** Raymond   **(c)** Nancy

(d) Bruce   **(e)** Sharon   **(f)** Alice

4.58 Certain words can be viewed whimsically as sequential combinations of symbols of the elements. For example, the given name Stephen is made up of the following chemical symbol sequence: S-Te-P-He-N. Analyze each of the following given names in a similar manner.

(a) Barbara   **(b)** Eugene   **(c)** Heather

(d) Monica   **(e)** Allan   **(f)** Bryce

Additional Problems

4.59 Carbon monoxide is a colorless, odorless gas that is toxic to humans. It combines with the metal nickel to form nickel carbonyl, a colorless liquid that boils at 43°C.

(a) List all physical properties of substances found in the preceding narrative.

(b) List all chemical properties of substances found in the preceding narrative.

4.60 A hard sugar cube is pulverized, and the resulting granules are heated in air until they discolor and then finally burst into flame and burn.

(a) List all physical changes to substances found in the preceding narrative.

(b) List all chemical changes to substances found in the preceding narrative.

4.61 Assign each of the following descriptions of matter to one of the following categories: element, compound, mixture.

(a) One substance present, one phase present, substance can be decomposed by chemical means

(b) Two substances present, one phase present

(c) One substance present, two elements present

(d) Two elements present, composition is variable

4.62 Assign each of the following descriptions of matter to one of the following categories: element, compound, mixture.

(a) One substance present, one phase present, substance cannot be decomposed by chemical means

(b) One substance present, three elements present

(c) Two substances present, two phases present

(d) Two elements present, composition is definite and constant

4.63 Indicate whether each of the following samples of matter is a heterogeneous mixture, a homogeneous mixture, a compound, or an element.

(a) a colorless single-phase liquid that when boiled away (evaporated) leaves behind a solid white residue

(b) a uniform red liquid with a boiling point of 59°C that cannot be broken down into simpler substances using chemical means

(c) a nonuniform, white crystalline substance, part of which dissolves in water and part of which does not

(d) a colorless single-phase liquid that completely evaporates without decomposition when heated and produces a gas that can be separated into simpler components using physical means

4.64 Indicate whether each of the following samples of matter is a heterogeneous mixture, a homogeneous mixture, a compound, or an element.

(a) a colorless gas, only part of which reacts with hot iron

(b) a "cloudy" liquid that separates into two layers upon standing for two hours

(c) a green solid, all of which melts at the same temperature to produce a liquid that decomposes upon further heating

(d) a colorless gas that cannot be separated into simpler substances using physical means and that reacts with copper to produce both a copper–nitrogen compound and a copper–oxygen compound

4.65 Classify the following as (1) heterogeneous mixture, (2) heterogeneous, but not a mixture, (3) homogeneous mixture, or (4) homogeneous, but not a mixture.

(a) an undissolved sugar cube in water

(b) a partially dissolved sugar cube in water

(c) a completely dissolved sugar cube in water

(d) an ice cube in water

4.66 Classify the following as (1) heterogeneous mixture, (2) heterogeneous, but not a mixture, (3) homogeneous mixture, or (4) homogeneous, but not a mixture.

(a) molten iron metal

(b) solid iron metal

(c) mix of molten and solid iron

(d) solid iron in water

4.67 In which of the following sequences of elements do all of the elements have two-letter symbols?

(a) magnesium, nitrogen, phosphorus

(b) bromine, iron, calcium

(c) aluminum, copper, chlorine

(d) boron, barium, beryllium

4.68 In which of the following sequences of elements do all of the elements have symbols that start with a letter not the first letter of the element's English name?

(a) silver, gold, mercury

(b) copper, helium, neon

(c) cobalt, chromium, sodium

(d) potassium, iron, lead

4.69 The chemical symbols Co and Hf when split into two capital letters produce the symbols of other elements: Co gives C and O and Hf gives H and F. With the help of Table 4.3, identify the other two-letter chemical symbols that when split into two separate capital letters produce the symbols of two other elements.

4.70 Reversal of the letters in the chemical symbols Ni and Ca produce the chemical symbols of other elements (In and Ac). With the help of Table 4.3, identify the other two-letter symbols for which this reversal process produces the symbol of another element.

Cumulative Problems

4.71 Specify the physical state of a pure substance at each of the following conditions or indicate that the state determination is not possible from the information given.

(a) 10°C below its freezing point

(b) 30°C above its melting point

(c) after sublimation has taken place

(d) at its boiling point

4.72 Specify the physical state of a pure substance at each of the following conditions or indicate that the state determination is not possible from the information given.

(a) 10°C below its melting point

(b) 30°C above its freezing point

(c) after decomposition has taken place

(d) after deposition has taken place

4.73 Calculate the following percents, expressing each percent to three significant figures.

(a) percent of the elements that are naturally occurring

(b) percent of the 110 chemical symbols that are one-letter symbols

(c) percent of the elements that have been discovered during the 1900s

(d) percent of elementary particles (atoms) in Earth's crust that are silicon, magnesium, or iron atoms

4.74 Calculate the following percents, expressing each percent to three significant figures.

(a) percent of the elements that are synthetic (laboratory produced)

(b) percent of the 110 chemical symbols in which the first two letters of the element's name are the symbol

(c) percent of the elements that were discovered during the 1800s

(d) percent of elementary particles (atoms) in Earth's crust that are aluminum, hydrogen, or calcium atoms

4.75 The following density determination data were obtained by three students analyzing unknown substances.

student I: mass = 4.32 g   volume = 3.78 mL
student II: mass = 5.73 g   volume = 5.02 mL
student III: mass = 1.52 g   volume = 1.33 mL

(a) Is it likely that the students were working with different unknowns or that they were working with the same substance?

(b) Is it possible to tell from the given data whether the unknowns were elements or compounds?

4.76 The following density determination data were obtained by three students analyzing unknown substances.

student I: mass = 27.2 g   volume = 23.6 mL
student II: mass = 30.3 g   volume = 28.3 mL
student III: mass = 55.6 g   volume = 42.5 mL

(a) Is it likely that the students were working with different unknowns or that they were working with the same substance?

(b) Is it possible to tell from the given data whether the unknowns were elements or compounds?

4.77 Three samples of a substance were subjected to analysis with each sample analyzed by a different technique. The results were:

technique I: 34.1% of Q and 65.9% of X
technique II: 34.12% of Q and 65.88% of X
technique III: 34.12497% of Q and 65.87503% of X

Is the substance that was analyzed likely an element, a compound, or a mixture?

4.78 Three samples of a substance were subjected to analysis with each sample analyzed by a different technique. The results were:

technique I: 34.2% of Z and 65.8% of D
technique II: 36.32% of Z and 63.68% of D
technique III: 37.2111% of Z and 62.7889% of D

Is the substance that was analyzed likely an element, a compound, or a mixture?

Answers to Practice Exercises

4.1 **(a)** physical **(b)** physical **(c)** physical **(d)** chemical

4.2 first box, mixture; second box, compound

5

Atoms, Molecules, Formulas, and Subatomic Particles

5.1 The Atom

Can a sample of a pure substance, say gold, be divided endlessly into smaller and smaller pieces of gold, or is there a limit to the subdivision process whereby a "smallest possible piece" of gold is obtained? In other words, is matter "continuous" or "discontinuous"? This is a concept that was debated by philosophers for many centuries without a conclusion being reached, although the "continuous" concept tended to be favored.

A definitive answer to this "continuous–discontinuous" question is now available. It came in the nineteenth century and is based on scientific experimentation rather than philosophical speculation. In a series of papers published in the period 1803–1807, the English chemist John Dalton (1766–1844; see "The Human Side of Chemistry 3") proposed that matter is *discontinuous;* that is, there is a limit to the process of subdividing matter into smaller and smaller particles. Dalton's proposal was based on data he and other scientists had collected concerning the amounts of different substances that react with each other. His data, inconsistent with the idea of infinitely divisible matter, were compatible with the concept that a limit to the process of physical subdivision of matter exists.

Dalton called these smallest particles of subdivision *atoms.* An **atom** *is the smallest particle of an element that can exist and still have the properties of the element.* Additional research, carried out by many scientists, has now validated Dalton's basic conclusion that the building blocks for all types of matter are atoms. Some of the details of Dalton's original proposals have had to be modified in the light of later, more sophisticated experiments, but the basic concept of atoms remains.

Today, among scientists, the concept that atoms are the building blocks for matter is a foregone conclusion. The large accumulated amount of supporting evidence for atoms is impressive. Key concepts about atoms, in terms of current knowledge, are found in what is known as the *atomic theory of matter*. The **atomic theory of matter** *is a set of five statements that summarizes modern-day scientific thought about atoms.* These five statements are:

1. All matter is made up of small particles called atoms, of which 113 different "types" are known, with each "type" corresponding to a different element.
2. All atoms of a given type are similar to one another and significantly different from all other types.
3. The relative number and arrangement of different types of atoms contained in a pure substance (its composition and structure) determine its identity.
4. Chemical change is a union, separation, or rearrangement of atoms to give new substances.
5. Only whole atoms can participate in or result from any chemical change, since atoms are considered indestructible during such changes.

Just how small is an atom? Atomic dimensions and masses, although not directly measurable, are known quantities obtained by calculation. The data used for the calculations come from measurements made on macroscopic amounts of pure substances.

> The word *atom* is derived from the Greek *atmos* meaning "uncut" or "indivisible." The Greek philosopher Democritus (460–370 B.C.), a proponent of the discontinuous matter concept, was the first to use this term.

The Human Side of Chemistry 3

John Dalton (1766–1844)

Born in Cumberland, England, in 1766, John Dalton was the second son of a poverty-stricken Quaker weaver. His formal education at the village school lasted until age 11. At age 12, Dalton himself was teaching in the village school. Shortly thereafter, he made his first attempts at scientific investigations, recording weather observations.

Throughout his life, Dalton had a particular interest in the study of weather. In 1787 he made his first entry in a notebook entitled "Observations on the Weather." He continued to record temperature, barometric pressure, rainfall, dew point, and so on, for the next 57 years; the last of over 200,000 observa-

tions was made the evening before his death.

In 1793, Dalton moved to Manchester, England, where he remained the rest of his life. He supported himself by private tutoring, which left him time to pursue his scientific investigations on an almost full-time basis. Some of his investigations involved color blindness, a personal affliction of his. He was the first person to describe color blindness.

His interest in meteorology was responsible for his greatest contribution to chemistry, the atomic theory of matter. From "weather" he turned his attention to the nature of the atmosphere and then to the study of gases in general. Dalton's atomic theory, first published in 1808, was based on his observations of the behavior of gases. He is also the formulator of the gas law now called Dalton's law of partial pressures.

Dalton remained a devout Quaker all his life. He was a very poor speaker and was not well received as a lecturer. He shunned honors and never found time for marriage. In later years, honors did come to him, including honorary doctor's degrees from Oxford and Cambridge Universities.

In 1832, some of his colleagues sought to present him to King William IV. Dalton objected because he did not want to wear the court dress. He finally went in the scarlet robes of Oxford University. Quakers do not wear scarlet, but Dalton, being color-blind to red, saw scarlet as gray. So he appeared before the king in scarlet but in gray to himself.

Upon his death in 1844, he was accorded a public funeral in Manchester. Over 40,000 persons passed by his casket, an appropriate tribute to a man who was so instrumental in revolutionizing the science of chemistry.

The diameter of an atom is about 10^{-10} m.

$$1 \text{ atom} \approx 10^{-10} \text{ m}$$

If one were to arrange atoms of this diameter in a straight line, it would take 1 million of them to extend across the dot that serves as a period at the end of this sentence.

The mass of an atom is about 10^{-23} g.

$$1 \text{ atom} \approx 10^{-23} \text{ g}$$

To produce a mass of 1 lb would require about 5×10^{25} such atoms. The number 5×10^{25} is so large that it is difficult to visualize its magnitude. The following comparison "hints" at this number's magnitude. If each of the 6 billion people on Earth were made a millionaire (receiving 1 million $1 bills), we would still need 8 billion other worlds, each inhabited by the same number of millionaires to have 5×10^{25} dollar bills in circulation.

Atoms are incredibly small particles. No one has seen or ever will see an atom with the naked eye. The question may thus be asked: "How can you be absolutely sure that something as minute as an atom really exists?" The achievements of twentieth-century scientific instrumentation have gone a long way toward removing any doubt about the existence of atoms. Electron microscopes, capable of producing magnification factors in the millions, have made it possible to photograph "images" of individual atoms. In 1976 physicists at the University of Chicago were successful in obtaining images of single atoms. One of these images is shown in Figure 5.1.

> **Atoms are very small.**
>
> - Imagine an apple enlarged to the size of this Earth. The atoms in the apple would then be about the size of cherries.
> - The diameter of an atom would have to be increased two hundred million times (2×10^8) to cause it to have the diameter of a penny.

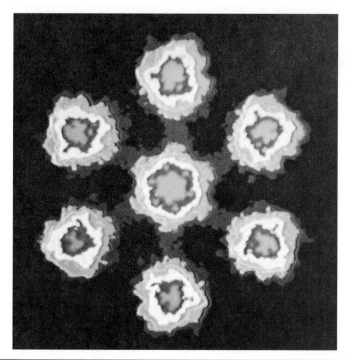

Figure 5.1

The bright spots in this photomicrograph are images of seven uranium atoms. The images were obtained using an electron microscope. *(Courtesy of M. Isaacson, Cornell University, and M. Ohtsuki, The University of Chicago)*

5.2 The Molecule

Free isolated atoms are rarely encountered in nature. Instead, under normal conditions of temperature and pressure, atoms are almost always found together in aggregates or clusters ranging in size from two atoms to numbers too large to count. When the group or cluster of atoms is relatively small and bound together tightly, the resulting entity is called a *molecule*. A **molecule** *is a group of two or more atoms that functions as a unit because the atoms are tightly bound together.* This resultant "package" of atoms behaves in many ways as a single, distinct particle would.

A **diatomic molecule** *is a molecule that contains two atoms.* It is the simplest type of molecule that can exist. Next in complexity are *triatomic* molecules. A **triatomic molecule** *is a molecule that contains three atoms.* Continuing on numerically, we have *tetratomic* molecules, *pentatomic* molecules, and so on.

The atoms contained in a molecule may all be of the same kind, or two or more kinds may be present. On the basis of this observation, molecules are classified into two categories: *homoatomic* and *heteroatomic*. A **homoatomic molecule** *is a molecule in which all atoms present are the same kind.* A pure substance containing homoatomic molecules must be an element. A **heteroatomic molecule** *is a molecule in which two or more different kinds of atoms are present.* Pure substances containing heteroatomic molecules must be compounds. Figure 5.2 shows general models for selected simple heteroatomic molecules.

The fact that homoatomic molecules exist indicates that individual atoms are not always the preferred structural unit for an element. The gaseous elements hydrogen, oxygen, nitrogen, fluorine, and chlorine exist in the form of diatomic molecules. There are four atoms present in a gaseous phosphorus molecule and eight atoms present in a gaseous sulfur molecule (see Fig 5.3). Some guidelines for determining which elements have individual atoms as their basic structural units and which exist in molecular form will be given in Section 7.2.

Reasons for the tendency of atoms to collect together into molecules and information on the binding forces involved are considered in Chapter 7. The important point at this time is that a molecule is a collection of atoms that functions as a single composite unit.

The Latin word *mole* means "a mass." The word *molecule* denotes a "little mass."

A generalized term for molecules containing several atoms is *polyatomic molecule*.

A diatomic molecule
containing one atom of
A and one atom of B

(a)

A triatomic molecule
containing two atoms of
A and one atom of B

(b)

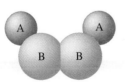

A tetratomic molecule
containing two atoms of
A and two atoms of B

(c)

A tetratomic molecule
containing three atoms of
A and one atom of B

(d)

Figure 5.2

General depictions of various simple heteroatomic molecules using models. Spheres of different sizes and colors represent different kinds of atoms.

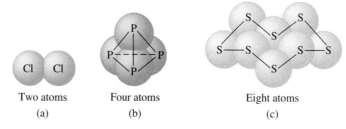

Two atoms   Four atoms   Eight atoms
(a)   (b)   (c)

Figure 5.3

Make-up of the homoatomic molecules present in the elements chlorine (a), phosphorus (b), and sulfur (c).

EXAMPLE 5.1

Classifying Molecules Based on Numbers of and Types of Atoms

Classify each of the following molecules as (1) *diatomic, triatomic,* etc., (2) *homoatomic* or *heteroatomic,* and (3) representing an *element* or a *compound*.

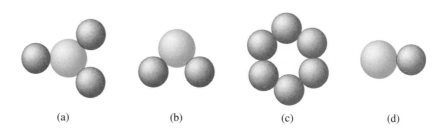

(a)   (b)   (c)   (d)

In Example 3.3 we noted that ozone (as in ozone layer) was a form of oxygen different from the oxygen we breathe. In ozone the molecules are triatomic, and in the oxygen we breathe the molecules are diatomic. In both cases the molecules are homoatomic because oxygen is an element.

SOLUTION

(a) Tetratomic (four atoms); heteroatomic (two kinds of atoms); a compound (two kinds of atoms)

(b) Triatomic (three atoms); heteroatomic (two kinds of atoms); a compound (two kinds of atoms)

(c) Hexatomic (six atoms); homoatomic (one kind of atom); an element (one kind of atom)

(d) Diatomic (two atoms); heteroatomic (two kinds of atoms); a compound (two kinds of atoms)

Practice Exercise 5.1

Classify each of the following molecules as (1) *diatomic, triatomic,* etc., (2) *homoatomic* or *heteroatomic,* and (3) representing an *element* or a *compound*.

Answers to practice exercises are located at the end of the chapter.

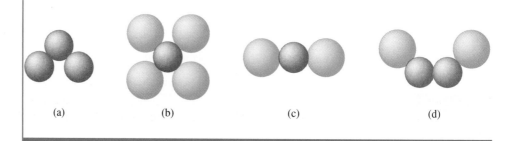

(a)   (b)   (c)   (d)

Many, but not all, compounds have heteroatomic molecules as their basic structural unit. Those compounds that do are called *molecular compounds.* Some compounds in the liquid and solid state, however, are not molecular; that is, the atoms present are not collected together into discrete heteroatomic molecules. These nonmolecular compounds still contain atoms of at least two kinds (a necessary requirement for a compound), but the form of aggregation is different. It involves an extended three-dimensional assembly of positively and negatively charged particles called *ions* (Sec. 7.4). Compounds that contain ions are called *ionic compounds.* The familiar substances sodium chloride (table salt) and calcium carbonate (limestone) are ionic compounds. The reasons some compounds have ionic rather than molecular structures are considered in Section 7.7.

For molecular compounds, the molecule is the smallest particle of the compound capable of a stable independent existence. It is the limit of physical subdivision for the compound. Consider the molecular compound sucrose (table sugar). Continued subdivision of a quantity of table sugar to yield smaller and smaller amounts would ultimately lead to the isolation of one single particle of table sugar—a molecule of table sugar. This molecule of sugar could not be broken down any further and still maintain the physical and chemical properties of table sugar. The sugar molecule could be broken down further by chemical (not physical) means to give atoms, but if that occurred we would no longer have sugar. The *molecule* is the limit of *physical* subdivision. The *atom* is the limit of *chemical* subdivision.

Every molecular compound has as its smallest characteristic unit a *unique* molecule. If two samples had the same molecule as a basic unit, both would have the same properties; thus, they would be one and the same compound. An alternative way of stating the same conclusion is: there is only one kind of molecule for any given molecular substance.

Since every molecule in a sample of a molecular compound is the same as every other molecule in the sample, it is commonly stated that molecular compounds are made up of a single kind of particle. Such terminology is correct as long as it is remembered that the particle referred to is the molecule. In a sample of a molecular compound there are at least two kinds of atoms present but only one kind of molecule.

The properties of molecules are very different from the properties of the atoms that make up the molecules. Molecules do not maintain the properties of their constituent elements. Table sugar is a white crystalline molecular compound with a sweet taste. None of the three elements present in table sugar (carbon, hydrogen, and oxygen) is a white solid or has a sweet taste. Carbon is a black solid, and hydrogen and oxygen are colorless gases.

Figure 5.4 summarizes the relationships between hetero- and homoatomic molecules and elements, compounds, and pure substances.

5.3 Natural and Synthetic Compounds

Approximately 9 million chemical compounds are now known, with more being characterized daily. No end appears to be in sight as to the number of compounds that can and will be prepared in the future. Approximately 10,000 new chemical substances are registered every week with Chemical Abstracts Service, a clearinghouse for new information concerning chemical substances.

Many compounds, perhaps the majority now known, are not naturally occurring substances. These synthetic (laboratory-produced) compounds are legitimate compounds and should not be considered "second class" or "unimportant" simply because they lack the distinction of being natural. Many of the plastics, synthetic fibers, and prescription drugs now in common use are synthetic materials produced through controlled chemical change carried out on an industrial scale.

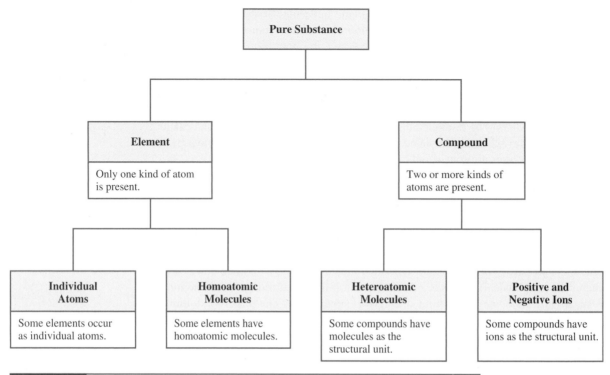

Figure 5.4

A comparison chart contrasting the basic structural units present in elements and compounds.

We have noted that chemists can produce compounds not found in nature. The reverse is also true. Nature is capable of making many compounds, especially those found in living systems, that chemists are not yet able to prepare in the laboratory.

There is a middle ground also. Many compounds that exist in nature can also be produced in the laboratory. A fallacy exists in the thinking of some people concerning these compounds that have "dual origins." A belief still persists that there is a difference between compounds prepared in the laboratory and samples of the same compounds found in nature. This is not true for pure samples of a compound. All pure samples of a compound, regardless of their origin, have the same composition. Since compositions are the same, properties will also be the same. There is no difference, for example, between a laboratory-prepared vitamin and a "natural" vitamin if both are pure samples of the same vitamin, despite frequent claims to the contrary.

> Two examples of compounds that were first discovered in nature and then later produced in a laboratory setting are the antibiotic penicillin G and the antirejection drug cyclosporin. The original source for penicillin G, the first antibiotic to be marketed, was a mold of the Penicillium family, from whence it got its name. Cyclosporin, a drug used to control rejection of a transplanted organ, such as a liver, by a patient's own immune system, was originally isolated from a type of soil fungus.

5.4 Chemical Formulas

A most important piece of information about a compound is its composition. *Chemical formulas* represent a concise means of specifying compound compositions. A **chemical formula** *is a notation made up of the chemical symbols of the elements present in a compound and numerical subscripts (located to the right of each chemical symbol) that indicate the number of atoms of each element present in a structural unit of the compound.*

The chemical formula for the compound we call aspirin is $C_9H_8O_4$. This formula provides us with the following information about an aspirin molecule: three elements are present—carbon (C), hydrogen (H), and oxygen (O)—and 21 atoms are present—9 carbon atoms, 8 hydrogen atoms, and 4 oxygen atoms.

When only one atom of a particular element is present in a molecule of a compound, the element's symbol is written without a numerical subscript in the formula of the compound. In the formula for rubbing alcohol, C_3H_6O, for example, the subscript 1 for the element oxygen is not written.

To write formulas correctly, it is necessary to follow strictly the capitalization rules for elemental symbols (Sec. 4.9). Making the error of capitalizing the second letter of an element's symbol can dramatically alter the meaning of a chemical formula. The formulas $CoCl_2$ and $COCl_2$ illustrate this point; the symbol Co stands for the element cobalt, whereas CO stands for one atom of carbon and one atom of oxygen. The properties of the compounds $CoCl_2$ and $COCl_2$ are dramatically different. The compound $CoCl_2$ is a blue crystalline solid with a melting point of 724°C and a boiling point of 1029°C. The compound $COCl_2$ is a highly toxic colorless gas with a melting point of -118°C and a boiling point of 8°C.

For molecular compounds, chemical formulas give the composition of the molecules making up the compounds. For ionic compounds, which have no molecules, a chemical formula gives the ion ratio found in the compound. For example, the ionic compound sodium oxide contains sodium ions and oxygen ions in a two-to-one ratio—twice as many sodium ions as oxygen ions. The formula of this compound is Na_2O, which expresses the ratio between the two types of ions present. The term *formula unit* is used to describe this smallest ratio between ions. The distinction between the formula unit of an ionic compound and the molecule of a molecular compound is graphically portrayed in Figure 5.5.

Sometimes chemical formulas contain parentheses, an example being $Al_2(SO_4)_3$. The interpretation of this formula is straightforward; in a formula unit there are present two aluminum (Al) ions and three SO_4 groups. The subscript following the parentheses always indicates the number of units in the formula of the polyatomic entity inside the parentheses. As another example, consider the compound $Pb(C_2H_5)_4$. Four units of C_2H_5 are present. In terms of atoms present, the formula $Pb(C_2H_5)_4$ represents 29 atoms: 1 lead (Pb) atom, $4 \times 2 = 8$ carbon (C) atoms, and $4 \times 5 = 20$ hydrogen (H) atoms. The formula could be

> A chemical symbol in a formula stands for one atom of the element. If more than one atom is to be indicated in a formula, a subscript number is used after the symbol.

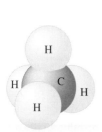

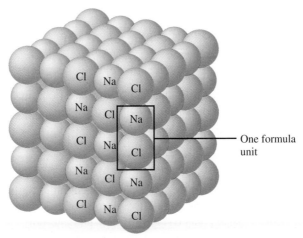

(a) A molecule of the molecular compound methane (CH_4)

(b) A formula unit of the ionic compound sodium chloride (NaCl)

One formula unit

Figure 5.5

A comparison of a formula unit of an ionic compound and a molecule of a molecular compound. A molecule can exist as a separate unit, whereas a formula unit is simply two or more ions "plucked" from a much larger array of ions.

(but is not) written as PbC_8H_{20}. Both versions of the formula convey the same information in terms of atoms present. However, $Pb(C_2H_5)_4$ gives the additional information that the C and H are present as C_2H_5 units and is therefore the preferred way of writing the formula. Further information concerning the use of parentheses (when and why) will be presented in Section 7.8. The important concern now is being able to interpret formulas that contain parentheses in terms of total atoms present. Example 5.2 deals with this skill in greater detail.

EXAMPLE 5.2

Interpreting Chemical Formulas in Terms of Atoms and Elements

Interpret each of the following chemical formulas in terms of how many atoms of each element are present in one structural unit of the substance.

(a) $C_8H_9O_2N$ (acetaminophen, the active ingredient in Tylenol).
(b) $(NH_4)_2C_2O_4$ (ammonium oxalate, used in the manufacture of explosives).
(c) $Ca_{10}(PO_4)_6(OH)_2$ (hydroxyapatite, present in tooth enamel).

SOLUTION

(a) We simply look at the subscripts following the symbols for the elements. This formula indicates that 8 carbon atoms, 9 hydrogen atoms, 2 oxygen atoms, and 1 nitrogen atom are present in one molecule of the compound.

(b) The subscript following the parenthesis, 2, indicates that two NH_4 units are present. Collectively, in these two units, we have 2 nitrogen atoms and $2 \times 4 = 8$ hydrogen atoms. In addition, 2 carbon atoms and 4 oxygen atoms are present.

(c) There are 10 calcium atoms. The amounts of phosphorus, hydrogen, and oxygen are affected by the subscripts outside the parentheses. There are 6 phosphorus atoms and 2 hydrogen atoms present. Oxygen atoms are present in two locations in the formula. There are a total of 26 oxygen atoms: 24 from the PO_4 subunits (6×4) and 2 from the OH subunits (2×1).

Practice Exercise 5.2

Interpret each of the following formulas in terms of how many atoms of each element are present in one structural unit of the substance.

(a) $C_8H_{10}N_4O_2$ (caffeine, an addictive central nervous system stimulant)
(b) $(NH_4)_3PO_4$ (ammonium phosphate, an ingredient in some lawn fertilizers)
(c) $Ca(NO_3)_2$ (calcium nitrate, used in fireworks to give a reddish color)

Figure 5.6 pictorially relates chemical formulas to the matter classifications of element, compound, and mixture. Note in the top third of the diagram that the formulas for molecules of an element contain only one type of atom and thus only one elemental symbol. In the middle third of the diagram we see that formulas for the compounds shown contain two types of atoms and thus two elemental symbols. Finally, in the bottom third of the diagram, we see that in mixtures different types of molecules, with different chemical formulas, must be present.

In addition to formulas, compounds have names. Naming compounds is not as simple as naming elements. Although the nomenclature of elements (Sec. 4.9) has been largely left up to the imagination of their discoverers, extensive sets of systematic rules exist for naming compounds. Rules must be used because of the large number of compounds that exist. Chapter 8 is devoted to compound nomenclature, and we will not worry about naming rules until then. For the time being, our focus will be on the meaning and significance of chemical

Elements

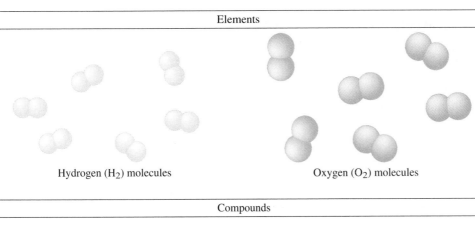

Hydrogen (H₂) molecules   Oxygen (O₂) molecules

Compounds

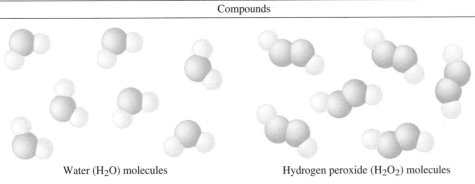

Water (H₂O) molecules   Hydrogen peroxide (H₂O₂) molecules

Mixtures

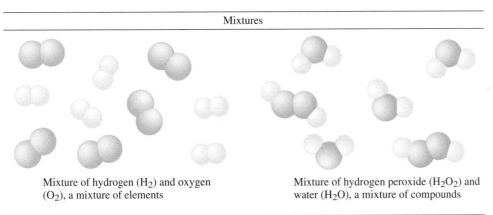

Mixture of hydrogen (H₂) and oxygen (O₂), a mixture of elements

Mixture of hydrogen peroxide (H₂O₂) and water (H₂O), a mixture of compounds

Figure 5.6

A contrast between the chemical formulas for molecules of elements and molecules of compounds and their pictorial representations.

formulas; knowing how to name the compounds that the formulas represent is not a prerequisite for understanding the meaning of formulas.

5.5 Subatomic Particles: Protons, Neutrons, and Electrons

Until the closing decades of the nineteenth century, scientists believed that atoms were solid indivisible spheres without substructure. Today this concept is known to be incorrect. Evidence from a variety of sources, some of which will be discussed in Section 5.9, indicates that atoms themselves are made up of smaller, more fundamental particles called *subatomic particles*.

Table 5.1	Charges and Masses of the Major Subatomic Particles		
	Electron	**Proton**	**Neutron**
Charge	-1	$+1$	0
Actual mass (g)	9.109×10^{-28}	1.673×10^{-24}	1.675×10^{-24}
Relative mass (based on the electron being one unit)	1	1837	1839
Relative mass (based on the neutron being one unit)	0 (1/1839)	1	1

Atoms of all 113 elements contain the same three types of subatomic particles. Different kinds of atoms differ only in the number of the various subatomic particles they contain.

A **subatomic particle** *is a very small particle that is a building block for atoms.* Three major types of subatomic particles exist: the *electron*, the *proton*, and the *neutron*. The key properties of *electrical charge* and *mass* for these three types of subatomic particles are given in Table 5.1. An **electron** *is a subatomic particle that possesses a negative (–) electrical charge.* Electrons were characterized in 1897 by the English physicist Joseph John Thomson (1856–1940). A **proton** *is a subatomic particle that possesses a positive (+) charge.* Protons were discovered in 1886 by the German physicist Eugen Goldstein (1850–1930). A **neutron** *is a subatomic particle that is neutral; that is, it has no charge.* The neutron, the last of the three subatomic particles to be identified, was characterized in 1932 by the English physicist James Chadwick (1891–1974).

Using the mass or relative mass values from Table 5.1, we see that the electron has the smallest mass of the three subatomic particles. Both protons and neutrons are very massive particles compared to the electron, being nearly 2000 times heavier. For most purposes the masses of protons and neutrons can be considered equal, although technically the neutron is slightly heavier (1839 versus 1837 on a relative mass scale where the electron has a value of 1; see Table 5.1).

Electrons and protons, the two types of *charged* subatomic particles, possess the same amount of electrical charge; the character of the charge is, however, opposite (negative versus positive). The fact that these subatomic particles are charged is most important because of the way in which charged particles interact. *Particles of opposite or unlike charge attract each other; particles of like charge repel each other.* This behavior of charged particles will be of major concern in many of the discussions in later portions of the text.

Arrangement of Subatomic Particles Within an Atom

The arrangement of subatomic particles within an atom is not haphazard. All protons and all neutrons are found at the center of an atom in a very small volume called the *nucleus* (Figure 5.7). A **nucleus** *is the small, dense, positively charged center of an atom; it contains an atom's protons and neutrons.* A nucleus always has a positive charge because of the positively charged protons that are present. Almost all of an atom's mass (over 99.9%) is found within its nucleus; all the heavy subatomic particles (protons and neutrons) are located there. The small size of the nucleus, coupled with its large amount of mass, causes nuclear material to be extremely dense.

Closely resembling the term *nucleus* is the term *nucleon*. A **nucleon** *is any subatomic particle found in the nucleus of an atom.* Thus, both protons and neutrons are nucleons and the nucleus can be regarded as containing a collection of nucleons (protons and neutrons).

The outer (extranuclear) region of an atom contains all of the electrons. It is an extremely large region compared to the nucleus. It is mostly empty space. It is a region in which the electrons move rapidly about the nucleus. The motion of the electrons in this extranuclear region determines the volume (size) of the atom in the same way as the blades of a fan determine a volume by their motion. The volume occupied by the electrons is sometimes referred

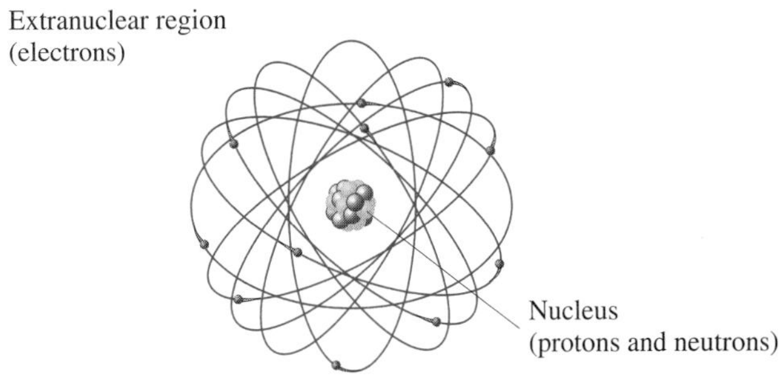

Extranuclear region
(electrons)

Nucleus
(protons and neutrons)

Figure 5.7

The protons and neutrons of an atom are found in the central nuclear region—the nucleus—and the electrons are found in an "electron cloud" outside the nucleus. Note that this figure is not drawn to scale; the correct scale would be comparable to a penny (the nucleus) in the center of a baseball field (the atom).

to as the *electron cloud.* Since electrons are negatively charged, the electron cloud is said to be negatively charged.

Figure 5.7 contrasts the nuclear and extranuclear regions of an atom.

Charge Neutrality of an Atom

An atom as a whole is neutral. How can it be that an entity possessing positive charge (the nuclear region) and negative charge (the extranuclear region or electron cloud) can end up neutral overall? For this to occur, the same amount of positive and negative charge must be present in the atom; equal numbers of positive and negative charges cancel each other. Atom neutrality thus requires that there be the same number of electrons and protons present in an atom, which is always the case for atoms.

$$\text{number of protons} = \text{number of electrons}$$

Size Relationships Within an Atom

The diameter of the nucleus of an atom is approximately 10^{-15} m, which is about 1/100,000 the 10^{-10} m diameter of an atom (see Fig. 5.8). As a help in visualizing this size contrast, imagine enlarging (magnifying) the nucleus until it is the size of a baseball (about 2.9 inches in diameter). If the nucleus were this large, the whole atom would have a diameter of approximately 2.5 miles. The electrons would still be smaller than the periods used to end sentences in this text, and they would move about within that 2.5-mile region.

The concentration of almost all the mass of an atom in the nucleus can also be illustrated by using our imagination. If a coin the same size as a copper penny contained copper nuclei (copper atoms stripped of their electrons) rather than copper atoms (which are mostly empty space), the coin would weigh 190,000,000 tons. Nuclei are indeed very dense matter.

Additional Subatomic Particles

Our just-completed discussion of the makeup of atoms in terms of subatomic particles is based on the existence of three types of subatomic particles: protons, neutrons, and electrons. This model of the atom is actually an oversimplification. In recent years, as the result of research carried out by nuclear physicists, the picture of the atom has lost its simplicity. Experimental evidence now available indicates that protons and neutrons themselves are made up of even smaller particles. Numerous other particles, with names such as leptons, mesons, and baryons, have been discovered. No simple theory is yet available that can explain all of these new discoveries relating to the complex nature of the nucleus.

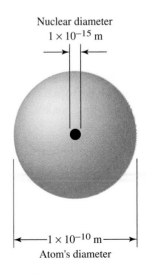

Nuclear diameter
1×10^{-15} m

1×10^{-10} m
Atom's diameter

Figure 5.8

Key atomic dimensions (not drawn to scale).

Despite the existence of other nuclear particles, we will continue to use the three-subatomic-particle model of the atom. It readily explains almost all chemical observations about atoms. We will have no occasion to deal with any of the recently discovered types of subatomic (or sub-subatomic) particles. Protons, neutrons, and electrons will meet all our needs.

We will also continue to use the concept that atoms are the fundamental building blocks for all types of matter (Sec. 5.1) despite the existence of protons, neutrons, and electrons. This is because under normal conditions subatomic particles do not lead an independent existence for any appreciable length of time. The only way they gain stability is by joining together to form an atom.

A significant collection of evidence is consistent with and supports the existence, nature, and arrangement of subatomic particles as given in this section. Several pieces of such evidence will be considered in Section 5.9.

5.6 Atomic Number and Mass Number

What determines whether a given atom is an atom of carbon or an atom of oxygen or an atom of gold? It is the number of protons present in the nucleus of the atom. The number of protons present determines atom identity. Every element has a characteristic number of protons associated with all of its atoms. For example:

All carbon atoms contain 6 protons.

All oxygen atoms contain 8 protons.

All gold atoms contain 79 protons.

If two atoms differ in the number of protons present, they must be atoms of two different elements. Conversely, if two atoms possess the same number of protons, they must be atoms of the same element.

Atomic Number

The characteristic number of protons associated with the atoms of a particular element is called the *atomic number* of the element. An **atomic number** *is the number of protons in the nucleus of an atom.*

> **Atomic numbers must be *whole* numbers since they are obtained by counting whole objects (protons). You cannot have $3\frac{1}{2}$ or $5\frac{1}{4}$ protons in a nucleus.**

Values of atomic numbers for the various elements, along with selected additional information, are printed on the inside front cover of this book. A check of the entries in the atomic number column of the data tabulation on the right shows that an entry exists for each of the numbers in the sequence 1 to 112 plus the number 114. The existence of an element that corresponds to each of the atomic numbers 1 through 112 is an indication of the order existing in nature. Scientists also interpret this continuous atomic number sequence 1 through 112 as evidence that there are no "missing elements" yet to be discovered in nature. The highest-atomic-numbered element that is naturally occurring is element 92 (uranium); elements 93 through 112 and 114 are all synthetic (Sec. 4.8).

The atomic numbers of the elements are also found on the left side of the inside front cover. The diagram there is a *periodic table,* a graphical presentation of selected characteristics of the elements. (The periodic table will be considered in detail in Chapter 6.) We note at this time that each box in the periodic table designates an element, with that element's symbol in the center of the box. The number above the symbol is the element's atomic number. The elements are arranged in the periodic table in order of increasing atomic number.

Previously, in Section 4.7, an element was defined as a pure substance that cannot be broken down into simpler substances by ordinary chemical means. Although this is a good historical definition for an element, we can now give a more rigorous definition using the concept of atomic number. An **element** *is a pure substance in which all atoms present have the same atomic number; that is, all atoms have the same number of protons.*

An atomic number, besides giving information about the number of protons present in an atom, also gives information about the number of electrons present. Because of charge neutrality (Sec. 5.5) an atom has the same number of protons and electrons. Thus,

$$\text{atomic number} = \text{number of protons} = \text{number of electrons}$$

The symbol Z is used as a general designation for atomic number.

Mass Number

Information about the number of neutrons present in an atom is not obtainable solely from an atomic number. A second number, a *mass number,* is needed in addition to the atomic number. A **mass number** *is the sum of the number of protons and the number of neutrons in the nucleus of an atom.*

$$\text{mass number} = \text{number of protons} + \text{number of neutrons}$$

The symbol A is used as a general designation for mass number. Since only protons and neutrons are present in a nucleus, the mass number gives the total number of subatomic particles present in the nucleus. The mass of an atom is almost totally accounted for by the protons and neutrons present (Sec. 5.5); hence the designation *mass* number.

> **Mass numbers, like atomic numbers, must always be *whole* numbers.**

Subatomic Particle Makeup of an Atom

Knowing the atomic number and mass number of an atom uniquely specifies the atom's makeup in terms of subatomic particles. The following equations show the relationship between subatomic particles and the two numbers.

$$\text{number of protons} = \text{atomic number} = Z$$

$$\text{number of electrons} = \text{atomic number} = Z$$

$$\text{number of neutrons} = \text{mass number} - \text{atomic number} = A-Z$$

> **The *sum* of the mass number and the atomic number for an atom (A + Z) has significance. It corresponds to the *total* number of subatomic particles present in the atom (protons + neutrons + electrons).**

Note that the neutron count is obtained through subtraction of the atomic number from the mass number.

Mass numbers are not tabulated in a manner similar to atomic numbers because, as we learn in the next section, most elements lack a unique mass number.

The mass and atomic numbers of a given atom are often specified using the notation

$$_{Z}^{A}\text{E}$$

Here E represents the symbol of the element being considered. The atomic number is placed as a subscript in front of the elemental symbol. The mass number is placed as a superscript in front of the elemental symbol. Examples of such notation for actual atoms include

$$_{9}^{19}\text{F} \quad _{11}^{23}\text{Na} \quad _{79}^{197}\text{Au}$$

The first of these notations specifies a fluorine atom that has an atomic number of 9 and a mass number of 19.

Examples 5.3 and 5.4 are sample calculations showing the interrelationships among atomic number, mass number, and the subatomic particle composition of atoms.

EXAMPLE 5.3

Determining the Subatomic Makeup of an Atom Given Its Atomic Number and Mass Number

Determine the following information for an atom that has an atomic number of 11 and a mass number of 23.

(a) The number of protons present

(b) The number of neutrons present

(c) The number of electrons present

SOLUTION

(a) There are 11 protons because the atomic number is always equal to the number of protons present.

(b) There are 12 neutrons because the number of neutrons is always obtained by subtracting the atomic number from the mass number ($23 - 11 = 12$).

$$\underbrace{(\text{Protons} + \text{neutrons})}_{\text{Mass number}} - \underbrace{\text{protons}}_{\substack{\text{Atomic} \\ \text{number}}} = \text{neutrons}$$

(c) There are 11 electrons because the number of protons and the number of electrons are always the same in an atom.

Practice Exercise 5.3

Determine the following information for an atom that has an atomic number of 79 and a mass number of 197.

(a) The number of protons present

(b) The number of electrons present

(c) The number of neutrons present

EXAMPLE 5.4

Determining the Atomic Number and Mass Number of an Atom Given Its Subatomic Makeup

Write a complete symbol ($^A_Z E$) for an atom whose nucleus contains 15 protons and 16 neutrons.

SOLUTION

Since 15 protons are present in the atom, the atomic number of the atom is 15. From the atomic number, the identity of the atom is determined using the information found inside the front cover. Using either tabulation there, we find that the atomic number 15 belongs to the element phosphorus (P).

The mass number of the atom is obtained by adding the numbers of protons and neutrons. The mass number is $15 + 16 = 31$.

The complete symbol of the atom is $^{31}_{15}P$. The atomic number is always the subscript and the mass number is always the superscript.

Practice Exercise 5.4

Write a complete symbol (^A_ZE) for an atom whose nucleus contains 53 protons and 74 neutrons.

5.7 Isotopes

Charge neutrality in an atom (Sec. 5.5) requires the presence of an equal number of protons and electrons. Because neutrons have no electrical charge, their numbers in atoms do not have to be the same as the number of protons or electrons. Most atoms contain more neutrons than either protons or electrons.

Studies of atoms of various elements show that the number of neutrons present in atoms of an element is usually not constant; it varies over a small range. This means that not all atoms of an element have to be identical. They must have the same number of protons and electrons, but they can differ in the number of neutrons. For example, three kinds of naturally occurring oxygen atoms exist. All oxygen atoms have eight protons and eight electrons. Most oxygen atoms also contain eight neutrons. Some oxygen atoms exist, however, that contain nine neutrons, and a few exist that contain ten neutrons. Designations for these three kinds of oxygen atoms are

$$^{16}_{8}\text{O} \quad ^{17}_{8}\text{O} \quad ^{18}_{8}\text{O}$$

Atoms of an element that differ in neutron count are called *isotopes*. Three oxygen isotopes exist. **Isotopes** *are atoms of an element that have the same number of protons and electrons but different numbers of neutrons.* Isotopes always have the same atomic number and different mass numbers.

The presence of one or more additional neutrons in the tiny nucleus of an atom has essentially no effect on the way it behaves chemically. Thus, isotopes of an element have the same chemical properties. Isotopes have the same number of electrons, and it is electrons that determine chemical properties. When two atoms interact chemically, the outer part (electrons) of one interacts with the outer part (electrons) of the other. The small nuclear centers never come in contact with each other during a chemical interaction between atoms.

Isotopes of an element can have slightly different physical properties because they have different numbers of neutrons and therefore different masses. Physical property differences are greatest for elements of low atomic number. For such elements, differences in mass between isotopes are relatively large when compared to the masses of the isotopes themselves. For example, $^2_1$H is twice as heavy as $^1_1$H, and the density of $^2_1$H is twice that of $^1_1$H (0.18 g/L versus 0.090 g/L.)

Most elements occurring naturally are mixtures of isotopes. The various isotopes of a given element are of varying abundance; usually one isotope is predominant. Typical of this situation is the element magnesium, which exists in nature in three isotopic forms: $^{24}_{12}$Mg, $^{25}_{12}$Mg, and $^{26}_{12}$Mg. The *percent abundances* for these three isotopes are, respectively, 78.70%, 10.13%, and 11.17%. A **percent abundance** *is the percent of atoms in a natural sample of a pure element that are a particular isotope of the element.* Percent abundances are number percents (number of atoms) rather than mass percents. A sample of 10,000 magnesium atoms would contain 7870 $^{24}_{12}$Mg atoms, 1013 $^{25}_{12}$Mg atoms, and 1117 $^{26}_{12}$Mg atoms. Table 5.2 gives natural isotopic abundances and isotopic masses for the elements with atomic numbers 1 through 12. The unit used for specifying the mass of the various isotopes, amu, will be discussed in Section 5.8.

The percent abundances of the isotopes of an element may vary slightly in samples obtained from different locations, but such variations are ordinarily extremely small. We will assume in this text that the isotopic composition of an element is a constant.

The word *isotope* comes from the Greek *iso*, meaning "equal," and *topos*, meaning "place." Isotopes occupy an equal place (location) in listings of elements because all isotopes of an element have the same atomic number.

Hydrogen isotopes are unique among isotopes in that each isotope has a different name.
$^1_1$H protium (symbol H)
$^2_1$H deuterium (symbol D)
$^3_1$H tritium (symbol T)

Table 5.2	Isotopic Data for Elements with Atomic Numbers 1 through 12. Information given for each isotope includes mass number, isotopic mass in amu, and percent abundance.

1	Hydrogen	2	Helium	3	Lithium
	$^1_1$H 1.008 amu 99.985% $^2_1$H 2.014 amu 0.015% $^3_1$H 3.016 amu trace		$^3_2$He 3.016 amu trace $^4_2$He 4.003 amu 100%		$^6_3$Li 6.015 amu 7.42% $^7_3$Li 7.016 amu 92.58%
4	Beryllium	5	Boron	6	Carbon
	$^9_4$Be 9.012 amu 100%		$^{10}_5$B 10.013 amu 19.6% $^{11}_5$B 11.009 amu 80.4%		$^{12}_6$C 12.000 amu 98.89% $^{13}_6$C 13.003 amu 1.11% $^{14}_6$C 14.003 amu trace
7	Nitrogen	8	Oxygen	9	Fluorine
	$^{14}_7$N 14.003 amu 99.63% $^{15}_7$N 15.000 amu 0.37%		$^{16}_8$O 15.995 amu 99.759% $^{17}_8$O 16.999 amu 0.037% $^{18}_8$O 17.999 amu 0.204%		$^{19}_9$F 18.998 amu 100%
10	Neon	11	Sodium	12	Magnesium
	$^{20}_{10}$Ne 19.992 amu 90.92% $^{21}_{10}$Ne 20.994 amu 0.26% $^{22}_{10}$Ne 21.991 amu 8.82%		$^{23}_{11}$Na 22.990 amu 100%		$^{24}_{12}$Mg 23.985 amu 78.70% $^{25}_{12}$Mg 24.986 amu 10.13% $^{26}_{12}$Mg 25.983 amu 11.17%

Isotopic masses, although not whole numbers, have values that are very close to whole numbers. This fact can be verified by looking at the isotopic masses in Table 5.2. If an isotopic mass is rounded off to the closest whole number, this value is the same as the mass number of the isotope. This statement can be verified using the data in Table 5.2.

Twenty-three elements have only one naturally occurring form; that is, they are "monoisotopic." For these elements, all atoms found in nature are identical to each other. Of the simpler elements (atomic numbers of 20 or less), those with only one form are

$$^9_4\text{Be} \quad ^{19}_9\text{F} \quad ^{23}_{11}\text{Na} \quad ^{27}_{13}\text{Al} \quad ^{31}_{15}\text{P}$$

The existence of isotopes adds clarification to the wording used in some of the statements of atomic theory (Sec. 5.1). Statement 1 reads: "All matter is made up of small particles called atoms, of which 113 different 'types' are known." It should now be apparent why the word "types" was put in quotation marks. Because of the existence of isotopes, atoms of each type are similar, but not identical. Atoms of a given element are similar in that they have the same atomic number, but not identical since they may have different mass numbers.

Statement 2 reads: "All atoms of a given type are similar to one another and significantly different from all other types." All atoms of an element are similar in chemical properties and differ significantly from atoms of other elements with different chemical properties.

It is possible for isotopes of two different elements to have the same mass number. For example, the element iron (atomic number 26) exists in nature in four isotopic forms, one of which is $^{58}_{26}$Fe. The element nickel, with an atomic number two units greater than that of iron, exists in nature in five isotopic forms, one of which is $^{58}_{28}Ni$. Thus, atoms of both iron and nickel exist with a mass number of 58. Thus, mass numbers are not unique for elements as are atomic numbers. Atoms of different elements that have the same mass number are called *isobars*; $^{58}_{26}Fe$ and $^{58}_{28}Ni$ are isobars. **Isobars** *are atoms that have the same mass number but different atomic numbers.* Even though atoms of two *different* elements can have the same mass number (isobars), they cannot have the same atomic number. All atoms of a given atomic number must necessarily be atoms of the same element.

There are 286 isotopes that occur naturally. In addition, over 2000 more isotopes have been synthesized in the laboratory (from the naturally occurring ones) using nuclear rather than chemical reactions. (Section 17.6 considers such nuclear reactions.) These unstable synthetic isotopes all have the common characteristic of being radioactive. Radioactive isotopes eventually revert back to naturally occurring isotopes. Many of these unstable isotopes, despite their instability, have important uses in chemical and biological research as well as in medicine.

EXAMPLE 5.5

Distinguishing Between Isotopes and Isobars

Indicate whether the members of each of the following pairs are isotopes, isobars, or neither.

(a) $^{42}_{20}X$ and $^{43}_{20}Q$

(b) $^{40}_{19}X$ and $^{40}_{20}Q$

(c) $^{44}_{20}X$ and $^{45}_{21}Q$

(d) an atom X with 20 protons and 21 neutrons and an atom Q with 19 protons and 21 neutrons

SOLUTION

(a) These atoms are isotopes. Both atoms have the same atomic number of 20. Isotopes differ from each other in neutron count, which is the case here. Atom X has 22 neutrons, and atom Q has 23 neutrons.

(b) These atoms are isobars. They have the same mass number (40) and different atomic numbers (19 and 20).

(c) These atoms are not isotopes or isobars. Isotopes must have the same atomic number and isobars must have the same mass number. Neither is the case here.

(d) These atoms are not isotopes or isobars. They are not isotopes because a differing number of protons means differing atomic numbers. They are not isobars because the mass numbers differ: 41 for atom X and 40 for atom Q. The two atoms contain the same number of neutrons. However, the definition for isobars is based on the same mass number rather than on the same neutron count.

Practice Exercise 5.5

Indicate whether the members of each of the following pairs are isotopes, isobars, or neither.

(a) $^{27}_{13}X$ and $^{28}_{14}Q$

(b) $^{46}_{20}X$ and $^{48}_{20}Q$

(c) $^{36}_{16}X$ and $^{36}_{18}Q$

(d) an atom X with eight protons and nine neutrons and an atom Q with nine protons and ten neutrons

An analogy involving isotopes and identical twins is helpful in understanding the fact that all atoms of an element need not have the same mass. Identical twins need not weigh the same even though they have identical "gene packages." They are identical twins by the gene criterion regardless of their masses. Likewise, isotopes, even though they have different masses, are atoms of the same element by atomic number criterion (same number of protons).

5.8 Atomic Masses

The existence of isotopes means that atoms of an element can have several different masses. For example, magnesium atoms can have any one of three masses because there are three magnesium isotopes. Which of these three magnesium isotopic masses is used in situations in which the mass of the element magnesium needs to be specified? The answer is none of them. Instead, a *weighted average mass* that takes into account the existence of isotopes and their relative abundances is used. These weighted average masses are called *atomic masses.*

Atomic mass values for the elements are found inside the front cover of this book. There are two such listings there. On the inside cover's right side (atomic number–atomic mass listing), atomic masses are given to the maximum number of significant figures possible, which varies from element to element. Typical atomic mass values include

14.0067 for the element nitrogen

32.065 for the element sulfur

126.90447 for the element iodine

On the inside cover's left side (periodic table; see Sec. 5.6), the atomic mass values are the numbers underneath the symbols of the elements. (The number above each element's symbol is its atomic number; see Sec. 5.6.) In this periodic table listing, the atomic masses are given to the hundredths decimal place. It is this "rounded form" of the atomic mass that is most often used in chemical calculations (Chapters 9 and 10).

Atomic mass values are not *mass numbers.* They cannot possibly be mass numbers because they are not whole numbers. Mass numbers, which are *counts* of the number of protons and neutrons present in nuclei (Sec. 5.6), must be whole numbers. Atomic masses are calculated numbers obtained from data on isotopic masses and isotopic abundances.

The starting point for understanding the origins of atomic mass values is a consideration of the formal definition for an atomic mass. An **atomic mass** *is the relative mass of an average atom of an element on a scale using the $^{12}_{6}C$ atom as the reference.* The meaning of two terms found within this definition, *relative mass* and *average atom,* is crucial to understanding the definition as a whole.

A detailed consideration of the periodic table, its value, significance, and use is the topic of a considerable portion of Chapter 6.

Relative Mass

The usual standards of mass, such as grams or pounds, are not convenient for use with atoms, because very small numbers are always encountered. For example, the mass in grams of a $^{238}_{92}U$ atom, one of the heaviest atoms known, is 3.95×10^{-22}. To avoid repeatedly encountering such small numbers scientists have chosen to work with relative rather than actual mass values.

A relative mass value for an atom is the mass of that atom relative to some standard rather than the actual mass value of the atom in grams. The term *relative* means "as compared to." The choice of the standard is arbitrary; this gives scientists control over the magnitude of the numbers on the relative scale, thus avoiding very small numbers.

For most purposes in chemistry, relative mass values serve just as well as actual mass values. Knowing how many times heavier one atom is than another, information obtainable from a relative mass scale, is just as useful as knowing the actual mass values of the atoms involved. Example 5.6 illustrates the procedures involved in constructing a relative mass scale and also points out some of the characteristics of such a scale.

EXAMPLE 5.6

Constructing a Relative Mass Scale

Construct a relative mass scale for the hypothetical atoms Q, X, and Z given the following information about them.

1. Atoms of Q are four times heavier than those of X.
2. Atoms of X are three times heavier than those of Z.

SOLUTION

Atoms of Z are the lightest of the three types of atoms. We will arbitrarily assign atoms of Z a mass value of one unit. The unit name can be anything we wish, and we shall choose "snick." On this basis, one atom of Z has a mass value of 1 snick. Atoms of Z will be our scale reference point. Atoms of X will have a mass value of 3 snicks (three times as heavy as Z) and Q atoms a value of 12 snicks (four times as heavy as X).

The name snick chosen for the mass unit was arbitrary. The assignment of the value 1 for the mass of Z, the reference point on the scale, was also arbitrary. What if we had chosen to call the unit a "smerge" and had chosen a value of 3 for the mass of an atom of Z? If this had been the case, the resulting relative scale would have appeared as

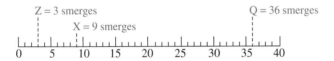

Which of the preceding relative scales is the "better" scale? The answer is that the scales are equivalent. The relationships between the masses of Q, X, and Z are the same on the two scales, even though the reference points and unit names differ. On the "snick" scale, for example, Q is four times heavier than X (12/3); on the "smerge" scale, Q is also four times heavier than X (36/9).

Notice that we did not need to know the actual masses of Q, X, and Z to set up either the "snick" or "smerge" scale. All that is needed to set up a relative scale is a set of interrelationships among quantities. One value—the reference point—is arbitrarily assigned, and all other values are determined by using the known interrelationships.

The information given at the start of this example is sufficient to set up an infinite number of relative mass scales. Each scale would differ from the others in choice of reference point and unit name. All the scales would, however, be equivalent to each other, and each scale would provide all of the mass relationships obtainable from an actual mass scale except for actual mass values.

Practice Exercise 5.6

Construct a relative mass scale for the hypothetical atoms Q, X, and Z given the following information about them.

1. Atoms of Q are two times as heavy as those of X.
2. Atoms of X are four times as heavy as those of Z.
3. The reference point for the scale is Z = 3.00 sloops.

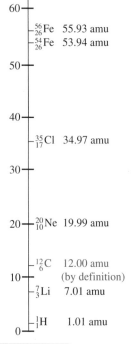

Figure 5.9

Relative masses of selected isotopes on the $^{12}_{6}C$ atomic mass scale.

A relative scale of atomic masses has been set up in a manner similar to that used in Example 5.6. The unit is called the *atomic mass unit,* abbreviated amu. The arbitrary reference point involves a particular isotope of carbon, $^{12}_{6}C$. The mass of this isotope is set at 12.00000 amu. The masses of all other atoms are then determined relative to that of $^{12}_{6}C$. For example, if an atom is twice as heavy as a $^{12}_{6}C$ atom, its mass is 24.00000 amu on the scale, and if an atom weighs half as much as a $^{12}_{6}C$ atom, its scale mass is 6.00000 amu.

The masses of all atoms have been determined relative to each other experimentally. Actual values for the masses of selected isotopes on the $^{12}_{6}C$ scale are given in Figure 5.9.

Reread the formal definition of atomic mass given earlier in this section. Note how $^{12}_{6}C$ is mentioned explicitly in the definition because of the central role it plays in the setting up of the relative atomic mass scale.

On the basis of the values given in Figure 5.9 it is possible to state, for example, that $^{238}_{92}U$ is 4.256 times as heavy as $^{56}_{26}Fe$ (238.05 amu/55.93 amu = 4.256) and $^{56}_{26}Fe$ is 2.798 times as heavy as $^{20}_{10}Ne$ (55.93 amu/19.99 amu = 2.798). We do not need to know the actual masses of the atoms involved to make such statements; relative masses are sufficient to calculate the information.

Average Atom

Since isotopes exist, the mass of an atom of a specific element can have one of several values. For example, oxygen atoms can have any one of three masses, since three isotopes exist: $^{16}_{8}O$, $^{17}_{8}O$, and $^{18}_{8}O$. Despite mass variances among isotopes, the atoms of an element are treated as if they all had a single common mass. The common mass value used is a *weighted average mass,* which takes into account the natural abundances and atomic masses of the isotopes of an element.

The validity of the weighted average mass concept rests on two points. First, extensive studies of naturally occurring elements have shown that the percent abundance of the isotopes of a given element is generally constant. No matter where the element sample is obtained on Earth, it generally contains the same percentage of each isotope. Because of these constant isotopic ratios, the mass of an "average atom" does not vary. Second, chemical operations are always carried out with very large numbers of atoms. The tiniest piece of matter visible to the eye contains more atoms than can be counted by a person in a lifetime. The numbers are so great that any collection of atoms a chemist works with will be representative of naturally occurring isotopic ratios.

Weighted Averages

Atomic masses are weighted averages calculated from the following three pieces of information:

1. The *number* of isotopes that exist for the element
2. The *isotopic mass* for each isotope, that is, the relative mass of each isotope on the $^{12}_{6}C$ scale
3. The *percent abundance* of each isotope

Table 5.2 gives these data for selected elements.

Examples 5.7 and 5.8 illustrate the operations needed to calculate weighted averages. Example 5.7 is a general exercise concerning weighted averages, and Example 5.8 illustrates the calculation of an atomic mass by the method of weighted averages.

EXAMPLE 5.7

Calculation of a Weighted Average

Sulfur oxides are air pollutants that arise primarily from the burning of coal. A student measures the sulfur oxide concentration in the atmosphere on five successive days with the following results:

Tuesday	24 ppb (parts per billion)
Wednesday	21 ppb
Thursday	21 ppb
Friday	24 ppb
Saturday	15 ppb

What is the average sulfur oxide concentration in the air over this time period, based on the student's measurements?

SOLUTION

Let us solve this problem using two different methods. The first method involves procedures familiar to you—the "normal" way of taking an average. By this method, the average is found by dividing the sum of the numbers by the number of values summed.

$$\frac{24 + 21 + 21 + 24 + 15}{5}\text{ppb} = 21\text{ ppb}\quad\text{(calculator answer)}$$

$$= 21.0\text{ ppb}\quad\text{(correct answer)}$$

Now let us solve this same problem again, this time treating it as a "weighted average" problem. To do this, we organize the given information in a different way. In our list of pollutant concentrations we have three different concentration values: 24 ppb, 21 ppb, and 15 ppb.

Two of the five values (40.0%) are 24 ppb.

Two of the five values (40.0%) are 21 ppb.

One of the five values (20.0%) is 15 ppb.

We will use the data in this "percent form" for our weighted average calculation.

To find the weighted average, we multiply each distinct value (24, 21, and 15) by its fractional abundance, that is, by its percentage expressed in decimal form, and then we sum the products from the multiplications.

$$0.400 \times 24\text{ ppb} =\ \ 9.6\text{ ppb}$$
$$0.400 \times 21\text{ ppb} =\ \ 8.4\text{ ppb}$$
$$0.200 \times 15\text{ ppb} =\ \ \underline{3.0\text{ ppb}}$$
$$\mathbf{21.0\ ppb}\ \text{(same average value as before)}$$

This averaging method, although it appears somewhat more involved than the "normal" method, is the one that must be used in calculating atomic masses because of the form in which data about isotopes are obtained. The percent abundances of isotopes are experimentally determinable quantities. The total number of atoms of various isotopes present in nature, a prerequisite for using the "normal" average method, is not easily determined. Therefore, we use the "percent" method.

CHEMICAL EXTENSION

Two oxides of sulfur are involved in sulfur oxide air pollution: sulfur dioxide (SO_2) and sulfur trioxide (SO_3). The sulfur dioxide forms when oxygen from the air reacts with sulfur compounds present in coal. Coal contains from 1% to 7% sulfur by mass. Once in the atmosphere, the sulfur dioxide so formed reacts with additional oxygen to form sulfur trioxide.

Most of the effects of sulfur oxide air pollution on human health are related to the irritation of the respiratory tract and eyes by sulfur dioxide. Of particular concern are the effects of SO_2 exposure on individuals who suffer from chronic respiratory diseases such as bronchitis and asthma. These individuals exhibit diminished lung functions at much lower SO_2 levels than healthy individuals.

Practice Exercise 5.7

Carbon monoxide is an air pollutant that arises primarily from the operation of automobiles. A student measures the carbon monoxide concentration in the atmosphere on five successive days with the following results:

Wednesday	11 ppm (parts per million)
Thursday	11 ppm
Friday	8 ppm
Saturday	8 ppm
Sunday	7 ppm

Using the "weighted average" method, calculate the average carbon monoxide concentration in the air over this time period based on the student's measurements.

EXAMPLE 5.8

Calculation of Atomic Mass from Isotopic Masses and Percent Abundances

Magnesium occurs in nature in three isotopic forms: $^{24}_{12}Mg$ (78.70% abundance), $^{25}_{12}Mg$ (10.13% abundance), and $^{26}_{12}Mg$ (11.17% abundance). The relative masses of these three isotopes, respectively, are 23.985, 24.986, and 25.983 amu. Calculate the atomic mass of magnesium from these data.

SOLUTION

The atomic mass of an element is calculated using the weighted average method illustrated in Example 5.7. Each of the isotopic masses is multiplied by the fractional abundance associated with that mass, and then the products are summed.

$$0.7870 \times 23.985 \text{ amu} = 18.876195 \text{ amu} = 18.88 \text{ amu}$$

$$0.1013 \times 24.986 \text{ amu} = 2.5310818 \text{ amu} = 2.531 \text{ amu}$$

$$0.1117 \times 25.983 \text{ amu} = 2.9023011 \text{ amu} = 2.902 \text{ amu}$$

Note that the method for converting percentages to fractional abundances is always the same. The decimal point in the percentage is moved two places to the left. For example,

$$78.70\% \text{ becomes } 0.7870$$

Significant figures are always an important part of an atomic mass calculation. Both isotopic masses and percent abundances are experimentally determined numbers.

Summing, to obtain the atomic mass, gives

$$(18.88 + 2.531 + 2.902) \text{ amu} = 24.313 \text{ amu} \quad \text{(calculator answer)}$$

$$= 24.31 \text{ amu} \quad \text{(correct answer)}$$

Since the number 18.88 is known only to the hundredths place, the answer can be expressed only to the hundredths place.

The above calculation involved an element that exists in three isotopic forms. An atomic mass calculation for an element having four isotopic forms would be carried out in an almost identical fashion. The only difference would be four products to calculate (instead of three) and four terms in the resulting sum.

CHEMICAL EXTENSION

The element magnesium is a light (low-density) silvery-white metal. At high temperatures, it burns with a brilliant white flame. Magnesium's major source, unlike that of copper and iron, is not ore deposits but rather salt brines and seawater. Magnesium compounds obtained from evaporation of such liquids are subjected to electrolysis to free the elemental magnesium.

Over one-half of the magnesium metal produced worldwide each year is used in aluminum–magnesium alloys. The attractiveness of these alloys is primarily their low density. Such alloys find use wherever their low density translates into significant energy savings: in aircraft, railroad passenger cars, rapid transit vehicles, and bus bodies. For a period of time in the 1970s, these alloys were used in the superstructure of warships because the lower mass of the ship allowed higher speeds. However, during the Falkland Islands War, the Royal Navy discovered a major disadvantage of such alloy use—their flammability when subjected to missile attack.

Practice Exercise 5.8

Silicon occurs in nature in three isotopic forms: $^{28}_{14}\text{Si}$ (92.21% abundance), $^{29}_{14}\text{Si}$ (4.70% abundance), and $^{30}_{14}\text{Si}$ (3.09% abundance). The relative masses of these three isotopes, respectively, are 27.977, 28.976, and 29.974 amu. Calculate the atomic mass of silicon from these data.

In Example 5.8 the atomic mass of magnesium was calculated to be 24.31 amu. How many magnesium atoms have a mass of 24.31 amu? The answer is none. Magnesium atoms have a mass of 23.985, 24.986, or 25.983 amu, depending on which isotope they are. The mass 24.31 amu is the mass of an "average" magnesium atom. It is this average mass that is used in calculations even though no magnesium atoms have masses equal to this average value. Only in the case where all atoms have the same mass will the isotopic mass and the atomic mass be the same.

Atomic masses are subject to change and they do change. Every two years an updated atomic mass listing is published by the International Union of Pure and Applied Chemistry (IUPAC). This update, produced by an international committee of chemists, takes into account all new research on isotopic abundances.

New atomic mass values often have less uncertainty than the older values they replace. This is a reflection of the increasingly sophisticated instrumentation available to current researchers, which enables them to make "better" measurements.

Illustrative of the atomic mass changes that do occur are those contained in the 2001 update report issued by the IUPAC. The atomic masses of three elements were changed:

zinc	65.39 becomes 65.409
krypton	83.80 becomes 83.798
dysprosium	162.50 becomes 162.500

Atomic mass revisions such as these explain why various textbooks (and periodic tables) often differ in a few atomic mass values. The differing values come from different IUPAC reports on atomic masses.

The uncertainty associated with atomic mass values varies from element to element. For example, we have

B	10.811	amu
F	18.9984032	amu
Si	28.0855	amu
Pb	207.2	amu

What causes such variance in uncertainty? The key factor is the constancy of isotopic percentage abundance measurements among various samples of an element. Although all elements have an essentially constant set of isotopic percentage abundances, there is slight variation among samples obtained from different sources. All variations are small; however, for some elements they are greater than for others. When only one form of an element occurs in nature, such as F, values with less uncertainty can be obtained for atomic mass.

An atomic mass cannot be calculated for all elements. Recall, from Section 4.8, that not all elements are naturally occurring substances. Twenty-five of the known elements are "synthetic," having been produced in the laboratory from naturally occurring elements. Obviously a weighted average atomic mass cannot be calculated for these laboratory-produced elements, since the amount of each isotope produced varies, depending on the laboratory experiment carried out.

Tabulations of atomic masses do contain entries for the synthetic elements. Such entries are the mass number of the most stable isotope of the synthetic element. (All isotopes of all synthetic elements are unstable.) Such mass numbers are always enclosed in parentheses to distinguish them from calculated atomic masses. Note the presence of such entries in both the atomic mass listing and the periodic table inside the front cover.

Before leaving the subject of atomic masses, we need to consider one additional question. How do scientists determine the abundances and masses of the various isotopes of an element? A device known as a *mass spectrometer* is the key to obtaining such information.

A schematic diagram of a mass spectrometer is given in Figure 5.10. The important components of this instrument are a source of charged particles of the element under investigation, an aligning system to create a narrow beam of the charged particles, a magnetic field to affect the path of the charged particles, and a means of detecting the charged particles (such as a photographic plate).

Suppose oxygen gas containing all three isotopes is admitted to the instrument. The gaseous molecules are bombarded with an energetic electron beam, producing positively charged oxygen species. The aligning system produces a narrow beam of these "charged atoms," which then enters the magnetic field. The most massive particles (heaviest isotope) are not deflected by the magnetic field as much as the less massive ones, so the charged atoms are divided into separate beams that strike the photographic plate at different points depending on their masses. The more abundant isotopes will create more intense lines on the plate. The relative intensities of the lines correlate exactly with the relative abundances of the isotopes.

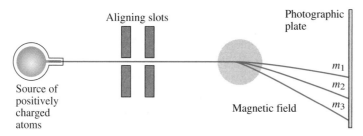

Figure 5.10

A simplified diagram of a mass spectrometer, an instrument used to obtain information about isotopic masses and abundances.

5.9 Evidence Supporting the Existence and Arrangement of Subatomic Particles

A significant collection of evidence is consistent with and supports the existence, nature, and arrangement of subatomic particles as given in Section 5.5. Two historically important types of experiments illustrate some of the sources of this evidence. *Discharge tube experiments* resulted in the original concept that the atom contained negatively and positively charged particles. *Metal foil experiments* provided evidence for the existence of a nucleus within the atom.

Discharge Tube Experiments

Neon signs, fluorescent lights, and television tubes are all basic components of our modern technological society. The forerunner for all three of these developments was the *gas discharge tube.* Gas discharge tubes also provided some of the first evidence that an atom consisted of still smaller particles (subatomic particles).

The principle behind the operation of a gas discharge tube—that gases at low pressure conduct electricity—was discovered in 1821 by the English chemist Humphry Davy (1778–1829). Subsequently, gas discharge tube studies were carried out by many scientists.

A simplified diagram of a gas discharge tube is shown in Figure 5.11. The apparatus consists of a sealed glass tube containing two metal disks called *electrodes.* The glass tube also has a side arm for attachment to a vacuum pump. During operation, the electrodes are connected to a source of electrical power. (The electrode attached to the positive side of the electrical power source is called the *anode*; the one attached to the negative side is known as the *cathode.*) Use of the vacuum pump allows the amount of gas within the tube to be varied. The smaller the amount of gas present, the lower the pressure within the tube.

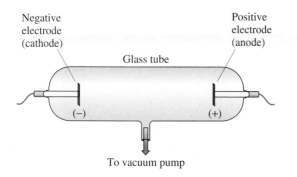

Figure 5.11

A simplified version of a gas discharge tube.

Early studies with gas discharge tubes showed that when the tube was almost evacuated (low pressure), electricity flowed from one electrode to the other and the residual gas became luminous (it glowed). Different gases in the tube gave different colors to the glow. After the pressure in the tube was reduced to still lower levels (very little gas remaining), it was found that the luminosity disappeared but the electrical conductance continued, as shown by a greenish glow given off by the tube's glass walls. This glow was the initial discovery of what became known as *cathode rays*. Their discovery marked the beginning of nearly 40 years of discharge tube experimentation that ultimately led to the characterization of both the electron and the proton.

The term *cathode rays* comes from the observation that when an obstacle is placed between the negative electrode (cathode) and the opposite glass wall, a sharp shadow the shape of the obstacle is cast on that wall. This indicates that the rays are coming from the cathode.

Further studies showed that these cathode rays caused certain minerals such as sphalerite (zinc sulfide) to glow. Glass plates were coated with sphalerite and observed under high magnification while being bombarded with cathode rays. The light emitted by the sphalerite coating consisted of many pinpoint flashes. This observation suggested that cathode rays were in reality a stream of extremely small particles.

Joseph John Thomson (1856–1940), an English physicist, provided many facts about the nature of cathode rays. Using a variety of materials as cathodes, he showed that cathode ray production was a general property of matter. By using a specially designed cathode ray tube (see Fig. 5.12), he also found that cathode rays could be deflected by charged plates or a magnetic field. The rays were repelled by the north pole or negative plate and attracted to the south pole or positive plate, thus indicating that they were negatively charged. In 1897, Thomson concluded that cathode rays were streams of negatively charged particles, which today we call *electrons.* Further experiments by others proved that his conclusions were correct.

In 1886, a German physicist, Eugen Goldstein (1850–1930), showed that positive particles were also present in discharge tubes. He used a discharge tube in which the cathode was a metal plate with a large number of holes drilled in it. The usual cathode rays were observed to stream from cathode to anode. In addition, rays of light appeared to stream from each of the holes in the cathode in a direction opposite to that of the cathode rays (see Fig. 5.13). Because these rays were observed streaming through the holes or channels in the cathode, Goldstein called them *canal rays*.

> Thomson's cathode ray tube experiments have evolved into today's televisions and computer monitors (still often called CRTs).

Figure 5.12

J.J. Thomson's cathode ray tube involved the use of both electrical and magnetic fields.

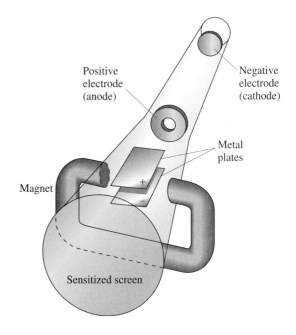

Positive electrode (anode)

Negative electrode (cathode)

Metal plates

Magnet

Sensitized screen

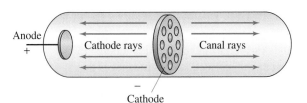

Figure 5.13

A gas discharge tube modified to detect canal rays.

Further research showed that canal rays were of many different types, in contrast to cathode rays, which are only one type, and that the particles making up canal rays were much heavier than those of cathode rays. The type of canal rays produced depended upon the gas in the tube. The simplest canal rays were eventually identified as the particles now called protons.

Canal rays are now known to be gas atoms that have lost one or more electrons. Their origin and behavior in a discharge tube can be understood as follows. Electrons (cathode rays) emitted from the cathode collide with residual gas molecules (air) on the way to the anode. Some of these electrons have enough energy to knock electrons away from the gas molecules, leaving behind a positive particle (the remainder of the gas molecule). These positive particles are attracted to the cathode, and some of them pass through the holes or channels. The fact that atoms, under certain conditions, can lose electrons will be discussed further in Section 7.4.

On the basis of discharge tube experiments, Thomson proposed in 1898 that the atom was composed of a sphere of positive electricity containing most of the mass, and that small

The Human Side of Chemistry 4

Ernest Rutherford (1871–1937)

Ernest Rutherford, born in 1871 on a farm in New Zealand, was the 4th in a family of 12 children. He left New Zealand for England at the age of 25 after having won a scholarship to Cambridge University. Rutherford's "gold foil experiment" is only one of many contributions that he made to the sciences of chemistry and physics. Indeed, three years prior to carrying out this famous experiment, in 1908, he received the Nobel Prize in chemistry for his investigations into the nature of radioactivity.

While in graduate school at Cambridge, his professor, J. J. Thomson, encouraged him to study the newly discovered phenomenon of radioactivity. His research in this area led to the discovery of the alpha and beta radiation associated with radioactivity.

In 1899, he moved to Canada, spending nine years at McGill University doing further research on alpha and beta particles. At McGill he found that alpha particles are helium nuclei and that beta particles are electrons. For this work he was awarded his Nobel Prize.

In 1907, he returned to England (Manchester University) where further studies on alpha particles led to his famous gold foil experiment, considered to be Rutherford's greatest and most fruitful contribution to scientific knowledge. Many years later Rutherford described the unexpected results of this experiment as follows: "It was about as credible as if you had fired a 15-inch shell at a piece of tissue paper and it came back and hit you."

World War I brought an abrupt change in direction, a switch from atoms to submarines. He studied underwater acoustics, supplying the government with much information needed to advance the technology of submarine detection.

Following the war, in 1919, Rutherford moved to Cambridge University, assuming the position formerly held by J. J. Thomson, the professor who guided him as a graduate student. Here, he again was on the "forefront" of scientific advances, this time discovering nuclear transformations, the process in which an atom of an element changes into another element (a topic to be discussed in Chapter 17 of this text).

Many of the students Rutherford guided as a professor went on to make major scientific discoveries of their own. Among his graduate students were ten future recipients of the Nobel Prize. He lived to see some of them receive their prizes.

Rutherford's research work was diverse—radioactivity at McGill, atomic physics at Manchester, and nuclear physics at Cambridge. His research was "world class" at all three institutions. He was a very talented researcher. Element 104, rutherfordium, carries Rutherford's name.

negative electrons were attached to the surface of the positive sphere. He postulated that a high voltage could pull off surface electrons to produce cathode rays. Thomson's model of the atom, sometimes referred to as the "raisin muffin" or "plum pudding" model—with the electrons as the raisins—is now known to be incorrect. Its significance is that it set the stage for an experiment, commonly called the gold foil experiment, that led to the currently accepted arrangement of protons and electrons in the atom.

Metal Foil Experiments

In 1911 Ernest Rutherford (1871–1937; see "The Human Side of Chemistry 4") designed an experiment to test the Thomson model of the atom. In this experiment thin sheets of metal foil were bombarded by alpha particles from a radioactive source. Alpha particles, which are positively charged, are ejected at high speeds from some radioactive materials. The phenomenon of radioactivity had been discovered in 1896 and gave further evidence that electrical charges existed within the atom. Gold was chosen as the target metal because it is easily hammered into very thin sheets. The experimental setup for Rutherford's experiment is shown in Figure 5.14. Alpha particles do not appreciably penetrate lead, so a lead plate with a slit was used to produce a narrow alpha particle beam. Each time an alpha particle hit the fluorescent screen, a flash of light was produced.

Rutherford expected that all the alpha particles, since they were so energetic, would pass straight through the thin gold foil. His reasoning was based on the Thomson model, in which the mass and positive charge of the gold atoms were distributed uniformly through each atom. As each positive alpha particle neared the foil, Rutherford assumed that it would be confronted by a uniform positive charge. All particles would be affected the same way (no deflection), which would support the Thomson model of the atom.

The results from the experiment were very surprising. Most of the particles—more than 99%—went straight through as expected. A few, however, were appreciably deflected by something that had to be much heavier than the alpha particles themselves. A very few particles were deflected almost directly back toward the alpha particle source. Similar results were obtained when elements other than gold were used as targets.

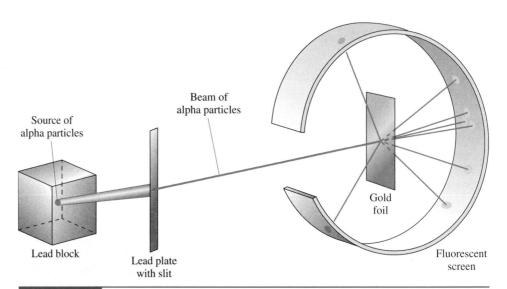

Source of alpha particles

Beam of alpha particles

Lead block

Lead plate with slit

Gold foil

Fluorescent screen

Figure 5.14

Rutherford's gold foil–alpha particle experiment. Most of the alpha particles went straight through the foil, but a few were deflected at large angles.

Extensive study of the results of his experiments led Rutherford to propose the following explanation.

1. A very dense, small nucleus exists in the center of the atom. This nucleus contains most of the mass of the atom and all of the positive charge.

2. Electrons occupy most of the total volume of the atom and are located outside the nucleus.

3. When an alpha particle scores a direct hit on a nucleus, it is deflected back along the incoming path.

4. A near miss of a nucleus by an alpha particle results in repulsion and deflection.

5. Most of the alpha particles pass through without any interference, because most of the atomic volume is empty space.

6. Electrons have so little mass that they do not deflect the much larger alpha particles (an alpha particle is almost 8000 times heavier than an electron).

Many other experiments have since verified Rutherford's conclusion that at the center of an atom there is a nucleus that is very small and very dense.

Summary

1. **Atoms and Molecules** An atom is the smallest particle of an element that can exist and still have the properties of the element. Free isolated atoms are rarely encountered in nature. Instead, atoms are almost always found together in aggregates or clusters. A molecule is a group of two or more atoms that functions as a unit because the atoms are tightly bound together.

2. **Types of Molecules** Molecules are of two types: homoatomic and heteroatomic. Homoatomic molecules are molecules in which all atoms present are of the same kind. A pure substance containing homoatomic molecules is an element. Heteroatomic molecules are molecules in which two or more different kinds of atoms are present. Pure substances that contain heteroatomic molecules must be compounds. The molecule is the limit of physical subdivision. The atom is the limit of chemical subdivision.

3. **Chemical Formulas** Chemical formulas are used to specify compound composition in a concise manner. They consist of the symbols of the elements present in the compound and numerical subscripts (located to the right of each symbol) that indicate the number of each element present in a molecule of the compound.

4. **Subatomic Particles** Subatomic particles, the very small building blocks from which atoms are made, are of three major types: electrons, protons, and neutrons. Electrons are negatively charged, protons are positively charged, and neutrons have no charge. Protons and neutrons have much larger masses than electrons. All neutrons and protons are found at the center of the atom in the nucleus. The electrons occupy the region about (around) the nucleus. Discharge tube experiments played a key role in the discovery of electrons and protons. Metal foil experiments were instrumental in establishing that an atom had a nucleus.

5. **Atomic Number and Mass Number** Each atom has a characteristic atomic number (Z) and mass number (A). The atomic number is equal to the number of protons in the nucleus of the atom. The mass number is equal to the total number of protons and neutrons in the nucleus. In designating various types of atoms, the atomic number is specified as a subscript to the left of an atom's elemental symbol, and the mass number is specified as a superscript to the left of the atom's elemental symbol.

6. **Isotopes and Isobars** Isotopes are atoms that have the same number of protons and electrons but have different numbers of neutrons. The isotopes of an element always have the same atomic number and different mass numbers. Isotopes of an element have the same chemical properties. Isobars are atoms of different elements that have the same mass number.

7. **Atomic Mass** The atomic mass of an element is the relative mass of an average atom of the element on a scale using the $^{12}_{6}C$ atom as the reference. An atomic mass depends on the percent abundances and masses of the naturally occurring isotopes of an element.

Key Terms

The new terms defined in this chapter are

atom   *Sec. 5.1*
atomic mass   *Sec. 5.8*
atomic number   *Sec. 5.6*
atomic theory of matter   *Sec. 5.1*
chemical formula   *Sec. 5.4*
diatomic molecule   *Sec. 5.2*
electron   *Sec. 5.5*
element   *Sec. 5.6*
heteroatomic molecule   *Sec. 5.2*
homoatomic molecule   *Sec. 5.2*

isobars   *Sec. 5.7*
isotopes   *Sec. 5.7*
mass number   *Sec. 5.6*
molecule   *Sec. 5.2*
neutron   *Sec. 5.5*
nucleon   *Sec. 5.5*
nucleus   *Sec. 5.5*
percent abundance   *Sec. 5.7*
proton   *Sec. 5.5*
subatomic particle   *Sec. 5.5*
triatomic molecule   *Sec. 5.2*

Practice Problems

Atoms and Molecules (Secs. 5.1 and 5.2)

5.1 Which of the following concepts are *not* consistent with the statements of modern-day atomic theory?

(a) Atoms are the basic building blocks for all kinds of matter.

(b) Different "types" of atoms exist.

(c) All atoms of a given "type" are identical.

(d) Atoms are considered indestructible during chemical change processes.

5.2 Which of the following concepts are *not* consistent with the statements of modern-day atomic theory?

(a) Only whole atoms can participate in chemical reactions.

(b) Atoms change identity during chemical change processes.

(c) 118 different "types" of atoms are known.

(d) Chemical change is a union, separation, or rearrangement of atoms to give new substances.

5.3 Which of the terms *heteroatomic, homoatomic, diatomic, triatomic, element,* and *compound* apply to each of the following molecules? (More than one term may apply in a given situation.)

(a)   **(b)** 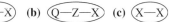 **(c)**   **(d)**

5.4 Which of the terms *heteroatomic, homoatomic, diatomic, triatomic, element,* and *compound* apply to each of the following molecules? (More than one term may apply in a given situation.)

(a)   **(b)**   **(c)**   **(d)**

5.5 Indicate whether each of the following statements is *true* or *false*. If a statement is false, change it to make it true. (Such a rewriting should involve more than merely converting the statement to the negative of itself.)

(a) Molecules must contain three or more atoms.

(b) The atom is the limit of chemical subdivision for an element.

(c) All compounds have molecules as their basic structural unit.

(d) The diameter of an atom is approximately 10^{-8} meters.

5.6 Indicate whether each of the following statements is *true* or *false*. If a statement is false, change it to make it true. (Such a rewriting should involve more than merely converting the statement to the negative of itself.)

(a) A molecule of an element may be homoatomic or heteroatomic depending on which element is involved.

(b) Heteroatomic molecules do not maintain the properties of their constituent elements.

(c) Only one kind of atom may be present in a homoatomic molecule.

(d) The mass of an atom is approximately 10^{-23} gram.

Chemical Formulas (Sec. 5.4)

5.7 What is the chemical formula for each of the following molecules?

(a)   **(b)**  **(c)**   **(d)**

5.8 What is the chemical formula for each of the following molecules?

(a)

5.9 On the basis of its formula, classify each of the following substances as an element or compound.

(a) $NaClO_2$ **(b)** CO

(c) S_8 **(d)** Al

5.10 On the basis of its formula, classify each of the following substances as an element or compound.

(a) AlN   **(b)** CO_2

(c) Co   **(d)** O_3

5.11 Write the chemical formulas for compounds in which the combining ratio of atoms is as follows:

(a) carbon:hydrogen, 20:30

(b) hydrogen:sulfur:oxygen, 2:1:4

(c) hydrogen:carbon:nitrogen, 1:1:1

(d) potassium:manganese:oxygen, 1:1:4

5.12 Write the chemical formulas for compounds in which the combining ratio of atoms is as follows:

(a) carbon:hydrogen:oxygen, 10:14:2

(b) carbon:hydrogen:oxygen, 2:6:1

(c) sodium:oxygen:hydrogen, 1:1:1

(d) potassium:nitrogen:oxygen, 1:1:3

5.13 In each of the following pairs of formulas indicate whether the first listed formula denotes *more total atoms,* the *same number of total atoms,* or *fewer total atoms* than the second listed formula.

(a) N_2O and NO_2

(b) KNO_3 and $Ca(OH)_2$

(c) $Ba(ClO)_2$ and $Ba(ClO_2)_2$

(d) $Al_2(SO_4)_3$ and $Ba_3(PO_4)_2$

5.14 In each of the following pairs of formulas, indicate whether the first listed formula denotes *more total atoms*, the *same number of total atoms*, or *fewer total atoms* than the second listed formula.

(a) HN_3 and NH_3 **(b)** $NaClO_3$ and $Be(CN)_2$

(c) $CaSO_4$ and $Mg(OH)_2$ **(d)** $Be_3(PO_4)_2$ and $Be(C_2H_3O_2)_2$

5.15 What difference in meaning, if any, is there in the following pairs of notations?

(a) NO and No   **(b)** Cs_2 and CS_2

(c) $CoBr_2$ and $COBr_2$ **(d)** H and H_2

5.16 What difference in meaning, if any, is there in the following pairs of notations?

(a) Hf and HF   **(b)** TiN and Tin

(c) $NICl_2$ and $NiCl_2$ **(d)** O_2 and O_3

5.17 The following molecular formulas are *incorrectly* written. Rewrite each formula in the correct manner.

(a) H3PO4

(b) $SICL_4$ (a silicon–chlorine compound)

(c) NOO

(d) 2HO (two H atoms and two O atoms)

5.18 The following molecular formulas are *incorrectly* written. Rewrite each formula in the correct manner.

(a) H2CO3

(b) $ALBR_3$ (an aluminum–bromine compound)

(c) HSH

(d) $2NO_2$ (two N atoms and four O atoms)

Subatomic Particles (Sec. 5.5)

5.19 Match the terms *proton, neutron,* and *electron* to each of the following subatomic particle descriptions. It is possible that more than one term may apply in a given situation.

(a) possesses a negative charge

(b) has a mass slightly less than that of a neutron

(c) can be called a nucleon

(d) is the heaviest of the three particles

5.20 Match the terms *proton, neutron,* and *electron* to each of the following subatomic particle descriptions. It is possible that more than one term may apply in a given situation.

(a) has no charge

(b) has a charge equal to but opposite in sign to that of an electron

(c) is not found in the nucleus

(d) has a positive charge

5.21 Indicate whether each of the following statements about the nucleus of an atom is *true* or *false.*

(a) The nucleus of an atom is neutral.

(b) The nucleus of an atom contains only neutrons.

(c) The number of nucleons present in the nucleus is always equal to the number of electrons present outside the nucleus.

(d) The nucleus accounts for almost all of the mass of an atom.

5.22 Indicate whether each of the following statements about the nucleus of an atom is *true* or *false.*

(a) The nucleus accounts for almost all the volume of an atom.

(b) The nucleus can be positively or negatively charged, depending on the identity of the atom.

(c) The nucleus of an atom contains an equal number of protons, neutrons, and electrons.

(d) The nucleus of an atom is always positively charged.

Atomic Number and Mass Number (Sec. 5.6)

5.23 What is the atomic number associated with the listed element or the element name associated with the listed atomic number?

(a) tin   **(b)** silver

(c) atomic number 28   **(d)** atomic number 53

5.24 What is the atomic number associated with the listed element or the element name associated with the listed atomic number?

(a) lead   **(b)** beryllium

(c) atomic number 18   **(d)** atomic number 56

5.25 For each of the following atoms, specify the atomic number and the mass number.

(a) $^{53}_{24}Cr$ **(b)** $^{103}_{44}Ru$ **(c)** $^{256}_{101}Md$ **(d)** $^{34}_{16}S$

5.26 For each of the following atoms, specify the atomic number and the mass number.

(a) $^{67}_{30}Zn$ **(b)** $^{9}_{4}Be$ **(c)** $^{40}_{20}Ca$ **(d)** $^{3}_{1}H$

5.27 Indicate whether (1) the atomic number, (2) the mass number, or (3) both the atomic number and the mass number are needed to determine the following.

(a) Number of protons in an atom

(b) Number of neutrons in an atom

(c) Number of nucleons in an atom

(d) Total number of subatomic particles in an atom

5.28 What information about the subatomic particles present in an atom is obtained from each of the following?

(a) atomic number

(b) mass number

(c) mass number – atomic number

(d) mass number + atomic number

5.29 What is the complete symbol (^A_ZE) for atoms composed of the following sets of subatomic particles?

(a) 5 protons, 5 electrons, and 6 neutrons

(b) 8 protons, 8 electrons, and 8 neutrons

(c) 13 protons, 13 electrons, and 14 neutrons

(d) 18 protons, 18 electrons, and 22 neutrons

5.30 What is the complete symbol (^A_ZE) for atoms composed of the following sets of subatomic particles?

(a) 4 protons, 4 electrons, and 5 neutrons

(b) 7 protons, 7 electrons, and 8 neutrons

(c) 15 protons, 15 electrons, and 16 neutrons

(d) 20 protons, 20 electrons, and 28 neutrons

5.31 Determine the number of protons, electrons, and neutrons in each of the following atoms.

(a) $^{32}_{16}$S   (b) $^{63}_{29}$Cu   (c) $^{50}_{22}$Ti   (d) $^{238}_{92}$U

5.32 Determine the number of protons, electrons, and neutrons in each of the following atoms.

(a) $^{35}_{17}$Cl   (b) $^{55}_{25}$Mn   (c) $^{127}_{53}$I   (d) $^{209}_{83}$Bi

5.33 Determine the number of protons, electrons, and neutrons present in atoms with the following characteristics.

(a) atomic number = 27 and mass number = 59

(b) mass number = 103 and Z = 45

(c) Z = 69 and A = 169

(d) atomic number = 9 and A = 19

5.34 Determine the number of protons, electrons, and neutrons present in atoms with the following characteristics.

(a) A = 103 and atomic number = 44

(b) Z = 41 and A = 93

(c) mass number = 59 and atomic number = 27

(d) Z = 59 and mass number = 141

5.35 For each of the atoms listed in Problem 5.33, what is the charge (magnitude and sign) on the atom's nucleus?

5.36 For each of the atoms listed in Problem 5.34, what is the charge (magnitude and sign) on the atom's nucleus?

5.37 Write complete symbols (^A_ZE) for atoms with the following characteristics.

(a) contains 15 electrons and 16 neutrons

(b) oxygen atom with 10 neutrons

(c) chromium atom with a mass number of 54

(d) gold atom that contains 276 subatomic particles

5.38 Write complete symbols (^A_ZE) for atoms with the following characteristics.

(a) contains 20 electrons and 24 neutrons

(b) radon atom with a mass number of 211

(c) silver atom that contains 157 subatomic particles

(d) beryllium atom that contains 9 nucleons

5.39 Characterize each of the following pairs of atoms as containing (1) the same number of neutrons, (2) the same number of electrons, or (3) the same total number of subatomic particles.

(a) $^{40}_{20}$Ca and $^{41}_{19}$K   (b) $^{30}_{14}$Si and $^{32}_{16}$S

(c) $^{23}_{11}$Na and $^{24}_{12}$Mg   (d) $^7_3$Li and $^6_3$Li

5.40 Characterize each of the following pairs of atoms as containing (1) the same number of neutrons, (2) the same number of electrons, or (3) the same total number of subatomic particles.

(a) $^{13}_6$C and $^{14}_7$N   (b) $^{18}_8$O and $^{19}_9$F

(c) $^{37}_{17}$Cl and $^{36}_{18}$Ar   (d) $^{35}_{17}$Cl and $^{37}_{17}$Cl

5.41 What are the names of the elements that each of the following atoms represents?

(a) mass number = 45, 24 neutrons are present

(b) Cl atom with two more neutrons than $^{35}_{17}$Cl

(c) atomic number = 38, 50 neutrons are present

(d) atom with two more protons and two more electrons than $^{70}_{31}$Ga

5.42 What are the names of the elements that each of the following atoms represents?

(a) mass number = 60, 32 neutrons are present

(b) Ca atom with three more neutrons than $^{40}_{20}$Ca

(c) atomic number = 14, 14 neutrons are present

(d) atom with one more proton and one more electron than $^{40}_{18}$Ar

5.43 Indicate whether each of the following pieces of information is sufficient to determine the element identity for an atom.

(a) number of neutrons

(b) atomic number

(c) number of nucleons

(d) nuclear charge

5.44 Indicate whether each of the following pieces of information is sufficient to determine the element identity for an atom.

(a) mass number

(b) number of protons

(c) total mass of the nucleus

(d) sum of the atomic number and mass number

Isotopes and Isobars (Sec. 5.7)

5.45 Write complete symbols (^A_ZE) for the four naturally occurring isotopes of iron whose mass numbers are 54, 56, 57, and 58.

5.46 Write complete symbols (^A_ZE) for the four naturally occurring isotopes of chromium whose mass numbers are 50, 52, 53, and 54.

5.47 Five naturally occurring isotopes of the element zirconium exist. Knowing that the heaviest isotope has a mass number of 96 and that the other isotopes have, respectively, 2, 4, 5, and 6 fewer neutrons, write the complete symbol ($_Z^A E$) for each of the five isotopes.

5.48 Four naturally occurring isotopes of the element strontium exist. Knowing that the lightest isotope has a mass number of 84 and that the other isotopes have, respectively, 2, 4, and 5 more neutrons, write the complete symbol ($_Z^A E$) for each of the four isotopes.

5.49 The following are selected properties for the most abundant isotope of a particular element. Which of these properties would also be the same for the second-most-abundant isotope of the element?

(a) Mass number is 70

(b) 31 electrons are present

(c) Isotopic mass is 69.92 amu

(d) Isotope reacts with chlorine to give a green compound

5.50 The following are selected properties for the most abundant isotope of a particular element. Which of these properties would also be the same for the second-most-abundant isotope of the element?

(a) Atomic number is 31

(b) Does not react with the element gold

(c) 40 neutrons are present

(d) Density is 1.0301 g/mL

5.51 Indicate whether the members of each of the following pairs of atoms are isotopes, isobars, or neither.

(a) $_{12}^{24}X$ and $_{12}^{26}Q$ **(b)** $_{31}^{71}X$ and $_{32}^{71}Q$

(c) $_{25}^{56}X$ and $_{24}^{54}Q$ **(d)** $_{27}^{57}X$ and $_{27}^{60}Q$

5.52 Indicate whether the members of each of the following pairs of atoms are isotopes, isobars, or neither.

(a) $_{30}^{64}X$ and $_{29}^{64}Q$ **(b)** $_{10}^{20}X$ and $_{10}^{22}Q$

(c) $_{16}^{36}X$ and $_{17}^{35}Q$ **(d)** $_{28}^{60}X$ and $_{26}^{60}Q$

5.53 Indicate whether the members of each of the following pairs of atoms, specified in terms of subatomic particle composition, are isotopes, isobars, or neither.

(a) (24p, 24e, 26n) and (24p, 24e, 28n)

(b) (24p, 24e, 28n) and (25p, 25e, 27n)

(c) (24p, 24e, 26n) and (25p, 25e, 26n)

(d) (24p, 24e, 26n) and (24p, 24e, 24n)

5.54 Indicate whether the members of each of the following pairs of atoms, specified in terms of subatomic particle composition, are isotopes, isobars, or neither.

(a) (30p, 30e, 39n) and (31p, 31e, 38n)

(b) (30p, 30e, 35n) and (29p, 29e, 36n)

(c) (30p, 30e, 34n) and (30p, 30e, 36n)

(d) (30p, 30e, 38n) and (32p, 32e, 38n)

5.55 Isobars with a mass number of 40 exist for Ar, K, and Ca. Write the complete symbol ($_Z^A E$) for each of these isobars.

5.56 Isobars with a mass number of 50 exist for Ti, V, and Cr. Write the complete symbol ($_Z^A E$) for each of these isobars.

Atomic Masses (Sec. 5.8)

5.57 What is the atomic mass associated with the listed element or the element name associated with the listed atomic mass?

(a) iron   **(b)** nitrogen   **(c)** 40.08 amu   **(d)** 126.90 amu

5.58 What is the atomic mass associated with the listed element or the element name associated with the listed atomic mass?

(a) phosphorus   **(b)** nickel   **(c)** 101.07 amu   **(d)** 20.18 amu

5.59 Construct a relative mass scale for the hypothetical atoms Q, X, and Z, given that atoms of Q are two times as heavy as those of X, atoms of X are two times as heavy as atoms of Z, and the reference point for the scale is X = 4.00 bebs.

5.60 Construct a relative mass scale for the hypothetical atoms Q, X, and Z, given that atoms of Q are three times as heavy as those of X, atoms of X are four times as heavy as atoms of Z, and the reference point for the scale is Z = 2.50 bobs.

5.61 The atoms of element Z each have an average mass three-fourths that of a $_6^{12}C$ atom. Another element, X, has atoms whose average mass is three times the mass of Z atoms. A third element, Q, has atoms with an average mass nine times that of $_6^{12}C$.

(a) Construct a relative mass scale, based on $_6^{12}C$ having a mass of 12 amu, for the elements Z, X, and Q.

(b) Based on atomic masses rounded off to whole numbers, what is the identity of elements Z, X, and Q?

5.62 The atoms of element Z each have an average mass one-third that of a $_6^{12}C$ atom. Another element, X, has atoms whose average mass is four times the mass of Z atoms. A third element, Q, has atoms whose average mass is twice the mass of X atoms.

(a) Construct a relative mass scale, based on $_6^{12}C$ having a mass of 12 amu, for the elements Z, X, and Q.

(b) Based on atomic masses rounded off to whole numbers, what is the identity of elements Z, X, and Q?

5.63 A football team has the following distribution of players: 22.0% are defensive linemen with an average mass of 271 lb, 19.0% are defensive backs with an average mass of 175 lb, 26.0% are offensive linemen with an average mass of 263 lb, 15.0% are offensive backs with an average mass of 182 lb, and 19.0% are specialty team members with an average mass of 191 lb. What is the average mass of a football player on this team?

5.64 The assortment of automobiles on a used-car lot is categorized as follows: 18.0% are 1 year old, 10.0% are 2 years old, 33.0% are 3 years old, 4.0% are 4 years old, 31.0% are 5 years old, and 4.0% are 6 years old. What is the average age of a car on this used-car lot?

5.65 Calculate the atomic mass of chlorine on the basis of the following percent composition and isotopic mass data for the naturally occurring isotopes.

chlorine–35 = 75.53% (34.9689 amu)

chlorine–37 = 24.47% (36.9659 amu)

5.66 Calculate the atomic mass of copper on the basis of the following percent composition and isotopic mass data for the naturally occurring isotopes.

copper–63 = 69.09% (62.9298 amu)

copper–65 = 30.91% (64.9278 amu)

5.67 Calculate the atomic mass of titanium on the basis of the following percent composition and isotopic mass data for the naturally occurring isotopes.

titanium–46 = 7.93% (45.95263 amu)

titanium–47 = 7.28% (46.9518 amu)

titanium–48 = 73.94% (47.94795 amu)

titanium–49 = 5.51% (48.94787 amu)

titanium–50 = 5.34% (49.9448 amu)

5.68 Calculate the atomic mass of sulfur on the basis of the following percent composition and isotopic mass data for the naturally occurring isotopes.

sulfur–32 = 95.0% (31.9721 amu)

sulfur–33 = 0.76% (32.9715 amu)

sulfur–34 = 4.22% (33.9679 amu)

sulfur–36 = 0.014% (35.9671 amu)

5.69 Each of the following elements has only two naturally occurring isotopes. Determine, in each case, which isotope is more abundant using only the atomic mass value for the element that is listed on the periodic table inside the front cover.

(a) $^{14}_{7}N$ and $^{15}_{7}N$ (b) $^{50}_{23}V$ and $^{51}_{23}V$

(c) $^{121}_{51}Sb$ and $^{123}_{51}Sb$ (d) $^{191}_{77}Ir$ and $^{193}_{77}Ir$

5.70 Each of the following elements has only two naturally occurring isotopes. Determine, in each case, which isotope is more abundant using only the atomic mass value for the element that is listed on the periodic table inside the front cover.

(a) $^{10}_{5}B$ and $^{11}_{5}B$ (b) $^{69}_{31}Ga$ and $^{71}_{31}Ga$

(c) $^{107}_{47}Ag$ and $^{109}_{47}Ag$ (d) $^{203}_{81}Tl$ and $^{205}_{81}Tl$

5.71 Based on atomic mass values, what is the mass ratio (to three significant figures) of the following?

(a) one atom of Cl to one atom of Rb

(b) two atoms of Cl to two atoms of Rb

(c) one atom of Ag to two atoms of Cu

(d) ten atoms of H to one atom of Be

5.72 Based on atomic mass values, what is the mass ratio (to three significant figures) of the following?

(a) one atom of Cr to one atom of Si

(b) one atom of Si to one atom of Cr

(c) three atoms of Ni to three atoms of Se

(d) one atom of Br to five atoms of Cs

5.73 What are the name and symbol of the element whose average atoms have a mass

(a) close to four times the mass of an average nitrogen atom?

(b) that is 81.2% of the mass of an average silver atom?

(c) that is three times the atomic number of lithium?

(d) close to one-fifth the mass of an average neon atom?

5.74 What are the name and symbol of the element whose average atoms have a mass

(a) close to three times the mass of an average beryllium atom?

(b) that is 37.1% of the mass of an average zinc atom?

(c) that is three times the atomic number of fluorine?

(d) close to one-half the mass of an average silicon atom?

5.75 The arbitrary standard for the atomic mass scale is the exact number 12 for the mass of $^{12}_{6}C$. Why, then, is the atomic mass of carbon listed as 12.01?

5.76 The atomic mass of fluorine is 19.00 amu and that of copper is 63.55 amu. All fluorine atoms have a mass of 19.00 amu, and not a single copper atom has a mass of 63.55 amu. Explain.

Evidence Supporting the Existence of Subatomic Particles (Sec. 5.9)

5.77 Indicate whether each of the following statements about discharge tube and metal foil experiments is *true* or *false*.

(a) Information obtained from discharge tube experiments led to the characterization of both protons and electrons.

(b) Canal rays are negatively charged particles produced in a discharge tube.

(c) Metal foil experiments led to the discovery of neutrons.

(d) Many different types of cathode rays are known.

5.78 Indicate whether each of the following statements about discharge tube and metal foil experiments is *true* or *false*.

(a) In metal foil experiments almost all of the bombarding particles were stopped by the metal foil.

(b) Metal foil experiments led to the concept that an atom has a nucleus.

(c) Cathode rays and canal rays move in opposite directions in a discharge tube.

(d) Many different types of canal rays have been observed but only one type of cathode ray is known.

Additional Problems

5.79 Based on the given information, determine the numerical value of the subscript x in each of the following chemical formulas.

(a) $Na_2S_xO_3$; formula unit contains 7 atoms.

(b) $Ba(ClO_x)_2$; formula unit contains 9 atoms.

(c) $Na_xP_xO_{10}$; formula unit contains 16 atoms.

(d) $C_xH_{2x}O_x$; formula unit contains 24 atoms.

5.80 Based on the given information, determine the numerical value of the subscript x in each of the following chemical formulas.

(a) BaS_2O_x; formula unit contains 6 atoms.

(b) $Al_2(SO_x)_3$; formula unit contains 17 atoms.

(c) SO_xCl_x; formula unit contains 5 atoms.

(d) $C_xH_{2x}Cl_x$; formula unit contains 8 atoms.

5.81 How many protons are present in seven molecules of the compound $C_6H_{12}O_6$ (glucose, blood sugar)?

5.82 How many electrons are present in nine molecules of the compound $C_{12}H_{22}O_{11}$ (sucrose, table sugar)?

5.83 A mixture contains the following five pure substances: O_2, N_2O, H_2O, CCl_4, and CH_2Br_2.

(a) How many different kinds of heteroatomic molecules are present in the mixture?

(b) How many different kinds of pentaatomic molecules are present in the mixture?

(c) How many different compounds are present in the mixture?

(d) How many different kinds of atoms are present in the mixture?

(e) How many total atoms are present in a mixture sample containing three molecules of each component?

5.84 A mixture contains the following five pure substances: N_2, N_2H_4, NH_3, CH_4, and CH_3Cl.

(a) How many different kinds of molecules that contain four or fewer atoms are present in the mixture?

(b) How many different kinds of homoatomic molecules are present in the mixture?

(c) How many different kinds of atoms are present in the mixture?

(d) How many total atoms are present in a mixture sample containing five molecules of each component?

(e) How many total hydrogen atoms are present in a mixture sample containing four molecules of each component?

5.85 Indicate whether each of the following statements concerning sodium isotopes is *true* or *false*.

(a) $^{23}_{11}Na$ has one more electron than does $^{24}_{11}Na$.

(b) $^{23}_{11}Na$ and $^{24}_{11}Na$ contain the same number of neutrons.

(c) $^{23}_{11}Na$ has one fewer subatomic particle than $^{24}_{11}Na$.

(d) $^{23}_{11}Na$ and $^{24}_{11}Na$ have the same atomic number.

5.86 Indicate whether each of the following statements concerning magnesium isotopes is *true* or *false*.

(a) $^{24}_{12}Mg$ has one more proton than $^{25}_{12}Mg$.

(b) $^{24}_{12}Mg$ and $^{25}_{12}Mg$ contain the same number of subatomic particles in their nucleus.

(c) $^{24}_{12}Mg$ has one fewer neutron than $^{25}_{12}Mg$.

(d) $^{24}_{12}Mg$ and $^{25}_{12}Mg$ have different mass numbers.

5.87 Arrange the five atoms $^{42}_{20}Ca$, $^{39}_{19}K$, $^{44}_{21}Sc$, $^{37}_{18}Ar$, and $^{43}_{22}Ti$ in order of

(a) increasing number of electrons.

(b) decreasing number of neutrons.

(c) increasing number of protons.

(d) decreasing mass.

5.88 Arrange the five atoms $^{92}_{40}Zr$, $^{89}_{39}Y$, $^{94}_{41}Nb$, $^{87}_{38}Sr$, and $^{93}_{42}Mo$ in order of

(a) decreasing number of electrons.

(b) increasing number of neutrons.

(c) increasing number of nucleons.

(d) decreasing number of subatomic particles.

5.89 Write the complete symbol for the isotope of boron with each of the following characteristics.

(a) contains two fewer neutrons than $^{10}_{5}B$

(b) contains three more subatomic particles than $^{9}_{5}B$

(c) contains the same number of neutrons as $^{14}_{7}N$

(d) contains the same number of subatomic particles as $^{14}_{7}N$

5.90 Write the complete symbol for the isotope of chromium with each of the following characteristics.

(a) contains two more neutrons than $^{55}_{24}Cr$

(b) contains two fewer subatomic particles than $^{52}_{24}Cr$

(c) contains the same number of neutrons as $^{60}_{29}Cu$

(d) contains the same number of subatomic particles as $^{60}_{29}Cu$

5.91 Copper consists of two naturally occurring isotopes with masses of 62.9298 amu and 64.9278 amu.

(a) How many protons are in the nucleus of each isotope?

(b) How many electrons are in an atom of each isotope?

(c) How many neutrons are in the nucleus of each isotope?

5.92 Silver consists of two naturally occurring isotopes with masses of 106.9041 amu and 108.9047 amu.

(a) How many protons are in the nucleus of each isotope?

(b) How many electrons are in an atom of each isotope?

(c) How many neutrons are in the nucleus of each isotope?

5.93 Naturally occurring aluminum has a single isotope. Determine the following for naturally occurring atoms of aluminum.

(a) atomic number

(b) mass number

(c) number of neutrons in the nucleus

(d) isotopic mass, in amu, to three significant figures

5.94 Naturally occurring sodium has a single isotope. Determine the following for naturally occurring atoms of sodium.

(a) atomic number

(b) mass number

(c) number of neutrons in the nucleus

(d) isotopic mass, in amu, to three significant figures

5.95 Identify the element with each of the following characteristics.

(a) has an atomic mass 1.679 times that of phosphorus

(b) has three times as many electrons as zinc

(c) has twenty more protons than beryllium

(d) has twice as many charged subatomic particles as calcium

5.96 Identify the element with each of the following characteristics.

(a) has a nuclear charge twice that of magnesium

(b) has 30 more electrons than potassium

(c) has an atomic mass 2.381 times that of bromine

(d) has three times as many protons as sulfur

5.97 Suppose it was decided to redefine the atomic mass scale by choosing as an arbitrary reference point a value of 20.000 amu to represent the naturally occurring mixture of nickel isotopes. What would be the atomic mass of the following elements on the new atomic mass scale?

(a) silver

(b) gold

5.98 Suppose it was decided to redefine the atomic mass scale by choosing as an arbitrary reference point a value of 40.000 amu to represent the mass of fluorine atoms. (All fluorine atoms are identical; that is, fluorine is monoisotopic.) What would be the atomic mass of the following elements on the new atomic mass scale?

(a) sulfur

(b) platinum

5.99 In a hypothetical molecule XZ_2, 43.2% of the mass is from X and the rest from Z. If a relative mass scale were established with the mass of Z assigned a value of exactly 80, what would be the relative atomic mass of X?

5.100 In a hypothetical molecule Q_2Z, 35.1% of the mass is from Q and the rest from Z. If a relative mass scale were established with the mass of Q assigned a value of exactly 40, what would be the relative atomic mass of Z?

5.101 The following isotopic mass ratios were determined experimentally:

$^{19}_{9}F/^{12}_{6}C = 1.5832$; $^{35}_{17}Cl/^{19}_{9}F = 1.8406$; $^{81}_{35}Br/^{37}_{17}Cl = 2.3140$

Based on this information, what is the mass of $^{81}_{35}Br$ in atomic mass units?

5.102 The following isotopic mass ratios were determined experimentally:

$^{17}_{8}O/^{12}_{6}C = 1.4166$; $^{32}_{16}S/^{17}_{8}O = 1.8808$; $^{80}_{34}Se/^{32}_{16}S = 2.4996$

Based on this information, what is the mass of $^{80}_{34}Se$ in atomic mass units?

5.103 Chlorine has two isotopes with mass numbers of 35 and 37. Combination of chlorine with bromine produces BrCl. An examination of the product BrCl molecules shows the presence of only three isotopically different types of molecules with approximate masses of 114, 116, and 118 amu. Based on this information, how many isotopes of Br exist and what are the mass numbers of the isotopes?

5.104 Chlorine has two isotopes with mass numbers of 35 and 37. Combination of copper with chlorine produces CuCl. An examination of the product CuCl molecules shows the presence of only three isotopically different types of molecules with approximate masses of 98, 100, and 102 amu. Based on this information, how many isotopes of Cu exist and what are the mass numbers of the isotopes?

5.105 Three naturally occurring isotopes of potassium exist: $^{39}_{19}K$, $^{40}_{19}K$, and $^{41}_{19}K$. The atomic mass of potassium is 39.102 amu. Which of the three potassium isotopes is most abundant? Explain your answer.

5.106 Naturally occurring boron is a mixture of two isotopes: $^{10}_{5}B$ and $^{11}_{5}B$. Given that the atomic mass of boron is 10.811 amu, estimate the abundances of the two isotopes to the nearest 10%.

Cumulative Problems

5.107 Assign each of the following descriptions of matter to one of the following categories: *element, compound,* or *mixture*.

(a) one substance present, two phases present, all molecules are heteroatomic

(b) two substances present, one phase present, all molecules are homoatomic

(c) one phase present, all molecules are homoatomic, all molecules are identical

(d) one phase present, both homoatomic and heteroatomic molecules present

5.108 Assign each of the following descriptions of matter to one of the following categories: *element, compound,* or *mixture*.

(a) one substance present, one phase present, one kind of homoatomic molecule present

(b) two substances present, two phases present, all molecules are heteroatomic

(c) one phase present, two kinds of homoatomic molecules present

(d) one phase present, all molecules are triatomic, all molecules are heteroatomic, all molecules are identical

5.109 What is the atomic number of the element of highest atomic mass whose symbol does not derive from the English name of the element?

5.110 What is the atomic number of the element of lowest atomic mass whose symbol does not derive from the English name of the element?

5.111 The density of gold is 19.3 g/cm^3, and the mass of a single gold atom is 3.27×10^{-22} g. How many gold atoms are present in a piece of gold whose volume is 3.22 cm^3?

5.112 The density of copper is 8.93 g/cm^3, and the mass of a single copper atom is 1.06×10^{-22} g. How many copper atoms are present in a copper bar whose dimensions are 2.00 cm $\times$ 3.00 cm $\times$ 5.00 cm?

5.113 In 1.00 g of fluorine atoms there are 3.17×10^{22} fluorine atoms. If you lined these atoms up side by side, how many miles long would the line of fluorine atoms be? The diameter of a fluorine atom is 1.44×10^{-8} cm.

5.114 In 1.00 g of fluorine atoms there are 3.17×10^{22} fluorine atoms. If you started counting these atoms at the rate of 10 per second, how many years would it take to count all the atoms in the 1.00-g sample?

5.115 An electron has a mass of 5.5×10^{-4} amu. What percentage of the total mass of a Pb atom with a mass of 207 amu is due to its electrons?

5.116 How many electrons, with a mass of 5.5×10^{-4} amu, would it take to equal the mass of one proton, which has a mass of 1.0073 amu?

5.117 The diameter of an atom is approximately 10^5 times as large as the diameter of the nucleus. If the nucleus were enlarged to the size of a Ping-Pong ball (1.5 in. in diameter), what would be the diameter of the atom, in miles?

5.118 The diameter of an atom is approximately 10^{-8} cm, and that of the nucleus is 10^{-13} cm. How many times larger is the volume of the atom than the volume of the nucleus? The formula for the volume of a sphere is $V = 0.524 \times d^3$ (where d is the diameter of the sphere).

Answers to Practice Exercises

5.1 **(a)** triatomic, homoatomic, element **(b)** pentatomic, heteroatomic, compound **(c)** triatomic, heteroatomic, compound **(d)** tetratomic, heteroatomic, compound

5.2 **(a)** 8 C, 10 H, 4 N, 2 O **(b)** 3 N, 12 H, 1 P, 4 O **(c)** 1 Ca, 2 N, 6 O

5.3 79 protons, 79 electrons, 118 neutrons

5.4 $^{127}_{53}I$

5.5 **(a)** neither **(b)** isotopes **(c)** isobars **(d)** neither

5.6 Z = 3.00 sloops, X = 12.0 sloops, and Q = 24.0 sloops

5.7 9.0 ppm

5.8 28.09 amu

6

Electronic Structure and Chemical Periodicity

6.1 The Periodic Law

During the early part of the nineteenth century, scientists began to look for order in the increasing amount of chemical information that had become available. They knew that certain elements had properties that were very similar to those of other elements, and they sought to use these similarities as a means for arranging or classifying the elements.

In 1869, these efforts culminated in the discovery of what is now called the *periodic law*. The **periodic law** *states that when elements are arranged in order of increasing atomic number, elements with similar chemical behavior occur at periodic (regularly recurring) intervals.* Proposed independently by both the Russian chemist Dmitri Ivanovich Mendeleev (1834–1907; see "The Human Side of Chemistry 5") and the German chemist Julius Lothar Meyer (1830–1895), the periodic law is one of the most important of all chemical laws.

The preceding statement of the periodic law is in modern-day language. It differs from the original 1869 statements in that the phrase *atomic number* has replaced "atom mass." The use of "atom mass" in the original statements reflected theories prevalent in 1869. According to these theories, the masses of atoms were their most important distinguishing properties—a knowledge of subatomic particles (Sec. 5.5) was still 30 years away. When the details of subatomic structure were finally discovered, it became obvious that the properties of atoms were related not to their masses but to the number and arrangement of their electrons. Thus, the periodic law was modified to reflect this new knowledge.

Figure 6.1 shows parts of the repeating pattern for chemical properties (periodic law) for the sequence of elements with atomic numbers 3 through 20. The elements shown in the same color have similar chemical properties. For the sake of simplicity, only two of the

The Human Side of Chemistry 5

Dmitri Ivanovich Mendeleev (1834–1907)

Dmitri Ivanovich Mendeleev, born in Siberia in 1834, the youngest of 17 children in his family, is Russia's most famous chemist. His periodic classification of the elements is heralded as his greatest achievement. Other areas of research for him included petroleum chemistry and pressure and temperature studies on gases.

Mendeleev's talents as a prolific writer and a popular teacher directly influenced his discovery of the periodic law. In 1867, two years prior to his great discovery, Mendeleev was appointed professor of general chemistry at the University of St. Petersburg. To help his students, he immediately began writing a new textbook. In preparation for writing some material about the elements and their properties, he wrote down the properties of each element on separate cards. While sorting through his "flash" cards trying to "organize things," he "discovered" the periodic repetition of properties of the elements. Within a month he had written a paper on the subject that was delivered before the Russian Chemical Society. The textbook he was writing, *General Chemistry*, became a classic, going through eight Russian editions and numerous English, French, and German editions.

Mendeleev's first periodic table listed the elements according to increasing atom mass and grouped them (in columns) according to chemical reactivity. In doing this he realized that several "holes" existed in the elemental arrangement, which he interpreted as evidence of missing (yet to be discovered) elements. Two of the vacant spaces were directly under aluminum and silicon. He named these undiscovered elements eka-aluminum and eka-silicon and predicted what their properties would be. Within 15 years, these elements—now known as gallium and germanium—were isolated from naturally occurring minerals. Their properties matched closely with those that Mendeleev had predicted.

Mendeleev remained a professor at St. Petersburg until 1890, when he resigned because of a controversy with the Ministry of Education concerning student political rights. He took the side of the students. In 1893, he became director of the Russian Bureau of Weights and Measures, a post he held until his death.

It should be pointed out that the German chemist Julius Lothar Meyer (1830–1895) discovered the periodic law simultaneously and independently of Mendeleev. Unknown to each other, both published papers on the subject in 1869. Because Mendeleev did more with the periodic law, once it was published, he is generally given more credit for its development.

In 1882, when selected to receive the Davy medal of the British Royal Society for his work concerning systematic classification of the elements, Mendeleev insisted that Meyer be co-recipient of the award, formally recognizing Meyer's independent conception of the periodic law.

Element 101, mendelevium, bears his name.

---

periodic relationships (repeating patterns) are shown; for the elements listed, similar properties are found in every eighth element.

The concept of a periodic variation or pattern should not be new to you because numerous everyday examples of this exist. Most, but not all, involve time. Mondays occur at regular intervals of seven days. Office workers get paid at regular intervals, such as every two weeks. The red–green–yellow light sequence for a traffic semaphore repeats itself in a periodic fashion.

To be useful, the relationships generated by the periodic law must be easily visualized. This is the purpose of *periodic tables*, the subject of Section 6.2.

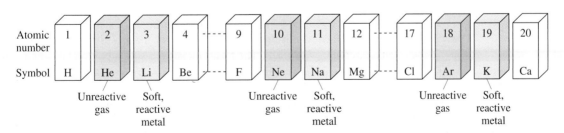

Figure 6.1

An illustration of the periodicity in properties that occurs when elements are arranged in order of increasing atomic number. He, Ne, and Ar have similar chemical properties as do Li, Na, and K.

6.2 The Periodic Table

A **periodic table** *is a tabular arrangement of the elements in order of increasing atomic number such that elements having similar chemical behavior are grouped in vertical columns.* It is a graphical portrayal of the periodic law.

The most commonly used form of the periodic table is the one shown in Figure 6.2 (and also inside the front cover of this book). Note the locations in this table for the elements He, Ne, and Ar and the elements Li, Na, and K—groups of elements with similar properties (Fig. 6.1). The elements He, Ne, and Ar are all found in the far right column of the table and the elements Li, Na, and K are all found in the far left column of the table. From further study of the far right column of this periodic table, we can deduce that the elements Kr, Xe, and Rn will have properties similar to those of He, Ne, and Ar since all of these elements are in the same column of the table.

Within the periodic table each element is represented by a square box. Within each box are given the symbol, atomic number, and atomic mass of the element, as shown in Figure 6.3. It should be noted that some periodic tables, but not the ones in this text, also give the element's name within the box.

> Using the information on a periodic table, you can quickly determine the number of protons and electrons for atoms of an element. However, no information concerning neutrons is available from a periodic table; mass numbers are not part of the information given, because they are not unique for an element.

Periods and Groups of Elements

Two terms integrally connected with the use of a periodic table are the terms *period* and *group*.

A **period** *is a horizontal row of elements in the periodic table.* For identification purposes, the periods are numbered sequentially, with Arabic numbers, starting at the top of the periodic table. (These period numbers are sometimes, but not usually, shown on the left-hand side of a periodic table.) Period 3 is the third row of elements, period 4 the fourth row of elements, and so forth. The elements Na, Mg, Al, Si, P, S, Cl, and Ar are all members of period 3 (see Fig. 6.2). Period 1 has only two elements—H and He.

A **group** *is a vertical column of elements in the periodic table.* The elements within a group have similar chemical properties. Three different labeling schemes for groups are in common use, two of which are shown in Figure 6.2 (located above each group). The bottom set of group labels in Figure 6.2, which involve Roman numerals and the letters A and B, is widely used in North America. Europeans use a similar convention (not shown) that numbers the groups consecutively as IA through VIIIA and then IB through VIIIB. North American and European group labels match for only four groups (IA, IIA, IB, and IIB). In an effort to eliminate "confusion" between the two systems, the International Union of Pure and Applied Chemistry (IUPAC) has proposed a new convention that numbers the groups simply using the numbers 1 through 18 (no A's or B's and no Roman numerals), as is shown in the top set of labels in Figure 6.2. We will use the traditional North American convention because the IUPAC system is still not widely used.

> The elements within a given periodic table group show numerous similarities in chemical properties, the degree of similarity varying from group to group. In no case are the group members "clones" of one another. Each element has some individual characteristics not found in other elements of the group. By analogy, the members of a human family often bear many resemblances to each other, but each member also has some (and often much) individuality.

Four groups of elements also have common (nonnumerical) names. On the extreme left side of the periodic table are found the *alkali metals* (Li, Na, K, Rb, Cs) and the *alkaline earth metals* (Be, Mg, Ca, Sr, Ba). **Alkali metal** *is a general name for any element in Group IA of the periodic table, excluding hydrogen.* The alkali metals are soft, shiny metals that readily react with water. **Alkaline earth metal** *is a general name for any element in group IIA of the periodic table.* The alkaline earth metals are soft, shiny metals also, but are only moderately reactive with water. On the extreme right side of the periodic table are found the *halogens* (F, Cl, Br, I, At) and the *noble gases* (He, Ne, Ar, Kr, Xe, Rn). **Halogen** *is a general name for any element in Group VIIA of the periodic table.* The halogens are very reactive colored substances that are gases at room temperature or become such at temperatures slightly above room temperature. **Noble gas** *is a general name for any element in Group VIIIA of the periodic table.* Noble gases are unreactive gases that undergo few, if any, chemical reactions.

1 Group IA	2 Group IIA	3 Group IIIB	4 Group IVB	5 Group VB	6 Group VIB	7 Group VIIB	8 Group	9 Group VIIIB	10 Group	11 Group IB	12 Group IIB	13 Group IIIA	14 Group IVA	15 Group VA	16 Group VIA	17 Group VIIA	18 Group VIIIA
1 H 1.01																	2 He 4.00
3 Li 6.94	4 Be 9.01											5 B 10.81	6 C 12.01	7 N 14.01	8 O 16.00	9 F 19.00	10 Ne 20.18
11 Na 22.99	12 Mg 24.30											13 Al 26.98	14 Si 28.09	15 P 30.97	16 S 32.06	17 Cl 35.45	18 Ar 39.95
19 K 39.10	20 Ca 40.08	21 Sc 44.96	22 Ti 47.87	23 V 50.94	24 Cr 52.00	25 Mn 54.94	26 Fe 55.84	27 Co 58.93	28 Ni 58.69	29 Cu 63.55	30 Zn 65.41	31 Ga 69.72	32 Ge 72.64	33 As 74.92	34 Se 78.96	35 Br 79.90	36 Kr 83.80
37 Rb 85.47	38 Sr 87.62	39 Y 88.91	40 Zr 91.22	41 Nb 92.91	42 Mo 95.94	43 Tc (98)	44 Ru 101.07	45 Rh 102.91	46 Pd 106.42	47 Ag 107.87	48 Cd 112.41	49 In 114.82	50 Sn 118.71	51 Sb 121.76	52 Te 127.60	53 I 126.90	54 Xe 131.29
55 Cs 132.91	56 Ba 137.33	57 La 138.91	72 Hf 178.49	73 Ta 180.95	74 W 183.84	75 Re 186.21	76 Os 190.23	77 Ir 192.22	78 Pt 195.08	79 Au 196.97	80 Hg 200.59	81 Tl 204.38	82 Pb 207.2	83 Bi 208.98	84 Po (209)	85 At (210)	86 Rn (222)
87 Fr (223)	88 Ra (226)	89 Ac (227)	104 Rf (261)	105 Db (262)	106 Sg (266)	107 Bh (264)	108 Hs (269)	109 Mt (268)	110 Ds (271)	111 (272)	112 (277)		114 (289)				

58 Ce 140.12	59 Pr 140.91	60 Nd 144.24	61 Pm (145)	62 Sm 150.36	63 Eu 151.96	64 Gd 157.25	65 Tb 158.93	66 Dy 162.50	67 Ho 164.93	68 Er 167.26	69 Tm 168.93	70 Yb 173.04	71 Lu 174.97
90 Th 232.04	91 Pa 231.04	92 U 238.03	93 Np (237)	94 Pu (242)	95 Am (243)	96 Cm (248)	97 Bk (247)	98 Cf (251)	99 Es (252)	100 Fm (257)	101 Md (260)	102 No (259)	103 Lr (262)

Metals ← → Non-metals

Figure 6.2

The most commonly used form of the periodic table.

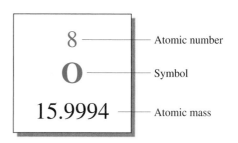

Figure 6.3

Arrangement of information about elements within "boxes" of the periodic table.

8 — Atomic number

O — Symbol

15.9994 — Atomic mass

The location of any element in the periodic table is specified by giving its group number and its period number. The element gold (Au), with an atomic number of 79, belongs to group IB and is in period 6. Nitrogen (N), with an atomic number of 7, belongs to group VA and is in period 2.

The Shape of the Periodic Table

There is one location in the periodic table of Figure 6.2 where the practice of arranging the elements according to increasing atomic number seems to be violated. This is the location of elements 57 and 89, which are both in group IIIB. Shown next to them, in group IVB, are elements 72 and 104, respectively. The missing elements, 58–71 and 90–103, are located in two rows at the bottom of the periodic table. Technically, these elements should be included in the body of the table, as shown in Figure 6.4a. However, to have a more compact table, they are placed in the position shown in Figure 6.4b. This arrangement should present no problems to the user of the periodic table as long as it is recognized for what it is—a space-saving device.

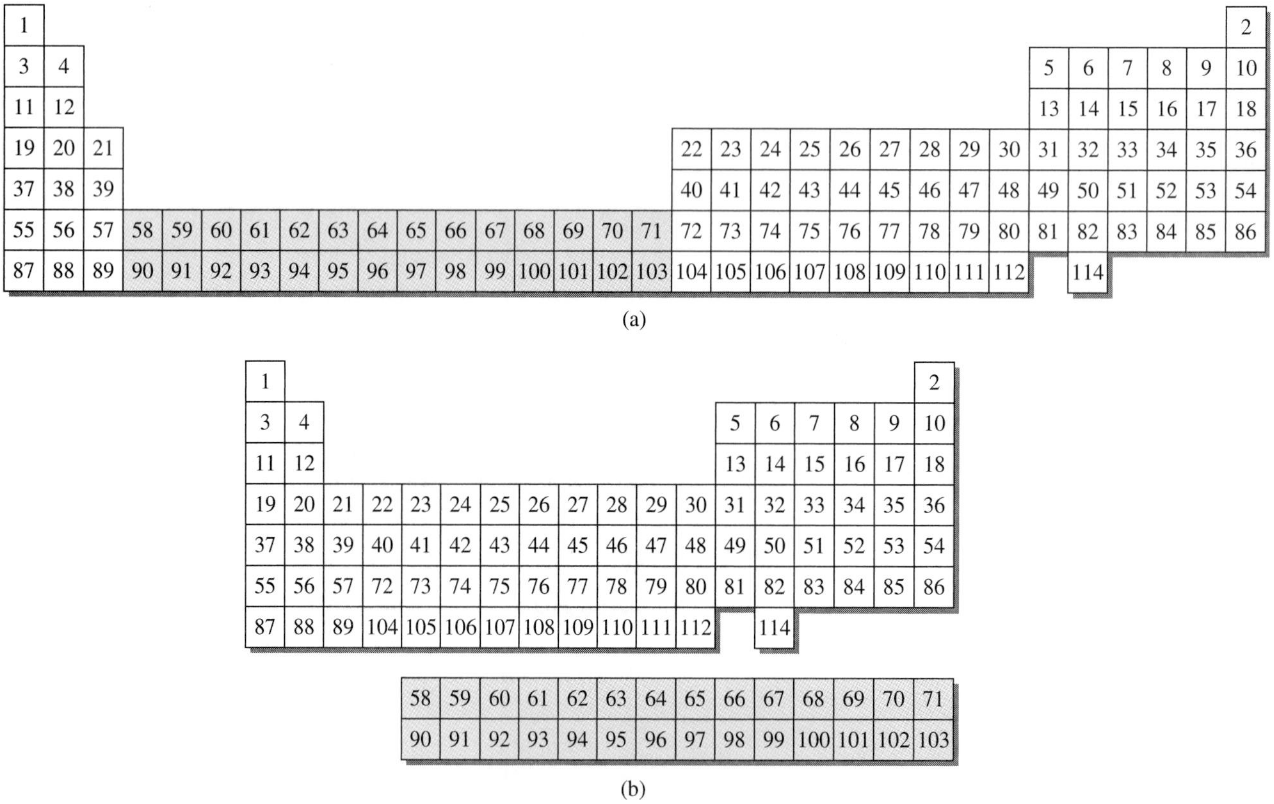

(a)

(b)

Figure 6.4

"Long" and "short" forms of the periodic table: (a) periodic table with elements 58–71 and 90–103 (in color) in their proper positions—the "long" form; (b) periodic table modified to conserve space by placing elements 58–71 and 90–103 (in color) below the rest of the elements—the "short" form.

Students often assume that a corollary of the fact that the elements are arranged in the periodic table in order of increasing atomic number is that they are also arranged in order of increasing atomic mass. This latter conclusion is correct most of the time; however, there are exceptions.

The following element sequences are exceptions to the generalization that atomic mass increases with increasing atomic number. In each triad, the middle element has the smaller atomic mass.

18	19	20
Ar	K	Ca
39.95	39.10	40.08

27	28	29
Co	Ni	Cu
58.93	58.69	63.55

52	53	54
Te	I	Xe
127.60	126.90	131.29

These inversions in atomic mass result from isotopic abundances (Sec. 5.7). Let us consider such data for the Ar–K–Ca triad.

^{36}Ar	0.34%	^{39}K	93.10%	^{40}Ca	96.97%
^{38}Ar	0.07%	^{40}K	0.01%	^{42}Ca	0.64%
^{40}Ar	99.60%	^{41}K	6.88%	^{43}Ca	0.14%
				^{44}Ca	2.06%
				^{46}Ca	0.003%
				^{48}Ca	0.18%

For argon the heaviest isotope $(A = 40)$ is most abundant, and for potassium the lightest isotope $(A = 39)$ is most abundant; this results in potassium having a lower atomic mass than argon. The atomic masses of argon and calcium are almost equal (39.95 and 40.08 amu, respectively); the dominant isotope for both elements has a mass number of 40.

Later in this chapter we will find that a periodic table conveys much more information about the elements than the symbols, atomic numbers, and atomic masses that are printed on it. Information about the arrangement of electrons in an atom is "coded" into the table, as is information concerning some physical and chemical property trends. Indeed, the periodic table is considered to be the single most useful study aid available for organizing information about the elements.

For many years after the formulation of the periodic law and the periodic table, both were considered to be empirical. The law worked and the table was very useful, but there was no explanation available for the law or for why the periodic table had the shape it had. It is now known that the theoretical basis for both the periodic law and the periodic table involves the arrangement of electrons in atoms. The properties of the elements repeat themselves in a periodic manner because the arrangement of electrons about the nucleus of an atom follows a periodic "pattern." Electron arrangements and an explanation of the periodic law and periodic table in terms of electronic theory are the subject matter for most of the remainder of this chapter.

6.3 The Energy of an Electron

In Section 5.5 we learned the following facts about electrons.

1. They are one of three fundamental subatomic particles.
2. They have very little mass in comparison to protons and neutrons.
3. They are located outside the nucleus of the atom.
4. They move rapidly about the nucleus in a volume that defines the size of the atom.

Much more must be known about electrons to understand the chemical behavior of the various elements, why some elements have similar chemical properties, and the theoretical basis for the periodic law.

More specific information concerning the behavior and arrangement of electrons within the extranuclear region of an atom is derived from a complex mathematical model for electron behavior called *quantum mechanics*. All the early work on this subject was done by physicists rather than by chemists.

During the early part of the twentieth century (1910–1930) a major revolution occurred in the field of physics, a revolution that profoundly affected chemistry. During this time it became clear that the "established" laws of physics, the laws that had been used for many years to predict the behavior of macroscopic objects, could not explain the behavior of extremely small objects such as atoms and electrons. The work of a number of European physicists led to this conclusion and also beyond it. These same physicists were able to formalize new laws that did apply to small objects. It is from these new laws, collectively called quantum mechanics, that information concerning the arrangement and behavior of electrons about a nucleus came.

A major force in the development of quantum mechanics was the Austrian physicist Erwin Schrödinger (1887–1961; see "The Human Side of Chemistry 6"). In 1926 he showed that the laws of quantum mechanics could be used to characterize the motion of electrons. Most of the concepts in the remainder of this chapter come from solutions to the equations developed by Schrödinger.

A formal discussion of quantum mechanics is beyond the scope of this course (and many other chemistry courses also) because of the rigorous mathematics involved. The answers obtained from quantum mechanics are, however, simple enough to be understood to a surprisingly large degree at the level of an introductory chemistry course. A consideration of these quantum-mechanical answers will enable us to develop a system for specifying electron arrangements around a nucleus and further to develop basic rules governing compound formation. This latter topic will occupy our attention in Chapter 7.

Present-day quantum mechanical theory describes the arrangement of an atom's electrons in terms of their energies. Indeed, the energy of an electron is the property that is most important to any consideration of its behavior about the nucleus.

The Human Side of Chemistry 6

Erwin Schrödinger (1887–1961)

Erwin Schrödinger, an only child, was born in 1887 in Vienna, Austria. He was taught at home until age 11, and science was always part of his environment—his father was a botanist, and his maternal grandfather a professor of chemistry. His formal education began in 1898 when he entered a gymnasium (high school), where he received a sound classical education. This early schooling in the "classics" influenced Schrödinger's actions throughout his life. The culmination of his formal education was a doctorate in theoretical physics, in 1910, from the University of Vienna.

After service as an artillery officer during World War I, Schrödinger held professorships in Stuttgart and Zürich. During his years in Switzerland his pioneering papers on the equations of quantum theory were formulated and published.

To have more contact with notable physicists (including Albert Einstein), he moved, in 1927, to Berlin. All went well until Hitler's new regime (which he could not accept) came into power. With the rise of Nazism, he accepted an invitation to be a guest professor at Oxford. In 1933, only a few weeks after leaving Germany for England, he received the Nobel Prize in physics for his quantum theory work.

Hitler's activities continued to affect his career. After a short stay in England, he returned to his native Austria despite an uncertain political situation there. When the Germans occupied Austria in 1938, Schrödinger fled to Italy and from there to the United States. Following a short stay at Princeton University, he became director of the School of Theoretical Physics in Dublin, Ireland. For the next 17 years he worked in Dublin. In 1956, he again returned to Austria, where he remained until his death.

In addition to his pioneering work in quantum theory, Schrödinger's research touched the areas of thermodynamics, specific heats of solids, theory of color, and physical aspects of the living cell.

Schrödinger was "many-sided." He loved poetry and wrote verse of his own. He was a sculptor and also widely acquainted with modern and classical art. He quoted Greek philosophers as he lectured on problems of theoretical physics. A colleague's tribute upon his death was: "The breadth of his knowledge was as marvelous as the power of his thought."

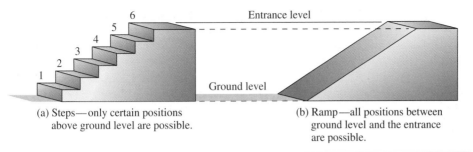

(a) Steps—only certain positions above ground level are possible.

(b) Ramp—all positions between ground level and the entrance are possible.

Figure 6.5

A stairway with quantized position levels (the steps) versus a ramp with continuous position levels.

The energy of an electron is manifested primarily in its velocity. The higher its energy, the higher its average velocity. The faster an electron travels, the farther it tends to move from the nucleus with which it is associated.

A most significant characteristic of an electron's energy is that it is a *quantized property*. A **quantized property** *is a property that can have only certain values, that is, not all values are allowed.* Since an electron's energy is quantized, an electron can have only certain specific energies.

Quantization is a phenomenon not commonly encountered in the macroscopic world. Somewhat analogous is the process of a person climbing a flight of stairs. In Figure 6.5a you see six steps between ground level and the entrance level. As a person climbs these stairs there are only six permanent positions he or she can occupy (with both feet together). Thus the person's position (height above ground level) is quantized; only certain positions are allowed. The opposite of quantized is continuousness. A person climbing a ramp up to the entrance (Fig. 6.5b) would be able to assume a continuous set of heights above ground level; all values are allowed.

The energy of an electron determines its behavior about the nucleus. Since electron energies are quantized, only certain behavior patterns are allowed. Descriptions of electron behaviors involve the use of the terms *shell, subshell,* and *orbital.* Sections 6.4–6.6 consider the meaning of these terms and the mathematical interrelationships between them.

> The floors of a building are quantized. An elevator will stop at the 5th or the 6th floor, but not at the 5.5 floor. The frets on the neck of a guitar make it quantized because the notes between the frets cannot be played. An instrument like a cello is continuous because it does not have frets; any note can be played.

6.4 Electron Shells

It was mentioned in Section 6.3 that electrons with higher energy will tend to be found farther from the nucleus than those with lower energy. Based on energy–distance-from-the-nucleus considerations, electrons can be grouped into *electron shells* or electron main energy levels. An **electron shell** *is a region of space about a nucleus that contains electrons that have approximately the same energy and that spend most of their time approximately the same distance from the nucleus.*

A shell number, *n*, is used to identify each electron shell. Electron shells are numbered 1, 2, 3, and so on, outward from the nucleus. Electron energy increases as the distance of the electron shell from the nucleus increases. An electron in shell 1 has the minimum amount of energy that an electron can have. No known atom has electrons beyond shell 7.

The maximum number of electrons possible in an electron shell varies; the higher the shell energy, the more electrons the shell can accommodate. The farther electrons are from the nucleus (a higher-energy shell), the greater the volume of space available for them; hence the more electrons there can be in the shell. (Conceptually, electron shells may be considered to be nested one inside another, somewhat like the layers of flavors inside a jawbreaker or similar type of candy.)

> Electrons that occupy shell 1 are closer to the nucleus and have a lower energy than do electrons in shell 2.

Table 6.1	Important Characteristics of Electron Shells	
Shell	**Number Designation (n)**	**Electron Capacity ($2n^2$)**
1st	1	$2 \times 1^2 = 2$
2nd	2	$2 \times 2^2 = 8$
3rd	3	$2 \times 3^2 = 18$
4th	4	$2 \times 4^2 = 32$
5th	5	$2 \times 5^2 = 50^*$
6th	6	$2 \times 6^2 = 72^*$
7th	7	$2 \times 7^2 = 98^*$

The maximum number of electrons in this shell has never been attained in any element now known.

The lowest-energy shell ($n = 1$) accommodates a maximum of 2 electrons. In the second, third, and fourth shells, 8, 18, and 32 electrons, respectively, are allowed. A very simple mathematical equation can be used to calculate the maximum number of electrons allowed in any given shell:

$$\text{shell electron capacity} = 2n^2, \text{ where } n = \text{shell number}$$

For example, when $n = 4$, the value $2n^2 = 2(4^2) = 32$, which is the number previously given for the number of electrons allowed in the fourth shell. Although there is a maximum electron occupancy for each shell or main energy level, a shell may hold fewer than the allowable number of electrons in a given situation.

Table 6.1 summarizes concepts presented in this section concerning electron shells or electron main energy levels.

6.5 Electron Subshells

All (both) electrons in the first shell have the same energy, but in higher shells all the electrons do not have the same energy. Their energies are all close to each other in magnitude, but they are not identical. The range of energies for electrons in a shell is due to the existence within an electron shell of *electron subshells* or electron energy sublevels. An **electron subshell** *is a region of space within an electron shell that contains electrons that have the same energy.*

The number of subshells within a shell varies. A shell contains the same number of subshells as its own shell number; that is,

$$\text{number of subshells in a shell} = n, \quad \text{where } n = \text{shell number}$$

Thus, each successive shell has one more subshell than the previous one. Shell 3 contains three subshells, shell 4 contains four subshells, and shell 5 contains five subshells.

Subshells are identified by a number and a letter. The number indicates the shell to which the subshell belongs. The letters are *s, p, d,* or *f* (all lowercase letters), which, in that order, denote subshells of increasing energy within a shell. The lowest energy subshell within a shell is always the *s* subshell, the next higher the *p* subshell, then the *d,* and finally the *f.* Shell 1 has only one subshell, the 1*s.* Shell 2 has two subshells, the 2*s* and 2*p.* The 3*s,* 3*p,* and 3*d* subshells are found in shell 3. Shell 4 contains four subshells, the 4*s,* 4*p,* 4*d,* and 4*f* subshells. Figure 6.6 gives information concerning subshells for the first seven shells. Note in that figure that number-and-letter designations are not given for all the subshells in shells 5, 6, and 7; some of these subshells—those denoted with a dash—are not needed to describe

> The early development of quantum mechanics was closely linked to the study of light (spectral) emissions produced by atoms. The letters *s, p, d,* and *f*—used to denote subshells—are derived, respectively, from old spectroscopic terminology that describes spectral lines as sharp, principal, diffuse, and fundamental.

Shell 1	1s					
Shell 2	2s	2p				
Shell 3	3s	3p	3d			
Shell 4	4s	4p	4d	4f		
Shell 5	5s	5p	5d	5f	—	
Shell 6	6s	6p	6d	—	—	—
Shell 7	7s	7p	—	—	—	—

Subshells

Figure 6.6

Subshells needed to describe the electron arrangements of the known elements. Note that not all subshells in shells 5, 6, and 7 are needed. Four of the five subshells in shell 5 are used, only three of the six subshells in shell 6 are used, and only two of the seven subshells in shell 7 are used.

the electron arrangements of the 113 known elements. Why they are not needed will become evident in Section 6.7. The important concept now is that we will never need subshell types other than *s, p, d,* and *f* in describing the electron arrangements of the known elements.

The maximum number of electrons that a subshell can hold varies from 2 to 14 depending on the type of subshell—*s, p, d,* or *f.* An *s* subshell can accommodate only 2 electrons. What shell the *s* subshell is located in does not affect the maximum electron occupancy figure; that is, the 1*s*, 2*s*, 3*s*, 4*s*, 5*s*, 6*s*, and 7*s* subshells all have a maximum electron occupancy of 2. Subshells of the *p, d,* and *f* types can accommodate maximums of 6, 10, and 14 electrons, respectively. Again the maximum numbers of electrons in these types of subshells depend only on the subshell types and are independent of shell number. Figure 6.7

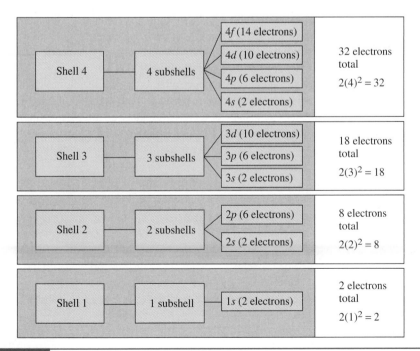

Figure 6.7

The distribution of electrons within shells and subshells for the first four electron shells.

summarizes the information presented in this section for subshells found in shells 1 through 4. Within a shell, the sum of the subshell electron occupancies is the same as the shell electron occupancy $(2n^2)$. For example, in shell 4, an *s* subshell containing 2 electrons, a *p* subshell containing 6 electrons, a *d* subshell containing 10 electrons, and an *f* subshell containing 14 electrons add up to a total of 32 electrons, which is the maximum occupancy of shell 4 as calculated by the $2n^2$ formula.

6.6 Electron Orbitals

The last and most basic of the three terms used in describing electron arrangements about nuclei is *electron orbital*. An **electron orbital** *is a region of space within an electron subshell where an electron with a specific energy is most likely to be found.*

An analogy for the relationship between shells, subshells, and orbitals can be found in the physical layout of a high-rise condominium complex. A shell is the counterpart of a floor of the condominium in our analogy. Just as each floor will contain apartments (of different sizes), a shell contains subshells (of different sizes). Further, just as apartments contain rooms, subshells contain orbitals. An apartment is a collection of rooms; a subshell is a collection of orbitals. A floor of a condominium building is a collection of apartments; a shell is a collection of subshells.

The characteristics of electron orbitals, the rooms in our "electron apartment house," include

1. The number of orbitals in a subshell varies, being one for an *s* subshell, three for a *p* subshell, five for a *d* subshell, and seven for an *f* subshell.
2. The maximum number of electrons in an orbital does not vary. It is always 2.
3. The notation used to designate orbitals is the same as that used for subshells. Thus, orbitals in the 4*f* subshell (there are seven of them) are called 4*f* orbitals.

We have already noted (Sec. 6.5) that all electrons in a subshell have the same energy. Thus all electrons in orbitals of the same subshell will have the same energy. This means that *shell and subshell designations are sufficient to specify the energy of an electron.* This statement will be of great importance in the discussions of Section 6.7.

Orbitals have a definite size and shape related to the type of subshell in which they are found. (Remember, an orbital is a region of space. We are not talking about the size and shape of an electron, but rather the size and shape of a region of space where an electron is found.) At any given time an electron can be at only one point in an orbital, but because of its rapid

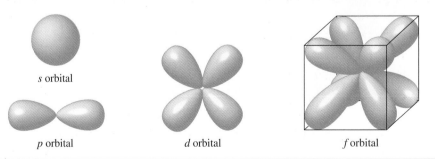

s orbital

p orbital   *d* orbital   *f* orbital

Figure 6.8

The shapes of atomic orbitals. To improve the perspective, the *f* orbital is shown within a cube with one lobe pointing to each corner of the cube. Only two electrons may occupy an orbital regardless of the number of lobes present in its shape. Some *d* and *f* orbitals have shapes related to, but not identical to, those shown.

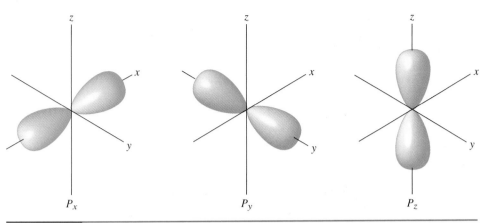

Figure 6.9

The orientation of the three orbitals in a *p* subshell.

1s   2s   3s

Figure 6.10

Size relationships among *s* orbitals in different shells.

movement throughout the orbital, it "occupies" the entire orbital. An analogy would be the "definite" volume occupied by rotating fan blades. Typical *s, p, d,* and *f* orbital shapes are given in Figure 6.8. Notice that the shapes increase in complexity in the order *s, p, d,* and *f.* Some of the *d* and *f* orbitals have shapes related, but not identical, to those shown in Figure 6.8.

Orbitals within the same subshell differ mainly in orientation. For example, the three 2*p* orbitals look the same but are aligned in different directions—along the *x, y,* and *z* axes in a Cartesian coordinate system (see Fig. 6.9).

Orbitals of the same type but in different shells (for example, 1*s,* 2*s,* and 3*s*) have the same general shape but differ in size (volume) (see Fig. 6.10).

EXAMPLE 6.1

Interrelationships Among Electron Shells, Electron Subshells, and Electron Orbitals

Determine the following information about electron shells, electron subshells, and electron orbitals.

(a) The number of electron subshells in shell 4

(b) The number of electron orbitals in a 3*d* subshell

(c) The maximum number of electrons that could be contained in a 2*p* subshell

(d) The maximum number of electrons that could be contained in a 2*p* orbital

SOLUTION

(a) The number of subshells in a shell is the same as the shell number. Thus, shell 4 will contain four subshells.

(b) The number of orbitals in a given type (*s, p, d, f*) subshell is independent of the shell number. Each *s* subshell (1*s,* 2*s,* 3*s,* etc.) contains one orbital, each *p* subshell contains three orbitals, each *d* subshell contains five orbitals, and each *f* subshell contains seven orbitals. Thus, a 3*d* subshell contains five orbitals.

(c) The maximum number of electrons in a subshell depends on the number of orbitals the subshell contains, with each orbital holding two electrons. In a *p* subshell there are three orbitals. Thus, a *p* subshell can accommodate six electrons (two per orbital).

(d) The maximum number of electrons in an orbital is two. It does not matter what type of orbital it is (2*s*, 2*p*, 3*d*, etc.). All orbitals hold a maximum of two electrons.

Practice Exercise 6.1

Determine the following information about electron shells, electron subshells, and electron orbitals.

(a) The number of electron subshells in shell 3
(b) The number of electron orbitals in a 2*p* subshell
(c) The maximum number of electrons that could be contained in a 4*d* subshell
(d) The maximum number of electrons that could be contained in a 1*s* orbital

Answers to practice exercises are located at the end of the chapter.

Electron Spin

A final feature of electron behavior is that electrons possess a property called *electron spin*. Experimental studies indicate that as an electron "moves about" within an orbital, it spins on its own axis in either a clockwise or counterclockwise direction. Furthermore, when two electrons are present in an orbital, they always have opposite spins; that is, one is spinning clockwise and the other counterclockwise. This situation of opposite spins is energetically the most favorable state for two electrons in the same orbital. **Electron spin** *is a property of an electron associated with its spinning on its own axis*. We will have more to say about electron spin when we discuss orbital diagrams in Section 6.8.

6.7 Electron Configurations

An **electron configuration** *is a statement of how many electrons an atom has in each of its subshells*. Since subshells group electrons according to energy (Sec. 6.5), electron configurations indicate how many electrons an atom has of various energies.

Electron configurations are not written out in words; a shorthand system with symbols is used. Subshells containing electrons, listed in order of increasing energy, are designated using number–letter combinations (1*s*, 2*s*, 2*p*, etc.). A superscript following each subshell designation indicates the number of electrons in that subshell. The electron configuration for oxygen using this shorthand notation is

$$1s^2 2s^2 2p^4 \quad \text{(read "one-}s\text{-two, two-}s\text{-two, two-}p\text{-four")}$$

An oxygen atom thus has an electron arrangement of two electrons in the 1*s* subshell, two electrons in the 2*s* subshell, and four electrons in the 2*p* subshell.

Aufbau Principle

To determine the electron configuration for an atom, a procedure called the *Aufbau principle* (German *aufbauen*, to build) is used. The **Aufbau principle** *states that electrons normally occupy electron subshells in an atom in order of increasing subshell energy*. This guideline brings order to what could be a very disorganized situation. Many subshells exist about the

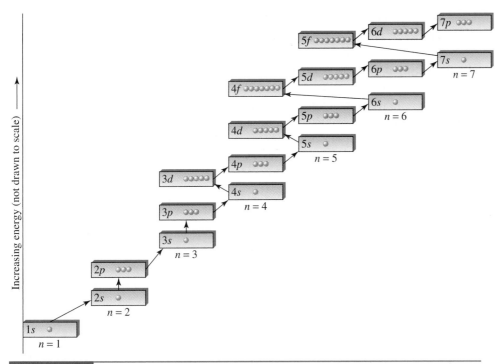

Figure 6.11

Relative energies and filling order for electron subshells.

nucleus of any given atom. Electrons do not occupy these subshells in a random, haphazard fashion; a very predictable pattern, governed by the Aufbau principle, exists for electron subshell occupancy. *Subshells are filled in order of increasing energy.*

Use of the Aufbau principle requires knowledge concerning the electron capacities of orbitals and subshells (which we already have; see Sec. 6.6) and knowledge concerning the relative energies of subshells (which we now consider).

The ordering of electron subshells in terms of increasing energy, which is experimentally determined, is more complex than might be expected. This is because the energies of subshells in different shells often "overlap," as shown in Figure 6.11. Beginning with shell 4, one or more lower-energy subshells of a specific shell have energies lower than the higher-energy subshells of a preceding shell and thus acquire electrons first. For example, the $4s$ subshell acquires electrons before the $3d$ subshell does (see Fig. 6.11). As another example, the s subshell of the sixth energy level fills before the d subshell of the fifth energy level or the f subshell of the fourth energy level (again refer to Fig. 6.11).

Aufbau Diagram

The sequence in which subshells acquire electrons must be learned before electron configurations can be written. A useful mnemonic (memory) device, called an *Aufbau diagram*, helps considerably with this learning process. As can be seen from Figure 6.12, an **Aufbau diagram** *is a listing of electron subshells in the order in which electrons occupy them.* All s subshells are located in column 1 of the diagram, all p subshells in column 2, and so on. Subshells belonging to the same shell are found in the same row. The order of subshell filling is given by following the diagonal arrows, starting with the top one. The $1s$ subshell fills first. The second arrow points to (goes through) the $2s$ subshell. It fills next. The third arrow points to both the $2p$ and $3s$ subshells. The $2p$ fills first, followed by the $3s$. Any time a single arrow points to more than one subshell, start at the tail of the arrow and work to its head to determine the

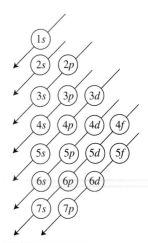

Figure 6.12

An Aufbau diagram is an aid to remembering subshell filling order.

proper filling sequence. The $3p$ subshell fills next, and so on. An Aufbau diagram is an easy way to catalog the information given in Figure 6.11.

Writing Electron Configurations

We are now ready to write electron configurations. Let us systematically consider electron configurations for the first few elements in the periodic table.

Hydrogen ($Z = 1$) has only one electron, which goes into the $1s$ subshell, which has the lowest energy of all subshells. Hydrogen's electron configuration is written as

$$1s^1$$

Helium ($Z = 2$) has two electrons, both of which occupy the $1s$ subshell. (Remember, an s subshell contains one orbital, and an orbital can accommodate two electrons.) Helium's electron configuration is

$$1s^2$$

For lithium ($Z = 3$), with three electrons, the third electron cannot enter the $1s$ subshell since its maximum capacity is two electrons. (All s subshells are completely filled with two electrons; see Sec. 6.6.) The third electron is placed in the next-highest energy subshell, the $2s$. The electron configuration for lithium is thus

$$1s^2 2s^1$$

With beryllium ($Z = 4$), the additional electron is placed in the $2s$ subshell, which is now completely filled, giving beryllium the electron configuration

$$1s^2 2s^2$$

In boron ($Z = 5$), the $2p$ subshell, the subshell of next highest energy (check Fig. 6.11 or 6.12 to be sure), becomes occupied for the first time. Boron's electron configuration is

$$1s^2 2s^2 2p^1$$

An *electron configuration* is a shorthand notation designating the *subshells* in an atom that are occupied by electrons. The sum of the superscripts in an electron configuration equals the total number of electrons present and hence equals the atomic number of the element.

A p subshell can accommodate six electrons since there are three orbitals within it (Sec. 6.6). The $2p$ subshell can thus accommodate the additional electrons found in C, N, O, F, and Ne. The electron configurations of these elements are

$$
\begin{aligned}
\text{C } (Z = 6): & \quad 1s^2 2s^2 2p^2 \\
\text{N } (Z = 7): & \quad 1s^2 2s^2 2p^3 \\
\text{O } (Z = 8): & \quad 1s^2 2s^2 2p^4 \\
\text{F } (Z = 9): & \quad 1s^2 2s^2 2p^5 \\
\text{Ne } (Z = 10): & \quad 1s^2 2s^2 2p^6
\end{aligned}
$$

With sodium ($Z = 11$), the $3s$ subshell acquires an electron for the first time.

$$\text{Na } (Z = 11): \quad 1s^2 2s^2 2p^6 3s^1$$

Note the pattern that is developing in the electron configurations we have written so far. Each element has an electron configuration the same as the one just before it with the addition of one electron.

Electron configurations for other elements are obtained by simply extending the principles we have just illustrated. Electrons are added to subshells, always filling one of lower energy before adding electrons to the next-highest subshell, until the correct number of electrons has been accommodated.

EXAMPLE 6.2

Writing Electron Configurations for Atoms

Write the electron configurations for (a) calcium ($Z = 20$) and (b) selenium ($Z = 34$).

SOLUTION

(a) The number of electrons in a calcium atom is 20. Remember, the atomic number (Z) gives the number of electrons (Sec. 5.6). We will need to fill subshells, in order of increasing energy, until 20 electrons have been accommodated.

The $1s$, $2s$, and $2p$ subshells fill first, accommodating a total of 10 electrons among them.

$$1s^2 2s^2 2p^6 \ldots$$

Next, according to Figure 6.11 or 6.12, the $3s$ fills and then the $3p$ subshell.

$$1s^2 2s^2 2p^6 3s^2 3p^6 \ldots$$

We have accommodated 18 electrons at this point. We will need to add two more electrons to get our desired number of 20.

These last two electrons are added to the $4s$ subshell.

$$1s^2 2s^2 2p^6 3s^2 3p^6 4s^2$$

These last two electrons completely fill the $4s$ subshell.

(b) To write the electron configuration for selenium, we continue along the same lines as in part (a), remembering that the maximum electron subshell populations are $s = 2$, $p = 6$, and $d = 10$.

The first 18 electrons, as with calcium (part a), will fill the $1s$, $2s$, $2p$, $3s$, and $3p$ subshells.

$$1s^2 2s^2 2p^6 3s^2 3p^6 \ldots$$

The $4s$ subshell fills next, accommodating 2 electrons, followed by the $3d$ subshell, which accommodates 10 electrons.

$$1s^2 2s^2 2p^6 3s^2 3p^6 4s^2 3d^{10} \ldots$$

We now have a total of 30 electrons.

Four more electrons are needed, which are added to the next higher subshell in energy, the $4p$.

$$1s^2 2s^2 2p^6 3s^2 3p^6 4s^2 3d^{10} 4p^4$$

The $4p$ subshell can accommodate 6 electrons, but we do not want it filled to capacity because that would give us too many electrons.

To double check that we have the correct number of electrons, 34, we add the superscripts in our final electron configuration.

$$2 + 2 + 6 + 2 + 6 + 2 + 10 + 4 = 34$$

The sum of the superscripts in any electron configuration should add up to the atomic number if the configuration is for a neutral atom.

Practice Exercise 6.2

Write the electron configuration for (a) chlorine ($Z = 17$) and (b) antimony ($Z = 51$).

It should be noted that for a few elements in the middle of the periodic table the actual distribution of electrons within subshells differs slightly from that obtained using the Aufbau principle and Aufbau diagram. These exceptions are caused by very small energy differences between some subshells and are not important in the uses we shall make of electronic configurations.

6.8 Orbital Diagrams

The arrangement of electrons about a nucleus can be specified in terms of *subshell* occupancy or *orbital* occupancy. The notation for specifying subshell occupancy (electron configurations) was considered in the previous section. In this section we consider electron arrangements at the orbital level.

Two principles are needed to specify orbital occupancy for electrons, one we have already considered and one that is new. The principles are

1. The Aufbau principle
2. Hund's rule

The Aufbau principle was used extensively in Section 6.7 in writing electron configurations. Hund's rule has not been encountered previously. The namesake for Hund's rule is the German physicist Friedrich Hund (1896–1997), an early worker in the field of quantum mechanics.

Hund's rule *states that when electrons are placed in a set of orbitals of equal energy (the orbitals of a subshell), the order of filling for the orbitals is such that each orbital will be occupied by one electron before any orbital receives a second electron.* Such a pattern of orbital filling minimizes repulsions between electrons. Numerous applications of Hund's rule are required in stating electron arrangements in terms of orbital occupancies, that is, in drawing *orbital diagrams*.

An **orbital diagram** *is a diagram that shows how many electrons an atom has in each of its occupied electron orbitals.* In drawing orbital diagrams, we use circles to represent orbitals and arrows with a single barb to denote electrons. The orbital diagram for hydrogen, with its one electron, is

> An *electron configuration* specifics *subshell* occupancy for electrons and an *orbital diagram* specifies *orbital* occupancy for electrons.

$$\text{H:} \quad \textcircled{\uparrow} \\ 1s$$

A helium atom contains two electrons, both of which occupy the $1s$ orbital. The orbital diagram for helium is

$$\text{He:} \quad \textcircled{\uparrow\downarrow} \\ 1s$$

The two electrons present are of opposite spin (see Sec. 6.6). Note the notation used to indicate this—one arrow points up and the other points down. The two electrons are said to be *paired*. **Paired electrons** *are two electrons of opposite spin present in the same orbital.* When only one electron is present in an orbital it is said to be *unpaired*. An **unpaired electron** *is a single electron in an orbital.* The electron configuration for hydrogen involves an unpaired electron.

Orbital diagrams for the next two elements, lithium and beryllium, are drawn according to reasoning similar to that followed for H and He. The two electrons in the $2s$ orbital of Be are paired.

Li: (⬆⬇) (⬆)
         1s   2s

Be: (⬆⬇) (⬆⬇)
         1s   2s

Boron has the electron configuration $1s^2 2s^2 2p^1$. The fifth electron in boron must enter a $2p$ orbital, since both the $1s$ and $2s$ orbitals are full. Boron's orbital diagram is

B: (⬆⬇) (⬆⬇) (⬆)( )( )
      1s   2s   2p

Note that it does not matter which one of the three $2p$ orbitals of boron is shown as containing an electron. All three orbitals have the same energy. Each of the following notations is equivalent.

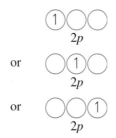

            2p

or ( )(⬆)( )
        2p

or ( )( )(⬆)
        2p

With carbon, element 6, we encounter the use of Hund's rule for the first time. Carbon has two electrons in the $2p$ subshell ($1s^2 2s^2 2p^2$). Do the two electrons go into the same orbital (paired), or do they go into separate equivalent orbitals (unpaired)? Hund's rule indicates that the latter is the case. The orbital diagram for carbon is

C: (⬆⬇) (⬆⬇) (⬆)(⬆)( )
      1s   2s   2p

Again, it does not matter which two of the three $2p$ orbitals are shown as containing electrons since all three orbitals have the same energy. However, the two $2p$ electrons must have the same spin. It will always be the case, for energy reasons, that unpaired electrons in orbitals of equal energy will have the same spin.

Nitrogen, with the electron configuration $1s^2 2s^2 2p^3$, contains three unpaired electrons, all with the same spin.

N: (⬆⬇) (⬆⬇) (⬆)(⬆)(⬆)
      1s   2s   2p

With oxygen ($1s^2 2s^2 2p^4$), two of the four $2p$ electrons must pair up, leaving two unpaired electrons. Fluorine, the next element in the periodic table, has only one unpaired electron. Finally, with neon, all of the electrons are paired up.

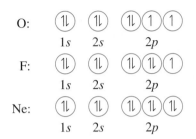

EXAMPLE 6.3

Writing Orbital Diagrams for the Electrons in Atoms

Draw an orbital diagram for the electrons in the element vanadium. The electron configuration for vanadium is

$$1s^2 2s^2 2p^6 3s^2 3p^6 4s^2 3d^3$$

SOLUTION

All the occupied subshells of vanadium are completely filled except for the $3d$. There are five orbitals in the $3d$ subshell, and there are three electrons present. Following Hund's rule, we will put one electron in each of three $3d$ orbitals, giving us three unpaired electrons. The three unpaired electrons will have the same spin.

Practice Exercise 6.3

Draw an orbital diagram for the electrons in the element sulfur. The electron configuration for sulfur is

$$1s^2 2s^2 2p^6 3s^2 3p^4$$

Atoms may be classified as *paramagnetic* or *diamagnetic* on the basis of unpaired electrons. A **paramagnetic atom** *is an atom that has an electron arrangement containing one or more unpaired electrons.* The presence of unpaired electrons causes paramagnetic substances to be slightly attracted to a magnet. Measurement of paramagnetism provides experimental verification of the presence of unpaired electrons. A **diamagnetic atom** *is an atom that has an electron arrangement in which all electrons are paired.*

6.9   Electron Configurations and the Periodic Law

A knowledge of electron configurations for the elements provides an explanation for the periodic law. Recall, from Section 6.1, that the periodic law points out that the properties of the elements repeat themselves in a regular manner when the elements are ordered in sequence of increasing atomic number. Those elements with similar chemical properties are placed one under another in vertical columns (groups) in a periodic table.

Groups of elements have similar chemical properties because of similarities that exist in the electron configurations of the elements of the group. *Chemical properties repeat themselves in a regular manner among the elements because electron configurations repeat themselves in a regular manner among the elements.*

To illustrate this correlation between similar chemical properties and similar electron configurations, let us look at the electron configurations of two groups of elements known to have similar chemical properties.

We begin with the elements lithium, sodium, potassium, and rubidium—all members of group IA of the periodic table. The electron configurations for these elements are

$$_3\text{Li:} \quad 1s^2 \overparen{2s^1}$$
$$_{11}\text{Na:} \quad 1s^2 2s^2 2p^6 \overparen{3s^1}$$
$$_{19}\text{K:} \quad 1s^2 2s^2 2p^6 3s^2 3p^6 \overparen{4s^1}$$
$$_{37}\text{Rb:} \quad 1s^2 2s^2 2p^6 3s^2 3p^6 4s^2 3d^{10} 4p^6 \overparen{5s^1}$$

We see that each of these elements has one outer s electron (shown in color), the last s electron added by the Aufbau principle. It is this similarity in outer shell electron arrangements that causes these elements to have similar chemical properties.

The elements fluorine, chlorine, bromine, and iodine of group VIIA of the periodic table have similar chemical properties. The electron configurations for these four elements are

$$_9\text{F:} \quad 1s^2 \overparen{2s^2 2p^5}$$
$$_{17}\text{Cl:} \quad 1s^2 2s^2 2p^6 \overparen{3s^2 3p^5}$$
$$_{35}\text{Br:} \quad 1s^2 2s^2 2p^6 3s^2 3p^6 \overparen{4s^2} 3d^{10} \overparen{4p^5}$$
$$_{53}\text{I:} \quad 1s^2 2s^2 2p^6 3s^2 3p^6 4s^2 3d^{10} 4p^6 \overparen{5s^2} 4d^{10} \overparen{5p^5}$$

Once again similarities in electron configurations are readily apparent. This time the repeating pattern is the seven electrons (in color) in the outermost s and p subshells.

Section 7.2 will consider in depth the fact that the electrons most important in controlling chemical properties are those found in the outermost shell of an atom.

6.10 Electron Configurations and the Periodic Table

One of the strongest pieces of supporting evidence for the assignment of electrons to shells, subshells, and orbitals is the periodic table itself. The basic shape and structure of this table, which were determined many years before electrons were even discovered, are consistent with and can be explained by electron configurations. Indeed, the specific location of an element in the periodic table can be used to obtain information about its electron configuration.

As the first step in linking electron configurations to the periodic table, let us analyze the general shape of the periodic table in terms of columns of elements. As shown in Figure 6.13, we have on the extreme left of the table 2 columns of elements, in the center an area containing 10 columns of elements, to the right a block of 6 columns of elements, and at the bottom of the table, in two rows, 14 columns of elements.

The numbers of columns of elements in the various regions of the periodic table—2, 6, 10, and 14—is the same as the maximum numbers of electrons that the various types of subshells can accommodate. We will see shortly that this is a very significant observation; the number matchup is no coincidence. The various columnar regions of the periodic table are called the s area (2 columns), the p area (6 columns), the d area (10 columns), and the f area (14 columns), as shown in Figure 6.14.

The concept of *distinguishing electrons* is the key to obtaining "electron configuration information" from the periodic table. For an element, the **distinguishing electron** *is the last electron added to the element's electron configuration when the configuration is written according to the Aufbau principle.* This last electron added is the one that causes an element's electron configuration to differ from that of the element immediately preceding it in the periodic table, hence the term distinguishing electron.

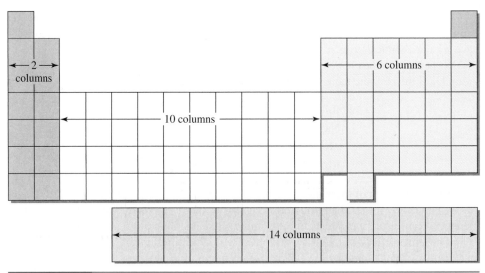

Figure 6.13

The periodic table can be divided into four areas—areas that are 2, 6, 10, and 14 columns wide.

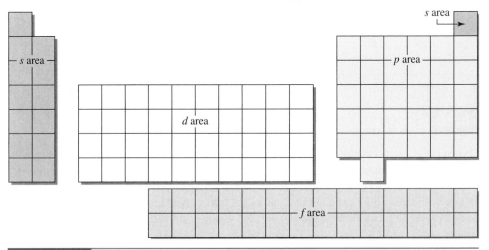

Figure 6.14

Within the four periodic table areas are elements whose distinguishing electron is located, respectively, in *s, p, d,* and *f* subshells.

For all elements located in the *s* area of the periodic table the distinguishing electron is always found in an *s* subshell. All *p* area elements have distinguishing electrons in *p* subshells. Similarly, elements in the *d* and *f* areas of the periodic table have, respectively, distinguishing electrons located in *d* and *f* subshells. Thus, the area location of an element in the periodic table can be used to determine the type of subshell that contains the distinguishing electron. Note that the element helium is considered to belong to the *s* rather than the *p* area of the periodic table even though its table position is on the right-hand side. (The reason for this placement of helium will be explained in Sec. 7.3.)

The extent of filling of the subshell containing an element's distinguishing electron can also be determined from the element's position in the periodic table. All elements in the first column of a specific area contain only one electron in the subshell, all elements in the

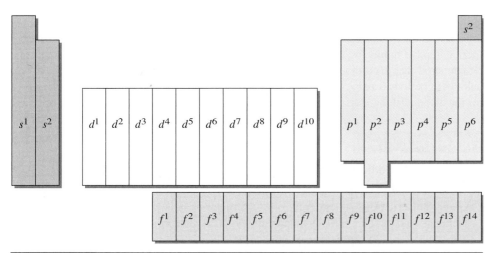

Figure 6.15

The extent of filling of the subshell containing the distinguishing electron can be determined from the element's position within a given area.

second column contain two electrons in the subshell, and so on. Thus all elements in the first column of the *p* area (group IIIA) have an electron configuration ending in p^1. Elements in the second column of the *p* area (group IVA) have electron configurations ending in p^2, and so forth. Similar relationships hold in other areas of the table, as shown in Figure 6.15. A few exceptions to these generalizations do exist in the *d* and *f* areas (because of irregular electron configurations; see Sec. 6.7), but we will not be concerned with them in this text.

We can also use the periodic table to determine the shell in which the distinguishing electron is located. The relationship used involves the number of the period in which the element is found. In the *s* and *p* areas, the period number gives the shell number directly. In the *d* area, the period number minus one is equal to the shell number. (Remember that the 3*d* subshell is filled during the fourth period.) For similar reasons, in the *f* area the period

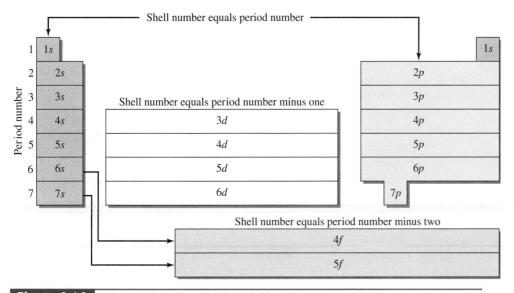

Figure 6.16

Relationship between the shell number for a distinguishing electron and the period number of the element's periodic table position.

number minus two equals the shell number. Thus, the subshell that contains the distinguishing electron for elements of period 6 may be the 6*s*, 6*p*, 5*d* (period number minus one), or 4*f* (period number minus two), depending on the location of the element in period 6. It must be remembered that even though the *f* area is located at the bottom of the table, it correctly belongs in periods 6 and 7. The complete matchup between period number and shell number for the distinguishing electron is given in the periodic table of Figure 6.16.

EXAMPLE 6.4

Using the Periodic Table as a Guide in Obtaining Information About Distinguishing Electrons

Using the periodic table and Figures 6.13 through 6.16, determine the following for the elements calcium ($Z = 20$), manganese ($Z = 25$), and tellurium ($Z = 52$).

(a) The type of subshell in which the distinguishing electron is found
(b) The extent of filling of the subshell containing the distinguishing electron
(c) The shell in which the subshell containing the distinguishing electron is found

SOLUTION

(a) Knowing the area of the periodic table in which an element is found is sufficient to determine the type of subshell in which the distinguishing electron is found.

Ca: Since this element is found in the *s* area of the periodic table, the distinguishing electron will be in an *s* subshell.

Mn: Since this element is found in the *d* area of the periodic table, the distinguishing electron will be in a *d* subshell.

Te: Since this element is found in the *p* area of the periodic table, the distinguishing electron will be in a *p* subshell.

(b) The extent of filling of the subshell containing the distinguishing electron is determined by noting the column in the area that the element occupies.

Ca: Since this element is in the second column of the *s* area, the *s* subshell involved contains two electrons (s^2).

Mn: Since this element is in the fifth column of the *d* area, the *d* subshell involved contains five electrons (d^5).

Te: Since this element is in the fourth column of the *p* area, the *p* subshell involved contains four electrons (p^4).

(c) The shell number of the subshell containing the distinguishing electron is obtained from the period number, sometimes directly and sometimes with modifications.

Ca: Since this element is in period 4, the *s* subshell involved is the 4*s*; therefore, the electron configuration for Ca ends in $4s^2$.

Mn: Since this element is in period 4, the *d* subshell involved is the 3*d* (period number minus one); therefore, the electron configuration for Mn ends in $3d^5$.

Te: Since this element is in period 5, the *p* subshell involved is the 5*p*; therefore, the electron configuration for Te ends in $5p^4$.

Practice Exercise 6.4

Using the periodic table and Figures 6.13 through 6.16, determine the following for the elements cesium ($Z = 55$) and silver ($Z = 47$).

(a) The type of subshell in which the distinguishing electron is found
(b) The extent of filling of the subshell containing the distinguishing electron
(c) The shell in which the subshell containing the distinguishing electron is found

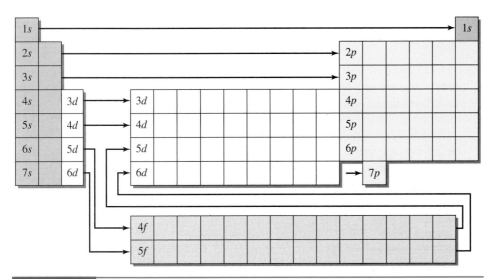

Figure 6.17

Using the periodic table as a guide to obtain the order in which subshells are occupied by electrons.

To write complete electron configurations, you must know the order in which the various electron subshells are filled. Up until now, we have obtained this filling order by using an Aufbau diagram (Fig. 6.12). We can also obtain this information directly from the periodic table. To obtain it, we merely follow a path of increasing atomic number through the table, noting the various subshells as we encounter them.

1. Begin with hydrogen and helium (the $1s$ elements, period 1).
2. Continue through the elements in the other periods, in order of increasing atomic number.
3. As you move across a period
   (a) Add electrons to the ns subshell as you pass through the s area.
   (b) Add electrons to the np subshell as you pass through the p area.
   (c) Add electrons to the $(n - 1)d$ subshell as you pass through the d area.
   (d) Add electrons to the $(n - 2)f$ subshell as you pass through the f area.
4. Continue until you reach the element whose electron configuration you are writing.

Figure 6.17 illustrates this way of using the periodic table. Note particularly how the f area of the periodic table is worked into the scheme.

EXAMPLE 6.5

Using the Periodic Table as a Guide in Writing an Electron Configuration

Write the complete electron configuration of $_{86}$Rn using the periodic table as your guide.

SOLUTION

The needed information will be obtained by "working our way" through the periodic table. We will start at hydrogen and go from box to box, in order of increasing atomic number, until we arrive at position 86, radon. Every time we traverse the s area we will fill an s subshell; the p area, a p subshell; and so on. We will remember that the shell number for s and p subshells is the period number; for d subshells, the period number minus one; and for f subshells, the period number minus two.

Let us begin our journey through the periodic table. As we cross period 1 we encounter H and He, both 1*s* elements. We thus add the 1*s* electrons.

$$\text{Rn:} \quad 1s^2 \ldots$$

In traversing period 2 we pass through the *s* area (elements 3 and 4) and the *p* area (elements 5–10) in that order. We add 2*s* and 2*p* electrons.

$$\text{Rn:} \quad 1s^2 2s^2 2p^6 \ldots$$

Our trip through period 3 is very similar to that of period 2. Only *s*-area elements (11 and 12) and *p*-area elements (13–18) are encountered. The message is to add *s* and *p* electrons—3*s* and 3*p* because we are in period 3.

$$\text{Rn:} \quad 1s^2 2s^2 2p^6 3s^2 3p^6 \ldots$$

In passing through period 4 we go through the *s* area (elements 19 and 20), the *d* area (elements 21–30), and the *p* area (elements 31–36). Electrons to be added are those in the 4*s*, 3*d* (period number minus one), and 4*p* subshells, in that order.

$$\text{Rn:} \quad 1s^2 2s^2 2p^6 3s^2 3p^6 4s^2 3d^{10} 4p^6 \ldots$$

In period 5 we encounter scenery similar to that in period 4—the *s* area (elements 37 and 38), the *d* area (elements 39–48), and the *p* area (elements 49–54). Hence, the 5*s*, 4*d* (period number minus one), and 5*p* subshells are filled in order.

$$\text{Rn:} \quad 1s^2 2s^2 2p^6 3s^2 3p^6 4s^2 3d^{10} 4p^6 5s^2 4d^{10} 5p^6 \ldots$$

Our journey ends in period 6, which we completely traverse. We go through the 6*s* area (elements 55 and 56), the 4*f* area (elements 58–71), the 5*d* area (elements 57 and 72–80), and finally the 6*p* area (elements 81–86). We will completely fill the 6*p* subshell since radon is in the last column of the *p* area.

$$\text{Rn:} \quad 1s^2 2s^2 2p^6 3s^2 3p^6 4s^2 3d^{10} 4p^6 5s^2 4d^{10} 5p^6 6s^2 4f^{14} 5d^{10} 6p^6$$

Practice Exercise 6.5

Write the complete electron configuration of $_{54}$Xe using the periodic table as a guide.

Again, let us mention that a few slightly irregular electronic configurations are encountered in the *d* and *f* areas of the periodic table. The generalizations in this section do not address this problem. We will be working mostly with *s*- and *p*-area elements in future chapters. There are no irregularities in electron configurations for these elements.

6.11 Classification Systems for the Elements

The elements can be classified in several ways. The two most common classification systems are

1. A system based on the electron configurations of the elements, in which elements are described as *noble gas, representative, transition,* or *inner transition* elements.
2. A system based on selected physical properties of the elements, in which elements are described as *metals* or *nonmetals.*

The classification scheme based on electron configurations of the elements is depicted in Figure 6.18. This type of classification system is used in numerous discussions in subsequent chapters.

> The electron configurations of the noble gases will be an important focal point when we consider chemical bonding in Chapter 7.

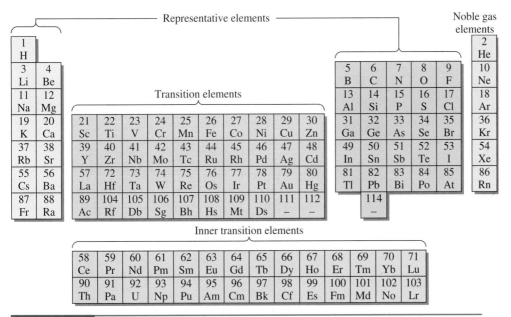

Figure 6.18

Elemental classification scheme based on the electron configurations of the elements.

A **noble-gas element** *is an element located in the far right column (Group VIIIA) of the periodic table.* Such elements are all gases at room temperature, and they have little tendency to form chemical compounds. With one exception, the distinguishing electron for a noble gas completes the *p* subshell. Therefore, they have electron configurations ending in p^6. The exception is helium, in which the distinguishing electron completes the first shell—a shell that has only two electrons. Helium's electron configuration is $1s^2$.

A **representative element** *is an element located in the s area or the first five columns of the p area of the periodic table.* The distinguishing electron for these elements partially or completely fills an *s* subshell or partially fills a *p* subshell. The representative elements include most of the more common elements.

A **transition element** *is an element located in the d area of the periodic table.* The common feature in the electronic configurations of the transition elements is the presence of the distinguishing electron in a *d* subshell.

An **inner-transition element** *is an element located in the f area of the periodic table.* The characteristic feature of the electronic configuration for such elements is the presence of the distinguishing electron in an *f* subshell.

On the basis of selected physical properties of the elements, the second of the two classification schemes divides the elements into the categories of metals and nonmetals. A **metal** *is an element that has the characteristic properties of luster, thermal conductivity, electrical conductivity, and malleability.* With the exception of mercury, all metals are solids at room temperature (25°C). Metals are good conductors of heat and electricity. Most metals are ductile (they can be drawn into wires) and malleable (they can be rolled into sheets). Most metals have high luster (shine), high density, and high melting points. Among the more familiar metals are the elements iron, aluminum, copper, silver, and gold.

A **nonmetal** *is an element characterized by the absence of the properties of luster, thermal conductivity, electrical conductivity, and malleability.* Many of the nonmetals, such as hydrogen, oxygen, nitrogen, and the noble gases, are gases at room temperature (25°C). The only nonmetal that is a liquid at room temperature is bromine. Solid nonmetals include carbon, sulfur, and phosphorus. In general, the nonmetals have lower densities and melting points than metals.

> The four most abundant elements in the human body (Sec. 4.8), which constitute more than 99% of all the atoms in the body, are representative elements—hydrogen, oxygen, carbon, and nitrogen.

> With two exceptions, all metals are silvery-white to silvery-gray to dull-gray in color. The two exceptions are copper and gold.

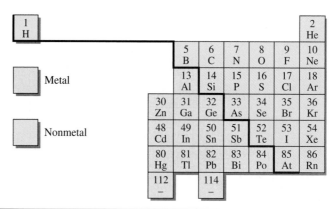

Figure 6.19

A portion of the periodic table showing the dividing line between metals and nonmetals. All elements that are not shown are metals.

The majority of the elements are metals: only 22 elements are nonmetals, the rest (91) are metals. It is not necessary to memorize which elements are nonmetals and which are metals. As can be seen from Figure 6.19, the location of an element in the periodic table correlates directly with its classification as a metal or nonmetal. The steplike heavy line that runs through the *p* area of the periodic table separates the metals from the nonmetals; metals are on the left and nonmetals on the right. Note that the element hydrogen is a nonmetal even though it is located on the left side of the periodic table.

The fact that the vast majority of elements are metals in no way indicates that metals are more important than nonmetals. Most nonmetals are relatively abundant and are found in many important compounds. For example, water (H_2O) is a compound involving two nonmetals. An analysis of the previously given abundances of the elements in the Earth's crust (Fig. 4.13) in terms of metals and nonmetals shows that three of the four most abundant elements, which account for 83% of all atoms, are nonmetals—oxygen, silicon, and hydrogen.

6.12 Chemical Periodicity

Chemical periodicity *is the variation in properties of elements as a function of their positions in the periodic table.* In this section we consider two properties of elements that exhibit chemical periodicity, metallic–nonmetallic character and atomic size (atomic radius). In Section 7.10 we consider a third property exhibiting chemical periodicity—electronegativity.

Metallic and Nonmetallic Character

The physical properties that distinguish metals and nonmetals were considered in Section 6.11. In general, these properties are opposites for the two classes of substances.

Not all metals possess metallic properties to the same extent; they have them, but to varying degrees. For example, some metals are better conductors of electricity than other metals. Similarly, not all nonmetals possess nonmetallic properties to the same extent.

Metallic and nonmetallic character for the various elements—the extent to which they possess metallic and nonmetallic properties—can be correlated with periodic table position for the elements.

1. Metallic character increases from right to left within a given period in the periodic table.
2. Metallic character increases from top to bottom within a group in the periodic table.

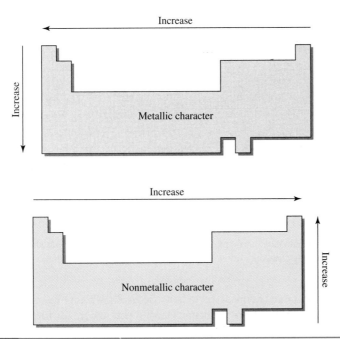

Figure 6.20

Chemical periodicity is associated with the properties of metallic character and nonmetallic character.

Thus, among the period-3 elements, Na is more metallic than Mg, which in turn is more metallic than Al. For the group IA elements, K is more metallic than Na, since it is farther down in group IA.

Similar, but opposite, trends exist for nonmetallic character.

1. Nonmetallic character increases from left to right within a given period in the periodic table.
2. Nonmetallic character increases from bottom to top within a group in the periodic table.

Figure 6.20 summarizes the chemical periodicity associated with the properties of metallic and nonmetallic character. The combined effect of these two trends is that the most nonmetallic elements are those in the upper right portion of the periodic table, a generalization consistent with Figure 6.19 in the preceding section of this chapter.

In Section 6.11 a "dividing line" was shown for classifying elements as metals or nonmetals (Fig. 6.19). Elements to the far left of this line have dominant metallic character. Elements to the far right of this line have dominant nonmetallic character. Most elements whose periodic table positions "touch" this metal–nonmetal dividing line actually show some properties that are characteristic of both metals and nonmetals. These elements are frequently given a classification of their own: *metalloids* (or *semimetals*). A **metalloid** *is an element with properties intermediate between those of metals and nonmetals.* Figure 6.21 identifies the metalloid elements. Several of the metalloids, including silicon, germanium, and antimony, are *semiconductors*. A **semiconductor** *is an element that does not conduct electrical current at room temperature but does so at higher temperatures.* Semiconductor elements are very important in the electronics industry.

Atomic Size

The quantum mechanical model for an atom (Sec. 6.3) suggests that the shape of an atom is approximately spherical and that the size of the atom can be expressed in terms of the radius of a sphere. The most commonly used unit for expressing atomic sizes (atomic radii) is the picometer (Sec. 3.1). In this unit most atomic radii fall in the range of 50–200 pm.

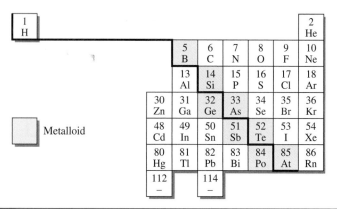

Figure 6.21

A portion of the periodic table showing the metalloids (color), elements that show some properties characteristic of metals and some other properties characteristic of nonmetals.

Atomic sizes exhibit chemical periodicity, as is shown in Figure 6.22, which gives atomic radii for the representative elements. General periodic trends for atomic radii are

1. Atomic radii tend to decrease from left to right within a period of the periodic table.
2. Atomic radii tend to increase from top to bottom within a periodic table group.

The trends in atomic radii values can be explained on the basis of the number of electrons present in the atom and their energies. In traversing a period in the periodic table (from left

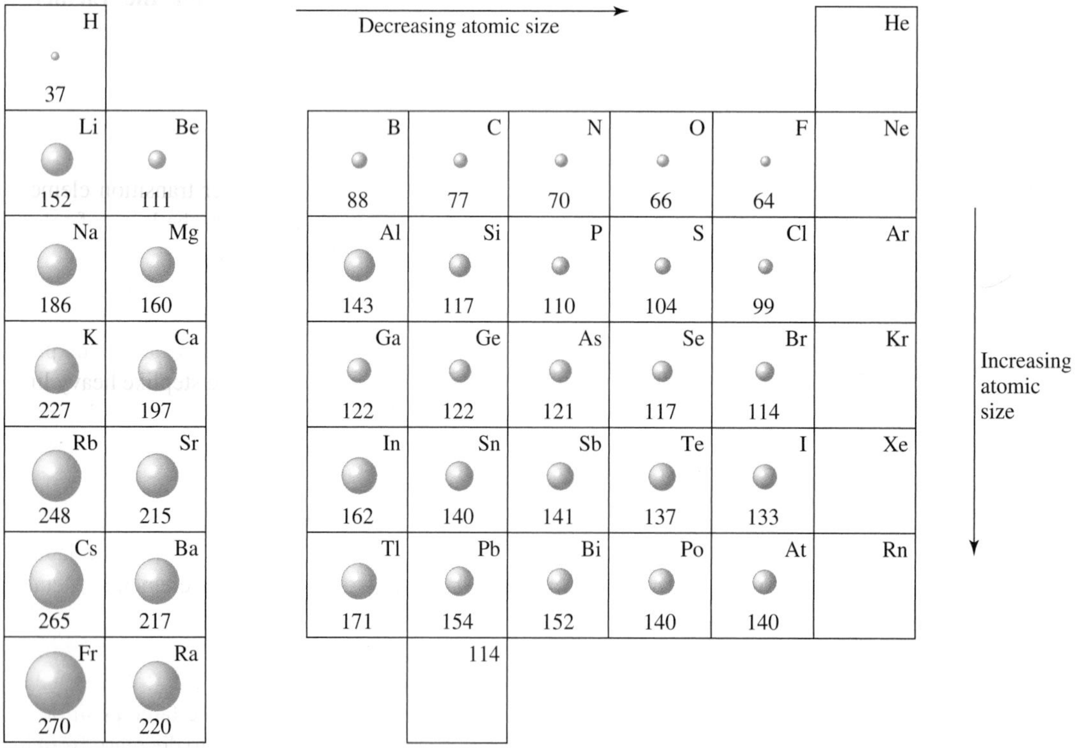

H			Decreasing atomic size →					He
37								
Li	Be		B	C	N	O	F	Ne
152	111		88	77	70	66	64	
Na	Mg		Al	Si	P	S	Cl	Ar
186	160		143	117	110	104	99	
K	Ca		Ga	Ge	As	Se	Br	Kr
227	197		122	122	121	117	114	
Rb	Sr		In	Sn	Sb	Te	I	Xe
248	215		162	140	141	137	133	
Cs	Ba		Tl	Pb	Bi	Po	At	Rn
265	217		171	154	152	140	140	
Fr	Ra		114					
270	220							

Increasing atomic size

Figure 6.22

Atomic radii, in picometers, of the representative elements.

to right), we note that (1) nuclear charge (atomic number) increases and (2) the added electrons enter the same shell, the outermost shell. The increased nuclear charge draws the electrons in this outermost shell closer to the nucleus, resulting in smaller atomic radii. Down a group in the periodic table, atomic radii increase because the electrons are added to shells with higher numbers. Recall, from Section 6.4, that the higher the shell number, the greater the average distances between the electrons and the nucleus.

Summary

1. **Periodic Law** The periodic law states that certain sets of physical and chemical properties occur at periodic (regularly occurring) intervals when the elements are arranged in order of increasing atomic number.

2. **Periodic Table** The periodic table is a tabular arrangement of the elements, in order of increasing atomic number, that places elements with similar properties in the same vertical columns of the table. It is a graphical portrayal of the periodic law. A group in the periodic table is a vertical column of elements. Roman numerals are used to distinguish the various periodic table groups. A period in the periodic table is a horizontal row of elements. Regular numbers are used to distinguish the various periodic table periods.

3. **Electron Shells** An electron shell is a region of space about a nucleus that contains electrons that have approximately the same energy and that spend most of their time approximately the same distance from the nucleus. The electron capacity of an electron shell is $2n^2$, where n is the shell number.

4. **Electron Subshells** An electron subshell is a region of space within an electron shell that contains electrons that have the same energy. The number of electron subshells in a particular electron shell is equal to the shell number. Each subshell can hold a specific maximum number of electrons. These values are 2, 6, 10, and 14 for s, p, d, and f subshells, respectively.

5. **Electron Orbitals** An electron orbital is a region of space within an electron subshell where an electron with a specific energy is most likely to be found. Each electron subshell contains one or more orbitals. For s, p, d, and f subshells there are 1, 3, 5, and 7 orbitals, respectively. No more than two electrons may occupy an orbital.

6. **Electron Configuration** An electron configuration is a statement of how many electrons an atom has in each of its electron subshells. The determination of an electron configuration is based on the Aufbau principle, which is that electrons normally occupy the lowest-energy subshell available for occupancy.

7. **Orbital Diagram** An orbital diagram is a statement of how many electrons an atom has in each of its orbitals. Electrons occupy the orbitals of a subshell such that each orbital within the subshell acquires one electron before any orbital acquires a second electron. All electrons in such singly occupied orbitals have the same spin.

8. **Electron Configurations and the Periodic Law** Chemical properties repeat themselves in a regular manner among the elements because electron configurations repeat themselves in a regular manner among the elements.

9. **Electron Configurations and the Periodic Table** The groups of the periodic table contain elements with similar electron configurations. Thus, the location of an element in the periodic table can be used to obtain information about its electron configuration.

10. **Classification Systems for the Elements** On the basis of electron configuration, elements are classified into four categories: noble gases (far right column of the periodic table), representative elements (s and p area of the periodic table, with the exception of the last p-area column), transition elements (d area of the periodic table), and inner transition elements (f area of the periodic table). On the basis of selected physical properties, elements are classified as metals or nonmetals. Metals exhibit luster, thermal conductivity, electrical conductivity, and malleability. Nonmetals are characterized by the absence of the properties associated with metals. The steplike heavy line that runs through the right third of the periodic table separates the metals on the left from the nonmetals on the right. The majority of elements are metals.

11. **Chemical Periodicity** Metallic and nonmetallic character of the elements can be correlated with periodic table position for the elements. Metallic character increases from right to left within a given period and increases from top to bottom within a group in the periodic table. Atom size (atomic radii) can also be correlated with periodic table position for the elements. Atomic radii tend to decrease from left to right within a period and increase from top to bottom within a periodic table group.

Key Terms

The new terms defined in this chapter are

alkali metal   *Sec. 6.2*
alkaline earth metal   *Sec. 6.2*
Aufbau diagram   *Sec. 6.7*
Aufbau principle   *Sec. 6.7*
chemical periodicity   *Sec. 6.12*
diamagnetic atom   *Sec. 6.8*
distinguishing electron   *Sec. 6.10*
electron configuration   *Sec. 6.7*
electron orbital   *Sec. 6.6*
electron shell   *Sec. 6.4*
electron spin   *Sec. 6.6*
electron subshell   *Sec. 6.5*
group   *Sec. 6.2*
halogen   *Sec. 6.2*
Hund's rule   *Sec. 6.8*

inner transition element   *Sec. 6.11*
metal   *Sec. 6.11*
metalloid   *Sec. 6.12*
noble gas   *Sec. 6.2 and 6.11*
nonmetal   *Sec. 6.11*
orbital diagram   *Sec. 6.8*
paired electrons   *Sec. 6.8*
paramagnetic atom   *Sec. 6.8*
period   *Sec. 6.2*
periodic law   *Sec. 6.1*
periodic table   *Sec. 6.2*
quantized property   *Sec. 6.3*
representative element   *Sec. 6.11*
semiconductor   *Sec. 6.12*
transition element   *Sec. 6.11*
unpaired electron   *Sec. 6.8*

Practice Problems

Periodic Law and Periodic Table (Secs. 6.1 and 6.2)

6.1 Give the symbol of the element that occupies each of the following positions in the periodic table.

(a) period 4, group IIIA   **(b)** period 5, group IVB
(c) group IA, period 2   **(d)** group VIIA, period 3

6.2 Give the symbol of the element that occupies each of the following positions in the periodic table.

(a) period 1, group IA   **(b)** period 6, group IB
(c) group IIIB, period 4   **(d)** group IVA, period 5

6.3 Based on their periodic table positions, characterize each of the following pairs of elements as belonging to (1) the same group, (2) the same period, or (3) neither the same group nor the same period.

(a) $_{28}Ni$ and $_{36}Kr$ **(b)** $_{12}Mg$ and $_{55}Cs$
(c) $_{8}O$ and $_{34}Se$ **(d)** $_{19}K$ and $_{29}Cu$

6.4 Based on their periodic table positions, characterize each of the following pairs of elements as belonging to (1) the same group, (2) the same period, or (3) neither the same group nor the same period.

(a) $_{49}In$ and $_{82}Pb$ **(b)** $_{42}Mo$ and $_{48}Cd$
(c) $_{13}Al$ and $_{81}Tl$ **(d)** $_{20}Ca$ and $_{21}Sc$

6.5 For each of the following sets of elements, choose the two that would be expected to have similar chemical properties.

(a) $_{19}K$, $_{29}Cu$, $_{37}Rb$, $_{41}Nb$ **(b)** $_{13}Al$, $_{14}Si$, $_{15}P$, $_{33}As$
(c) $_{9}F$, $_{40}Zr$, $_{50}Sn$, $_{53}I$ **(d)** $_{11}Na$, $_{12}Mg$, $_{54}Xe$, $_{55}Cs$

6.6 For each of the following sets of elements, choose the two that would be expected to have similar chemical properties.

(a) $_{11}Na$, $_{14}Si$, $_{23}V$, $_{55}Cs$ **(b)** $_{13}Al$, $_{19}K$, $_{32}Ge$, $_{50}Sn$
(c) $_{37}Rb$, $_{38}Sr$, $_{54}Xe$, $_{56}Ba$ **(d)** $_{2}He$, $_{6}C$, $_{8}O$, $_{10}Ne$

6.7 The following statements either define or are closely related to the terms *periodic law, period*, or *group*. Match the terms to the appropriate statements.

(a) This is a vertical arrangement of elements in the periodic table.
(b) The properties of the elements repeat in a regular way as the atomic numbers increase.
(c) The chemical properties of elements 12, 20, and 38 demonstrate this principle.
(d) Elements 24 and 33 belong to this arrangement.

6.8 The following statements either define or are closely related to the terms *periodic law, period*, and *group*. Match the terms to the appropriate statements.

(a) This is a horizontal arrangement of elements in the periodic table.
(b) Element 19 begins this arrangement in the periodic table.
(c) The element carbon is the first member of this arrangement.
(d) Elements 10, 18, 36, and 54 belong to this arrangement.

6.9 Identify each of the following elements by name.

(a) period 2 halogen
(b) period 3 alkali metal
(c) period 4 noble gas
(d) period 5 alkaline earth metal

6.10 Identify each of the following elements by name.

(a) period 2 alkali metal
(b) period 3 noble gas

(c) period 4 alkaline earth metal

(d) period 5 halogen

6.11 How many elements exist with an atomic number less than 40 that are

(a) halogens   **(b)** noble gases

(c) alkali metals   **(d)** alkaline earth metals

6.12 How many elements exist with an atomic number greater than 20 that are

(a) halogens   **(b)** noble gases

(c) alkali metals   **(d)** alkaline earth metals

Terminology Associated with Electron Arrangements (Secs. 6.3–6.6)

6.13 Give the maximum number of electrons that can occupy each of the following electron subshells.

(a) $5s$ **(b)** $3d$ **(c)** $2p$ **(d)** $4f$

6.14 Give the maximum number of electrons that can occupy each of the following electron subshells.

(a) $6p$ **(b)** $1s$ **(c)** $5f$ **(d)** $4d$

6.15 Give the maximum number of electrons that can occupy each of the following electron orbitals.

(a) $3s$ **(b)** $3p$ **(c)** $4d$ **(d)** $5f$

6.16 Give the maximum number of electrons that can occupy each of the following electron orbitals.

(a) $1s$ **(b)** $3d$ **(c)** $5p$ **(d)** $4d$

6.17 In each of the following pairs of items identify the item that can accommodate the most electrons.

(a) $3d$ subshell, second shell

(b) shell with $n = 1$, $2p$ subshell

(c) $3p$ orbital, $3p$ subshell

(d) $4f$ subshell, third shell

6.18 In each of the following pairs of items identify the item that can accommodate the most electrons.

(a) first shell, third shell

(b) $4f$ subshell, $4d$ subshell

(c) second shell, $5f$ subshell

(d) $3d$ orbital, $3d$ subshell

6.19 Indicate whether each of the following statements is *true* or *false*.

(a) An orbital has a definite size and shape, which are related to the energy of the electrons it could contain.

(b) All shells accommodate the same number of electrons.

(c) All of the orbitals in a subshell have the same energy.

(d) A $2p$ and a $3p$ subshell would contain the same number of orbitals.

(e) The fourth shell is made up of six subshells.

6.20 Indicate whether each of the following statements is *true* or *false*.

(a) All the subshells in a shell have the same energy.

(b) A d subshell always contains five orbitals.

(c) An s orbital is shaped something like a four-leaf clover.

(d) The $n = 3$ shell can accommodate a maximum of 18 electrons.

(e) All subshells accommodate the same number of electrons.

6.21 Describe the general shape of each of the following orbitals.

(a) $4s$ **(b)** $4p$ **(c)** $4d$ **(d)** $6s$

6.22 Describe the general shape of each of the following orbitals.

(a) $1s$ **(b)** $2p$ **(c)** $3d$ **(d)** $5p$

6.23 Which of the following electron subshell and electron orbital designations is not allowed?

(a) $4s$ subshell   **(b)** $2d$ orbital

(c) $1p$ subshell   **(d)** $4f$ orbital

6.24 Which of the following electron subshell and electron orbital designations is not allowed?

(a) $2d$ subshell   **(b)** $4s$ orbital

(c) $3p$ subshell   **(d)** $2f$ orbital

Electron Configurations (Sec. 6.7)

6.25 In which member of each of the following pairs of subshells would an electron have the higher energy?

(a) $2s$ or $3s$ **(b)** $3p$ or $3d$

(c) $5p$ or $7s$ **(d)** $4s$ or $4p$

6.26 In which member of each of the following pairs of subshells would an electron have the higher energy?

(a) $3d$ or $4d$ **(b)** $4f$ or $4s$

(c) $6s$ or $6d$ **(d)** $5d$ or $6p$

6.27 With the help of an Aufbau diagram, write the complete electron configuration for each of the following atoms.

(a) $_{13}Al$ **(b)** $_{7}N$

(c) $_{18}Ar$ **(d)** $_{12}Mg$

6.28 With the help of an Aufbau diagram, write the complete electron configuration for each of the following atoms.

(a) $_{20}Ca$ **(b)** $_{10}Ne$

(c) $_{6}C$ **(d)** $_{15}P$

6.29 With the help of an Aufbau diagram, write the complete electron configuration for each of the following atoms.

(a) $_{26}Fe$ **(b)** $_{37}Rb$

(c) $_{53}I$ **(d)** $_{86}Rn$

6.30 With the help of an Aufbau diagram, write the complete electron configuration for each of the following atoms.

(a) $_{31}Ga$ (b) $_{38}Sr$
(c) $_{48}Cd$ (d) $_{88}Ra$

6.31 Based on total number of electrons present, identify the element represented by each of the following electron configurations.

(a) $1s^22s^22p^6$
(b) $1s^22s^22p^63s^23p^64s^1$
(c) $1s^22s^22p^63s^23p^64s^23d^2$
(d) $1s^22s^22p^63s^23p^64s^23d^{10}$

6.32 Based on total number of electrons present, identify the element represented by each of the following electron configurations.

(a) $1s^22s^22p^2$
(b) $1s^22s^22p^63s^23p^3$
(c) $1s^22s^22p^63s^23p^64s^2$
(d) $1s^22s^22p^63s^23p^64s^23d^6$

Orbital Diagrams (Sec. 6.8)

6.33 Draw the electron orbital diagram associated with each of the following electron configurations.

(a) $1s^22s^1$ (b) $1s^22s^22p^5$
(c) $1s^22s^22p^63s^23p^3$ (d) $1s^22s^22p^63s^23p^64s^23d^8$

6.34 Draw the electron orbital diagram associated with each of the following electron configurations.

(a) $1s^22s^22p^3$ (b) $1s^22s^22p^63s^2$
(c) $1s^22s^22p^63s^23p^5$ (d) $1s^22s^22p^63s^23p^64s^23d^7$

6.35 Draw electron orbital diagrams for the following elements.

(a) $_6C$ (b) $_{10}Ne$
(c) $_{11}Na$ (d) $_{15}P$

6.36 Draw electron orbital diagrams for the following elements.

(a) $_7N$ (b) $_9F$
(c) $_{12}Mg$ (d) $_{16}S$

6.37 How many unpaired electrons are there in an atom of the following elements?

(a) boron (b) chlorine
(c) potassium (d) zinc

6.38 How many unpaired electrons are there in an atom of the following elements?

(a) lithium (b) aluminum
(c) calcium (d) bromine

6.39 Indicate whether atoms of each of the elements in Problem 6.37 are paramagnetic or diamagnetic.

6.40 Indicate whether atoms of each of the elements in Problem 6.38 are paramagnetic or diamagnetic.

Electron Configurations and the Periodic Law (Sec. 6.9)

6.41 Indicate whether or not the elements represented by the given pairs of electron configurations have similar chemical properties.

(a) $1s^22s^1$ and $1s^22s^2$
(b) $1s^22s^22p^6$ and $1s^22s^22p^63s^23p^6$
(c) $1s^22s^22p^3$ and $1s^22s^22p^63s^23p^64s^23d^3$
(d) $1s^22s^22p^63s^23p^6$ and $1s^22s^22p^63s^23p^64s^23d^{10}4p^6$

6.42 Indicate whether or not the elements represented by the given pairs of electron configurations have similar chemical properties.

(a) $1s^22s^22p^4$ and $1s^22s^22p^5$
(b) $1s^22s^2$ and $1s^22s^22p^2$
(c) $1s^22s^1$ and $1s^22s^22p^63s^23p^64s^1$
(d) $1s^22s^22p^6$ and $1s^22s^22p^63s^23p^64s^23d^6$

Electron Configurations and the Periodic Table (Sec. 6.10)

6.43 Indicate the position in the periodic table, in terms of s, p, d, or f area, of each of the following elements.

(a) $_{23}V$ (b) $_{34}Se$
(c) $_{56}Ba$ (d) $_{62}Sm$

6.44 Indicate the position in the periodic table, in terms of s, p, d, or f area, of each of the following elements.

(a) $_{50}Sn$ (b) $_{59}Pr$
(c) $_{78}Pt$ (d) $_{87}Fr$

6.45 For each of the following elements, identify the subshell ($2s$, $3p$, $4f$, etc.) that contains the distinguishing electron.

(a) $_{20}Ca$ (b) $_{31}Ga$
(c) $_{40}Zr$ (d) $_{79}Au$

6.46 For each of the following elements, identify the subshell ($2s$, $3p$, $4f$, etc.) that contains the distinguishing electron.

(a) $_{16}S$ (b) $_{38}Sr$
(c) $_{39}Y$ (d) $_{61}Pm$

6.47 Identify the element by name whose electron configuration, after the distinguishing electron has been added, ends in

(a) $4p^5$ (b) $7s^1$
(c) $4d^2$ (d) $3d^3$

6.48 Identify the element by name whose electron configuration, after the distinguishing electron has been added, ends in

(a) $3p^1$ (b) $5s^2$
(c) $3d^1$ (d) $5d^3$

6.49 For each of the following elements, give the number of electrons present in the electron subshell that contains the distinguishing electron.

(a) carbon   (b) calcium

(c) arsenic   (d) tungsten

6.50 For each of the following elements, give the number of electrons present in the electron subshell that contains the distinguishing electron.

(a) aluminum   (b) titanium

(c) bromine   (d) xenon

6.51 Indicate the position in the periodic table where each of the following occurs by giving the symbol of the element.

(a) The $3p$ subshell begins filling.

(b) The $2s$ subshell begins filling.

(c) The $5d$ subshell begins filling.

(d) The $3d$ subshell begins filling.

6.52 Indicate the position in the periodic table where each of the following occurs by giving the symbol of the element.

(a) The $4p$ subshell begins filling.

(b) The $5f$ subshell begins filling.

(c) The $3s$ subshell begins filling.

(d) The $7s$ subshell begins filling.

6.53 Indicate the position in the periodic table where each of the following occurs by giving the symbol of the element.

(a) The $4p$ subshell becomes completely filled.

(b) The $2s$ subshell becomes half-filled.

(c) The fourth shell begins filling.

(d) The fourth shell becomes completely filled.

6.54 Indicate the position in the periodic table where each of the following occurs by giving the symbol of the element.

(a) The $3d$ subshell becomes completely filled.

(b) The $4p$ subshell becomes half-filled.

(c) The third shell becomes half-filled.

(d) The third shell becomes completely filled.

6.55 Using only the periodic table, determine the complete electron configuration for each of the following elements.

(a) $_{33}$As   (b) $_{51}$Sb

(c) $_{37}$Rb   (d) $_{48}$Cd

6.56 Using only the periodic table, determine the complete electron configuration for each of the following elements.

(a) $_{22}$Ti   (b) $_{56}$Ba

(c) $_{35}$Br   (d) $_{19}$K

6.57 How many $3d$ electrons are found in atoms of each of the following elements?

(a) titanium   (b) nickel

(c) selenium   (d) palladium

6.58 How many $4d$ electrons are found in atoms of each of the following elements?

(a) zinc   (b) yttrium

(c) silver   (d) iodine

6.59 Using the periodic table as a guide, indicate the number of

(a) $3p$ electrons in a $_{16}$S atom

(b) $3d$ electrons in a $_{29}$Cu atom

(c) $4s$ electrons in a $_{37}$Rb atom

(d) $4d$ electrons in a $_{30}$Zn atom

6.60 Using the periodic table as a guide, indicate the number of

(a) $3s$ electrons in a $_{12}$Mg atom

(b) $4p$ electrons in a $_{32}$Ge atom

(c) $3d$ electrons in a $_{47}$Ag atom

(d) $4p$ electrons in a $_{15}$P atom

Classification Systems for the Elements (Sec. 6.11)

6.61 Classify each of the following elements as a noble gas, representative element, transition element, or inner transition element.

(a) $_{29}$Cu   (b) $_{32}$Ge

(c) $_{36}$Kr   (d) $_{63}$Eu

6.62 Classify each of the following elements as a noble gas, representative element, transition element, or inner transition element.

(a) $_{3}$Li   (b) $_{79}$Au

(c) $_{54}$Xe   (d) $_{72}$Hf

6.63 Classify each of the following general properties as characteristic of metallic elements or of nonmetallic elements.

(a) ductile

(b) low electrical conductivity

(c) high thermal conductivity

(d) good heat insulator

6.64 Classify each of the following general properties as characteristic of metallic elements or of nonmetallic elements.

(a) nonmalleable

(b) high luster

(c) low thermal conductivity

(d) brittle

6.65 Indicate whether each of the following groups in the periodic table contains (1) more metals than nonmetals or (2) more nonmetals than metals.

(a) group IA   (b) group IIIA

(c) group VIA   (d) group VIIIA

6.66 Indicate whether each of the following groups in the periodic table contains (1) more metals than nonmetals or (2) more nonmetals than metals.

(a) group IIA   (b) group IIIB

(c) group IVA   (d) group VIA

6.67 Identify the nonmetal in each of the following sets of elements.

(a) S, Na, K   **(b)** Cu, Li, P

(c) Be, I, Ca   **(d)** Fe, Cl, Ga

6.68 Identify the nonmetal in each of the following sets of elements.

(a) Al, H, Mg   **(b)** C, Sn, Pb

(c) Ti, V, F   **(d)** Sr, Se, Sm

6.69 Identify the metal in each of the following sets of elements.

(a) H, He, Li   **(b)** S, Cl, K

(c) N, Fe, O   **(d)** Hg, Ne, F

6.70 Identify the metal in each of the following sets of elements.

(a) C, Br, Pb   **(b)** Ar, Kr, Na

(c) P, Ga, Se   **(d)** Zn, I, Xe

6.71 Identify the lowest-atomic-numbered element that is

(a) a nonmetal

(b) a noble gas

(c) a representative metal

(d) an inner transition element

6.72 Identify the lowest-atomic-numbered element that is

(a) a transition element

(b) a representative nonmetal

(c) an inner transition metal

(d) a metal

Chemical Periodicity (Sec. 6.12)

6.73 Based on general periodic table trends, indicate whether each of the following statements is *true* or *false*.

(a) The metallic character for a period of elements increases in proceeding from left to right in the periodic table.

(b) The nonmetallic character for a group of elements decreases in proceeding down a group in the periodic table.

(c) The atomic radius for a period of elements increases in proceeding from left to right in the periodic table.

6.74 Based on general periodic table trends, indicate whether each of the following statements is *true* or *false*.

(a) The nonmetallic character for a period of elements decreases in proceeding from left to right in the periodic table.

(b) The atomic radius for a group of elements decreases in proceeding down a group in the periodic table.

(c) The metallic character for a group of elements increases in proceeding down a group in the periodic table.

6.75 Identify the metalloid in each of the following sets of elements.

(a) Ge, Ga, Zn   **(b)** B, Al, Ga

(c) Pb, Po, P   **(d)** Te, I, Xe

6.76 Identify the metalloid in each of the following sets of elements.

(a) Se, As, Br   **(b)** Br, I, At

(c) Sn, Sb, Sm   **(d)** Al, Si, P

6.77 Using the periodic table, indicate which member of each pair is more metallic.

(a) $_{12}$Mg or $_{14}$Si   **(b)** $_{47}$Ag or $_{79}$Au

(c) $_{16}$S or $_{17}$Cl   **(d)** $_{4}$Be or $_{37}$Rb

6.78 Using the periodic table, indicate which member of each pair is more metallic.

(a) $_{11}$Na or $_{19}$K   **(b)** $_{22}$Ti or $_{29}$Cu

(c) $_{32}$Ge or $_{34}$Se   **(d)** $_{30}$Zn or $_{56}$Ba

6.79 Using the periodic table, indicate which member of each pair is more nonmetallic.

(a) $_{9}$F or $_{35}$Br   **(b)** $_{14}$Si or $_{15}$P

(c) $_{25}$Mn or $_{30}$Zn   **(d)** $_{17}$Cl or $_{33}$As

6.80 Using the periodic table, indicate which member of each pair is more nonmetallic.

(a) $_{6}$C or $_{8}$O   **(b)** $_{12}$Mg or $_{20}$Ca

(c) $_{50}$Sn or $_{52}$Te   **(d)** $_{53}$I or $_{81}$Tl

6.81 Indicate which member of each of the following pairs of elements has the larger atomic radius.

(a) $_{7}$N or $_{8}$O   **(b)** $_{17}$Cl or $_{31}$Ga

(c) $_{19}$K or $_{35}$Br   **(d)** $_{19}$K or $_{37}$Rb

6.82 Indicate which member of each of the following pairs of elements has the larger atomic radius.

(a) $_{15}$P or $_{16}$S   **(b)** $_{15}$P or $_{33}$As

(c) $_{35}$Br or $_{52}$Te   **(d)** $_{37}$Rb or $_{53}$I

Additional Problems

6.83 What is wrong with each of the following attempts to write an electron configuration?

(a) $1s^2 2s^3$ **(b)** $1s^2 2s^2 2p^2 3s^2$

(c) $1s^2 2s^2 3s^2$ **(d)** $1s^2 2s^2 2p^6 3s^2 3d^{10}$

6.84 What is wrong with each of the following attempts to write an electron configuration?

(a) $1s^2 1p^6$ **(b)** $1s^2 2s^4$

(c) $1s^2 2s^2 2p^4 3s^2$ **(d)** $1s^2 2s^2 2p^6 3s^2 3p^6 3d^{10}$

6.85 In what period and group in the periodic table is an element with each of the following electron configurations located?

(a) $1s^2 2s^2 2p^6 3s^1$

(b) $1s^2 2s^2 2p^6 3s^2 3p^1$

(c) $1s^2 2s^2 2p^6 3s^2 3p^6 4s^2 3d^1$

(d) $1s^2 2s^2 2p^6 3s^2 3p^6 4s^2 3d^{10} 4p^5$

6.86 In what period and group in the periodic table is an element with each of the following electron configurations located?

(a) $1s^2 2s^2 2p^2$

(b) $1s^2 2s^2 2p^6 3s^2$

(c) $1s^2 2s^2 2p^6 3s^2 3p^6 4s^2 3d^2$

(d) $1s^2 2s^2 2p^6 3s^2 3p^6 4s^2 3d^{10} 4p^6 5s^2$

6.87 Assign values to x and y in each of the following electron configurations.

(a) Ca: $1s^2 2s^2 2p^6 3s^2 3p^x 4s^y$

(b) Al: $1s^2 2s^2 2p^6 3s^x 3p^y$

(c) Zn: $1s^2 2s^2 2p^6 3s^2 3p^6 4s^x 3d^y$

(d) Kr: $1s^2 2s^2 2p^6 3s^2 3p^6 4s^x 3d^{10} 4p^y$

6.88 Assign values to x and y in each of the following electron configurations.

(a) Ti: $1s^2 2s^2 2p^6 3s^2 3p^x 4s^y 3d^2$

(b) Ar: $1s^2 2s^2 2p^6 3s^x 3p^y$

(c) Cl: $1s^2 2s^2 2p^x 3s^2 3p^y$

(d) Se: $1s^2 2s^2 2p^6 3s^2 3p^6 4s^2 3d^x 4p^y$

6.89 Write electron configurations for the following elements.

(a) the group IVA element in the same period as $_{15}P$

(b) the period 2 element in the same group as $_{50}Sn$

(c) the lowest-atomic-numbered nonmetal in period VA

(d) the period 2 element that has three unpaired electrons

6.90 Write electron configurations for the following elements.

(a) the group IIIA element in the same period as $_4Be$

(b) the period 3 element in the same group as $_5B$

(c) the lowest-atomic-numbered metal in group IA

(d) the period 3 element that has three unpaired electrons

6.91 Referring only to the periodic table, determine the element of lowest atomic number whose electron configuration contains the following.

(a) two completely filled orbitals

(b) two completely filled subshells

(c) two completely filled shells

(d) two completely filled p subshells

6.92 Referring only to the periodic table, determine the element of lowest atomic number whose electron configuration contains the following.

(a) three completely filled orbitals

(b) three completely filled subshells

(c) three completely filled shells

(d) three completely filled s subshells

6.93 How many electrons are in each of the following?

(a) the outermost p subshell of a group IVA element

(b) the outermost p subshell of a group VIA element

(c) the outermost s subshell of a group IA element

(d) the outermost s subshell of a group IVB element

6.94 How many electrons are in each of the following?

(a) the outermost s subshell of a group IIA element

(b) the outermost d subshell of a group IVB element

(c) the outermost p subshell of a group VA element

(d) the outermost s subshell of a group VIIA element

6.95 Determine how many elements there are whose electron configurations end in the following.

(a) ns^2 　　　　　　　　　**(b)** $ns^2 np^4$

(c) $ns^2 (n-1)d^1$ 　　　　**(d)** $ns^2 (n-1)d^{10} np^3$

6.96 Determine how many elements there are whose electron configurations end in the following.

(a) ns^1 　　　　　　　　　**(b)** $ns^2 np^6$

(c) $ns^2 (n-1)d^2$ 　　　　**(d)** $ns^2 (n-1)d^{10} np^2$

6.97 In what group(s) in the periodic table would you expect to find each of the following?

(a) a representative element with two unpaired electrons

(b) a transition element with three unpaired electrons

(c) an element with two unpaired d electrons

(d) an element with one unpaired s electron

6.98 In what group(s) in the periodic table would you expect to find each of the following?

(a) a transition element with no unpaired electrons

(b) a nonmetal with no unpaired electrons

(c) an element with three unpaired p electrons

(d) an element with one unpaired d electron

6.99 Indicate whether the first listed element in each of the following pairs of elements contains (1) more unpaired electrons, (2) fewer unpaired electrons, or (3) the same number of unpaired electrons than the second listed element.

(a) $_{13}Al$, $_{14}Si$ 　　　**(b)** $_7N$, $_{15}P$

(c) $_{32}Ge$, $_{34}Se$ 　　　**(d)** $_{23}V$, $_{35}Br$

6.100 Indicate whether the first listed element in each of the following pairs of elements contains (1) more unpaired electrons, (2) fewer unpaired electrons, or (3) the same number of unpaired electrons than the second listed element.

(a) $_{15}P$, $_{16}S$ 　　　　**(b)** $_8O$, $_{34}Se$

(c) $_{49}In$, $_{53}I$ 　　　　**(d)** $_{22}Ti$, $_{52}Te$

6.101 Indicate which element or elements have the electron characteristics below. In those cases where a series of elements have the indicated characteristics, do not write all the symbols but rather write the atomic numbers of the first and last elements in the series, for example, elements 70–83.

(a) a total of 84 electrons

(b) only four $3d$ electrons

(c) two $7s$ electrons (note that more than one element qualifies)

(d) a total of six s electrons (note that more than one element qualifies)

6.102 Indicate which element or elements have the electron characteristics below. In those cases where a series of elements have the indicated characteristics, do not write all the symbols but rather write the atomic numbers of the first and last elements in the series, for example, elements 70–83.

(a) only one electron in shell 7

(b) a total of fifteen p electrons (they are not all in the same subshell)

(c) twelve electrons in shell 3

(d) six $5p$ electrons (note that more than one element qualifies)

6.103 Referring only to the periodic table, determine the element of lowest atomic number whose electron configuration contains the following.

(a) more p electrons than s electrons

(b) more d electrons than p electrons

6.104 Referring only to the periodic table, determine the element of lowest atomic number whose electron configuration contains the following.

(a) more f electrons than s electrons

(b) more d electrons than s electrons

6.105 If an orbital could hold three electrons instead of the actual two electrons, what would be the electron configurations of the following elements? Assume that all other relationships associated with electron configurations remain unchanged.

(a) aluminum   **(b)** bromine

(c) sulfur   **(d)** iron

6.106 If an orbital could hold three electrons instead of the actual two electrons, what would be the electron configurations of the following elements? Assume that all other relationships associated with electron configurations remain unchanged.

(a) chlorine   **(b)** magnesium

(c) krypton   **(d)** zinc

Cumulative Problems

6.107 Which of the six elements oxygen, lithium, helium, boron, strontium, and potassium belongs in each of the following classifications?

(a) has an atomic number less than 20

(b) is not ductile and malleable

(c) all atoms contain an equal number of protons and electrons

(d) belongs to period 4 of the periodic table

6.108 Which of the six elements nitrogen, beryllium, argon, aluminum, silver, and gold belongs in each of the following classifications?

(a) period and group numbers are numerically equal

(b) readily conducts electricity and heat

(c) has an atomic mass greater than its atomic number

(d) all atoms have a nuclear charge greater than +20

6.109 The electron configuration of the isotope $^{12}_{6}C$ is $1s^2 2s^2 2p^2$. What is the electron configuration for the isotope $^{13}_{6}C$?

6.110 The electron configuration of the isotope $^{16}_{8}O$ is $1s^2 2s^2 2p^4$. What is the electron configuration for the isotope $^{18}_{8}O$?

6.111 How many subatomic particles are present in an atom whose isotopic mass is 36.96590 amu and whose electron configuration is $1s^2 2s^2 2p^6 3s^2 3p^5$?

6.112 How many subatomic particles are present in an atom whose isotopic mass is 29.97376 amu and whose electron configuration is $1s^2 2s^2 2p^6 3s^2 3p^2$?

6.113 An element that is a member of group IIA has atoms that possess a total of 18 p electrons. What is the nuclear charge for atoms of this element?

6.114 An element that is a member of group VIA has atoms that possess a total of 10 s electrons. What is the nuclear charge for atoms of this element?

6.115 In one molecule of the compound carbon dioxide (CO_2),

(a) how many total electrons are present?

(b) what percentage of total electrons present came from oxygen atoms?

(c) what percentage of total electrons present are s electrons?

(d) what percentage of total charged subatomic particles present are electrons?

6.116 In one molecule of the compound ammonia (NH_3),

(a) how many total electrons are present?

(b) what percentage of total electrons present came from the nitrogen atom?

(c) what percentage of total electrons present are p electrons?

(d) what percentage of total charged subatomic particles present are protons?

6.117 Make a list of elements that have all four of the following characteristics: (1) neither a metal nor metalloid, (2) a representative element, (3) atomic number is less than 40, and (4) nuclear charge is greater than 10.

6.118 Make a list of elements that have all four of the following characteristics: (1) neither a nonmetal nor metalloid, (2) a *non-transition* element, (3) an atomic mass less than 150 amu, and (4) a period number greater than 4.

Answers to Practice Exercises

6.1 (a) 3   **(b)** 3   **(c)** 10   **(d)** 2

6.2 (a) $1s^2 2s^2 2p^6 3s^2 3p^5$

    (b) $1s^2 2s^2 2p^6 3s^2 3p^6 4s^2 3d^{10} 4p^6 5s^2 4d^{10} 5p^3$

6.3 ① ① ①①① ① ①①①
    1s   2s   2p   3s   3p

6.4 (a) s for Cs, d for Ag

    (b) s^1 for Cs, d^9 for Ag

    (c) $6s^1$ for Cs, $4d^9$ for Ag

6.5 $1s^2 2s^2 2p^6 3s^2 3p^6 4s^2 3d^{10} 4p^6 5s^2 4d^{10} 5p^6$

7

Chemical Bonds

7.1 Types of Chemical Bonds

In Section 5.2 we considered the fact that chemical compounds are conveniently divided into two broad classes called *ionic compounds* and *molecular compounds*. Ionic and molecular compounds can be distinguished from each other on the basis of general physical properties. Ionic compounds tend to have high melting points (500–2000°C) and are good conductors of electricity when they are in a molten (liquid) state. Molecular compounds, on the other hand, generally have much lower melting points and tend to be gases, liquids, or low-melting solids. They do not conduct electricity in the molten state. Ionic compounds, unlike molecular compounds, do not have molecules as their basic structural unit. Instead, an extended array of positively and negatively charged particles called *ions* is present.

Some combinations of elements produce ionic compounds, whereas other combinations of elements form molecular compounds. What determines whether the interaction of two elements produces ions (an ionic compound) or molecules (a molecular compound)? An answer to this question requires information concerning the nature of *chemical bonds*. A **chemical bond** *is the attractive force that holds two atoms together in a more complex unit.* Chemical bonds form as the result of interactions between electrons found in the combining atoms. Thus, chemical bond considerations are closely linked to electron configurations (Sec. 6.7).

Corresponding to the two broad categories of chemical compounds are two types of chemical attractive forces (chemical bonds): *ionic bonds* and *covalent bonds*. An **ionic bond** *is a chemical bond formed through the transfer of one or more electrons from one atom or group of atoms to another.* As suggested by its name, the ionic bond model (electron transfer) is used

Another designation for a *molecular compound* is *covalent compound*. The modifier *molecular* draws attention to the basic structural unit present (the molecule) and the modifier *covalent* focuses on the mode of bond formation (electron sharing).

in describing the attractive forces in ionic compounds. A **covalent bond** *is a chemical bond formed through the sharing of one or more pairs of electrons between two atoms.* The covalent bond model (electron sharing) is used in describing the attractions between atoms in molecular compounds.

It is important to emphasize, even before we consider the details of these two bond models, that the concepts of ionic and covalent bonds are actually "convenience concepts." Most bonds are not 100% ionic or 100% covalent. Instead, most bonds have some degree of both ionic and covalent character, that is, some degree of both the transfer and sharing of electrons. However, it is easiest to understand these intermediate bonds (the real bonds) by relating them to the pure or ideal bond types called ionic and covalent.

There are two fundamental concepts that are common to and necessary for understanding both the ionic and covalent bonding models. These concepts are

Purely ionic bonds involve complete transfer of electrons from one atom to another. *Purely* covalent bonds involve equal sharing of electrons. Experimentally, it is found that *most* actual bonds have some degree of both ionic and covalent character. The exception is bonds between identical atoms; here, the bonding is purely covalent.

1. Not all electrons in an atom participate in bonding. Those that are available are called *valence electrons.*

2. Certain arrangements of electrons are more stable than other arrangements of electrons. The *octet rule* addresses this situation.

Section 7.2 deals with the concept of valence electrons and Section 7.3 discusses the octet rule.

7.2 Valence Electrons and Lewis Symbols

Certain electrons, called *valence electrons*, are particularly important in determining the bonding characteristics of a given atom. A **valence electron** *is an electron in the outermost electron shell of a representative element or noble gas element.* Valence electrons are always found in either *s* or *p* subshells or both. Note the restriction on the use of this definition; it applies only for representative and noble gas elements (Sec. 6.11). Many commonly encountered elements are representative elements; hence the definition finds much use. (We will not consider in this text the more complicated valence electron definitions for transition or inner-transition elements (Sec. 6.11); the presence of incompletely filled *inner d* or *f* subshells is the complicating factor in definitions for these elements.)

The number of valence electrons in an atom of a representative element can be determined from the atom's electron configuration, as is illustrated in Example 7.1.

The term *valence* is derived from the Latin *valentia,* meaning "capacity" (to form bonds).

EXAMPLE 7.1

Determining the Number of Valence Electrons an Atom Possesses

Determine the number of valence electrons present in atoms of each of the following elements.

(a) $_{12}Mg$ **(b)** $_{17}Cl$ **(c)** $_{34}Se$

SOLUTION

(a) The element magnesium has two valence electrons, as can be seen by examining its electron configuration.

┌─── Number of valence electrons
$1s^2 2s^2 2p^6 3s^2$
└─── Highest value of the electron shell number

The highest value of the electron shell number is $n = 3$. Only two electrons are found in shell 3, two electrons in the $3s$ subshell.

(b) The element chlorine has seven valence electrons.

$$1s^2 2s^2 2p^6 \underset{\text{Highest value of the}}{\underbrace{3s^2 3p^5}}$$

Total of seven valence electrons

Highest value of the electron shell number

Electrons in two different subshells can simultaneously be valence electrons. The highest shell number is 3, and both the $3s$ and $3p$ subshells belong to shell number 3. Hence, all electrons in both subshells are valence electrons.

(c) The element selenium has six valence electrons.

$$1s^2 2s^2 2p^6 3s^2 3p^6 4s^2 3d^{10} 4p^4$$

Total of six valence electrons

Highest value of the electron shell number

The $3d$ electrons are not counted as valence electrons because the $3d$ subshell is in shell 3 and shell 3 is not the shell with maximum n value. Shell 4 is the outermost shell, the shell with maximum n value.

Practice Exercise 7.1

Determine the number of valence electrons present in atoms of the following elements.

(a) $_{20}$Ca   **(b)** $_{16}$S   **(c)** $_{35}$Br

Answers to practice exercises are located at the end of the chapter.

It seems reasonable that the outermost electrons of atoms are those involved in bonding when you remember that when atoms collide—an event that is necessary before atoms can combine—it will be the outermost parts of the atoms that come into contact with each other. Also, since the outermost electrons are located the farthest from the nucleus, they are therefore the least tightly bound (attraction to the nucleus decreases with distance) and thus the most susceptible to change (transfer or sharing).

Chemists have developed a shorthand system for designating the number of valence electrons present in an atom. This system uses *Lewis symbols*. A **Lewis symbol** *is the chemical symbol of an element surrounded by dots equal in number to the number of valence electrons present in atoms of the element*. Lewis symbols for the first 20 elements, arranged as in the periodic table, are given in Figure 7.1. Lewis symbols, named in honor of the American chemist Gilbert Newton Lewis, an early contributor to chemical bonding theory and the person who first used them (see "The Human Side of Chemistry 7"), are also frequently called *electron-dot structures*.

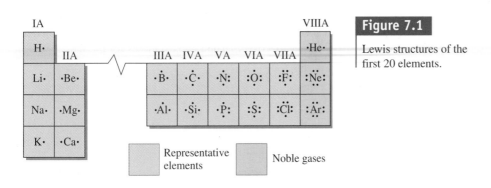

Figure 7.1

Lewis structures of the first 20 elements.

> **In a Lewis symbol, the chemical symbol represents the nucleus and all of the nonvalence electrons. The valence electrons are then shown as dots.**

The general practice when writing Lewis symbols is to place the first four dots separately on the four sides of the chemical symbol and then begin pairing the dots as further dots are added. It makes no difference on which side of the symbol the process of adding dots begins. The following notations for the Lewis symbol of the element magnesium are all equivalent.

$$\overset{\cdot}{Mg}\cdot \quad \underset{\cdot}{Mg}\cdot \quad \cdot\underset{\cdot}{Mg} \quad \cdot\overset{\cdot}{Mg} \quad \overset{\cdot}{\underset{\cdot}{Mg}} \quad \cdot\overset{\cdot}{Mg}\cdot$$

Three important generalizations about valence electrons can be drawn from a study of the Lewis symbols in Figure 7.1.

1. *Representative elements in the same group of the periodic table have the same number of valence electrons.* This should not be surprising to you. Elements in the same group in the periodic table have similar chemical properties as a result of having similar outer shell electron configurations (Sec. 6.9). The electrons in the outermost shell are the valence electrons.

2. *The number of valence electrons for representative elements in a group is the same as the periodic table group number.* For example, the Lewis symbols for O and S, both members of group VIA, show six dots. Similarly, the Lewis symbols of H, Li, Na, and K, all members of group IA, show one dot.

3. *The maximum number of valence electrons for any element is eight.* Only the noble gases (Sec. 6.11), beginning with Ne, have the maximum number of eight electrons. Helium, with only two valence electrons, is the exception in the noble gas family; obviously an element with a grand total of two electrons cannot have eight valence electrons. Although shells with *n* greater than 2 are capable of holding more than eight electrons, they do so only when they are no longer the outermost shell and thus are not the valence shell. For example, selenium (Example 7.1) has 18 electrons in its third shell; however, shell 4 is the valence shell in selenium.

EXAMPLE 7.2

Writing Lewis Symbols for Elements

Write Lewis symbols for the following elements.

(a) O, S, and Se **(b)** B, C, and N

SOLUTION

(a) These elements are all group VIA elements and thus possess the same number of valence electrons, which is six. (The number of valence electrons and the periodic-table group number will always be the same for representative elements.) The Lewis symbols, which all have six dots, are

$$\cdot\overset{\cdot\cdot}{\underset{\cdot\cdot}{O}}: \quad \cdot\overset{\cdot\cdot}{\underset{\cdot\cdot}{S}}: \quad \cdot\overset{\cdot\cdot}{\underset{\cdot\cdot}{Se}}:$$

(b) These elements are sequential elements in period 2 of the periodic table; B is in group IIIA (three valence electrons), C is in group IVA (four valence electrons) and N is in group VA (five valence electrons). The Lewis symbols for these elements are

$$\cdot\overset{\cdot}{B}\cdot \quad \cdot\overset{\cdot}{C}\cdot \quad :\overset{\cdot}{N}\cdot$$

Practice Exercise 7.2

Write Lewis symbols for the following elements.

(a) Be, Mg, and Ca **(b)** P, S, and Cl

The Human Side of Chemistry 7

Gilbert Newton Lewis (1875–1946)

Gilbert Newton Lewis, an American chemist, is recognized as one of the foremost chemists of the 20th century. The son of a lawyer, he received his primary education at home. He read at age 3 and was intellectually precocious.

Lewis was born in Weymouth, Massachusetts (1875), but was raised in Nebraska where his family moved in 1884. In 1889, at age 14, he received his first formal education, attending the University of Nebraska preparatory school. He studied two years at the University of Nebraska and then in 1893 transferred to Harvard University where he obtained a Ph.D. in chemistry in 1899.

After two years of study in Germany and a stint as a government chemist in the Philippines, he became a professor of chemistry at Massachusetts Institute of Technology (1905–1912). From 1912 to his death in 1946, the University of California at Berkeley was his home.

In 1916 he published a paper proposing that a chemical bond involved a pair of electrons shared or held jointly by two atoms. This proved to be one of the most fruitful ideas in the history of chemistry. Within a few years of his 1916 paper, with contributions from other scientists, a formal theory of chemical bonding based on sharing of electron pairs had been developed. Lewis is also the one who developed the symbolism now called Lewis symbols or electron-dot structures. His memoirs indicate that he

first had the idea of shared electrons while he was lecturing to an introductory chemistry class.

Lewis also made significant contributions in many other areas of chemistry besides bonding theory. Thermodynamic studies were a major interest during most of his career. In 1938 a generalized theory for acids and bases (now called Lewis acid–base theory) came from his mind and pen. Still later, he was the first to isolate "heavy" hydrogen ($^2_1$H).

Lewis was not only a profound researcher but also a teacher of renown. He strongly felt that students learn by doing. Among his teaching techniques was the use of large problem sets to reinforce concepts taught in lecture. This problem set approach is still a part of most chemistry classes today.

7.3 The Octet Rule

A key concept in modern bonding theory is that certain arrangements of valence electrons are more stable than others. The term *stable* as used here refers to the idea that a system (in this case an arrangement of electrons) does not easily undergo spontaneous change.

The valence electron configurations possessed by the noble gases (He, Ne, Ar, Kr, Xe, and Rn; see Sec. 6.11) are considered to be the *most stable of all valence electron configurations*. For helium, this most stable electron configuration involves two outer shell electrons. The rest of the noble gases have eight outer shell electrons.

He: $\boxed{1s^2}$

Ne: $1s^2 \boxed{2s^2 2p^6}$

Ar: $1s^2 2s^2 2p^6 \boxed{3s^2 3p^6}$

Kr: $1s^2 2s^2 2p^6 3s^2 3p^6 \boxed{4s^2} 3d^{10} \boxed{4p^6}$

Xe: $1s^2 2s^2 2p^6 3s^2 3p^6 4s^2 3d^{10} 4p^6 \boxed{5s^2} 4d^{10} \boxed{5p^6}$

Rn: $1s^2 2s^2 2p^6 3s^2 3p^6 4s^2 3d^{10} 4p^6 5s^2 4d^{10} 5p^6 \boxed{6s^2} 4f^{14} 5d^{10} \boxed{6p^6}$

The common feature among these noble gas electron configurations is *completely filled* outermost *s* and *p* subshells.

The conclusion that a noble gas configuration is the most stable of all outer shell electron configurations is based on the chemical properties of the noble gases. These elements are the *most unreactive* of all the elements. They are the only elemental gases found in nature in the form of individual uncombined atoms. There are no known compounds of He and Ne, and only a few compounds of Ar, Kr, Xe, and Rn. The noble gases appear to be "happy" the way they are. They have little or no "desire" to form bonds to other atoms.

The *outermost* electron shell of an atom is also called the *valence* electron shell.

The majority of noble gas compounds contain the element xenon. The first was synthesized in 1962. The three most studied xenon compounds are all xenon fluorides—XeF_2, XeF_4, and XeF_6. All three xenon fluorides are white solids at room temperature that react rapidly with water, even traces of moisture.

Atoms of many elements that lack the very stable outer shell electron configuration of the noble gases tend to attain it in chemical reactions that result in compound formation. This observation has become known as the *octet rule* because of the eight outer shell electrons possessed by five of the six noble gases. A formal statement of the **octet rule** is: *In forming compounds, atoms of elements lose, gain, or share electrons in such a way as to produce a noble gas electron configuration for each of the atoms involved.*

Application of the octet rule to many different systems has shown that it has value in predicting correctly the observed combining ratios of atoms. For example, it explains why two hydrogen atoms rather than some other number are bonded to one oxygen atom in the molecular compound water. It explains why the formula of the ionic compound sodium chloride is NaCl rather than $NaCl_2$, $NaCl_3$, or Na_2Cl.

There are exceptions to the octet rule, but it is still used because of the large amount of information that it is able to correlate. It is particularly effective in explaining compound formation involving only representative elements. Often complications arise with transition and inner-transition elements because of the involvement of *d* and *f* electrons in the bonding.

7.4 The Ionic Bond Model

> The pronunciation for the word *ion* is "eye-on."

Electron transfer between two or more atoms is the basic premise for the ionic bond model. This electron transfer produces charged particles called *ions*. An **ion** *is an atom (or group of atoms) that is electrically charged as the result of loss or gain of electrons.* An atom is neutral only when the number of its protons (positive charges) is equal to the number of its electrons (negative charges). Loss or gain of electrons destroys this proton–electron balance and leaves a net charge on the atom.

> An atom's nucleus *never* changes during the process of ion formation. The numbers of neutrons and protons remain constant.

If one or more electrons are gained by an atom, a negatively charged ion is produced; excess negative charge is present because electrons now outnumber protons. The loss of one or more electrons by an atom results in the formation of a positively charged ion; more protons than electrons are now present, resulting in excess positive charge. Note that the excess positive charge associated with a positive ion is never caused by proton gain but always by electron loss. If the number of protons remains constant and the number of electrons decreases, the result is net positive charge. The number of protons, which determines the identity of the element (Sec. 5.7), never changes during ion formation.

The terms *anion* and *cation* describe ions of opposite charge. An **anion** *is a negatively charged ion.* A **cation** *is a positively charged ion.*

> The term *anion* is pronounced "an-eye-on," and the term *cation* is pronounced "cat-eye-on."

The charge on an ion is directly correlated with the number of electrons lost or gained. Loss of one, two, or three electrons gives cations with +1, +2, or +3 charges, respectively. Similarly, a gain of one, two, or three electrons gives anions with −1, −2, or −3 charges, respectively. (Atoms that have lost or gained more than three electrons are very seldom encountered.)

The notation for charges on ions is a superscript placed to the right of the elemental symbol. Some examples of ion symbols are

$$\text{positive ions (cations)} \quad Na^+, K^+, Ca^{2+}, Mg^{2+}, Al^{3+}$$

$$\text{negative ions (anions)} \quad Cl^-, Br^-, O^{2-}, S^{2-}, N^{3-}$$

Note that a single plus or minus sign is used to denote a charge of one, instead of using the notation 1+ or 1−. Also note that in multicharged ions the number precedes the charge sign; that is, the correct notation for a charge of plus two is 2+ rather than +2.

The chemical properties of a particle (atom or ion) depend on the particle's electron arrangement. Because an ion has a different electron configuration (fewer or more electrons) than the atom from which it was formed, it has different chemical properties as well. For example, water solutions containing Na^+ ions are very stable even though the element sodium (neutral Na) reacts vigorously with water.

EXAMPLE 7.3

Writing Symbols for Ions

Give the symbol for each of the following ions.

(a) the ion formed when an aluminum atom loses three electrons

(b) the ion formed when a sulfur atom gains two electrons

SOLUTION

(a) A neutral aluminum atom contains 13 protons and 13 electrons, since the atomic number of aluminum is 13 (obtained from the periodic table). The aluminum ion formed by the loss of three electrons would still contain 13 protons but would have only 10 electrons because three electrons were lost.

$$13 \text{ protons} = 13 + \text{charges}$$
$$10 \text{ electrons} = \underline{10 - \text{charges}}$$
$$\text{net charge} = 3 +$$

The symbol for the aluminum ion is Al^{3+}.

(b) The atomic number of sulfur is 16. Thus, 16 protons and 16 electrons are present in a neutral sulfur atom. A gain of two electrons raises the electron count to 18.

$$16 \text{ protons} = 16 + \text{charges}$$
$$18 \text{ electrons} = \underline{18 - \text{charges}}$$
$$\text{net charge} = 2 -$$

The symbol for the sulfur ion is thus S^{2-}.

Practice Exercise 7.3

Give the symbol for each of the following ions.

(a) the ion formed when a calcium atom loses two electrons

(b) the ion formed when a nitrogen atom gains three electrons

EXAMPLE 7.4

Determining the Number of Protons and Electrons in Ions

Determine the number of protons and electrons present in the following ions.

(a) P^{3-} (b) Mg^{2+}

SOLUTION

(a) The number of protons present is the same as in a neutral atom and is therefore given by the atomic number, which is 15 for phosphorus.

$$\text{number of protons} = \text{atomic number} = 15$$

The number of electrons in a neutral phosphorus atom is 15. The charge of -3 on the ion indicates the gain of three electrons.

$$\text{number of electrons} = 15 + 3 = 18$$

(b) The atomic number of magnesium is 12.

$$\text{number of protons} = \text{atomic number} = 12$$

The number of electrons in a neutral Mg atom is 12. The charge of +2 on the ion indicates the loss of two electrons.

$$\text{number of electrons} = 12 - 2 = 10$$

Practice Exercise 7.4

Determine the number of protons and electrons present in the following ions.

(a) K^+ **(b)** S^{2-}

So far our discussion about electron transfer and ion formation has focused on the loss or gain of electrons by isolated individual atoms. During ionic bond formation, ion formation occurs only when atoms of two elements are present—an element that can lose electrons and an element that can gain electrons. The total number of electrons lost by atoms of the one element is the same as the total number gained by atoms of the other element. Thus, positive and negative ions must always be formed at the same time (see Fig. 7.2).

The mutual attraction between the positive and negative ions that results from electron transfer constitutes the force that holds the ions together as an ionic compound. This force is referred to as an *ionic bond*. An **ionic bond** *is the chemical bond resulting from the attraction of positive and negative ions for each other.*

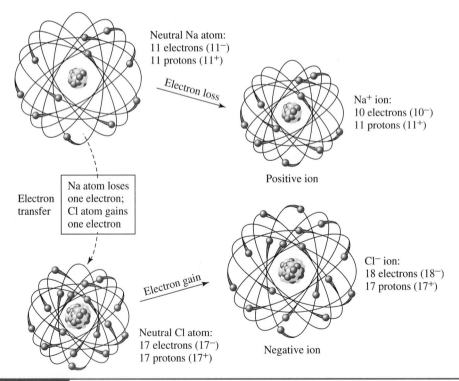

Neutral Na atom:
11 electrons (11⁻)
11 protons (11⁺)

Electron loss

Na⁺ ion:
10 electrons (10⁻)
11 protons (11⁺)

Positive ion

Electron transfer

Na atom loses one electron; Cl atom gains one electron

Electron gain

Neutral Cl atom:
17 electrons (17⁻)
17 protons (17⁺)

Cl⁻ ion:
18 electrons (18⁻)
17 protons (17⁺)

Negative ion

Figure 7.2

Electron transfer always produces positive and negative ions at the same time. You cannot form positive ions without also forming negative ions. Proton count does not change during ion formation. It is electrons that are transferred.

7.5 The Sign and Magnitude of Ionic Charge

Figure 7.2 illustrates the ion formation process that occurs when the elements sodium and chlorine interact to form the compound NaCl. Sodium atoms lose (transfer) one electron to chlorine atoms, producing Na^+ and Cl^- ions. These ions combine in a one-to-one ratio to give the compound NaCl.

Why do sodium atoms form Na^+ and not Na^{2+} or Na^- ions? Why do chlorine atoms form Cl^- ions rather than Cl^{2-} or Cl^+ ions? In general, what determines the specific number of electrons lost or gained in electron transfer processes?

The octet rule (Sec. 7.3) provides a very simple and straightforward explanation for the charge magnitude associated with ions of the representative elements. *Atoms tend to gain or lose electrons until they have obtained an electron configuration that is the same as that of a noble gas.*

Consider the element sodium, which has the electron configuration

$$1s^2 2s^2 2p^6 3s^1$$

It can attain a noble gas configuration by losing one electron (to give it the electron configuration of neon) or by gaining seven electrons (to give it the electron configuration of argon).

$$Na \ (1s^2 2s^2 2p^6 3s^1) \quad \overset{\text{Loss of } 1e^-}{\underset{\text{Gain of } 7e^-}{\diagup\diagdown}}$$

$Na^+ \quad (1s^2 2s^2 2p^6)$
Electron configuration of neon

$Na^{7-} \quad (1s^2 2s^2 2p^6 3s^2 3p^6)$
Electron configuration of argon

The first process, the loss of one electron, is more energetically favorable than the gain of seven electrons and is the process that occurs. The process that involves the fewer number of electrons will always be the more energetically favorable process and will be the process that occurs.

Consider the element chlorine, which has the electron configuration

$$1s^2 2s^2 2p^6 3s^2 3p^5$$

It can attain a noble gas configuration by losing seven electrons (to give it the electron configuration of neon) or by gaining one electron (to give it the electron configuration of argon). The latter occurs for the reason cited previously.

$$Cl \ (1s^2 2s^2 2p^6 3s^2 3p^5) \quad \overset{\text{Loss of } 7e^-}{\underset{\text{Gain of } 1e^-}{\diagup\diagdown}}$$

$Cl^{7+} \quad (1s^2 2s^2 2p^6)$
Electron configuration of neon

$Cl^- \quad (1s^2 2s^2 2p^6 3s^2 3p^6)$
Electron configuration of argon

The type of consideration we have just used for the elements sodium and chlorine leads to the following generalizations.

1. Metal atoms containing one, two, or three valence electrons (the metals in groups IA, IIA, and IIIA of the periodic table) tend to lose electrons to acquire a noble gas electron configuration. The noble gas involved is the one preceding the metal in the periodic table.

Group IA metals form $+1$ ions.

Group IIA metals form $+2$ ions.

Group IIIA metals form $+3$ ions.

Group IA metals are all located one periodic table position past a noble gas. Thus, they will each have one more electron than the preceding noble gas. This electron must be lost if a noble gas configuration is to be obtained. Group IIA and IIIA metals are two and three periodic table positions, respectively, beyond a noble gas. Consequently, two and three electrons, respectively, must be lost for these metals to attain a noble gas electron configuration.

> The positive charge on metal ions from groups IA, IIA, and IIIA has a magnitude equal to the metal's periodic table group number.

2. Nonmetal atoms containing five, six, or seven valence electrons (the nonmetals in groups VA, VIA, and VIIA of the periodic table) tend to gain electrons to acquire a noble gas configuration. The noble gas involved is the one following the nonmetal in the periodic table.

Group VIIA nonmetals form -1 ions.

Group VIA nonmetals form -2 ions.

Group VA nonmetals form -3 ions.

> Nonmetals from groups VA, VIA, and VIIA form negative ions whose charge is equal to the group number minus 8. For example, S, in group VIA, forms S^{2-} ions ($6 - 8 = -2$).

The nonmetal ionic charge guidelines can be explained by reasoning similar to that for the metals, only this time the periodic table positions are those immediately preceding the noble gases. Consequently, electrons must be gained to attain noble gas configurations.

3. Elements in group IVA occupy unique positions relative to the noble gases. They are located equidistant between two noble gases. For example, the element carbon is four positions beyond helium and four positions before neon. Theoretically, ions with charges of $+4$ or -4 could be formed by elements in this group, but in most cases the bonding that results is more adequately described by the covalent bond model to be discussed in Section 7.10.

Isoelectronic Species

An ion formed in the preceding manner with an electronic configuration the same as that of a noble gas is said to be *isoelectronic* with the noble gas. **Isoelectronic species** *are ions, or an atom and ions, having the same number and configuration of electrons.* An atom and an ion or two ions may be isoelectronic. Numerous ions that are isoelectronic with a given noble gas exist, as can be seen from the entries in Table 7.1.

It should be emphasized that an ion that is isoelectronic with a noble gas does not have the properties of the noble gas. It has not been converted into the noble gas. The number of protons in the nucleus of the isoelectronic ion is different from that in the noble gas. These points are emphasized by the comparison in Table 7.2 between Mg^{2+} and Ne, the noble gas with which Mg^{2+} is isoelectronic.

| Table 7.1 | Ions Isoelectronic with Selected Noble Gases | | | | | | |

Electron Configuration	Anions			Noble Gas	Cations		
$1s^2$			H^-	He	Li^+	Be^{2+}	
$1s^2 2s^2 2p^6$	N^{3-}	O^{2-}	F^-	Ne	Na^+	Mg^{2+}	Al^{3+}
$1s^2 2s^2 2p^6 3s^2 3p^6$	P^{3-}	S^{2-}	Cl^-	Ar	K^+	Ca^{2+}	Sc^{3+}

Table 7.2	Comparison of the Structure of a Mg^{2+} Ion and a Ne Atom, the Noble Gas Atom Isoelectronic with the Ion	
	Ne Atom	**Mg^{2+} Ion**
Protons (in the nucleus)	10	12
Electrons (around the nucleus)	10	10
Atomic number	10	12
Charge	0	2+

7.6 Ionic Compound Formation

The use of *Lewis structures* is helpful in visualizing the formation of simple ionic compounds. A **Lewis structure** *is a grouping of Lewis symbols that shows either the transfer of electrons or the sharing of electrons in chemical bonds.* Lewis symbols involve atoms of individual elements. Lewis structures involve compounds. The reaction between sodium (with one valence electron) and chlorine (with seven valence electrons) is represented as follows with a Lewis structure.

$$Na \cdot + \cdot \overset{..}{\underset{..}{Cl}}: \longrightarrow Na^+ [:\overset{..}{\underset{..}{Cl}}:]^- \longrightarrow NaCl$$

The loss of an electron by sodium empties its valence shell. The next inner shell, which contains eight electrons (a noble gas configuration), then becomes the valence shell. After the valence shell of chlorine gains one electron, it then has the "desired" eight valence electrons.

When sodium, which has one valence electron, combines with oxygen, which has six valence electrons, two sodium atoms are needed to meet the "electron requirements" of one oxygen atom.

$$\begin{matrix} Na \cdot \\ \\ Na \cdot \end{matrix} \overset{..}{\underset{..}{O}}: \longrightarrow \begin{matrix} Na^+ \\ Na^+ \end{matrix} [:\overset{..}{\underset{..}{O}}:]^{2-} \longrightarrow Na_2O$$

Note how oxygen's need for two additional electrons dictates that each of the two sodium atoms supply one electron to the oxygen atom; hence the formula is Na_2O.

An opposite situation to that for Na_2O occurs in the reaction between calcium, which has two valence electrons, and chlorine, which has seven valence electrons. Here, two chlorine atoms are required to accommodate electrons transferred from one calcium atom because a chlorine atom can accept only one electron. (It has seven valence electrons and needs only eight.)

$$\begin{matrix} & \cdot \overset{..}{\underset{..}{Cl}}: \\ Ca \cdot & \\ & \cdot \overset{..}{\underset{..}{Cl}}: \end{matrix} \longrightarrow Ca^{2+} \begin{matrix} [:\overset{..}{\underset{..}{Cl}}:]^- \\ [:\overset{..}{\underset{..}{Cl}}:]^- \end{matrix} \longrightarrow CaCl_2$$

We can usually tell whether a compound is ionic (consisting of ions) or covalent (consisting of molecules) from its composition. In general, metals tend to form cations (positive ions) and nonmetals tend to form anions (negative ions). Therefore, *ionic compounds are generally compounds in which both a metal and nonmetal are present*, as in NaCl, Na_2O, and $CaCl_2$. In contrast, *covalent compounds are generally compounds in which only nonmetals are present*, as in H_2O (see Sec. 7.10).

EXAMPLE 7.5

Using Lewis Structures to Depict Ionic Compound Formation

Show the formation of the following ionic compounds using Lewis structures.

(a) K_3P (b) NaF   (c) Al_2O_3

SOLUTION

(a) Potassium (a group IA element) has one valence electron, which it would "like" to lose. Phosphorus (a group VA element) has five valence electrons and would thus "like" to acquire three more. Three potassium atoms will be required to supply enough electrons for one phosphorus atom.

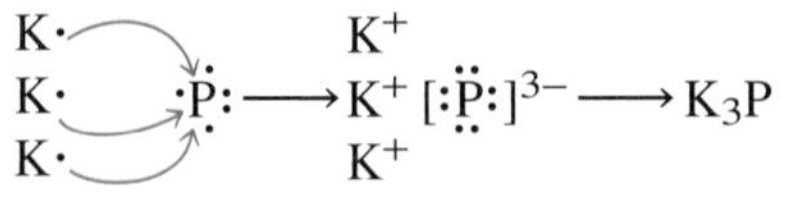

(b) Sodium (a group IA element) has one valence electron, and fluorine (a group VIIA element) has seven valence electrons. The transfer of the one sodium valence electron to a fluorine atom will result in each atom having a noble gas electron configuration. Thus, these two elements combine in a one-to-one ratio.

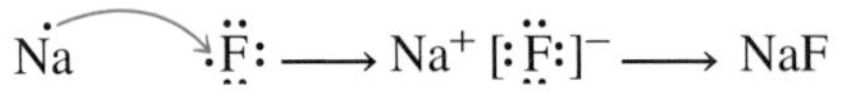

(c) Aluminum (a group IIIA element) has three valence electrons, all of which need to be lost through electron transfer. Oxygen (a group VIA element) has six valence electrons and thus needs to acquire two more. Three oxygen atoms are needed to accommodate the electrons given up by two aluminum atoms.

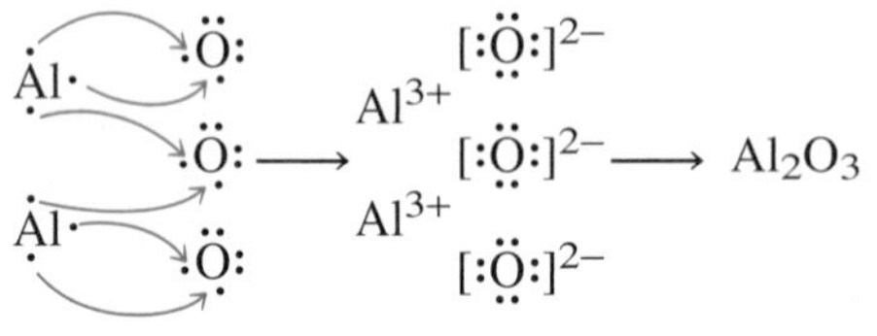

Practice Exercise 7.5

Show the formation of the following ionic compounds using Lewis structures.

(a) KF   (b) Li_2O (c) Ca_3P_2

7.7 Formulas for Ionic Compounds

Since total electron loss always equals total electron gain in an electron transfer process, ionic compounds are always neutral; no net charge is present. The total positive charge on the ions that have lost electrons is always exactly counterbalanced by the total negative charge on the ions that have gained electrons. Thus, *the ratio in which positive and negative ions combine is the ratio that achieves charge neutrality for the resulting compound.* This generalization can be used instead of Lewis structures to determine ionic compound formulas. Ions are combined in the ratio that causes the positive and negative charges to add to zero.

The correct combining ratio when K^+ and S^{2-} ions combine is two to one. Two K^+ ions (each of $+1$ charge) will be required to balance the charge on a single S^{2-} ion.

$$2(K^+): \quad (2\text{ ions}) \times (\text{charge of } +1) = +2$$
$$S^{2-}: \quad (1\text{ ion}) \times \underline{(\text{charge of } -2) = -2}$$
$$\text{net charge} = 0$$

Hence, the formula is K_2S.

Example 7.6 gives further illustration of the procedures needed to determine correct combining ratios between ions and to write correct ionic formulas from the combining ratios. Written ionic formulas are consistent with the following guidelines.

> **In any ionic compound the total positive charge (from the cations) plus the total negative charge (from the anions) must add up to zero. Ionic compounds cannot have a net charge.**

1. The symbol for the positive ion is always written first.
2. The charges on the ions that are present are *not* shown in the formula. Knowledge of charges is necessary to determine the formula, but once it is determined, the charges are not explicitly written.
3. The numbers in the formula (the subscripts) give the combining ratio for the ions.

EXAMPLE 7.6

Using Ionic Charges to Determine the Formula of an Ionic Compound

Determine the formula for the compound that is formed when each of the following types of ions interact.

(a) Ba^{2+} and Cl^- (b) Ba^{2+} and S^{2-} (c) Ba^{2+} and N^{3-}

SOLUTION

(a) Ba^{2+} and Cl^- ions will combine in a one-to-two ratio because this combination will cause the total charge to add up to zero. One Ba^{2+} ion gives a total positive charge of 2. Two Cl^- ions give a total negative charge of 2. Thus, the formula of the compound is $BaCl_2$.

(b) The formula of this compound is simply BaS (a one-to-one ratio between ions). One Ba^{2+} ion contributes two units of positive charge, and that is counterbalanced by two units of negative charge from the S^{2-} ion.

(c) The numbers in the charges for these ions are 2 and 3. The lowest common multiple of 2 and 3 is 6 ($2 \times 3 = 6$). Thus, we will need six units of positive charge and six units of negative charge. Three Ba^{2+} ions are needed to give the six units of positive charge, and two N^{3-} ions are needed to give the six units of negative charge. The combining ratio of ions is three to two, and the formula is Ba_3N_2. The strategy used in determining this formula, finding the lowest common multiple in the charges of the ions, will always work.

Practice Exercise 7.6

Determine the formula of the compound that is formed when each of the following types of ions interact.

(a) Na^+ and P^{3-} (b) Be^{2+} and P^{3-} (c) Al^{3+} and P^{3-}

Before leaving the subject of ions, ionic bonds, and formulas for ionic compounds, let us quickly review the key principles about ionic bonding that have been presented.

Table 7.3	General Formulas for Ionic Compounds as a Function of Periodic Table Position of the Metal and Nonmetal		
	Nonmetals (X)		
Metals (M)	VIIA (-1 ions)	VIA (-2 ions)	VA (-3 ions)
IA ($+1$ ions)	MX	M_2X	M_3X
IIA ($+2$ ions)	MX_2	MX	M_3X_2
IIIA ($+3$ ions)	MX_3	M_2X_3	MX

1. Ionic compounds usually contain both a metallic and a nonmetallic element.
2. The metallic element atoms lose electrons to produce positive ions, and the nonmetallic element atoms gain electrons to produce negative ions.
3. The electrons lost by the metal atoms are the same ones that are gained by the nonmetal atoms. Electron loss must always equal electron gain.
4. The ratio in which positive metal ions and negative nonmetal ions combine is the lowest whole number ratio that achieves charge neutrality for the resulting compound.
5. Metals from groups IA, IIA, and IIIA of the periodic table form ions with charges of $+1$, $+2$, and $+3$, respectively. Nonmetals of groups VIIA, VIA, and VA of the periodic table form ions with charges of -1, -2, and -3, respectively. Table 7.3 lists, in general terms, all of the possible metal–nonmetal combinations from these periodic table groups that result in the formation of ionic compounds.

7.8 Structure of Ionic Compounds

Ionic solids consist of positive and negative ions arranged in such a way that each ion is surrounded by nearest neighbors of the opposite charge. Any given ion is bonded by electrostatic (positive–negative) attractions to all of the other ions of opposite charge immediately surrounding it. Figure 7.3 gives two three-dimensional depictions of the arrangement of ions for the ionic compound NaCl (table salt).

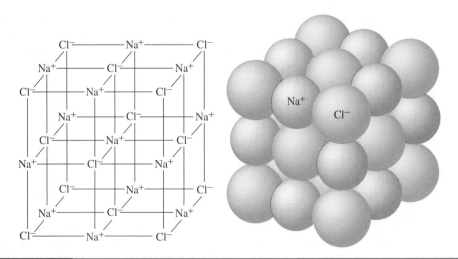

Figure 7.3

The arrangement of ions in sodium chloride. Each of the Na^+ ions is surrounded on all six sides by Cl^- ions, and each of the Cl^- ions is surrounded on all sides by Na^+ ions.

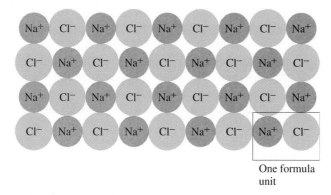

One formula
unit

Figure 7.4

Two-dimensional cross-section of an ionic solid (NaCl). No molecule can be distinguished in this structure. Instead, we can recognize a basic formula unit that is repeated indefinitely.

The alternating array of positive and negative ions present in an ionic compound means that discrete molecules do not exist in such compounds (Sec. 5.4). Therefore, the formulas of ionic compounds cannot represent the composition of molecules of these substances. Instead, such formulas represent the simplest combining ratio for the ions present. The formula for sodium chloride, NaCl, indicates that sodium and chloride ions are present in a one-to-one ratio in this compound. Chemists use the term *formula unit*, rather than molecule, to refer to the smallest unit of an ionic compound. A **formula unit** *is the smallest whole-number repeating ratio of ions present in an ionic compound that results in charge neutrality*. A formula unit is "hypothetic," because it does not exist as a separate entity; it is only "a part" of the extended array of ions that constitute an ionic solid (see Fig. 7.4).

Although the formulas for ionic compounds represent only ratios, they are used in equations and chemical calculations in the same way as the formulas for molecular species. Remember, however, that they cannot be interpreted as indicating that molecules exist for these substances. They represent the simplest ratio of ions.

The ions present in an ionic solid adopt an arrangement that maximizes attractions between ions of opposite charge and minimizes repulsions between ions of like charge. The specific arrangement that is adopted depends on ion sizes and on the ratio between positive and negative ions. Arrangements are usually very symmetrical and result in crystalline solids—that is, solids with highly regular shapes. Crystalline solids usually have flat surfaces or faces that make definite angles with one another.

Looked at closely, salt grains reveal the cubic shape of the NaCl crystal. The intricate shapes of many gems such as rubies and emeralds reflect the arrangement of their microscopic ionic arrays.

7.9 Polyatomic Ions

To this point in this chapter all references to and comments about ions have involved *monoatomic ions*. A **monoatomic ion** *is an ion formed from a single atom through loss or gain of electrons*. Such ions are very common and very important. Another large and important category of ions, called *polyatomic ions*, exists. A **polyatomic ion** *is an ion formed from a group of atoms (held together by covalent bonds) through loss or gain of electrons*. Numerous ionic compounds exist in which the positive or negative ion (sometimes both) contains more than one atom. Polyatomic ions are very stable species, generally maintaining their identity during chemical reactions.

An example of a polyatomic ion is the sulfate ion, SO_4^{2-}. This ion contains four oxygen atoms and one sulfur atom, and the whole group of five atoms has acquired a -2 charge. The whole sulfate group is the ion rather than any one atom within the group. Covalent bonding, discussed in Section 7.10, holds the sulfur and oxygen atoms together.

Polyatomic ions are not molecules. They never occur alone as molecules do. Instead, they are always found associated with ions of opposite charge. Polyatomic ions are *pieces* of compounds, not compounds. Ionic compounds require the presence of both positive and negative ions and are neutral overall. Polyatomic ions are always charged species.

Formulas for ionic compounds containing polyatomic ions are determined in the same way as those for ionic compounds containing monoatomic ions (Sec. 7.7). The basic rule is the same: the total positive and negative charge present must add up to zero.

Two conventions not encountered previously in formula writing often arise when writing formulas with polyatomic ions. They are

1. When more than one polyatomic ion of a given kind is required in a formula, the polyatomic ion is enclosed in parentheses, and a subscript is placed outside the parentheses to indicate the number of polyatomic ions needed.

2. To preserve the identity of polyatomic ions, the same elemental symbol may be used more than once in a formula.

Example 7.7 contains examples illustrating the use of both of these new conventions. Besides the sulfate ion, four other polyatomic ions are involved in this example: OH^- (hydroxide ion), NO_3^- (nitrate ion), NH_4^+ (ammonium ion), and CN^- (cyanide ion). The formulas for numerous other polyatomic ions are considered in Section 8.4.

EXAMPLE 7.7

Writing Formulas for Ionic Compounds Containing Polyatomic Ions

Determine the formulas for the ionic compounds containing the following pairs of ions.

(a) K^+ and SO_4^{2-} (b) Na^+ and NO_3^-
(c) Ca^{2+} and OH^- (d) NH_4^+ and CN^-

SOLUTION

(a) To equalize the total positive and negative charges, we need two K^+ ions for each SO_4^{2-} ion. We indicate the presence of the two K^+ ions with the subscript 2 following the symbol of the ion. The formula of the compound is K_2SO_4. The convention that the positive ion is always written first in the formula still holds when polyatomic ions are present.

(b) Since both of these ions possess a charge of one, combining them in a one-to-one ratio will balance the charge. The formula of the compound is $NaNO_3$.

(c) Two OH^- ions are needed to balance the charge on one Ca^{2+} ion. Since more than one polyatomic ion is needed, the formula will contain parentheses: $Ca(OH)_2$. The subscript 2 outside the parentheses indicates two of what is inside the parentheses. If parentheses were not used, the formula would appear to be $CaOH_2$, which is not intended and which actually conveys false information. The formula $Ca(OH)_2$ indicates a formula unit containing one Ca atom, two O atoms, and two H atoms (Sec. 5.4); the formula $CaOH_2$ would indicate a formula unit containing one Ca atom, one O atom, and two H atoms. Verbally the correct formula, $Ca(OH)_2$, would be read as "C-A" (pause) "O-H-taken-twice."

(d) In this compound both ions are polyatomic, a perfectly legal situation. Since the ions have equal but opposite charges, they will combine in a one-to-one ratio. The formula is thus NH_4CN. No parentheses are needed because we need only one polyatomic ion of each type in a formula unit. Parentheses are used only when there are two or more polyatomic ions of a given kind in a formula unit. What is different about this formula is the appearance of the symbol for the element nitrogen (N) at two locations in the formula. This could be prevented by combining the two nitrogens, giving the formula N_2H_4C. However, combining is not done in situations like this because the identity of the polyatomic ions present is lost in the resulting combined formula. The formula N_2H_4C does not convey the message that NH_4^+ and CN^- ions are present; the formula NH_4CN does. Thus, in writing formulas that contain polyatomic ions, we always maintain the identities of these ions even if it means having the same elemental symbol at more than one location in the formula.

Practice Exercise 7.7

Determine the formulas for the ionic compounds containing the following pairs of ions.

(a) K^+ and OH^- **(b)** Na^+ and CN^-

(c) Ca^{2+} and NO_3^- **(d)** NH_4^+ and SO_4^{2-}

Figure 7.5 gives pictorial representations of the ionic makeup of the four compounds whose formulas were determined in Example 7.7.

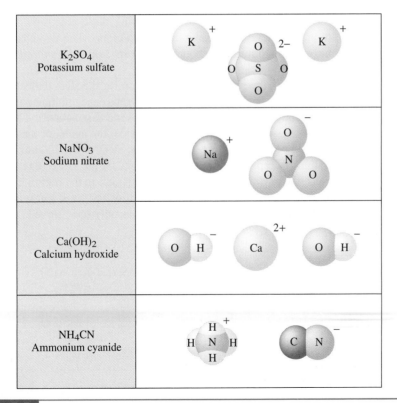

Figure 7.5

Ionic components of selected compounds in which polyatomic ions are present.

We begin our discussion of covalent bonding and the molecular compounds that result from such bonding by listing several key differences between ionic bonding (Sec. 7.4) and covalent bonding and the resulting ionic and molecular compounds.

1. Ionic bonds form between atoms of a metal and a nonmetal. Covalent bond formation occurs between two nonmetal atoms. The two nonmetal atoms can be identical but need not be so.

2. *Electron transfer* is the mechanism by which ionic bond formation occurs. Covalent bond formation involves *electron sharing*.

3. In an ionic compound, discrete molecules do not exist since such compounds involve an extended array of alternating positive and negative ions. In covalently bonded compounds the basic structural unit is the molecule. Indeed, such compounds are called *molecular* compounds.

4. All ionic compounds are solids at room temperature. Molecular compounds may be solids (glucose), liquids (water), or gases (carbon dioxide).

5. An ionic solid, if soluble in water, forms an aqueous solution that conducts electricity. The electrical conductance is related to the presence of ions (charged particles) in the solution. A molecule compound, if soluble in water, usually produces a nonconducting aqueous solution.

A **covalent bond** *is a chemical bond resulting from two nuclei attracting the same shared electrons*. A consideration of the simple hydrogen (H_2) molecule provides initial insights into the nature of the covalent bond. When two hydrogen atoms, each with a single electron, are brought together, the orbitals containing these electrons *overlap* (as shown in Fig. 7.6) to produce an orbital common to both atoms. The two electrons, one from each H atom, move throughout this new orbital and are said to be *shared* by the two nuclei.

Once two orbitals overlap, the most favorable location for the shared electrons is the "area" directly between the two nuclei. Here the two electrons can simultaneously interact with (be attracted to) both nuclei, a situation that produces increased stability. This simple analogy illustrates the "increased stability" concept. Consider the nuclei of the two hydrogen atoms in H_2 to be "old potbellied stoves" and the two electrons to be running around each of the stoves trying to keep warm. When the two nuclei are close together (an H_2 molecule), the electrons have two sources of heat. In particular, in the region between the nuclei (the overlap region) the electrons can keep both front and back warm at the same time. This is a better situation than when each electron has only one "stove" (nucleus) as a source of heat.

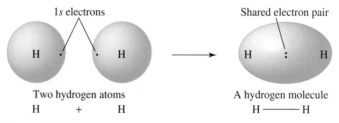

Figure 7.6

Electron sharing can occur only when electron orbitals from two atoms overlap. In the H_2 molecule it is the $1s$ orbitals that overlap.

In terms of Lewis structures, this sharing of electrons by the two hydrogen atoms is diagrammed as

Shared electron pair

$$H \cdot \overset{\frown}{} \cdot H \longrightarrow H{:}H$$

The two shared electrons do double duty, helping each of the hydrogen atoms achieve a helium noble gas configuration.

7.11 Lewis Structures for Molecular Compounds

Using the octet rule, which applies to both electron transfer and electron sharing (Sec. 7.3) and Lewis structures (Sec. 7.6), let us now consider the formation of selected simple covalently bonded molecules containing the element chlorine. Chlorine, located in group VIIA of the periodic table, has seven valence electrons. Its Lewis structure is

$$\cdot \ddot{\underset{\cdot\cdot}{C}}l\colon$$

Chlorine needs one additional electron to achieve the octet of electrons that makes it isoelectronic with the noble gas argon. In ionic compounds, where it bonds to metals, the Cl receives the needed electron via electron transfer. When Cl combines with another nonmetal, a common situation, the octet of electrons is completed via electron sharing. Representative of the situation where chlorine obtains its eighth valence electron through an electron-sharing process are the molecules HCl, Cl_2, and BrCl, whose Lewis structures are as follows.

> In covalent bonding, hydrogen atoms violate the "octet part" of the octet rule—they need only two electrons—but they do not violate the "noble gas part" of the octet rule—after sharing electrons they are isoelectronic with helium.

$$H \cdot \overset{\frown}{} \cdot \ddot{\underset{\cdot\cdot}{C}}l\colon \longrightarrow H{:}\ddot{\underset{\cdot\cdot}{C}}l\colon$$

$$\colon\!\ddot{\underset{\cdot\cdot}{C}}l \cdot \overset{\frown}{} \cdot \ddot{\underset{\cdot\cdot}{C}}l\colon \longrightarrow \colon\!\ddot{\underset{\cdot\cdot}{C}}l{:}\ddot{\underset{\cdot\cdot}{C}}l\colon$$

$$\colon\!\ddot{\underset{\cdot\cdot}{B}}r \cdot \overset{\frown}{} \cdot \ddot{\underset{\cdot\cdot}{C}}l\colon \longrightarrow \colon\!\ddot{\underset{\cdot\cdot}{B}}r{:}\ddot{\underset{\cdot\cdot}{C}}l\colon$$

The HCl and BrCl molecules illustrate the point that the two atoms involved in a covalent bond need not be identical (as in H_2 or Cl_2).

A common practice when writing Lewis structures for covalently bonded molecules is to represent the *shared* electron pairs with dashes. Using this notation, the previously discussed H_2, HCl, Cl_2, and BrCl molecules are written as

$$H{-}H \qquad H{-}\ddot{\underset{\cdot\cdot}{C}}l\colon \qquad \colon\!\ddot{\underset{\cdot\cdot}{C}}l{-}\ddot{\underset{\cdot\cdot}{C}}l\colon \qquad \colon\!\ddot{\underset{\cdot\cdot}{B}}r{-}\ddot{\underset{\cdot\cdot}{C}}l\colon$$

The atoms in covalently bonded molecules often possess both *bonding* and *nonbonding* electrons. **Bonding electrons** *are pairs of valence electrons that are shared between atoms in a covalent bond.* Each of the chlorine atoms in the molecules HCl, Cl_2, and BrCl possess one pair of bonding electrons. **Nonbonding electrons** *are pairs of valence electrons about an atom that are not involved in electron sharing.* Each of the chlorine atoms in HCl, Cl_2, and BrCl possesses three pairs of nonbonding electrons, as does the bromine atom in BrCl.

> Nonbonding electron pairs are often also referred to as *unshared electron pairs* or *lone electron pairs* (or simply *lone pairs*).

$$H:H \qquad H:\overset{..}{\underset{..}{F}}: \qquad :\overset{..}{\underset{..}{F}}:\overset{..}{\underset{..}{F}}: \qquad :\overset{..}{\underset{..}{Br}}:\overset{..}{\underset{..}{F}}:$$

Nonbonding electrons

Bonding electrons

Bonding electrons (black)
Nonbonding electrons (blue)

In Section 7.17 we will learn that nonbonding electron pairs play an important role in determining the shape (geometry) of molecules in which three or more atoms are present.

The number of covalent bonds that an atom forms is equal to the number of electrons it needs to achieve a noble gas configuration. Note that the chlorine atoms in HCl, Cl$_2$, and BrCl all formed one covalent bond. For chlorine, seven valence electrons plus one electron acquired by electron sharing (one bond) give the eight valence electrons needed for a noble gas electronic configuration. The elements oxygen, nitrogen, and carbon have, respectively, six, five, and four valence electrons. Therefore, these elements form, respectively, two, three, and four covalent bonds. The number of covalent bonds these three elements form is reflected in the formulas of their simplest hydrogen compounds H$_2$O, NH$_3$, and CH$_4$. Lewis structures for these three molecules are as follows.

Oxygen has six valence electrons and gains two more through sharing

Nitrogen has five valence electrons and gains three more through sharing

Carbon has four valence electrons and gains four more through sharing

Thus, we see that just as the octet rule was useful in determining the ratio of ions in ionic compounds, we can use it to predict formulas in molecular compounds. Example 7.8 gives additional illustrations of the use of the octet rule to determine formulas for molecular compounds.

EXAMPLE 7.8

Writing Lewis Structures for Simple Molecular Compounds

Write Lewis structures for the simplest molecular compound formed from the following pairs of nonmetals.

(a) phosphorus and hydrogen

(b) sulfur and fluorine

(c) oxygen and chlorine

SOLUTION

(a) Phosphorus is in group VA of the periodic table and thus has five valence electrons. It will form three covalent bonds which through electron sharing will give it eight valence electrons (noble gas configuration). Hydrogen, in group IA of the periodic table, has one valence electron and will form only one covalent bond. For H an "octet" is two electrons; the noble gas that hydrogen "mimics" is helium, which has only two valence electrons. Using Lewis structures, we obtain a compound with the formula PH_3.

(b) Sulfur has six valence electrons and fluorine has seven valence electrons. Thus sulfur will form two covalent bonds $(6 + 2 = 8)$, and fluorine will form one covalent bond $(7 + 1 = 8)$. The formula of the compound is SF_2.

(c) Oxygen, with six valence electrons, will form two covalent bonds, and chlorine, with seven valence electrons, will form only one covalent bond. The formula of the compound is thus Cl_2O, which has the following Lewis structure.

Practice Exercise 7.8

Write Lewis structures for the simplest molecular compound formed from the following pairs of nonmetals.

(a) nitrogen and fluorine   **(b)** carbon and chlorine   **(c)** sulfur and bromine

7.12 Single, Double, and Triple Covalent Bonds

A **single covalent bond** *is a covalent bond in which two atoms share one pair of valence electrons*. All bonds in all of the molecules discussed in the previous section were *single* covalent bonds.

Single covalent bonds are not adequate to explain covalent bonding in all molecules. Sometimes two atoms must share two or three pairs of electrons in order to provide a complete octet of electrons for each atom involved in the bonding. Such bonds are called *double* covalent bonds and *triple* covalent bonds. A **double covalent bond** *is a covalent bond in which two atoms share two pairs of valence electrons*. A double covalent bond between two atoms is approximately twice as strong as a single covalent bond between the same two atoms; that is, it takes approximately twice as much energy to break the double bond as it takes to break the single bond. A **triple covalent bond** *is a covalent bond in which two atoms share three pairs of valence electrons.* A triple covalent bond is approximately three times as strong as a single covalent bond between the same two atoms. The term *multiple covalent bond* is a designation that applies collectively to both double and triple covalent bonds. This designation is often shortened to simply "multiple bond."

One of the simplest molecules possessing a multiple covalent bond is the N_2 molecule; a triple covalent bond is present. A nitrogen atom has five valence electrons and needs three additional electrons to complete its octet.

$$\cdot \ddot{N} \cdot$$

In a N_2 molecule the only sharing that can take place is between the two nitrogen atoms. They are the only atoms present. Thus, to acquire a noble gas electron configuration, each nitrogen atom must share three of its electrons with the other nitrogen atom.

$$:\dot{N}\cdot \longrightarrow \longleftarrow \cdot\dot{N}: \longrightarrow :N:::N: \quad \text{or} \quad :N\equiv N:$$

Notice how all three shared electron pairs are placed in the area between the two nitrogen atoms in the above bonding diagrams. Note also that three lines are used to denote a triple covalent bond, paralleling the use of one line to denote a single covalent bond.

In "bookkeeping" electrons in a Lewis structure, to make sure that all atoms in the molecule have achieved their octet of electrons, *all* electrons in a multiple covalent bond are considered to "belong" to *both* of the atoms involved in that bond. The "bookkeeping" for the N_2 molecule would be

Each of the circles about a N atom contains eight valence electrons. Again, all the electrons in a multiple covalent bond are considered to belong to each of the atoms in the bond. Circles are never drawn to include just some of the electrons in a multiple covalent bond.

A slightly more complicated molecule containing a triple covalent bond is the molecule C_2H_2 (acetylene). A carbon–carbon triple covalent bond is present as well as two carbon–hydrogen single covalent bonds. The arrangement of valence electrons in C_2H_2 is as follows.

$$H:\dot{C}\cdot \longrightarrow \longleftarrow \cdot\dot{C}:H \longrightarrow H:C:::C:H \quad \text{or} \quad H-C\equiv C-H$$

The two atoms in a triple covalent bond are commonly the same element. However, they do not have to be. The molecule HCN (hydrogen cyanide) contains a heteroatomic triple covalent bond.

$$H:C:::N: \quad \text{or} \quad H-C\equiv N:$$

Double covalent bonds are found in numerous molecules. A very common molecule that contains bonding of this type is carbon dioxide (CO_2). In fact, there are two carbon–oxygen double covalent bonds present in CO_2.

$$:\ddot{O}\cdot\cdot\ddot{C}\cdot\cdot\ddot{O}: \longrightarrow :\ddot{O}::C::\ddot{O}: \quad \text{or} \quad :\ddot{O}=C=\ddot{O}:$$

Note in the following diagram how the circles are drawn for the octet of electrons about each of the atoms in CO_2.

> **The triple bond that holds the N_2 molecule together is very strong, and thus N_2 is a very unreactive substance. The nitrogen that makes up most of our atmosphere does not react with most living things. Many manufacturers package their products in nitrogen to keep them fresh.**

> **A single line (dash) is used to denote a single covalent bond, two lines to denote a double covalent bond, and three lines to denote a triple covalent bond.**

7.13 Valence Electron Count and Number of Covalent Bonds Formed

Not all elements can form multiple covalent bonds. There must be at least two vacancies in an atom's valence electron shell prior to bond formation if it is to participate in a multiple covalent bond. This requirement eliminates group VIIA elements (F, Cl, Br, I) and hydrogen from participating in such bonds. The group VIIA elements have seven valence electrons and one vacancy, and hydrogen has one valence electron and one vacancy. All bonds formed by these elements are single covalent bonds.

Double bonding becomes possible for elements needing two electrons to complete their octet, and triple bonding becomes possible when three or more electrons are needed to complete an octet. Note that the word *possible* was used twice in the previous sentence. Multiple bonding does not have to occur when an element has two or three or four vacancies in its octet; single covalent bonds can be formed instead. The "bonding behavior" of an element, when more than one behavior is possible, is determined by what other element or elements it is bonded to.

Let us consider the possible "bonding behaviors" for O (six valence electrons; two octet vacancies), N (five valence electrons; three octet vacancies), and C (four valence electrons; four octet vacancies).

To complete its octet by electron sharing, an oxygen atom can form either two single bonds or one double bond.

$:\!\overset{\textstyle |}{\underset{\textstyle ..}{O}}\!-$	$:\!\overset{..}{\underset{..}{O}}\!=\!=$
Two single bonds	One double bond

Nitrogen is a very versatile element with respect to bonding. It can form single, double, or triple covalent bonds as dictated by the other atoms present in a molecule.

$-\overset{..}{\underset{\textstyle |}{N}}-$	$-\overset{..}{N}\!=\!=$	$:\!N\!\equiv\!\equiv$
Three single bonds	One single and one double bond	One triple bond

Note that in each of these bonding situations a nitrogen atom forms three bonds. A double bond counts as two bonds, and a triple bond as three bonds. Since nitrogen has only five valence electrons, it must form three covalent bonds to complete its octet.

Carbon is an even more versatile element than nitrogen with respect to variety of types of bonding as illustrated by the following possibilities for bonding.

$-\overset{\textstyle |}{\underset{\textstyle |}{C}}-$	$-\overset{\textstyle |}{C}\!=\!=$	$=\!=\!C\!=\!=$	$-C\!\equiv\!\equiv$
Four single bonds	Two single bonds and one double bond	Two double bonds	One single bond and one triple bond

7.14 Coordinate Covalent Bonds

In the covalent bonds considered so far (single, double, and triple), each of the participating atoms in the bond contributed an equal number of electrons to the bond. There is another *less common* way in which a covalent bond can form. It is possible for both electrons in a shared electron pair to come from the same atom; that is, one atom supplies two electrons and the other atom none. Such a covalent bond is called a *coordinate covalent bond*.

A **coordinate covalent bond** *is a covalent bond in which both electrons of a shared electron pair come from one of the two atoms involved in the bond.* Coordinate covalent bonding allows an atom that has two (or more) vacancies in its valence shell to share a pair of nonbonding electrons located on another atom.

The ammonium ion, NH_4^+, is an example of a species containing a coordinate covalent bond. The formation of an NH_4^+ ion can be viewed as resulting from the reaction of a hydrogen ion, H^+, with an ammonia molecule, NH_3. Doing the "bookkeeping" on all the valence electrons involved in this reaction, using **x**'s for nitrogen electrons and dots for hydrogen electrons, we get

$$H{:}\overset{\overset{\textstyle H}{\times\times}}{\underset{\underset{\textstyle H}{\times\times}}{N}}{\times} + H^+ \longrightarrow \left[H{:}\overset{\overset{\textstyle H}{\times\times}}{\underset{\underset{\textstyle H}{}}{N}}{\times}H \right]^+$$

Coordinate covalent bond

An H^+ ion has no electrons, hydrogen having lost its only electron when it became an ion. The H^+ ion has two vacancies in its valence shell; that is, it needs two electrons to become isoelectronic with the noble gas helium. The nitrogen in NH_3 possesses a pair of nonbonding electrons. These electrons are used in forming the new nitrogen–hydrogen bond. The new species formed, the NH_4^+ ion, is charged, since the H^+ was charged. The $+1$ charge on the NH_4^+ ion is dispersed over the *entire* molecule; it is not localized on the "new" hydrogen atom.

The element oxygen quite often forms coordinate covalent bonds. Consider the Lewis structures of the molecules HOCl (hypochlorous acid) and $HClO_2$ (chlorous acid).

$$H{:}\overset{..}{\underset{..}{O}}{:}\overset{..}{\underset{..}{Cl}}{:} \qquad H{:}\overset{..}{\underset{..}{O}}{:}\overset{..}{\underset{..}{Cl}}{:}\overset{\times\times}{\underset{\times\times}{O}}{\times}$$

In the first structure, all the bonds are "ordinary" covalent bonds. In the second structure, which differs from the first in that a second oxygen atom is present, the "new" chlorine–oxygen bond is a coordinate covalent bond. The second oxygen atom with six valence electrons (denoted by **x**'s) needs two more electrons for an octet. It shares one of the nonbonding electron pairs present on the chlorine atom. (The chlorine atom does not need any of the oxygen's electrons since it already has an octet.)

Atoms participating in coordinate covalent bonds generally deviate from the common bonding pattern (Sec. 7.13) expected for that type of atom. For example, oxygen normally forms two bonds; yet in the molecules N_2O (dinitrogen monoxide) and CO (carbon monoxide), which contain coordinate covalent bonds, oxygen forms one and three bonds, respectively.

$$:N{:::}N{:}\overset{..}{\underset{..}{O}}{:} \qquad \text{or} \qquad :N{\equiv}N{-}\overset{..}{\underset{..}{O}}{:}$$

$$:C{::}\overset{\times\times}{O}{\times} \qquad \text{or} \qquad :C{\equiv}O:$$

Once a coordinate covalent bond is formed, there is no way to distinguish it from any of the other covalent bonds in a molecule; all electrons are identical regardless of their source. The main use of the concept of coordinate covalency is to help rationalize the existence of certain molecules and polyatomic ions whose bonding electron arrangement would otherwise present problems. Figure 7.7 contrasts the formation of a "regular" covalent bond with that of a coordinate covalent bond.

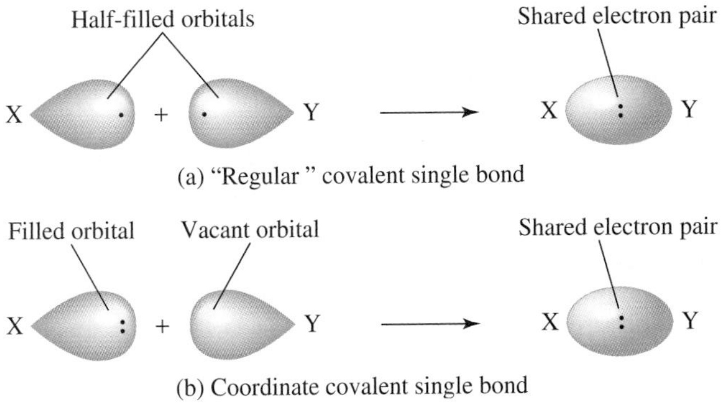

Half-filled orbitals

Shared electron pair

X + Y ⟶ X Y

(a) "Regular" covalent single bond

Filled orbital Vacant orbital

Shared electron pair

X + Y ⟶ X Y

(b) Coordinate covalent single bond

Figure 7.7

(a) A regular covalent single bond results from the overlap of two half-filled orbitals. (b) A coordinate covalent single bond results from the overlap of a filled and an empty orbital.

7.15 Resonance Structures

In Section 7.12 it was noted that, in general, triple covalent bonds are stronger than double covalent bonds, which in turn are stronger than single covalent bonds. **Bond strength** *is a measure of the energy it takes to break a covalent bond; that is, to separate bonded atoms to give neutral particles.* It can be determined experimentally.

Another experimentally determinable parameter of bonds is *bond length.* **Bond length** *is the distance between the nuclei of covalently bonded atoms* (see Fig. 7.8). A definite relationship exists between bond strength and bond length. It is found that as bond strength increases, bond length decreases; that is, the stronger the bond, the shorter the distance between the nuclei of the atoms of the bond. Thus, in general, triple covalent bonds are shorter than double covalent bonds, which are shorter than single covalent bonds.

Most Lewis structures for molecules give bonding pictures that are consistent with available experimental information on bond strength and bond length. However, there are some molecules for which no single Lewis structure that is consistent with such information can be written.

The molecule sulfur dioxide (SO_2) is an example of a situation in which a single Lewis structure does not adequately describe bonding. A plausible Lewis structure for SO_2, in which the octet rule is satisfied for all three atoms, is

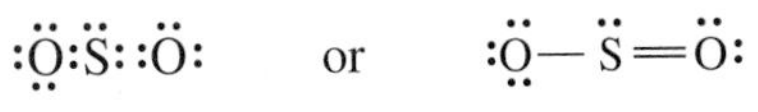

$$:\!\ddot{O}\!:\!\ddot{S}\!:\!:\!\ddot{O}\!: \quad \text{or} \quad :\!\ddot{O}\!-\!\ddot{S}\!=\!\ddot{O}\!:$$

However, this Lewis structure suggests that one sulfur–oxygen bond, the double bond, should be stronger and shorter than the other sulfur–oxygen bond, the single bond. Experiment shows that this is not the case; both sulfur–oxygen bonds are equivalent, with both bond length and bond strength characteristics intermediate between those for sulfur–oxygen single and double bonds. A Lewis structure depicting this intermediate situation cannot be written.

The solution to the phenomenon in which no single Lewis structure adequately describes bonding involves the use of two or more Lewis structures, known as *resonance structures*, to represent the bonding in the molecule. **Resonance structures** *are two or more Lewis structures for a molecule or polyatomic ion that have the same arrangement of atoms, contain the same number of electrons, and differ only in the location of the electrons.*

Two resonance structures exist for an SO_2 molecule.

$$:\!\ddot{O}\!:\!\ddot{S}\!:\!:\!\ddot{O}\!: \longleftrightarrow :\!\ddot{O}\!:\!:\!\ddot{S}\!:\!\ddot{O}\!: \quad \text{or} \quad :\!\ddot{O}\!-\!\ddot{S}\!=\!\ddot{O}\!: \longleftrightarrow :\!\ddot{O}\!=\!\ddot{S}\!-\!\ddot{O}\!:$$

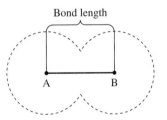

Bond length

A   B

Figure 7.8

The length of a covalent bond between two atoms is the distance separating the nuclei of the atoms.

A double-headed arrow is used to connect resonance structures. The only difference between the two SO_2 resonance structures is in the location of one pair of electrons. The positioning of this pair of electrons determines whether the oxygen atom on the right or the left is the oxygen atom involved in the double bond.

The actual bonding in a SO_2 molecule is said to be a *resonance hybrid* of the two contributing resonance structures. Beginning chemistry students frequently misinterpret the concept of a resonance hybrid. They incorrectly envision that a molecule—SO_2 in this case—is constantly changing (resonating) between various resonance structure forms. This is not the case. For example, SO_2 is not a mixture of two kinds of molecules, nor does a single type of molecule flip-flop back and forth between the two resonance forms. There is only one kind of SO_2 molecule, and the bonding in it is an average of that depicted by the resonance structures. SO_2 molecules exist "full time" in this average state. A mule, the offspring of a donkey and horse, can be considered a hybrid of a donkey and a horse. However, it is not a horse at one instant and a donkey at another; it is always a mule. Likewise, a molecule has only one real structure, which is different from any of the resonance structures; it has characteristics of each one but does not match any one of them exactly.

Sometimes three, four, or even more resonance structures can be drawn for a molecule. Again, such resonance structures must all contain the same number of electrons and have the same arrangement of atoms; the structures may differ only in the location of the electrons about the atoms.

7.16 Systematic Procedures for Drawing Lewis Structures

The task of constructing Lewis structures for molecules and polyatomic ions containing many electrons, or for which several resonance hybrids must be used to describe the bonding, can become quite frustrating if it is approached in a nonsystematic trial-and-error manner. Use of a systematic approach to writing Lewis structures will enable a student to avoid most of this frustration. The following guidelines make the drawing of a Lewis structure for any molecule or polyatomic ion that obeys the octet rule, even a very complicated structure, a straightforward procedure.

STEP 1 Determine the total number of valence electrons present in the molecule, that is, the total number of dots that must appear in the Lewis structure.

The total number of valence electrons is found by adding up the number of valence electrons each atom in the molecule or ion possesses. If the species is a polyatomic ion, add one electron for each unit of negative charge present or subtract one electron for each positive charge. Do not worry about keeping track of which electrons come from which atoms. Only their total number is needed.

STEP 2 Write the symbols of the atoms in the molecule, arranged in the way in which they are bonded to each other, and then place a single covalent bond (two electrons) between each pair of bonded atoms. Either a pair of dots or a dash can be used in denoting the single bond(s).

Determining which atom is the *central atom*, that is, which atom has the most other atoms bonded to it, is the key to determining the arrangement of atoms in a molecule or ion. Most other atoms present will be bonded to the central atom. For most molecular compounds containing just two elements, the molecular formula is of help in deciding the identity of the central atom. The central atom is the atom that appears only once in the formula; for example, S is the central atom in SO_3, O is the central atom in H_2O, and P is the central atom in PF_3. In a molecular compound containing hydrogen, oxygen, and an additional element, it is the additional element that is the central atom; for example, N is the central atom in HNO_3, and S is the central atom in H_2SO_4. In compounds of this type the oxygen atoms are bonded to the central atom and the hydrogen atoms are bonded to the oxygens. Carbon is the central atom in almost all carbon-containing compounds. Hydrogen and fluorine are never the central atom.

STEP 3 Add nonbonding electron pairs to the molecular structure such that each atom bonded to the central atom has an octet of electrons.

Remember that for hydrogen an "octet" of electrons is only two electrons. The noble gas electron configuration acquired by hydrogen is that of helium, and helium has only two electrons.

STEP 4 Place any remaining electrons on the central atom in the structure.

The number of remaining electrons is obtained by subtracting from the total number of valence electrons (step 1) the number of electrons used in steps 2 and 3.

STEP 5 If there are not enough electrons to give the central atom an octet, form multiple covalent bonds by shifting nonbonding electron pairs from surrounding atoms into bonding locations.

The elements C, N, and O are the elements most frequently involved in multiple bonding situations. The elements H, F, Cl, Br, and I in terminal atom positions do not participate in multiple bond formation.

STEP 6 Count the total number of electrons in the completed Lewis structure to make sure it is equal to the total number of valence electrons available for bonding, as calculated in step 1.

This step serves as a "double check" on the correctness of the Lewis structure.

The three examples that follow illustrate the use of the preceding procedures for drawing Lewis structures. In each part of each example the outlined procedure is followed step by step so that you will become familiar with it.

EXAMPLE 7.9

Drawing a Lewis Structure for a Molecular Compound Using Systematic Procedures

The acid present in the greatest amount in *acid rain* is sulfuric acid, a compound with the chemical formula H_2SO_4. Using systematic procedures, draw the Lewis structure for H_2SO_4.

SOLUTION

STEP 1 Both sulfur and oxygen atoms have six valence electrons and hydrogen has one valence electron. The total number of valence electrons present in this molecule is 32.

$$
\begin{aligned}
1\ S{:}\quad 1 \times 6 &= 6 \text{ valence electrons} \\
4\ O{:}\quad 4 \times 6 &= 24 \text{ valence electrons} \\
2\ H{:}\quad 2 \times 1 &= \underline{2 \text{ valence electrons}} \\
& \ 32 \text{ valence electrons}
\end{aligned}
$$

STEP 2 In compounds containing H, O, and an additional element, the additional element (S in this case) is the central atom. The oxygen atoms are attached to this central $\overset{\cdot\cdot}{S}$ atom, and the hydrogen atoms are attached to the oxygen atoms. Drawing this atomic arrangement with single covalent bonds (two electrons) placed between all bonded atoms gives

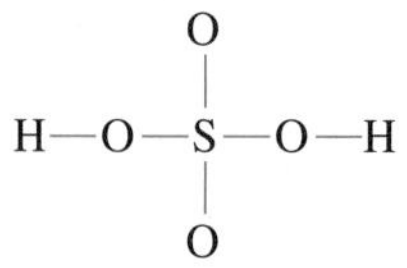

$$
\begin{array}{c}
\text{O} \\
| \\
\text{H—O—S—O—H} \\
| \\
\text{O}
\end{array}
$$

STEP 3 Adding nonbonding electrons to the structure to complete the octets of all atoms bonded to the central atom gives

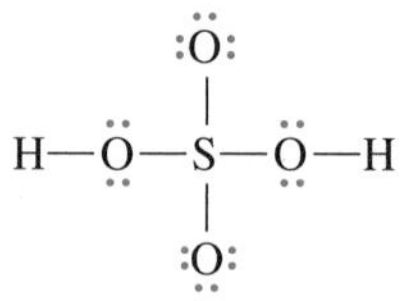

$$
\begin{array}{c}
:\overset{\cdot\cdot}{\text{O}}: \\
| \\
\text{H—}\overset{\cdot\cdot}{\underset{\cdot\cdot}{\text{O}}}\text{—S—}\overset{\cdot\cdot}{\underset{\cdot\cdot}{\text{O}}}\text{—H} \\
| \\
:\underset{\cdot\cdot}{\text{O}}:
\end{array}
$$

STEP 4 We started out with 32 valence electrons (step 1). Twelve were used in step 2 and 20 in step 3. This accounts for all of the available electrons. None are available to add to the central atom. This is fine because the central sulfur atom already has an octet of electrons; it is participating in four single bonds (8 electrons).

STEP 5 No double or triple bonds are needed since all atoms have an octet of electrons with only single bonds present.

STEP 6 There are 32 electrons in the Lewis structure, the same number of electrons as calculated in step 1.

CHEMICAL EXTENSION

All rainfall, even that in the most pristine (unpolluted) areas, is acidic. This acidity results from the presence of carbon dioxide in the atmosphere, which reacts with water to produce carbonic acid. (The same reaction also occurs in a carbonated beverage.) Acid rain is rain with an acidity greater than that which normally occurs.

The cause of acid rain is sulfur oxide and nitrogen oxide air pollution. After being discharged into the atmosphere, these pollutants are chemically converted into sulfuric acid and nitric acid.

The most observable effect of acid rain is the slow, but definite, corrosion of carbonate-based building materials; this phenomenon is sometimes called "stone leprosy." Objects, including statues, made of limestone and marble, which are forms of calcium carbonate ($CaCO_3$), are the most severely affected materials.

Practice Exercise 7.9

Write the Lewis structure for the molecule $HClO_3$.

EXAMPLE 7.10

Drawing a Lewis Structure for a Polyatomic Ion Using Systematic Procedures

The compound sodium sulfite (Na_2SO_3), which contains the polyatomic sulfite ion (SO_3^{2-}), is used in many processed foods as an antioxidant. Using systematic procedures, draw the Lewis structure of the sulfite ion.

SOLUTION

STEP 1 The sulfur atom has six valence electrons and each oxygen atom also has six valence electrons. (Sulfur and oxygen are in the same group in the periodic table.) There are two additional valence electrons present because of the -2 charge associated with this polyatomic ion.

$$1 \text{ S:} \quad 1 \times 6 = \quad 6 \text{ valence electrons}$$

$$3 \text{ O:} \quad 3 \times 6 = 18 \text{ valence electrons}$$

$$\text{charge of } -2 = \underline{\quad 2 \text{ valence electrons}}$$

$$26 \text{ valence electrons}$$

If the polyatomic ion had had a positive charge instead of a negative one, we would have had to subtract valence electrons from the total instead of adding. A positive charge would have denoted loss of electrons, and the electrons lost would have been valence electrons.

STEP 2 The sulfur atom is the central atom with all three oxygen atoms individually attached to it. Drawing this atomic arrangement with single covalent bonds (two electrons) placed between all bonded atoms gives

$$\text{O—S—O}$$
$$|$$
$$\text{O}$$

STEP 3 Adding nonbonding electrons to the structure to complete the octets of the oxygen atoms gives

$$:\ddot{\text{O}}\text{—S—}\ddot{\text{O}}:$$
$$|$$
$$:\ddot{\text{O}}:$$

STEP 4 We started out with 26 electrons (step 1). Six electrons were used in step 2 (single bonds) and 18 electrons in step 3 (nonbonding electron pairs). This leaves 2 electrons not yet used. These 2 remaining electrons are available for placement on the central sulfur atom, which needs 2 more electrons to complete its octet.

$$:\ddot{\text{O}}\text{—}\ddot{\text{S}}\text{—}\ddot{\text{O}}:$$
$$|$$
$$:\ddot{\text{O}}:$$

STEP 5 No double or triple bonds are needed since the action in step 4 causes the central sulfur atom to have an octet of electrons.

We must remember, however, that the Lewis structure we have just generated is that of a polyatomic ion. We therefore need to enclose it in large brackets and place the ionic charge for it outside the right bracket in a superscript position.

$$\left[:\ddot{\text{O}}\text{—}\ddot{\text{S}}\text{—}\ddot{\text{O}}: \atop \quad | \atop \quad :\ddot{\text{O}}: \right]^{2-}$$

STEP 6 There are 26 electrons in the Lewis structure, the same number of electrons as calculated in step 1.

In this example we have shown the bonding within a polyatomic ion. This polyatomic ion, as well as all other polyatomic ions, is not a stable entity that exists alone. Polyatomic ions are *parts* of ionic compounds. The ion SO_3^{2-} would be found in ionic compounds such as Na_2SO_3, K_2SO_3, and $(NH_4)_2SO_3$. Ionic compounds containing polyatomic ions offer an interesting combination of both ionic and covalent bonds: covalent bonding *within* the polyatomic ion and ionic bonding *between* it and ions of opposite charge.

CHEMICAL EXTENSION

Exposure to oxygen causes many foods to undergo changes in color and flavor. Often these changes, called oxidation, present no hazard to health but simply change the foods' appearance. The color changes that occur in sliced apples and peeled potatoes over time are representative of such changes.

Food additives called antioxidants prevent oxidation changes. Among the antioxidants approved for use in foods are vitamin C and vitamin E. Another group of approved antioxidants, which cost less than vitamins, are the sulfites (SO_3^{2-}). Sulfites are used to prevent oxidation in many processed foods, alcoholic beverages (especially wine), and drugs. They were formerly used in restaurant salad bars to keep raw fruits and vegetables looking fresh, but this practice is now banned after some people experienced allergic reactions. For most people, sulfites pose no problem in the amounts used in food products.

Practice Exercise 7.10

Draw the Lewis structure of the phosphite ion, PO_3^{3-}.

EXAMPLE 7.11

Drawing a Lewis Structure for a Molecule That Has Resonance Forms

Ozone (O_3), the triatomic form of oxygen, is the main "irritant" in photochemical smog, the type of smog that requires sunlight for its production. Draw the Lewis structures for O_3; there is more than one structure because of resonance.

SOLUTION

STEP 1   There is a total of 18 valence electrons present, 6 from each oxygen atom.

STEP 2   One of the oxygen atoms may be considered the central atom, and the other two oxygen atoms are attached to it. Drawing this atomic arrangement with single covalent bonds (two electrons) placed between all bonded atoms gives

$$O—O—O$$

STEP 3   Adding six nonbonding electrons to each of the terminal oxygen atoms to complete their octets gives

$$:\overset{..}{\underset{..}{O}}—O—\overset{..}{\underset{..}{O}}:$$

STEP 4   We started out with 18 electrons (step 1). Four electrons were used in step 2 (single bonds) and 12 in step 3 (nonbonding electron pairs). This leaves 2 electrons available for placement on the central oxygen atom.

$$:\overset{..}{\underset{..}{O}}—\overset{..}{O}—\overset{..}{\underset{..}{O}}:$$

STEP 5   The addition of the two nonbonding electrons to the central oxygen atom is not sufficient to give this oxygen an octet of electrons. It still lacks two electrons. This problem is solved by moving a nonbonding pair of electrons from one of the terminal oxygen atoms into the oxygen–oxygen bonding region.

This action, which produces a double bond, gives the central oxygen atom an octet of electrons.

$$:\ddot{O}-\ddot{O}\overset{\frown}{\underset{\cdot\cdot}{O}}:$$

There are two choices for the terminal oxygen atom that is involved in double bond formation. The ramification of this situation is that resonance structures exist. The availability of choices for multiple bond formation (to give a central atom an octet of electrons) is always a signal for the existence of resonance structures. In this case, the resonance structures are

$$:\ddot{O}-\ddot{O}=\underset{\cdot\cdot}{O}: \longleftrightarrow :\underset{\cdot\cdot}{O}=\ddot{O}-\ddot{O}:$$

STEP 6 There are 18 electrons in each of the resonance Lewis structures, the same number of electrons as calculated in step 1.

CHEMICAL EXTENSION

Air pollutants are of two types: primary and secondary. Primary air pollutants are directly emitted into the atmosphere from natural and human-related sources. Secondary air pollutants are produced within the atmosphere from the interaction of primary pollutants. Ozone is a secondary air pollutant. It is formed in a two-step process, within the atmosphere, from the interaction of NO_2 (nitrogen dioxide, a primary pollutant), O_2 (normal atmospheric oxygen), and sunlight.

Atmospheric ozone levels are significantly influenced by meteorological conditions. Concentration levels are usually higher during the summer months, when there is more sunlight than during winter months. High ozone levels can, however, occur during winter months as a result of air stagnation (temperature inversion).

The secondary air pollutant ozone affects the eyes and the mucous membranes of the nose and throat. Its effects are more severe in individuals with chronic lung disease.

Practice Exercise 7.11

Draw the Lewis structure of the SO_3 molecule.

The systematic approach to drawing Lewis structures for molecules and polyatomic ions illustrated in Examples 7.9 to 7.11 does not take into account the origin of the electrons in a chemical bond, that is, which atoms contribute which electrons to the bond. Thus, no distinction between normal covalent bonds and coordinate covalent bonds is made by the procedures of this system. This is acceptable. Each electron in a bond belongs to the bond as a whole. Electrons do not have labels of genealogy.

7.17 Molecular Geometry

Lewis structures give the number and types of bonds present in molecules. They do not, however, convey any information about molecular shape, that is, *molecular geometry*. **Molecular geometry** *is a description of the three-dimensional arrangement of atoms within a molecule*, Indeed, Lewis structures falsely imply that all molecules have flat, two-dimensional shapes.

Molecular geometry is an important factor in determining the physical and chemical properties of substances. Dramatic relationships between geometry and properties are often observed in research associated with the development of prescription drugs. A small change in overall molecular geometry, caused by addition or removal of atoms, can enhance drug effectiveness and/or decrease drug side effects. Studies also show that the human senses of taste and smell depend in part on the shapes of molecules.

For simple molecules (only a few atoms), molecular geometry can be predicted using the information present in a molecule's Lewis structure and a procedure called *valence-shell-electron-pair-repulsion theory (VSEPR theory)*. **VSEPR theory** *is a set of procedures for predicting the geometry of a molecule from the information contained in the molecule's Lewis structure.*

The central concept of VSEPR theory is that electron pairs in the valence shell of an atom adopt an arrangement in space about the nucleus that minimizes the repulsions between the like-charged (all negative) electron pairs. Minimization occurs when the electron pairs are as far away from each other as possible. Repulsion-minimizing arrangements about a nucleus for two, three, and four electron pairs are as follows.

> The acronym VSEPR is pronounced "vesper."

1. Two electron pairs, to be as far apart as possible from each other, will be found on opposite sides of a nucleus, that is, at 180° angles to each other. Such an electron pair arrangement is said to be *linear*.

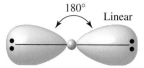

> The preferred arrangement of a given number of valence electron pairs about a central atom is the one that maximizes the separation among them. Such an arrangement minimizes repulsions between electron pairs.

2. Three electron pairs are as far apart as possible when they are found at the corners of an equilateral triangle. In such an arrangement, they are separated by angles of 120°, giving a *trigonal planar* arrangement of the electron pairs.

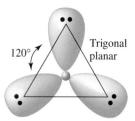

3. A *tetrahedral* arrangement of electron pairs minimizes repulsions between four sets of electron pairs. A tetrahedron is a four-sided geometrical figure, all four sides being identical equilateral triangles. The angle between any two electron pairs is 109°.

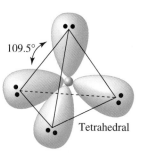

Figure 7.9 shows these three electron-pair arrangements—linear, trigonal planar, and tetrahedral—using balloons that have their ends tied together. When two, three, and four like-sized balloons are tied together they naturally assume the shapes we are talking about.

Most simple molecules have a central atom to which all other atoms are bonded. The first step in applying VSEPR theory to such molecules involves determining the number of

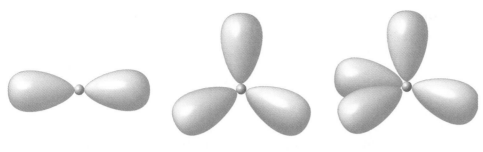

Two balloons, linear   Three balloons, trigonal planar   Four balloons, tetrahedral

Figure 7.9

When balloons of the same size and shape are tied together, they will assume positions in space similar to those taken by pairs of valence electrons around a central atom.

electron pairs present on the *central atom*. This count is obtained from the molecule's Lewis structure using the following VSEPR theory conventions.

Electron Groups

Before we use VSEPR theory to predict molecular geometry, an expansion of the concept of an "electron pair" to that of an "electron group" is needed. This will enable us to extend VSEPR theory to molecules in which double covalent bonds and triple covalent bonds are present. A **VSEPR electron group** *is a group of valence electrons present in a localized region about an atom in a molecule.* An electron group may contain two electrons (a single covalent bond), four electrons (a double covalent bond) or six electrons (a triple covalent bond). Electron groups that contain four and six electrons repel other electron groups in the same way electron pairs do. This makes sense. The four electrons in a double covalent bond or the six electrons in a triple covalent bond are localized in the region between two bonded atoms in a manner similar to the two electrons in a single covalent bond.

Based on these conventions, the VSEPR electron group count about the *central atom* in molecules with the following Lewis structures is as indicated.

:N≡N—Ö: Central atom has two VSEPR electron groups (single bond and triple bond)

:Ö—S̈=Ö: Central atom has three VSEPR electron groups (single bond, double bond, and nonbonding electron pair)

H—Ö—H   Central atom has four VSEPR electron groups (two single bonds and two nonbonding electron pairs)

Let us now apply VSEPR theory to molecules in which two, three, and four VSEPR electron groups are present about a central atom. Our operational rules will be

1. Draw a Lewis structure for the molecule and identify the specific atom for which geometrical information is desired. (This atom will usually be the central atom in the molecule.)
2. Determine the number of VSEPR electron groups present about the central atom. The following conventions govern this determination.
   (a) No distinction is made between bonding and nonbonding electron groups. Both are counted.
   (b) Single, double, and triple covalent bonds are all counted equally as "one electron group" because each takes up only one region of space about a central atom.
3. Predict the VSEPR electron group arrangement about the atom by assuming that the electron groups orient themselves in a manner that minimizes repulsions (Fig. 7.9).

Molecules with Two VSEPR Electron Groups

All molecules with two VSEPR electron groups are linear. Two common molecules in this category are carbon dioxide (CO_2) and hydrogen cyanide (HCN), whose Lewis structures are

$$:\ddot{O}=C=\ddot{O}: \qquad H-C\equiv N:$$

In CO_2 the central carbon atom's two VSEPR groups are the two double bonds. In HCN the central carbon atom's two VSEPR groups are a single bond and a triple bond. In both molecules the VSEPR electron groups arrange themselves on opposite sides of the carbon atom to produce a linear molecule.

Molecules with Three VSEPR Electron Groups

Two molecular geometries are associated with molecules that have three VSEPR electron groups: *trigonal planar* and *angular* (or *bent*) (see Fig. 7.10). The former results when all three VSEPR groups are bonding and the latter when one of the three VSEPR groups is nonbonding. Illustrative of these two situations are the molecules H_2CO (formaldehyde) and SO_2 (sulfur dioxide), whose Lewis structures are

---

Two VSEPR electron groups

Two double bonds	One triple bond and one single bond

Linear	Linear
CO_2	HCN

Three VSEPR electron groups

One double bond and two single bonds	One double bond, one single bond, and one nonbonding pair
Trigonal planar	Angular or bent
H_2CO, $COCl_2$	SO_2, S_2O

Four VSEPR electron groups

Four single bonds	Three single bonds and one nonbonding pair	Two single bonds and two nonbonding pairs
Tetrahedral	Trigonal pyramidal	Angular or bent
CH_4, $SiCl_4$	NH_3, PF_3	H_2O, OF_2

Note: The figure does not consider cases where only one bond is present — for example, one single bond and three nonbonding pairs. If only one bond is present, the molecule is diatomic. All diatomic molecules have the same geometry: The two atoms lie along a straight line.

Figure 7.10

Molecular geometries associated with various combinations of bonding and nonbonding electrons about a central atom that obeys the octet rule.

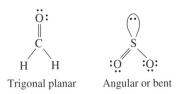

Trigonal planar   Angular or bent

The shape of the SO_2 molecule is described as *angular* rather than *trigonal planar* because molecular geometry describes only *atom positions*. The positions of nonbonding electron pairs are "ignored" when coining words to describe molecular geometry. Do not interpret this to mean that nonbonding electron pairs are unimportant in molecular geometry determinations; indeed, in the case of SO_2, it is the presence of the nonbonding electron pair that makes the molecule angular rather than linear.

Molecules with Four VSEPR Electron Groups

Three molecular geometries are possible for molecules with four VSEPR electron groups: *tetrahedral* (no nonbonding electron pairs present), *trigonal pyramidal* (one nonbonding electron pair present), and *angular* (two nonbonding electron pairs present). The molecules CH_4 (methane), NH_3 (ammonia), and H_2O (water) illustrate this sequence of molecular geometries.

Again, note how the word that is used to describe the geometry of a molecule does not take into account the positions of nonbonding electron pairs.

Molecular Formula	Lewis Structure and VSEPR Electron-Pair Analysis		Molecular Geometry
C_2H_2 (acetylene)	H—C≡C—H		H—C≡C—H Straight chain of four atoms—linear
	Two VSEPR electron groups; linear C center	Two VSEPR electron groups; linear C center	
HN_3 (hydrogen azide)	H—N̈=N=N̈:		N̈=N=N̈: / H Chain of four atoms with one bend
	Three VSEPR electron groups; angular N center	Two VSEPR electron groups; linear N center	
H_2O_2 (hydrogen peroxide)	H—Ö—Ö—H		Ö—Ö / H H Chain of four atoms with two bends
	Four VSEPR electron groups; angular O center	Four VSEPR electron groups; angular O center	

Figure 7.11

Using VSEPR theory to predict molecular geometry for molecules containing more than one central atom.

VSEPR electron-group arrangement and molecular geometry are not the same when a central atom possesses nonbonding electron pairs. The word used to describe the geometry in such cases does not include the positions of the nonbonding electron pairs.

Figure 7.10 summarizes the relationships among the number of VSEPR electron groups about a central atom, molecular geometry, and terminology used to describe molecular geometry.

VSEPR theory can also be used to predict the molecular geometry of molecules that contain more than one central atom. In such situations, each central atom is considered separately and then the results of the separate analyses are combined to obtain the overall molecular geometry. Figure 7.11 shows the results of such procedures for selected molecules containing more than one central atom.

EXAMPLE 7.12

Using VSEPR Theory to Predict Molecular Geometry

Using VSEPR theory, predict the molecular geometry of each of the following molecules given their Lewis structures.

(a)
$$:\!\overset{..}{\underset{}{C}l}\!:$$
$$| $$
$$H\!-\!C\!-\!H$$
$$| $$
$$:\!\overset{}{\underset{..}{C}l}\!:$$

(b)
$$:\!\overset{..}{\underset{..}{F}}\!-\!\overset{..}{\underset{}{P}}\!-\!\overset{..}{\underset{..}{F}}\!:$$
$$| $$
$$:\!\overset{}{\underset{..}{F}}\!:$$

(c)
$$\overset{\displaystyle H\ \ H}{\underset{\displaystyle }{|\ \ |}}$$
$$H\!-\!C\!=\!C\!-\!H$$

SOLUTION

(a) The Lewis structure for this molecule shows that the central carbon atom has four VSEPR electron groups present (four single bonds). No nonbonding valence shell electrons are present on the carbon atom.

   The arrangement of the four electron groups about the central atom is tetrahedral.

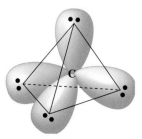

Since each electron group is a bonding group, the molecular shape will also be tetrahedral.

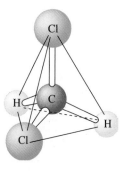

(b) Electrons are present in four locations about the central atom—three electron groups involved in single bonds and one electron pair that is nonbonding.

The arrangement of four electron groups about a central atom is always tetrahedral. It does not matter whether the groups are bonding or nonbonding. Arrangement depends only on the number of groups and not on how the groups function.

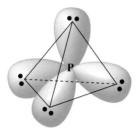

However, the molecular geometry will be different from the electron group geometry since a nonbonding electron pair is present. The molecular geometry is trigonal pyramidal.

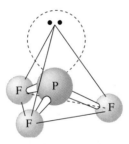

Note that the vacant corner of the tetrahedron (the corner that has a nonbonding electron pair instead of an atom) is not considered when coining a word to describe the molecular geometry. Molecular geometry describes the arrangement of *atoms only.*

(c) This molecule has two central atoms, the two carbon atoms. A VSEPR theory analysis needs to be carried out for each of the carbon atoms and then the results of the two analyses combined.

Each of the two carbon atoms has three VSEPR electron groups about it (two single bonds and a double bond). The geometry associated with three VSEPR electron groups is always trigonal planar.

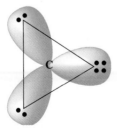

Since all positions about the carbon atoms are bonding positions (no nonbonding electron pairs are present on the carbon atoms) there will also be a trigonal planar arrangement of atoms about each carbon atom. The geometry of the molecule is therefore the following.

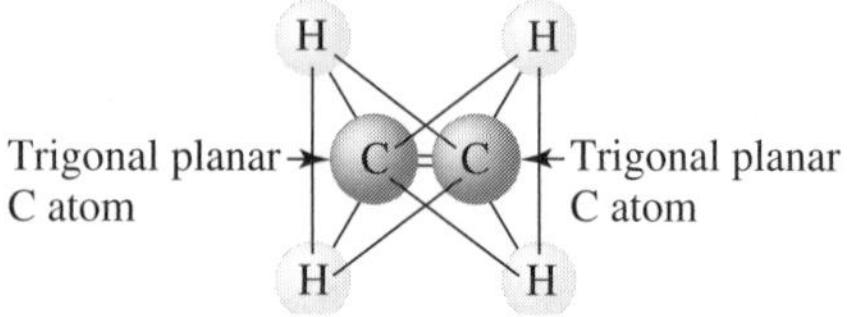

A simpler representation of the geometry of this molecule is the following

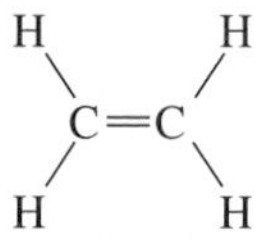

Practice Exercise 7.12

Using VSEPR theory, predict the molecular geometry of each of the following molecules given their Lewis structures.

(a)

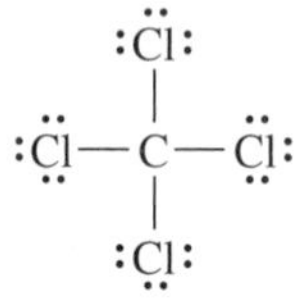

(b)

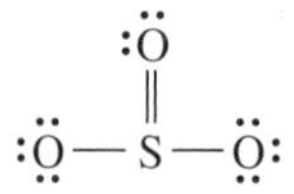

(c)

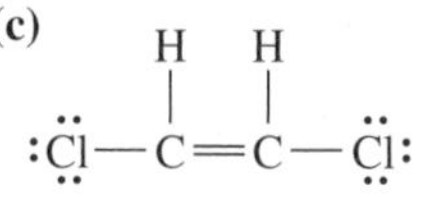

7.18 Electronegativity

The ionic and covalent bonding models we have developed in this chapter seem to represent two very distinct forms of bonding. Actually, the two models are closely related to each other; they are the extremes of a broad continuum of bonding patterns. The close relationship between the two bonding models becomes apparent when the concepts of *electronegativity* (discussed in this section) and *bond polarity* (discussed in the next section) are considered.

The electronegativity concept has its origins in the fact that the nuclei of various elements have differing abilities to attract shared electrons (in a bond) to themselves. Different electron-attracting abilities result from differences in size, nuclear charge, and number of nonvalence electrons present for atoms of various elements. As a result of these factors, some elements are better electron attractors than other elements.

Electronegativity *is a measure of the relative attraction that an atom has for the shared electrons in a bond.* Electronegativity values are unitless numbers on a relative scale that are obtained from bond energies and other related experimental data. Electronegativities cannot be directly measured in the laboratory. A number of electronegativity scales exist. The most widely used one, a scale developed by the American chemist Linus Carl Pauling (1901–1994; see "The Human Side of Chemistry 8") is given in the format of a periodic table in Figure 7.12. On this scale, fluorine, the most electronegative of all elements, has arbitrarily been assigned a value of 4.0 (the maximum on the scale) and serves as the reference element. *The higher the electronegativity of an element, the greater the electron-attracting ability of atoms of that element.*

Patterns and trends in the electronegativity values given in Figure 7.12 include the following.

1. For representative elements (Sec. 6.11), electronegativity values generally increase from left to right within a period of the periodic table.

2. For the period 2 elements this left-to-right increase is regular, being 0.5 units.

Li	Be	B	C	N	O	F
1.0	1.5	2.0	2.5	3.0	3.5	4.0

Before VSEPR theory (Sec. 7.17) can be applied to a molecule one must know which atom is the central atom in the molecule. In general, the *least* electronegative of the atoms present occupies the position of central atom.

H 2.1																	He –
Li 1.0	Be 1.5											B 2.0	C 2.5	N 3.0	O 3.5	F 4.0	Ne –
Na 0.9	Mg 1.2											Al 1.5	Si 1.8	P 2.1	S 2.5	Cl 3.0	Ar –
K 0.8	Ca 1.0	Sc 1.3	Ti 1.5	V 1.6	Cr 1.6	Mn 1.5	Fe 1.8	Co 1.8	Ni 1.8	Cu 1.8	Zn 1.6	Ga 1.6	Ge 1.8	As 2.0	Se 2.4	Br 2.8	Kr –
Rb 0.8	Sr 1.0	Y 1.2	Zr 1.4	Nb 1.6	Mo 1.8	Tc 1.9	Ru 2.2	Rh 2.2	Pd 2.2	Ag 1.9	Cd 1.7	In 1.7	Sn 1.8	Sb 1.9	Te 2.1	I 2.5	Xe –
Cs 0.7	Ba 0.9	57–71 1.1–1.2	Hf 1.3	Ta 1.5	W 1.7	Re 1.9	Os 2.2	Ir 2.2	Pt 2.2	Au 2.4	Hg 1.9	Tl 1.8	Pb 1.8	Bi 1.9	Po 2.0	At 2.2	Rn –
Fr 0.7	Ra 0.9																

Figure 7.12

Electronegativities of the elements.

3. It is important to note how the electronegativity of hydrogen (period 1) compares with that of the period 2 elements. Hydrogen's value of electronegativity, 2.1, is between that of B and C.
4. For the period 3 elements the left-to-right increase is 0.3 units until the last two elements are reached, where it is 0.4 and 0.5.

Na	Mg	Al	Si	P	S	Cl
0.9	1.2	1.5	1.8	2.1	2.5	3.0

The Human Side of Chemistry 8

Linus Carl Pauling (1901–1994)

In 1960, an undergraduate chemistry student asked of Linus Carl Pauling the question: "Professor Pauling, I'd like to know what goes through your mind when you decide that you understand something." Pauling's answer was: "When I say 'Aha! So that's how it works!' But it isn't long before I realize that what I thought I had gotten right begged a number of questions. I realize that I hadn't really understood; that there's more to the problem than I had thought. So I puzzle over the problem some more until I again say, 'Aha! So that's how it *really* works.' Of course, that isn't the end of it, because still more questions arise, still more 'Aha's!' "

Pauling, who died in 1994 at age 93, is considered one of the greatest chemists of all time. Born in Portland, Oregon, the son of a pharmacist, he graduated from Oregon Agricultural College (now Oregon State University) in 1922 (as a chemical engineer) and then three years later obtained a Ph.D. in chemistry from California Institute of Technology (Cal Tech). After two years of further study in Europe he returned to Cal Tech as a faculty member.

During the 1930s, Pauling made numerous key contributions to chemistry including a scale of electronegativity that could be used to determine the ionic and covalent character of chemical bonds, the concept of bond orbital hybridization (the organization of orbitals of atoms in configurations that favor bonding), and the theory of resonance (distribution of electrons between two or more possible Lewis structures for a molecule). During this time period, the chairman of Cal Tech's chemistry department is said to have remarked: "Were all the rest of the chemistry department wiped away except Pauling, it would still be one of the most important departments of chemistry in the world."

In the 1940s, Pauling's interests included biological chemistry, where he is considered one of the founders of molecular biology. In this area he made important contributions concerning the three-dimensional structures of proteins and how enzymes and antibodies function.

In later years, he was concerned with what he called "orthomolecular medicine," in particular with the concept that large doses of ascorbic acid (vitamin C) could ward off common colds and prevent cancer.

Pauling also became an ardent advocate for peace and for a ban on the testing of nuclear weapons. He set in motion a petition that was signed by more than 11,000 scientists worldwide calling for such a ban. He also took vocal antiwar stands.

Pauling is the only person to receive two unshared Nobel Prizes. In 1954, for his contributions to chemistry, especially his work on chemical bonding theory, Pauling received the Nobel Prize in Chemistry. In 1962, Pauling received the Nobel Peace Prize in recognition of his efforts to end nuclear weapons testing.

5. Electronegativity values generally increase from bottom to top within a periodic table group.

The group and period trends in electronegativity values result in nonmetals generally having higher electronegativities than metals. This fact is consistent with our previous generalization (Sec. 7.5) that metals tend to lose electrons and nonmetals tend to gain electrons when an ionic bond is formed. Metals (low electronegativities, poor electron attractors) will give up electrons to nonmetals (high electronegativities, good electron attractors).

<table>
<tr><td>7.19</td><td>Bond Polarity</td></tr>
</table>

Different types of sharing are seen in everyday life. Partners in a business can own it equally, or one partner can have more than the other. It can be owned 60:40, 70:30, and so on.

When two atoms of *equal* electronegativity share one or more pairs of electrons, each atom exerts the same attraction for the electrons, which results in the electrons being *equally* shared. This type of bond is called a *nonpolar covalent bond.* A **nonpolar covalent bond** *is a covalent bond in which there is an equal sharing of electrons between two atoms.*

When the two atoms involved in a covalent bond have *different* electronegativities, the electron-sharing situation is not equal. The atom that has the higher electronegativity will attract the electrons more strongly than the other atom; this results in an unequal sharing of electrons. This type of covalent bond is called a *polar covalent bond.* A **polar covalent bond** *is a covalent bond in which there is an unequal sharing of electrons between two atoms.* Figure 7.13 pictorially contrasts a nonpolar covalent bond and a polar covalent bond using the molecules H_2 and HCl.

The significance of unequal sharing of electrons in a covalent bond is that it creates partial (fractional) positive and negative charges on the atoms involved in a bond. Although each atom involved in a polar covalent bond is initially uncharged, the unequal sharing of electrons means the electrons spend more time near the more electronegative atom of the bond (producing a partial negative charge). The presence of such partial charges on atoms within a molecule often significantly affects molecular properties (Sec. 7.20).

The unequal sharing of electrons in a covalent bond is often indicated by a notation that uses the lowercase Greek letter δ (delta). A δ− symbol, meaning a "partial negative charge," is placed above the relatively negative atom of the bond, and a δ+ symbol, meaning a "partial positive charge," is placed above the relatively positive atom.

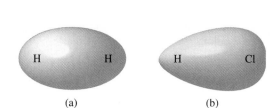

(a)   (b)

<table>
<tr><td>**Figure 7.13**</td></tr>
</table>

(a) In the nonpolar covalent bond present in H_2 (H—H), there is a symmetrical distribution of electron density between the two atoms; that is, equal sharing of electrons occurs.
(b) In the polar covalent bond present in HCl (H—Cl), electron density is displaced toward the Cl atom because of its greater electronegativity; that is, unequal sharing of electrons occurs.

$M^+ X:^-$	$\overset{\delta+\ \ \delta-}{Y:X}$	$X:X$
Ionic (full charges)	Polar covalent (partial charges)	Nonpolar covalent (no charges)
Electron transfer	Unequal sharing of electrons	Equal sharing of electrons

Figure 7.14

Notation used to denote electron transfer, unequal sharing of electrons, and equal sharing of electrons in chemical bonds.

With "delta notation," the bond in hydrogen chloride (HCl) would be depicted as

$$\overset{\delta+\ \ \ \delta-}{H-Cl}$$

Chlorine has an electronegativity of 3.0, and hydrogen's electronegativity is 2.1 (see Fig. 7.12). Since Cl is the more electronegative of the two elements, it dominates the electron-sharing process and draws the electrons closer to it. Hence, the Cl end of the bond has the $\delta-$ designation. Again, the $\delta-$ over the Cl atom indicates a partial negative charge; that is, the chlorine end of the molecule is negative with respect to the hydrogen end. The meaning of the $\delta+$ over the H is that the H end of the molecule is positive with respect to the Cl end. Partial charges are always charges less than $+1$ or -1. Charges of $+1$ and -1, full charges, would result when an electron is transferred from one atom to another. With partial charges we are talking about an intermediate charge state between 0 and 1. Figure 7.14 contrasts the use of formulas to denote electron transfer, unequal sharing of electrons, and equal sharing of electrons.

The direction of polarity of a polar covalent bond, that is, which atom of the bond bears the partial negative charge, can also be designated using an arrow with a cross at its other end ($\longmapsto$). The crossed end of the arrow is placed near the atom bearing the partial positive charge and the arrow's head is placed near the atom bearing the partial negative charge. Using this notation, the bond in the molecule HCl would be denoted as:

$$\overset{\longmapsto}{H-Cl}$$

Bond polarity *is a measure of the degree of inequality in the sharing of electrons in a chemical bond.* The absolute value of the difference in electronegativity of two bonded atoms gives a *rough* measure of the polarity to be expected in a bond. The greater the numerical difference in electronegativities, the greater the inequality of electron sharing and the greater the polarity of the bond. Note that it is the magnitude of the electronegativity *difference,* not the actual electronegativities of the bonded atoms, that determines the extent of polarity of a bond.

As the polarity of a bond increases, the bond is becoming increasingly ionic; that is, the *percent ionic character* of the bond is increasing. **Percent ionic character of a bond** *is a measure of how the actual charge separation (partial charge) in a bond due to electronegativity difference of the bonded atoms compares to the complete charge separation associated with ions.* Percent ionic character values are obtained from experiments involving the behavior of molecules in electric fields.

> **All bonds, except those between identical atoms, have some degree of polarity associated with them.**

A rough scale relating the difference in electronegativity to percent ionic character of a bond is as follows:

Electronegativity Difference	Percent Ionic Character
1.0	20%
1.5	40%
2.0	60%
2.5	80%

> **No known compound has a percent ionic character of 100%.**

A bond that has 50% ionic character correlates with an electronegativity difference of 1.7.

The message of the preceding discussion of bond polarity is that there is no natural boundary between ionic and covalent bonding. Most bonds are a mixture of pure ionic and pure covalent bonds; that is, unequal sharing of electrons occurs. Most bonds have both ionic and covalent character. Nevertheless, it is still convenient to use the terms ionic and covalent in describing chemical bonds, based on the following general, but arbitrary, guidelines, which relate to electronegativity differences.

> **The answer to the question "Is the bond ionic or covalent?" is usually "both." A better question is "How ionic or covalent is the bond?" The existence of partial charges on atoms because of unequal sharing of electrons means that all polar covalent bonds have partial ionic character.**

1. Bonds with the same or very similar electronegativities are called *nonpolar covalent bonds.* "Similar" is taken to mean an electronegativity difference of 0.5 or less (less than 10% ionic character).

   Technically, the only purely nonpolar covalent bonds are those between identical atoms. However, bonds with less than 10% ionic character behave very similarly to the purely nonpolar bonds.

2. Bonds where the electronegativity difference is greater than 0.5 but less than 2.0 are called *polar covalent bonds.*

3. Bonds where the difference in electronegativity between atoms is 2.0 or greater are called *ionic bonds.*

Figure 7.15 summarizes the interrelationships among the concepts of electronegativity difference between bonded atoms, bond type, a bond's covalent character, and a bond's ionic character.

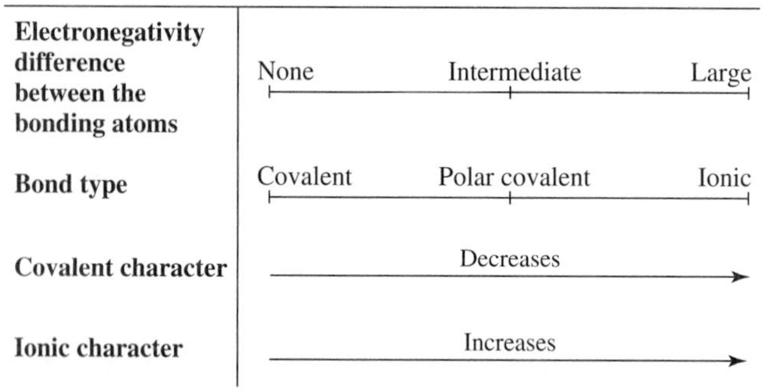

Electronegativity difference between the bonding atoms	None	Intermediate	Large
Bond type	Covalent	Polar covalent	Ionic
Covalent character		Decreases →	
Ionic character		Increases →	

Figure 7.15

The relationships among the ionic and covalent characters of a chemical bond and electronegativity difference between the bonded atoms.

EXAMPLE 7.13

Predicting the Polarity Characteristics of Bonds

(a) Rank the following bonds in order of increasing polarity.

$$\text{Cl—Cl, Ca—F, C—O, B—H, P—Br}$$

(b) Determine the direction of polarity for each bond.
(c) Classify each bond as nonpolar covalent, polar covalent, or ionic.

SOLUTION

Let us first calculate the electronegativity difference for each of the bonds using the electronegativity values of Figure 7.12.

$$\begin{aligned}
\text{Cl—Cl:} &\quad 3.0 - 3.0 = 0.0 \\
\text{Ca—F:} &\quad 4.0 - 1.0 = 3.0 \\
\text{C—O:} &\quad 3.5 - 2.5 = 1.0 \\
\text{B—H:} &\quad 2.1 - 2.0 = 0.1 \\
\text{P—Br:} &\quad 2.8 - 2.1 = 0.7
\end{aligned}$$

(a) Bond polarity increases as electronegativity difference increases. The ranking in terms of increasing bond polarity is thus

$$\text{Cl—Cl} < \text{B—H} < \text{P—Br} < \text{C—O} < \text{Ca—F}$$
$$0.0 \qquad 0.1 \qquad 0.7 \qquad 1.0 \qquad 3.0$$

(b) The direction of bond polarity is from the least electronegative atom to the most electronegative atom. The more electronegative atom will bear the partial negative charge (δ^-).

$$\overset{\longrightarrow}{\text{Cl—Cl}} \quad \overset{\longrightarrow}{\text{B—H}} \quad \overset{\longrightarrow}{\text{P—Br}} \quad \overset{\longrightarrow}{\text{C—O}} \quad \overset{\longrightarrow}{\text{Ca—F}}$$

(c) Nonpolar covalent bonds require a difference in electronegativity of 0.5 or less and ionic bonds an electronegativity difference of 2.0 or greater. The in-between region corresponds to polar covalent bonds.

nonpolar covalent:	Cl—Cl, B—H
polar covalent:	P—Br, C—O
ionic:	Ca—F

Practice Exercise 7.13

(a) Rank the following bonds in order of increasing polarity.

$$\text{N—S, H—H, Na—F, K—Cl, and F—Cl}$$

(b) Determine the direction of polarity for each bond.
(c) Classify each bond as nonpolar covalent, polar covalent, or ionic.

7.20 Molecular Polarity

Molecules, as well as bonds (Sec. 7.19), can have polarity. **Molecular polarity** *is a measure of the degree of inequality in the attraction of bonding electrons to various locations within a molecule.* In terms of electron attraction, if one part of a molecule is favored over other parts then the molecule is *polar*. A **polar molecule** *is a molecule in which there is an* *unsymmetrical distribution of electronic charge*. In a polar molecule, bonding electrons are more attracted to one part of the molecule than to other parts. A **nonpolar molecule** *is a molecule in which there is a symmetrical distribution of electronic charge*. Attraction for bonding electrons is the same in all parts of the molecule.

Molecular polarity depends on two factors: (1) bond polarities and (2) molecular geometry (Sec. 7.17). The presence of polar bonds within a molecule does not automatically mean that the molecule as a whole is polar. In molecules that are symmetrical the effects of polar bonds may cancel each other, resulting in a molecule as a whole having no polarity.

> A prerequisite for determining molecular polarity is a knowledge of molecular geometry.

Determining the molecular polarity of a diatomic molecule is simple because only one bond is present. If that bond is nonpolar, the molecule is nonpolar; if the bond is polar, the molecule is polar.

Molecular polarity determination in triatomic molecules is more complicated. Two different molecular geometries are possible in triatomic molecules: linear and angular. In addition, the symmetrical or unsymmetrical nature of the molecule must be considered. Let us consider the polarities of three specific triatomic molecules: CO_2 (linear), H_2O (angular), and HCN (linear).

Both bonds in the symmetrical linear CO_2 are polar. For both bonds the direction of polarity is toward oxygen since oxygen is more electronegative than carbon.

> Molecules in which all bonds are polar can, as a whole, be nonpolar if the bonds are so oriented in direction that the polarity effects cancel each other.

$$\overset{\longleftarrow \ \ \longrightarrow}{O=C=O}$$

Despite the presence of the two polar bonds, CO_2 molecules are *nonpolar*. The effects of the two polar bonds cancel. The shift of electronic charge toward one oxygen atom is exactly compensated for by the shift of electronic charge toward the other oxygen atom. Thus, one end of the molecule is not negatively charged relative to the other end (a requirement for polarity), and the molecule is nonpolar.

The nonlinear (angular) triatomic H_2O molecule is polar. The bond polarities associated with the two hydrogen–oxygen bonds do not cancel each other because of the nonlinearity of the molecule.

> It is the polarity of water molecules that makes cooking in microwave ovens possible. The electromagnetic field of the microwave interacts with the polar water molecules causing them to shake. The food is cooked by the heat generated by the motion of the water molecules.

$$\underset{H \qquad H}{\overset{\diagdown \ O \ \diagup}{}}$$

As a result of their orientation, both bonds contribute to an accumulation of negative charge on the oxygen atom. The two bond polarities are equal in magnitude but are not opposite in their direction.

The generalization that linear triatomic molecules are nonpolar and nonlinear triatomic molecules are polar, which is consistent with our discussion of CO_2 and H_2O molecular polarities, is not valid. The linear molecular HCN, which is polar, invalidates this generalization. Both bond polarities contribute to nitrogen's acquiring a partial negative charge relative to hydrogen in HCN.

$$\overset{\longrightarrow \ \ \longrightarrow}{H-C\equiv N}$$

(Note that the two polarity arrows were drawn in the same direction because nitrogen is more electronegative than carbon and carbon is more electronegative than hydrogen.)

Commonly encountered geometries for molecules containing four and five atoms are, respectively, trigonal planar and tetrahedral. Molecules with those geometries in which all of the atoms attached to the central atom are identical, such as SO_3 (trigonal planar) and CH_4 (tetrahedral), are *nonpolar*. The individual bond polarities cancel as the result of the highly symmetrical arrangement of atoms around the central atom (see Fig. 7.16).

If two or more kinds of atoms are attached to the central atom in a trigonal planar or tetrahedral molecule, the molecule will be polar. The high degree of symmetry required for cancellation of the individual bond polarities is no longer present. For example, if one

	Type	Cancellation of Polar Bonds	Example
Linear molecules with two identical bonds	B—A—B	←+ +→	CO_2
Trigonal planar molecules with three identical bonds	B | A ／ ＼ B B		SO_3
Tetrahedral molecules with four identical bonds	B | A ／ ＼ B B | B		CH_4

Figure 7.16

Examples of nonpolar molecules in which all bonds present are polar.

of the hydrogen atoms in CH_4 (a nonpolar molecule) is replaced by a chlorine atom, a polar molecule results, even though the resulting CH_3Cl is still a tetrahedral molecule. A carbon–chlorine bond has a greater polarity than a carbon–hydrogen bond; chlorine has an electronegativity of 3.0 and hydrogen has an electronegativity of only 2.1. Figure 7.17 contrasts the polar CH_3Cl and nonpolar CH_4 molecules. Note that the direction of polarity of the carbon–chlorine bond is opposite to that of the carbon–hydrogen bonds.

Figure 7.17

(a) Methane (CH_4) is a nonpolar tetrahedral molecule; (b) methyl chloride (CH_3Cl) is a polar tetrahedral molecule. Bond polarities cancel in the first case but not in the second one.

(a) CH_4, a nonpolar molecule   (b) CH_3Cl, a polar molecule

EXAMPLE 7.14

Predicting the Polarity of Molecules Given Their Molecular Geometry

Predict the polarity of each of the following molecules.

(a) NH_3 (trigonal pyramidal)

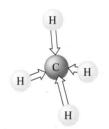

(b) H_2S (angular)

:S:
／ ＼
H   H

(c) N_2O (linear)

(d) C_2H_2 (linear)

SOLUTION

Knowledge of molecular geometry, which is given for each molecule in this example, is a prerequisite for predicting molecular polarity.

(a) Noncancellation of the individual bond polarities in the trigonal pyramidal NH_3 molecule results in it being a *polar* molecule.

Net polarity
toward nitrogen

The bond polarity arrows all point toward the nitrogen atom because nitrogen is more electronegative than hydrogen.

(b) For the bent H_2S molecule, the shift in electron density in the polar sulfur–hydrogen bonds will be toward the sulfur atom, because sulfur is more electronegative than hydrogen. The H_2S molecule as a whole is *polar* because of the noncancellation of the individual sulfur–hydrogen bond polarities.

Net polarity
toward sulfur

(c) The structure of the linear N_2O molecule is unsymmetrical; a nitrogen atom rather than the oxygen atom is the central atom. The nitrogen–nitrogen bond is nonpolar; the nitrogen–oxygen bond is polar. The molecule as a whole is *polar* because of the polarity of the nitrogen–oxygen bond.

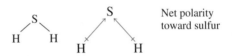

Net polarity
toward oxygen

(d) The carbon–carbon bond is nonpolar. The two carbon–hydrogen bonds are polar and are "equal but opposite" in terms of effect; that is, they cancel. The molecule as a whole is thus *nonpolar.*

$$H—C≡C—H$$

No net polarity

Practice Exercise 7.14

Predict the polarity of the following molecules.

(a) CCl_4 (tetrahedral)

(b) OF_2 (angular)

(c) H_2CO (trigonal planar)

(d) HN_3 (angular)

Summary

1. **Chemical Bonds** Chemical bonds are the attractive forces that hold atoms together in more complex units. Chemical bonds result from the transfer of electrons between atoms (ionic bond) or from the sharing of electrons between atoms (covalent bond).

2. **Valence Electrons** Valence electrons, for representative elements, are the electrons in the outermost electron shell, which is the shell with the highest shell number. Valence electrons are particularly important in determining the bonding characteristics of a given atom.

3. **Lewis Structure** A Lewis structure for an atom is a notation for designating the number of valence electrons present in the atom. It consists of an element's symbol with one dot for each valence electron placed about the elemental symbol.

4. **Octet Rule** In compound formation, atoms of elements lose, gain, or share electrons in such a way as to produce a noble gas electron configuration for each of the atoms involved.

5. **Ions** An ion is an atom (or group of atoms) that is electrically charged as the result of loss or gain of electrons. The notation for charges on ions is a superscript placed to the right of the elemental symbol. An anion is a negatively charged ion. A cation is a positively charged ion.

6. **Ionic Compound** Ionic compounds usually contain a metallic element and a nonmetallic element. Metal atoms lose one or more electrons, producing cations. Nonmetal atoms acquire the electrons lost by the metal atoms, producing anions. The oppositely charged ions attract one another, creating ionic bonds.

7. **Charge Magnitude for Ions** Metal atoms containing one, two, or three valence electrons tend to lose such electrons, producing ions of +1, +2, and +3 charge, respectively. Nonmetal atoms containing five, six, or seven valence electrons tend to gain electrons, producing ions of −3, −2, and −1 charge, respectively.

8. **Formulas for Ionic Compounds** The ratio in which positive and negative ions combine is the ratio that causes the total amount of positive and negative charges to add up to zero.

9. **Structure of Ionic Compounds** Ionic solids consist of positive and negative ions arranged in such a way that each ion is surrounded by ions of the opposite charge. Individual molecules are not present in an ionic compound.

10. **Polyatomic Ions** A polyatomic ion is a group of covalently bonded atoms that has acquired a charge through the loss or gain of electrons. Polyatomic ions are very stable entities that generally maintain their identity during chemical reactions. Polyatomic ions never occur alone as do molecules; they are a part of an ionic compound.

11. **Molecular Compound** Molecular compounds usually involve two or more nonmetals and covalent bonds. Covalent bond formation occurs through electron sharing between nonmetal atoms. The covalent bond results from the common attraction of two nuclei for the shared electrons.

12. **Bonding and Nonbonding Electron Pairs** Bonding electrons are pairs of valence electrons that are shared between atoms in a covalent bond. Nonbonding electrons are pairs of valence electrons about an atom that are not involved in electron sharing.

13. **Types of Covalent Bonds** One shared pair of electrons constitutes a single covalent bond. Two or three pairs of electrons may be shared between atoms to give double and triple covalent bonds. Most often, both atoms of the bond contribute an equal number of electrons to the bond. In a few cases, both electrons of a shared electron pair come from the same atom; this is a coordinate covalent bond.

14. **Number of Covalent Bonds Formed** There is a strong tendency for nonmetals to form a particular number of covalent bonds. The number of valence electrons the nonmetal has and the number of covalent bonds it forms gives a sum of eight.

15. **Resonance Structures** Resonance structures are two or more Lewis structures for a molecular compound that have the same arrangement of atoms, contain the same number of electrons, and differ only in the location of the electrons.

16. **Molecular Geometry and VSEPR Theory** Molecular geometry describes the way atoms in a molecule are arranged in space relative to one another. VSEPR theory is a set of procedures used to predict molecular geometry from a compound's Lewis structure. VSEPR theory is based on the concept that valence shell electron groups about an atom (bonding and nonbonding) orient themselves as far away from one another as possible (to minimize repulsions).

17. **Electronegativity** Electronegativity is a measure of the relative attraction that an atom has for the shared

Summary (*Continued*)

electrons in a bond. Electronegativity values are useful in predicting the polarity of a bond.

18. **Bond Polarity** When atoms of like electronegativity participate in a bond, the bonding electrons are equally shared and the bond is nonpolar. When atoms of differing electronegativity participate in a bond, the bonding electrons are unequally shared, and the bond is polar. In a polar bond, the more electronegative bond dominates the sharing process. The greater the

electronegativity difference between the two bonded atoms, the greater the polarity of the bond.

19. **Molecular Polarity** Molecules as a whole can have polarity. If individual bond polarities do not cancel because of the symmetrical nature of a molecule, then the molecule as a whole is polar. Molecular polarity is a measure of the degree of inequality in the attraction of bonding electrons to various locations within a molecule.

Key Terms

The new terms defined in this chapter are

anion   *Sec. 7.4*

bond length   *Sec. 7.15*

bond polarity   *Sec. 7.19*

bond strength   *Sec. 7.15*

bonding electrons   *Sec. 7.11*

cation   *Sec. 7.4*

chemical bond   *Sec. 7.1*

coordinate covalent bond   *Sec. 7.14*

covalent bond   *Sec. 7.1 and 7.10*

double covalent bond   *Sec. 7.12*

electronegativity   *Sec. 7.18*

formula unit   *Sec. 7.8*

ion   *Sec. 7.4*

ionic bond   *Sec. 7.1 and 7.4*

isoelectronic species   *Sec. 7.5*

Lewis structure   *Sec. 7.6*

Lewis symbol   *Sec. 7.2*

molecular geometry   *Sec. 7.17*

molecular polarity   *Sec. 7.20*

monoatomic ion   *Sec. 7.9*

nonbonding electrons   *Sec. 7.11*

nonpolar covalent bond   *Sec. 7.19*

nonpolar molecule   *Sec. 7.20*

octet rule   *Sec. 7.3*

percent ionic character of a bond   *Sec. 7.19*

polar covalent bond   *Sec. 7.19*

polar molecule   *Sec. 7.20*

polyatomic ion   *Sec. 7.9*

resonance structures   *Sec. 7.15*

single covalent bond   *Sec. 7.12*

triple covalent bond   *Sec. 7.12*

valence electron   *Sec. 7.2*

VSEPR electron group   *Sec. 7.17*

VSEPR theory   *Sec. 7.17*

Practice Problems

Valence Electrons (Sec. 7.2)

7.1 How many valence electrons do atoms with the following electron configurations have?

(a) $1s^2 2s^2 2p^1$

(b) $1s^2 2s^2 2p^5$

(c) $1s^2 2s^2 2p^6 3s^1$

(d) $1s^2 2s^2 2p^6 3s^2 3p^6 4s^2 3d^{10} 4p^1$

7.2 How many valence electrons do atoms with the following electron configurations have?

(a) $1s^2 2s^2 2p^3$

(b) $1s^2 2s^2 2p^6 3s^2 3p^2$

(c) $1s^2 2s^2 2p^6 3s^2 3p^6 4s^1$

(d) $1s^2 2s^2 2p^6 3s^2 3p^6 4s^2 3d^{10} 4p^2$

7.3 For each of the following pairs of representative elements, indicate whether the first listed element has (1) more valence electrons, (2) fewer valence electrons, or (3) the same number of valence electrons compared to the second listed element.

(a) Be, Li   **(b)** N, P   **(c)** F, O   **(d)** Al, Cl

7.4 For each of the following pairs of representative elements, indicate whether the first listed element has (1) more valence electrons, (2) fewer valence electrons, or (3) the same number of valence electrons compared to the second listed element.

(a) K, Ca   **(b)** S, Se   **(c)** N, B   **(d)** Si, Na

7.5 Give the periodic table group number and the number of valence electrons present for each of the following representative elements.

(a) lithium   **(b)** neon

(c) calcium   **(d)** iodine

7.6 Give the periodic table group number and the number of valence electrons present for each of the following representative elements.

(a) magnesium   **(b)** potassium

(c) phosphorus   **(d)** bromine

7.7 Identify the element with each of the following characteristics.

(a) period 2 element with four valence electrons

(b) period 2 element with seven valence electrons

(c) period 3 element with two valence electrons

(d) period 3 element with five valence electrons

7.8 Identify the element with each of the following characteristics.

(a) period 2 element with one valence electron

(b) period 2 element with three valence electrons

(c) period 3 element with three valence electrons

(d) period 3 element with six valence electrons

Lewis Structures for Atoms (Sec. 7.2)

7.9 Draw Lewis structures for atoms of the following elements.

(a) $_4$Be   **(b)** $_8$O   **(c)** $_{17}$Cl   **(d)** $_{34}$Se

7.10 Draw Lewis structures for atoms of the following elements.

(a) $_5$B   **(b)** $_9$F   **(c)** $_{11}$Na   **(d)** $_{32}$Ge

7.11 Each of the following Lewis structures represents a period 2 representative element. Determine the element's identity in each case.

(a) $\cdot \dot{X} \cdot$ **(b)** $\cdot \dot{\underset{\cdot}{X}} \cdot$ **(c)** $\cdot \dot{\underset{\cdot \cdot}{X}} :$ **(d)** $X \cdot$

7.12 Each of the following Lewis structures represents a period 3 representative element. Determine the element's identity in each case.

(a) $X \cdot$ **(b)** $\cdot \underset{\cdot \cdot}{\overset{\cdot \cdot}{X}} :$ **(c)** $\underset{\cdot \cdot}{\overset{\cdot \cdot}{X}} \cdot$ **(d)** $\cdot \dot{\underset{\cdot \cdot}{X}} \cdot$

7.13 For each of the following pairs of elements, indicate whether the Lewis symbol for the first listed element contains (1) more dots, (2) fewer dots, or (3) the same number of dots compared to the Lewis symbol for the second listed element.

(a) Li, Be   **(b)** Mg, Ca   **(c)** Al, K   **(d)** Ar, Ne

7.14 For each of the following pairs of elements, indicate whether the Lewis symbol for the first listed element contains (1) more dots, (2) fewer dots, or (3) the same number of dots compared to the Lewis symbol for the second listed element.

(a) Si, P   **(b)** O, S   **(c)** F, N   **(d)** Cs, Br

Notation for Ions (Sec. 7.4)

7.15 Give the symbol for each of the following ions.

(a) a lithium atom that has lost one electron

(b) a phosphorus atom that has gained three electrons

(c) a bromine atom that has gained one electron

(d) a barium atom that has lost two electrons

7.16 Give the symbol for each of the following ions.

(a) an iodine atom that has gained one electron

(b) a zinc atom that has lost two electrons

(c) a gallium atom that has lost three electrons

(d) a sodium atom that has lost one electron

7.17 Calculate the number of protons and electrons present in each of the following ions.

(a) Li^+ **(b)** N^{3-} **(c)** Ca^{2+} **(d)** Cl^-

7.18 Calculate the number of protons and electrons present in each of the following ions.

(a) Al^{3+} **(b)** O^{2-} **(c)** Be^{2+} **(d)** F^-

7.19 What would be the symbol for an ion with each of the following characteristics?

(a) an aluminum ion with ten electrons

(b) an oxygen ion with ten electrons

(c) a magnesium ion with two more protons than electrons

(d) a beryllium ion with two fewer electrons than protons

7.20 What would be the symbol for an ion with each of the following characteristics?

(a) a sodium ion with ten electrons

(b) a fluorine ion with ten electrons

(c) a sulfur ion with two fewer protons than electrons

(d) a calcium ion with two more protons than electrons

7.21 Classify each of the following species as (1) a cation, (2) an anion, or (3) not an ion.

(a) Cl^- **(b)** Na^+ **(c)** F   **(d)** N^{3-}

7.22 Classify each of the following species as (1) a cation, (2) an anion, or (3) not an ion.

(a) O^{2-} **(b)** Mg^{2+} **(c)** K   **(d)** F^-

The Sign and Magnitude of Ionic Charge (Sec. 7.5)

7.23 Use the periodic table to predict whether each of the following elements forms a cation or an anion.

(a) beryllium   **(b)** selenium

(c) rubidium   **(d)** strontium

7.24 Use the periodic table to predict whether each of the following elements forms a cation or an anion.

(a) nitrogen   **(b)** magnesium

(c) chlorine   **(d)** iodine

8.2   Types of Binary Ionic Compounds

The term *binary* means "two." A **binary compound** *is a compound in which only two elements are present.* The compounds NaCl, CO_2, NH_3, and P_4O_{10} are all binary compounds. Any number of atoms of the two elements may be present in a molecule or formula unit of a binary compound, but only two elements may be present. A **binary ionic compound** *is an ionic compound in which monoatomic positively charged metal ions and monoatomic negatively charged nonmetal ions are present.* A consideration of nomenclature rules for binary ionic compounds will be our entry point into the realm of chemical nomenclature.

Binary ionic compounds, which are the simplest type of ionic compound, may be divided into two categories based on the identity of the metal ion that is present.

1. *Fixed-charge* binary ionic compounds contain a metal that can form only *one type* of positive ion.
2. *Variable-charge* binary ionic compounds contain a metal that can form *more than one type* of positive ion.

In general, all metals lose electrons when forming ions. Fixed-charge metals always exhibit the same behavior in ion formation; that is, they always lose the same number of electrons. A **fixed-charge metal** *is a metal that forms only one type of positive ion, which always has the same charge magnitude.*

Variable-charge metals do not always lose the same number of electrons upon ion formation. For example, the variable-charge metal iron sometimes forms an Fe^{2+} ion and at other times an Fe^{3+} ion. A **variable-charge metal** *is a metal that forms more than one type of positive ion, with the ion types differing in charge magnitude.*

Which metals are fixed-charge metals and which metals are variable-charge metals? This information is a prerequisite for naming binary ionic compounds. Fixed-charge binary ionic compounds are named in one way, and variable-charge binary ionic compounds are named in a slightly different way.

A limited number of fixed-charge metals exist. If these are learned, then all other metals will be variable-charge metals. The fixed-charge metals are the group IA and IIA metals plus Al, Ga, Zn, Cd, and Ag. Figure 8.1 shows the periodic table positions for these 15 fixed-charge metals. The charge magnitude for these metals can be related to their position in the periodic table. Looking at Figure 8.1, we see that all group IA elements form +1 ions. The charges for ions of elements in groups IIA and IIIA are +2 and +3, respectively. Group numbers and charge also directly correlate for Zn, Cd, and Ag, the other fixed-charge metallic ions. The reason for this charge–periodic table correlation, the octet rule, was considered in Section 7.3.

> **All the inner transition metals (*f* area of the periodic table), most of the transition metals (*d* area), and a few representative metals (*p* area) exhibit variable-charge behavior (see Fig. 8.1).**

Figure 8.1

Periodic table showing the fixed-charge metallic ions. For these metals, ionic charge correlates with periodic table group number.

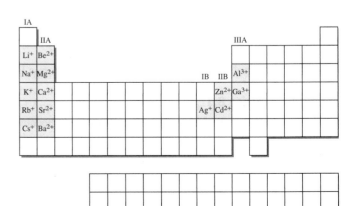

Table 8.1	Ionic Charges Associated with Ions of the More Common Variable-Charge Metals
Element	**Ions Formed**
Chromium	Cr^{2+} and Cr^{3+}
Cobalt	Co^{2+} and Co^{3+}
Copper	Cu^{+} and Cu^{2+}
Gold	Au^{+} and Au^{3+}
Iron	Fe^{2+} and Fe^{3+}
Lead	Pb^{2+} and Pb^{4+}
Manganese	Mn^{2+} and Mn^{3+}
Tin	Sn^{2+} and Sn^{4+}

The vast majority of metals, all except the 15 fixed-charge metals, are variable-charge metals. Charge magnitude for ions of such metals cannot be easily related to periodic table position. (The presence of *d* or *f* electrons in most of these metals complicates octet rule considerations in a manner beyond what we consider in this book.) Table 8.1 gives the charges for selected commonly encountered variable-charge metal ions. Note that the charges of +2 and +3 are a common combination. However, there are also other combinations: two +2 and +4 pairings (Pb and Sn), a +1 and +2 pairing (Cu), and a +1 and +3 pairing (Au) are listed in Table 8.1.

8.3 Nomenclature for Binary Ionic Compounds

The names of binary ionic compounds are based on the names of the ions of which they are composed. Ion names are obtained as follows.

1. Fixed-charge metal ions take the full name of the element plus the word "ion."

Na^{+} sodium ion
Ca^{2+} calcium ion
Al^{3+} aluminum ion

> A fixed-charge metal cation has the same name as its parent element.

2. Variable-charge metal ions take the full name of the element followed by a Roman numeral (in parentheses) that gives the ionic charge plus the word "ion."

Fe^{2+} iron(II) ion
Fe^{3+} iron(III) ion
Cu^{2+} copper(II) ion

3. Nonmetal ions take the *stem* of the name of the element followed by the suffix *-ide* plus the word "ion."

Cl^{-} chloride ion
S^{2-} sulfide ion
O^{2-} oxide ion

The stem of the name of the nonmetal is always the first few letters of the nonmetal's name, that is, the name of the nonmetal with its ending chopped off. Table 8.2 gives the stem part of the name for the more common nonmetallic ions.

Table 8.2	Names for the More Common Nonmetal Ions		
Element	**Stem**	**Name of Ion**	**Formula**
Bromine	brom-	bromide ion	Br^-
Carbon	carb-	carbide ion	C^{4-}
Chlorine	chlor-	chloride ion	Cl^-
Fluorine	fluor-	fluoride ion	F^-
Hydrogen	hydr-	hydride ion	H^-
Iodine	iod-	iodide ion	I^-
Nitrogen	nitr-	nitride ion	N^{3-}
Oxygen	ox-	oxide ion	O^{2-}
Phosphorus	phosph-	phosphide ion	P^{3-}
Sulfur	sulf-	sulfide ion	S^{2-}

Fixed-Charge Binary Ionic Compounds

Names for fixed-charge binary ionic compounds are assigned using the following rule:

The full name of the metallic element is given first, followed by a separate word consisting of the stem of the nonmetallic element name and the suffix -ide.

Thus, to name the compound NaF we start with the name of the metal (sodium), follow it with the stem of the name of the nonmetal (fluor-), and then add the suffix -ide. The name becomes *sodium fluoride.*

> **Fixed-charge binary ionic compounds contain a metal that can form only one type of cation.**

The name of the metal ion present is always exactly the same as the name of the metal itself. The metal's name is never shortened as are nonmetal names. Example 8.2 illustrates further the use of the preceding rule in naming fixed-charge binary ionic compounds.

EXAMPLE 8.2

Naming Fixed-Charge Binary Ionic Compounds

Name the following fixed-charge binary ionic compounds.

(a) KCl   **(b)** MgF_2 **(c)** Na_2O **(d)** Be_3N_2

SOLUTION

The general pattern for naming compounds of this type is

> First word: name of metal
> Second word: stem of name of nonmetal + -ide

(a) The metal is potassium, and the nonmetal is chlorine. Thus, the compound's name is potassium <u>chlor</u>ide (the stem of the nonmetal name is underlined).

(b) The metal is magnesium, and the nonmetal is fluorine. Hence, the name is magnesium <u>fluor</u>ide. Note that no mention is made of the subscript 2 found after the symbol for fluorine in the formula. *The name of an ionic compound never contains any reference to formula subscript numbers.* Since magnesium is a fixed-charge metal, there is only one ratio in which magnesium and fluorine may combine. Thus, just telling the name of the elements present in the compound is adequate nomenclature.

(c) Sodium (Na) and oxygen (O) are present in the compound. Its name is sodium oxide.

(d) This compound is named beryllium nitride.

Practice Exercise 8.2

Name the following fixed-charge binary ionic compounds.

(a) $BaCl_2$ (b) AlP (c) K_2O (d) Ca_3N_2

Variable-Charge Binary Ionic Compounds

Names for variable-charge binary ionic compounds are assigned using the following rule:

The full name of the metallic element with a Roman numeral appended to it is given first, followed by a separate word consisting of the stem of the nonmetallic element name and the suffix -ide.

Two different iron chlorides are known: one contains Fe^{2+} ions ($FeCl_2$) and the other contains Fe^{3+} ions ($FeCl_3$). The names of these two chlorides are, respectively, iron(II) chloride and iron(III) chloride. Without the use of Roman numerals (or some equivalent system), these two compounds would have the same name (iron chloride), an unacceptable situation.

In naming variable-charge binary ionic compounds, the Roman numeral to be used in the name is the metal ion charge, which can be calculated using the charge on the nonmetal ion and the principle of charge neutrality. The charge neutrality principle is that the total positive charge and the total negative charge must add to zero (Sec. 7.7). The calculation of metal ion charge (Roman numeral) using this technique, as well as the use of Roman numerals in the names of variable-charge binary ionic compounds, is illustrated in Example 8.3. As you work through Example 8.3, be sure to note that Roman numerals are *never* a part of the *formula* of a variable-charge binary ionic compound but *always* a part of the *name* of a variable-charge binary ionic compound.

> Variable-charge binary ionic compounds contain a metal that can form more than one type of cation.

> The cation name precedes the anion name in the names of *all* binary ionic compounds (both fixed-charge and variable-charge).

EXAMPLE 8.3

Naming Variable-Charge Binary Ionic Compounds

Name the following variable-charge binary ionic compounds.

(a) CuO (b) Cu_2O (c) Mn_2S_3 (d) $AuCl_3$

SOLUTION

The general pattern for naming compounds of this type is

First word: name of metal + Roman numeral
Second word: stem of name of nonmetal + -ide

(a) The metal ion charge in this compound, a quantity needed to determine the Roman numeral to be used in the name, is easily calculated using the following procedure.

$$\text{copper charge} + \text{oxygen charge} = 0$$

The oxide ion has a -2 charge (Sec. 7.5). Therefore,

$$\text{copper charge} + (-2) = 0$$

Solving, we get

$$\text{copper charge} = +2$$

Therefore, the copper ions present are Cu^{2+}, and the name of the compound is copper(II) oxide.

(b) For charge balance in this compound we have the equation

$$2(\text{copper charge}) + \text{oxygen charge} = 0$$

Note that we have to take into account the number of copper ions present, two in this case. The oxide ion carries a -2 charge (Sec. 7.5). Therefore,

$$2(\text{copper charge}) + (-2) = 0$$

$$2(\text{copper charge}) = +2$$

$$\text{copper charge} = +1$$

Here we note that we are interested in the charge on a *single* copper ion $(+1)$ and not in the total positive charge present $(+2)$. Since Cu^+ ions are present, the compound is named copper(I) oxide. As is the case for all ionic compounds, the name does not contain any reference to the numerical subscripts in the compound's formula.

(c) The charge balance equation is

$$2(\text{manganese charge}) + 3(\text{sulfur charge}) = 0$$

Substituting a charge of -2 into the equation for sulfur (Sec. 7.5), we get

$$2(\text{manganese}) + 3(-2) = 0$$

$$2(\text{manganese}) = +6$$

$$\text{manganese} = +3$$

The compound is thus named manganese(III) sulfide.

(d) This compound is gold(III) chloride. Chloride ions carry a -1 charge (Sec. 7.5). Since the compound contains three chloride ions (-3 total charge), the single gold ion must bear a $+3$ charge to counterbalance the -3 charge of the chlorides. Hence, Au^{3+} ions are present.

Practice Exercise 8.3

Name the following variable-charge binary ionic compounds.

(a) FeO   **(b)** Fe_2O_3 **(c)** PbO_2 **(d)** AuCl

An older method for indicating the charge on metal ions uses suffixes rather than Roman numerals. This system is more complicated and less precise than the Roman numeral system and fortunately is being abandoned. It is mentioned here because it is still encountered (especially on the labels of bottles of chemicals; see Fig. 8.2). In this system, when a metal has two common ionic charges, the suffix *-ous* is used for the ion of lower charge and the suffix *-ic* for the ion of higher charge. Table 8.3 compares the two systems for the metals where the old system is most often encountered. Note that in the old system the Latin names for metals are used (Sec. 4.9).

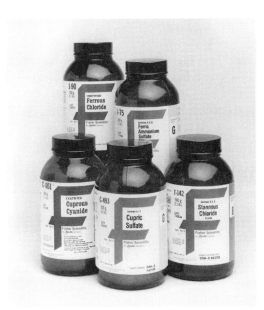

Figure 8.2

An older method of naming ionic compounds containing a variable-charge metal is still used on the labels of many laboratory chemicals. (Fisher Scientific)

The ruddy rust that forms on iron is ferric oxide (Fe_2O_3). This same compound gives red bricks and red clay sewer pipes their color. It also is the pigment of burnt sienna in artists' paints. Rouge, one of the most ancient of cosmetics, gets its color from ferric oxide as well. Barns were traditionally painted red because farmers used red ferric oxide paint as a way to preserve the wood. People still paint their barns red without knowing why.

Table 8.3	Comparison of IUPAC and Old System Names for Selected Metal Ions		

Element	Ions	Preferred Name	Old System Name
Copper	Cu^+	copper(I) ion	cuprous ion
	Cu^{2+}	copper(II) ion	cupric ion
Iron	Fe^{2+}	iron(II) ion	ferrous ion
	Fe^{3+}	iron(III) ion	ferric ion
Tin	Sn^{2+}	tin(II) ion	stannous ion
	Sn^{4+}	tin(IV) ion	stannic ion
Lead	Pb^{2+}	lead(II) ion	plumbous ion
	Pb^{4+}	lead(IV) ion	plumbic ion
Gold	Au^+	gold(I) ion	aurous ion
	Au^{3+}	gold(III) ion	auric ion

EXAMPLE 8.4

Naming Variable-Charge Binary Ionic Compounds Using the "-ic,-ous" System

Two systems exist for indicating the charge on the metal ion present in variable-charge binary ionic compounds: (1) a system that uses Roman numerals and (2) a system that uses the suffixes -ic and -ous. Change the following compound names from one system to the other system.

(a) iron(III) chloride   **(b)** copper(I) fluoride
(c) stannic bromide   **(d)** plumbous fluoride

SOLUTION

The information found in Table 8.3 is needed to effect the desired name changes.

(a) The given name indicates that Fe^{3+} ion is present. The alternative name for Fe^{3+} ion, from Table 8.3, is ferric ion. The name iron(III) chloride thus becomes ferric chloride.

(b) The alternative name for the copper(I) ion, Cu^+, is cuprous ion. Therefore, copper(I) fluoride becomes cuprous fluoride.

(c) The stannic ion, from Table 8.3, is Sn^{4+} ion; the alternative name for this ion is tin(IV). Therefore, the changed (and preferred) name for this compound is tin(IV) bromide.

(d) Since the plumbous ion and Pb^{2+} ion are equivalent (Table 8.3), the names plumbous fluoride and lead(II) fluoride are equivalent.

Practice Exercise 8.4

Change each of the following variable-charge binary ionic compound names from the "Roman numeral" to the "-ic, -ous" system, or vice versa.

(a) copper(II) chloride   **(b)** tin(II) fluoride

(c) ferrous bromide   **(d)** auric oxide

The overall strategy for naming binary ionic compounds, as discussed in this section, is summarized in the "nomenclature decision tree" of Figure 8.3.

8.4   Nomenclature for Ionic Compounds Containing Polyatomic Ions

In Section 7.9 the topic of polyatomic ions was considered, and at that time a limited number of such ions were introduced in the context of formula writing. We now consider some additional facts about polyatomic ions.

Hundreds of different polyatomic ions exist. Table 8.4 gives the names and formulas for the most common ones. Note that almost all the polyatomic ions listed in Table 8.4 contain

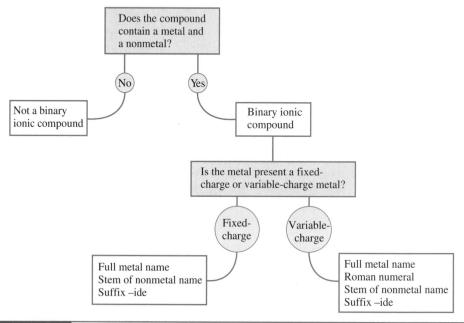

Figure 8.3

A "nomenclature decision tree" for naming binary ionic compounds.

oxygen atoms. The names, but not necessarily the formulas, of some of these common poly-atomic ions should be familiar to you. Many of these ions are found in commercial products. Examples are fertilizers (phosphates, sulfates, nitrates), baking soda and baking powder (bicarbonates), and building materials (carbonates, sulfates).

There is no easy way to learn the formulas and names for all the common polyatomic ions. Memorization is required. The charges and formulas for the various polyatomic ions cannot be related easily to the periodic table as was the case for many of the monoatomic ions. In Table 8.4 the most frequently encountered polyatomic ions are in color. Their formulas and names should definitely be memorized. Your instructor may want you to memorize others, too. The inability to recognize the presence of polyatomic ions (both by name and by formula) in a compound is a major stumbling block for many chemistry students. It requires some effort to overcome this obstacle.

> The green corrosion on bronze statues is $Cu_2(OH)_2CO_3$, a substance that contains two different types of polyatomic ions. It forms from copper reacting with water, oxygen, and carbon dioxide in the air.

Table 8.4 Formulas and Names of Some Common Polyatomic Ions

Key Element Present	Formula	Name of Ion
Nitrogen	NO_3^-	nitrate ion
	NO_2^-	nitrite ion
	NH_4^+	ammonium ion
	N_3^-	azide ion
Sulfur	SO_4^{2-}	sulfate ion
	HSO_4^-	hydrogen sulfate (bisulfate ion)**
	SO_3^{2-}	sulfite ion
	HSO_3^-	hydrogen sulfite (bisulfite ion)**
	$S_2O_3^{2-}$	thiosulfate ion
Phosphorus	PO_4^{3-}	phosphate ion
	HPO_4^{2-}	hydrogen phosphate ion
	$H_2PO_4^-$	dihydrogen phosphate ion
	PO_3^{3-}	phosphite ion
Carbon	CO_3^{2-}	carbonate ion
	HCO_3^-	hydrogen carbonate (bicarbonate ion)**
	$C_2O_4^{2-}$	oxalate ion
	$C_2H_3O_2^-$	acetate ion
	CN^-	cyanide ion
	OCN^-	cyanate ion
	SCN^-	thiocyanate ion
Chlorine	ClO_4^-	perchlorate ion
	ClO_3^-	chlorate ion
	ClO_2^-	chlorite ion
	ClO^-	hypochlorite ion
Oxygen	O_2^{2-}	peroxide ion
Boron	BO_3^{3-}	borate ion
Hydrogen	H_3O^+	hydronium ion*
	OH^-	hydroxide ion
Metals	MnO_4^-	permanganate ion
	CrO_4^{2-}	chromate ion
	$Cr_2O_7^{2-}$	dichromate ion

> Polyatomic-ion-containing compounds found in the home:
>
> $NaHCO_3$ Baking soda
> $MgSO_4$ Epsom salt
> Na_3PO_4 Cleaning agent
> $Al(OH)_3$ Antacid
> $NaOH$ Drain cleaner
> $MgCO_3$ Plaster of Paris
> $NaClO$ Clorox
> $CaCO_3$ Chalk

*This ion is encountered only in aqueous solution.

**These alternate names are often encountered.

The following interrelationships among polyatomic ion names and formulas should be helpful as you become more familiar with the information in Table 8.4.

1. Most of the ions have a negative charge, which can vary from -1 to -3. Only two positive ions are listed in the table: NH_4^+ (ammonium) and H_3O^+ (hydronium).

2. Four of the polyatomic ions have names ending in -ide: OH^- (hydroxide), CN^- (cyanide), N_3^- (azide), and O_2^{2-} (peroxide). These names represent exceptions to the rule that the suffix -ide be reserved for use in naming monoatomic ions.

3. A number of -ate, -ite pairs of ions exist, for example, SO_4^{2-} (sulfate) and SO_3^{2-} (sulfite). The ion in the pair with the higher number of oxygens is always the *-ate* ion. The *-ite* ion always contains one fewer oxygen than the *-ate* ion.

4. A number of pairs of ions exist where one member of the pair differs from the other by having a hydrogen atom present, for example, CO_3^{2-} (carbonate) and HCO_3^- (hydrogen carbonate or bicarbonate). In such pairs, the charge on the hydrogen-containing ion is always one less than the charge on the other ion.

5. Two pairs of ions exist in which the difference between pair members is that a sulfur atom has replaced an oxygen atom in one member of the pair: $(SO_4^{2-}, S_2O_3^{2-})$ and (OCN^-, SCN^-). The prefix thio- is used to denote this replacement of oxygen by sulfur. The names of these pairs of ions, respectively, are sulfate–thiosulfate and cyanate–thiocyanate.

> The prefix *bi-* in polyatomic ion names means *hydrogen* rather than the number *two*.

The names of ionic compounds containing polyatomic ions are derived in the same way as those of binary ionic compounds (Sec. 8.3). Recall that the rule for naming binary ionic compounds is: Give the name of the metallic element first (including, when needed, a Roman numeral indicating ion charge), and then as a separate word give the stem of the nonmetallic element name to which the suffix -ide is appended.

For our present situation, *if the polyatomic ion is positive, its name is substituted for that of the metal. If the polyatomic ion is negative, its name is substituted for the nonmetal stem plus -ide.* In the case where both positive and negative ions are polyatomic, dual substitution occurs and the resulting name includes just the names of the polyatomic ions. Example 8.5 illustrates the use of these rules.

> Note that the word "ion" is never used in the name of an ionic compound. Ions are charged particles. When cations and anions combine to form an ionic compound, which is neutral, the word ion is no longer applicable.

EXAMPLE 8.5

Naming Polyatomic-Ion-Containing Compounds

Name the following polyatomic-ion-containing compounds.

(a) Na_3PO_4 (b) $Fe(NO_3)_3$ (c) Cu_2SO_4 (d) NH_4CN

SOLUTION

(a) The positive ion present is the sodium ion (Na^+). The negative ion is the polyatomic phosphate ion (PO_4^{3-}). The name of the compound is sodium phosphate. No Roman numeral is needed in the name since sodium is a fixed-charge metal. As in naming binary ionic compounds (Sec. 8.3), subscripts in the formula are not incorporated into the name.

(b) The positive ion present is iron, and the negative ion is the nitrate ion (NO_3^-). Since iron is a variable-charge metal, a Roman numeral must be used to indicate ionic charge. In this case, the Roman numeral is III. The fact that iron ions carry a $+3$ is deduced by noting that there are three nitrate ions present, each of which carries a -1 charge. The charge on the single iron ion present must be a $+3$ in order to counterbalance the total negative charge of -3. The name of the compound, therefore, is iron(III) nitrate.

(c) The positive ion present is Cu(I) ion. The negative ion is the polyatomic sulfate ion (SO_4^{2-}). The name of the compound is copper(I) sulfate. The determination that copper (a variable-charge metal) is present as copper(I) involves the following calculation dealing with charge balance.

$$2(\text{copper charge}) + (\text{sulfate charge}) = 0$$
$$2(\text{copper charge}) + (-2) = 0$$
$$2(\text{copper charge}) = +2$$
$$\text{copper charge} = +1$$

(d) Both the positive and negative ions in this compound are polyatomic—the ammonium ion (NH_4^+) and the cyanide ion (CN^-). The name of the compound is simply the combination of the names of the two polyatomic ions: ammonium cyanide.

Practice Exercise 8.5

Name the following polyatomic-ion-containing compounds.

(a) K_2CO_3 **(b)** $Co(NO_3)_2$ **(c)** $Fe_2(SO_4)_3$ **(d)** $(NH_4)_3PO_4$

Polyatomic-ion-containing compounds are not binary ionic compounds; more than two elements are present. Most are *ternary compounds*. A **ternary compound** *is a compound containing three different elements*. Compounds containing more than three elements do exist, and we will encounter a few [such as ammonium sulfate, $(NH_4)_2SO_4$], but we will not worry about specific classification terms for such compounds. Most compounds considered in this text are either binary compounds or ternary compounds.

The overall strategy for naming ionic compounds—both binary (Sec. 8.3) and polyatomic-ion-containing (Sec. 8.4)—is summarized in the "nomenclature decision tree" of Figure 8.4.

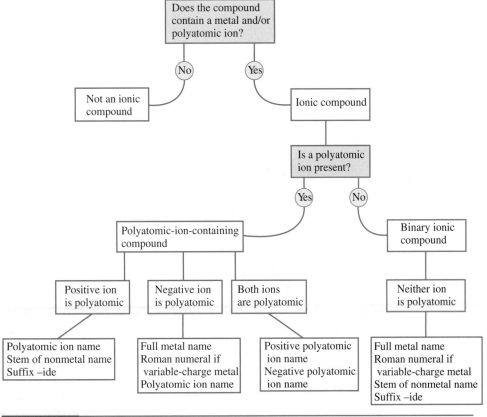

Figure 8.4

A "nomenclature decision tree" for naming binary and polyatomic-ion-containing ionic compounds.

8.5 Nomenclature for Binary Molecular Compounds

A **binary molecular compound** *is a molecular compound in which only two nonmetallic elements are present.* Such compounds are named in the following manner. The two nonmetals present are named in the order in which they appear in the formula. [The least electronegative nonmetal (Sec. 7.18) is usually written first in the formula.] *The name of the first nonmetal is used in full, followed by a separate word containing the stem of the name of the second nonmetal and the suffix -ide. Numerical prefixes, giving numbers of atoms present, precede the names of both nonmetals.* The use of these prefixes is in direct contrast to the procedures used for naming ionic compounds. For ionic compounds, formula subscripts are not mentioned in the name.

Prefixes are needed in naming binary molecular compounds because numerous different compounds exist for many pairs of nonmetallic elements. For example, all of the following nitrogen–oxygen molecular compounds exist: NO, NO_2, N_2O, N_2O_3, N_2O_4, and N_2O_5. The prefixes used are the standard Greek numerical prefixes, which are given in Table 8.5 for the numbers 1 through 10, and which are used in words such as *mono*nucleosis, *di*chromatic, *tri*angle, *tetra*gram, *penta*gon, *hexa*pod, *hepta*thlon, *octa*ve, *ennea*hedron, and *dec*ade. Example 8.6 shows how these prefixes are used in naming binary molecular compounds.

Parallels and differences between naming binary molecular compounds and binary ionic compounds are:

1. For binary molecular compounds, the *less electronegative element* is named first (just as with the metal in a binary ionic compound); in both cases the element's name is used in unmodified form.
2. For binary molecular compounds, the *more electronegative element* is named second and takes the suffix *-ide* (just as with the nonmetal in a binary ionic compound).
3. For binary molecular compounds, a numerical prefix indicates the number of atoms of each nonmetal present (an opposite situation to that for binary ionic compounds, where numerical prefixes are never used).

> You cannot predict the formulas of most molecular compounds in the same way that you can predict the formulas of ionic compounds. That is why we name them using prefixes that explicitly indicate their composition.

Table 8.5	Greek Numerical Prefixes from 1 to 10
Greek Prefix	**Number**
Mono-	1
Di-	2
Tri-	3
Tetra-	4
Penta-	5
Hexa-	6
Hepta-	7
Octa-	8
Ennea-*	9
Deca-	10

The prefix ennea- is preferred to the Latin nona- by IUPAC, but nona- is still frequently used.

EXAMPLE 8.6

Naming Binary Molecular Compounds

Name the following binary molecular compounds.

(a) N_2O_5　　(b) PF_3　　(c) S_4N_4　　(d) $SiCl_4$

SOLUTION

The name of each of these compounds will consist of two words with the following general formats.

> First word:　numerical prefix $+$ full name of the first nonmetal
> Second word:　numerical prefix $+$ stem of name of second nonmetal $+$ -ide

(a) The elements present are nitrogen and oxygen. The two portions of the name before adding Greek numerical prefixes are *nitrogen* and *oxide*. Adding the prefixes gives *dinitrogen* (two nitrogen atoms are present) and *pentoxide* (five oxygen atoms are present). (When an element name begins with an *a* or *o*, the *a* or *o* at the end of the Greek prefix is dropped for ease of pronunciation—pentoxide instead of pentaoxide.) The name of this compound is dinitrogen pentoxide.

(b) When there is only one atom of the first nonmetal present, it is standard procedure to omit the prefix *mono-* for the element. Following this guideline, we have for the name of the compound phosphorus trifluoride. The prefix *mono-* is never used in the first position in the name of a binary molecular compound.

(c) The prefix for four atoms is *tetra-*. This compound is, therefore, tetrasulfur tetranitride.

(d) Omitting the initial *mono-* (see part b), we name this compound silicon tetrachloride.

Practice Exercise 8.6

Name the following binary molecular compounds.

(a) S_2O　　(b) NCl_3　　(c) P_4S_6　　(d) CF_4

There is one exception to the use of Greek numerical prefixes in naming binary molecular compounds. Binary compounds with hydrogen listed as the first element in the formula are named without prefix use. Thus, the compounds H_2S and HCl are named hydrogen sulfide and hydrogen chloride, respectively.

All procedures for naming compounds so far presented in this chapter have been based on IUPAC rules (Sec. 8.1) and have produced *systematic names*. A **systematic name** *is a name for a compound based on IUPAC rules that conveys information about the composition of the compound.* A few binary molecular compounds have names completely unrelated to the IUPAC rules for naming such compounds. They have "common" or "trivial" names, which were coined before the development of the systematic rules. A **common name** *is a name for a compound not based on IUPAC rules that does not convey information about the composition of the compound.* At one time, in the early history of chemistry, all compounds had common names. With the advent of systematic nomenclature, most common names were discontinued. A few, however, have persisted and are now officially accepted. The most "famous" example of this is the compound H_2O, which has the systematic name hydrogen oxide. This name is never used; H_2O is known

> **Numerical prefixes are not used in naming molecular compounds that have hydrogen as the first element in the chemical formula, such as HCl, HBr, and H_2S.**

Table 8.6	Some Binary Molecular Compounds That Have Common Names	
Compound Formula	**Accepted Common Name**	
H_2O	water	
H_2O_2	hydrogen peroxide	
NH_3	ammonia	
N_2H_4	hydrazine	
CH_4	methane	
C_2H_6	ethane	
PH_3	phosphine	
AsH_3	arsine	

as *water*, a name that is not going to be changed. Another very common example of common nomenclature involves the compound NH_3, which is *ammonia*. Table 8.6 gives additional examples of compounds for which common names are used in preference to systematic names. Such exceptions are actually very few compared to the total number of compounds named by systematic rules.

The "nomenclature decision tree" in Figure 8.5 contrasts the strategy used in naming binary molecular compounds and binary ionic compounds.

Writing formulas for binary molecular compounds given their names is a very easy task. The Greek prefixes in the names of such compounds tell you exactly how many atoms of each kind are present. For example, the compounds dinitrogen pentoxide and carbon dioxide have, respectively, the formulas N_2O_5 (di- and penta-) and CO_2 (mono- and di-). Remember, the prefix mono- is always dropped at the start of a name, as in carbon dioxide, as was noted in Example 8.5.

Figure 8.5

A "nomenclature decision tree" for naming binary molecular compounds and binary ionic compounds.

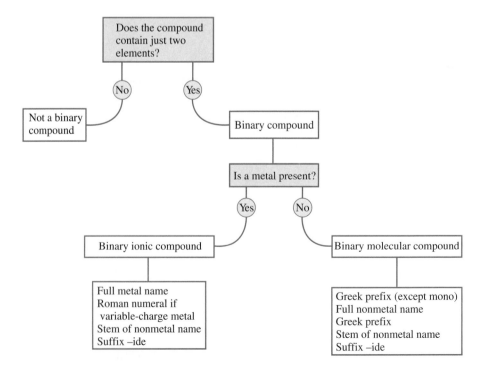

8.6 Nomenclature for Acids

The rules for naming binary molecular compounds do not carry over to naming ternary molecular compounds in a manner similar to that in ionic compound nomenclature. Different sets of new rules exist for naming various "families" of ternary molecular compounds. We will consider in this section one set of such rules—those for naming ternary *acids*, the most commonly encountered family of ternary molecular compounds.

Many hydrogen-containing molecular compounds dissolve in water to give solutions with properties markedly different from those of the compounds that were dissolved. These solutions, which we will discuss in detail in Chapter 14, are called *acids*.

Not all hydrogen-containing molecular compounds are acids. Those that are can be recognized by their formulas, which have hydrogen written as the first element in the formula.

$$\text{Acids:} \quad HCl, H_2S, H_2SO_4, HNO_3$$
$$\text{Nonacids:} \quad NH_3, CH_4, PH_3, SiH_4$$

Water, which is the solvent for most acidic solutions, is generally not considered to be an acid. Despite this, H is written first in its formula (H_2O).

Acids are molecular compounds (no ions are present). When acids are dissolved in water, the interaction with the water produces ions. The acid molecules can be considered to break apart to form ions in the presence of water. (No ion formation occurs if the water is not present.)

An **acid** *is a hydrogen-containing molecular compound whose molecules yield hydrogen ions* (H^+) *when dissolved in water*. The production of the positive H^+ ion, which is the species that gives acids their characteristic properties, is common to all acids. The negative ion produced varies from acid to acid. For example,

$$HCl, \text{ in water, produces } H^+ \text{ and } Cl^- \text{ ions.}$$
$$HNO_3, \text{ in water, produces } H^+ \text{ and } NO_3^- \text{ ions.}$$
$$HCN, \text{ in water, produces } H^+ \text{ and } CN^- \text{ ions.}$$

Because of differences in the properties of acids and the anhydrous (without water) compounds from which they are produced, the acids and the anhydrous compounds are given different names. Acid nomenclature is derived from the names of the parent anhydrous compounds (which we considered in Sec. 8.5).

The names of acids are derived from the names of the *negative ions* produced from the acids' interaction with water. There are three nomenclature rules based on whether the name of the negative ion ends in *-ide*, *-ate*, or *-ite*.

1. *-ide rule*: When the name of the negative ion produced from the acid ends in *-ide* the acid name has four parts:
   1. the prefix *hydro-*
   2. the stem of the name of the negative ion
   3. the suffix *-ic*
   4. the word *acid*

   Examples of the application of this rule are

   HCl (chloride) is named *hydro*chlor*ic acid*.

   HCN (cyanide) is named *hydro*cyan*ic acid*.

Acids, when dissolved in water, have a sour taste. (The word acid comes from the Latin word *acidus*, which means "sour.") Acids cause the dye litmus to change from blue to red. (Litmus is a naturally occurring vegetable dye obtained from lichens.) When certain metals, such as zinc and iron, are placed in acids, they dissolve, liberating hydrogen gas.

Although it was done in earlier times, the tasting of compounds (such as acids) as part of the identification process is not generally considered a good idea these days.

Not all acids are ternary molecular compounds. The acids HCl and HBr (hydrochloric acid and hydrobromic acid) are examples of binary molecular compounds.

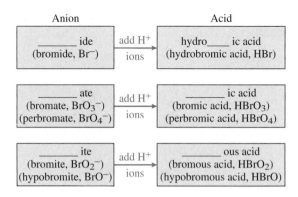

Figure 8.6

Summary of the ways in which anion names and acid names are related.

2. *-ate rule*: When the name of the negative ion produced from the acid ends in *-ate* the acid name has three parts:
   1. the name of the negative ion less the *-ate* ending
   2. the suffix *-ic*
   3. the word *acid*

   Examples of the application of this rule are

   HNO_3 (nitrate) is named nitr*ic acid*.

   $HClO_4$ (perchlorate) is named perchlor*ic acid*.

3. *-ite rule*: When the name of the negative ion produced from the acid ends in *-ite* the acid name has three parts:
   1. the name of the negative ion less the *-ite* ending
   2. the suffix *-ous*
   3. the word *acid*

   Examples of the application of this rule are

   HNO_2 (nitrite) is named nitr*ous acid*.

   $HClO$ (hypochlorite) is named hypochlor*ous acid*.

Figure 8.6 summarizes the three ways in which anion names and acid names are related, and Examples 8.7 and 8.8 illustrate the use of these relationships in naming acids.

EXAMPLE 8.7

Naming Hydrogen-Containing Compounds as Acids

Name the following compounds as acids.

(a) H_2CO_3 (b) HF   (c) $HClO_2$ (d) H_2SO_4

SOLUTION

(a) The negative ion present in solutions of this acid is the carbonate ion (CO_3^{2-}). Removing the *-ate* ending from the word carbonate and replacing it with the suffix *-ic* (rule 2) and then adding the word *acid* gives the name *carbonic acid*.

(b) The negative ion present in solutions of this acid is the fluoride ion (F^-). This is a rule 1 (*-ide*) situation. The name of the acid is formed from using the prefix *hydro-*, the suffix *-ic*, and the word *acid*. The acid name is *hydrofluoric acid*.

(c) The negative ion present in solutions of this acid is the chlorite ion (ClO_2^-). (If you are unsure of the identity of a particular negative ion, you can consult Table 8.4.) This is a rule 3 (-ite) situation. Removing the *-ite* ending from the word chlorite and replacing it with the suffix *-ous* and then adding the word *acid* produces the name *chlorous acid.*

(d) The negative ion present in solutions of this acid is the sulfate ion (SO_4^{2-}). Changing the *-ate* ending to *-ic* (rule 2) and adding the word acid gives the name *sulfuric acid.* (For acids involving sulfur, "ur" from sulfur is reinserted into the acid name for phonetic reasons.)

Practice Exercise 8.7

Name the following compounds as acids.

(a) H_2SO_3 **(b)** HBr   **(c)** H_3PO_4 **(d)** $HC_2H_3O_2$

EXAMPLE 8.8

Writing Chemical Formulas for Acids Given Their Names

Give the chemical formulas of each of the following acids.

(a) chlorous acid   **(b)** hydrocyanic acid

SOLUTION

In this problem we will go through a reasoning process that is the reverse of that in Example 8.7. Students often find this "reverse" process to be more challenging than the "forward" process.

(a) The *-ous* ending in the acid's name indicates that the negative ion it forms in solution is an *-ite* ion. It is the *chlorite* ion. From Table 8.4, the chlorite ion has the formula ClO_2^-. The hydrogen present in the acid, for formula-writing purposes, is considered to be H^+ ion. Combining these two ions, H^+ and ClO_2^-, in a one-to-one ratio will produce the neutral compound $HClO_2$ (chlorous acid).

(b) Since this is a hydro_____ic acid, the negative ion present in acid solution will be an *-ide* ion. It is the cyanide ion, whose formula is CN^- (Table 8.4). Combining the H^+ and CN^- ions in the ratio that produces neutrality, one to one, gives the formula HCN for *hydrocyanic acid.*

Practice Exercise 8.8

Give the chemical formulas of each of the following acids.

(a) phosphorous acid   **(b)** perchloric acid

The overall strategy for identifying and naming acids is summarized in the "nomenclature decision tree" of Figure 8.7.

Acids are often subclassified into the categories *nonoxyacid* and *oxyacid*. A **nonoxyacid** *is a molecular compound composed of hydrogen and one or more nonmetals other than oxygen that produce* H^+ *ions in aqueous solution.* All common nonoxyacids, with one exception, HCN, are binary compounds. An **oxyacid** *is a molecular compound composed of hydrogen, oxygen, and one or more other elements that produces* H^+ *ions in aqueous solution.*

Figure 8.7

A "nomenclature decision tree" for naming acids.

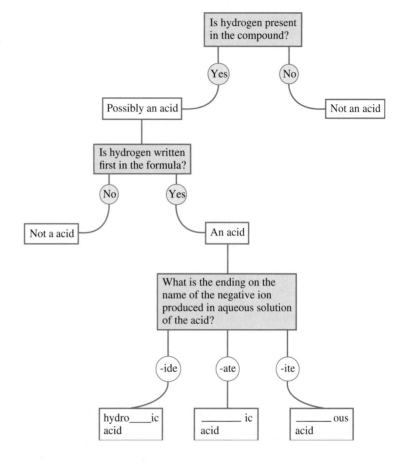

Nonoxyacids, but not oxyacids, exist as both pure compounds and water solutions. Conventions for distinguishing between the two situations are as follows, using HCl (a gas in the pure state) as the example.

HCl in the pure state: HCl(g) hydrogen chloride

HCl in aqueous solution: HCl(aq) hydrochloric acid

Table 8.7 gives other examples of this dual naming system for nonoxyacids. Since oxyacids are encountered only as aqueous solutions, a dual naming system is not needed for them.

Table 8.7	The Dual Naming System for Molecular Compounds Containing Hydrogen and a Nonmetal Other Than Oxygen	
Formula	**Name of Pure Compound**	**Name of Water Solution**
HF	hydrogen fluoride	hydrofluoric acid
HBr	hydrogen bromide	hydrobromic acid
HI	hydrogen iodide	hydroiodic acid
H_2S	hydrogen sulfide	hydrosulfuric acid*

*For acids involving sulfur, "ur" from sulfur is reinserted in the acid name for pronunciation reasons.

For several nonmetals a series of oxyacids exist. The formulas of the members of the series differ from each other only in oxygen content. For example, there are four chlorine-containing oxyacids whose formulas are

$$HClO_4, HClO_3, HClO_2, \text{ and } HClO$$

In such a series of oxyacids, one of the acids is the "-ic" acid, the acid whose name has no prefix and ends in -ic. The names of the other acids in the series are related, in a constant manner, to the "-ic" acid by prefixes and suffixes as follows.

1. An oxyacid containing *one less oxygen atom* than the "-ic" acid is the "-ous" acid.
2. An oxyacid containing *two less oxygen atoms* than the "-ic" acid is the "hypo-____-ous" acid.
3. An oxyacid containing *one more oxygen atom* than the "-ic" acid is the "per-____-ic" acid.

These relationships between names for a series of oxyacids can be summarized as follows, where n represents the number of oxygen atoms in the "-ic" acid.

Number of Oxygen Atoms	Name
$n + 1$	per____ic acid
n	____ic acid
$n - 1$	____ous acid
$n - 2$	hypo____ous acid

Most nonmetals do not form a "complete" series of four simple oxyacids as does chlorine. Bromine is the only other nonmetal for which such a series of oxyacids exists. For both sulfur and nitrogen only the "-ic" and "-ous" acids are known. For both phosphorus and iodine there are three oxyacids; there is no $n + 1$ acid (per____ic acid) for phosphorus and no $n - 1$ acid (____ous acid) for iodine. Hypofluorous acid, HOF, is the only known fluorine-containing oxyacid.

8.7 Nomenclature Rules—A Summary

Within this chapter various sets of nomenclature rules have been presented. Students sometimes have problems deciding which set of rules to use in a given situation. This dilemma can be avoided if, when confronted with the request to name a compound, the following reasoning pattern is used in this order.

1. Decide, first, whether the compound is ionic or molecular. If a metal or polyatomic ion is present, the compound is ionic. If not, the compound is molecular.
2. If the compound is ionic, then classify it as binary ionic or polyatomic-ion-containing ionic and use the rules appropriate for that classification.
3. If the compound is molecular, classify it as an acid or nonacid. An acid must have the element hydrogen present (written first in the chemical formula), and the compound must be in water solution. If the compound is a nonacid, name it according to the rules for binary molecular compounds.

Figure 8.8 portrays the preceding information in the format of a "nomenclature decision tree."

Figure 8.8

A "nomenclature decision tree" for deciding the classification of a compound that is to be named.

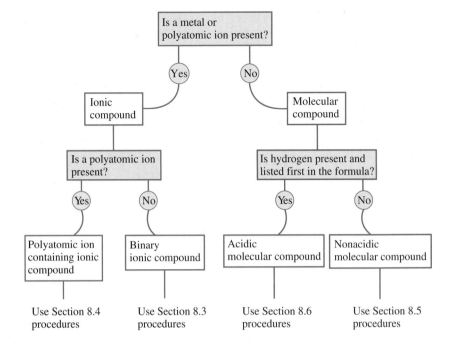

EXAMPLE 8.9

General Exercise in Chemical Nomenclature

Name the following compounds.

(a) N_2O_4 (b) $Cu(NO_3)_2$ (c) $H_2S(aq)$ (d) Na_3N

SOLUTION

(a) Neither a metal nor a polyatomic ion is present. Thus, the compound is not ionic. If it is not ionic, it must be molecular. Since no hydrogen is present, the compound is a nonacidic molecular compound. Such compounds are named using Greek numerical prefixes to denote the number of atoms present. The name of the compound is dinitrogen tetroxide.

(b) This compound meets both tests for an ionic compound. Both a metal and a polyatomic ion are present. The metal present is a variable-charge metal, which means the name of the compound will have a Roman numeral in it. The metal ion present is copper(II) and the polyatomic ion is the nitrate ion. The name of the compound is copper(II) nitrate.

(c) This compound does not meet either test for an ionic compound. It is thus a molecular compound. Hydrogen is present and it is listed first in the formula. We have an acid. The (aq) means the acid is in solution rather than the pure state. The negative ion produced from the acid is sulf*ide* ion. Using the *-ide* rule for naming acids, the name of the acid becomes hydrosulfuric acid.

(d) A metal is present but no polyatomic ion is present. Thus, we have a binary ionic compound. The metal present is a fixed-charge metal, meaning that no Roman numeral is needed in the name. The name of the compound is sodium nitride. Greek numerical prefixes are never used in naming ionic compounds. (The name *trisodium nitride* would be wrong.)

Practice Exercise 8.9

Name the following compounds.

(a) $(NH_4)_2S$ (b) $AlCl_3$ (c) P_4O_6 (d) HNO_2

Example 8.9, as well as most of the other examples in this chapter, have involved naming compounds given their formulas. It is important to be able to do the reverse of this also—to write chemical formulas when given chemical names. Often students who can go "forward" without difficulty have problems with this "reverse" process. Example 8.10 involves writing chemical formulas given the names of the compounds.

EXAMPLE 8.10

General Exercise in Using Chemical Names to Obtain Chemical Formulas

Write chemical formulas for the following compounds.

(a) magnesium sulfide   (b) beryllium hydroxide
(c) copper(II) carbonate   (d) heptasulfur dioxide

SOLUTION

(a) Magnesium is a metal, and sulfur is a nonmetal. We have a binary ionic compound. The lack of a Roman numeral in the name indicates that the metal is a fixed-charge metal. Magnesium forms +2 ions (Fig. 8.1). The sulfide ion has a charge of -2 (Sec. 7.5). Combining the ions Mg^{2+} and S^{2-} in a one-to-one ratio, the ratio that will cause their charges to add to zero, gives the formula MgS.

(b) This is also an ionic compound. Both a metal (beryllium) and a polyatomic ion (hydroxide) are present. Beryllium is a fixed-charge metal, forming ions with a +2 charge (Sec. 7.5). The formula of the hydroxide ion is OH^- (Table 8.4). Combining these ions, Be^{2+} and OH^-, in the ratio that causes their charges to add to zero gives the formula $Be(OH)_2$. Remember that any time more than one polyatomic ion is needed in a formula, parentheses must be used. (Review Examples 7.5 and 7.6 if you are still having trouble determining the ratio in which ions combine.)

(c) This is another ionic compound containing a polyatomic ion. The Roman numeral associated with the word copper indicates that the copper ions present are Cu^{2+} ions. The Roman numeral gives the charge on the ions. The formula of the polyatomic carbonate ion is CO_3^{2-} (Table 8.4). Combining Cu^{2+} and CO_3^{2-} ions in a one-to-one ratio gives the formula $CuCO_3$. Note, again, that a Roman numeral in the name of a compound never becomes part of the formula of the compound. The Roman numeral's purpose is to give information about the ionic charge of a variable-charge metal.

(d) The presence of only nonmetals and the presence of Greek numerical prefixes both indicate that this is a molecular compound. The Greek numerical prefixes give directly the number of each type of atom present. The chemical formula is S_7O_2.

Practice Exercise 8.10

Write chemical formulas for the following compounds.

(a) calcium perchlorate   (b) iron(II) nitrate
(c) lithium oxide   (d) sulfur trioxide

Summary

1. **Chemical Nomenclature** Chemical nomenclature is the system of names used to distinguish compounds from each other and the rules needed to devise these names.

2. **IUPAC Nomenclature Rules** IUPAC nomenclature rules are a systematic set of rules for naming compounds based on the premise that a compound's name should convey information about the composition of the compound.

3. **Binary Ionic Compounds** Binary ionic compounds contain positively charged monoatomic metal ions and negatively charged monoatomic nonmetal ions. They are divided into two categories: those containing *fixed-charge* metal ions and those containing *variable-charge* metal ions. Binary ionic compounds are named by giving the full name of the metallic element first, followed by a separate word containing the stem of the nonmetallic element name and the suffix *-ide*. A Roman numeral (in parentheses) specifying ionic charge is appended to the name of the metallic element if it is variable-charge metal.

4. **Polyatomic-Ion-Containing Ionic Compounds** Poly-atomic-ion-containing ionic compounds are named in a way similar to binary ionic compounds, with the following modifications. If the polyatomic ion is positive, its name is substituted for that of the metal. If the polyatomic ion is negative, its name is substituted for the nonmetal stem plus *-ide*. When both positive and negative ions are polyatomic, dual substitution occurs and the resulting name includes just the names of the polyatomic ions.

5. **Binary Molecular Compounds** Binary molecular compounds are molecular compounds in which just two nonmetallic elements are present. The two nonmetals are named in the order they occur in the formula of the compound. The name of the first nonmetal is used in full. The name of the second nonmetal is treated as was the non-metal in binary ionic compounds; the stem of the name is given and the suffix *-ide* is added. Numerical prefixes, giving numbers of atoms, precede the names of both nonmetals. A few binary molecular compounds have common names—names not based on systematic rules.

6. **Acids** An acid is a hydrogen-containing molecular compound whose molecules yield H^+ ions when dissolved in water. Acids are subclassified into the categories nonoxyacids and oxyacids. Nonoxyacids have names that contain the prefix *hydro-* and suffix *-ic acid*. Oxyacids that produce a negative ion in solution that has a name ending in *-ate* are named *-ic acids*. Oxyacids that produce a negative ion in solution that has a name ending in *-ite* are named *-ous acids*.

Key Terms

The new terms defined in this chapter are

acid   *Sec. 8.6*
binary compound   *Sec. 8.2*
binary ionic compound   *Sec. 8.2*
binary molecular compound   *Sec. 8.5*
chemical nomenclature   *Sec. 8.1*
common name   *Sec. 8.5*

fixed-charge metal   *Sec. 8.2*
nonoxyacid   *Sec. 8.6*
oxyacid   *Sec. 8.6*
systematic name   *Sec. 8.5*
ternary compound   *Sec. 8.4*
variable-charge metal   *Sec. 8.2*

Practice Problems

Nomenclature Classifications for Compounds (Sec. 8.1)

8.1 In which of the following pairs of compounds are both members of the pair binary compounds?

(a) NO and NO_2 **(b)** H_2O and $HClO$
(c) HNO_3 and $HClO_3$ **(d)** N_4S_4 and P_4O_{10}

8.2 In which of the following pairs of compounds are both members of the pair not binary compounds?

(a) H_2O and H_2O_2 **(b)** $KClO_2$ and $KClO_3$
(c) HCN and KCN **(d)** $NaBr$ and $NaBrO$

8.3 In which of the following pairs of compounds are both members of the pair ionic compounds?

(a) KNO_3 and K_3N **(b)** $NaCl$ and HCl
(c) SO_2 and S_2O **(d)** K_3N and K_2O

8.4 In which of the following pairs of compounds are both members of the pair molecular compounds?

(a) BrF and BrCl   **(b)** $AlBr_3$ and NaBr

(c) P_4O_6 and HNO_3 **(d)** H_2CO_3 and CO

8.5 In which of the following pairs of compounds is one member of the pair an ionic compound and the other member of the pair a molecular compound?

(a) AlN and HF   **(b)** $CuCO_3$ and H_2CO_3

(c) SO_3 and H_2SO_4 **(d)** NaCl and $Mg(OH)_2$

8.6 In which of the following pairs of compounds is one member of the pair an ionic compound and the other member of the pair a molecular compound?

(a) CO and CO_2 **(b)** Be_3N_2 and Cl_2O

(c) HNO_3 and KNO_3 **(d)** MgO and CaS

Types of Binary Ionic Compounds (Sec. 8.2)

8.7 Classify each of the following metals as a fixed-charge metal or a variable-charge metal.

(a) Mg   **(b)** Ni   **(c)** Ca   **(d)** Au

8.8 Classify each of the following metals as a fixed-charge metal or a variable-charge metal.

(a) Ba   **(b)** Ag   **(c)** Cr   **(d)** Li

8.9 In which of the following pairs of fixed-charge metals do both members of the pair form ions with the same charge?

(a) Be and Mg   **(b)** Ag and Zn

(c) Al and K   **(d)** Na and Li

8.10 In which of the following pairs of fixed-charge metals do both members of the pair form ions with the same charge?

(a) Ag and Na   **(b)** Al and Ga

(c) Be and Cd   **(d)** Ca and Zn

8.11 Classify each of the following compounds as a fixed-charge binary ionic compound or a variable-charge binary ionic compound.

(a) FeO   **(b)** $NiCl_2$ **(c)** K_2O **(d)** Al_2S_3

8.12 Classify each of the following compounds as a fixed-charge binary ionic compound or a variable-charge binary ionic compound.

(a) Cu_2S **(b)** AgCl   **(c)** AuCl   **(d)** Na_3N

Nomenclature of Binary Ionic Compounds (Sec. 8.3)

8.13 Name the following metal cations.

(a) K^+ **(b)** Mg^{2+} **(c)** Cu^+ **(d)** Pb^{4+}

8.14 Name the following metal cations.

(a) Li^+ **(b)** Be^{2+} **(c)** Au^{3+} **(d)** Sn^{2+}

8.15 Name the following nonmetal anions.

(a) Br^- **(b)** N^{3-} **(c)** S^{2-} **(d)** Se^{2-}

8.16 Name the following nonmetal anions.

(a) I^- **(b)** P^{3-} **(c)** O^{2-} **(d)** Te^{2-}

8.17 What is the chemical formula of each of the following monoatomic ions?

(a) zinc   **(b)** lead(II)   **(c)** calcium   **(d)** nitride

8.18 What is the chemical formula of each of the following monoatomic ions?

(a) silver   **(b)** gold(I)   **(c)** selenide   **(d)** bromide

8.19 Name each of the following fixed-charge binary ionic compounds.

(a) MgO   **(b)** Li_2S **(c)** AgCl   **(d)** $ZnBr_2$

8.20 Name each of the following fixed-charge binary ionic compounds.

(a) BeS   **(b)** $GaCl_3$ **(c)** CaO   **(d)** Cd_3P_2

8.21 Indicate whether or not a Roman numeral is required in the name of each of the following binary ionic compounds.

(a) ZnO   **(b)** Ag_2S **(c)** Cu_2O **(d)** $FeBr_3$

8.22 Indicate whether or not a Roman numeral is required in the name of each of the following binary ionic compounds.

(a) Al_2O_3 **(b)** CoF_3 **(c)** Ag_3N **(d)** BaS

8.23 What is the charge on the metal ion in each of the following variable-charge binary ionic compounds?

(a) FeO   **(b)** $NiCl_2$ **(c)** Au_2O_3 **(d)** Co_3N_2

8.24 What is the charge on the metal ion in each of the following variable-charge binary ionic compounds?

(a) SnO   **(b)** PbS_2 **(c)** Cu_2S **(d)** Fe_2O_3

8.25 Name each compound in the following pairs of variable-charge binary ionic compounds.

(a) $FeBr_2$ and $FeBr_3$ **(b)** Cu_2O and CuO

(c) SnS and SnS_2 **(d)** NiO and Ni_2O_3

8.26 Name each compound in the following pairs of variable-charge binary ionic compounds.

(a) $SnCl_4$ and $SnCl_2$ **(b)** FeS and Fe_2S_3

(c) Cu_3N and Cu_3N_2 **(d)** NiI_2 and NiI_3

8.27 Name each of the following binary ionic compounds.

(a) $AlCl_3$ **(b)** $NiCl_3$ **(c)** ZnO   **(d)** CoO

8.28 Name each of the following binary ionic compounds.

(a) Au_2O **(b)** Ag_2O **(c)** CuCl   **(d)** KCl

8.29 Change each of the following compound names from the Roman numeral to the "-ic, -ous" system or vice versa.

(a) lead(IV) oxide   **(b)** gold(III) chloride

(c) ferric iodide   **(d)** stannous bromide

8.30 Change each of the following compound names from the Roman numeral to the "-ic, -ous" system or vice versa.

(a) cuprous chloride　　　　**(b)** ferrous sulfide

(c) tin(IV) nitride　　　　　**(d)** lead(II) oxide

8.31 Write chemical formulas for the following binary ionic compounds.

(a) iron(II) sulfide　　　　**(b)** tin(IV) sulfide

(c) lithium sulfide　　　　　**(d)** zinc sulfide

8.32 Write chemical formulas for the following binary ionic compounds.

(a) cobalt(III) sulfide　　　　**(b)** manganese(II) sulfide

(c) calcium sulfide　　　　　**(d)** aluminum sulfide

8.33 In which of the following pairs of compound names do the two names in the pair denote the same compound?

(a) cupric chloride and copper(I) chloride

(b) stannic bromide and tin(IV) bromide

(c) ferrous oxide and iron(II) oxide

(d) plumbous sulfide and lead(IV) sulfide

8.34 In which of the following pairs of compound names do the two names in the pair denote the same compound?

(a) auric chloride and gold(III) chloride

(b) ferric bromide and iron(III) bromide

(c) cuprous oxide and copper(I) oxide

(d) stannous sulfide and tin(II) sulfide

8.35 Each of the following correct names is paired with an incorrect chemical formula. Change the incorrect chemical formula to a correct chemical formula.

(a) calcium chloride—$CaCl$

(b) sodium oxide—NaO

(c) iron(III) bromide—Fe_3Br

(d) copper(I) sulfide—CuS

8.36 Each of the following correct names is paired with an incorrect chemical formula. Change the incorrect chemical formula to a correct chemical formula.

(a) potassium bromide—KBr_2

(b) magnesium nitride—MgN

(c) copper(II) chloride—Cu_2Cl

(d) iron(III) oxide—Fe_3O

Nomenclature for Ionic Compounds Containing Polyatomic Ions (Sec. 8.4)

8.37 Write the chemical formulas, including ionic charge, of the following polyatomic ions.

(a) phosphate　　　**(b)** permanganate

(c) nitrate　　　　**(d)** cyanide

8.38 Write the chemical formulas, including ionic charge, of the following polyatomic ions.

(a) hydroxide　　　**(b)** sulfate

(c) carbonate　　　**(d)** ammonium

8.39 With the help of Table 8.4, give the names for the following polyatomic ions.

(a) O_2^{2-}　　**(b)** $S_2O_3^{2-}$　　**(c)** $C_2O_4^{2-}$　　**(d)** ClO_3^-

8.40 With the help of Table 8.4, give the names for the following polyatomic ions.

(a) N_3^-　　**(b)** BO_3^{3-}　　**(c)** SCN^-　　**(d)** CrO_4^{2-}

8.41 Write chemical formulas for both ions in each of the following pairs of polyatomic ions.

(a) sulfate and sulfite

(b) phosphate and hydrogen phosphate

(c) hydroxide and peroxide

(d) chromate and dichromate

8.42 Write chemical formulas for both ions in each of the following pairs of polyatomic ions.

(a) nitrate and nitrite

(b) chlorate and perchlorate

(c) cyanide and azide

(d) cyanate and thiocyanate

8.43 In which of the following pairs of compounds are polyatomic ions present in both members of the pair?

(a) NO_2 and KNO_2

(b) $ZnSO_4$ and NH_4Cl

(c) $Fe(NO_3)_3$ and $AlNO_3$

(d) $Ca_3(PO_4)_2$ and Ca_3N_2

8.44 In which of the following pairs of compounds are polyatomic ions present in both members of the pair?

(a) SO_3 and $CaSO_3$

(b) NH_4Br and $KClO$

(c) Cu_2CO_3 and $CuCO_3$

(d) Na_2SO_4 and Na_2S

8.45 Classify each of the following ionic compounds as a fixed-charge ionic compound or a variable-charge ionic compound.

(a) $ZnSO_4$　　　　**(b)** $Ba(OH)_2$

(c) $Fe(NO_3)_3$　　**(d)** $CuCO_3$

8.46 Classify each of the following ionic compounds as a fixed-charge ionic compound or a variable-charge ionic compound.

(a) $AgNO_3$　　　**(b)** $PbSO_4$

(c) $Sn(CO_3)_2$　　**(d)** K_3PO_4

8.47 What is the charge on the variable-charge metal ion present in each of the following polyatomic-ion-containing compounds?

(a) $FeSO_4$　　　**(b)** $Pb(SO_4)_2$

(c) $Ni_2(SO_4)_3$　　**(d)** Cu_2SO_4

8.48 What is the charge on the variable-charge metal ion present in each of the following polyatomic-ion-containing compounds?

(a) $PbCO_3$　　　**(b)** $Sn(CO_3)_2$

(c) $Fe_2(CO_3)_3$　　**(d)** Cu_2CO_3

8.49 What is the name of each of the compounds in Problem 8.45?

8.50 What is the name of each of the compounds in Problem 8.46?

8.51 Name each compound in the following pairs of polyatomic-ion-containing compounds.

(a) $Fe_2(CO_3)_3$ and $FeCO_3$
(b) Au_2SO_4 and $Au_2(SO_4)_3$
(c) $Sn(OH)_2$ and $Sn(OH)_4$
(d) $Cr(C_2H_3O_2)_3$ and $Cr(C_2H_3O_2)_2$

8.52 Name each compound in the following pairs of polyatomic-ion-containing compounds.

(a) $CuNO_3$ and $Cu(NO_3)_2$
(b) $Pb_3(PO_4)_2$ and $Pb_3(PO_4)_4$
(c) $Mn(CN)_3$ and $Mn(CN)_2$
(d) $Co(ClO_3)_2$ and $Co(ClO_3)_3$

8.53 Name each of the following polyatomic-ion-containing compounds.

(a) NH_4NO_3 (b) NH_4Cl
(c) Na_3PO_4 (d) Cu_3PO_4

8.54 Name each of the following polyatomic-ion-containing compounds.

(a) NH_4CN (b) $(NH_4)_2SO_4$
(c) $AgNO_3$ (d) $AuNO_3$

8.55 Write chemical formulas for the following compounds containing polyatomic ions.

(a) silver carbonate   (b) gold(I) nitrate
(c) chromium(III) sulfate   (d) ammonium acetate

8.56 Write chemical formulas for the following compounds containing polyatomic ions.

(a) copper(II) sulfate
(b) manganese(III) hydroxide
(c) ammonium nitrate
(d) magnesium phosphate

8.57 Write chemical formulas for the following compounds containing polyatomic ions.

(a) ferric sulfate   (b) cuprous cyanide
(c) stannic carbonate   (d) plumbous hydroxide

8.58 Write chemical formulas for the following compounds containing polyatomic ions.

(a) auric cyanide   (b) ferrous nitrate
(c) plumbous phosphate   (d) cupric chlorate

Nomenclature for Binary Molecular Compounds (Sec. 8.5)

8.59 Write the number that corresponds to each of the following prefixes.

(a) hepta-   (b) penta-   (c) tri-   (d) deca-

8.60 Write the number that corresponds to each of the following prefixes.

(a) tetra-   (b) octa-   (c) hexa-   (d) ennea-

8.61 Name the following binary molecular compounds.

(a) P_4O_{10} (b) SF_4 (c) CBr_4 (d) ClO_2

8.62 Name the following binary molecular compounds.

(a) S_4N_2 (b) SO_3 (c) IF_7 (d) N_2O_4

8.63 Write chemical formulas for the following binary molecular compounds.

(a) iodine monochloride
(b) nitrogen trichloride
(c) sulfur hexafluoride
(d) oxygen difluoride

8.64 Write chemical formulas for the following binary molecular compounds.

(a) disulfur monoxide
(b) tetraphosphorus hexoxide
(c) carbon dioxide
(d) silicon tetrachloride

8.65 Name the following binary molecular compounds.

(a) H_2S (b) HF   (c) NH_3 (d) CH_4

8.66 Name the following binary molecular compounds.

(a) HCl   (b) H_2Se (c) N_2H_4 (d) H_2O_2

8.67 Write chemical formulas for the following binary molecular compounds.

(a) phosphine   (b) hydrogen bromide
(c) ethane   (d) hydrogen telluride

8.68 Write chemical formulas for the following binary molecular compounds.

(a) hydrogen iodide   (b) hydrazine
(c) hydrogen peroxide   (d) arsine

8.69 The chemical formula for a compound consisting of two nitrogen atoms and one oxygen atom is written as N_2O, and the compound's name is dinitrogen monoxide. The chemical formula is not written as ON_2 and the compound is not named oxygen dinitride. Explain why.

8.70 The chemical formula for a compound consisting of one nitrogen atom and one oxygen atom is written as NO and the compound's name is nitrogen monoxide. The chemical formula is not written as ON and the compound is not named oxygen nitride. Explain why.

8.71 The correct name for the compound $NaNO_3$ is not sodium nitrogen trioxide. Explain why.

8.72 The correct name for the compound $BaSO_4$ is not barium sulfur tetroxide. Explain why.

Nomenclature for Acids (Sec. 8.6)

8.73 Indicate whether or not each of the following hydrogen-containing compounds forms an acid in aqueous solution.

(a) CH_4 **(b)** H_2S **(c)** HCN   **(d)** NH_3

8.74 Indicate whether or not each of the following hydrogen-containing compounds forms an acid in aqueous solution.

(a) HCl   **(b)** HClO   **(c)** SiH_4 **(d)** CH_4

8.75 Name the acids that produce each of the following negative ions in aqueous solution.

(a) CN^- **(b)** SO_4^{2-} **(c)** NO_2^- **(d)** BO_3^{3-}

8.76 Name the acids that produce each of the following negative ions in aqueous solution.

(a) NO_3^- **(b)** I^- **(c)** PO_3^{3-} **(d)** $C_2O_4^{2-}$

8.77 What is the chemical formula for each of the acids in Problem 8.75?

8.78 What is the chemical formula for each of the acids in Problem 8.76?

8.79 Name each of the following compounds as acids.

(a) HNO_3 **(b)** HI   **(c)** HClO   **(d)** $HC_2H_3O_2$

8.80 Name each of the following compounds as acids.

(a) $HClO_3$ **(b)** $HClO_4$ **(c)** H_2S **(d)** HCl

8.81 Give the name and the chemical formula of the anion produced by each of the following acids.

(a) hydrochloric acid   **(b)** chlorous acid
(c) perchloric acid   **(d)** sulfuric acid

8.82 Give the name and the chemical formula of the anion produced by each of the following acids.

(a) hydrocyanic acid   **(b)** nitrous acid
(c) hypochlorous acid   **(d)** phosphoric acid

8.83 Supply the missing name in each of the following pairs of name–formula combinations.

(a) H_3AsO_4 (arsenic acid); H_3AsO_3 (?)
(b) HIO_3 (iodic acid); HIO_4 (?)
(c) H_3PO_3 (phosphorus acid); H_3PO_2 (?)
(d) HBrO (hypobromous acid); $HBrO_2$ (?)

8.84 Supply the missing name in each of the following pairs of name–formula combinations.

(a) HIO_3 (iodic acid); HIO (?)
(b) H_2SeO_4 (selenic acid); H_2SeO_3 (?)
(c) HBrO (hypobromous acid); $HBrO_4$ (?)
(d) HNO_2 (nitrous acid); HNO_3 (?)

8.85 Name each of the following compounds.

(a) HBr(g)   **(b)** HCN(aq)   **(c)** H_2S(g)   **(d)** HI(aq)

8.86 Name each of the following compounds.

(a) HBr(aq)   **(b)** HCN(g)   **(c)** H_2S(aq)   **(d)** HI(g)

8.87 Write chemical formulas for the following acids.

(a) chloric acid   **(b)** nitrous acid
(c) hydrofluoric acid   **(d)** acetic acid

8.88 Write chemical formulas for the following acids.

(a) hypochlorous acid   **(b)** sulfurous acid
(c) nitric acid   **(d)** oxalic acid

Additional Problems

8.89 Determine the value of x in each of the following "incomplete" chemical formulas for ionic compounds.

(a) Na_xSO_4 **(b)** KCl_x **(c)** $Ba_3(PO_4)_x$ **(d)** $Al_x(CO_3)_3$

8.90 Determine the value of x in each of the following "incomplete" chemical formulas for ionic compounds.

(a) K_xN **(b)** $Ca(NO_3)_x$ **(c)** $Mg_x(CN)_2$ **(d)** Zn_xSO_4

8.91 In which of the following pairs of ionic compounds do both members of the pair have positive ions with the same charge?

(a) BaO and NaCl   **(b)** Fe_2O_3 and $NiCl_3$
(c) $MgCl_2$ and BeO   **(d)** AlN and KF

8.92 In which of the following pairs of ionic compounds do both members of the pair have positive ions with the same charge?

(a) Co_2O_3 and $CoCl_3$ **(b)** Cu_2O and CuO
(c) K_2O and Al_2O_3 **(d)** MgS and NaI

8.93 In which of the following pairs of compounds would both members of the pair have names that contain Roman numerals?

(a) NaCl and CuCl   **(b)** Al_2O_3 and Fe_2O_3
(c) $Cu(NO_3)_2$ and NiO   **(d)** Ag_2SO_4 and $CuSO_4$

8.94 In which of the following pairs of compounds would both members of the pair have names that contain Roman numerals?

(a) $AuCl_3$ and $FeCl_3$ **(b)** K_3N and AlN
(c) FeO and BaO   **(d)** NiS and $Cu_3(PO_4)_2$

8.95 In which of the following pairs of compounds would both members of the pair have names that contain Greek numerical prefixes?

(a) SO_3 and N_2O **(b)** AlN and CO
(c) HCl and H_2S **(d)** PCl_3 and NF_3

8.96 In which of the following pairs of compounds would both members of the pair have names that contain Greek numerical prefixes?

(a) CO and CO_2 **(b)** N_2O and K_2O
(c) OF_2 and BaF_2 **(d)** HBr and H_2Se

8.97 How many ions are present in one formula unit of each of the following compounds?

(a) K_3N **(b)** KN_3 **(c)** Na_2O **(d)** Na_2O_2

8.98 How many ions are present in one formula unit of each of the following compounds?

(a) NaCN   **(b)** NaSCN   **(c)** K_2O **(d)** K_2O_2

8.99 In each of the following pairs of anions, select the one that contains the greatest number of oxygen atoms.

(a) sulfate and sulfite   **(b)** chlorate and perchlorate

(c) hydroxide and peroxide   **(d)** chromate and dichromate

8.100 In each of the following pairs of anions, select the one that contains the greatest number of oxygen atoms.

(a) nitride and nitrate   **(b)** phosphate and phosphite

(c) chlorite and hypochlorite   **(d)** acetate and cyanide

8.101 Write chemical formulas for the following compounds.

(a) calcium nitride   **(b)** calcium nitrate

(c) calcium nitrite   **(d)** calcium cyanide

8.102 Write chemical formulas for the following compounds.

(a) sodium sulfide   **(b)** sodium sulfate

(c) sodium sulfite   **(d)** sodium thiosulfate

8.103 Write chemical formulas for the following compounds.

(a) potassium phosphide

(b) potassium phosphate

(c) potassium hydrogen phosphate

(d) potassium dihydrogen phosphate

8.104 Write chemical formulas for the following compounds.

(a) magnesium carbide

(b) magnesium bicarbonate

(c) magnesium carbonate

(d) magnesium hydrogen carbonate

8.105 Indicate which compounds in each of the following groups have names that contain the prefix *di-*.

(a) N_2O, Li_2O, CO_2, K_2S

(b) K_2O, K_2CO_3, NO_2, SO_2

(c) BeF_2, $BeCl_2$, SF_2, SCl_2

(d) Au_2O_3, Fe_2O_3, Al_2O_3, N_2O_3

8.106 Indicate which compounds in each of the following groups have names that contain the suffix *-ide*.

(a) CaS, CO, SO_3, Be_3N_2

(b) K_3N, KNO_3, KNO_2, $KClO$

(c) MgO, AlN, KF, KOH

(d) $NaCN$, $NaOH$, Na_2CO_3, NaF

8.107 Indicate which compounds in each of the following groups have names that contain the suffix *-ous* or *-ate*.

(a) $CaCO_3$, H_2CO_3, $Ca(NO_2)_2$, HNO_2

(b) $NaClO_4$, $NaClO_3$, $NaClO_2$, $NaClO$

(c) $HClO_4$, $HClO_3$, $HClO_2$, $HClO$

(d) $LiOH$, $LiCN$, Li_2CO_3, Li_3PO_4

8.108 Indicate which compounds in each of the following groups have names that contain the suffix *-ic* or *-ite*.

(a) $HBr(g)$, $HBr(aq)$, $HCN(g)$, $HCN(aq)$

(b) K_2SO_4, KCN, $KMnO_4$, K_3PO_3

(c) NH_4Cl, NH_4CN, $(NH_4)_2SO_4$, NH_4NO_2

(d) Li_3N, LiN_3, $LiNO_3$, $LiNO_2$

8.109 Indicate which compounds in each of the following groups possess molecules or formula units that are pentatomic.

(a) sodium cyanide, sodium thiocyanate, sodium hypochlorite, sodium nitrate

(b) aluminum sulfide, magnesium nitride, beryllium phosphide, potassium hydroxide

(c) beryllium oxide, iron(II) oxide, iron(III) oxide, sulfur dioxide

(d) gold(I) cyanide, gold(III) cyanide, gold(I) chlorate, gold(III) chlorate

8.110 Indicate which compounds in each of the following groups possess molecules or formula units that are heptatomic.

(a) magnesium cyanide, magnesium oxalate, magnesium chloride, magnesium perchlorate

(b) aluminum oxide, calcium thiosulfate, beryllium thiocyanate, dinitrogen pentoxide

(c) iron(III) sulfide, iron(III) nitride, iron(III) sulfate, iron(III) hypochlorite

(d) dichlorine heptoxide, sulfur trioxide, sulfur hexafluoride, ammonium sulfide

8.111 The chemical formula of a hydroxide of nickel is $Ni(OH)_3$. What are the chemical formulas of the following nickel compounds in which nickel has the same ionic charge as in $Ni(OH)_3$?

(a) nickel sulfate   **(b)** nickel oxide

(c) nickel oxalate   **(d)** nickel nitrate

8.112 The chemical formula of a phosphate of vanadium is VPO_4. What are the chemical formulas of the following vanadium compounds in which vanadium has the same ionic charge as in VPO_4?

(a) vanadium sulfate   **(b)** vanadium hydroxide

(c) vanadium oxide   **(d)** vanadium nitrate

8.113 The chemical formula of the ionic compound potassium superoxide is KO_2. The chemical formula of the ionic compound nitronium perchlorate is NO_2ClO_4. What is the chemical formula for the ionic compound nitronium superoxide?

8.114 The chemical formula of the ionic compound calcium perrhenate is $Ca(ReO_4)_2$. The chemical formula of the ionic compound nitrosonium hydrogen sulfate is $NOHSO_4$. What is the chemical formula for the ionic compound nitrosonium perrhenate?

8.115 Each of the following correct chemical formulas is paired with an incorrect chemical name for the compound represented by the chemical formula. Change the incorrect chemical name to a correct chemical name.

(a) BeO—beryllium(II) oxide

(b) $MgCl_2$—magnesium dichloride

(c) Na_2CO_3—disodium carbon trioxide

(d) $(NH_4)_2SO_4$—diammonium sulfate

8.116 Each of the following correct chemical formulas is paired with an incorrect chemical name for the compound represented by the chemical formula. Change the incorrect chemical name to a correct chemical name.

(a) NO_2—mononitrogen dioxide
(b) H_2SO_4—dihydrogen sulfur tetroxide
(c) CuS—copper sulfide
(d) K_2S—potassium(II) sulfide

8.117 Write the chemical formula for each of the substances mentioned in the following word descriptions.

(a) Sulfur trioxide reacts with water to form sulfuric acid.
(b) Perchloric acid reacts with cadmium to produce cadmium perchlorate.
(c) Zinc carbonate decomposes to form zinc oxide and carbon dioxide.
(d) Magnesium reacts with iron(II) chloride to produce magnesium chloride and iron.

8.118 Write the chemical formula for each of the substances mentioned in the following word descriptions.

(a) Cadmium sulfide reacts with sulfuric acid to produce hydrogen sulfide gas.
(b) Silicon dioxide reacts with hydrofluoric acid to produce silicon tetrafluoride.
(c) Lead(II) nitrate reacts with potassium iodide to produce lead(II) iodide.
(d) Sodium phosphate reacts with silver nitrate to produce silver phosphate and sodium nitrate.

8.119 Classify each of the following compounds as (1) binary ionic, (2) ternary ionic, (3) binary molecular, or (4) ternary molecular.

(a) hydrocyanic acid   **(b)** hydrobromic acid
(c) sodium peroxide   **(d)** hydrogen peroxide

8.120 Classify each of the following compounds as (1) binary ionic, (2) ternary ionic, (3) binary molecular, or (4) ternary molecular.

(a) sodium azide   **(b)** ammonium azide
(c) beryllium nitride   **(d)** beryllium nitrate

Cumulative Problems

8.121 Give the name of the simplest binary compound that forms between elements with the following electron configurations.

(a) $1s^2 2s^2 2p^6 3s^2 3p^5$ and $1s^2 2s^2 2p^6 3s^2$
(b) $1s^2 2s^2 2p^4$ and $1s^2 2s^2 2p^5$

8.122 Give the name of the simplest binary compound that forms between elements with the following electron configurations.

(a) $1s^2 2s^2 2p^3$ and $1s^2 2s^2 2p^6 3s^2 3p^6 4s^1$
(b) $1s^2 2s^2 2p^6 3s^2 3p^4$ and $1s^2 2s^2 2p^6 3s^2 3p^5$

8.123 After determining the value of x in each of the following chemical formulas, name the compounds.

(a) $SiCl_x$ **(b)** $MgCl_x$ **(c)** K_xN **(d)** NCl_x

8.124 After determining the value of x in each of the following chemical formulas, name the compounds.

(a) CCl_x **(b)** BeO_x **(c)** NF_x **(d)** $AlBr_x$

8.125 A compound has the chemical formula $M(XO_3)_2$, where M is a metal and X is a nonmetal. Assign a name to this compound given the following information.

(1) M is a metal with two valence electrons.
(2) X forms a monoatomic ion with a charge of -1.
(3) X has an electronegativity between 2.6 and 2.9.
(4) The sum of the periodic table period numbers for X and M is 6.

8.126 A compound has the chemical formula M_3XO_3, where M is a metal and X is a nonmetal. Assign a name to this compound given the following information.

(1) M is a metal whose electron configuration "ends" in s^1.
(2) Both M and X are found in period 2 of the periodic table.
(3) There are 38 protons in one formula unit of M_3XO_3.

8.127 Determine the name for a binary ionic compound that has the following characteristics.

(1) Positive and negative ions are present in a one-to-one ratio.
(2) All ions present have the electron configuration $1s^2 2s^2 2p^6$.
(3) One of the elements present has an atomic number that is six less than the atomic number of the other element.

8.128 Determine the name for a binary ionic compound that has the following characteristics.

(1) Positive and negative ions are present in a one-to-two ratio.
(2) All ions present have the electron configuration $1s^2 2s^2 2p^6 3s^2 3p^6$.
(3) Neutral atoms of the more electronegative element present have 17 electrons.

8.129 Determine the name for a binary molecular compound that has the following characteristics.

(1) Its molecules are triatomic.
(2) There are twice as many atoms of the more electronegative element per molecule as there are of the less electronegative element.
(3) Both elements are in period 2 of the periodic table.
(4) Atoms of one element contain four valence electrons, and atoms of the other element contain six valence electrons.

8.130 Determine the name for a binary molecular compound that has the following characteristics.

(1) Its molecules are hexatomic.
(2) There are twice as many atoms of the more electronegative element per molecule as there are of the less electronegative element.

(3) The two elements occupy adjacent positions in period 2 of the periodic table.

(4) The sum of the valence electrons for an atom of each element is 11.

8.131 Determine the name for a compound that has the following characteristics.

(1) Monoatomic and polyatomic ions are present.

(2) The ratio between positive and negative ions is one to two.

(3) A group IIA element that loses one half of its total electrons upon ion formation is present.

(4) The polyatomic ion contains equal numbers of atoms of two nonmetallic elements.

(5) The sum of the atomic numbers for the two elements involved in the polyatomic ion is 13.

(6) One of the elements present in the compound forms a monoatomic ion with a charge of -3 that is isoelectronic with Ne.

8.132 Determine the name for a compound that has the following characteristics.

(1) Monoatomic and polyatomic ions are present.

(2) The ratio between positive and negative ions is one to three.

(3) A group IIIA element whose ion electron configuration is isoelectronic with Ne is present.

(4) The polyatomic ion contains equal numbers of atoms of two nonmetallic elements.

(5) The sum of the atomic numbers for the two elements involved in the polyatomic ion is 9.

Answers to Practice Exercises

8.1 (a) ionic **(b)** ionic **(c)** ionic **(d)** molecular

8.2 (a) barium chloride **(b)** aluminum phosphide **(c)** potassium oxide **(d)** calcium nitride

8.3 (a) iron(II) oxide **(b)** iron(III) oxide **(c)** lead(IV) oxide **(d)** gold(I) chloride

8.4 (a) cupric chloride **(b)** stannous fluoride **(c)** iron(II) bromide **(d)** gold(III) oxide

8.5 (a) potassium carbonate **(b)** cobalt(II) nitrate **(c)** iron(III) sulfate **(d)** ammonium phosphate

8.6 (a) disulfur monoxide **(b)** nitrogen trichloride **(c)** tetraphosphorus hexasulfide **(d)** carbon tetrafluoride

8.7 (a) sulfurous acid **(b)** hydrobromic acid **(c)** phosphoric acid **(d)** acetic acid

8.8 (a) H_3PO_3 **(b)** $HClO_4$

8.9 (a) ammonium sulfide **(b)** aluminum chloride **(c)** tetraphosphorus hexoxide **(d)** nitrous acid

8.10 (a) $Ca(ClO_4)_2$ **(b)** $Fe(NO_3)_2$ **(c)** Li_2O **(d)** SO_3

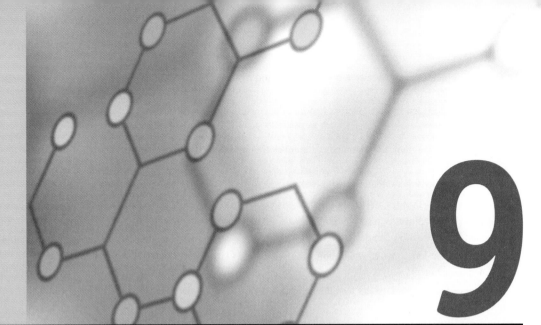

Chemical Calculations: The Mole Concept and Chemical Formulas

9.1 The Law of Definite Proportions

This chapter is the first of two chapters dealing with "chemical arithmetic," that is, with the quantitative relationships between elements and compounds. The emphasis in this chapter will be on quantitative relationships that involve chemical formulas. In Chapter 10, the emphasis will be on quantitative relationships that involve chemical equations.

In Section 4.7 we saw that compounds are pure substances with the following characteristics.

1. They are chemical combinations of two or more elements.
2. They can be broken down into constituent elements by chemical, but not physical, means.
3. They have a definite, constant elemental composition.

In this section we consider further the fact that compounds have a definite composition.

The composition of a compound can be determined by decomposing a weighed amount of the compound into its elements and then determining the masses of the individual elements. Alternatively, determining the mass of a compound formed by the combination of known masses of elements will also allow the calculation of its composition.

Studies of composition data for many compounds have led to the conclusion that the percentage of each element present in a given compound does not vary. This conclusion has been formalized into a statement known as the **law of definite proportions:** *In a pure compound, the elements are always present in the same definite proportion by mass.*

The Human Side of Chemistry 9

Joseph-Louis Proust (1754–1826)

Joseph-Louis Proust (pronounced Proost), born in Angers, France, began his chemistry studies at home in his father's apothecary shop. Later he studied in Paris. It is Proust's research on the constancy of composition for various compounds that brought the law of definite proportions into focus.

Proust was a dedicated analyst, with little interest in theory. His most important work, during the period 1797–1807, involved copious experiments on the composition of various compounds. Other research involved studies on starch, camphor, preparation of mercury from cinnabar ore, and isolation of grape sugar (glucose) from grapes.

Almost all of Proust's work, including that related to the law of definite proportions, was carried out in Spain. At the invitation of the Spanish government, in 1791, he became director of the Royal Laboratory at Madrid. This laboratory, financed by the government, was elegantly equipped. Almost all the vessels, even those in common use, were made of platinum.

His research work in Madrid was stopped abruptly in 1808 when his laboratory was destroyed by French troops occupying Spain. Reduced to poverty, and plagued with ill health, he still refused an offer of 100,000 francs from Napoleon to come back to France and supervise the manufacture of glucose from grapes (a most abundant crop in both Spain and France).

Later, in 1817, after Louis XVIII had come to power, Proust returned to France, and in 1820, he took over the apothecary shop of his brother Joachim, who was in still poorer health. Proust remained in France until his death in 1826.

The French chemist Joseph-Louis Proust (1754–1826; see "The Human Side of Chemistry 9") is responsible for the work that established this law as one of the fundamentals of chemistry.

Let us consider how composition data obtained from decomposing a compound are used to illustrate the law of definite proportions. Samples of the compound of *known masses* are decomposed. The masses of the constituent elements present in each sample are then obtained. This experimental information (elemental masses and the original sample mass) is then used to calculate the percent composition of the samples. Within the limits of experimental error, the calculated percentages for any given element present turn out to be the same, validating the law of definite proportions. Example 9.1 gives actual decomposition data for a compound and shows how they are used to verify the law of definite proportions.

EXAMPLE 9.1

Using Decomposition Data to Illustrate the Law of Definite Proportions

Two samples of ammonia (NH_3) of differing mass and from different sources are individually decomposed to yield ammonia's constituent elements (nitrogen and hydrogen). The results of the decomposition experiments are as follows.

	Sample Mass Before Decomposition (g)	Mass of Nitrogen Produced (g)	Mass of Hydrogen Produced (g)
Sample 1	1.840	1.513	0.327
Sample 2	2.000	1.644	0.356

Show that these data are consistent with the law of definite proportions.

> The law of definite proportions is also called the *law of constant composition*.

SOLUTION

Calculating the percent nitrogen in each sample will be sufficient to show whether the data are consistent with the law. Both nitrogen percentages should come out the same if the law is obeyed.

$$\text{percent nitrogen} = \frac{\text{mass of nitrogen obtained}}{\text{total sample mass}} \times 100$$

SAMPLE 1

$$\% \text{ N} = \frac{1.513 \text{ g}}{1.840 \text{ g}} \times 100 = 82.22826\% \quad \text{(calculator answer)}$$

$$= 82.23\% \quad \textbf{(correct answer)}$$

SAMPLE 2

$$\% \text{ N} = \frac{1.644 \text{ g}}{2.000 \text{ g}} \times 100 = 82.2\% \quad \text{(calculator answer)}$$

$$= 82.20\% \quad \textbf{(correct answer)}$$

Note that the percentages are close to being equal but are not identical. This is due to measuring uncertainties and rounding in the original experimental data. The two percentages are the same to three significant figures. The difference lies in the fourth significant digit. Recall from Section 2.4 that the last of the significant digits in a number (the fourth one here) has uncertainty in it. The two percentages above are considered to be the same within experimental error.

An alternative way of treating the given data to illustrate the law of definite proportions involves calculating the mass ratio between nitrogen and hydrogen. This ratio should be the same for each sample; otherwise the two samples are not the same compound.

SAMPLE 1

$$\frac{\text{mass of N}}{\text{mass of H}} = \frac{1.513 \text{ g}}{0.327 \text{ g}} = 4.6269113 \quad \text{(calculator answer)}$$

$$= 4.63 \quad \textbf{(correct answer)}$$

SAMPLE 2

$$\frac{\text{mass of N}}{\text{mass of H}} = \frac{1.644 \text{ g}}{0.356 \text{ g}} = 4.6179775 \quad \text{(calculator answer)}$$

$$= 4.62 \quad \textbf{(correct answer)}$$

Again, slight differences in the ratios are caused by measuring errors and rounding in the original data.

CHEMICAL EXTENSION

At room temperature and pressure, ammonia is a colorless gas. Its pungent odor can be detected at a level in air of about 50 parts per million (ppm). At levels of 100–200 ppm, ammonia sharply irritates the eyes and the air passages in the lungs.

Ammonia is among the "top five" chemical compounds produced industrially in the United States in terms of mass—in excess of 35 billion pounds is produced annually.

Approximately 80% of the ammonia produced is used as chemical fertilizer—either directly or indirectly. Under increased pressure, ammonia is easily liquefied. In liquid form it can be directly used as a fertilizer. In addition, ammonia can be converted to solid derivatives, which include ammonium nitrate (NH_4NO_3), ammonium sulfate [$(NH_4)_2SO_4$], and urea [$(NH_2)_2CO$]. Urea, which is often made at the site of the ammonia synthesis plant, has a greater nitrogen content by mass than any other solid fertilizer.

Practice Exercise 9.1

Two samples of dichlorine heptoxide (Cl_2O_7) of differing mass and from different sources are individually decomposed to yield the elements chlorine and oxygen. The results of the decomposition experiments are as follows.

	Sample Mass Before Decomposition (g)	Mass of Chlorine Produced (g)	Mass of Oxygen Produced (g)
Sample 1	2.724	1.056	1.668
Sample 2	8.367	3.243	5.124

Show that these data are consistent with the law of definite proportions.

Answers to practice exercises are located at the end of the chapter.

The constancy of composition for compounds can also be examined by considering the mass ratios in which elements combine to form compounds. Let us consider the reaction between the elements calcium and sulfur to produce the compound calcium sulfide. Suppose an attempt is made to combine various masses of sulfur with a fixed mass of calcium. A set of possible experimental data for this attempt is given in the first four lines of Table 9.1. Note that, regardless of the mass of S present, only a certain amount, 44.4 g, reacts with the 55.6 g of Ca. The excess S is left over in an unreacted form. The data therefore illustrate that Ca and S will react in only one fixed mass ratio ($55.6/44.4 = 1.25$) to form CaS. This fact is consistent with the law of definite proportions. Note also that if the amount of Ca used is doubled (line 5 of Table 9.1), the amount of S with which it reacts also doubles (compare lines 1 and 5 of the table). Nevertheless, the ratio in which the substances react ($111.2/88.8$) still remains 1.25.

Figure 9.1 relates the law of definite proportions to Dalton's atomic theory (Sec. 5.1). This figure shows pictorially the formation of the compound tin(II) sulfide, SnS, from atoms of tin and sulfur. Note that a formula unit of SnS must always contain one atom of tin and one atom of sulfur.

9.2 Calculation of Formula Masses

Formula masses play a role in almost all chemical calculations and will be used extensively in later sections of this chapter and in succeeding chapters. A **formula mass** *is the sum of the atomic masses of all atoms present in one formula unit of a substance, expressed in atomic mass units.* Formula masses, like the atomic masses from which they are calculated, are relative masses based on the $^{12}_{6}C$ relative mass scale (Sec. 5.9). They can be calculated for compounds (both molecular and ionic) and for elements that exist in molecular form.

Table 9.1 Data Illustrating the Law of Definite Proportions

Mass of Ca Used (g)	Mass of S Used (g)	Mass of CaS Formed (g)	Mass of Excess Unreacted Sulfur (g)	Ratio in Which Substances React
55.6	44.4	100.0	none	1.25
55.6	50.0	100.0	5.6	1.25
55.6	100.0	100.0	55.6	1.25
55.6	200.0	100.0	155.6	1.25
111.2	88.8	200.0	none	1.25

Figure 9.1

Illustration of the law of definite proportions at the level of atoms.

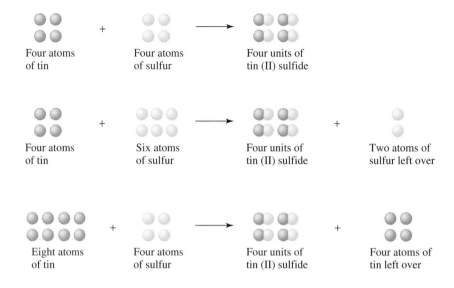

Four atoms of tin + Four atoms of sulfur → Four units of tin (II) sulfide

Four atoms of tin + Six atoms of sulfur → Four units of tin (II) sulfide + Two atoms of sulfur left over

Eight atoms of tin + Four atoms of sulfur → Four units of tin (II) sulfide + Four atoms of tin left over

The term *molecular mass* is used interchangeably with *formula mass* by many chemists in referring to substances that contain discrete molecules. It is incorrect, however, to use the term *molecular mass* when talking about ionic substances, since molecules are not their basic structural unit (Sec. 7.8).

Some chemistry books use the terms *formula weight* and *molecular weight* rather than *formula mass* and *molecular mass*. This practice is not followed in this book because mass is technically more correct (Sec. 3.3).

Once the formula of a substance has been established, its formula mass is calculated by adding together the atomic masses of all the atoms in the formula. If more than one atom of any element is present, that element's atomic mass must be added as many times as there are atoms of the element present.

Example 9.2 contains sample formula mass calculations. Significant figures (Sec. 2.5) are a consideration because atomic masses are obtained from experimentally determined numbers (Sec. 5.8).

EXAMPLE 9.2

Using Chemical Formulas and Atomic Masses to Calculate Formula Masses

Calculate the formula masses for the following compounds.

(a) N_2H_4 (hydrazine, a rocket fuel)
(b) $ClNO_3$ (chlorine nitrate, a substance involved in the Antarctic ozone hole phenomenon)
(c) CHF_2Cl (chlorodifluoromethane, a Freon replacement)
(d) $C_9H_8O_4$ (aspirin, a mild pain reliever)

SOLUTION

Formula masses are obtained simply by adding the atomic masses of the constituent elements, counting each atomic mass as many times as the symbol for the element occurs in the formula.

(a) A molecule of N_2H_4 contains six atoms: two atoms of N and four atoms of H. The formula mass, the collective mass of these six atoms, is calculated as follows:

$$2 \text{ atom N} \times \frac{14.0067 \text{ amu}}{1 \text{ atom N}} = 28.0134 \text{ amu}$$

$$4 \text{ atom H} \times \frac{1.00794 \text{ amu}}{1 \text{ atom H}} = 4.03176 \text{ amu}$$

$$\text{Formula mass} = 32.04516 \text{ amu} \quad \text{(calculator answer)}$$

$$= 32.0452 \text{ amu} \quad \textbf{(correct answer)}$$

The conversion factors in the calculation were derived from the atomic mass listing on the right side of the inside front cover of the text. The atomic mass, the mass of an average atom, is 14.0067 amu for N and 1.00794 amu for H. The calculator answer, 32.04516 amu, had to be rounded down one significant figure, since the atomic mass of N is known only to the fourth decimal place. (Recall the rules for significant figures in addition; Sec. 2.4.)

Often, conversion factors are not explicitly shown in a formula mass calculation as just shown; the calculation is simplified as follows:

N: 2 × 14.0067 amu = 28.0134 amu

H: 4 × 1.00794 amu = $\underline{\text{4.03176 amu}}$

formula mass = 32.04516 amu (calculator answer)

= 32.0452 amu (**correct answer**)

(b) Using the simplified calculation method, we calculate the formula mass for $ClNO_3$ as

Cl: 1 × 35.453 amu = 35.4527 amu

N: 1 × 14.0067 amu = 14.0067 amu

O: 3 × 15.9994 amu = $\underline{\text{47.9982 amu}}$

formula mass = 97.4576 amu (calculator and correct answer)

(c) Using the simplified calculation method, we calculate the formula mass for CHF_2Cl as

C: 1 × 12.0107 amu   = 12.0107 amu

H: 1 × 1.00794 amu   = 1.00794 amu

F: 2 × 18.9984032 amu = 37.9968064 amu

Cl: 1 × 35.453 amu   = $\underline{\text{35.4527 amu}}$

formula mass = 86.4681464 amu (calculator answer)

= 86.4681 amu (**correct answer**)

Note how the atomic masses of C and Cl limit the precision of the formula mass to ten-thousandths. Why does the atomic mass listing not give a more precise atomic mass for Cl, you may ask. A more fundamental question is "Why does the uncertainty in the atomic masses of the elements vary as much as it does?" Some elements have atomic masses known to 0.00001 amu; others are known to only 0.1 amu. A major factor in atomic mass uncertainty is the constancy of the abundance percentages for the isotopes of an element. Abundance percentages fluctuate more for some elements than for others; the less the fluctuation, the less the atomic mass uncertainty. (The fluctuation is a relative matter; none of the fluctuations is large on an absolute scale.) The atomic masses given in the atomic mass listing on the right side of the inside front cover of the text are stated to current uncertainty limits.

(d) There are 21 atoms in a molecule of this compound. Its formula mass is

C: 9 × 12.0107 amu = 108.0963 amu

H: 8 × 1.00794 amu =   8.06352 amu

O: 4 × 15.9994 amu =   63.9976 amu

formula mass = 180.15742 amu   (calculator answer)

= 180.1574 amu   (**correct answer**)

An important point concerning significant figures in formula mass calculations is formally encountered in the carbon part of this calculation. Here, we have the multiplication (9 × 12.0107). Nine is a counted (exact) number and 12.0107 has six significant figures. The answer shown, 108.0963, contains seven significant figures rather than six allowed by the multiplication rule. Why? A multiplication involving a small whole number, 9 in this case, can be viewed as an addition. We are repeatedly adding 12.0107 to itself. As an addition, an uncertainty in the ten-thousandths place is allowed in the answer (108.0963).

Situations similar to that occurring here for carbon will be encountered repeatedly in future formula mass calculations. Our operational rule, for significant figure purposes, in all such situations will be to consider the calculation to be a pure addition problem rather than a combined multiplication–addition problem. Using this approach, we will often be able to justify one more significant figure in the calculated formula mass.

Practice Exercise 9.2

Calculate the formula masses for the following substances.

(a) H_2SO_4 (sulfuric acid, an industrial acid)

(b) $C_{16}H_{18}N_2O_5S$ (penicillin V, an important antibiotic)

In Example 9.2, all the formula masses were calculated purposely to the maximum number of significant figures possible. This was done to emphasize the fact that the number of significant figures in "input" atomic masses determines the uncertainty in calculated formula masses.

Formula masses, relative masses on the $^{12}_6C$ relative mass scale, can be used for mass comparisons. For example, using the formula masses calculated in Example 9.2, we can make the statement that one molecule of $C_9H_8O_4$ is 1.84856 times as heavy as one molecule of $ClNO_3$ (180.1575 amu/97.4576 amu = 1.84856).

9.3 Significant Figures and Atomic Mass

Many of the numerical problems in this and later chapters require conversion factors that involve formula masses. To ensure that these formula-mass–based conversion factors are never the limiting factor in significant figure considerations, we will use formula masses that usually contain one more significant figure than required by the data in the problem. The uncertainty in calculated answers should always reflect uncertainty in the data provided rather than uncertainty in formula masses.

The operational rules we will follow throughout the remainder of the text relative to calculating formula masses are:

1. Atomic masses rounded to the hundredths place will be used as the input numbers for the formula mass determination. (This is why the atomic masses in the periodic table inside the front cover are atomic masses rounded to the hundredths place.)

2. The formula mass calculation will be considered a pure addition problem rather than a combination multiplication–addition problem for significant figure purposes (see Example 9.2, part d). Thus, the addition rule will always govern significant figure determinations.

Example 9.3 illustrates the use of these operational rules in formula mass calculations.

EXAMPLE 9.3

Calculation of Formula Masses Using Specific Operational Rules

Calculate the formula masses for the following substances using the operational rules given in the text.

(a) $C_7H_6O_2$ (benzoic acid, a food preservative)

(b) $PtCl_2(NH_3)_2$ (cis-Platin, a chemotherapy agent)

SOLUTION

(a) Atomic masses, rounded to hundredths, are the starting point for the calculation. From the periodic table (inside front cover) we obtain

C: 12.01 amu   H: 1.01 amu   O: 16.00 amu

We calculate the formula's mass, a sum of atomic masses, as

C: 7×12.01 amu $=$ 84.07 amu

H: 6×1.01 amu $=$ 6.06 amu

O: 2×16.00 amu $=$ 32.00 amu

        formula mass $=$ 122.13 amu   (calculator and **correct answer**)

Since all of the input atomic masses have an uncertainty of hundredths, the formula mass uncertainty will also be hundredths (addition rule for significant figures).

(b) From the periodic table, we obtain the needed atomic masses rounded to hundredths.

Pt: 195.08 amu   Cl: 35.45 amu   N: 14.01 amu   H: 1.01 amu

Summing the atomic masses, using each an appropriate number of times, we get

Pt: 1×195.08 amu $=$ 195.08 amu

Cl: $2 \times$ 35.45 amu $=$ 70.90 amu

N: $2 \times$ 14.01 amu $=$ 28.02 amu

H: $6 \times$ 1.01 amu $=$ 6.06 amu

        formula mass $=$ 300.06 amu (calculator and correct answer)

Based on the addition rule for significant figures, the uncertainty in the answer is hundredths since all of the input atomic masses had uncertainties of hundredths.

Practice Exercise 9.3

Calculate the formula masses for the following substances using the operational rules given in the text.

(a) CCl_4 (carbon tetrachloride, an industrial solvent)

(b) $C_7H_{14}O_2$ (isopentyl acetate, banana-like flavoring agent)

9.4 Percent Composition

A useful piece of information about a compound is its *percent composition*. **Percent composition** *is the percent by mass of each element present in a compound.* For instance, the percent composition of water is 88.81% oxygen and 11.19% hydrogen.

Percent compositions are frequently used to compare compound compositions. The compounds gold (III) iodide (AuI_3), gold (III) nitrate [$Au(NO_3)_3$], and gold (I) cyanide (AuCN) contain, respectively, 34.10%, 51.43%, and 88.33% gold by mass. If you were given the choice of receiving a gift of 1 lb of one of these three gold compounds, which one would you choose?

The percent composition of a compound can be calculated from experimental decomposition data as was done in Example 9.1. Such a calculation can be carried out even if the formula or the identity of the compound is unknown. A compound's chemical formula can also be used to obtain percentage composition information, as is illustrated in Example 9.4.

EXAMPLE 9.4

Using a Compound's Formula to Calculate Percent Composition

Calculate the percent composition (using atomic masses to two decimal places) of the pain reliever acetaminophen. Its chemical formula is $C_8H_9O_2N$.

SOLUTION

First, we calculate the formula mass of $C_8H_9O_2N$, using atomic masses rounded to the hundredths decimal place.

$$
\begin{aligned}
\text{C:} &\quad 8 \times 12.01 \text{ amu} = 96.08 \text{ amu} \\
\text{H:} &\quad 9 \times 1.01 \text{ amu} = 9.09 \text{ amu} \\
\text{O:} &\quad 2 \times 16.00 \text{ amu} = 32.00 \text{ amu} \\
\text{N:} &\quad 1 \times 14.01 \text{ amu} = \underline{14.01 \text{ amu}} \\
&\quad \text{formula mass} = 151.18 \text{ amu} \quad \text{(calculator and \textbf{correct answer})}
\end{aligned}
$$

The mass percent of each element in the compound is found by dividing the mass contribution of each element, in amu, by the total mass (formula mass), in amu, and multiplying by 100.

$$\% \text{ element} = \frac{\text{mass of element in one formula unit}}{\text{formula mass}} \times 100$$

Finding percentages, we have

$$\% \text{ C} = \frac{96.08 \text{ amu}}{151.18 \text{ amu}} \times 100 = 63.55338\% \quad \text{(calculator answer)}$$
$$= 63.55\% \quad \text{(\textbf{correct answer})}$$

$$\% \text{ H} = \frac{9.09 \text{ amu}}{151.18 \text{ amu}} \times 100 = 6.0127\% \quad \text{(calculator answer)}$$
$$= 6.01\% \quad \text{(\textbf{correct answer})}$$

$$\% \text{ O} = \frac{32.00 \text{ amu}}{151.18 \text{ amu}} \times 100 = 21.166821\% \quad \text{(calculator answer)}$$
$$= 21.17\% \quad \text{(\textbf{correct answer})}$$

$$\% \text{ N} = \frac{14.01 \text{ amu}}{151.18 \text{ amu}} \times 100 = 9.2670988\% \quad \text{(calculator answer)}$$

$$= 9.267\% \qquad \textbf{(correct answer)}$$

To check our work we can add the percentages of all the parts. They, of course, have to total 100. (On occasion, rounding errors may not cancel, and totals such as 99.99% or 100.01% may be obtained.)

$$63.55\% + 6.01\% + 21.17\% + 9.267\% = 99.997\% \quad \text{(calculator answer)}$$

$$= 100.00\% \quad \textbf{(correct answer)}$$

CHEMICAL EXTENSION

Acetaminophen has replaced aspirin as the most widely used of all nonprescription pain relievers. Its use now accounts for over half of the over-the-counter pain reliever market. It is marketed generically and under trade names such as Tylenol, Datril, Tempra, and Anacin-3. Acetaminophen's availability in a liquid form makes it an ideal medication for small children and other patients who have difficulty taking solid tablets.

Acetaminophen, often called the "aspirin substitute," has no irritating effect on the intestinal tract as does aspirin and yet has comparable pain-relieving and fever-reducing effects. However, it is not effective against inflammation as is aspirin and is thus of limited use for the aches and pains of arthritis. Also, acetaminophen does not have any blood thinning properties, as does aspirin, and therefore is not useful as an aid in reducing the risk of stroke and heart attack.

Practice Exercise 9.4

What is the percent composition (using atomic masses to two decimal places) of vitamin C, a compound necessary in small amounts for the normal growth of humans, whose formula is $C_6H_8O_6$?

Percent compositions can also be calculated from mass data obtained from compound synthesis or compound decomposition experiments. Example 9.5 shows how synthesis data are treated to yield percent composition.

EXAMPLE 9.5

Using Synthesis Data to Calculate Percent Composition

The production of a 13.50-g sample of the hormone adrenaline requires 7.96 g of C, 0.96 g of H, 3.54 g of O, and 1.04 g of N. What is the percent composition of this compound?

SOLUTION

The total mass of the compound sample is given as 13.50 g. We divide the mass of each element present by this total mass (13.50 g) and multiply by 100 to give percentage.

$$\% \text{ element} = \frac{\text{mass of element}}{\text{total sample mass}} \times 100$$

Finding the percentages, we have

$$\% \text{ C}: \frac{7.96 \text{ g}}{13.50 \text{ g}} \times 100 = 58.962963\% \quad \text{(calculator answer)}$$

$$= 59.0\% \qquad \textbf{(correct answer)}$$

$$\% \text{ H:} \quad \frac{0.96 \text{ g}}{13.50 \text{ g}} \times 100 = 7.1111111\% \quad \text{(calculator answer)}$$

$$= 7.1\% \quad \text{(correct answer)}$$

$$\% \text{ O:} \quad \frac{3.54 \text{ g}}{13.50 \text{ g}} \times 100 = 26.222222\% \quad \text{(calculator answer)}$$

$$= 26.2\% \quad \text{(correct answer)}$$

$$\% \text{ N:} \quad \frac{1.04 \text{ g}}{13.50 \text{ g}} \times 100 = 7.7037037\% \quad \text{(calculator answer)}$$

$$= 7.70\% \quad \text{(correct answer)}$$

Checking our work, we see that the percentages do add up correctly.

$$59.0\% + 7.1\% + 26.2\% + 7.70\% = 100.0\%$$

CHEMICAL EXTENSION

Adrenaline, also known as epinephrine, has the chemical formula $C_9H_{13}O_3N$. It is a hormone released from the adrenal glands into the bloodstream, usually in response to pain, excitement, anger, or fear, that increases the level of glucose in the blood. The extra glucose, a key source of energy for body processes, in turn increases rate and force of heart contraction, muscular strength, and blood pressure. These changes cause the body to function at a "higher" level. For this reason, adrenaline is often called the "fight or flight" hormone.

Adrenaline-caused increased blood pressure results from increased heart action accompanied by constriction of peripheral blood vessels. Injectable local anesthetics usually contain adrenaline because it constricts blood vessels in the vicinity of the injection. This prevents the blood from rapidly distributing the anesthetic and prolongs the anesthetic effect in the target tissue. Adrenaline is also used to reduce hemorrhage.

Practice Exercise 9.5

The sour taste of vinegar is caused by the compound acetic acid. Calculate the percent composition of acetic acid, knowing that the synthesis of 24.03 g of this compound requires 9.61 g of C, 1.62 g of H, and 12.80 g of O.

9.5 The Mole: The Chemist's Counting Unit

Two common methods exist for specifying the quantity of material in a sample of a substance: (1) in terms of units of *mass* and (2) in terms of units of *amount*. We measure *mass* by using a balance (Sec. 3.3). Common mass units are gram, kilogram, and pound. For substances that consist of discrete units, we can specify the *amount* of substance present by indicating the number of units present—12, or 27, or 113, and so on.

We all use both units of mass and units of amount on a daily basis. We work well with this dual system. Sometimes it does not matter which type of unit is used; at other times one system is preferred over the other. When buying potatoes at the grocery store we can decide on quantity in either mass units (10-lb bag, 20-lb bag, etc.) or amount units (9 potatoes, 15 potatoes, etc.). When buying eggs, amount units are used almost exclusively—12 eggs (1 dozen), 24 eggs (2 dozen), and so on. On the other hand, peanuts and grapes are almost always purchased in weighed quantities. It is impractical to count the number of grapes in a bunch. Very few people go to the store with the idea of buying 117 grapes.

In chemistry, as in everyday life, both the mass and amount methods of specifying quantity find use. Again, the specific situation dictates the method used. In laboratory work, practicality dictates working with quantities of known mass (12.3 g, 0.1365 g, etc.). (Counting out a given number of atoms for a laboratory experiment is impossible, since we cannot see individual atoms.)

In performing chemical calculations, after the laboratory work has been done, it is often useful (even necessary) to think of quantities of substances present in terms of atoms or formula units. A problem exists when this is done—very, very large numbers are always encountered. Any macroscopic sample of a chemical substance contains many trillions of atoms or formula units.

To cope with this "large number problem" chemists have found it convenient to use a special counting unit. Employment of such a unit should not surprise you, as specialized counting units are used in many areas. The two most common counting units are *dozen* and *pair*. Other more specialized counting units exist. For example, at an office supply store, paper is sold by the *ream* (500 sheets), and pencils by the *gross* (144 pencils). (See Fig. 9.2.)

The chemist's counting unit is called a *mole*. What is unusual about the mole is its magnitude. A **mole** *is a counting unit based on the number* 6.022×10^{23}. This extremely large number is necessitated by the extremely small size of atoms, molecules, and ions. The use of a traditional counting unit, such as a dozen, would be, at best, only a slight improvement over counting atoms singly.

> A more technical definition for a *mole* will be given in Section 9.7.

$$6.022 \times 10^{23} \text{ atoms} = 5.018 \times 10^{22} \text{ dozen atoms} = 1 \text{ mole atoms}$$

Note how the use of the mole counting unit decreases very significantly the magnitude of numbers encountered. The number 1 represents 6.022×10^{23} objects, the number 2, double that number of objects. (Why the number 6.022×10^{23} was chosen as the counting unit rather than some other number will be discussed in Section 9.6.)

The number 6.022×10^{23} also has a special name. **Avogadro's number** *is the name given to the numerical value* 6.022×10^{23}*, the number of objects in a mole*. This designation honors the Italian physicist Lorenzo Romano Amedeo Carlo Avogadro (1776–1856; see "The Human Side of Chemistry 10"), whose pioneering work on gases later proved to be valuable in determining the number of particles present in a given volume of a substance.

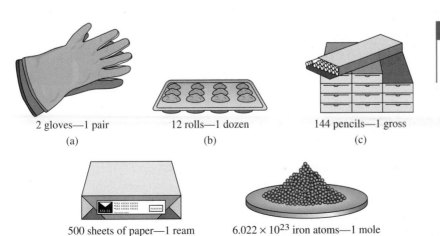

Figure 9.2
Some counting units used to denote quantities in terms of amount.

2 gloves—1 pair
(a)

12 rolls—1 dozen
(b)

144 pencils—1 gross
(c)

500 sheets of paper—1 ream
(d)

6.022×10^{23} iron atoms—1 mole
(e)

The Human Side of Chemistry 10

Lorenzo Romano Amedeo Carlo Avogadro (1776–1856)

Amedeo Avogadro, born in Turin, Italy, followed in the footsteps of his father, obtaining a law degree at age 16 and for some years engaging in the practice of law. At age 24, he began to study, privately, mathematics and physics. Abandoning the law profession at age 30, he spent most of his later life as a professor of mathematical physics at the University of Turin.

Avogadro was the first scientist to distinguish between atoms and molecules. His most important paper, published in 1811, suggested that all gases (at a given temperature) contained the same number of particles per unit volume and that these particles need not be individual atoms but might be combinations of atoms. The thrust of this paper, known now as Avogadro's law, received little attention from his contemporaries. Turin, where Avogadro was professor, was outside the mainstream of scientific activity in 1811.

It was not until after his death that his "genius" was recognized. In 1858, two years after his death, Stanislao Cannizzaro (1826–1910), one of his countrymen, pointed out the full significance of Avogadro's law at an international gathering of scientists. The "time was right," and acceptance was rapid. One chemist, after reading Cannizzaro's paper, wrote "The scales seemed to fall from my eyes. Doubts disappeared and a feeling of quiet certainty took their place." In 1911, in Turin, in commemoration of the 100th anniversary of the first publication of his law, a monument honoring Avogadro was unveiled by the king in one of the greatest posthumous tributes to a scientist in history.

Avogadro's name is attached to a number he never determined. During the early 1900s, chemical theory developed to the point that estimates could be made of the number of molecules in a given volume of gas. That equal volumes of all gases really contain the same number of molecules (Avogadro's law) was a provable reality. Jean-Baptiste Perrin (1870–1942), in 1909, was the first to use the phrase "Avogadro's number" to denote the calculated number of molecules in a specific volume of gas. The honor is appropriate since Avogadro's law was the "springboard" for Perrin's work as well as that of many other scientists working in the same research area.

One mole of a chemical substance contains Avogadro's number of "particles." The word *particles,* as used in the preceding sentence, can have a number of meanings, including

1. the number of *molecules* of a molecular compound
2. the number of *formula units* of an ionic compound
3. the number of *atoms* of an element

In solving mathematical problems dealing with the number of atoms, molecules, or ions present in a given amount of material, Avogadro's number becomes part of the conversion factor used to relate number of particles present to moles present.

From the definition

$$1 \text{ mole} = 6.022 \times 10^{23} \text{ objects}$$

two conversion factors can be derived.

$$\frac{1 \text{ mole}}{6.022 \times 10^{23} \text{ objects}} \quad \text{and} \quad \frac{6.022 \times 10^{23} \text{ objects}}{1 \text{ mole}}$$

Example 9.6 illustrates the use of particle-to-mole conversion factors.

EXAMPLE 9.6

Calculating the Number of Objects in a Molar Quantity

How many objects are there in each of the following quantities?

(a) 1.20 moles of carbon monoxide (CO) molecules

(b) 2.53 moles of silver (Ag) atoms

(c) 0.025 mole of magnesium sulfate ($MgSO_4$) formula units

(d) 2.25 moles of watermelons

SOLUTION

We will use dimensional analysis (Sec. 3.7) in solving each part of this problem. All of the parts are similar in that we are given a certain number of moles of substance and want to find the number of particles contained in the given number of moles. All parts can be classified as moles-to-particles problems, and each solution will involve the use of Avogadro's number.

$$\boxed{\text{Moles of substance}} \xrightarrow[\text{number}]{\text{Avogadro's}} \boxed{\text{Particles of substance}}$$

(a) The given quantity is 1.20 moles of CO molecules, and the desired quantity is number of CO molecules.

$$1.20 \text{ moles CO} = ? \text{ CO molecules}$$

The setup, by dimensional analysis, involves only one conversion factor.

$$1.20 \text{ moles CO} \times \frac{6.022 \times 10^{23} \text{ CO molecules}}{1 \text{ mole CO}}$$

$$= 7.2264 \times 10^{23} \text{ CO molecules} \quad \text{(calculator answer)}$$

$$= 7.23 \times 10^{23} \text{ CO molecules} \quad \textbf{(correct answer)}$$

(b) The given quantity is 2.53 moles of silver atoms, and the desired quantity is the actual number of silver atoms present.

$$2.53 \text{ moles Ag} = ? \text{ Ag atoms}$$

The setup, with the same conversion factor as in part (a), is

$$2.53 \text{ moles Ag} \times \frac{6.022 \times 10^{23} \text{ Ag atoms}}{1 \text{ mole Ag}}$$

$$= 1.523566 \times 10^{24} \text{ Ag atoms} \quad \text{(calculator answer)}$$

$$= 1.52 \times 10^{24} \text{ Ag atoms} \quad \textbf{(correct answer)}$$

(c) The fact that we are dealing with formula units here (an ionic compound), rather than atoms or molecules, does not change the way the problem is solved.

$$0.025 \text{ mole MgSO}_4 = ? \text{ formula units MgSO}_4$$

The conversion factor setup is

$$0.025 \text{ mole MgSO}_4 \times \frac{6.022 \times 10^{23} \text{ formula units MgSO}_4}{1 \text{ mole MgSO}_4}$$

$$= 1.5055 \times 10^{22} \text{ MgSO}_4 \text{ formula units} \quad \text{(calculator answer)}$$

$$= 1.5 \times 10^{22} \text{ MgSO}_4 \text{ formula units} \quad \textbf{(correct answer)}$$

(d) Use of the mole as a counting unit is usually found only in a chemical context. Technically, however, any type of object can be counted in units of moles. One mole denotes 6.02×10^{23} objects; it does not matter what the objects are—even watermelons. Just as we can talk about dozens of watermelons, we can talk about moles of watermelons, although the latter involves a *very large* watermelon patch.

$$2.25 \text{ moles watermelons} \times \frac{6.022 \times 10^{23} \text{ watermelons}}{1 \text{ mole watermelons}}$$

$$= 1.35495 \times 10^{24} \text{ watermelons} \quad (\text{calculator answer})$$

$$= 1.35 \times 10^{24} \text{ watermelons} \quad (\textbf{correct answer})$$

Practice Exercise 9.6

How many objects are there in each of the following quantities?

(a) 2.67 moles of carbon dioxide (CO_2) molecules

(b) 1.45 moles of sodium chloride (NaCl) formula units

It is somewhat unfortunate, because of its similarity to the word molecule, that the name mole was selected as the name for the chemist's counting unit. Students often think that *mole* is an abbreviated form of the word *molecule*. That is not the case. The word mole comes from the Latin *moles*, which means "heap or pile." A mole is a macroscopic amount, a heap or pile of objects, that can easily be seen. A molecule is a particle too small to be seen with the naked eye.

In Example 9.6 we calculated the number of objects present in samples ranging in size from 0.025 mole to 2.53 moles. Our answers were numbers carrying the exponents 10^{22}, 10^{23}, or 10^{24}. Numbers with these exponents are inconceivably large. The magnitude of Avogadro's number itself is so large that it is almost incomprehensible. There is nothing in our experience to relate to it. (When chemists count, they count in really big jumps.) Many attempts have been made to create word pictures of the vast size of Avogadro's number. Such pictures, however, really only hint at its magnitude, since other large numbers must be used in the word pictures. Several such word pictures are given in Figure 9.3.

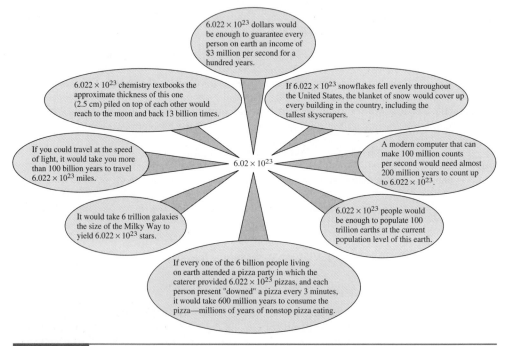

Figure 9.3

"Word pictures" of how big Avogadro's number is.

9.6 The Mass of a Mole

How much does a mole weigh; that is, what is its mass? Consider a similar (but more familiar) question first: "How much does a dozen weigh?" Your response is now immediate. "A dozen what?" you reply. The mass of a dozen identical objects obviously depends on the identity of the object. For example, the mass of a dozen elephants will be "somewhat greater" than the mass of a dozen marshmallows. The mole, like the dozen, is a counting unit. Similarly, the mass of a mole of objects will depend on the identity of the object. Thus, the mass of a mole, *molar mass*, is not one set number; it varies, being different for each different chemical substance. This is in direct contrast to the *molar number*, Avogadro's number, which is the same for all chemical substances.

The **molar mass of an element is** *a mass in grams numerically equal to the atomic mass of the element when the element is present in atomic form.* Thus, if we know the atomic mass of an element, we also know the mass of one mole of atoms of the element. The two quantities are numerically the same, differing only in units. For the elements carbon, oxygen, and sodium we can write the following numerical relationships.

$$\text{mass of 1 carbon atom} = 12.01 \text{ amu} \quad \text{(atomic mass)}$$

$$\text{mass of 1 mole of carbon atoms} = 12.01 \text{ g} \quad \textbf{(molar mass)}$$

$$\text{mass of 1 oxygen atom} = 16.00 \text{ amu} \quad \text{(atomic mass)}$$

$$\text{mass of 1 mole of oxygen atoms} = 16.00 \text{ g} \quad \textbf{(molar mass)}$$

$$\text{mass of 1 sodium atom} = 22.99 \text{ amu} \quad \text{(atomic mass)}$$

$$\text{mass of 1 mole of sodium atoms} = 22.99 \text{ g} \quad \textbf{(molar mass)}$$

It is not a coincidence that the mass in grams of a mole of atoms of an element and the element's atomic mass are numerically equal. Avogadro's number has the value that it has in order to cause this relationship to exist. Experimentally it was determined that when 6.022×10^{23} atoms of an element are present, molar masses (in grams) and atomic masses (in amu) are numerically equal. Again, only when the chemist's counting unit has the value 6.022×10^{23} does this relationship hold. Avogadro's number is thus the "connecting link" between the chemist's microscopic mass scale (amu) and macroscopic mass scale (grams). The use of the mass unit grams in specifying the molar mass of an element is a must. The use of other mass units would require a different counting unit than 6.022×10^{23} for a numerical match between atomic mass and molar mass.

The molecular form of an element will have a different molar mass than its atomic form. Consider the element chlorine, which is found in nature in the form of diatomic molecules (Cl_2). The mass of one mole of chlorine atoms (Cl) is different from the mass of one mole of chlorine molecules (Cl_2). Since there are two atoms in each molecule of chlorine, the molar mass of molecular chlorine is twice the molar mass of atomic chlorine. The following relationships hold for chlorine, whose atomic mass is 35.45 amu.

$$6.022 \times 10^{23} \text{ Cl atoms} = 1 \text{ mole Cl atoms} = 35.45 \text{ g Cl}$$

$$6.022 \times 10^{23} \text{ Cl}_2 \text{ molecules} = 1 \text{ mole Cl}_2 \text{ molecules} = 70.90 \text{ g Cl}_2$$

Note that one mole of molecular chlorine contains twice as many atoms as one mole of atomic chlorine; however, the number of discrete particles present is the same (Avogadro's number) in both cases. For atomic chlorine, atoms are considered to be the object counted; for molecular chlorine, molecules are considered to be the discrete particle counted. There is the same number of atoms in the former case as there is of molecules in the latter case. This atomic–molecular chlorine situation is analogous to the difference between a dozen shoes and a dozen pairs of shoes. Both the mass and the actual number of shoes for the dozen pairs of shoes are double those for the dozen shoes.

The existence of some elements in molecular form becomes a source of error in chemical calculations if care is not taken to distinguish properly between atomic and molecular forms of the element. The phrase "one mole of chlorine" is an ambiguous term. Does it mean one mole of chlorine atoms (Cl), or does it mean one mole of chlorine molecules (Cl_2)? To avoid ambiguity, it is important to always state explicitly the chemical form of an element being discussed. Using the chemical formula Cl_2 or referring to the substance as molecular chlorine avoids ambiguity.

The **molar mass of a compound** *is a mass in grams that is numerically equal to the formula mass of the compound.* Thus, for compounds a numerical equivalence exists between molar mass and formula mass if the molar mass is specified in grams. When we add atomic masses to get the formula mass (in amu) of a compound, we are simultaneously finding the mass of one mole of compound (in grams). The molar mass–formula mass relationships for the compounds water (H_2O), ammonia (NH_3), and barium chloride ($BaCl_2$) are

$$\text{mass of 1 } H_2O \text{ molecule} = 18.02 \text{ amu} \quad \text{(formula mass)}$$

$$\text{mass of 1 mole of } H_2O \text{ molecules} = 18.02 \text{ g} \quad \textbf{(molar mass)}$$

$$\text{mass of 1 } NH_3 \text{ molecule} = 17.04 \text{ amu} \quad \text{(formula mass)}$$

$$\text{mass of 1 mole of } NH_3 \text{ molecules} = 17.04 \text{ g} \quad \textbf{(molar mass)}$$

$$\text{mass of 1 } BaCl_2 \text{ formula unit} = 208.23 \text{ amu} \quad \text{(formula mass)}$$

$$\text{mass of 1 mole of } BaCl_2 \text{ formula units} = 208.23 \text{ g} \quad \textbf{(molar mass)}$$

Figure 9.4 pictures molar quantities of a number of common substances. Note again how the mass of a mole varies in numerical value. On the other hand, all of the pictured amounts of substances contain the same number of units—6.022×10^{23} atoms, molecules, or formula units.

A generalized definition for the *molar mass* of a substance, which is independent of whether the substance is an element or a compound, can be given. A **molar mass** *is the mass, in grams, of one mole of atoms, molecules, or formula units of a substance.*

It should now be very evident to you why the chemist's counting unit, the mole, has the value it does—6.022×10^{23}. *Avogadro's number represents the experimentally determined number of atoms, molecules, or formula units contained in a sample of a pure substance with a mass in grams numerically equal to the atomic mass or formula mass of the pure substance.*

The numerical match between molar mass and atomic or formula mass makes the calculation of the mass of any given number of moles of a substance a very simple procedure. In solving problems of this type, the numerical value of the molar mass becomes part of the conversion factor used to convert from moles to grams.

$$\boxed{\text{Moles of substance}} \xrightarrow[\text{mass}]{\text{Molar}} \boxed{\text{Grams of substance}}$$

For example, for the compound CO_2, which has a formula mass of 44.01 amu, we can write the equality

$$44.01 \text{ g } CO_2 = 1 \text{ mole } CO_2$$

From this statement two conversion factors can be written.

$$\frac{44.01 \text{ g } CO_2}{1 \text{ mole } CO_2} \quad \text{and} \quad \frac{1 \text{ mole } CO_2}{44.01 \text{ g } CO_2}$$

Example 9.7 illustrates the use of gram-to-mole conversion factors like these.

58.44 g table salt

18.02 g water

180.17 g aspirin

200.59 g mercury

63.55 g copper

342.34 g table sugar

253.80 g iodine

Figure 9.4

One mole each of seven different substances: the elements copper (Cu), mercury (Hg), and iodine (I_2) and the compounds water (H_2O), table salt (NaCl), table sugar ($C_{12}H_{22}O_{11}$), and aspirin ($C_9H_8O_4$).

EXAMPLE 9.7

Calculating the Mass, in Grams, of Molar Quantities

Calculate the mass, in grams, of each of the following quantities of matter.

(a) 1.50 moles of CH_4 molecules

(b) 2.50 moles of NaCl formula units

(c) 1.68 moles of N_2 molecules

(d) 1.68 moles of N atoms

SOLUTION

We will use dimensional analysis to solve each of these problems. The relationship between molar mass and atomic or formula mass will serve as a conversion factor in the setup of each problem.

(a) The given quantity is 1.50 moles of CH_4 molecules, and the desired quantity is grams of CH_4. Thus,

$$1.50 \text{ moles } CH_4 = ? \text{ g } CH_4$$

The formula mass of CH_4 is calculated to be 16.05 amu.

$$
\begin{array}{lll}
\text{C:} & 1 \times 12.01 \text{ amu} = & 12.01 \text{ amu} \\
4 \text{ H:} & 4 \times 1.01 \text{ amu} = & \underline{4.04 \text{ amu}} \\
& \text{formula mass} = & 16.05 \text{ amu}
\end{array}
$$

Using the formula mass, we can write the equality

$$16.05 \text{ g } CH_4 = 1 \text{ mole } CH_4$$

The dimensional analysis setup for the problem with the gram-to-mole equation used as a conversion factor is

$$1.50 \text{ moles } \cancel{CH_4} \times \frac{16.05 \text{ g } CH_4}{1 \text{ mole } \cancel{CH_4}} = 24.075 \text{ g } CH_4 \quad \text{(calculator answer)}$$

$$= 24.1 \text{ g } CH_4 \quad \textbf{(correct answer)}$$

(b) The given quantity is 2.50 moles of NaCl formula units and the desired quantity is grams of NaCl.

$$2.50 \text{ moles NaCl} = ? \text{ g NaCl}$$

The calculated formula mass of NaCl is 58.44 amu. Thus,

$$58.44 \text{ g NaCl} = 1 \text{ mole NaCl}$$

With this relationship as a conversion factor, the setup for the problem becomes

$$2.50 \text{ moles } \cancel{NaCl} \times \frac{58.44 \text{ g NaCl}}{1 \text{ mole } \cancel{NaCl}} = 146.1 \text{ g NaCl} \quad \text{(calculator answer)}$$

$$= 146 \text{ g NaCl} \quad \textbf{(correct answer)}$$

(c) The given quantity is 1.68 moles of N_2 molecules. The desired quantity is grams of N_2 molecules. Thus,

$$1.68 \text{ moles } N_2 = ? \text{ g } N_2$$

We are dealing here with diatomic nitrogen molecules (N_2) and not nitrogen atoms. Thus, 28.02 amu, twice the atomic mass of nitrogen, is the formula mass used in the mole-to-gram statement.

$$28.02 \text{ g } N_2 = 1 \text{ mole } N_2$$

With this relationship as a conversion factor, the setup becomes

$$1.68 \text{ moles } \cancel{N_2} \times \frac{28.02 \text{ g } N_2}{1 \text{ mole } \cancel{N_2}} = 47.0736 \text{ g } N_2 \quad \text{(calculator answer)}$$

$$= 47.1 \text{ g } N_2 \quad \textbf{(correct answer)}$$

(d) The given quantity is 1.68 moles of N atoms, and the desired quantity is grams of N atoms. Thus,

$$1.68 \text{ moles N} = ? \text{ g N}$$

This problem differs from the previous one in that atoms, rather than molecules, of nitrogen are being counted. The atomic mass of nitrogen is 14.01 amu, and the mole-to-gram equality statement is

$$14.01 \text{ g N} = 1 \text{ mole N}$$

With this relationship as a conversion factor, the setup becomes

$$1.68 \text{ moles } \cancel{N} \times \frac{14.01 \text{ g N}}{1 \text{ mole } \cancel{N}} = 23.5368 \text{ g N} \quad \text{(calculator answer)}$$

$$= 23.5 \text{ g N} \quad \textbf{(correct answer)}$$

Notice that the mass of 1.68 moles of N_2 (part c) is twice the mass of 1.68 moles of N (part d). This is as expected; there are twice as many atoms of N in 1.68 moles of N_2 as in 1.68 moles of N.

Practice Exercise 9.7

Calculate the mass, in grams, of each of the following molar quantities of matter.

(a) 1.81 moles of ClF_5 molecules

(b) 0.622 mole of O_3 molecules

9.7 Significant Figures and Avogadro's Number

Now that the mass of a molar amount of a substance has been considered (Sec. 9.6), let us revisit the concept of Avogadro's number.

In Section 9.5 we defined the mole simply as

$$1 \text{ mole} = 6.022 \times 10^{23} \text{ objects}$$

Although this statement conveys correct information (the value of Avogadro's number to four significant figures is 6.022×10^{23}), it is not the officially accepted definition for Avogadro's number. The official definition, which is mass based, is: The **mole** *is the amount of a substance that contains as many particles (atoms, molecules, or formula units) as there are* $^{12}_{6}C$ *atoms in 12.000000 grams of* $^{12}_{6}C$. The value of Avogadro's number is an experimentally determined quantity (the number of atoms in 12.000000 g of $^{12}_{6}C$ atoms) rather than an exactly defined quantity. Its value is not even mentioned in the definition. The most up-to-date experimental value for Avogadro's number is 6.0221367×10^{23}, which is consistent with our previous definition. In calculations we will never need a value with as many significant figures as the experimentally determined one.

In problem solving, conversion factors that involve Avogadro's number should never be the "limiting factor" in significant figure considerations. To ensure that this is the case, our approach will be to carry more digits than required by the data in "Avogadro's number" conversion factors. [We used a similar approach for "atomic mass" conversion factors (Sec. 9.3).] Our operational rule will be to use the value 6.022×10^{23} for Avogadro's number in problem-solving situations. This will suffice in most situations. However, remember that more significant figures are available for use if the data ever require such.

> One mole of a substance represents a *fixed number* of chemical entities and has a *fixed mass*. It is more than just a counting unit (Sec. 9.5) that specifies the number of objects. The official definition of the mole specifies the number of objects in a *fixed mass* of a substance.

9.8 Relationship Between Atomic Mass Units and Gram Units

The atomic mass unit (amu) and the grams unit (g) are related to one another through Avogadro's number.

$$6.022 \times 10^{23} \text{ amu} = 1.000 \text{ g}$$

That this is the case can be deduced from the following equalities:

atomic mass of N = mass of 1 N atom = 14.01 amu

molar mass of N = mass of 6.022×10^{23} N atoms = 14.01 g

Because the second equality involves 6.022×10^{23} times as many atoms as the first equality and the masses come out numerically equal, the gram unit must be 6.022×10^{23} times larger than the amu unit.

Using a dimensional analysis set-up, the relationship

$$6.022 \times 10^{23} \text{ amu} = 1.000 \text{ g}$$

can be derived from the preceding data on nitrogen in the following manner:

$$\frac{6.022 \times 10^{23} \text{ atoms N}}{1 \text{ mole N}} \times \frac{1 \text{ mole N}}{14.01 \text{ g N}} \times \frac{14.01 \text{ amu}}{1 \text{ atom N}}$$

$$= 6.022 \times 10^{23} \text{ amu/g} \quad \text{(calculator and \textbf{correct answer})}$$

The first conversion factor is based on the number of atoms in 1 mole of N, the second on the molar mass of N, and the third on the atomic mass of N.

EXAMPLE 9.8

Using the Relationship Between Grams and Atomic Mass Units as a Conversion Factor

What is the mass, in grams, of a molecule whose mass on the amu scale is 104.00 amu?

SOLUTION

This is a one-step problem based on the conversion factor

$$6.022 \times 10^{23} \text{ amu} = 1.000 \text{ g}$$

The dimensional analysis setup is

$$104.00 \text{ amu} \times \frac{1.000 \text{ g}}{6.022 \times 10^{23} \text{ amu}} = 1.7270009 \times 10^{-22} \text{ g} \quad \text{(calculator answer)}$$

$$= 1.727 \times 10^{-22} \text{ g} \quad \text{(\textbf{correct answer})}$$

Practice Exercise 9.8

What is the mass in grams of a molecule whose mass on the amu scale is 169.00 amu?

9.9 Counting Particles by Weighing

In a laboratory situation it is often necessary to work with equal numbers of atoms of two different substances or twice as many atoms of one type than of another, and so forth. Atoms are so small that it is impossible to count them one by one. How does the chemist resolve this requirement of "equal or proportional numbers of atoms"? The problem is solved by means of the concept of "counting by weighing," a concept closely related to molar mass (Sec. 9.6).

To illustrate the "counting-by-weighing" concept, let us compare the masses of two kinds of atoms—oxygen and nitrogen—when varying but equal numbers of atoms are present. (The choice of elements in the comparison is arbitrary; any two could be used.) From a table of atomic masses we determine that the mass of a single oxygen atom (on the average) is 16.00 amu and that of a single nitrogen atom (on the average) is 14.01 amu. Using these single atom masses and varying the equal numbers of atoms present from 1 to 2 to 5 to 100 to 1 billion to Avogadro's number generates the mass comparisons given in Table 9.2. Note that the mass ratio is identical in every case; also note that the numerical value of this ratio is the same as the ratio of the atomic masses of oxygen and nitrogen. What Table 9.2 establishes is that *the mass ratio of equal numbers of atoms of two elements is the same as the ratio of the atomic masses of the two elements.* This statement is true regardless of the elements involved in the comparison. The ratio between two other

Table 9.2	Mass Ratios for Varying Equal Numbers of Oxygen and Nitrogen Atoms

Atomic Ratio	Mass Ratio of Equal Numbers of Atoms
$\dfrac{1 \text{ atom O}}{1 \text{ atom N}}$	$\dfrac{1 \times 16.00 \text{ amu}}{1 \times 14.01 \text{ amu}} = \dfrac{16.00}{14.01} = 1.142$
$\dfrac{2 \text{ atoms O}}{2 \text{ atoms N}}$	$\dfrac{2 \times 16.00 \text{ amu}}{2 \times 14.01 \text{ amu}} = \dfrac{32.00}{28.02} = 1.142$
$\dfrac{5 \text{ atoms O}}{5 \text{ atoms N}}$	$\dfrac{5 \times 16.00 \text{ amu}}{5 \times 14.01 \text{ amu}} = \dfrac{80.00}{70.05} = 1.142$
$\dfrac{100 \text{ atoms O}}{100 \text{ atoms N}}$	$\dfrac{100 \times 16.00 \text{ amu}}{100 \times 14.01 \text{ amu}} = \dfrac{1.600 \times 10^3}{1.401 \times 10^3} = 1.142$
$\dfrac{1.000 \times 10^9 \text{ atoms O}}{1.000 \times 10^9 \text{ atoms N}}$	$\dfrac{1.000 \times 10^9 \times 16.00 \text{ amu}}{1.000 \times 10^9 \times 14.01 \text{ amu}} = \dfrac{1.600 \times 10^{10}}{1.401 \times 10^{10}} = 1.142$
$\dfrac{6.022 \times 10^{23} \text{ atoms O}}{6.022 \times 10^{23} \text{ atoms N}}$	$\dfrac{6.022 \times 10^{23} \times 16.00 \text{ amu}}{6.022 \times 10^{23} \times 14.01 \text{ amu}} = \dfrac{9.635 \times 10^{24}}{8.437 \times 10^{24}} = 1.142$

elements will not be 1.142 (as was the case for O and N); each pair of elements will have its own unique mass ratio determined by the atomic masses of the elements involved.

Turning the statement of the conclusion derived from Table 9.2 around (reversing it) gives an even more useful generalization (from a laboratory viewpoint): *If samples of two or more elements have mass ratios equal to their atomic mass ratios, they must contain equal numbers of atoms.* Thus, for example, since the atomic mass ratio between Au and Cu is 3.099 (196.97/63.55), any time samples of these elements are weighed out in a 3.099-to-1.000 ratio the samples will contain equal numbers of atoms. A 3.099-g sample of Au contains the same number of atoms as a 1.000-g sample of Cu. Similarly, 3.099-lb and 3.099-ton samples of Au contain, respectively, the same numbers of atoms as 1.000-lb and 1.000-ton samples of Cu. Any time, regardless of units (g, lb, kg, ton, etc.), Au and Cu are present in a 3.099-to-1.000 mass ratio, equal numbers of atoms are present.

Our comparisons (and generalizations) so far have involved atoms and elements. Parallel comparisons (and generalizations) exist for molecules (or formula units) and compounds. The generalizations for compounds differ from those for elements only in that "formula mass" has replaced atomic mass and "molecules" (or formula units) has replaced atoms. The substitution of formula masses for atomic masses is valid because both are based on the same $^{12}_6\text{C}$ scale (Sec. 9.2). Therefore, the following samples all contain the same number of molecules.

$$18.02 \text{ g } H_2O \qquad (\text{formula mass of } H_2O = 18.02 \text{ amu})$$

$$44.01 \text{ g } CO_2 \qquad (\text{formula mass of } CO_2 = 44.01 \text{ amu})$$

$$64.06 \text{ g } SO_2 \qquad (\text{formula mass of } SO_2 = 64.06 \text{ amu})$$

We can now make the following generalization, which is applicable to both elements and compounds concerning *counting by weighing:* samples of two or more pure substances (elements or compounds) found to have mass ratios equal to the ratios of their atomic or formula masses must contain identical numbers of particles (atoms, molecules, or formula units). The following samples therefore contain the same numbers of particles.

$$26.98 \text{ g Al} \qquad (\text{atomic mass} = 26.98 \text{ amu})$$

$$18.02 \text{ g } H_2O \quad (\text{formula mass} = 18.02 \text{ amu})$$

The particles for water are molecules, and the particles for aluminum are atoms.

We see, therefore, that to obtain samples of elements or compounds containing equal numbers of particles, we merely weigh out quantities (in any units) whose mass ratio is numerically equal to the ratio of the substance's atomic or formula masses.

EXAMPLE 9.9

Using the Concept of "Counting by Weighing"

Two different hydrogen–oxygen compounds are known: water (H_2O) and hydrogen peroxide (H_2O_2). Using the concept of "counting particles by weighing," determine whether equal numbers of molecules of these two compounds are present in the following pairs of samples.

(a) 14.00 g H_2O_2 and 8.00 g H_2O

(b) 3.512 g H_2O_2 and 1.860 g H_2O

SOLUTION

The formula mass of H_2O_2 is 34.02 amu, and that of H_2O 18.02 amu. The ratio of formula masses for these two compounds is

$$\frac{\text{formula mass } H_2O_2}{\text{formula mass } H_2O} = \frac{34.02 \text{ amu}}{18.02 \text{ amu}} = 1.8879023 \quad \text{(calculator answer)}$$

$$= 1.888 \quad \textbf{(correct answer)}$$

The ratio of masses for the two samples in each pair must equal this number, 1.888, if equal numbers of molecules are present.

(a) The ratio of the masses of the two samples is

$$\frac{\text{mass } H_2O_2}{\text{mass } H_2O} = \frac{14.00 \text{ g}}{8.00 \text{ g}} = 1.75 \quad \text{(calculator and } \textbf{correct answer)}$$

Since this ratio is not equal to 1.888, equal numbers of molecules are not present. The fact that the ratio is less than 1.888 indicates that there are fewer H_2O_2 molecules (the numerator of the ratio) than H_2O molecules.

(b) The ratio of the masses of the two samples is

$$\frac{\text{mass } H_2O_2}{\text{mass } H_2O} = \frac{3.512 \text{ g}}{1.860 \text{ g}} = 1.888172 \quad \text{(calculator answer)}$$

$$= 1.888 \quad \textbf{(correct answer)}$$

This time the formula mass ratio and the mass ratio are the same (to four significant figures). Thus, equal numbers of molecules are present (to four significant figures).

CHEMICAL EXTENSION

Pure hydrogen peroxide (H_2O_2) is an almost colorless (slightly bluish), viscous liquid. It is an extremely corrosive substance that must be handled with great care.

After storage for a long time, H_2O_2 decomposes to water and oxygen (O_2). Decomposition is accelerated by traces of metals (including Cu and Fe), dirt, heat, light, and biological materials such as blood. Because decomposition is accelerated by light, H_2O_2 is stored in dark-colored bottles. Hydrogen peroxide has found some use as a component of rocket fuel.

Hydrogen peroxide is miscible with water in all proportions, and aqueous solutions of H_2O_2 have numerous uses. A 30% by mass aqueous solution finds use as a bleaching agent for textiles and furs. When used as a human hair bleach (as a 60% solution), it reacts with the pigments in the hair. Dilute (3%) H_2O_2 solutions, available in drugstores, are used as mild antiseptics. If you have ever used dilute H_2O_2 to disinfect a wound, you probably noticed some "foaming" occur. Blood accelerates the decomposition of the H_2O_2 to water and oxygen, and it is the oxygen that kills bacteria.

Practice Exercise 9.9

Carbon monoxide (CO) and nitrogen monoxide (NO) are air pollutants that enter the atmosphere in automobile exhaust. Using the concept of "counting particles by weighing," determine whether equal numbers of molecules of these two pollutants are present in the following pairs of samples.

(a) 10.00 g CO and 11.00 g NO

(b) 7.011 g CO and 7.511 g NO

9.10 The Mole and Chemical Formulas

A chemical formula has two meanings or interpretations: (1) a microscopic level interpretation and (2) a macroscopic level interpretation.

The first of these two interpretations was discussed in Section 5.4. At the *microscopic level* a chemical formula indicates the number of atoms of each element present in one molecule or formula unit of a substance. The subscripts in the formula are interpreted to mean the numbers of atoms of the various elements present in one unit of the substance. The formula C_2H_6, interpreted at the microscopic level, conveys the information that two atoms of C and six atoms of H are present in one molecule of C_2H_6.

Now that the mole concept has been introduced, a macroscopic interpretation of formulas is possible. At the *macroscopic level* a chemical formula indicates the number of moles of atoms of each element present in one mole of a substance. The subscripts in the formula are interpreted to mean the numbers of moles of atoms of the various elements present in one mole of the substance. The designation "macroscopic" is given to this molar interpretation, since moles are "laboratory-sized" quantities of atoms. The formula C_2H_6, interpreted at the macroscopic level, conveys the information that 2 moles of C atoms and 6 moles of H atoms are present in 1 mole of C_2H_6.

It is now evident, then, that the subscripts in a formula always carry a dual meaning: "atom" at the microscopic level and "moles of atoms" at the macroscopic level.

The validity of the molar interpretation for subscripts in a formula derives from the following line of reasoning. In x molecules of C_2H_6, where x is any number, there are $2x$ atoms of C and $6x$ atoms of H. Regardless of the value of x, there must always be two times as many C atoms as molecules and six times as many H atoms as molecules; that is,

$$\text{number of } C_2H_6 \text{ molecules} = x$$

$$\text{number of C atoms} = 2x$$

$$\text{number of H atoms} = 6x$$

Now let x equal 6.022×10^{23}, the value of Avogadro's number. With this x value, the following statements are true.

$$\text{number of } C_2H_6 \text{ molecules} = 6.022 \times 10^{23}$$

$$\text{number of C atoms} = 2 \times 6.022 \times 10^{23} = 1.2044 \times 10^{24} \quad \text{(calculator answer)}$$

$$= 1.204 \times 10^{24} \quad \textbf{(correct answer)}$$

$$\text{number of H atoms} = 6 \times 6.022 \times 10^{23} = 3.6132 \times 10^{24} \quad \text{(calculator answer)}$$
$$= 3.613 \times 10^{24} \quad \textbf{(correct answer)}$$

Since 6.022×10^{23} is equal to 1 mole, 1.2044×10^{24} to 2 moles, and 3.6132×10^{24} to 6 moles, these statements may be changed to read

$$\text{number of } C_2H_6 \text{ molecules} = 1 \text{ mole}$$
$$\text{number of C atoms} = 2 \text{ moles}$$
$$\text{number of H atoms} = 6 \text{ moles}$$

Thus, the mole ratio is the same as the subscript ratio: 2 to 6.

In calculations where the moles of a particular element within a compound are asked for, the subscript of that particular element in the chemical formula of the compound becomes part of the conversion factor used to convert from moles of compound to moles of element within the compound.

$$\boxed{\text{Moles of compound}} \xrightarrow[\text{subscript}]{\text{Formula}} \boxed{\text{Moles of element within compound}}$$

For example, again using C_2H_6 as our chemical formula, we can write the following conversion factors.

$$\text{For C:} \quad \frac{2 \text{ moles C atoms}}{1 \text{ mole } C_2H_6 \text{ molecules}} \quad \text{or} \quad \frac{1 \text{ mole } C_2H_6 \text{ molecules}}{2 \text{ moles C atoms}}$$

$$\text{For H:} \quad \frac{6 \text{ moles H atoms}}{1 \text{ mole } C_2H_6 \text{ molecules}} \quad \text{or} \quad \frac{1 \text{ mole } C_2H_6 \text{ molecules}}{6 \text{ moles H atoms}}$$

Example 9.10 illustrates the use of conversion factors of this type in a problem-solving context.

EXAMPLE 9.10

Calculating Molar Quantities of Compound Components

The characteristic odor of pineapple is due to ethyl butyrate, a compound with the formula $C_6H_{12}O_2$. How many moles of each type of atom present in $C_6H_{12}O_2$ are contained in a 2.65-mole sample of this compound?

SOLUTION

The formula $C_6H_{12}O_2$ specifies that 1 mole of this substance will contain 6 moles of carbon atoms, 12 moles of hydrogen atoms, and 2 moles of oxygen atoms. This information leads to the following conversion factors:

$$\frac{6 \text{ moles C atoms}}{1 \text{ mole } C_6H_{12}O_2} \quad \frac{12 \text{ moles H atoms}}{1 \text{ mole } C_6H_{12}O_2} \quad \frac{2 \text{ moles O atoms}}{1 \text{ mole } C_6H_{12}O_2}$$

Using the first of these conversion factors, the moles of C atoms present are calculated as follows:

$$2.65 \text{ moles } C_6H_{12}O_2 \times \frac{6 \text{ moles C atoms}}{1 \text{ mole } C_6H_{12}O_2} = 15.9 \text{ moles C atoms}$$

(calculator and **correct answer**)

Similarly, using the second conversion factor, the moles of H atoms present are calculated.

$$2.65 \text{ moles } C_6H_{12}O_2 \times \frac{12 \text{ moles H atoms}}{1 \text{ mole } C_6H_{12}O_2} = 31.8 \text{ moles H atoms}$$

(calculator and **correct answer**)

Finally, using the third conversion factor, we obtain the moles of O atoms present.

$$2.65 \text{ moles } C_6H_{12}O_2 \times \frac{2 \text{ moles O atoms}}{1 \text{ mole } C_6H_{12}O_2} = 5.3 \text{ moles O atoms} \quad \text{(calculator answer)}$$

$$= 5.30 \text{ moles O atoms} \quad \textbf{(correct answer)}$$

If the question "How many total moles of atoms are present?" were asked, we could obtain the answer by adding the moles of C, H, and O atoms just calculated.

$$(15.9 + 31.8 + 5.30) \text{ moles atoms} = 53 \text{ moles atoms} \quad \text{(calculator answer)}$$

$$= 53.0 \text{ moles atoms} \quad \textbf{(correct answer)}$$

Alternatively, by noting that there are a total of 20 moles of atoms present in 1 mole of $C_6H_{12}O_2$ (the sum of the subscripts in the formula), we could calculate the total moles of atoms present using the following setup:

$$2.65 \text{ moles } C_6H_{12}O_2 \times \frac{20 \text{ moles of atoms}}{1 \text{ mole } C_6H_{12}O_2} = 53 \text{ moles of atoms} \quad \text{(calculator answer)}$$

$$= 53.0 \text{ moles of atoms} \quad \textbf{(correct answer)}$$

CHEMICAL EXTENSION

Compounds called *esters* are largely responsible for the flavor and fragrance of fruits and flowers. Generally, a natural flavor or odor is caused by a mixture of esters, with one particular compound being dominant. The synthetic production of these "dominant" compounds is the basis for the flavoring agents used in ice cream, gelatins, soft drinks, and so on.

The general structure for an ester is

Pineapple flavor (ethyl butyrate) involves the ester with three carbon atoms on the left and two carbon atoms on the right.

What is surprising about ester flavoring agents is how closely some of them resemble each other in structure. For example, apple and pineapple flavors differ by one carbon atom (on the right); a 5-carbon-atom chain versus an 8-carbon-atom chain makes the difference between banana and orange flavors.

Practice Exercise 9.10

How many moles of each type of atom are present in each of the following molar quantities?

(a) 0.753 mole of CO_2 molecules

(b) 1.31 moles of P_4O_{10} molecules

Table 9.3 serves as a summary of the mole relationships we have considered up to this point in the chapter. Note, through comparison of the third and fourth entries in Table 9.3, that the same-size molar quantity of a given type of atom and its monoatomic ion are considered to have the same mass. This is because the mass of electrons is so very small (Sec. 5.5).

Table 9.3	Mole Relationships			
Name	Formula	Formula Mass (amu)	Mass of 1 Mole Formula Units (g)	Number and Kind of Particles in 1 Mole
Atomic nitrogen	N	14.01	14.01	6.022×10^{23} N atoms
Molecular nitrogen	N_2	28.02	28.02	$\begin{cases} 6.022 \times 10^{23} \ N_2 \ \text{molecules} \\ 2(6.022 \times 10^{23}) \ \text{N atoms} \end{cases}$
Zinc	Zn	65.41	65.41	6.022×10^{23} Zn atoms
Zinc ions	Zn^{2+}	65.41*	65.41	$6.022 \times 10^{23} \ Zn^{2+}$ ions
Calcium chloride	$CaCl_2$	110.98	110.98	$\begin{cases} 6.022 \times 10^{23} \ CaCl_2 \ \text{units} \\ 6.022 \times 10^{23} \ Ca^{2+} \ \text{ions} \\ 2(6.022 \times 10^{23}) \ Cl^- \ \text{ions} \end{cases}$

Recall that the electron has negligible mass; thus ions and atoms have essentially the same mass.

9.11   The Mole and Chemical Calculations

In this section we combine the major points we have learned about moles in previous sections to produce a general approach to problem solving that is applicable to a variety of types of chemical calculations.

The three quantities most often calculated in chemical problems are

1. The number of *particles* of a substance, that is, the number of atoms, molecules, or formula units.
2. The number of *moles* of a substance.
3. The number of *grams* (mass) of a substance.

These quantities are interrelated. The conversion factors dealing with these relationships, as previously noted, involve the concepts of (1) Avogadro's number, (2) molar mass, and (3) molar interpretation of chemical formula subscripts.

1. Avogadro's number (Sec. 9.5) provides a relationship between the number of particles of a substance and the number of moles of the same substance.

Particles of substance	Avogadro's number	Moles of substance

2. Molar mass (Sec. 9.6) provides a relationship between the number of grams of a substance and the number of moles of the same substance.

Grams of substance	Molar mass	Moles of substance

3. Molar interpretation of chemical formula subscripts (Sec. 9.10) provides a relationship between the number of moles of a substance and the number of moles of its component parts.

Moles of compound	Formula subscript	Moles of element within compound

The preceding three concepts can be combined into a single diagram that is very useful in problem solving. This diagram, Figure 9.5, can be viewed as a "road map" from which conversion factor sequences (pathways) can be obtained. It gives all the needed relationships for solving two general types of problems.

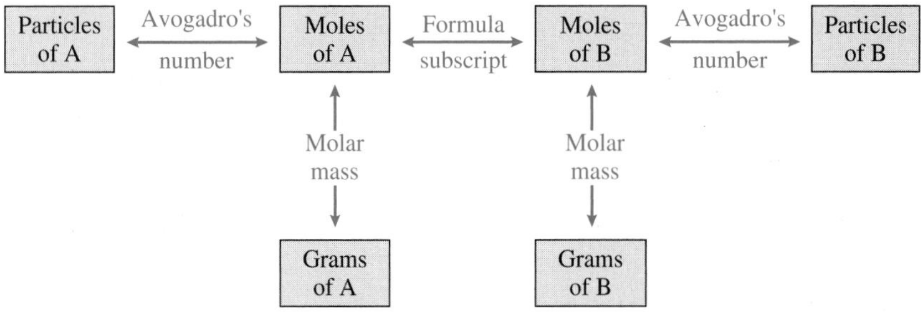

Figure 9.5

Useful relationships for solving chemical-formula–based problems.

1. Calculations for which information (moles, grams, particles) is given about a particular substance, and additional information (moles, grams, particles) is needed concerning the *same* substance.

2. Calculations for which information (moles, grams, particles) is given about a particular substance, and information (moles, grams, particles) is needed concerning a *component* of that same substance.

For the first type of problem, only the left side of Figure 9.5 (the A boxes) is needed. For problems of the second type, both sides of the diagram (both A and B boxes) are used.

The thinking pattern needed to use Figure 9.5 is very simple.

1. Determine which box in the diagram represents the *given* quantity in the problem.

2. Next, locate the box that represents the *desired* quantity.

3. Finally, follow the indicated pathway that takes you from the *given* quantity to the *desired* quantity. This involves simply following the arrows. There will always be only one pathway possible for the needed transition.

Examples 9.11 to 9.15 illustrate a few of the types of problems that can be solved using the relationships shown in Figure 9.5. In the first three examples we will need only the A side of the diagram; the following two examples make use of both the A and B sides of the diagram.

EXAMPLE 9.11

Calculating the Number of Particles in a Given Mass of Substance

Nicotine, the second most widely used central nervous system stimulant in our society (caffeine is first), occurs naturally in tobacco leaves. Its chemical formula is $C_{10}H_{14}N_2$. How many nicotine molecules are present in a 0.0015-g sample of nicotine (a typical amount in a cigarette)?

SOLUTION

We will solve this problem by using the three steps of dimensional analysis (Sec. 3.7) and Figure 9.5.

STEP 1 The given quantity is 0.0015 g of $C_{10}H_{14}N_2$, and the desired quantity is molecules of $C_{10}H_{14}N_2$.

$$0.0015 \text{ g } C_{10}H_{14}N_2 = ? \text{ molecules } C_{10}H_{14}N_2$$

In terms of Figure 9.5, this is a "grams of A" to "particles of A" problem. We are given grams of substance A and desire to find particles (molecules) of that same substance.

STEP 2 Figure 9.5 gives us the "pathway" (sequence of conversion factors) needed to work the problem. We want to start in the "grams of A" box and end up in the "particles of A" box. The pathway is

$$\boxed{\text{Grams of A}} \xrightarrow[\text{mass}]{\text{Molar}} \boxed{\text{Moles of A}} \xrightarrow[\text{number}]{\text{Avogadro's}} \boxed{\text{Particles of A}}$$

Using dimensional analysis, the setup for this sequence of conversion factors is

$$0.0015 \text{ g } C_{10}H_{14}N_2 \times \frac{1 \text{ mole } C_{10}H_{14}N_2}{162.26 \text{ g } C_{10}H_{14}N_2} \times \frac{6.022 \times 10^{23} \text{ molecules } C_{10}H_{14}N_2}{1 \text{ mole } C_{10}H_{14}N_2}$$

$$\text{grams A} \longrightarrow \text{moles A} \longrightarrow \text{particles A}$$

The number 162.26 that was used in the first conversion factor is the formula mass of $C_{10}H_{14}N_2$. It is not given in the problem, but had to be calculated using atomic masses and the operational rules given in Section 9.3.

STEP 3 The solution to the problem, obtained by doing the arithmetic, is

$$\frac{0.0015 \times 1 \times 6.022 \times 10^{23}}{162.26 \times 1} \text{ molecules } C_{10}H_{14}N_2$$

$$= 5.5669912 \times 10^{18} \text{ molecules } C_{10}H_{14}N_2 \quad \text{(calculator answer)}$$

$$= 5.6 \times 10^{18} \text{ molecules } C_{10}H_{14}N_2 \quad \textbf{(correct answer)}$$

CHEMICAL EXTENSION

Nicotine is readily and completely absorbed from the stomach after oral administration and from the lungs upon inhalation. Its mild effect on the central nervous system is rather transient. After an initial response, depression follows. Most cigarettes contain between 0.5 mg and 2.0 mg of nicotine, of which approximately 20% (between 0.1 mg and 0.4 mg) will actually be inhaled and absorbed into the bloodstream.

Nicotine is quickly distributed throughout the body, rapidly penetrating the brain, all body organs, and, in general, all body fluids. There is now strong evidence that cigarette smoking affects a developing fetus.

Nicotine can influence the action of prescription drugs that a smoker is taking concurrently. The carcinogenic effects associated with cigarette smoking are not caused by nicotine but by other compounds present in tobacco.

Practice Exercise 9.11

The koala bear feeds exclusively on eucalyptus leaves. Its digestive system detoxifies the eucalyptus oil, a poison to other animals. The predominant compound in eucalyptus oil is eucalyptol, which has the formula of $C_{10}H_{18}O$. How many eucalyptol molecules are present in a 2.00-g sample of this compound?

EXAMPLE 9.12

Calculating the Mass of a Given Number of Particles of a Substance

The "starting material" for the production of photochemical smog is the air pollutant nitrogen dioxide (NO_2). Its interaction with ultraviolet light from the sun produces the chemical species needed for smog production. What would be the mass, in grams, of an NO_2 sample in which 100 billion (1.000×10^{11}) molecules are present?

SOLUTION

STEP 1 The given quantity is 1.000×10^{11} NO_2 molecules. The desired quantity is grams of NO_2.

$$1.000 \times 10^{11} \ NO_2 \ \text{molecules} = ? \ \text{g} \ NO_2$$

In terms of Figure 9.5, this is a "particles of A" to "grams of A" problem.

STEP 2 The pathway for this problem is the exact reverse of the one used in the previous example. We are given particles and asked to find grams of the same substance.

$$\boxed{\text{Particles of A}} \xrightarrow[\text{number}]{\text{Avogadro's}} \boxed{\text{Moles of A}} \xrightarrow[\text{mass}]{\text{Molar}} \boxed{\text{Grams of A}}$$

Using dimensional analysis, the setup is

$$1.000 \times 10^{11} \ \overline{NO_2 \ \text{molecules}} \quad \times \quad \frac{1 \ \overline{\text{mole } NO_2}}{6.022 \times 10^{23} \ \overline{NO_2 \ \text{molecules}}} \quad \times \quad \frac{46.01 \ \text{g} \ NO_2}{1 \ \overline{\text{mole } NO_2}}$$

$$\underbrace{\text{particles A}} \quad \longrightarrow \quad \underbrace{\text{moles A}} \quad \longrightarrow \quad \underbrace{\text{grams A}}$$

The molar mass of NO_2, 46.01 g, which was used in the second conversion factor, was calculated from the atomic masses of nitrogen and oxygen. Avogadro's number in the first conversion factor, specified to four significant figures, is consistent with the given number 1.000×10^{11} having four significant figures.

STEP 3 The final answer is obtained by doing the arithmetic.

$$\frac{1.000 \times 10^{11} \times 1 \times 46.01}{6.022 \times 10^{23} \times 1} \text{g} \ NO_2$$

$$= 7.6403118 \times 10^{-12} \ \text{g} \ NO_2 \quad \text{(calculator answer)}$$

$$= 7.640 \times 10^{-12} \ \text{g} \ NO_2 \quad \textbf{(correct answer)}$$

CHEMICAL EXTENSION

The molecule nitrogen dioxide (NO_2) is an example of a simple molecule whose bonding does not conform to the standard rules for bonding (Sec. 7.16). The presence of an odd number of valence electrons in nitrogen dioxide (17; 5 from nitrogen and 6 from each oxygen) makes it impossible to write a Lewis structure in which all electrons are paired as required by the octet rule. Thus, an unpaired electron is present in the Lewis structure of NO_2 (on the N atom).

$$:\ddot{O}::\dot{N}:\ddot{O}:$$

Despite the "nonconforming" nature of the bonding in nitrogen dioxide, this nonflammable reddish-brown colored gas is a naturally occurring compound present in the atmosphere in low concentrations. The unpaired electron present causes NO_2 to be very reactive; it is the precursor molecule for the formation of ozone in the lower atmosphere. Certain wavelengths of radiation from the sun cause NO_2 molecules to break apart into NO and O (atomic oxygen). The very reactive atomic oxygen reacts with "normal" oxygen (O_2) to produce ozone (O_3). Ozone is the most abundant irritant present in photochemical smog.

Practice Exercise 9.12

The artificial sweetener aspartame (Nutra-Sweet), with the chemical formula $C_{14}H_{18}N_2O_5$, is the sugar substitute used in most diet soft drinks. What would be the mass in grams of a pure sample of this sweetener that contains 1 billion (1.000×10^9) molecules?

EXAMPLE 9.13

Calculating the Mass of an Atom or a Molecule

Calculate the mass, in grams, of each of the following chemical entities.

(a) a single atom of Cu **(b)** a single molecule of CO_2

SOLUTION

(a)

STEP 1 The given quantity is 1 atom of Cu, and the desired quantity is grams of Cu.

$$1 \text{ atom Cu} = ? \text{ g Cu}$$

In terms of Figure 9.5 this is a "particles of A" to "grams of A" problem.

STEP 2 The pathway for this problem is identical to that used in the previous example. We are given particles (one particle in this case), and asked to find grams of the same substance.

$$\boxed{\text{Particles of A}} \xrightarrow[\text{number}]{\text{Avogadro's}} \boxed{\text{Moles of A}} \xrightarrow[\text{mass}]{\text{Molar}} \boxed{\text{Grams of A}}$$

Using dimensional analysis, the setup is

$$1 \text{ atom Cu} \quad \times \quad \frac{1 \text{ mole Cu}}{6.022 \times 10^{23} \text{ atoms Cu}} \quad \times \quad \frac{63.55 \text{ g Cu}}{1 \text{ mole Cu}}$$

$$\text{particles A} \longrightarrow \qquad \text{moles A} \qquad \longrightarrow \quad \text{grams A}$$

The second conversion factor contains the atomic mass of Cu, which was obtained from the periodic table.

STEP 3 The solution to the problem, obtained by doing the arithmetic, is

$$\frac{1 \times 1 \times 63.55}{6.022 \times 10^{23} \times 1} \text{ g Cu} = 1.0552972 \times 10^{-22} \text{ g Cu} \quad \text{(calculator answer)}$$

$$= 1.055 \times 10^{-22} \text{ g Cu} \qquad \textbf{(correct answer)}$$

The mass of a single atom is, indeed, a very small mass in the unit of grams — 0.0000000000000000000001055 g. Commonly used analytical balances are capable of weighing to ±0.0001 g. The mass of a single atom can be calculated but, obviously, it cannot be determined using an analytical balance.

(b)

STEP 1 The given quantity is 1 molecule of CO_2 and the desired quantity is grams of CO_2.

$$1 \text{ molecule } CO_2 = ? \text{ g } CO_2$$

In terms of Figure 9.5, this problem, like part (a), is a "particles of A" to "grams of A" problem.

STEP 2 We are dealing with an identical setup to that in part (a), except that a molecule rather than an atom is the particle of concern.

$$\boxed{\text{Particles of A}} \xrightarrow[\text{number}]{\text{Avogadro's}} \boxed{\text{Moles of A}} \xrightarrow[\text{mass}]{\text{Molar}} \boxed{\text{Grams of A}}$$

Using dimensional analysis, the conversion factor setup becomes

$$1 \text{ molecule } CO_2 \times \frac{1 \text{ mole } CO_2}{6.022 \times 10^{23} \text{ molecules } CO_2} \times \frac{44.01 \text{ g } CO_2}{1 \text{ mole } CO_2}$$

particles A $\longrightarrow$ moles A $\longrightarrow$ grams A

The second conversion factor involves the molar mass of CO_2, which is calculated using the atomic masses of carbon and oxygen obtained from the periodic table.

STEP 3 Collecting the conversion factors and doing the arithmetic gives

$$\frac{1 \times 1 \times 44.01}{6.022 \times 10^{23} \times 1} \text{ g } CO_2 = 7.3082032 \times 10^{-23} \text{ g } CO_2 \quad \text{(calculator answer)}$$

$$= 7.308 \times 10^{-23} \text{ g } CO_2 \quad \textbf{(correct answer)}$$

Note that a Cu atom [part (a)] weighs more than a CO_2 molecule [part (b)]. The atoms present in a CO_2 molecule are small atoms when compared to a larger Cu atom.

Practice Exercise 9.13

Calculate the mass, in grams, of each of the following chemical entities.

(a) a single atom of Fe   **(b)** a single molecule of CH_4

EXAMPLE 9.14

Calculating the Mass of an Element Present in a Given Mass of Compound

Caffeine, the stimulant in coffee and tea, has the formula $C_8H_{10}N_4O_2$. How many grams of nitrogen are present in a 50.0-g sample of caffeine?

SOLUTION

STEP 1 There is an important difference between this problem and the preceding three; here we are dealing with not one but two substances, caffeine and nitrogen. The given quantity is grams of caffeine (substance A), and we are asked to find the grams of nitrogen (substance B). This is a "grams of A" to "grams of B" problem.

$$50.0 \text{ g } C_8H_{10}N_4O_2 = ? \text{ g N}$$

STEP 2 The appropriate set of conversions for a "grams of A" to "grams of B" problem, from Figure 9.5, is

| Grams of A | $\xrightarrow{\text{Molar mass}}$ | Moles of A | $\xrightarrow{\text{Formula subscript}}$ | Moles of B | $\xrightarrow{\text{Molar mass}}$ | Grams of B |

The mathematical setup involving conversion factors is

$$50.0 \text{ g } C_8H_{10}N_4O_2 \times \frac{1 \text{ mole } C_8H_{10}N_4O_2}{194.22 \text{ g } C_8H_{10}N_4O_2} \times \frac{4 \text{ moles N}}{1 \text{ mole } C_8H_{10}N_4O_2} \times \frac{14.01 \text{ g N}}{1 \text{ mole N}}$$

grams A $\longrightarrow$ moles A $\longrightarrow$ moles B $\longrightarrow$ grams B

The number 194.22 that is used in the first conversion factor is the formula mass for caffeine. The conversion from "moles of A" to "moles of B" (the second conversion factor) is derived using the information contained in the formula for caffeine. One mole of caffeine contains 4 moles of N because the subscript for N in the formula is 4. The number 14.01 in the final conversion factor is the atomic mass of nitrogen.

STEP 3 Collecting the numbers from the various conversion factors together and doing the arithmetic gives us our answer.

$$\frac{50.0 \times 1 \times 4 \times 14.01}{194.22 \times 1 \times 1} \text{ g N} = 14.426938 \text{ g N} \quad \text{(calculator answer)}$$

$$= 14.4 \text{ g N} \quad \textbf{(correct answer)}$$

CHEMICAL EXTENSION

Caffeine, as well as nicotine (Example 9.11), are examples of compounds that are called *alkaloids*. Alkaloids are physiologically active compounds produced by plants.

People in various parts of the world have known for centuries that physiological effects can be obtained by eating or chewing the leaves, roots, or bark of certain plants. Over 5000 different compounds that are physiologically active have been isolated from such plants.

Besides caffeine (coffee beans and tea leaves) and nicotine (tobacco leaves), other alkaloids include cocaine (coca plant), quinine, atropine, and opium. The latter three alkaloids are currently used in medicine.

Quinine, which occurs in cinchona bark, is used to treat malaria. Atropine, which is isolated from the belladonna plant, is used to dilate the pupil of the eye in patients undergoing eye examinations. Opium, the dried juice of the oriental poppy plant, has been used for centuries as a pain-killing drug. It contains numerous alkaloids, including morphine and codeine.

Practice Exercise 9.14

Glucose, $C_6H_{12}O_6$, is one of the main sources of energy used by living organisms. How many grams of oxygen are present in 10.0 g of glucose?

EXAMPLE 9.15

Calculating the Number of Particles of an Element Present in a Given Mass of a Compound

The compound cholesterol has the chemical formula $C_{27}H_{46}O$. How many hydrogen atoms would be present in a 2.000-g sample of cholesterol?

SOLUTION

STEP 1 The given quantity is 2.000 g of $C_{27}H_{46}O$ (cholesterol) and the desired quantity is atoms of hydrogen.

$$2.000 \text{ g } C_{27}H_{46}O = ? \text{ atoms H}$$

In the jargon of Figure 9.5, this is a "grams of A" to "particles of B" problem.

STEP 2 From Figure 9.5, the appropriate pathway for solving this problem is

| Grams of A | $\xrightarrow[\text{mass}]{\text{Molar}}$ | Moles of A | $\xrightarrow[\text{subscript}]{\text{Formula}}$ | Moles of B | $\xrightarrow[\text{number}]{\text{Avogadro's}}$ | Particles of B |

Using dimensional analysis, the conversion factor setup becomes

$$2.000 \text{ g } C_{27}H_{46}O \times \frac{1 \text{ mole } C_{27}H_{46}O}{386.73 \text{ g } C_{27}H_{46}O} \times \frac{46 \text{ moles H}}{1 \text{ mole } C_{27}H_{46}O} \times \frac{6.022 \times 10^{23} \text{ atoms H}}{1 \text{ mole H}}$$

grams A $\longrightarrow$ moles A $\longrightarrow$ moles B $\longrightarrow$ particles B

The numbers in the second conversion factor (46 and 1) were obtained from the chemical formula for cholesterol; they are, thus, exact numbers. One mole of $C_{27}H_{46}O$ contains 46 moles of hydrogen; the hydrogen subscript is 46.

STEP 3 The solution to the problem, obtained by doing the arithmetic, is

$$\frac{2.000 \times 1 \times 46 \times 6.022 \times 10^{23}}{386.73 \times 1 \times 1} \text{ atoms H} = 1.432586 \times 10^{23} \text{ atoms H}$$

(calculator answer)

$$= 1.433 \times 10^{23} \text{ atoms H}$$

(**correct answer**)

CHEMICAL EXTENSION

The compound cholesterol is absolutely necessary for the proper functioning of the human body. Within the human body, cholesterol is an important component of cell membranes (up to 25% by mass), nerve tissue, and brain tissue (about 10% by dry mass).

Cholesterol is the precursor (starting material) for the synthesis of many other compounds that the body needs. These other compounds include bile salts (necessary for the digestion of certain types of foods), male and female sex hormones, and vitamin D.

The human body, mainly within the liver, synthesizes about 1 gram of cholesterol each day, an amount sufficient to meet the body's biosynthetic needs. Therefore, cholesterol is not necessary in the diet. When we ingest cholesterol, the amount synthesized by the body is reduced. However, the reduction is less than the amount ingested, resulting in an increase in total body cholesterol levels. Medical science now considers such higher-than-normal cholesterol levels a risk factor for cardiovascular disease.

Practice Exercise 9.15

Ethylene glycol, the major ingredient in most automobile antifreeze solutions, has the formula $C_2H_6O_2$. How many atoms of oxygen are present in a 2.000-g sample of ethylene glycol?

9.12 Purity of Samples

Most substances used in chemical laboratories and most substances encountered in everyday situations contain impurities, that is, they are not 100% pure. Sometimes the impurities are naturally present substances, and other times impurities have been added on purpose to give a desired effect.

Drinking water contains naturally present minerals (impurities) that give drinking water its taste; distilled water, which lacks such minerals, has a very "flat" taste. Table salt contains potassium iodide as an additive (an impurity). This additive serves as an iodine source for the human body; iodine is necessary for the proper functioning of the thyroid gland.

The amount of impurities present in a sample of a substance is specified using *percent purity*. **Percent purity** *is the percent by mass of a specified substance in an impure sample of the substance.* For example, the purity of laboratory-grade sodium hydroxide (NaOH) is 98.2% by mass.

Percent purity values can be used as conversion factors in several types of problem-solving situations. The information that the purity of a sodium chloride (NaCl) sample is 99.23% by mass is a source for six conversion factors. Three of the conversion factors are

$$\frac{99.23 \text{ g NaCl}}{100 \text{ g impure NaCl}} \qquad \frac{0.77 \text{ g impurities}}{100 \text{ g impure NaCl}} \qquad \frac{0.77 \text{ g impurities}}{99.23 \text{ g NaCl}}$$

The reciprocals of these conversion factors are the other three of the six relationships.

> A general discussion of the use of percent as a conversion factor has been previously given. It is found in Section 3.10.

EXAMPLE 9.16

Calculating Masses of Substances in an Impure Sample

A 32.00-g sample of nitric acid (HNO_3) has a purity of 96.20% by mass. Calculate the following for this sample of nitric acid.

(a) the mass, in grams, of HNO_3 present
(b) the mass, in grams, of impurities present

SOLUTION

(a) From the given percent purity, the following needed conversion factor is obtained.

$$\frac{96.20 \text{ g HNO}_3}{100 \text{ g impure HNO}_3}$$

The dimensional analysis setup, using this conversion factor, is

$$32.00 \text{ g impure HNO}_3 \times \frac{96.20 \text{ g HNO}_3}{100 \text{ g impure HNO}_3}$$

Canceling units and doing the arithmetic gives

$$\frac{32.00 \times 96.20}{100} \text{ g HNO}_3 = 30.784 \text{ g HNO}_3 \quad \text{(calculator answer)}$$

$$= 30.78 \text{ g HNO}_3 \quad \text{(correct answer)}$$

(b) Similarly, we have, for the amount of impurities present, the dimensional analysis setup

$$32.00 \text{ g impure HNO}_3 \times \frac{3.80 \text{ g impurities}}{100 \text{ g impure HNO}_3}$$

$$= 1.216 \text{ g impurities} \quad \text{(calculator answer)}$$

$$= 1.22 \text{ g impurities} \quad \text{(correct answer)}$$

CHEMICAL EXTENSION

Pure nitric acid, HNO_3, is a colorless liquid that boils at 83°C. Light or heat causes it to decompose into NO_2, O_2, and H_2O. The presence of the NO_2 in partially decomposed nitric acid (or aqueous solutions of nitric acid) produces a yellow or brown tinge. Nitric acid is extremely soluble in water; 16 moles will dissolve in a liter of water. The nitric acid encountered in a chemical laboratory always involves aqueous solutions of the acid.

Nitric acid was once called "aqua fortis," Latin for "strong water." It dissolves several metals, including silver and copper, which are not attacked by other acids. Nitric acid stains proteins, including woolen fabrics and human skin, a bright yellow color.

The largest use for nitric acid is in nitrate fertilizer production; it is the source for nitrate ions. Nitric acid is also important in the production of explosives, including nitroglycerin and trinitrotoluene (TNT).

Practice Exercise 9.16

A 90.00-g sample of sulfuric acid (H_2SO_4) has a purity of 98.45% by mass. Calculate the following for this sample of sulfuric acid.

(a) the mass, in grams, of H_2SO_4 present

(b) the mass, in grams, of impurities present

Percent purity, when appropriate, can be incorporated into "grams to moles to particles" problems like those previously considered in Section 9.11. Example 9.17 illustrates how this is done.

EXAMPLE 9.17

Calculating Atoms of a Substance in an Impure Sample

How many iron (Fe) atoms are present in a 25.00-g sample of iron (iron ore) that has a purity of 87.70% by mass iron? Assume that there are no iron-containing impurities present.

SOLUTION

If the iron sample were 100% pure, this calculation would be a standard "grams of A" to "particles of A" problem, in terms of Figure 9.5. The pathway would be

$$\boxed{\text{Grams of A}} \xrightarrow[\text{mass}]{\text{Molar}} \boxed{\text{Moles of A}} \xrightarrow[\text{number}]{\text{Avogadro's}} \boxed{\text{Particles of A}}$$

To take into account the fact that the sample is impure, we add a new step at the start of the pathway that involves "grams of impure A" to "grams of A". The enhanced pathway is

$$\boxed{\text{Grams of Impure A}} \xrightarrow[\text{purity}]{\text{Percent}} \boxed{\text{Grams of A}} \xrightarrow[\text{mass}]{\text{Molar}} \boxed{\text{Moles of A}} \xrightarrow[\text{number}]{\text{Avogadro's}} \boxed{\text{Particles of A}}$$

The dimensional analysis setup consistent with the new pathway is

$$25.00 \text{ g impure Fe} \times \frac{87.70 \text{ g Fe}}{100 \text{ g impure Fe}} \times \frac{1 \text{ mole Fe}}{55.84 \text{ g Fe}} \times \frac{6.022 \times 10^{23} \text{ atoms Fe}}{1 \text{ mole Fe}}$$

Collecting the numbers from the various conversion factors together and doing the arithmetic gives us our answer.

$$\frac{25.00 \times 87.70 \times 1 \times 6.022 \times 10^{23}}{100 \times 55.84 \times 1} \text{ atoms Fe} = 2.3644762 \times 10^{23} \text{ atoms Fe}$$

$$\text{(calculator answer)}$$

$$= 2.364 \times 10^{23} \text{ atoms}$$
$$\text{(correct answer)}$$

CHEMICAL EXTENSION

Iron is the most used of all metals. Both worldwide and in the United States, more iron is consumed than all other metals combined. Iron consumption is not, however, in the form of elemental iron. Rather, the end product form for almost all iron consumption is that of steel, an iron alloy. An alloy is a solid material with metallic properties that is produced by melting together two or more elements, at least one of which is a metal.

There are two major types of steel: (1) carbon steel, and (2) alloy steel. Both types are metal alloys containing iron and carbon. Alloy steels contain one or more other metals in addition to iron. Approximately 90% of all steel produced in the United States is carbon steel.

The properties of steel depend to a large extent on the percentage of carbon present. Low-carbon steel (less than 0.25% C by mass) is very ductile and is used in such products as steel cans, wire, nails, and sheet metal for automobile bodies. Medium-carbon steel (0.25% to 0.7% C by mass) is less ductile and much stronger and is used in railroad rails, machine axles, beams, and girders. High-carbon steel (0.7% to 1.5% C by mass) is very hard and is used for springs, razor blades, surgical instruments, and tools.

In alloy steels, particular metals improve specific characteristics of the steel. Chromium improves hardness and resistance to corrosion. Molybdenum and tungsten increase heat resistance. Nickel increases toughness. Vanadium adds springiness. Manganese improves resistance to wear. The best-known alloy steel is *stainless steel*. This corrosion-resistant steel contains 14–18% Cr and 7–9% nickel.

Practice Exercise 9.17

How many carbon dioxide (CO_2) molecules are present in a 37.75-g sample of carbon dioxide that has a purity of 99.37% by mass?

9.13 Empirical and Molecular Formulas

Chemical formulas provide a great deal of useful information about the substances they represent and, as seen in previous sections of this chapter, are key entities in many types of calculations. How are formulas themselves determined? They are determined by calculation using experimentally obtained information.

Depending on the amount of experimental information available, two types of chemical formulas may be obtained: an *empirical formula* or a *molecular formula*. The **empirical formula** (*or simplest formula*) *is a chemical formula that gives the* smallest *whole number ratio of atoms present in a formula unit of a compound*. In empirical formulas the subscripts in the formula cannot be reduced to a simpler set of numbers by division with a small integer. The **molecular formula** (*or true formula*) *is a chemical formula that gives the* actual *number of atoms present in a formula unit of a compound*.

For ionic compounds the empirical and molecular formulas are almost always the same; that is, the actual ratio of ions present in a formula unit and the smallest ratio of ions present are the same. For molecular compounds the two types of formulas may be the same, but frequently they are not. When they are not the same, the molecular formula is a multiple of the empirical formula.

molecular formula = whole number multiplier × empirical formula

Table 9.4 contrasts the differences between the two types of formulas for selected compounds.

Table 9.4	A Comparison of Empirical and Molecular Formulas for Selected Compounds		
Compound	Empirical Formula	Molecular Formula	Whole Number Multiplier
Dinitrogen tetrafluoride	NF_2	N_2F_4	2
Hydrogen peroxide	HO	H_2O_2	2
Sodium chloride	$NaCl$	$NaCl$	1
Benzene	CH	C_6H_6	6

EXAMPLE 9.18

Obtaining Empirical Formulas from Molecular Formulas

Write the empirical formula for each of the following compounds.

(a) C_2H_4 (ethene, used to make polyethylene)
(b) C_8H_{18} (octane, a component of gasoline)
(c) CH_4O (methanol, an industrial solvent)
(d) $C_6H_{12}O_6$ (glucose, often called blood sugar)

SOLUTION

(a) Each of the formula subscripts is divided by 2 to give the empirical formula CH_2.
(b) Each of the formula subscripts is divided by 2 to give the empirical formula C_4H_9.
(c) There is no small number that will divide evenly into each of the formula subscripts other than 1. Thus, the empirical formula is the same as the given formula: CH_4O.
(d) Each of the formula subscripts is divided by 6 to give the empirical formula CH_2O.

Practice Exercise 9.18

Write the empirical formulas for each of the following compounds.

(a) P_4O_{10} (b) $C_3H_{10}N_2$ (c) C_2H_6O (d) $B_3N_3H_6$

9.14 Determination of Empirical Formulas

The determination of a compound's chemical formula from experimental data is usually carried out in two calculational steps: (1) elemental composition data are used to determine the compound's empirical formula (simplest ratio of atoms present—Sec. 9.13), and (2) this empirical formula and molecular mass data are used to determine the compound's molecular formula (actual ratio of atoms present—Sec. 9.13).

In this section we consider the first of these two steps, and then in Section 9.15 we learn the additional procedures needed to go from an empirical formula to a molecular formula.

Elemental composition data are sufficient for an empirical formula calculation. Such data may be in the form of percent composition or simply the mass of each element present in a known mass of compound. Example 9.19 shows the steps used in obtaining an empirical formula from percent composition data. Example 9.20 uses the mass of the elements present in a compound sample of known mass to obtain an empirical formula.

EXAMPLE 9.19

Determining an Empirical Formula from Percent Composition Data

Percent composition data for the chlorofluorocarbon Freon-12 by mass are: 9.933% carbon, 58.63% chlorine, and 31.44% fluorine. Based on these data, what is the empirical formula of Freon-12?

SOLUTION

The problem of calculating the empirical formula of a compound from percent composition data can be broken down into three steps.

1. Determine the number of grams of each element in a sample of the compound.
2. Convert the grams of each element to moles of element.
3. Express the mole ratio between the elements in terms of *small whole numbers*.

STEP 1 Mass percentage values are independent of sample size; that is, they apply to samples of all sizes. In working with mass percentages in empirical formula calculations, it is convenient to assume a 100.0-g sample size. When this is done, mass percentages translate directly into gram amounts. The mass of each element in the 100.0-g sample is numerically equal to the percentage value.

$$C: \quad 9.933\% \text{ of } 100.0 \text{ g} = \quad 9.933 \text{ g}$$

$$Cl: \quad 58.63\% \text{ of } 100.0 \text{ g} = 58.63 \text{ g}$$

$$F: \quad 31.44\% \text{ of } 100.0 \text{ g} = 31.44 \text{ g}$$

STEP 2 We next convert the grams (from step 1) to moles. We need moles information to determine the subscripts in the formula of the compound. Formula subscripts give the ratio of the number of moles of each element present in a compound (Sec. 9.9).

$$9.933 \text{ g C} \times \frac{1 \text{ mole C}}{12.01 \text{ g C}} = 0.82706078 \text{ mole C} \quad \text{(calculator answer)}$$

$$= 0.8271 \text{ mole C} \qquad \textbf{(correct answer)}$$

$$58.63 \text{ g Cl} \times \frac{1 \text{ mole Cl}}{35.45 \text{ g Cl}} = 1.6538787 \text{ moles Cl} \quad \text{(calculator answer)}$$

$$= 1.654 \text{ moles Cl} \qquad \textbf{(correct answer)}$$

$$31.44 \text{ g F} \times \frac{1 \text{ mole F}}{19.00 \text{ g F}} = 1.6547368 \text{ moles F} \quad \text{(calculator answer)}$$

$$= 1.655 \text{ moles F} \qquad \textbf{(correct answer)}$$

Thus, in 100.0 g of compound there are 0.8271 mole of C, 1.654 moles of Cl, and 1.655 moles of F.

STEP 3 The subscripts in a formula are expressed as whole numbers, not as decimals. To obtain whole numbers from the decimals of step 2, each of the numbers is divided by the smallest of the numbers.

$$C: \quad \frac{0.8271 \text{ mole}}{0.8271 \text{ mole}} = 1 \quad \text{(calculator answer)}$$

$$= 1.000 \qquad \textbf{(correct answer)}$$

$$\text{Cl:} \quad \frac{1.654 \text{ moles}}{0.8271 \text{ mole}} = 1.9997581 \quad \text{(calculator answer)}$$

$$= 2.000 \quad \textbf{(correct answer)}$$

$$\text{F:} \quad \frac{1.655 \text{ moles}}{0.8271 \text{ mole}} = 2.0009672 \quad \text{(calculator answer)}$$

$$= 2.001 \quad \textbf{(correct answer)}$$

Thus the ratio of carbon to chlorine to fluorine in this compound is 1 to 2 to 2, and the empirical formula is CCl_2F_2.

The calculated value for one of the three formula subscripts (2.001) has a nonzero digit to the right of the decimal place. This is because of experimental error in the original mass percentages. This calculated value is, however, sufficiently close to a whole number that we easily recognize what the whole number is.

CHEMICAL EXTENSION

Chlorofluorocarbons (CFCs) are compounds composed of the elements chlorine, fluorine, and carbon. CFCs are synthetic compounds that have been developed primarily for use as refrigerants. The most widely used of the CFCs are marketed under the trade name *Freon*.

Freons possess ideal properties for use as refrigerant gases. They are inert, nontoxic, and easily compressible. Prior to their development, ammonia was used in refrigeration. Ammonia is toxic, and leaking ammonia-based refrigeration units have been fatal.

Freon use in refrigeration and air conditioning has stopped because of an unanticipated environmental problem associated with their use. Once released into the atmosphere, CFCs persist for long periods without reaction. Consequently, they slowly drift upward in the atmosphere, finally reaching the stratosphere, the location of the "ozone layer."

It is in the stratosphere that environmental problems occur. At these high altitudes, the CFCs are exposed to ultraviolet light from the sun, which activates them. In an activated form, they upset the "ozone–oxygen balance" present there, decreasing the amount of ozone present.

Practice Exercise 9.19

Acetone is the chemical solvent present in fingernail polish remover. Its percent composition is 62.0% carbon, 10.4% hydrogen, and 27.5% oxygen. What is the empirical formula for the compound?

EXAMPLE 9.20

Determining an Empirical Formula Using Mass Data

Analysis of a sample of ibuprofen, the active ingredient in Advil, shows that the sample contains 7.568 g of carbon, 0.881 g of hydrogen, and 1.551 g of oxygen. Use these data to calculate the empirical formula of ibuprofen.

SOLUTION

Having data given in grams instead of as percent composition (Example 9.19) simplifies the calculation of an empirical formula. The grams are changed to moles, and we are ready to find the smallest whole-number ratio between elements.

STEP 1 Each of the given gram amounts is converted to moles using dimensional analysis.

$$7.568 \text{ g C} \times \frac{1 \text{ mole C}}{12.01 \text{ g C}} = 0.63014154 \text{ mole C} \quad \text{(calculator answer)}$$

$$= 0.6301 \text{ mole C} \quad \textbf{(correct answer)}$$

$$0.881 \text{ g H} \times \frac{1 \text{ mole H}}{1.01 \text{ g H}} = 0.87227722 \text{ mole H} \quad \text{(calculator answer)}$$

$$= 0.872 \text{ mole H} \quad \textbf{(correct answer)}$$

$$1.551 \text{ g O} \times \frac{1 \text{ mole O}}{16.00 \text{ g O}} = 0.0969375 \text{ mole O} \quad \text{(calculator answer)}$$

$$= 0.09694 \text{ mole O} \quad \textbf{(correct answer)}$$

STEP 2 Dividing the mole quantities from step 1 by the *smallest* of the three numbers gives

$$\text{C:} \quad \frac{0.6301 \text{ mole}}{0.09694 \text{ mole}} = 6.4998968 \quad \text{(calculator answer)}$$

$$= 6.500 \quad \textbf{(correct answer)}$$

$$\text{H:} \quad \frac{0.872 \text{ mole}}{0.09694 \text{ mole}} = 8.9952547 \quad \text{(calculator answer)}$$

$$= 9.00 \quad \textbf{(correct answer)}$$

$$\text{O:} \quad \frac{0.09694 \text{ mole}}{0.09694 \text{ mole}} = 1 \quad \text{(calculator answer)}$$

$$= 1.000 \quad \textbf{(correct answer)}$$

Frequently all whole numbers or near-whole numbers are obtained at this point (as was the case in Example 9.19), and the calculation is finished. Sometimes, however, as is the case in this example, we obtain one or more numbers that are not even close to being whole numbers. The number we obtained for carbon, 6.50, is such a number. What do we do?

The number 6.50 is recognizable as being close to a simple fraction; it is $6\frac{1}{2}$. When we obtain simple fractions, we clear them by multiplying *all of the numbers* in our set of numbers by a common factor. We multiply all numbers by 2 if we are dealing with halves, 3 if we are dealing with thirds, 4 if we are dealing with fourths, and so on. Following this procedure, we obtain for the situation in this example the following:

$$\text{C:} \quad 6\tfrac{1}{2} \times 2 = 13$$
$$\text{H:} \quad 9 \times 2 = 18$$
$$\text{O:} \quad 1 \times 2 = 2$$

We now have a whole-number ratio. The empirical formula of ibuprofen is $C_{13}H_{18}O_2$. The fractions that most commonly occur in empirical formula calculations are fourths, 0.25 $\left(\frac{1}{4}\right)$ and 0.75 $\left(\frac{3}{4}\right)$, which are multiplied by 4 to clear; thirds, 0.33 $\left(\frac{1}{3}\right)$ and 0.67 $\left(\frac{2}{3}\right)$, which are multiplied by 3 to clear; and half, 0.50 $\left(\frac{1}{2}\right)$, which is multiplied by 2 to clear.

CHEMICAL EXTENSION

Consumers are faced with shelves filled with choices when looking for an over-the-counter medicine to treat aches, pains, and fever. The multitude of brands available, however, represent only five chemical formulations. Besides the long-available aspirin and acetaminophen (Example 9.4), consumers can now purchase ibuprofen, naproxen, or ketoprofen.

Ibuprofen, marketed under the brand names Advil, Motrin-IB, and Nuprin, was cleared by the FDA in 1984 for nonprescription sales. Ibuprofen is more effective than either aspirin or acetaminophen in reducing dental and menstrual pain. However, it is more expensive.

Naproxen, marketed under the brand names Aleve and Anaprox, was cleared for nonprescription sales by the FDA in 1994. The effects of naproxen last longer in the body (8–12 hours per dose) than the effects of ibuprofen (4–6 hours) and of aspirin and acetaminophen (4 hours).

Ketoprofen, marketed under the brand names Orudis KT and Actron, was approved by the FDA in 1996 for over-the-counter sales. Consumers take ketoprofen in 12.5-mg doses, compared with the dosages of 200 to 500 mg for other over-the-counter pain relievers. The easy-to-swallow size of this product is touted by its manufacturers.

Practice Exercise 9.20

The tip of the head of a strike-anywhere match contains a phosphorus–sulfur compound that readily ignites when drawn over a rough surface. What is the empirical formula of this ignitable compound given that a sample of the compound contains 0.5629 g of P and 0.4371 g of S?

Figure 9.6 summarizes the procedures used in Examples 9.19 and 9.20 for calculating empirical formulas. If mass percents are the starting point for the calculation (Example 9.19), you start in box 1. If mass data is the starting point (Example 9.20), the procedure is shortened by one step as you start in box 2.

Examples 9.19 and 9.20 suggest that compounds are always broken down completely into their elements when they are analyzed. This is not always the case. Elemental analysis often involves changing the compound to be analyzed into other compounds that contain the elements present in the original compound. A common procedure of this type is *combustion analysis*. This procedure is used extensively for compounds containing carbon, hydrogen, and oxygen (or just carbon and hydrogen). **Combustion analysis** *is a method used to measure the amounts of carbon and hydrogen present in a combustible compound that contains these two elements (and perhaps other elements) when that compound is burned in pure O_2.* The basic idea in combustion analysis is as follows. A sample of the compound is burned completely in pure oxygen. The products of the reaction are CO_2 and H_2O. All carbon atoms originally present end up in CO_2 molecules, and all hydrogen atoms originally present end up in H_2O molecules (see Fig. 9.7). From the mass of CO_2 and H_2O produced, it is possible

(1)		(2)		(3)		(4)
Mass % of each element	Assume 100 g sample →	Grams of each element	Use atomic masses →	Moles of each element	Calculate mole ratios →	Empirical formula

Figure 9.6

Outline of the procedure used to calculate the empirical formula of a compound from (a) percentage composition data (start in box 1) and (b) mass data (start in box 2).

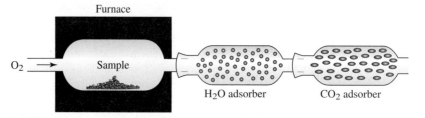

Furnace

O_2 → Sample H_2O adsorber CO_2 adsorber

Figure 9.7

Combustion analysis method for determining the percentages of carbon and hydrogen in a compound. A weighed sample of the compound is burned in a stream of oxygen, producing gaseous CO_2 and H_2O. These gases then pass through a series of tubes. One tube contains a substance that adsorbs H_2O; the substance in the other tube adsorbs CO_2. By comparing the masses of these tubes before and after the reaction, the analyst can determine the masses of hydrogen and carbon present in the compound that was burned.

to calculate the empirical formula of the original compound. Example 9.21 shows how this is done for a carbon–hydrogen compound, and Example 9.22 deals with the same situation for a carbon–hydrogen–oxygen compound.

EXAMPLE 9.21

Determining an Empirical Formula Using Combustion Analysis Data

Ethylene, a compound that contains only carbon and hydrogen, is commercially used as a fruit-ripening agent. A sample of this compound is burned in a combustion analysis apparatus and 3.14 g of CO_2 and 1.29 g of H_2O are produced. What is the empirical formula of ethylene?

SOLUTION

Every carbon atom in the CO_2 and every hydrogen atom in the H_2O came from ethylene. We will need to calculate the number of moles of carbon present in the CO_2 and then the number of moles of hydrogen present in the H_2O. Both these calculations are "grams of A" to "moles of B" problems in terms of the jargon of Figure 9.5.

The dimensional analysis setup from which we obtain the moles of carbon is

$$3.14 \text{ g } CO_2 \times \frac{1 \text{ mole } CO_2}{44.01 \text{ g } CO_2} \times \frac{1 \text{ mole C}}{1 \text{ mole } CO_2}$$

$$= 0.07134742 \text{ mole C} \quad \text{(calculator answer)}$$
$$= 0.0713 \text{ mole C} \quad \text{(correct answer)}$$

The last conversion factor used in this calculation comes from the formula CO_2. One molecule of CO_2 contains one atom of carbon. Thus, one mole of CO_2 will contain one mole of carbon.

The dimensional analysis setup from which we obtain the moles of hydrogen is

$$1.29 \text{ g } H_2O \times \frac{1 \text{ mole } H_2O}{18.02 \text{ g } H_2O} \times \frac{2 \text{ moles H}}{1 \text{ mole } H_2O}$$

$$= 0.14317425 \text{ mole H} \quad \text{(calculator answer)}$$
$$= 0.143 \text{ mole H} \quad \text{(correct answer)}$$

The last conversion factor in this setup is obtained from the information in the formula H_2O. There are 2 moles of H in 1 mole of H_2O.

With the moles of C and of H both known, we can now calculate the whole-number mole ratio by dividing each mole amount by the smallest of these amounts.

$$\text{C:} \quad \frac{0.0713 \text{ mole}}{0.0713 \text{ mole}} = 1 \quad \text{(calculator answer)}$$

$$= 1.00 \quad \textbf{(correct answer)}$$

$$\text{H:} \quad \frac{0.143 \text{ mole}}{0.0713 \text{ mole}} = 2.00561 \quad \text{(calculator answer)}$$

$$= 2.01 \quad \textbf{(correct answer)}$$

Carbon and hydrogen are present in the compound in a one-to-two molar ratio. The empirical formula of the compound is CH_2.

CHEMICAL EXTENSION

Ethylene, a colorless, flammable gas with a slightly sweet odor, is both a natural product and an important industrial chemical. Its molecular formula is C_2H_4.

Ethylene occurs naturally in small amounts in plants where it functions as a plant hormone. Many fruits and vegetables, especially citrus fruit and tomatoes, release ethylene as they ripen. The released ethylene triggers further ripening.

The commercial fruit industry uses to advantage this ripening function for ethylene. Crops, such as bananas and tomatoes, are picked green to prevent spoiling and bruising during transportation to markets. At its destination, the fruit is exposed to ethylene gas, which stimulates the ripening process; 1 kg of tomatoes can be ripened by exposure to 0.1 mg of ethylene.

Industrially, U.S. chemical manufacturers produce more ethylene each year than any other compound except sulfuric acid. The ethylene does not reach consumers directly; most of it goes into the production of plastics and synthetic fibers. Almost half of the ethylene produced is consumed in the production of the well-known plastic called polyethylene.

Practice Exercise 9.21

A compound containing only carbon and hydrogen is subjected to combustion analysis. From a sample of the compound, 67.5 g of CO_2 and 13.8 g of H_2O are obtained. What is the empirical formula of the compound?

When the compound to be analyzed using combustion analysis contains oxygen in addition to carbon and hydrogen, the calculation of the amount (or percent) of oxygen in the compound proceeds in a different manner than that shown in Example 9.19. Only part of the oxygen involved in the production of CO_2 and H_2O comes from the compound itself, and part comes from the O_2 used to support combustion of the compound. In this situation, the amount of hydrogen and carbon present is calculated as before. Then the sum of the hydrogen and carbon masses is subtracted from the original compound mass to obtain the mass of oxygen present in the original sample of compound.

EXAMPLE 9.22

Determining an Empirical Formula Using Combustion Analysis Data

The compound ascorbic acid (vitamin C) is a carbon–hydrogen–oxygen compound. This water-soluble vitamin cannot be stored in the body and thus must continuously be supplied by the diet. Combustion analysis of a 3.08-g ascorbic acid sample yields 6.17 g of CO_2 and 2.52 g of H_2O. What is the empirical formula for ascorbic acid?

SOLUTION

First, we will calculate the grams of C in the original sample from the grams of CO_2 and the grams of H in the original sample from the grams of H_2O. The grams of O present in the original sample can then be calculated by difference: the difference between the original sample mass and the masses of C and H. We cannot calculate the grams of oxygen present in the original sample from the CO_2 and H_2O masses because the oxygen atoms in these compounds came from two sources: (1) ascorbic acid and (2) the oxygen in air.

With masses of the individual elements known, we then obtain moles of the individual elements. Dividing these molar amounts by the smallest number of moles of an element present leads to the subscripts for the empirical formula. The following flowchart outlines the overall calculational process.

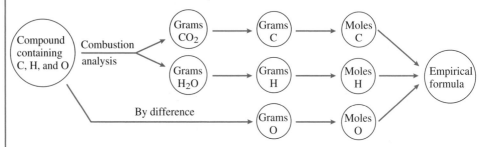

Converting the grams of CO_2 and grams of H_2O to grams of C and grams of H, respectively, involves "grams of A" to "grams of B" calculations.

$$6.17 \text{ g } CO_2 \times \frac{1 \text{ mole } CO_2}{44.01 \text{ g } CO_2} \times \frac{1 \text{ mole C}}{1 \text{ mole } CO_2} \times \frac{12.01 \text{ g C}}{1 \text{ mole C}}$$

$$= 1.6837468 \text{ g C} \quad \text{(calculator answer)}$$

$$= 1.68 \text{ g C} \quad \textbf{(correct answer)}$$

$$2.52 \text{ g } H_2O \times \frac{1 \text{ mole } H_2O}{18.02 \text{ g } H_2O} \times \frac{2 \text{ moles H}}{1 \text{ mole } H_2O} \times \frac{1.01 \text{ g H}}{1 \text{ mole H}}$$

$$= 0.28248612 \text{ g H} \quad \text{(calculator answer)}$$

$$= 0.282 \text{ g H} \quad \textbf{(correct answer)}$$

The next-to-last conversion factor in the CO_2 calculation comes from the formula CO_2. One CO_2 molecule contains one C atom. Likewise, the formula H_2O is the key to obtaining information about hydrogen; one H_2O molecule contains two H atoms.

The difference between the mass of the sample and the masses of C and H in the sample is the mass of O in the sample.

$$\text{grams O} = \text{grams of sample} - \text{grams C} - \text{grams H}$$

$$= 3.08 \text{ g} - 1.68 \text{ g} - 0.282 \text{ g}$$

$$= 1.118 \text{ g} \quad \text{(calculator answer)}$$

$$= 1.12 \text{ g} \quad \textbf{(correct answer)}$$

The next step is to go from grams of element to moles of element. These are simple one-step "grams of A" to "moles of A" calculations.

$$1.68 \text{ g C} \times \frac{1 \text{ mole C}}{12.01 \text{ g C}} = 0.13988343 \text{ mole C} \quad \text{(calculator answer)}$$

$$= 0.140 \text{ mole C} \quad \textbf{(correct answer)}$$

$$0.282 \text{ g H} \times \frac{1 \text{ mole H}}{1.01 \text{ g H}} = 0.27920792 \text{ mole H} \quad \text{(calculator answer)}$$

$$= 0.279 \text{ mole H} \quad \textbf{(correct answer)}$$

$$1.12 \text{ g O} \times \frac{1 \text{ mole O}}{16.00 \text{ g O}} = 0.07 \text{ mole O} \quad \text{(calculator answer)}$$

$$= 0.0700 \text{ mole O} \quad \textbf{(correct answer)}$$

Dividing each of these molar values by the smallest among them, 0.0700 mole O, gives us the small whole numbers that lead to the empirical formula of the compound.

$$\text{C:} \quad \frac{0.140 \text{ mole}}{0.0700 \text{ mole}} = 2 \quad \text{(calculator answer)}$$

$$= 2.00 \quad \textbf{(correct answer)}$$

$$\text{H:} \quad \frac{0.279 \text{ mole}}{0.0700 \text{ mole}} = 3.9857142 \quad \text{(calculator answer)}$$

$$= 3.99 \quad \textbf{(correct answer)}$$

$$\text{O:} \quad \frac{0.0700 \text{ mole}}{0.0700 \text{ mole}} = 1 \quad \text{(calculator answer)}$$

$$= 1.00 \quad \textbf{(correct answer)}$$

The empirical formula of ascorbic acid is C_2H_4O.

CHEMICAL EXTENSION

Vitamins are organic compounds that must be obtained from dietary sources and that are essential in trace amounts for the proper functioning of the human body. They must be obtained from the diet, because the human body cannot synthesize them in the required amounts.

There are 13 known vitamins; scientists believe that the discovery of additional vitamins is unlikely. Strong evidence that the vitamin family is complete comes from the fact that many people have lived for years being fed, intravenously, solutions containing the known vitamins and nutrients, and they have not developed any known vitamin-deficiency disease.

Solubility characteristics divide the vitamins into two classes: the water-soluble vitamins and the fat-soluble vitamins. Water-soluble vitamins (C and eight B vitamins) must be constantly replenished in the body, because they are rapidly eliminated from the body in the urine. Fat-soluble vitamins (A, D, E, and K) are stored in fat tissues, are needed in periodic doses, and are more likely to be toxic when consumed in excess.

Practice Exercise 9.22

The food preservative butylated hydroxytoluene (BHT) is known to contain only the elements C, H, and O. A sample of this compound is burned in a combustion analysis apparatus. The sample size is 2.20 g, and 6.60 g of CO_2 and 2.16 g of H_2O are produced. What is the empirical formula of BHT?

9.15 Determination of Molecular Formulas

To determine the molecular formula of a compound we need additional information besides its empirical formula. That additional information is the compound's molecular mass. A number of experimental methods exist, whose details are beyond the scope of this book, for obtaining such molecular mass information.

A compound's molecular formula will always be a *whole-number multiple* of its empirical formula (Sec. 9.13).

$$\text{molecular formula} = (\text{empirical formula})_x \quad \text{where } x = \text{whole number}$$

The value of x is calculated by dividing the compound's molecular mass by the compound's empirical formula mass.

$$x = \frac{\text{molecular mass (experimentally determined quantity)}}{\text{empirical formula mass (calculated from atomic masses)}}$$

If the value of x was found to be 5.00 and the compound's empirical formula was CH_2, then the compound's molecular formula would be

$$(CH_2)_5 = C_5H_{10}$$

Example 9.23 illustrates further the mechanics involved in obtaining a molecular formula from empirical formula and molecular mass data.

EXAMPLE 9.23

Calculating a Molecular Formula from an Empirical Formula and Molecular Mass Data

Determine the molecular formula of each of the following compounds from the given empirical formula and molecular mass information.

(a) empirical formula = CH; molecular mass = 78.12 amu

(b) empirical formula = NH_2; molecular mass = 32.06 amu

(c) empirical formula = CO; molecular mass = 28.01 amu

(d) empirical formula = C_4H_9; molecular mass = 114.26 amu

SOLUTION

In each case we will determine the whole-number multiplier (x) that relates empirical and molecular formulas to each other.

$$x = \frac{\text{molecular mass (MM)}}{\text{empirical formula mass (EFM)}}$$

(a) First, we calculate the empirical formula mass of CH from atomic masses. The atomic mass of C is 12.01 amu and that of H is 1.01 amu.

$$\text{EFM of CH} = 12.01 \text{ amu} + 1.01 \text{ amu} = 13.02 \text{ amu}$$

The whole-number multiplier is

$$x = \frac{\text{MM}}{\text{EFM}} = \frac{78.12 \text{ amu}}{13.02 \text{ amu}} = 6.000$$

Therefore, the molecular formula is

$$(CH)_6 = C_6H_6$$

(b) The empirical formula mass of NH_2 is 16.03 amu. Nitrogen has an atomic mass of 14.01 amu and each hydrogen has an atomic mass of 1.01 amu. The whole-number multiplier, x, is

$$x = \frac{\text{MM}}{\text{EFM}} = \frac{32.06 \text{ amu}}{16.03 \text{ amu}} = 2.000$$

The molecular formula is $(NH_2)_2 = N_2H_4$

(c) The empirical formula mass is 28.01 amu. The whole-number multiplier, x, is

$$x = \frac{MM}{EFM} = \frac{28.01 \text{ amu}}{28.01 \text{ amu}} = 1.000$$

An x value of 1.000 means that the empirical formula and the molecular formula are one and the same. Both are CO.

(d) The empirical formula mass is 57.13 amu, the value of x is 2.000, and the molecular formula is $(C_4H_9)_2 = C_8H_{18}$.

Practice Exercise 9.23

Determine the molecular formula of each of the following compounds from the given empirical formula and molecular mass information.

(a) empirical formula = HO; molecular mass = 34.02 amu

(b) empirical formula = SN; molecular mass = 184.28 amu

Example 9.23 showed how a molecular formula is determined when the empirical formula is already known. Example 9.24 illustrates a "complete" molecular formula determination, one in which we begin at the beginning—with percent composition data. There are two different approaches to such a calculation, and both approaches are illustrated in Example 9.24. The two approaches differ in the selection of the basis (amount of compound) for the calculation. The two approaches are:

1. *Select 100.0 g of compound as the basis for the calculation*, determine the empirical formula, and then use the empirical formula and the molecular mass to determine the molecular formula.

2. *Select 1.000 mole of compound as the basis for the calculation*, and directly determine the molecular formula.

In the first approach, the molecular mass data are used at the end of the problem. In the second approach, the molecular mass data are used at the beginning of the calculation. Either way, the answer is the same.

EXAMPLE 9.24

Calculating a Molecular Formula from Percent Composition and Molecular Mass Data

Butane, a compound containing only carbon and hydrogen, is the fuel used in many camp stoves. Butane has a percent composition by mass of 82.63% C and 17.37% H and a molecular mass of 58.14 amu. Determine the molecular formula of butane using

(a) 100.0 g of butane as the basis for the calculation.

(b) 1.000 mole of butane as the basis for the calculation.

SOLUTION

(a) Taking 100.0 g of compound as our basis, we will have the following amounts of carbon and hydrogen present.

C: 82.63% of 100.0 g = 82.63 g

H: 17.37% of 100.0 g = 17.37 g

We next convert these gram amounts of C and H to moles of the same.

$$82.63 \text{ g C} \times \frac{1 \text{ mole C}}{12.01 \text{ g C}} = 6.8800999 \text{ moles C} \quad \text{(calculator answer)}$$
$$= 6.880 \text{ moles C} \qquad \text{(\textbf{correct answer})}$$

$$17.37 \text{ g H} \times \frac{1 \text{ mole H}}{1.008 \text{ g H}} = 17.232142 \text{ moles H} \quad \text{(calculator answer)}$$
$$= 17.23 \text{ moles H} \qquad \text{(\textbf{correct answer})}$$

Note that the atomic mass of hydrogen is specified to thousandths (1.008) rather than the "usual" hundredths (1.01). The given amount for hydrogen (17.37 g) requires an atomic mass with at least four significant figures.

Dividing each of these molar values by the smallest one (6.880) gives the following results.

$$\text{C:} \quad \frac{6.880 \text{ mole}}{6.880 \text{ mole}} = 1 \qquad \text{(calculator answer)}$$
$$= 1.000 \quad \text{(\textbf{correct answer})}$$

$$\text{H:} \quad \frac{17.23 \text{ mole}}{6.880 \text{ mole}} = 2.5043604 \quad \text{(calculator answer)}$$
$$= 2.504 \qquad \text{(\textbf{correct answer})}$$

Our result for hydrogen is not a whole number or a near-whole number that can be rounded to a whole number. For hydrogen, we should recognize that the number we are dealing with is $2\frac{1}{2}$. Multiplication of both molar ratios by 2 will clear the fraction. This gives us our needed whole-number ratio.

$$\text{C:} \quad 1 \times 2 = 2$$
$$\text{H:} \quad 2\frac{1}{2} \times 2 = 5$$

The empirical formula of butane is C_2H_5.

To make the transition from empirical formula to molecular formula, we first determine the formula mass for the empirical formula.

$$\text{C:} \quad 2 \times 12.01 \text{ amu} = 24.02 \text{ amu}$$
$$\text{H:} \quad 5 \times 1.01 \text{ amu} = \underline{5.05 \text{ amu}}$$
$$29.07 \text{ amu}$$

The molecular mass of butane, given in the problem statement, is 58.14 amu. We next determine how many times larger the molecular mass is than the empirical formula mass. This is done by dividing the molecular mass by the empirical formula mass.

$$\frac{58.14 \text{ amu}}{29.07 \text{ amu}} = 2.000$$

We have just determined the multiplication factor that converts the empirical formula into a molecular formula. Each of the subscripts in the empirical formula is multiplied by 2.

$$(C_2H_5)_2 = C_4H_{10}$$

The molecular formula of butane is C_4H_{10}.

(b) We will go through this same calculation again, this time using 1.000 mole of butane as our basis. The mass of 1.000 mole of butane is 58.14 g. We know this because it was given in the problem statement; the molecular mass of butane is 58.14 amu.

We first determine the number of grams of each element present in our basis amount of 58.14 g of butane.

C: 82.63% of 58.14 g = 48.041082 g   (calculator answer)

   = 48.04 g   **(correct answer)**

H: 17.37% of 58.14 g = 10.098918 g   (calculator answer)

   = 10.10 g   **(correct answer)**

We next change grams of element to moles of element. Formula subscripts are always determined from molar information.

$$48.04 \text{ g } C \times \frac{1 \text{ mole C}}{12.01 \text{ g } C} = 4 \text{ moles C} \quad \text{(calculator answer)}$$

   = 4.000 moles C   **(correct answer)**

$$10.10 \text{ g } H \times \frac{1 \text{ mole H}}{1.008 \text{ g } H} = 10.019841 \text{ moles H} \quad \text{(calculator answer)}$$

   = 10.02 moles H   **(correct answer)**

At this stage in the calculation, because the basis for the calculation is one mole of compound, we will always obtain whole-number molar amounts or near-whole-number molar amounts that can be rounded to whole numbers. (If we do not, we have a mistake somewhere in our calculation.) These whole-number molar amounts are the subscripts in the molecular formula. Thus, the molecular formula of butane is C_4H_{10}.

What caused this formula, C_4H_{10}, to be a molecular formula rather than an empirical formula? Again, it is the fact that the basis for the calculation was 1 mole of compound.

This second method for calculating a molecular formula is usually shorter than the first method.

CHEMICAL EXTENSION

Raw natural gas, natural gas as it is obtained from a well, is a mixture of the hydro-carbons methane (CH_4), ethane (C_2H_6), propane (C_3H_8), and butane (C_4H_{10}). Methane dominates in the mixture (50–90%), with lesser amounts of ethane (5–20%), propane (3–18%), and butane (1–7%).

Processing of the raw natural gas involves removal of the propane and butane as well as other substances (impurities) present. The processed mixture (methane and ethane) is the commercial natural gas used to heat homes. The removed hydrocarbons (propane and butane) are marketed as liquid petroleum (LP) gas.

Both propane and butane are camping stove fuels. Both are gases at normal temper-atures, but butane condenses to a liquid at 0°C and so cannot be used in camping under cold conditions. Thus, butane fuel is used in the southern United States and propane (which condenses to a liquid at −42°C) in colder northern climates. Cigarette lighters use butane as a fuel; it is present as a liquid in a small pressurized compartment.

Practice Exercise 9.24

The compound oxalic acid, which is used in commercial laundries to remove rust stains, has a molecular mass of 90.04 amu and a percent composition of 26.68% C, 2.24% H, and 71.08% O. Determine the molecular formula of oxalic acid using

(a) 100.0 g of oxalic acid as the basis for the calculation.

(b) 1.000 mole of oxalic acid as the basis for the calculation.

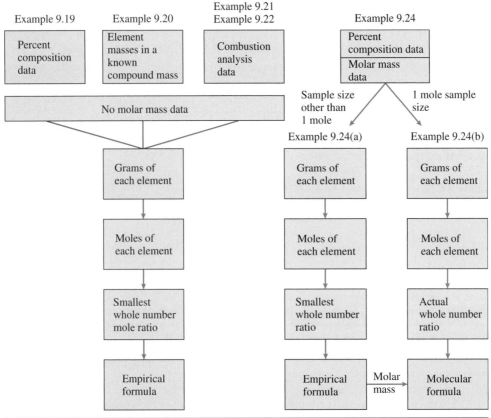

Figure 9.8

A summary of how empirical formulas and molecular formulas are calculated from various types of data.

Figure 9.8 summarizes the methods used in this section and the preceding one to obtain empirical and molecular formulas. It contrasts each of the example calculations we carried out in these sections.

Summary

1. **Law of Definite Proportions** Study of composition data for many compounds led to the law of definite proportions: In a pure compound, the elements are always present in the same definite proportion by mass.

2. **Formula Mass** The formula mass of a substance is the sum of the atomic masses of the atoms present in one formula unit of the substance. An operational rule, in this text, for calculation of formula masses is that all atomic masses are rounded to the hundredths place before they are used as input numbers in a formula mass calculation.

3. **Percent Composition** The percent composition of a compound specifies the percent by mass of each

element present in the compound. The formula mass and chemical formula of a compound can be used to calculate percent composition.

4. **The Mole** The mole is the chemist's counting unit. One mole of any substance—element or compound—consists of 6.022×10^{23} formula units of the substance.

5. **Avogadro's Number** Avogadro's number is the name given to the numerical value 6.022×10^{23}.

6. **Molar Mass** The molar mass of a substance is the mass in grams of the substance that is numerically equal to the substance's formula mass. Molar mass is not a set number; it is different for each chemical substance.

7. **Counting Particles by Weighing** The mass ratio of equal numbers of atoms of two elements is the same as the ratio of the atomic masses of the two elements. Thus, if samples of two or more elements have mass ratios equal to their atomic mass ratios, they must contain equal numbers of atoms. Similar parallel comparisons exist for molecules and compounds.

8. **The Mole and Chemical Formulas** The numerical subscripts in a chemical formula give the number of moles of atoms of the various elements present in 1 mole of the substance.

9. **The Mole and Chemical Calculations** The three quantities most often calculated in chemical problems are the number of particles present (atoms or molecules), the number of moles present, and the number of grams present. These quantities are interrelated. The conversion factors needed relate to the concepts of

(1) Avogadro's number, (2) molar mass, and (3) molar interpretation of chemical formula subscripts.

10. **Percent Purity** Most substance are not 100% pure. In precise work, account must be taken of the impurities present. Percent purity is the mass percentage of a specified substance in an impure sample of the substance.

11. **Empirical and Molecular Formulas** An empirical formula gives the *smallest* whole-number ratio of atoms present in a formula unit of a compound. A molecular formula gives the *actual* ratio of atoms present in a formula unit of a compound. The empirical formula of a compound can be determined from percent composition data or combustion analysis data. The molecular formula can be determined from the empirical formula if the molar mass of the substance is known.

Key Terms

The new terms defined in this chapter are

Avogadro's number   *Sec. 9.5*

combustion analysis   *Sec. 9.14*

empirical formula   *Sec. 9.13*

formula mass   *Sec. 9.2*

law of definite proportions   *Sec. 9.1*

molar mass   *Sec. 9.6*

molar mass of a compound   *Sec. 9.6*

molar mass of an element   *Sec. 9.6*

mole   *Sec. 9.5 and 9.7*

molecular formula   *Sec. 9.13*

percent composition   *Sec. 9.4*

percent purity   *Sec. 9.12*

Practice Problems

Law of Definite Proportions (Sec. 9.1)

9.1 The air pollutant SO_2 is produced when coal is burned. A 1.00-g sample of SO_2 is found to contain 33.4% S and 66.6% O. What would be the percent composition of a 5.00-g sample of this compound?

9.2 The air pollutant NO is a component of automobile exhaust. A 5.00-g sample of NO is found to contain 46.7% N and 53.3% O. What would be the percent composition of a 1.00-g sample of this compound?

9.3 Two different samples of a pure compound (containing elements A and D) were analyzed with the following results.

Sample I: 17.35 g of compound yielded 10.03 g of A and 7.32 g of D.

Sample II: 22.78 g of compound yielded 13.17 g of A and 9.61 g of D.

Show that these data are consistent with the law of definite proportions.

9.4 Two different samples of a pure compound (containing elements E and D) were analyzed with the following results.

Sample I: 11.21 g of compound yielded 5.965 g of E and 5.245 g of D.

Sample II: 13.65 g of compound yielded 7.263 g of E and 6.387 g of D.

Show that these data are consistent with the law of definite proportions.

9.5 It has been found experimentally that the elements X and Q react to produce two different compounds, depending on conditions. Some sample experimental data are as follows.

Experiment	Grams of X	Grams of Q	Grams of Compound
1	3.37	8.90	12.27
2	0.561	1.711	2.272
3	26.9	71.0	97.9

Which two of the three experiments produced the same compound?

9.6 It has been found experimentally that the elements R and T react to produce two different compounds, depending upon conditions. Some sample experimental data are as follows.

Experiment	Grams of R	Grams of T	Grams of Compound
1	1.926	1.074	3.000
2	3.440	1.560	5.000
3	4.494	2.506	7.000

Which two of the three experiments produced the same compound?

9.7 The elemental composition, by mass, of carbon monoxide (CO) is 42.9% C and 57.1% O. What is the maximum amount of CO that could be formed from 42.9 g of C and 80.0 g of O?

9.8 The elemental composition, by mass, of carbon dioxide (CO_2) is 27.3% C and 72.7% O. What is the maximum amount of CO_2 that could be formed from 30.0 g of C and 72.7 g of O?

Formula Masses (Sec. 9.2)

9.9 Calculate the formula mass of each of the following substances using the operational rules given in Section 9.3.

(a) Na_4SiO_4 (sodium silicate, a flame retardant)

(b) H_3BO_3 (boric acid, a mild antiseptic)

(c) $C_{12}H_{11}NO_2$ (Sevin, an insecticide)

(d) $C_{22}H_{30}ClNO_2$ (Darvon, a prescription pain killer)

9.10 Calculate the formula mass of each of the following substances using the operational rules given in Section 9.3.

(a) $NaHCO_3$ (sodium bicarbonate, baking soda)

(b) Tl_2SO_4 (thallium (I) sulfate, a rat and ant poison)

(c) $C_5H_8NO_4Na$ (MSG, a flavor enhancer used in Chinese cooking)

(d) $C_{20}H_{24}N_2O_2$ (quinine, an antimalarial drug)

9.11 Calculate the formula mass of each of the following substances using the operational rules given in Section 9.3.

(a) $C_2H_4(OH)_2$ (ethylene glycol, an automotive antifreeze component)

(b) $(H_2N)_2CO$ (urea, a component of urine)

(c) $Mg_3(Si_2O_5)_2(OH)_2$ (talc, a mineral from which talcum powder comes)

(d) $Al_2Si_2O_5(OH)_2$ (kaolinite, a form of clay)

9.12 Calculate the formula mass of each of the following substances using the operational rules given in Section 9.3.

(a) $Al(OH)_3$ (aluminum hydroxide, a water purification chemical)

(b) $C_3H_5(OH)_3$ (glycerin, a substance derived from fats)

(c) $KLi_2Al(Si_2O_5)_2(OH)_2$ (a form of the mineral mica)

(d) $Ca_3Be(OH)_2(Si_3O_{10})$ (a form of the mineral aminoffite)

9.13 The compound 2-butene-1-thiol, which is responsible in part for the characteristic odor of skunks, has a formula mass of 88.19 amu and the formula C_yH_8S. What number does y stand for in this formula?

9.14 The compound 1-propanethiol, which is the eye irritant that is released when fresh onions are chopped up, has a formula mass of 76.18 amu and the formula C_3H_yS. What number does y stand for in this formula?

9.15 A compound that is 51.5% oxygen by mass contains two oxygen atoms per molecule. What is the formula mass of the compound?

9.16 A compound that is 52.1% oxygen by mass contains three oxygen atoms per molecule. What is the formula mass of the compound?

Percent Composition (Sec. 9.4)

9.17 Calculate the percent composition for each of the following compounds. (Round off all atomic masses to 0.01 amu before using them.)

(a) SnF_2 (tin (II) fluoride, a toothpaste additive)

(b) $FeSO_4$ (iron (II) sulfate, used in treatment of iron deficiency)

(c) $C_{12}H_{22}O_{11}$ (sucrose, table sugar)

(d) $C_{14}H_9Cl_5$ (DDT, an insecticide)

9.18 Calculate the percent composition for each of the following compounds. (Round off all atomic masses to 0.01 amu before using them.)

(a) $C_{10}H_8$ (naphthalene, ingredient in some mothballs)

(b) NaCN (sodium cyanide, used to extract gold from ores)

(c) C_7H_{16} (heptane, a component of gasoline)

(d) $C_{55}H_{72}MgN_4O_5$ (chlorophyll, present in all green plants)

9.19 Calculate the percent composition of each of the following compounds from the given information.

(a) 1.271 g of Cu and 0.320 g of O completely react to produce a sample of the compound.

(b) Decomposition of a 49.31-g sample of the compound yields 15.96 g of Na, 11.13 g of S, and 22.22 g of O.

(c) A 6.48-g sample of N is reacted with oxygen to give 25.00 g of compound.

(d) The reaction of 3.34 g of S with 10.00 g of O produces 10.00 g of compound and 3.34 g of unreacted (leftover) O.

9.20 Calculate the percent composition of each of the following compounds from the given information.

(a) 1.12 g of Fe and 0.48 g of O completely react to produce a sample of the compound.

(b) Decomposition of an 18.03-g sample of the compound yields 4.79 g of K, 6.38 g of Cr, and 6.86 g of O.

(c) A 5.76-g sample of P is reacted with oxygen to give 13.20 g of compound.

(d) The reaction of 4.67 g of N with 10.00 g of O produces 10.00 g of compound and 4.67 g of unreacted (leftover) O.

9.21 In which of the following pairs of compounds do both members of the pair have a mass percent of oxygen that exceeds 50%?

(a) CO and CO_2
(b) ClO and ClO_2
(c) $LiNO_3$ and $RbNO_3$
(d) Na_3PO_4 and $Be_3(PO_4)_2$

9.22 In which of the following pairs of compounds do both members of the pair have a mass percent of oxygen that exceeds 50%?

(a) NO and NO_2
(b) H_2O and H_2O_2
(c) BeO and MgO
(d) $BaSO_4$ and $Al_2(SO_4)_3$

9.23 Calculate the percent compositions of acetylene (C_2H_2) and benzene (C_6H_6). Explain your results.

9.24 Calculate the percent compositions of propene (C_3H_6) and cyclobutane (C_4H_8). Explain your results.

9.25 A sample of the compound hydrazine, N_2H_4, contains 52.34 g of H. How many grams of N does it contain?

9.26 A sample of the compound sodium azide, NaN_3, contains 57.08 g of Na. How many grams of N does it contain?

The Mole as a Counting Unit (Sec. 9.5)

9.27 How many particles (atoms, molecules, formula units, or ions) are present in 1.00 mole of each of the following?

(a) silver (Ag) atoms
(b) water (H_2O) molecules
(c) sodium nitrate $(NaNO_3)$ formula units
(d) sulfate (SO_4^{2-}) ions

9.28 How many particles (atoms, molecules, formula units, or ions) are present in 1.00 mole of each of the following?

(a) copper (Cu) atoms
(b) ammonia (NH_3) molecules
(c) potassium carbonate (K_2CO_3) formula units
(d) phosphate (PO_4^{3-}) ions

9.29 How many carbon atoms are present in each of the following molar quantities of carbon?

(a) 2.50 moles   (b) 3.25 moles
(c) 0.23 mole   (d) 0.3114 mole

9.30 How many boron atoms are present in each of the following molar quantities of boron?

(a) 1.50 moles   (b) 4.21 moles
(c) 0.97 mole   (d) 1.079 moles

9.31 Calculate the number of molecules present in each of the following samples of molecular compounds.

(a) 1.50 moles CO_2 (b) 0.500 mole NH_3
(c) 2.33 moles PF_3 (d) 1.115 moles N_2H_4

9.32 Calculate the number of molecules present in each of the following samples of molecular compounds.

(a) 4.69 moles CO   (b) 0.433 mole SO_3
(c) 1.44 moles P_2H_4 (d) 2.307 moles H_2O_2

Molar Mass (Sec. 9.6)

9.33 What is the mass, in grams, of 1.00 mole of each of the following elements?

(a) Cu   (b) Ba   (c) Si   (d) U

9.34 What is the mass, in grams, of 1.00 mole of each of the following elements?

(a) Ag   (b) K   (c) Al   (d) I

9.35 What is the molar mass, to hundredths, of each of the following compounds?

(a) $Al(OH)_3$ (b) Mg_3N_2
(c) $Cu(NO_3)_2$ (d) La_2O_3

9.36 What is the molar mass, to hundredths, of each of the following compounds?

(a) NH_4Cl (b) $HClO_4$
(c) $Ca(ClO_4)_2$ (d) HfO_2

9.37 Calculate the mass, in grams, of 1.357 moles of each of the following compounds.

(a) NaCl   (b) Na_2S (c) $NaNO_3$ (d) Na_3PO_4

9.38 Calculate the mass, in grams, of 0.981 mole of each of the following compounds.

(a) SO_2 (b) SO_2Cl_2 (c) S_4N_4 (d) $Li_2S_2O_3$

9.39 In each of the following pairs of molar-sized quantities, select the quantity that has the greater mass, in grams.

(a) 2.00 moles of Cu and 2.00 moles of O
(b) 1.00 mole of Br and 5.00 moles of Be
(c) 2.00 moles of CO and 1.50 moles of N_2O
(d) 4.87 moles of B_2H_6 and 0.35 mole of U

9.40 In each of the following pairs of molar-sized quantities, select the quantity that has the greater mass, in grams.

(a) 3.00 moles of Na and 3.00 moles of Al
(b) 1.00 mole of I and 9.00 moles of Li
(c) 2.00 moles of CO_2 and 1.00 mole of SO_3
(d) 9.00 moles of Be and 1.00 mole of HCl

Relationship Between Atomic Mass Units and Gram Units (Sec. 9.8)

9.41 A 0.123-mole sample of a pure substance has a mass of 5.904 g. What is the molar mass of the substance?

9.42 A 0.571-mole sample of a pure substance has a mass of 36.60 g. What is the molar mass of the substance?

9.43 What is the mass, in grams, of an atom whose mass on the amu scale is 19.0 amu?

9.44 What is the mass, in grams, of an atom whose mass on the amu scale is 23.0 amu?

9.45 Identify the element that contains atoms with an average mass of 5.143×10^{-23} g.

9.46 Identify the element that contains atoms with an average mass of 2.326×10^{-23} g.

9.47 How many Al atoms of mass 26.98 amu are present in 26.98 g of Al?

9.48 How many P atoms of mass 30.97 amu are present in 30.97 g of P?

Counting Particles by Weighing (Sec. 9.9)

9.49 What is the Ag–nonmetal mass ratio needed to obtain an equal number of atoms of Ag and each of the following nonmetals?

(a) Cl   **(b)** H   **(c)** O   **(d)** P

9.50 What is the Cu–nonmetal mass ratio needed to obtain an equal number of atoms of Cu and each of the following nonmetals?

(a) F   **(b)** He   **(c)** N   **(d)** S

9.51 What is the Al–Be mass ratio needed to obtain the following?

(a) an equal number of both types of atoms

(b) twice as many Al atoms as Be atoms

(c) twice as many Be atoms as Al atoms

(d) 80% Al atoms and 20% Be atoms

9.52 What is the Si–N mass ratio needed to obtain the following?

(a) an equal number of both types of atoms

(b) three times as many Si atoms as N atoms

(c) three times as many N atoms as Si atoms

(d) 75% Si atoms and 25% N atoms

9.53 Using the concept of counting by weighing, determine whether or not each of the following pairs of samples contains the same number of particles (atoms or molecules).

(a) 28.086 g of Si and 35.453 g of Cl

(b) 72.59 g of Ge and 74.72 g of Ga

(c) 1.427 g of NH_3 and 2.685 g of N_2H_4

(d) 64.14 g of S and 128.14 g of SO_2

9.54 Using the concept of counting by weighing, determine whether or not each of the following pairs of samples contains the same number of particles (atoms or molecules).

(a) 3.813 g of B and 2.314 g of Li

(b) 2.0020 g of C and 2.3347 g of N

(c) 9.01 g of H_2O and 15.78 g of H_2O_2

(d) 41.4 g of Au and 44.6 g of AuCN

9.55 Using the concept of counting by weighing, indicate whether each of the following samples contains more atoms, the same number of atoms, or fewer atoms than 23.6 g of copper (Cu).

(a) 24.3 g of Fe   **(b)** 19.6 g of Ni

(c) 11.52 g of P   **(d)** 11.2 g of Al

9.56 Using the concept of counting by weighing, indicate whether each of the following samples contains more atoms, the same number of atoms, or fewer atoms than 14.0 g of sodium (Na).

(a) 77.3 g of I   **(b)** 126 g of Pb

(c) 20.2 g of K   **(d)** 9.00 g of N

9.57 Using the concept of counting by weighing, determine the number of grams of each of the following elements that will contain three times as many atoms as 10.00 g of P.

(a) S   **(b)** Be   **(c)** U   **(d)** Si

9.58 Using the concept of counting by weighing, determine the number of grams of each of the following elements that will contain twice as many atoms as 20.00 g of Zn.

(a) Li   **(b)** Al   **(c)** Cu   **(d)** Ba

The Mole and Chemical Formulas (Sec. 9.10)

9.59 Write the six mole-to-mole conversion factors that can be derived from the formula Na_3PO_4.

9.60 Write the six mole-to-mole conversion factors that can be derived from the formula K_2SO_4.

9.61 In which of the following pairs of compound amounts do both members of the pair contain the same number of moles of sulfur atoms?

(a) 1.0 mole Na_2SO_4 and 0.50 mole $Na_2S_2O_3$

(b) 2.00 moles S_3Cl_2 and 1.50 moles S_2O

(c) 3.00 moles $H_2S_2O_5$ and 6.00 moles H_2SO_4

(d) 1.00 mole $Na_3Ag(S_2O_3)_2$ and 2.00 moles S_2F_{10}

9.62 In which of the following pairs of compound amounts do both members of the pair contain the same number of moles of nitrogen atoms?

(a) 0.50 mole N_2O_5 and 1.0 mole of N_2O_4

(b) 2.00 moles HNO_3 and 2.00 moles HNO_2

(c) 3.00 moles NH_3 and 1.00 mole HN_3

(d) 1.50 moles $(NH_4)_2SO_4$ and 1.00 mole $(NH_4)_3PO_4$

9.63 Which amount in each of the following pairs of amounts contains the greater number of total moles of atoms?

(a) 2.00 moles $NaAuBr_4$ and 2.00 moles Au_2Te_3

(b) 1.00 mole $C_2H_2Cl_4$ and 1.00 mole CCl_4

(c) 3.00 moles $Ba(NO_3)_2$ and 3.00 moles $BaSO_4$

(d) 1.20 moles NH_4CN and 1.30 moles NH_4Cl

9.64 Which amount in each of the following pairs of amounts contains the greater number of total moles of atoms?

(a) 3.00 moles Cl_2O and 3.00 moles Cl_2O_3

(b) 1.00 mole C_2H_6 and 1.00 mole $SOCl_2$

(c) 2.00 moles $CaSO_4$ and 2.00 moles NH_4Br

(d) 1.00 mole $(NH_4)_2CO_3$ and 3.00 moles $Ba(OH)_2$

The Mole and Chemical Calculations (Sec. 9.11)

9.65 Calculate the number of atoms present in a 7.500-g sample of each of the following elements.

(a) S   (b) Be   (c) Ba   (d) Au

9.66 Calculate the number of atoms present in a 3.752-g sample of each of the following elements.

(a) Li   (b) V   (c) Hg   (d) Pb

9.67 Calculate the number of molecules present in a 25.0-g sample of each of the following compounds.

(a) HF   (b) N_2H_4 (c) SO_2 (d) H_2SO_4

9.68 Calculate the number of molecules present in a 52.0-g sample of each of the following compounds.

(a) HClO   (b) $HClO_2$ (c) $HClO_3$ (d) $HClO_4$

9.69 What is the mass, in grams, of each of the following quantities of chemical substance?

(a) 6.022×10^{23} atoms of Ag
(b) 1.750×10^{23} atoms of N
(c) 6.5×10^{30} molecules of CO_2
(d) 2431 molecules of CO

9.70 What is the mass, in grams, of each of the following quantities of chemical substance?

(a) 6.022×10^{23} atoms of Xe
(b) 3.125×10^{22} atoms of As
(c) 3.125×10^{22} molecules of H_3AsO_4
(d) 989 molecules of H_2O

9.71 What is the mass, in grams (to four significant figures), of a single atom (for elements) or single molecule (for compounds) of the following substances?

(a) Na   (b) Mg   (c) C_4H_{10} (d) C_6H_6

9.72 What is the mass, in grams (to four significant figures), of a single atom (for elements) or single molecule (for compounds) of the following substances?

(a) Cu   (b) B   (c) C_2H_6 (d) $C_{10}H_{20}$

9.73 Determine the number of grams of Cl present in each of the following amounts of chlorine-containing compounds.

(a) 100.0 g of NaCl   (b) 980.0 g of CCl_4
(c) 10.0 g of HCl   (d) 50.0 g of $BaCl_2$

9.74 Determine the number of grams of N present in each of the following amounts of nitrogen-containing compounds.

(a) 100.0 g of $NaNO_3$ (b) 980.0 g of N_2H_4
(c) 10.0 g of Na_3N (d) 50.0 g of KCN

9.75 Determine the number of phosphorus atoms present in 25.0 g of each of the following substances.

(a) PF_3 (b) Be_3P_2
(c) $POCl_3$ (d) $Na_5P_3O_{10}$

9.76 Determine the number of sulfur atoms present in 35.0 g of each of the following substances.

(a) CS_2 (b) S_4N_4
(c) SF_6 (d) $Al_2(SO_4)_3$

9.77 Determine the number of grams of O present in each of the following chemical samples.

(a) 2.0×10^{25} molecules of P_4O_{10} (b) 50.00 g of SO_3
(c) 3.50 moles of KNO_3 (d) 475 g of $Ca_3(PO_4)_2$

9.78 Determine the number of grams of N present in each of the following chemical samples.

(a) 6.3×10^{23} molecules of N_2O_5 (b) 26.00 g of HN_3
(c) 25.0 moles of NH_4Cl (d) 75.7 g of $Al(NO_3)_3$

9.79 What amount or mass of each of the following substances would be needed to obtain 1.000 g of S?

(a) moles of H_2S (b) molecules of S_8
(c) grams of $S_3N_3O_3Cl_3$ (d) atoms of S

9.80 What amount or mass of each of the following substances would be needed to obtain 1.000 g of Si?

(a) moles of SiH_4 (b) molecules of SiO_2
(c) grams of $(CH_3)_3SiCl$ (d) atoms of Si

9.81 What would be the mass, in grams, of a CO_2 sample in which each of the following were present?

(a) 2.70×10^{25} atoms of oxygen
(b) 2.70×10^{25} atoms of carbon
(c) 2.70×10^{25} total atoms
(d) 2.70×10^{25} molecules

9.82 What would be the mass, in grams, of an SO_3 sample in which each of the following were present?

(a) 3.80×10^{24} atoms of oxygen
(b) 3.80×10^{24} atoms of sulfur
(c) 3.80×10^{24} total atoms
(d) 3.80×10^{24} molecules

9.83 Progesterone, a female hormone, has the formula $C_{21}H_{30}O_2$. In a 25.00-g sample of this compound,

(a) how many moles of atoms are present?
(b) how many atoms of C are present?
(c) how many grams of O are present?
(d) how many progesterone molecules are present?

9.84 Testosterone, a male hormone, has the formula $C_{19}H_{28}O_2$. In a 30.00-g sample of this compound,

(a) how many moles of atoms are present?
(b) how many atoms of H are present?
(c) how many grams of C are present?
(d) how many testosterone molecules are present?

Purity of Samples (Sec. 9.12)

9.85 Calculate the following for a 325-g sample of 95.4% pure Fe_2S_3.

(a) mass, in grams, of Fe_2S_3 present

(b) mass, in grams, of impurities present

9.86 Calculate the following for a 25.4-g sample of 88.7% pure Cu_2S.

(a) mass, in grams, of Cu_2S present

(b) mass, in grams, of impurities present

9.87 A 62.03% pure sample of FeS contains 6.00 g of impurities. What is the mass, in grams, of the sample?

9.88 A 87.23% pure sample of NiS contains 0.83 g of impurities. What is the mass, in grams, of the sample?

9.89 What is the percent purity by mass of an NaCl sample in which 0.55 g of impurities are present per 67.21 g of NaCl?

9.90 What is the percent purity by mass of a KBr sample in which 2.30 g of impurities are present per 32.00 g of KBr?

9.91 How many copper atoms are present in a 35.00-g sample of impure copper that has a purity of 92.35% copper by mass? Assume that there are no copper-containing impurities present in the sample.

9.92 How many silver atoms are present in a 50.00-g sample of impure silver that has a purity of 89.98% silver by mass? Assume that there are no silver-containing impurities present in the sample.

9.93 What mass, in grams, of chromium is present in a 25.00-g sample of chromium ore that is 64.0% Cr_2O_3? Assume that impurities present do not contain any chromium.

9.94 What mass, in grams, of nickel is present in a 25.00-g sample of nickel ore that is 77.25% NiS? Assume that impurities present do not contain any nickel.

Empirical and Molecular Formulas (Secs. 9.13–9.15)

9.95 Each of the following is a correctly written molecular formula. In each case write the empirical formula for the substance.

(a) H_2O_2 **(b)** C_8H_{14} **(c)** C_3H_8 **(d)** S_4N_4

9.96 Each of the following is a correctly written molecular formula. In each case write the empirical formula for the substance.

(a) P_4O_{10} **(b)** Pb_3S_4 **(c)** C_4H_8 **(d)** C_5H_{12}

9.97 Given the following percent compositions, determine the empirical formula.

(a) 58.91% Na and 41.09% S

(b) 24.74% K, 34.76% Mn, and 40.50% O

(c) 2.06% H, 32.69% S, and 65.25% O

(d) 19.84% C, 2.50% H, 66.08% O, and 11.57% N

9.98 Given the following percent compositions, determine the empirical formula.

(a) 47.26% Cu and 52.74% Cl

(b) 40.27% K, 26.78% Cr, and 32.96% O

(c) 40.04% Ca, 12.00% C, and 47.96% O

(d) 28.03% Na, 29.28% C, 3.69% H, and 39.01% O

9.99 Convert each of the following molar ratios between elements in a compound to a whole-number molar ratio.

(a) 1.00-to-1.33   **(b)** 1.50-to-2.00

(c) 1.75-to-2.00-to-2.25   **(d)** 1.33-to-2.33-to-2.00

9.100 Convert each of the following molar ratios between elements in a compound to a whole-number molar ratio.

(a) 1.00-to-1.25   **(b)** 1.50-to-2.50

(c) 1.67-to-2.33-to-3.00   **(d)** 1.50-to-1.00-to-1.33

9.101 Determine the empirical formula for substances with each of the following percent compositions.

(a) 43.64% P and 56.36% O

(b) 72.24% Mg and 27.76% N

(c) 29.08% Na, 40.56% S, and 30.36% O

(d) 21.85% Mg, 27.83% P, and 50.32% O

9.102 Determine the empirical formula for substances with each of the following percent compositions.

(a) 54.88% Cr and 45.12% S

(b) 38.76% Cl and 61.24% O

(c) 59.99% C, 4.485% H, and 35.52% O

(d) 26.58% K, 35.35% Cr, and 38.06% O

9.103 Nickel forms two chlorides. One has 45.3% Ni by mass, and the other 35.5% Ni. What are the empirical formulas of these two chlorides?

9.104 Copper forms two fluorides. One has 77.0% Cu by mass, and the other 62.6% Cu. What are the empirical formulas of these two fluorides?

9.105 Ethyl mercaptan is an odorous substance added to natural gas to make leaks easily detectable. Analysis of a sample of ethyl mercaptan indicates that 5.798 g of C, 1.46 g of H, and 7.740 g of S are present. What is the empirical formula of this compound?

9.106 Hydroquinone is a compound used in developing photographic film. Analysis of a sample of hydroquinone indicates that 16.36 g of C, 1.38 g of H, and 7.265 g of O are present. What is the empirical formula of this compound?

9.107 A 2.00-g sample of beryllium metal is burned in an oxygen atmosphere to produce 5.55 g of a beryllium–oxygen compound. Determine the compound's empirical formula.

9.108 A 2.00-g sample of lithium metal is burned in an oxygen atmosphere to produce 4.31 g of a lithium–oxygen compound. Determine the compound's empirical formula.

9.109 Determine the empirical formula of the carbon–hydrogen compound that, upon combustion in a combustion analysis apparatus, generates each of the following sets of CO_2–H_2O data.

(a) 0.338 g of CO_2 and 0.277 g of H_2O

(b) 0.303 g of CO_2 and 0.0621 g of H_2O

(c) 0.225 g of CO_2 and 0.115 g of H_2O

(d) 0.314 g of CO_2 and 0.192 g of H_2O

9.110 Determine the empirical formula of the carbon–hydrogen compound that, upon combustion in a combustion analysis apparatus, generates each of the following sets of CO_2–H_2O data.

(a) 0.269 g of CO_2 and 0.221 g of H_2O

(b) 0.294 g of CO_2 and 0.120 g of H_2O

(c) 0.600 g of CO_2 and 0.184 g of H_2O

(d) 0.471 g of CO_2 and 0.0963 g of H_2O

9.111 Determine the empirical formula of the carbon–hydrogen compound that, upon combustion of a 0.7420-mg sample in a combustion analysis apparatus, generates 2.328 mg of CO_2.

9.112 Determine the empirical formula of the carbon–hydrogen compound that, upon combustion of a 0.4244-mg sample in a combustion analysis apparatus, generates 1.164 mg of CO_2.

9.113 A 3.750-g sample of the compound responsible for the odor of cloves (containing only C, H, and O) is burned in a combustion analysis apparatus. The mass of CO_2 produced is 10.05 g, and the mass of H_2O produced is 2.47 g. What is the empirical formula of the compound?

9.114 A 0.8640-g sample of the compound responsible for the pungent odor of rancid butter (containing only C, H, and O) is burned in a combustion analysis apparatus. The mass of CO_2 produced is 1.727 g, and the mass of H_2O produced is 0.7068 g. What is the empirical formula of the compound?

9.115 Determine the molecular formulas of compounds with the following empirical formulas and molecular masses.

(a) P_2O_5, 284 amu   (b) SN, 184 amu

(c) $C_3H_6O_2$, 74 amu   (d) BNH_2, 80.4 amu

9.116 Determine the molecular formulas of compounds with the following empirical formulas and molecular masses.

(a) NH_2, 32 amu   (b) P_2O_3, 220 amu

(c) $C_5H_{10}O_2$, 102 amu   (d) $SNCl_2$, 351 amu

9.117 Methyl benzoate, a compound used in the manufacture of perfumes, has a molecular mass of 136 amu, and its percent composition by mass is 70.57% C, 5.93% H, and 23.49% O. Determine the molecular formula of methyl benzoate using

(a) 100.0 g of methyl benzoate as the basis for the calculation.

(b) 1.00 mole of methyl benzoate as the basis for the calculation.

9.118 Adipic acid, a compound used as a raw material for the manufacture of nylon, has a molecular mass of 146 amu, and its percent composition by mass is 49.30% C, 6.91% H, and 43.79% O. Determine the molecular formula of adipic acid using

(a) 100.0 g of adipic acid as the basis for the calculation.

(b) 1.00 mole of adipic acid as the basis for the calculation.

9.119 Lactic acid, the substance that builds up in muscles and causes them to "hurt" when they are worked hard, has a molecular mass of 90.0 amu and a percent composition by mass of 40.0% C,

6.71% H, and 53.3% O. Determine the molecular formula of lactic acid using

(a) 100.0 g of lactic acid as the basis for the calculation.

(b) 1.00 mole of lactic acid as the basis for the calculation.

9.120 Citric acid, a flavoring agent in many carbonated beverages, has a molecular mass of 192 amu and a percent composition by mass of 37.50% C, 4.21% H, and 58.29% O. Determine the molecular formula of citric acid using

(a) 100.0 g of citric acid as the basis for the calculation.

(b) 1.00 mole of citric acid as the basis for the calculation.

Additional Problems

9.121 Select the quantity that has the greater mass in each of the following pairs of quantities. Make your selection using the periodic table but without performing an actual calculation.

(a) 1.00 mole of silver or 1.00 mole of gold

(b) 1.00 mole of sulfur or 6.022×10^{23} atoms of carbon

(c) 1.00 mole of Cl atoms or 1.00 mole of Cl_2 molecules

(d) 8.00 g of He or 6.022×10^{23} atoms of Ne

9.122 Select the quantity that has the greater mass in each of the following pairs of quantities. Make your selection using the periodic table but without performing an actual calculation.

(a) 1.00 mole of tin or 1.00 mole of lead

(b) 1.00 mole of phosphorus or 6.022×10^{23} atoms of boron

(c) 1.00 mole of O atoms or 1.00 mole of O_2 molecules

(d) 7.5 g of Be or 6.022×10^{23} atoms of Li

9.123 Select the quantity that has the greater number of atoms in each of the following pairs of quantities. Make your selection using the periodic table but without performing an actual calculation.

(a) 1.00 mole of P or 1.00 mole of P_4

(b) 21.0 g of Na or 1.00 mole of Na

(c) 63.5 g of Cu or 8.0 g of B

(d) 1.00 g of K or 6.022×10^{23} atoms of Be

9.124 Select the quantity that has the greater number of atoms in each of the following pairs of quantities. Make your selection using the periodic table but without performing an actual calculation.

(a) 1.00 mole of S or 1.00 mole of S_8

(b) 28.0 g of Al or 1.00 mole of Al

(c) 28.1 g of Si or 30.0 g of Mg

(d) 2.00 g of Na or 6.022×10^{23} atoms of He

9.125 Indicate whether each of the following statements concerning O_2 and O_3 molecules is *true* or *false*.

(a) The mass of one mole of each of these molecules is the same.

(b) The number of molecules in one mole of each of these molecules is the same.

(c) The mass of oxygen in one mole of each of these molecules is the same.

(d) The number of atoms of oxygen in one mole of each of these molecules is the same.

9.126 Indicate whether each of the following statements concerning S_2 and S_8 molecules is *true* or *false*.

(a) The mass of one mole of each of these molecules is the same.

(b) The number of molecules in one mole of each of these molecules is the same.

(c) The mass of sulfur in one mole of each of these molecules is the same.

(d) The number of atoms of sulfur in one mole of each of these molecules is the same.

9.127 How many grams of potassium and sulfur are theoretically needed to make 4.000 g of K_2S?

9.128 How many grams of beryllium and nitrogen are theoretically needed to make 3.00 g of Be_3N_2?

9.129 How many grams of B would contain the same number of atoms as there are in 3.50 moles of Xe?

9.130 How many grams of Si would contain the same number of atoms as there are in 2.10 moles of Ar?

9.131 How many grams of glucose, $C_6H_{12}O_6$, would contain the same mass of carbon as there is in 3.44 g of ethanol, C_2H_6O?

9.132 How many grams of ethanol, C_2H_6O, would contain the same mass of oxygen as there is in 7.08 g of glucose, $C_6H_{12}O_6$?

9.133 Individual samples of the three compounds Cr_2O_3, CrO_2, and CrO_3 are found to contain identical masses of Cr.

(a) Which sample has the largest total mass?

(b) Which sample contains the greatest number of oxygen atoms?

9.134 Individual samples of the three compounds V_2O_5, V_2O_3, and VO_2 are found to contain identical masses of V.

(a) Which sample has the largest total mass?

(b) Which sample contains the greatest number of oxygen atoms?

9.135 A certain alloy of Au, Cu, and Ni contains these elements in the atomic proportions 3:2:1, respectively. What is the mass, in grams, of a sample of this alloy containing a total of 1.00×10^{24} atoms?

9.136 A certain alloy of Sn, Pb, and Bi contains these elements in the atomic proportions 2:4:3, respectively. What is the mass, in grams, of a sample of this alloy containing Avogadro's number of atoms?

9.137 Identify each of the following elements.

(a) Atoms of this element have an average mass of 1.792×10^{-22} g.

(b) A sample of this element with a mass of 0.05232 g contains 4.840×10^{-3} mole of atoms.

9.138 Identify each of the following elements.

(a) Atoms of this element have an average mass of 6.647×10^{-24} g.

(b) A sample of this element with a mass of 0.2346 g contains 3.380×10^{-2} mole of atoms.

9.139 For the compound $(CH_3)_3SiCl$, calculate the

(a) mass percent of H present.

(b) atom percent of H present.

(c) mole percent of H present.

9.140 For the compound $(CH_3)_2SiCl_2$, calculate the

(a) mass percent of H present.

(b) atom percent of H present.

(c) mole percent of H present.

9.141 What are the empirical formulas for the compounds that contain each of the following?

(a) 9.0×10^{23} atoms of Na, 3.0×10^{23} atoms of Al, and 1.8×10^{24} atoms of F

(b) 3.2 g of S and 1.20×10^{23} atoms of O

(c) 0.36 mole of Ba, 0.36 mole of C, and 17.2 g of O

(d) 1.81×10^{23} atoms of H, 10.65 g of Cl, and 0.30 mole of O atoms

9.142 What are the empirical formulas for the compounds that contain each of the following?

(a) 3.0×10^{30} atoms of Fe, 3.0×10^{30} atoms of Cr, and 1.2×10^{31} atoms of O

(b) 0.0023 g of N and 2.0×10^{20} atoms of O

(c) 0.40 mole of Li, 6.4 g of S, and 0.80 mole of O

(d) 0.15 mole of S, 1.8×10^{23} atoms of O, and 5.7 g of F

9.143 Calculate the number of carbon atoms in 5.25 g of a compound that contains 92.26% C and 7.74% H by mass.

9.144 Calculate the number of nitrogen atoms in 5.25 g of a compound that contains 87.39% N and 12.61% H by mass.

9.145 A compound has an empirical formula of C_2H_3O. Calculate the molecular formula of the compound for each of the following cases.

(a) The compound's molecular mass is twice the compound's empirical formula mass.

(b) Molecules of the compound contain 18 atoms.

(c) The sum of the carbon and oxygen atoms in a molecule of the compound is 18.

(d) The mass of 0.010 mole of the compound is 0.86 g.

9.146 A compound has an empirical formula of C_3H_5O. Calculate the molecular formula of the compound for each of the following cases.

(a) The compound's molecular mass and empirical formula mass differ by a factor of 2.

(b) Molecules of the compound contain 18 atoms.

(c) Molecules of the compound contain more than 20 atoms but fewer than 30 atoms.

(d) The mass of 0.010 mole of the compound is 0.57 g.

9.147 A sample of a compound containing only C and H is burned in oxygen, and 13.75 g of CO_2 and 11.25 g of H_2O are obtained. What was the mass of the sample, in grams, that was burned?

9.148 A sample of a compound containing only C and H is burned in oxygen, and 14.66 g of CO_2 and 9.00 g of H_2O are obtained. What was the mass of the sample, in grams, that was burned?

9.149 A sample of a compound containing only C, H, and S was burned in oxygen, and 6.60 g of CO_2, 5.41 g of H_2O, and 9.61 g of SO_2 were obtained.

(a) What is the empirical formula of the compound?

(b) What was the mass, in grams, of the sample that was burned?

9.150 A sample of a compound containing only C, H, and N was burned in oxygen, and 6.60 g of CO_2, 6.76 g of H_2O, and 4.50 g of NO were obtained.

(a) What is the empirical formula of the compound?

(b) What was the mass, in grams, of the sample that was burned?

9.151 A sample containing NaF, Na_2SO_4, and $NaNO_3$ gives the following elemental analysis by mass: 18.1% F and 6.60% N. Calculate the mass percent of each compound in the mixture.

9.152 A sample containing NaF, Na_2SO_4, and $NaNO_3$ gives the following elemental analysis by mass: 11.3% F and 5.65% S. Calculate the mass percent of each compound in the mixture.

9.153 By analysis, a compound with the formula $KClO_x$ is found to contain 28.9% chlorine by mass. What is the value of the integer x?

9.154 By analysis, a compound with the formula H_3AsO_x is found to contain 52.78% arsenic by mass. What is the value of the integer x?

9.155 A 7.503-g sample of metal is reacted with excess oxygen to yield 10.498 g of the oxide MO. Calculate the molar mass of the element M.

9.156 A 11.17-g sample of metal is reacted with excess oxygen to yield 15.97 g of the oxide M_2O_3. Calculate the molar mass of the element M.

9.157 A certain compound contains only carbon, hydrogen, and oxygen. If it contains 47.4% carbon by mass, and if there is one oxygen atom present for every four hydrogen atoms, what is its empirical formula?

9.158 A certain compound contains only lead, carbon, and hydrogen. If it contains 64.07% lead by mass, and if there are two carbon atoms present for every five hydrogen atoms, what is its empirical formula?

Cumulative Problems

9.159 Calculate the molecular mass, in amu, of the compound H_3AsO_4 to the following number of significant figures.

(a) three **(b)** four **(c)** five **(d)** six

9.160 Calculate the molecular mass, in amu, of the compound H_3AsO_3 to the following number of significant figures.

(a) three **(b)** four **(c)** five **(d)** six

9.161 Calculate the density of the metal lead, in g/cm^3, given that 0.422 mole of lead occupies a volume of 7.74 cm^3.

9.162 Calculate the density of the metal potassium, in g/cm^3, given that 0.674 mole of potassium occupies a volume of 30.8 cm^3.

9.163 At a certain temperature, the density of water is 1.00 g/mL and the density of ethyl alcohol (C_2H_6O) is 0.789 g/mL. At this temperature, what volume of water contains the same number of molecules as are present in 225 mL of ethyl alcohol?

9.164 At a certain temperature, the density of water is 1.00 g/mL and the density of ethyl alcohol (C_2H_6O) is 0.789 g/mL. At this temperature, what volume of ethyl alcohol contains the same number of molecules as are present in 122 mL of water?

9.165 Calculate the volume of 2.50 moles of zinc, given that the density of the metal is 7.14 g/cm^3.

9.166 Calculate the volume of 1.25 moles of magnesium, given that the density of the metal is 1.74 g/cm^3.

9.167 Calculate the molar mass, to four significant figures, of each of the following compounds.

(a) magnesium nitrate

(b) sodium azide

(c) nickel(II) fluoride

(d) ammonium perchlorate

9.168 Calculate the molar mass, to four significant figures, of each of the following compounds.

(a) sodium thiosulfate

(b) copper(II) hydroxide

(c) ammonium chlorate

(d) aluminum hydrogen phosphate

9.169 The percent natural abundance of $^{40}_{19}K$ is 0.012%. How many $^{40}_{19}K$ atoms does a person ingest by drinking one cup of whole milk containing 371 mg of K per cup?

9.170 The percent natural abundance of $^{41}_{19}K$ is 6.88%. How many $^{41}_{19}K$ atoms does a person ingest by drinking one cup of whole milk containing 392 mg of K?

9.171 A 2.33-mole sample of the ionic compound $Al_2(SO_4)_3$ contains how many of the following?

(a) $Al_2(SO_4)_3$ formula units **(b)** Al^{3+} ions
(c) SO_4^{2-} ions **(d)** total ions

9.172 A 3.50-mole sample of the ionic compound $(NH_4)_3PO_4$ contains how many of the following?

(a) $(NH_4)_3PO_4$ formula units **(b)** NH_4^+ ions
(c) PO_4^{3-} ions **(d)** total ions

9.173 What mass of KCl would contain the same *total* number of *ions* as 32.5 g of $CaCl_2$?

9.174 What mass of LiF would contain the same *total* number of *ions* as 32.5 g of $MgCl_2$?

9.175 Calculate the percent by mass of sulfur in each of the following compounds.

(a) sulfur dioxide   **(b)** hydrogen sulfide
(c) sulfuric acid   **(d)** sulfurous acid

9.176 Calculate the percent by mass of nitrogen in each of the following compounds.

(a) ammonia   **(b)** sodium nitride
(c) sodium azide   **(d)** nitric acid

9.177 A mixture consists of 42.0% NaCl and 58.0% $CaCl_2$ by mass. What is the total number of chloride ions (Cl^-) present in 425 g of mixture?

9.178 A mixture consists of 22.0% $Cu(NO_3)_2$ and 78.0% $Fe(NO_3)_3$ by mass. What is the total number of nitrate ions (NO_3^-) present in 25.00 g of mixture?

9.179 The mass percent composition for an alloy that has a density of 8.31 g/cm^3 is 56.0% copper, 43.0% nickel, and 1.0% manganese. How many nickel atoms are in a block of this alloy measuring 10.0 cm × 30.0 cm × 62.0 cm?

9.180 The mass percent composition for an alloy that has a density of 8.28 g/cm^3 is 64.0% iron, 12.0% cobalt, and 24.0% molybdenum. How many molybdenum atoms are in a block of this alloy measuring 2.0 cm × 1.5 cm × 6.7 cm?

9.181 What volume, in milliliters, of a NaOH solution that is 12.0% NaOH by mass contains 0.275 mole of NaOH? The density of the solution is 1.131 g/mL.

9.182 What volume, in milliliters, of a H_3PO_4 solution that is 85.5% H_3PO_4 by mass contains 0.100 mole of H_3PO_4? The density of the solution is 1.70 g/mL.

9.183 How many total atoms are present in 15.0 mL of an ethyl alcohol–water solution that is 85.0% ethanol by mass? The formula of ethyl alcohol is C_2H_6O, and the density of the solution is 0.807 g/mL.

9.184 How many total atoms are present in 500.0 mL of an ethylene glycol–water solution that is 56.0% ethylene glycol by mass? The formula for ethylene glycol is $C_2H_6O_2$, and the density of the solution is 1.072 g/mL.

9.185 When copper is selling for $0.645 per pound, how many copper atoms could you buy for 1 cent?

9.186 When gold is selling for $390.00 per ounce, how many gold atoms could you buy for 1 cent?

Answers to Practice Exercises

9.1 The percents by mass of chlorine in the two samples are 38.77% and 38.76%.

9.2 **(a)** 98.078 amu **(b)** 350.390 amu

9.3 **(a)** 153.81 amu **(b)** 130.21 amu

9.4 40.91% C; 4.59% H; 54.50% O

9.5 40.0% C; 6.74% H; 53.27% O

9.6 **(a)** 1.61×10^{24} CO_2 molecules
(b) 8.73×10^{23} NaCl formula units

9.7 **(a)** 236 g ClF_5 **(b)** 29.9 g O_3

9.8 2.806×10^{-22} g

9.9 **(a)** different numbers of molecules present
(b) same number of molecules present

9.10 **(a)** 0.753 mole of C atoms and 1.51 moles of O atoms
(b) 5.24 moles of P atoms and 13.1 moles of O atoms

9.11 7.81×10^{21} molecules of $C_{10}H_{18}O$

9.12 4.888×10^{-13} g $C_{14}H_{18}N_2O_5$

9.13 **(a)** 9.273×10^{-23} g **(b)** 2.665×10^{-23} g

9.14 5.33 g O

9.15 3.880×10^{22} atoms of O

9.16 **(a)** 88.60 g **(b)** 1.40 g

9.17 5.133×10^{23} CO_2 molecules

9.18 **(a)** P_2O_5 **(b)** $C_3H_{10}N_2$ **(c)** C_2H_6O
(d) BNH_2

9.19 C_3H_6O

9.20 P_4S_3

9.21 CH

9.22 $C_{15}H_{24}O$

9.23 **(a)** H_2O_2 **(b)** S_4N_4

9.24 **(a)** $H_2C_2O_4$ **(b)** $H_2C_2O_4$

10 Chemical Calculations Involving Chemical Equations

10.1 The Law of Conservation of Mass

In an earlier consideration of chemical change (Sec. 4.4), it was noted that chemical changes are referred to as chemical reactions. As we learned there, a *chemical reaction* is a process in which at least one new substance is produced as a result of chemical change. It is usually easy to see that a chemical reaction has occurred. Color change, emission of heat and/or light, gas evolution, and solid formation are an indication that a chemical reaction has taken place.

The starting materials for a chemical reaction are known as *reactants*. **Reactants** *are the starting substances that undergo change in a chemical reaction.* As a chemical reaction proceeds, reactants are consumed (used up) and new materials with new chemical properties are produced. **Products** *are the substances produced as a result of a chemical reaction.*

From a molecular viewpoint, a chemical reaction (chemical change) involves the union, separation, or rearrangement of atoms to produce new substances. Figure 10.1 shows the rearrangement of atoms that occurs when methane (CH_4) reacts with oxygen (O_2) to produce carbon dioxide (CO_2) and water (H_2O). Hydrogen atoms originally associated with carbon atoms (CH_4) became associated with oxygen atoms (H_2O) as the result of the chemical reaction.

Studies of countless chemical reactions over a period of 200 years have shown that there is no detectable change in the quantity of matter present during an ordinary chemical reaction. This generalization concerning chemical reactions has been formalized into a statement known as the **law of conservation of mass:** *mass is neither created nor destroyed in any ordinary chemical reaction.* To demonstrate the validity of this law, the masses of all

345

Figure 10.1

Rearrangement of atoms that occurs when methane (CH_4) reacts with oxygen (O_2). The products are carbon dioxide (CO_2) and water (H_2O).

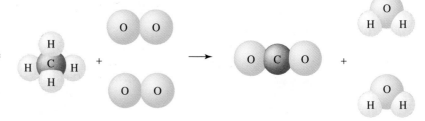

reactants (substances that react together) and all products (substances formed) in a chemical reaction are carefully determined. It is found that the sum of the masses of the products is always the same as the sum of the masses of the reactants. The French chemist Antoine-Laurent Lavoisier (1743–1794; see "The Human Side of Chemistry 11") is given credit for being the first to state this important relationship between the reactants and products of a chemical reaction.

Consider, as an illustrative example of this law, the reaction of known masses of the elements beryllium (Be) and oxygen (O) to form the compound beryllium oxide (BeO). Experimentally it is found that 36.03 g of Be will react with *exactly* 63.97 g of O. After the reaction no Be or O remains in elemental form; the only substance present is the product BeO, combined Be and O. When this product is weighed, its mass is found to be 100.00 g, which is the sum of the masses of the reactants (36.03 g + 63.97 g = 100.00 g). It is also found that when the 100.00 g of product BeO is heated to a high temperature in the absence of air, the BeO decomposes into Be and O, producing 36.03 g of Be and 63.97 g of O. Once again, no detectable mass change is observed; the mass of the reactants is equal to the mass of the products.

The Human Side of Chemistry 11

Antoine-Laurent Lavoisier (1743–1794)

Antoine-Laurent Lavoisier (pronounced Lav-wazy-ay), the son of a wealthy Parisian lawyer, is often called the "father of 'modern' chemistry." It is he who first appreciated the importance of carrying out very accurate (quantitative) measurements of chemical change. His work was performed on balances he had specially designed that were more accurate than any then known. From his "balance work" came the discovery of the law of conservation of mass. He was also the first to make quantitative measurements on the heat produced during chemical reactions.

Originally trained as a lawyer, Lavoisier entered the field of science through geology and from there became interested

in chemistry. This change in focus took place during his early twenties.

Lavoisier's studies on combustion are considered his major work. He was the first to realize that combustion involves the reaction of oxygen from air with the substance that is burned. He gave the name oxygen to the gas then known as "dephlogisticated air." In later years, he became interested in physiological chemistry.

Lavoisier's *Elementary Treatise on Chemistry*, published in 1789, was the first textbook based on quantitative experiments. He was also one of the first to use systematic nomenclature for elements and a few compounds.

In 1790 he was appointed secretary and treasurer of a commission established to standardize weights and measures used throughout France. The outgrowth of this commission's work was the metric system.

Early in life he invested money in a private tax collecting firm and married the daughter of one of the company executives. Firms of this type were licensed by the state to collect taxes and keep a portion of the proceeds. Lavoisier used his earnings from this endeavor (about 100,000 francs a year) to support his scientific work. This connection with tax collecting proved fatal.

During the French revolution all associated with this type of activity, known as "tax-farming," were denounced, arrested, and guillotined. Lavoisier's death, at age 51, came just two months before the end of the revolution. On the day of his death, one of Lavoisier's scientific colleagues gave this tribute: "It took but a moment to cut off that head; perhaps a hundred years will be required to produce another like it."

The law of conservation of mass is consistent with the statements of atomic theory (Sec. 5.1). Since all reacting chemical substances are made up of atoms (statement 1), each with its unique identity (statement 2), and these atoms can be neither created nor destroyed in a chemical reaction but merely rearranged (statement 4), it follows that the total mass after the reaction must equal the total mass before the reaction. We have the same number of atoms of each kind after the reaction as we started out with. An alternative way of stating the law of conservation of mass is: *The total mass of reactants and the total mass of products in a chemical reaction are always equal.*

The law of conservation of mass applies to all ordinary chemical reactions, and there is no known case of a *measurable* change in total mass during an *ordinary* chemical reaction.* This law will be a "guiding principle" for the discussion that follows about chemical equations and their use.

10.2 Writing Chemical Equations

A **chemical equation** *is a representation for a chemical reaction that uses chemical symbols and chemical formulas instead of words to describe the changes that occur in a chemical reaction.* The following example shows the contrast between a word description of a chemical reaction and a chemical equation for the same reaction.

> *Word description:* Magnesium oxide reacts with carbon to produce carbon monoxide and magnesium.
>
> *Chemical equation:* $MgO + C \longrightarrow CO + Mg$

In the same way that chemical symbols are considered the *letters* of chemical language, and formulas the *words* of the language, chemical equations can be considered the *sentences* of chemical language.

The conventions used in writing chemical equations are

1. The correct formulas of the *reactants* are always written on the *left* side of the equation.

$$MgO + C \longrightarrow CO + Mg$$

2. The correct formulas of the *products* are always written on the *right* side of the equation.

$$MgO + C \longrightarrow CO + Mg$$

3. The reactants and products are separated by an arrow pointing toward the products.

$$MgO + C \longrightarrow CO + Mg$$

4. Plus signs are used to separate different reactants or different products from each other.

$$MgO + C \longrightarrow CO + Mg$$

In reading chemical equations, plus signs on the reactant side of the equation are taken to mean "reacts with"; the arrow, "to produce"; and plus signs on the product side, "and."

A catchy, informal way of defining a chemical equation is to say that it gives the "before and after" picture of a chemical reaction. *Before* the reaction starts, only reactants are

In a chemical equation, the *reactants* (starting materials in a chemical reaction) are always written on the left side of the equation, and the *products* (substances produced in a chemical reaction) are always written on the right side of the equation.

*In the last half-century the law of conservation of mass has had to be qualified. Certain types of reactions that involve radioactive processes have been found to deviate from this law. In these processes, there is a conversion of a small amount of matter into energy rather than into another form of matter. A more general law incorporates this apparent discrepancy—the law of conservation of mass and energy. This law takes into account the fact that matter and energy are interconvertible. Note that the statement of the law of conservation of mass as given at the start of this discussion contains the phrase "ordinary chemical reaction." Radioactive processes are not considered to be ordinary chemical reactions.

present—the left side of the equation. *After* the reaction is completed, products are present—the right side of the equation.

For a chemical equation to be *valid*, it must satisfy two conditions.

1. *It must be consistent with experimental facts.* Only the reactants and products actually involved in a reaction are shown in an equation. The accurate formula must be used for each of these substances. For compounds, molecular rather than empirical formulas (Sec. 9.11) are always used. Elements in the solid and liquid states are represented in equations by the chemical symbol for the element. Elements that are gases at room temperature are represented by the molecular form in which they actually occur in nature. Monoatomic, diatomic, and tetratomic elemental gases are known.

> **The diatomic elemental gases are the elements whose names end in** *-gen* **(hydrogen, oxygen, and nitrogen) or** *-ine* **(fluorine, chlorine, bromine, and iodine).**

monoatomic:	He, Ne, Ar, Kr, Xe, Rn
diatomic:	H_2, O_2, N_2, F_2, Cl_2, Br_2 (vapor)*, I_2 (vapor)*
tetratomic:	P_4 (vapor)*, As_4 (vapor)*

2. *It must be consistent with the law of conservation of mass* (Sec. 10.1). There must be the same number of product atoms of each kind as there are reactant atoms of each kind, since atoms are neither created nor destroyed in an ordinary chemical reaction. Equations that satisfy the conditions of this law are said to be *balanced*. Using the four conventions previously listed for writing equations does not guarantee a balanced equation. Section 10.3 considers the steps that must be taken to ensure that an equation is balanced.

10.3 Balancing Chemical Equations

A **balanced chemical equation** *is a chemical equation that has the same number of atoms of each element involved in the reaction on each side of the equation.* It is therefore an equation consistent with the law of conservation of mass (Sec. 10.1).

An unbalanced equation is brought into balance by adding coefficients to the equation; such coefficients adjust the number of reactant and/or product molecules (or formula units) present. An **equation coefficient** *is a number placed to the left of a chemical formula in a chemical equation that changes the amount, but not the identity, of a substance.* In the notation $2 H_2O$, the 2 on the left is a coefficient; $2 H_2O$ means two molecules of H_2O, and $3 H_2O$ means three molecules of H_2O. Coefficients tell how many formula units of a given substance are present.

The following is a balanced chemical equation with the equation coefficients shown in color.

$$3\,Cu + 8\,HNO_3 \longrightarrow 3\,Cu(NO_3)_2 + 2\,NO + 4\,H_2O$$

The message of this balanced equation is "three Cu atoms react with eight HNO_3 molecules to produce three $Cu(NO_3)_2$ formula units, two NO molecules, and four H_2O molecules." A coefficient of 1 in a balanced equation is not explicitly written; it is considered to be understood. Both PCl_3 and H_3PO_3 have understood coefficients of 1 in the following balanced equation.

$$PCl_3 + 3\,H_2O \longrightarrow H_3PO_3 + 3\,HCl$$

---

*The four elements listed as vapors are not gases at room temperature but vaporize at slightly higher temperatures. The resultant vapors contain molecules with the formulas indicated. Even if these elements do not vaporize, they are still represented with these formulas.

The distinction between an *equation coefficient* placed in front of a chemical formula and a *subscript* in a chemical formula is a very important difference. An equation coefficient placed in front of a formula applies to the whole formula. In contrast, subscripts, also present in formulas, affect only parts of a formula.

Coefficient (affects both the H and O)

$$2\ H_2O$$

Subscript (affects only H)

The above notation denotes two molecules of H_2O; it also denotes a total of four H atoms and two O atoms.

Changing a chemical formula subscript affects the *identity* of the substance. Changing the subscript 2 in the formula SO_2 to a 3 to give SO_3 produces an identity change. The substances SO_2 (sulfur dioxide) and SO_3 (sulfur trioxide) are different compounds with distinctly different properties. *Subscripts can never be changed in the process of balancing a chemical equation.* In contrast, placing a coefficient in front of a chemical formula does not change *identity* but rather changes *amount*. The notation $2\ SO_2$ means two molecules of SO_2 and the notation $3\ SO_2$ means three molecules of SO_2.

We now proceed to the method for determining the proper coefficients to balance a given equation. It is introduced in the context of actually balancing two equations (Examples 10.1 and 10.2). Both of these examples should be studied carefully, because each includes detailed commentary concerning the "ins and outs" of balancing equations.

EXAMPLE 10.1

Balancing a Chemical Equation

Balance the equation $Fe_3O_4 + H_2 \longrightarrow Fe + H_2O$.

SOLUTION

STEP 1 *Examine the equation, and pick one element to balance first.* It is often convenient to identify the most complex substance first, that is, the substance with the greatest number of atoms per formula unit. For this most complex substance, whether a reactant or product, "key in" on the element within it that is present in the greatest amount (greatest number of atoms). Using this guideline, we select Fe_3O_4 and the element oxygen.

We note that there are four oxygen atoms on the left side of the equation (in Fe_3O_4) and only one oxygen atom on the right side (in H_2O). For the oxygen atoms to balance we will need four on each side. To obtain four atoms of oxygen on each side of the equation we place the coefficient 1 in front of Fe_3O_4 and the coefficient 4 in front of H_2O.

$$1\ Fe_3O_4 + H_2 \longrightarrow Fe + 4\ H_2O$$

The coefficient 1 (in front of Fe_3O_4) has been explicitly shown in the preceding equation to remind us that the Fe_3O_4 coefficient has been determined. (In the final balanced equation the 1 need not be shown.) We now have four oxygen atoms on each side of the equation.

$$1\ Fe_3O_4: \quad 1 \times 4 = 4$$
$$4\ H_2O: \quad 4 \times 1 = 4$$

> In balancing a chemical equation, chemical formula subscripts are *never changed*. You must leave the chemical formulas just as they are given. The only thing you can do is place coefficients in front of the chemical formulas.

STEP 2 *Now pick a second element to balance.* We will balance the element Fe next. (In this particular equation it does not matter whether we balance Fe or H second.) The number of Fe atoms on the left side of the equation is three; the coefficient 1 in front of Fe_3O_4 sets the Fe atom number at three. We will need three Fe atoms on the product side. This is accomplished by placing the coefficient 3 in front of Fe.

$$1\,Fe_3O_4 + H_2 \longrightarrow 3\,Fe + 4\,H_2O$$

Now there are three Fe atoms on each side of the equation.

STEP 3 *Now pick a third element to balance.* The only element left to balance is H. There are two H atoms on the left and eight H atoms on the right ($4\,H_2O$ involves 8 H atoms). Placing the coefficient 4 in front of H_2 on the left side gives 8 H atoms on that side.

$$1\,Fe_3O_4 + 4\,H_2 \longrightarrow 3\,Fe + 4\,H_2O$$

STEP 4 *As a final check on the correctness of the balancing procedure, count atoms on each side of the equation.* The following table can be constructed from our balanced equation.

$$Fe_3O_4 + 4\,H_2 \longrightarrow 3\,Fe + 4\,H_2O$$

Atom	Left Side	Right Side
Fe	$1 \times 3 = 3$	$3 \times 1 = 3$
O	$1 \times 4 = 4$	$4 \times 1 = 4$
H	$4 \times 2 = 8$	$4 \times 2 = 8$

All elements are in balance: three Fe atoms on each side, four O atoms on each side, and eight H atoms on each side.

Practice Exercise 10.1

Balance the equation $Fe_2O_3 + C \longrightarrow Fe + CO_2$.

Answers to practice exercises are located at the end of the chapter.

EXAMPLE 10.2

Balancing a Chemical Equation

Balance the equation $C_4H_{10} + O_2 \longrightarrow CO_2 + H_2O$.

SOLUTION

STEP 1 *Examine the equation, and pick one element to balance first.* The formula containing the most atoms is C_4H_{10}. We will balance the element H first. We have ten H atoms on the left and two H atoms on the right. The two sides are brought into balance by placing the coefficient 5 in front of H_2O on the right side. We now have ten H atoms on each side.

$$1\,C_4H_{10}: \quad 1 \times 10 = 10$$
$$5\,H_2O: \quad 5 \times 2 = 10$$

Our equation now has the following appearance.

$$1\,C_4H_{10} + O_2 \longrightarrow CO_2 + 5\,H_2O$$

In setting the H balance at ten atoms we are setting the coefficient in front of C_4H_{10} at 1. The 1 has been explicitly shown in the above equation to remind us that the C_4H_{10} coefficient has been determined. (In the final balanced equation the 1 should not be shown.)

STEP 2 *Now pick a second element to balance.* We will balance C next. You always balance the elements that appear in only one reactant and one product before trying to balance any elements appearing in several formulas on one side of the equation. Oxygen, our other choice for an element to balance at this stage, appears in two places on the product side of the equation. The number of carbon atoms is already set at four on the left side of the equation.

$$1\,C_4H_{10}\ \ 1 \times 4 = 4$$

We obtain a balance of four carbon atoms on each side of the equation by placing the coefficient 4 in front of CO_2.

$$1\,C_4H_{10} + O_2 \longrightarrow 4\,CO_2 + 5\,H_2O$$

STEP 3 *Now pick a third element to balance.* Only one element is left to balance—oxygen. The number of oxygen atoms on the right side of the equation is already set at 13: 8 O atoms from the CO_2 and 5 O atoms from the H_2O.

$$4\,CO_2:\ 4 \times 2 = 8$$

$$5\,H_2O:\ 5 \times 1 = 5$$

To obtain 13 O atoms on the left side of the equation, we need a fractional coefficient, $6\frac{1}{2}$.

$$6\tfrac{1}{2}\,O_2:\ 6\tfrac{1}{2} \times 2 = 13$$

The coefficient 6 gives 12 atoms and the coefficient 7 gives 14 atoms. The only way we can get 13 atoms is by using $6\frac{1}{2}$.

All the coefficients in the equation have now been determined.

$$1\,C_2H_6 + 6\tfrac{1}{2}\,O_2 \longrightarrow 4\,CO_2 + 5\,H_2O$$

Equations containing fractional coefficients are not considered to be written in their most conventional form. Although such equations are "mathematically" correct, they have some problems "chemically." The above equation indicates the need for $6\frac{1}{2}$ O_2 molecules among the reactants, but half an O_2 molecule does not exist as such. Step 4 shows how to take care of this "problem."

STEP 4 *After all coefficients have been determined, clear any fractional coefficients that are present.* We can clear the fractional coefficient present in this equation, $6\frac{1}{2}$, by multiplying *each* of the coefficients in the equation by the factor 2.

$$2\,C_4H_{10} + 13\,O_2 \longrightarrow 8\,CO_2 + 10\,H_2O$$

Now we have the equation in its conventional form. Note that *all the coefficients* had to be multiplied by 2, not just the fractional one. It will always be the case that whatever is done to a fractional coefficient to make it a whole number must also be carried out on all of the other coefficients.

If a coefficient involving $\frac{1}{3}$ had been present in the equation, we would have multiplied by 3 instead of by 2.

STEP 5 *As a final check on the correctness of the balancing procedure, count atoms on each side of the equation.* The following table can be constructed from our balanced equation.

$$2\,C_4H_{10} + 13\,O_2 \longrightarrow 8\,CO_2 + 10\,H_2O$$

Atom	Left Side	Right Side
C	$2 \times 4 = 8$	$8 \times 1 = 8$
H	$2 \times 10 = 20$	$10 \times 2 = 20$
O	$13 \times 2 = 26$	$(8 \times 2) + (10 \times 1) = 26$

All atom counts balance. We have accomplished our task.

Practice Exercise 10.2

Balance the equation $C_2H_6 + O_2 \longrightarrow CO_2 + H_2O$.

Some additional comments and guidelines concerning chemical equations and the process of balancing them are in order.

It is important to remember that the ultimate source of the information conveyed by a chemical equation is experimental data. Before you can write a chemical equation for a reaction, the identity and formulas of the reactants and products must be determined *by experiment*. Once you know these things, then you can write the balanced equation.

There are four additional guidelines concerning chemical equations and the balancing process:

1. *The coefficients in a balanced equation are always the smallest set of whole numbers that will balance the equation.* This is significant because more than one set of coefficients will balance an equation. Consider the following three equations.

$$2\,H_2 + O_2 \longrightarrow 2\,H_2O$$

$$4\,H_2 + 2\,O_2 \longrightarrow 4\,H_2O$$

$$8\,H_2 + 4\,O_2 \longrightarrow 8\,H_2O$$

All three of these equations are mathematically correct; there are equal numbers of H and O atoms on each side of the equation. The first equation, however, is considered the conventional form, because the coefficients used there are the smallest set of whole numbers that will balance the equation. The coefficients in the second equation are double those in the first, and the third equation has coefficients four times those of the first equation.

> The only way to learn to balance equations is through practice. The problems at the end of this chapter contain numerous equations for your practice.

2. *It is useful to consider polyatomic ions as single entities in balancing an equation, provided they maintain their identities in the chemical reaction.* Polyatomic ions have maintained their identity in a chemical reaction if they appear on both sides of the equation. For example, in an equation where sulfate units are present (SO_4^{2-}), both as reactants and products, balance them as a unit rather than trying to balance S and O separately. The reasoning would be: "We have two sulfates on this side, so we need two sulfates on that side," and so on.

3. *Subscripts in the chemical formula of a reactant or product should never be altered (changed) during the balancing process.* A student might try to balance the atoms of C in CO_2 at two by using the notation C_2O_4 instead of $2\,CO_2$. This is incorrect. The notation C_2O_4 denotes a molecule containing four atoms, whereas the notation $2\,CO_2$ denotes two molecules, each containing two atoms. The experimental fact is that a molecule of CO_2 contains two rather than four atoms. *The coefficient deals with the*

number of formula units of a substance, and the subscript deals with the composition of the substance. Subscripts illustrate the law of definite proportions (Sec. 9.1); coefficients relate to the law of conservation of mass (Sec. 10.1).

4. *Knowing the procedures for balancing equations does not enable you to predict what the products of a chemical reaction will be.* You are not expected, at this point, to be able to write down the products for a chemical reaction given what the reactants are. After learning how to balance equations, students sometimes get the mistaken idea that they ought to be able to write down equations from "scratch." This is not so. At this stage, you should be able to balance simple equations given *all* of the reactants and *all* of the products.

10.4 Special Symbols Used in Chemical Equations

In addition to the essential plus sign and arrow notation used in chemical equations, a number of optional symbols convey more information about a chemical reaction than just the chemical species involved. In particular, it is often useful to know the physical state of the substances involved in a chemical reaction. The optional symbols listed in Table 10.1 are used to specify the physical state of reactants and products.

The equations we balanced in Section 10.3 (Examples 10.1 and 10.2) are written as follows when the optional symbols are included.

$$Fe_3O_4(s) + 4\,H_2(g) \longrightarrow 3\,Fe(s) + 4\,H_2O(l)$$

$$2\,C_4H_{10}(g) + 13\,O_2(g) \longrightarrow 8\,CO_2(g) + 10\,H_2O(g)$$

Two more examples of the use of optional symbols are

$$NaCl(aq) + AgNO_3(aq) \longrightarrow AgCl(s) + NaNO_3(aq)$$

$$NaOH(aq) + HCl(aq) \longrightarrow NaCl(aq) + H_2O(l)$$

The optional symbols in these latter two equations indicate that both reactions take place in aqueous solution. In the first reaction, one of the products, AgCl, is insoluble, being present in the mixture as a solid. In the second reaction, the product NaCl is soluble and thus remains in solution.

Also, note that in the last equation the reactant HCl is functioning as an acid and must be designated as such (Sec. 8.6). The notation HCl(g) indicates hydrogen chloride in its gaseous state; the notation HCl(aq), hydrogen chloride dissolved in water, denotes hydrochloric acid.

> Some equations are much more difficult to balance than those you encounter in this chapter's examples and end-of-chapter problems. The procedures discussed here simply are not adequate for these more difficult equations. In Chapter 15 a more systematic method for balancing equations, specifically designed for these more difficult situations, will be presented.

Table 10.1	Symbols Used in Equations
Symbol	**Meaning**
Essential	
$\longrightarrow$	"to produce"
+	"reacts with" or "and"
Optional	
(s)	solid
(l)	liquid
(g)	gas
(aq)	aqueous solution (a substance dissolved in water)

10.5 Classes of Chemical Reactions

An almost inconceivable number of chemical reactions are possible. The problems associated with organizing our knowledge about them are diminished considerably by grouping the reactions into classes based on common characteristics. In this section we examine five classes of chemical reactions. These classes are

1. Synthesis reactions
2. Decomposition reactions
3. Single-replacement reactions
4. Double-replacement reactions
5. Combustion reactions

Although there are other categories of reactions besides these five, most of the reactions discussed in this text fall into one of these five categories.

Synthesis Reactions

The first of the five categories of reactions is the *synthesis reaction*. A **synthesis reaction** *is a chemical reaction in which a single product is produced from two (or more) reactants.*

$$X + Y \longrightarrow XY$$

Synthesis reactions always involve simpler substances being combined into a more complex substance. The reactants X and Y may be elements or compounds or an element and a compound. The product of the reaction, XY, is always a compound (see Fig. 10.2).

Some representative synthesis reactions with elements as the reactants are

$$H_2 + Cl_2 \longrightarrow 2\,HCl$$
$$S + O_2 \longrightarrow SO_2$$
$$Ni + S \longrightarrow NiS$$

Some examples of synthesis reactions in which compounds are involved as reactants are

$$SO_3 + H_2O \longrightarrow H_2SO_4$$
$$K_2O + H_2O \longrightarrow 2\,KOH$$
$$2\,NO + O_2 \longrightarrow 2\,NO_2$$
$$2\,NO_2 + H_2O_2 \longrightarrow 2\,HNO_3$$

Decomposition Reactions

A **decomposition reaction** *is a chemical reaction in which a single reactant is converted into two or more simpler substances.* It is thus the exact opposite of a synthesis reaction. The general equation for a decomposition reaction is

$$XY \longrightarrow X + Y$$

Heat or light is the common stimulus used to effect a decomposition reaction.

At sufficiently high temperatures all compounds can be broken down (decomposed) into their constituent elements. Such reactions include

$$2\,CuO \longrightarrow 2\,Cu + O_2$$
$$2\,H_2O \longrightarrow 2\,H_2 + O_2$$

Examples of decomposition reactions that result in at least one compound as a product are

$$NH_4NO_3 \longrightarrow N_2O + 2\,H_2O$$

$$2\,KClO_3 \longrightarrow 2\,KCl + 3\,O_2$$

A most common type of decomposition reaction is that which involves metal carbonates. Metal carbonates, when heated to a high temperature (the temperature needed varies with the metal), break down, releasing carbon dioxide gas and producing the metal oxide. Typical carbonate decomposition equations include the following.

$$Na_2CO_3 \longrightarrow Na_2O + CO_2$$

$$MgCO_3 \longrightarrow MgO + CO_2$$

$$Al_2(CO_3)_3 \longrightarrow Al_2O_3 + 3\,CO_2$$

Single-Replacement Reactions

A **single-replacement reaction** *is a chemical reaction in which one element within a compound is replaced by another element.* In this type of reaction there are always two reactants, an element and a compound, and two products, also an element and a compound. The general equation for a single-replacement reaction is

$$X + YZ \longrightarrow Y + XZ$$

Both metals and nonmetals can be replaced in this manner. Such reactions usually involve aqueous solutions. Examples include

$$Zn + H_2SO_4 \longrightarrow H_2 + ZnSO_4$$

$$Ni + 2\,HCl \longrightarrow H_2 + NiCl_2$$

$$Fe + CuSO_4 \longrightarrow Cu + FeSO_4$$

$$Mg + Ni(NO_3)_2 \longrightarrow Ni + Mg(NO_3)_2$$

Double-Replacement Reactions

A **double-replacement reaction** *is a chemical reaction in which two compounds exchange parts with each other and form two new compounds.* The general equation for such a reaction is

$$AX + BY \longrightarrow AY + BX$$

Double-replacement reactions generally involve ionic compounds in aqueous solution. Most often the positive ion from one compound exchanges with the positive ion of the other. The process may be thought of as "partner swapping," since each negative ion ends up paired with a new partner (positive ion). Such reactions include

$$AgNO_3 + NaCl \longrightarrow NaNO_3 + AgCl$$

$$NaF + HCl \longrightarrow NaCl + HF$$

$$AgNO_3 + HCl \longrightarrow AgCl + HNO_3$$

Figure 10.3 summarizes in pictorial form the four classes of chemical reactions we have so far considered. Example 10.3 is an exercise in classifying reactions into these four general types.

> When hydrogen peroxide is poured on a wound, it decomposes into water and oxygen:
>
> $$2\,H_2O_2 \longrightarrow 2\,H_2O + O_2$$
>
> The bubbles that form are oxygen gas.

> The decomposition reaction $CaCO_3 \longrightarrow CaO + CO_2$ occurs when limestone ($CaCO_3$) is heated in cement factories to produce calcium oxide, one of the components of cement.

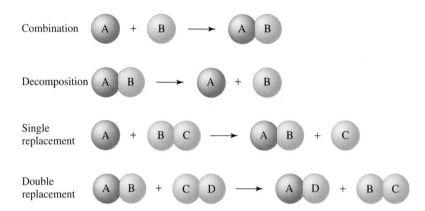

Figure 10.3

Schematic diagrams for synthesis, decomposition, single-replacement, and double-replacement types of chemical reactions.

EXAMPLE 10.3

Classification of Chemical Reactions into the Categories Synthesis, Decomposition, Single Replacement, and Double Replacement

Classify each of the following reactions as synthesis, decomposition, single replacement, or double replacement.

(a) $CuCO_3 \longrightarrow CuO + CO_2$

(b) $Fe + Cu(NO_3)_2 \longrightarrow Cu + Fe(NO_3)_2$

(c) $3\,Mg + N_2 \longrightarrow Mg_3N_2$

(d) $NH_4Cl + AgNO_3 \longrightarrow NH_4NO_3 + AgCl$

SOLUTION

(a) Since two substances are produced from a single substance, this reaction is a decomposition reaction.

(b) Having an element and a compound as reactants and an element and a compound as products is a characteristic of a single-replacement reaction.

(c) Two substances combine to form a single substance; hence, this reaction is classified as a synthesis reaction.

(d) This reaction is a double-replacement reaction. Ammonium ion and silver ion are changing places, that is, "swapping partners."

Practice Exercise 10.3

Classify each of the following reactions as synthesis, decomposition, single replacement, or double replacement.

(a) $2\,C + O_2 \longrightarrow 2\,CO$

(b) $2\,KNO_3 \longrightarrow 2\,KNO_2 + O_2$

(c) $Zn + 2\,AgNO_3 \longrightarrow Zn(NO_3)_2 + 2\,Ag$

(d) $Ni(NO_3)_2 + 2\,NaOH \longrightarrow Ni(OH)_2 + 2\,NaNO_3$

Combustion Reactions

Combustion reactions are a most common type of reaction. A **combustion reaction** *is a chemical reaction in which a substance reacts with oxygen (usually from air) that proceeds with evolution of heat and usually also a flame*. Hydrocarbons, binary compounds of carbon and hydrogen (of which many exist), are the most common type of compound that undergoes combustion. In hydrocarbon combustion, the carbon of the hydrocarbon combines with oxygen to produce carbon dioxide (CO_2). The hydrogen of the hydrocarbon also interacts with oxygen of air to give water (H_2O) as a product. The relative amounts of CO_2 and H_2O produced depends on the composition of the hydrocarbon.

$$2\,C_2H_2 + 5\,O_2 \longrightarrow 4\,CO_2 + 2\,H_2O$$

$$C_3H_8 + 5\,O_2 \longrightarrow 3\,CO_2 + 4\,H_2O$$

$$C_4H_8 + 6\,O_2 \longrightarrow 4\,CO_2 + 4\,H_2O$$

Note that the equation that was balanced in Example 10.2 was a combustion reaction.

Combustion of compounds containing oxygen as well as carbon and hydrogen (for example, CH_4O or C_3H_8O) also produce CO_2 and H_2O as products.

$$2\,CH_4O + 3\,O_2 \longrightarrow 2\,CO_2 + 4\,H_2O$$

$$2\,C_3H_8O + 9\,O_2 \longrightarrow 6\,CO_2 + 8\,H_2O$$

Example 10.4 is an equation-balancing exercise involving a reaction of this type.

> Hydrocarbon combustion reactions are the basis of industrial society, making possible the burning of gasoline in cars, of natural gas in homes, and of coal in factories. Gasoline, natural gas, and coal all contain hydrocarbons.
>
> Unlike most other chemical reactions, hydrocarbon combustion reactions are carried out for the energy they produce rather than for the material products.

EXAMPLE 10.4

Writing and Balancing an Equation for a Combustion Reaction

Diethyl ether, $C_4H_{10}O$, a very flammable compound, was one of the first general anesthetics. Its use for this purpose was first demonstrated by the Boston dentist William Morton, in 1846. Write the balanced chemical equation for the reaction that occurs when diethyl ether burns, that is, reacts with the oxygen (O_2) in air.

SOLUTION

The C atoms in the ether will end up in product CO_2 and the H atoms of the ether in product H_2O. There are two sources for the oxygen present in the product CO_2 and H_2O. Most of it comes from the O_2 of air; however, the reactant $C_4H_{10}O$ is also an oxygen source. The unbalanced equation for this reaction is

$$C_4H_{10}O + O_2 \longrightarrow CO_2 + H_2O$$

The equation is balanced using the procedures of Section 10.3.

STEP 1 *Balancing of H atoms*: There are ten H atoms on the left and only two H atoms on the right. Placing the coefficient 5 in front of H_2O balances the H atoms at ten on each side.

$$1\,C_4H_{10}O + O_2 \longrightarrow CO_2 + 5\,H_2O$$

STEP 2 *Balancing of C atoms*: An effect of balancing the H atoms at ten (step 1) is the setting of the C atoms on the left side of the equation at 4; the coefficient in front of $C_4H_{10}O$ is 1. Placing the coefficient 4 in front of CO_2 causes the carbon atoms to balance at four on each side of the equation.

$$1\,C_4H_{10}O + O_2 \longrightarrow 4\,CO_2 + 5\,H_2O$$

STEP 3 *Balancing of O atoms*: The oxygen content of the right side of the equation is set at 13 atoms; 8 oxygen atoms from $4\,CO_2$ and 5 oxygen atoms from $5\,H_2O$. To obtain 13 oxygen atoms on the left side of the equation, the coefficient 6 is placed in front of O_2; $6 O_2$ gives 12 oxygen atoms, and there is an additional O in $1 C_2H_6O$. Note that the element oxygen is present in all four formulas in the equation.

$$1 C_4H_{10}O + 6 O_2 \longrightarrow 4 CO_2 + 5 H_2O$$

STEP 4 *Final check*: The equation is balanced. There are 4 carbon atoms, 10 hydrogen atoms, and 13 oxygen atoms on each side of the equation.

$$C_4H_{10}O + 6 O_2 \longrightarrow 4 CO_2 + 5 H_2O$$

CHEMICAL EXTENSION

Diethyl ether is no longer used as a general anesthetic because of two drawbacks: (1) it causes nausea and irritation of the respiratory passage, and (2) it is a highly flammable substance, forming explosive mixtures with air, which can be set off by a spark.

In the 1930s nonether anesthetics were developed that solved the nausea–irritation problem. These compounds were, however, still extremely flammable. The hydrocarbon cyclopropane (C_3H_6) was the most used of these newer compounds.

It was not until the late 1950s and early 1960s that nonflammable general anesthetics became available. Most of the new compounds were halogenated ethers. Presence of halogen atoms (F, Cl, Br, I) in ethers decreases their flammability. Halogenated ethers in use today as general anesthetics include enflurane $(C_3H_2OF_5Cl)$ and methoxyflurane $(C_3H_4OF_2Cl_2)$.

Practice Exercise 10.4

Ethyl alcohol, C_2H_6O, is a component of many oxygenated gasolines that are used during winter time. Write the balanced chemical equation for the reaction that occurs when ethyl alcohol burns in air (O_2).

> Glucose sugar $(C_6H_{12}O_6)$ is "burned" in the body by the reaction $C_6H_{12}O_6 + 6 O_2 \longrightarrow 6 CO_2 + 6 H_2O$. This is accomplished with the help of enzymes rather than an open flame.

Many, but not all, reactions fall into one of the five categories we have discussed in this section. Even though this classification system is not all-inclusive, it is still very useful because of the many reactions it does help correlate.

In later chapters, two additional categories of reactions are considered. *Acid–base reactions* are the topic of Chapter 14, and *oxidation–reduction reactions* are considered in Chapter 15.

10.6 Chemical Equations and the Mole Concept

The coefficients in a balanced chemical equation, like the subscripts in a chemical formula (Sec. 9.9), have two levels of interpretation—a microscopic level of meaning and a macroscopic level of meaning.

The first of these two interpretations, the microscopic level, has been used in the previous sections of this chapter. At the *microscopic* level of interpretation the coefficients in a balanced chemical equation give directly the numerical relationships among formula units consumed (used up) and/or produced in the chemical reaction. Interpreted at the microscopic level, the chemical equation

$$4 NH_3 + 5 O_2 \longrightarrow 4 NO + 6 H_2O$$

conveys the information that four molecules of NH_3 react with five molecules of O_2 to produce four molecules of NO and six molecules of H_2O.

At the *macroscopic* level of interpretation, the coefficients in a balanced chemical equation give the fixed molar ratios between substances consumed and/or produced in the chemical reaction. Interpreted at the macroscopic level, the chemical equation

$$4\,NH_3 + 5\,O_2 \longrightarrow 4\,NO + 6\,H_2O$$

conveys the information that four moles of NH_3 react with five moles of O_2 to produce four moles of NO and six moles of H_2O.

The validity of the molar (macroscopic) interpretation of coefficients in an equation can be derived very straightforwardly from the microscopic level of interpretation. A balanced chemical equation remains valid (mathematically correct) when all of its coefficients are multiplied by the same number. (If molecules react in a 3-to-1 ratio, they will also react in a 6-to-2 or 9-to-3 ratio.) Multiplying the previous equation by y, where y is any number, we have

$$4y\,NH_3 + 5y\,O_2 \longrightarrow 4y\,NO + 6y\,H_2O$$

The situation for $y = 6.022 \times 10^{23}$ is of particular interest because $6.022 \times 10^{23} = 1$ mole.

Using $y = 1$ mole, we have by substitution

$$4 \text{ moles } NH_3 + 5 \text{ moles } O_2 \longrightarrow 4 \text{ moles } NO + 6 \text{ moles } H_2O$$

Thus, as with the subscripts in chemical formulas (Sec. 9.9), the coefficients in chemical equations carry a dual meaning. "Number of formula units" is the microscopic-level interpretation for equation coefficients and "moles of formula units" is the macroscopic-level interpretation.

In Section 10.3 it was noted that fractional equation coefficients are often obtained in the equation-balancing process. We can now further note that such fractional coefficients do have valid meaning for the macroscopic-level interpretation of a chemical equation ($3\frac{1}{2}$ moles, and so on), whereas they are totally unacceptable for the microscopic level of interpretation ($3\frac{1}{2}$ molecules, and so on).

The coefficients in a balanced chemical equation may be used to generate conversion factors used in problem solving. Numerous conversion factors are obtainable from a single balanced equation. Consider the balanced equation

$$P_4O_{10} + 6\,H_2O \longrightarrow 4\,H_3PO_4$$

Three mole-to-mole relationships can be obtained from this equation.

> One mole of P_4O_{10} produces four moles of H_3PO_4.

> Six moles of H_2O produce four moles of H_3PO_4.

> One mole of P_4O_{10} reacts with six moles of H_2O.

From these three macroscopic-level relationships, six conversion factors can be written.

From the first relationship,

$$\frac{1 \text{ mole } P_4O_{10}}{4 \text{ moles } H_3PO_4} \quad \text{and} \quad \frac{4 \text{ moles } H_3PO_4}{1 \text{ mole } P_4O_{10}}$$

From the second relationship,

$$\frac{6 \text{ moles } H_2O}{4 \text{ moles } H_3PO_4} \quad \text{and} \quad \frac{4 \text{ moles } H_3PO_4}{6 \text{ moles } H_2O}$$

From the third relationship,

$$\frac{1 \text{ mole } P_4O_{10}}{6 \text{ moles } H_2O} \quad \text{and} \quad \frac{6 \text{ moles } H_2O}{1 \text{ mole } P_4O_{10}}$$

To a chemist a chemical equation is a recipe. Just as the (very simple) recipe 3 cups flour + 1 cup milk = 2 cakes says that if 3 cups of flour are mixed with 1 cup of milk, 2 cakes can be made. The equation $3\,H_2 + N_2 \longrightarrow 2\,NH_3$ tells a chemist that when 3 moles of H_2 are reacted with 1 mole of nitrogen, 2 moles of NH_3 are formed.

Any chemical equation can be the source of numerous conversion factors. The more reactants and products there are in the equation, the greater the number of conversion factors.

Conversion factors obtained from equations are used in several different types of calculations. Example 10.5 illustrates some very simple applications of their use. In Section 10.9 we explore their use in more complicated problem-solving situations.

EXAMPLE 10.5

Calculating Molar Quantities Using a Balanced Chemical Equation

Two air pollutants present in automobile exhaust are carbon monoxide (CO) and nitrogen monoxide (NO). Within an automobile's catalytic converter, these two pollutants react with each other to produce carbon dioxide (CO_2) and nitrogen (N_2). The equation for the reaction is

$$2 CO + 2 NO \longrightarrow 2 CO_2 + N_2$$

(a) How many moles of N_2 are produced when 3.50 moles of CO reacts?

(b) How many moles of NO are needed to react with 2.31 moles of CO?

SOLUTION

Both parts of this problem are one-step mole-to-mole calculations. In each case the needed conversion factor is derived from the coefficients of the chemical equation.

$$\boxed{\text{Moles of substance A}} \xrightarrow[\text{coefficients}]{\text{Equation}} \boxed{\text{Moles of substance B}}$$

(a)

STEP **1** The given quantity is 3.50 moles of CO and the desired quantity is moles of N_2.

$$3.50 \text{ moles CO} = ? \text{ moles } N_2$$

STEP **2** The conversion factor needed to convert from moles of CO to moles of N_2 is derived from the coefficients of CO and N_2 in the balanced equation. The equation tells us that 2 moles of CO produces 1 mole of N_2. From this relationship two conversion factors are obtainable:

$$\frac{2 \text{ moles CO}}{1 \text{ mole } N_2} \quad \text{and} \quad \frac{1 \text{ mole } N_2}{2 \text{ moles CO}}$$

We will use the second of these conversion factors in solving the problem. The setup is

$$3.50 \text{ moles } \cancel{CO} \times \frac{1 \text{ mole } N_2}{2 \text{ moles } \cancel{CO}}$$

We used the second of the two conversion factors because it had moles of CO in the denominator, a requirement for the unit moles of CO to cancel.

STEP **3** Collecting numerical terms, after cancellation of units, gives

$$\frac{3.50 \times 1}{2} \text{ moles } N_2 = 1.75 \text{ moles } N_2 \text{ (calculator and \textbf{correct answer})}$$

Note that the coefficients in the equation enter directly into the numerical calculation. Having a correctly balanced equation is therefore of vital importance.

Using an unbalanced or misbalanced equation as a source of a conversion factor will lead to a wrong numerical answer.

(b)

STEP 1 Both the given species, CO, and the desired species, NO, are reactants.

$$2.31 \text{ moles CO} = ? \text{ moles NO}$$

STEP 2 Molar relationships obtained from an equation are not required to always involve one reactant and one product, as was the case in part (a). Molar relationships involving only reactants or only products are often needed and used. In this problem we will need the molar relationship between the two reactants, CO and NO, which is two to two. From this ratio the conversion factor

$$\frac{2 \text{ moles NO}}{2 \text{ moles CO}}$$

can be constructed, which is used in the setup of the problem as follows.

$$2.31 \text{ moles CO} \times \frac{2 \text{ moles NO}}{2 \text{ moles CO}}$$

STEP 3 Collecting numerical terms, after cancellation of units, gives

$$\frac{2.31 \times 2}{2} \text{ moles NO} = 2.31 \text{ moles NO} \quad \text{(calculator and \textbf{correct answer})}$$

In terms of significant figures, the numbers in conversion factors obtained from equation coefficients are considered exact numbers. Thus, since 2.31 contains three significant figures, the answer to this problem should also contain three significant figures.

CHEMICAL EXTENSION

Catalytic converters significantly decrease, but do not totally eliminate, the emission of carbon monoxide into the atmosphere by automobiles. What is the fate of the carbon monoxide that does enter the atmosphere? It is not converted to carbon dioxide as one might suppose; this reaction does not readily occur at atmospheric temperatures.

Because CO is not appreciably soluble in water, very little of it is removed in rainfall. The major removal mechanism for CO involves soil microorganisms. A number of types of soil fungi have the ability to metabolize CO.

The capacity for CO uptake by the soils of the United States far exceeds the amount of CO present in the air. Nevertheless, CO is still found at significant concentrations in the air. Why is this so? Soil and CO are not distributed uniformly. The largest CO-producing areas are large cities with their large numbers of automobiles, areas which contain much concrete and asphalt but little soil.

Practice Exercise 10.5

The natural gas burned to provide heat in many homes is predominantly methane (CH_4). Methane burns (reacts with O_2) as shown by the equation

$$CH_4 + 2\,O_2 \longrightarrow CO_2 + 2\,H_2O$$

(a) How many moles of H_2O are produced when 1.23 moles of CH_4 burns?

(b) How many moles of O_2 are needed to react with 3.61 moles of CH_4?

<table>
<tr><td>**10.7**</td><td>**Balanced Chemical Equations and the Law of Conservation of Mass**</td></tr>
</table>

We began this chapter with a discussion of the law of conservation of mass (Sec. 10.1), which became the basis for the concepts involved in balancing chemical equations (Sec. 10.3). Now that the relationship between chemical equation coefficients and moles has been addressed (Sec. 10.6), it is time to revisit the law of conservation of mass.

A balanced chemical equation can be used to verify the law of conservation of mass. Let us consider this verification process using the balanced equation

$$4\,NH_3 + 5\,O_2 \longrightarrow 4\,NO + 6\,H_2O$$

According to the coefficients in this balanced equation, the molar ratio among the reactants and product is 4 to 5 to 4 to 6, that is,

$$4\text{ moles }NH_3 + 5\text{ moles }O_2 \longrightarrow 4\text{ moles }NO + 6\text{ moles }H_2O$$

Substituting molar masses (Sec. 9.6) into this equation, which are, respectively, 17.04 g/mole for NH_3, 32.00 g/mole for O_2, 30.01 g/mole for NO, and 18.02 g/mole for H_2O, we obtain

$$4\,(17.04\text{ g}) + 5\,(32.00\text{ g}) = 4\,(30.01\text{ g}) + 6\,(18.02\text{ g})$$

Simplifying, we get

$$68.16\text{ g} + 160.00\text{ g} = 120.04\text{ g} + 108.12\text{ g}$$
$$228.16\text{ g} = 228.16\text{ g}$$

Thus, based on the molar interpretation of equation coefficients, the sum of the reactant masses is equal to the sum of the product masses, as it should be according to the law of conservation of mass.

Example 10.6 further illustrates the relationship between the coefficients in a balanced equation and the law of conservation of mass.

EXAMPLE 10.6

Verifying the Law of Conservation of Mass Using a Balanced Chemical Equation

Verify the law of conservation of mass using the balanced chemical equation

$$3\,FeO + 2\,Al \longrightarrow 3\,Fe + Al_2O_3$$

SOLUTION

The given balanced equation, interpreted in terms of moles, is

$$3\text{ moles }FeO + 2\text{ moles }Al \longrightarrow 3\text{ moles }Fe + 1\text{ mole }Al_2O_3$$

The mole masses of the substances involved in this reaction are

$$FeO:\ 1\text{ mole} = 71.84\text{ g}$$
$$Al:\ 1\text{ mole} = 26.98\text{ g}$$
$$Fe:\ 1\text{ mole} = 55.84\text{ g}$$
$$Al_2O_3:\ 1\text{ mole} = 101.96\text{ g}$$

Substituting these molar masses into the given equation in place of moles gives

$$3\,(71.84\text{ g}) + 2\,(26.98\text{ g}) = 3\,(55.84\text{ g}) + 1\,(101.96\text{ g})$$

Simplifying, we obtain

$$215.52 \text{ g} + 53.96 \text{ g} = 167.52 \text{ g} + 101.96 \text{ g}$$

$$269.48 \text{ g} = 269.48 \text{ g}$$

The sum of the reactant masses is equal to the sum of the product masses, a verification of the law of conservation of mass.

Practice Exercise 10.6

Verify the law of conservation of mass using the balanced chemical equation

$$2 \text{ C}_2\text{H}_2 + 5 \text{ O}_2 \longrightarrow 4 \text{ CO}_2 + 2 \text{ H}_2\text{O}$$

10.8 Calculations Based on Chemical Equations—Stoichiometry

A major area of concern for chemists is the quantities of materials consumed and produced in chemical reactions. This area of study is called *chemical stoichiometry*. **Chemical stoichiometry** *is the study of the quantitative relationships among reactants and products in a chemical reaction.* The word *stoichiometry*, pronounced stoy-kee-OM-eh-tree, is derived from the Greek *stoicheion* (element) and *metron* (measure). The stoichiometry of a chemical reaction always involves the *molar relationships* between reactants and products (Sec. 10.6) and thus is given by the coefficients in the balanced equation for the chemical reaction.

In a typical stoichiometric calculation, information is given about one reactant or product of a reaction (number of grams, moles, or particles), and similar information is requested concerning another reactant or product of the same reaction. The substances involved in such a calculation may both be reactants, may both be products, or may be one of each.

The conversion-factor relationships needed to solve problems of the above general type are given in Figure 10.4. This diagram should seem very familiar to you, for it is almost identical to Figure 9.5, with which you have worked repeatedly. There is only one difference between the two. In the Chapter 9 diagram, the subscripts in a chemical formula were listed as the basis for relating moles of given and desired substances to each other. In this new diagram, these same two quantities are related using the coefficients of a balanced chemical equation.

The most common type of stoichiometric calculation is a mass-to-mass (gram-to-gram) problem. In such problems the mass of one substance involved in a chemical reaction (either reactant or product) is given, and information is requested about the mass of another of the substances involved in the reaction (either a reactant or product). Situations requiring the solution of problems of this type are frequently encountered in laboratory settings. For

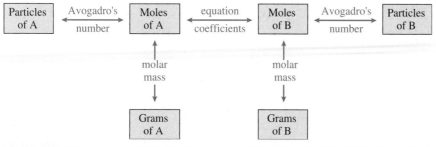

Figure 10.4

Conversion factor relationships needed for solving chemical-equation-based problems.

example, a chemist has available so many grams of a certain chemical and wants to know how many grams of another substance can be produced from it, or how many grams of a third substance are needed to react with it. Examples 10.7 and 10.8 are both problem-solving situations of the gram-to-gram type.

EXAMPLE 10.7

Calculating the Needed Mass of a Reactant in a Chemical Reaction

A mixture of hydrazine (N_2H_4) and hydrogen peroxide (H_2O_2) is used as a fuel for rocket engines. These two substances react as shown by the equation

$$N_2H_4(l) + 2\,H_2O_2(l) \longrightarrow N_2(g) + 4\,H_2O(g)$$

How many grams of H_2O_2 are needed to completely react with 50.0 g of N_2H_4?

SOLUTION

STEP 1 Here we are given information about one reactant (50.0 g of N_2H_4) and asked to calculate information about the other reactant (H_2O_2).

$$50.0 \text{ g } N_2H_4 = ? \text{ g } H_2O_2$$

STEP 2 This problem is of the "grams-of-A" to "grams-of-B" type. The pathway to be used in solving this type of problem, in terms of Figure 10.4, is

$$\boxed{\begin{array}{c}\text{Grams}\\\text{of A}\end{array}} \xrightarrow[\text{mass}]{\text{Molar}} \boxed{\begin{array}{c}\text{Moles}\\\text{of A}\end{array}} \xrightarrow[\text{coefficients}]{\text{Equation}} \boxed{\begin{array}{c}\text{Moles}\\\text{of B}\end{array}} \xrightarrow[\text{mass}]{\text{Molar}} \boxed{\begin{array}{c}\text{Grams}\\\text{of B}\end{array}}$$

STEP 3 The dimensional analysis setup for this pathway is

$$50.0 \text{ g } \cancel{N_2H_4} \times \frac{1 \text{ mole } \cancel{N_2H_4}}{32.06 \text{ g } \cancel{N_2H_4}} \times \frac{2 \text{ moles } \cancel{H_2O_2}}{1 \text{ mole } \cancel{N_2H_4}} \times \frac{34.02 \text{ g } H_2O_2}{1 \text{ mole } \cancel{H_2O_2}}$$

$$\text{grams A} \longrightarrow \text{moles A} \longrightarrow \text{moles B} \longrightarrow \text{grams B}$$

The number 32.06 in the first conversion factor is the molar mass of N_2H_4; the 2 and 1 in the second conversion factor are the coefficients of H_2O_2 and N_2H_4, respectively, in the balanced chemical equation; and the number 34.02 in the last conversion factor is the molar mass of H_2O_2.

STEP 4 The solution, obtained from combining all of the numerical factors, is

$$\frac{50.0 \times 1 \times 2 \times 34.02}{32.06 \times 1 \times 1} \text{g } H_2O_2 = 106.11353 \text{ g } H_2O_2 \quad \text{(calculator answer)}$$

$$= 106 \text{ g } H_2O_2 \qquad \textbf{(correct answer)}$$

CHEMICAL EXTENSION

Pure hydrazine is a colorless liquid having an ammonia-like odor that decomposes, often violently, upon heating. In a laboratory, hydrazine is normally used in an aqueous solution, where it can be handled safely.

Structurally, hydrazine (N_2H_4) bears the same relationship to ammonia (NH_3) that hydrogen peroxide (H_2O_2) does to water (H_2O).

Hydrazine as well as substituted hydrazines, where one or more of the hydrogen atoms are replaced by other groups, are useful rocket fuels. The methyl ($—CH_3$) and di-methyl derivatives of hydrazine, in combination with dinitrogen tetroxide (N_2O_4), were used in the lunar excursion module (LEM) for both landing and reblast-off. The reaction between these substances is *hypergolic*, that is, self-igniting. Astronauts control the reaction by opening and closing separate tanks containing the reacting substances.

Practice Exercise 10.7

The active ingredient in some antacid formulations is magnesium hydroxide [$Mg(OH)_2$], which reacts with stomach acid (HCl) to produce magnesium chloride ($MgCl_2$) and water. The equation for the reaction is

$$Mg(OH)_2(s) + 2\ HCl(aq) \longrightarrow MgCl_2(aq) + 2\ H_2O(l)$$

How many grams of $Mg(OH)_2$ are needed to react with 1.00 g of HCl?

EXAMPLE 10.8

Calculating the Mass of a Product in a Chemical Reaction

When baking soda ($NaHCO_3$) is heated it decomposes, producing carbon dioxide gas (CO_2). This carbon dioxide is responsible for the rising of bread, donuts, and cookies. The equation for the baking soda decomposition reaction is

$$2\ NaHCO_3 \longrightarrow Na_2CO_3 + CO_2 + H_2O$$

How many grams of CO_2 are produced when 1.00 g of $NaHCO_3$ decomposes?

SOLUTION

STEP 1 Here we are given information about the reactant (1.00 g of $NaHCO_3$) and asked to calculate information about one of the products (CO_2).

$$1.00\ g\ NaHCO_3 = ?\ g\ CO_2$$

STEP 2 This problem, like Example 10.7, is a "grams-of-A" to "grams-of-B" problem. The pathway used in solving it will be the same, which, in terms of Figure 10.4, is

$$\boxed{\begin{array}{c}\text{Grams}\\\text{of A}\end{array}} \xrightarrow[\text{mass}]{\text{Molar}} \boxed{\begin{array}{c}\text{Moles}\\\text{of A}\end{array}} \xrightarrow[\text{coefficients}]{\text{Equation}} \boxed{\begin{array}{c}\text{Moles}\\\text{of B}\end{array}} \xrightarrow[\text{mass}]{\text{Molar}} \boxed{\begin{array}{c}\text{Grams}\\\text{of B}\end{array}}$$

STEP 3 The dimensional analysis setup is

$$1.00\ \cancel{g\ NaHCO_3} \times \frac{1\ \cancel{\text{mole } NaHCO_3}}{84.01\ \cancel{g\ NaHCO_3}} \times \frac{1\ \cancel{\text{mole } CO_2}}{2\ \cancel{\text{moles } NaHCO_3}} \times \frac{44.01\ g\ CO_2}{1\ \cancel{\text{mole } CO_2}}$$

$$\text{grams A} \longrightarrow \text{moles A} \longrightarrow \text{moles B} \longrightarrow \text{grams B}$$

The chemical equation is the "bridge" that enables us to go from $NaHCO_3$ to CO_2. The numbers in the second conversion factor, the "bridge factor," are coefficients from this equation.

STEP 4 The solution, obtained from combining all of the numerical factors in the setup, is

$$\frac{1.00 \times 1 \times 1 \times 44.01}{84.01 \times 2 \times 1}\ g\ CO_2 = 0.2619331\ g\ CO_2 \quad \text{(calculator answer)}$$

$$= 0.262\ g\ CO_2 \quad \textbf{(correct answer)}$$

CHEMICAL EXTENSION

A problem with the use of baking soda in a dough (or batter) is the soapy taste of the sodium carbonate (Na_2CO_3) produced from the decomposition of the baking soda. The presence of an acidic ingredient in the batter combats this problem; a compound other than Na_2CO_3 is formed from the reaction with the acid. Most of these acid-produced products do not have objectionable flavor. Often-used acidic ingredients include buttermilk, sour milk, molasses, vinegar, and citrus fruit juices.

> Most recipes call for baking powder instead of baking soda. Baking powders are more convenient to use. They contain one or more dry acids or acidic compounds as part of the formulation of the powder, in addition to baking soda and starch (added to help stabilize the mixture and prevent the components (dry acid and baking soda) from reacting prematurely. The two active ingredients in baking powder react with each other at room temperature as soon as liquid is added to the batter.

Practice Exercise 10.8

The chemical equation for the photosynthesis reaction in plants is

$$6\,CO_2 + 6\,H_2O \longrightarrow C_6H_{12}O_6 + 6\,O_2$$

How many grams of oxygen (O_2) are produced by a plant from the consumption of 25.0 g of carbon dioxide (CO_2)?

"Grams-of-A" to "grams-of-B" problems (Examples 10.7 and 10.8) are not the only type of problem for which the coefficients in a balanced equation can be used to relate quantities of two substances. As further examples of the use of equation coefficients in problem solving, consider Example 10.9 (a "grams-of-A" to "moles-of-B" problem) and Example 10.10 (a "particles-of-A" to "grams-of-B" problem).

EXAMPLE 10.9

Calculating Moles of Product Produced in a Chemical Reaction

Automotive airbags inflate when sodium azide, NaN_3, rapidly decomposes to its constituent elements. The equation for the chemical reaction is

$$2\,NaN_3(s) \longrightarrow 2\,Na(s) + 3\,N_2(g)$$

How many moles of N_2 are produced when 1.000 g of NaN_3 decomposes?

SOLUTION

STEP 1 The given quantity is 1.000 g of NaN_3, and the desired quantity is moles of N_2.

$$1.000 \text{ g } NaN_3 = ? \text{ moles } N_2$$

STEP 2 This is a "grams-of-A" to "moles-of-B" problem. The pathway used to solve such a problem is, according to Figure 10.4,

$$\boxed{\text{Grams of A}} \xrightarrow[\text{mass}]{\text{Molar}} \boxed{\text{Moles of A}} \xrightarrow[\text{coefficients}]{\text{Equation}} \boxed{\text{Moles of B}}$$

STEP 3 The dimensional analysis setup is

$$1.000 \text{ g } \cancel{NaN_3} \times \frac{1 \cancel{\text{ mole } NaN_3}}{65.02 \text{ g } \cancel{NaN_3}} \times \frac{3 \text{ moles } N_2}{2 \cancel{\text{ moles } NaN_3}}$$

$$\text{grams A} \longrightarrow \text{moles A} \longrightarrow \text{moles B}$$

The number 65.02 in the first conversion factor is the molar mass of NaN_3.

STEP 4 The solution, obtained from combining all of the numbers in the manner indicated in the setup, is

$$\frac{1.000 \times 1 \times 3}{65.02 \times 2} \text{ moles } N_2 = 0.023069824 \text{ mole } N_2 \quad \text{(calculator answer)}$$

$$= 0.02307 \text{ mole } N_2 \quad \textbf{(correct answer)}$$

CHEMICAL EXTENSION

An increasing number of automobiles are equipped with airbags that rapidly inflate on collision, before car occupants can be thrown forward after impact. The chemistry principle involved in the inflation process involves a controlled chemical explosion that produces a large volume of gas (which inflates the bag).

Nitrogen gas (N_2) is the inflatant in most airbags. Airbags are activated when an impact causes a steel ball to compress a spring and electrically ignite a detonator cap, which in turn, causes sodium azide (NaN_3) to decompose, forming nitrogen gas and metallic sodium. (100 g of NaN_3 produces 56 L of N_2 gas at 25°C.) The sodium metal produced, an undesirable decomposition product because of its reactivity, instantaneously reacts with iron(III) oxide, which is included in the reaction container, to produce sodium oxide (Na_2O) and iron (Fe).

$$6\,Na + Fe_2O_3 \longrightarrow 3\,Na_2O + 2\,Fe$$

Practice Exercise 10.9

The thermal decomposition of solid potassium chlorate ($KClO_3$) serves as a convenient laboratory source of small amounts of oxygen gas. The reaction is

$$2\,KClO_3(s) \longrightarrow 2\,KCl(s) + 3\,O_2(g)$$

How many moles of $KClO_3$ must be decomposed to produce 5.00 g of O_2?

EXAMPLE 10.10

Calculating the Amount of a Substance Produced in a Chemical Reaction

The reaction of sulfuric acid (H_2SO_4) with elemental copper (Cu) produces three products—sulfur dioxide (SO_2), water (H_2O), and copper(II) sulfate ($CuSO_4$). How many grams of water will be produced at the same time that 5 billion (5.00×10^9) sulfur dioxide molecules are produced?

SOLUTION

Although a calculation of this type will not have a lot of practical significance, it will test your understanding of the problem-solving relationships under discussion in this section of the text.

The specifics of the chemical reaction of concern to us in this problem were given in "word" rather than "equation" form in the problem statement. These "words" must be translated into an equation before we can proceed with the problem solving. The equation is

$$H_2SO_4 + Cu \longrightarrow SO_2 + H_2O + CuSO_4$$

Having a chemical equation is not enough. It must be a *balanced* chemical equation. Using the balancing procedures of Section 10.3, the preceding equation, in balanced form, becomes

$$2\,H_2SO_4 + Cu \longrightarrow SO_2 + 2\,H_2O + CuSO_4$$

Now we are ready to proceed with the solving of our problem.

STEP 1 We are given a certain number of particles (molecules) and asked to find the number of grams of a related substance.

$$5.00 \times 10^9 \text{ molecules } SO_2 = ?\text{ g } H_2O$$

STEP 2 This is a "particles-of-A" to "grams-of-B" problem. Even though SO_2 and H_2O are both products, we can still work this problem in a manner similar to previous problems. The coefficients in a balanced equation relate reactants to products, reactants to reactants, and *products to products*. The pathway for this problem (see Fig. 10.4) is

$$\boxed{\begin{matrix}\text{Particles}\\\text{of A}\end{matrix}} \xrightarrow[\text{number}]{\text{Avogadro's}} \boxed{\begin{matrix}\text{Moles}\\\text{of A}\end{matrix}} \xrightarrow[\text{coefficients}]{\text{Equation}} \boxed{\begin{matrix}\text{Moles}\\\text{of B}\end{matrix}} \xrightarrow[\text{mass}]{\text{Molar}} \boxed{\begin{matrix}\text{Grams}\\\text{of B}\end{matrix}}$$

STEP 3 The dimensional analysis setup is

$$5.00 \times 10^9 \text{ molecules } SO_2 \times \frac{1 \text{ mole } SO_2}{6.022 \times 10^{23} \text{ molecules } SO_2} \times \frac{2 \text{ moles } H_2O}{1 \text{ mole } SO_2} \times \frac{18.02 \text{ g } H_2O}{1 \text{ mole } H_2O}$$

$$\text{particles A} \longrightarrow \text{moles A} \longrightarrow \text{moles B} \longrightarrow \text{grams B}$$

STEP 4 The solution, obtained by combining all of the numerical factors in the setup, is

$$\frac{5.00 \times 10^9 \times 1 \times 2 \times 18.02}{6.022 \times 10^{23} \times 1 \times 1} \text{ g } H_2O = 2.9923613 \times 10^{-13} \text{ g } H_2O$$

$$\text{(calculator answer)}$$

$$= 2.99 \times 10^{-13} \text{ g } H_2O$$

$$\text{(correct answer)}$$

CHEMICAL EXTENSION

Which chemical substance is industrially produced in the greatest amount in the United States? It is sulfuric acid, a runaway winner. Sulfuric acid has an almost two-to-one production lead over any other chemical.

Sulfuric acid, in the pure state, is a colorless, corrosive, oily liquid. It is usually marketed as a concentrated aqueous solution. It has been known since "ancient" times, when it was called *oil of vitriol*.

People rarely have direct contact with sulfuric acid because it is seldom part of finished consumer products. The closest encounter most people have with this acid (other than in a chemical laboratory) is as the acid present in an automobile battery.

Approximately 70% of sulfuric acid production is used in the manufacture of chemical fertilizers that contain phosphorus. The raw material for phosphate fertilizers is highly insoluble phosphate rock. Treatment of phosphate rock with H_2SO_4 produces phosphoric acid (H_3PO_4), which is then used to produce soluble phosphate- containing fertilizers.

Practice Exercise 10.10

When silver carbonate (Ag_2CO_3) is decomposed by heating, three products are produced: metallic silver (Ag), carbon dioxide (CO_2), and oxygen (O_2). How many grams of O_2 will be produced when 100 billion (1.00×10^{11}) Ag_2CO_3 formula units decompose?

10.9 The Limiting Reactant Concept

When a chemical reaction is carried out in a laboratory or industrial setting, the reactants are not usually present in the exact molar ratios specified in the balanced chemical equation for the reaction. Most often, on purpose, excess quantities of one or more of the reactants are present.

Numerous reasons exist for having some reactants present in excess. Sometimes such a procedure will cause a reaction to occur more rapidly. For example, large amounts of oxygen make combustible materials burn faster. Sometimes an excess of one reactant will ensure that another reactant, perhaps a very expensive one, is completely consumed.

When reactants are not present in the exact molar ratios specified by the balanced chemical equation for the reaction, the reaction proceeds only until one of the reactants is depleted, that is, completely used up. At this point, the reaction stops. This "reaction-stopping" reactant, which has been entirely consumed, is called the *limiting reactant*. A **limiting reactant** *is the reactant in a chemical reaction that is entirely consumed when the reaction goes to completion (stops)*. The limiting reactant "limits" the amount of product(s) formed. Other reactants present are often called *excess reactants;* they are not entirely consumed. For excess reactants, the excess remains unreacted because there is not enough of the limiting reactant present to react with the excess.

The concept of a limiting reactant plays a major role in chemical calculations of certain types. It must be thoroughly understood. Let us consider some simple but analogous nonchemical examples of a "limiting reactant" before we go on to limiting reactant calculations.

Suppose we have a vending machine that contains 40 50-cent candy bars and we have 30 quarters. In this case we can purchase only 15 candy bars. The quarters are the limiting reactant. The candy bars are present in excess. Suppose we have 10 slices of cheese and 18 slices of bread and we want to make as many cheese sandwiches as possible using 1 slice of cheese and 2 slices of bread per sandwich. The 18 slices of bread limit us to 9 sandwiches; 1 slice of cheese is left over. The bread is the limiting reactant in this case even though initially there was more bread (18 slices) than cheese (10 slices) present. The bread is still limiting because it is used up first.

An additional limiting reactant analogy that comes closer to the realm of molecules and atoms involves nuts and bolts. Assume we have ten identical nuts and ten identical bolts. From this collection we can make ten "one-nut, one-bolt" entities by screwing a nut on each bolt. This situation is depicted in Figure 10.5a.

Next, let us make "two-nut, one-bolt" entities from our same collection of nuts and bolts. This time we can make only five combinations, and we will have five bolts left over, as is shown in Figure 10.5b. We run out of nuts before all the bolts are used up. In chemical jargon, we say that the nuts are the limiting reactant.

Finally, let us consider making "one-nut, two-bolt" combinations. As is shown in Figure 10.5c, this time we do not have enough bolts; we can make five combinations, and we will have five nuts left over. The bolts are the limiting reactant.

Now let us consider, as is presented in Example 10.11, an extension of our nut–bolt discussion to a situation that cannot easily be reasoned out in one's head. Instead, a calculation must be performed.

> In combustion reactions, oxygen is not usually a limiting reactant because it is so plentiful in the air. But if a jar is put over a burning candle, oxygen becomes a limiting reactant. As soon as all the oxygen in the jar is used up, the candle sputters out.

EXAMPLE 10.11

Determining a Limiting Reactant in a Nonchemical Context

What will be the limiting reactant in the production of "two-nut, three-bolt" combinations from a collection of 284 nuts and 414 bolts?

SOLUTION

There are three possible answers to this problem.

1. We will run out of bolts first; bolts are the limiting reactant.
2. We will run out of nuts first; nuts are the limiting reactant.
3. The ratio of nuts to bolts is such that we run out of both at the same time; both are limiting reactants.

Figure 10.5

Starting with ten nuts and ten bolts, we can make (a) ten "one-nut, one-bolt" combinations, (b) five "two-nut, one-bolt" combinations with five bolts left over (the nuts are the limiting reactant); and (c) five "one-nut, two-bolt" combinations with 5 nuts left over (the bolts are the limiting reactant).

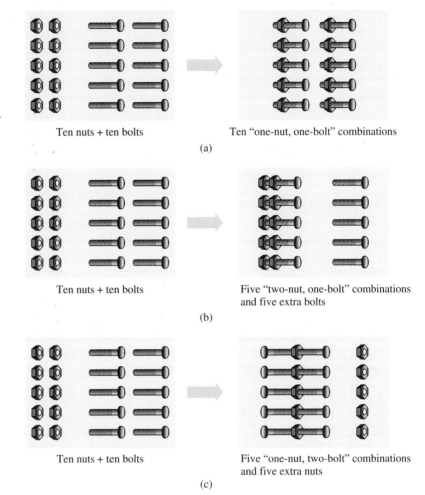

Ten nuts + ten bolts   Ten "one-nut, one-bolt" combinations

(a)

Ten nuts + ten bolts   Five "two-nut, one-bolt" combinations and five extra bolts

(b)

Ten nuts + ten bolts   Five "one-nut, two-bolt" combinations and five extra nuts

(c)

We determine which of these answers is the correct one by calculating how many nut–bolt combinations can be made from each of the "ingredients," assuming an excess of the other.

$$284 \text{ nuts} \times \frac{1 \text{ combination}}{2 \text{ nuts}} = 142 \text{ combinations} \quad \text{(calculator and correct answer)}$$

$$414 \text{ bolts} \times \frac{1 \text{ combination}}{3 \text{ bolts}} = 138 \text{ combinations} \quad \text{(calculator and correct answer)}$$

Because fewer combinations can be made from the bolts, the bolts are the limiting reactant.

Practice Exercise 10.11

What is the limiting reactant in the production of "three-nut, two-bolt" combinations from a collection of 261 nuts and 176 bolts?

Now let us proceed to chemical calculations that involve a limiting reactant. *Whenever the quantities of two or more reactants in a chemical reaction are given, it is necessary to determine which of the given quantities is the limiting reactant.*

Determining the limiting reactant can be accomplished by the following procedure.

1. Determine the number of *moles* of each of the reactants present.
2. Calculate the number of moles of product *each* of the molar amounts of reactant would produce if it were the only reactant amount given. If more than one product is formed in the reaction, you need to do this mole calculation for only one of the products.
3. The reactant that produces the *lesser number* of moles of product is the limiting reactant.

EXAMPLE 10.12

Determining the Limiting Reactant from Given Reactant Amounts

Silver and silver-plated objects tarnish in the presence of hydrogen sulfide (H_2S), a gas that originates from the decay of food, because of the reaction

$$4\,Ag + 2\,H_2S + O_2 \longrightarrow 2\,Ag_2S + 2\,H_2O$$

The black product Ag_2S is the tarnish. If 25.00 g of Ag, 5.00 g of H_2S, and 4.00 g of O_2 are present in a reaction mixture, which is the limiting reactant for tarnish formation?

SOLUTION

To determine the limiting reactant, we determine how many moles of product each of the reactants can form. In this particular problem there are two products: Ag_2S and H_2O. It is sufficient to calculate how many moles of either Ag_2S or H_2O are formed. The decision as to which product to use is arbitrary; we will choose H_2O.

The calculation type will be "grams of A" to "moles of B." We start with a given number of grams of reactant and desire to calculate moles of product. The pathway for the calculation, in terms of Figure 10.4, is

$$\boxed{\text{Grams of A}} \xrightarrow[\text{mass}]{\text{Molar}} \boxed{\text{Moles of A}} \xrightarrow[\text{coefficients}]{\text{Equation}} \boxed{\text{Moles of B}}$$

Note that we will have to go through this type of calculation three times because we have three reactants: once for Ag, once for H_2S and once for O_2.

For Ag:

$$25.00 \text{ g Ag} \times \frac{1 \text{ mole Ag}}{107.87 \text{ g Ag}} \times \frac{2 \text{ moles } H_2O}{4 \text{ moles Ag}} = 0.11588022 \text{ mole } H_2O$$

$$\text{(calculator answer)}$$

$$= 0.1159 \text{ mole } H_2O$$

$$\text{(correct answer)}$$

For H_2S:

$$5.00 \text{ g } H_2S \times \frac{1 \text{ mole } H_2S}{34.08 \text{ g } H_2S} \times \frac{2 \text{ moles } H_2O}{2 \text{ moles } H_2S} = 0.14671361 \text{ mole } H_2O$$

$$\text{(calculator answer)}$$

$$= 0.147 \text{ mole } H_2O$$

$$\text{(correct answer)}$$

For O_2:

$$4.00 \text{ g } O_2 \times \frac{1 \text{ mole } O_2}{32.00 \text{ g } O_2} \times \frac{2 \text{ moles } H_2O}{1 \text{ mole } O_2} = 0.25 \text{ mole } H_2O$$

(calculator answer)

$$= 0.250 \text{ mole } H_2O$$

(correct answer)

The limiting reactant is the reactant that will produce the fewest number of moles of H_2O. Looking at the numbers just calculated, we see that Ag will be the limiting reactant. Once the limiting reactant has been determined, the amount of that reactant present becomes the starting point for any further calculations about the chemical reaction under consideration. Example 10.13 illustrates this point.

CHEMICAL EXTENSION

The hydrogen sulfide molecule is the sulfur analog of water, with a central sulfur atom in place of the central oxygen atom.

Hydrogen sulfide is a gas at room temperature with an odor that resembles rotten eggs. Natural sources of it include bacterial decomposition of proteins and volcanic eruptions.

Nasal detection of hydrogen sulfide is possible at concentrations as low as 0.02 parts per million. Despite its easy detection, its presence can be dangerous because it exerts an anesthetic effect on nasal passages, causing the nose to rapidly lose its ability to detect its presence. Hydrogen sulfide is lethal at a concentration of 100 parts per million.

Decomposition of albumin, a sulfur-rich protein present in eggs, produces hydrogen sulfide. The pale green color often found at the interface of the yolk and white portion of a cooked egg is caused by the presence of hydrogen sulfide. Egg yolk is rich in iron-containing proteins, and H_2S produced from albumin during cooking reacts with these proteins to form iron sulfide, the source of the green color.

Practice Exercise 10.12

Old oil paintings are darkened by PbS, which forms by the reaction of the Pb in paint (a coloring agent) with H_2S in air.

$$Pb + H_2S \longrightarrow PbS + H_2$$

If 5.00 g of Pb and 1.00 g of H_2S are present in a reaction mixture, which is the limiting reactant for the "darkening" reaction?

EXAMPLE 10.13

Calculating the Mass of Product from Masses of Reactants

Aluminum metal, in the form of a powder, reacts vigorously with iron(III) oxide when the two substances are heated. The equation for the reaction is

$$2 \text{ Al}(s) + Fe_2O_3(s) \longrightarrow Al_2O_3(s) + 2 \text{ Fe}(l)$$

How many grams of molten Fe can be formed from a reaction mixture containing 30.0 g Al and 90.0 g of Fe_2O_3?

SOLUTION

First, we must determine the limiting reactant since specific amounts of both reactants are given in the problem. This determination involves two "grams-of-A" to "moles-of-B" calculations—one for the reactant Al and one for the reactant Fe_2O_3. The product we "key in" on is Fe because our final goal is the mass of iron produced.

$$\boxed{\text{Grams of A}} \xrightarrow[\text{mass}]{\text{Molar}} \boxed{\text{Moles of A}} \xrightarrow[\text{coefficients}]{\text{Equation}} \boxed{\text{Moles of B}}$$

For Al:

$$30.0 \text{ g Al} \times \frac{1 \text{ mole Al}}{26.98 \text{ g Al}} \times \frac{2 \text{ moles Fe}}{2 \text{ moles Al}} = 1.1119347 \text{ moles Fe} \quad \text{(calculator answer)}$$

$$= 1.11 \text{ moles Fe} \quad \textbf{(correct answer)}$$

For Fe_2O_3:

$$90.0 \text{ g Fe}_2\text{O}_3 \times \frac{1 \text{ mole Fe}_2\text{O}_3}{159.68 \text{ g Fe}_2\text{O}_3} \times \frac{2 \text{ moles Fe}}{1 \text{ mole Fe}_2\text{O}_3} = 1.1272545 \text{ moles Fe}$$

$$\text{(calculator answer)}$$

$$= 1.13 \text{ moles Fe}$$

$$\textbf{(correct answer)}$$

Thus, Al is the limiting reactant since fewer moles of Fe can be produced from it (1.11 moles) than from the Fe_2O_3 (1.13 moles).

We can now calculate the grams of Fe formed in the reaction using the 1.11 moles of Fe (formed from our limiting reactant) as our starting factor. The calculation will be a simple one-step "moles-of-A" to "grams-of-A" conversion.

$$\boxed{\text{Moles of Fe}} \xrightarrow[\text{mass}]{\text{Molar}} \boxed{\text{Grams of Fe}}$$

$$1.11 \text{ moles Fe} \times \frac{55.84 \text{ g Fe}}{1 \text{ mole Fe}} = 61.9824 \text{ g Fe} \quad \text{(calculator answer)}$$

$$= 62.0 \text{ g Fe} \quad \textbf{(correct answer)}$$

CHEMICAL EXTENSION

Freshly cut aluminum metal has a bright silvery appearance. The cut surface, however, quickly acquires a "duller" appearance as the aluminum reacts with O_2 from the air to form aluminum oxide, Al_2O_3. This oxide coating, which is very impermeable ("tough"), protects the aluminum from further reactions. When household aluminum objects are cleaned with scouring pads, the oxide coating is usually removed, giving the aluminum a shinier appearance. The cleaning, however, is in vain; a new oxide coating quickly forms.

Aluminum is a good conductor of heat, hence its use in kitchen cookware. The metal copper is a still better conductor of heat. This is why higher-priced pans have a copper-coated bottom.

The electrical conductivity of aluminum is 63.5% that of an equal volume of copper. However, when aluminum's lower density is considered, its conductivity is 2.1 times that of copper on a pound-for-pound basis. Ninety percent of all overhead electrical transmission lines in the United States are aluminum alloys.

Practice Exercise 10.13

Ammonia (NH_3) reacts with oxygen as shown in the following equation.

$$4\,NH_3 + 3\,O_2 \longrightarrow 2\,N_2 + 6\,H_2O$$

How many grams of H_2O can be formed from a reaction mixture containing 35.0 g of NH_3 and 50.0 g of O_2?

10.10 Yields: Theoretical, Actual, and Percent

When stoichiometric relationships (Secs. 10.8 and 10.9) are used to calculate the amount of a product that will be produced in a chemical reaction from given amounts of reactants, the answer obtained represents a *theoretical yield*. A **theoretical yield** *is the maximum amount of a product that can be obtained from given amounts of reactants in a chemical reaction if no losses or inefficiencies of any kind occur.* Examples 10.8 and 10.13 are theoretical yield calculations, although this fact was not noted at the time they were presented.

In most chemical reactions the amount of a given product isolated from the reaction mixture is less than that theoretically possible. Why is this so? Two major factors contribute to this situation.

1. Some product is almost always lost in the process of its isolation and purification and in such mechanical operations as transferring materials from one container to another.

2. Often a particular set of reactants undergoes two or more reactions simultaneously, forming undesired products (in small amounts) as well as the desired products. Reactants consumed in these side reactions obviously will not end up in the form of the desired products.

The net effect of these factors is that the actual quantities of product isolated—that is, the *actual yield*—are usually less, sometimes far less, than the theoretically possible amount. An **actual yield** *is the amount of a product actually obtained from a chemical reaction.* Actual yield is always an experimentally determined number; it cannot be calculated.

Product loss is specified in terms of *percent yield*. The **percent yield** *is the ratio of the actual (experimental) yield of a product in a chemical reaction to its theoretical (calculated) yield multiplied by 100 (to give percent).* The mathematical equation for percent yield is

$$\text{percent yield} = \frac{\text{actual yield}}{\text{theoretical yield}} \times 100$$

If the theoretical yield of a product for a reaction is calculated to be 17.9 g and the amount of product actually obtained (the actual yield) is 15.8 g, the percent yield is 88.3%.

$$\text{percent yield} = \frac{15.8\ \text{g}}{17.9\ \text{g}} \times 100 = 88.268156\%\ (\text{calculator answer})$$

$$= 88.3\%\ (\textbf{correct answer})$$

Note that percent yield cannot exceed 100%; you cannot obtain a yield greater than that which is theoretically possible.

EXAMPLE 10.14

Calculating the Theoretical Yield and Percent Yield for a Chemical Reaction

Mixing aluminum oxide with sulfuric acid produces aluminum sulfate, a compound that is used in the paper industry to strengthen paper and make it water resistant. The chemical equation for this reaction is

$$Al_2O_3(s) + 3 H_2SO_4(aq) \longrightarrow Al_2(SO_4)_3(aq) + 3 H_2O(l)$$

(a) What is the theoretical yield of $Al_2(SO_4)_3$ that can be obtained from a reaction mixture containing 75.0 g of Al_2O_3 and 150.0 g of H_2SO_4?

(b) If the actual yield of $Al_2(SO_4)_3$ for the reaction mixture in part (a) is 136 g, what is the percent yield of $Al_2(SO_4)_3$ for the reaction?

SOLUTION

(a) The limiting reactant must be determined before the theoretical yield can be calculated. Recalling the procedures of Examples 10.12 and 10.13 for determining the limiting reactant, we calculate the number of moles of $Al_2(SO_4)_3$ that can be produced from each individual reactant amount using a "grams-of-A" to "moles-of-B" type of calculations.

$$\boxed{\text{Grams of A}} \xrightarrow[\text{mass}]{\text{Molar}} \boxed{\text{Moles of A}} \xrightarrow[\text{coefficients}]{\text{Equation}} \boxed{\text{Moles of B}}$$

For Al_2O_3:

$$75.0 \text{ g Al}_2\text{O}_3 \times \frac{1 \text{ mole Al}_2\text{O}_3}{101.96 \text{ g Al}_2\text{O}_3} \times \frac{1 \text{ mole Al}_2(SO_4)_3}{1 \text{ mole Al}_2\text{O}_3} = 0.73558258 \text{ mole Al}_2(SO_4)_3$$

$$\text{(calculator answer)}$$
$$= 0.736 \text{ mole Al}_2(SO_4)_3$$
$$\text{(correct answer)}$$

For H_2SO_4:

$$150.0 \text{ g H}_2\text{SO}_4 \times \frac{1 \text{ mole H}_2\text{SO}_4}{98.08 \text{ g H}_2\text{SO}_4} \times \frac{1 \text{ mole Al}_2(SO_4)_3}{3 \text{ moles H}_2\text{SO}_4}$$

$$= 0.50978792 \text{ mole Al}_2(SO_4)_3$$
$$\text{(calculator answer)}$$
$$= 0.5098 \text{ mole Al}_2(SO_4)_3$$
$$\text{(correct answer)}$$

The calculations show that H_2SO_4 is the limiting reactant.

The maximum number of grams of $Al_2(SO_4)_3$ obtainable from the limiting reactant, that is, the theoretical yield, can now be calculated. It is done using a one-step "moles-of-A" to "grams-of-A" setup.

$$\boxed{\text{Moles of Al}_2(SO_4)_3} \xrightarrow[\text{mass}]{\text{Molar}} \boxed{\text{Grams of Al}_2(SO_4)_3}$$

$$0.5098 \text{ mole Al}_2(SO_4)_3 \times \frac{342.14 \text{ g Al}_2(SO_4)_3}{1 \text{ mole Al}_2(SO_4)_3} = 174.42297 \text{ g Al}_2(SO_4)_3$$

$$\text{(calculator answer)}$$
$$= 174.4 \text{ g Al}_2(SO_4)_3$$
$$\text{(correct answer)}$$

(b) The percent yield is obtained by dividing the actual yield by the theoretical yield and multiplying by 100.

$$\text{percent yield} = \frac{\text{actual yield}}{\text{theoretical yield}} \times 100 = \frac{136 \text{ g}}{174.4 \text{ g}} \times 100 = 77.981651\%$$

$$\text{(calculator answer)}$$

$$= 78.0\%$$

$$\text{(correct answer)}$$

CHEMICAL EXTENSION

Pure aluminum oxide is a solid that exists in a variety of *polymorphic* forms, that is, in a variety of different crystalline forms. One naturally occurring form, called *corundum*, has extremely hard crystals. Finely divided corundum is used as an abrasive in sandpaper.

Several familiar gemstones are corundum (Al_2O_3) crystals naturally colored with transition-metal impurities. The most familiar of these gemstones are rubies and sapphires. A ruby is Al_2O_3 with chromium as an impurity. Blue sapphires contain small amounts of iron and titanium, green ones cobalt, and yellow ones nickel and magnesium. Another aluminum-oxide-containing gemstone is the emerald. Its composition is more complex because beryllium oxide and silicon dioxide are also present; an approximate formula is $3 \text{ BeO} \cdot Al_2O_3 \cdot 6 \text{ SiO}_2$.

Artificial rubies and sapphires are made by melting Al_2O_3 with carefully measured amounts of transition metal oxides and then cooling the melt in such a way that large crystals are produced.

Practice Exercise 10.14

A mixture of 80.0 g of chromium(III) oxide (Cr_2O_3) and 8.00 g of carbon (C) is used to produce elemental chromium (Cr) by the reaction

$$Cr_2O_3 + 3 \text{ C} \longrightarrow 2 \text{ Cr} + 3 \text{ CO}$$

(a) What is the theoretical yield of Cr that can be obtained from the reaction mixture?

(b) The actual yield is 21.7 g Cr. What is the percent yield for the reaction?

10.11 Simultaneous and Sequential Chemical Reactions

The concepts presented so far in this chapter can easily be adapted to problem-solving situations that involve two or more chemical reactions. In some cases the two or more chemical reactions occur simultaneously, and in other cases they occur sequentially (one right after the other). Example 10.15 deals with a pair of simultaneous reactions, and Example 10.16 deals with three sequential reactions.

EXAMPLE 10.15

A Calculation Based on Simultaneous Chemical Reactions

A mixture contains 47.3% by mass magnesium carbonate ($MgCO_3$) and 52.7% by mass calcium carbonate ($CaCO_3$). The mixture is heated until both carbonates completely decompose as shown by the following equations.

$$MgCO_3 \longrightarrow MgO + CO_2$$

$$CaCO_3 \longrightarrow CaO + CO_2$$

How many grams of CO_2 are produced from decomposition of 78.3 g of the mixture?

SOLUTION

In solving this problem we will need to carry out two parallel calculations. In the one calculation we will determine the grams of CO_2 produced from the $MgCO_3$ component of the mixture and in the other the grams of CO_2 produced from the $CaCO_3$ component. Then we will add together the answers from the two parallel calculations to get our final answer, the total grams of CO_2 produced. The sequence of conversion factors for each setup is derived from the following pathway.

The number of grams of CO_2 produced from the $MgCO_3$ is given by the following setup.

$$78.3 \text{ g mixture} \times \frac{47.3 \text{ g } MgCO_3}{100.0 \text{ g mixture}} \times \frac{1 \text{ mole } MgCO_3}{84.31 \text{ g } MgCO_3} \times \frac{1 \text{ mole } CO_2}{1 \text{ mole } MgCO_3} \times \frac{44.01 \text{ g } CO_2}{1 \text{ mole } CO_2}$$

$$= 19.332818 \text{ g } CO_2 \quad \text{(calculator answer)}$$

$$= 19.3 \text{ g } CO_2 \quad \textbf{(correct answer)}$$

The number of grams of CO_2 produced from the $CaCO_3$ is given by the following setup.

$$78.3 \text{ g mixture} \times \frac{52.7 \text{ g } CaCO_3}{100.0 \text{ g mixture}} \times \frac{1 \text{ mole } CaCO_3}{100.09 \text{ g } CaCO_3} \times \frac{1 \text{ mole } CO_2}{1 \text{ mole } CaCO_3} \times \frac{44.01 \text{ g } CO_2}{1 \text{ mole } CO_2}$$

$$= 18.144 \text{ g } CO_2 \quad \text{(calculator answer)}$$

$$= 18.1 \text{ g } CO_2 \quad \textbf{(correct answer)}$$

The first conversion factor in each setup is derived from the given percentage of that compound in the mixture. The use of percentages as conversion factors was covered in Section 3.9.

The total number of grams of CO_2 produced is the sum of the grams of CO_2 in the individual reactions.

$$(19.3 + 18.1) \text{ g } CO_2 = 37.4 \text{ g } CO_2 \quad \text{(calculator and } \textbf{correct answer)}$$

CHEMICAL EXTENSION

Limestone is an impure form of calcium carbonate $(CaCO_3)$. When limestone is strongly heated, to temperatures of about 900°C, the $CaCO_3$ decomposes to produce carbon dioxide (CO_2) and a residue of calcium oxide (CaO). Industrially, calcium oxide is called *lime*.

Lime ranks among the top five chemicals produced in the United States in terms of amount produced. The dominant outlet for lime production is steelmaking. In the steelmaking process, lime is added to a molten charge of iron that contains impurities (mainly silicon dioxide and silicates). The lime reacts with the impurities to form slag, a glassy waste material that floats to the top of the molten iron and is removed.

A very small amount of lime, less than 1% of total production, is consumed in building construction. Mortar used in bricklaying is made from lime, sand, and water. Initially, the mortar sets up because of the formation of solid $Ca(OH)_2$. With time, the $Ca(OH)_2$ is converted to $CaCO_3$ by reaction with CO_2 in the air.

Practice Exercise 10.15

A mixture of gaseous fuel has the composition 83.1% methane (CH_4) and 16.9% ethane (C_2H_6). The combustion of this mixture produces CO_2 and H_2O as the only products. The combustion equations are

$$CH_4 + 2\,O_2 \longrightarrow CO_2 + 2\,H_2O$$

$$2\,C_2H_6 + 7\,O_2 \longrightarrow 4\,CO_2 + 6\,H_2O$$

How many moles of CO_2 would be produced from the combustion of 72.0 g of this gaseous fuel mixture?

Often, particularly in industrial processes, more than one chemical reaction is needed to change starting materials into desired products; a series of sequential chemical reactions is required. The product amount from each of the chemical reactions in the sequence becomes the starting material for the next chemical reaction in the sequence.

EXAMPLE 10.16

A Calculation Based on Sequential Chemical Reactions

In steelmaking, a series of three chemical reactions is needed to convert Fe_2O_3 (the iron-containing component of iron ore) to molten iron.

$$\text{reaction (1):} \quad 3\,Fe_2O_3 + CO \longrightarrow 2\,Fe_3O_4 + CO_2$$
$$\text{reaction (2):} \quad Fe_3O_4 + CO \longrightarrow 3\,FeO + CO_2$$
$$\text{reaction (3):} \quad FeO + CO \longrightarrow Fe + CO_2$$

Assuming that the reactant CO is present in excess, how many grams of Fe can be produced from 125 g of Fe_2O_3?

SOLUTION

The key substances in this set of reactions, from a calculational point of view, are the iron-containing species: Fe_2O_3, Fe_3O_4, FeO, and Fe. Note that the iron-containing species produced in the first and second reactions (Fe_3O_4 and FeO, respectively) are the reactants for the second and third reactions, respectively.

$$Fe_2O_3 \xrightarrow{\text{reaction(1)}} Fe_3O_4 \xrightarrow{\text{reaction(2)}} FeO \xrightarrow{\text{reaction(3)}} Fe$$

We can solve this problem using a single multiple-step setup. The sequence of conversion factors needed is that for a gram-to-gram problem with two additional intermediate mole-to-mole steps added.

Grams of Fe_2O_3	Molar mass →	Moles of Fe_2O_3	Eq. (1) coeff. →	Moles of Fe_3O_4	Eq. (2) coeff. →

Moles of FeO	Eq. (3) coeff. →	Moles of Fe	Molar mass →	Grams of Fe

The dimensional analysis setup is

$$125\ \text{g Fe}_2\text{O}_3 \times \frac{1\ \text{mole Fe}_2\text{O}_3}{159.68\ \text{g Fe}_2\text{O}_3} \times \frac{2\ \text{moles Fe}_3\text{O}_4}{3\ \text{moles Fe}_2\text{O}_3} \times \frac{3\ \text{moles FeO}}{1\ \text{mole Fe}_3\text{O}_4}$$

$$\times \frac{1\ \text{mole Fe}}{1\ \text{mole FeO}} \times \frac{55.84\ \text{g Fe}}{1\ \text{mole Fe}}$$

$$= 87.424849 \text{ g Fe} \quad \text{(calculator answer)}$$
$$= 87.4 \text{ g Fe} \quad \text{(correct answer)}$$

An alternative approach to solving this problem would involve setting up a separate calculation for each equation. As a first step, the number of moles of Fe_3O_4 produced in the first reaction would be calculated. In the second step, one would determine the moles of FeO obtained if all the Fe_3O_4 produced in the first reaction entered into the second reaction. In the final step, one would determine the grams of Fe derivable from the FeO produced in the second reaction. The answer obtained from this three-setup method is the same as that obtained from the one-setup method.

CHEMICAL EXTENSION

Iron is the most used of all metals. Indeed, 90% of all metal consumed today is iron, but primarily as steel, not pure iron. All steels are iron alloys.

There are two major types of steel: (1) carbon steels and (2) alloy steels. Both types are metal alloys containing iron and carbon. Alloy steels contain one or more other metals in addition to iron. Most steel produced is carbon steel.

Addition of specific metals improve the properties of steel. Chromium improves hardness and resistance to corrosion. Molybdenum and tungsten increase heat resistance. Nickel increases toughness. Vanadium adds springiness. Manganese improves resistance to wear.

The best-known alloy steel is *stainless steel*. This corrosion-resistant steel contains 14–18% chromium and 7–9% nickel. Stainless steel finds use in cutlery and instruments.

Practice Exercise 10.16

An older method for preparation of nitric acid (HNO_3) involves the following sequence of three chemical reactions.

$$\text{reaction (1):} \quad 4\,NH_3 + 5\,O_2 \longrightarrow 4\,NO + 6\,H_2O$$
$$\text{reaction (2):} \quad 2\,NO + O_2 \longrightarrow 2\,NO_2$$
$$\text{reaction (3):} \quad 3\,NO_2 + H_2O \longrightarrow 2\,HNO_3 + NO$$

Assuming an excess of O_2 and H_2O as reactants, how many grams of HNO_3 can be produced from 244 g of NH_3?

Summary

1. **Reactants and Products** The starting materials in a chemical reaction are called reactants and the substances produced as a result of a chemical reaction are called products.

2. **Law of Conservation of Mass** Mass is neither created nor destroyed in any ordinary chemical reaction. In a chemical reaction the sum of the masses of the products is always equal to the sum of the masses of the reactants.

3. **Chemical Equation** A chemical equation is a representation for a chemical reaction that uses chemical symbols and chemical formulas for the reactants and products involved in the chemical reaction. The formulas of the reactants are always written on the left side of the chemical equation and the formulas of the products are always found on the right side of the equation.

4. **Balanced Chemical Equation** A balanced chemical equation has the same number of atoms of each element involved in the reaction on each side of the equation. An unbalanced equation is brought into balance through the use of coefficients. A coefficient is a number that is placed to the left of the formula of a substance and that changes the amount, but not the identity, of the substance.

Summary (*Continued*)

5. **Classes of Chemical Reactions** Chemical reactions are grouped into classes based on common characteristics. A synthesis reaction is a reaction in which a single product is produced from two (or more) reactants. A decomposition reaction is a reaction in which a single reactant is converted into two or more simpler products. A single-replacement reaction is a reaction in which one element within a compound is replaced by another element. A double-replacement reaction is a reaction in which two compounds exchange parts with each other and form two different compounds. A combustion reaction involves the reaction of a substance with oxygen (usually from air) that proceeds with evolution of heat and usually also a flame.

6. **The Mole and Chemical Equations** The coefficients in a balanced chemical equation give the molar ratios between substances consumed or produced in the chemical reaction described by the equation.

7. **Chemical Stoichiometry** Chemical stoichiometry is the study of the quantitative relationships among reactants and products in a chemical reaction. The stoichiometry of a chemical reaction always involves the molar relationships (coefficients) among reactants and products.

8. **Limiting Reactant** A limiting reactant is completely consumed in a reaction. When it is used up, the reaction stops, thus limiting the quantities of products formed. Other reactants present are often called excess reactants; they are not entirely consumed.

9. **Theoretical Yield, Actual Yield, and Percent Yield** The theoretical yield is the maximum amount of a product that can be obtained from given amounts of reactants in a chemical reaction if no losses or inefficiencies of any kind occur. In most chemical reactions the amount of a given product isolated from the reaction mixture, the actual yield, is less than that theoretically possible. The percent yield compares the actual and theoretical yields.

Key Terms

The new terms defined in this chapter are

actual yield   *Sec. 10.10*

balanced chemical equation   *Sec. 10.3*

chemical equation   *Sec. 10.2*

chemical stoichiometry   *Sec. 10.8*

combustion reaction   *Sec. 10.5*

decomposition reaction   *Sec. 10.5*

double-replacement reaction   *Sec. 10.5*

equation coefficient   *Sec. 10.3*

law of conservation of mass   *Sec. 10.1*

limiting reactant   *Sec. 10.9*

percent yield   *Sec. 10.10*

products   *Sec. 10.1*

reactants   *Sec. 10.1*

single-replacement reaction   *Sec. 10.5*

synthesis reaction   *Sec. 10.5*

theoretical yield   *Sec. 10.10*

Practice Problems

The Law of Conservation of Mass (Sec. 10.1)

10.1 Based on the following description of a chemical reaction, calculate the numerical value of *x*.

$$(5.85 \text{ g NaCl}) + (16.98 \text{ g AgNO}_3) \rightarrow (x \text{ g AgCl}) + (8.50 \text{ g NaNO}_3)$$

10.2 Based on the following description of a chemical reaction, calculate the numerical value of *x*.

$$(7.62 \text{ g CS}_2) + (x \text{ g O}_2) \rightarrow (4.40 \text{ g CO}_2) + (12.82 \text{ g SO}_2)$$

10.3 A 4.2-g sample of sodium hydrogen carbonate is added to a solution of acetic acid weighing 10.0 g. The two substances react, releasing carbon dioxide gas to the atmosphere. After the reaction,

the contents of the reaction vessel weigh 12.0 g. What is the mass of carbon dioxide given off during the reaction?

10.4 A 1.00-g sample of solid calcium carbonate is added to a reaction flask containing 10.00 g of hydrochloric acid solution. The calcium carbonate slowly dissolves in the acid solution as evidenced by the generation of carbon dioxide gas. After 5 min of reaction, 0.21 g of carbon dioxide gas has been given off. At that time, what is the mass, in grams, of the reaction flask contents?

Chemical Equation Notation (Secs. 10.2 and 10.4)

10.5 In which of the following pairs of symbols and/or formulas for gaseous elements are both members of the pair written appropriately for use in chemical equations?

(a) N_2 and O_2

(b) Xe and Cl

(c) H_2 and He_2

(d) F_2 and I_2 (vapor)

10.6 In which of the following pairs of symbols and/or formulas for gaseous elements are both members of the pair written appropriately for use in chemical equations?

(a) Br_2 (vapor) and He

(b) H_2 and F_2

(c) Ar and Kr

(d) O_2 and N

10.7 What do the symbols in parentheses stand for in the following equations?

(a) $CaCO_3(s) \rightarrow CaO(s) + CO_2(g)$

(b) $SO_2(g) + H_2O(l) \rightarrow H_2SO_3(aq)$

10.8 What do the symbols in parentheses stand for in the following equations?

(a) $PCl_3(l) + Cl_2(g) \rightarrow PCl_5(s)$

(b) $NaCl(aq) + AgNO_3(aq) \rightarrow AgCl(s) + NaNO_3(aq)$

Balancing Chemical Equations (Sec. 10.3)

10.9 Indicate whether each of the following equations is *balanced* or *unbalanced*.

(a) $SO_3 + H_2O \rightarrow H_2SO_4$

(b) $CuO + H_2 \rightarrow Cu + H_2O$

(c) $CS_2 + O_2 \rightarrow CO_2 + SO_2$

(d) $AgNO_3 + KCl \rightarrow KNO_3 + AgCl$

10.10 Indicate whether each of the following equations is *balanced* or *unbalanced*.

(a) $BaCO_3 \rightarrow BaO + CO_2$

(b) $KCl + O_2 \rightarrow KClO_3$

(c) $Fe_2O_3 + CO \rightarrow Fe + CO_2$

(d) $H_2SO_4 + CaCO_3 \rightarrow H_2CO_3 + CaSO_4$

10.11 For each of the following balanced equations, indicate how many atoms of each element are present on the reactant and product sides of the chemical equation.

(a) $2 N_2 + 3 O_2 \rightarrow 2 N_2O_3$

(b) $4 NH_3 + 6 NO \rightarrow 5 N_2 + 6 H_2O$

(c) $PCl_3 + 3 H_2 \rightarrow PH_3 + 3 HCl$

(d) $Al_2O_3 + 6 HCl \rightarrow 2 AlCl_3 + 3 H_2O$

10.12 For each of the following balanced equations, indicate how many atoms of each element are present on the reactant and product sides of the chemical equation.

(a) $4 Al + 3 O_2 \rightarrow 2 Al_2O_3$

(b) $2 Na + 2 H_2O \rightarrow 2 NaOH + H_2$

(c) $2 Co + 3 HgCl_2 \rightarrow 2 CoCl_3 + 3 Hg$

(d) $H_2SO_4 + 2 NH_3 \rightarrow (NH_4)_2SO_4$

10.13 Balance the following chemical equations.

(a) $Cu + O_2 \rightarrow CuO$

(b) $H_2O \rightarrow H_2 + O_2$

(c) $BaCl_2 + Na_2S \rightarrow BaS + NaCl$

(d) $Mg + HBr \rightarrow MgBr_2 + H_2$

10.14 Balance the following chemical equations.

(a) $Fe + O_2 \rightarrow Fe_2O_3$

(b) $NaClO_3 \rightarrow NaCl + O_2$

(c) $Au_2S_3 + H_2 \rightarrow H_2S + Au$

(d) $NH_3 + O_2 \rightarrow N_2O + H_2O$

10.15 Balance the following chemical equations.

(a) $PbO + NH_3 \rightarrow Pb + N_2 + H_2O$

(b) $NaHCO_3 + H_2SO_4 \rightarrow Na_2SO_4 + H_2O + CO_2$

(c) $TiO_2 + C + Cl_2 \rightarrow TiCl_4 + CO_2$

(d) $NBr_3 + NaOH \rightarrow N_2 + NaBr + HBrO$

10.16 Balance the following chemical equations.

(a) $NH_3 + O_2 + CH_4 \rightarrow HCN + H_2O$

(b) $KClO_3 + HCl \rightarrow KCl + Cl_2 + H_2O$

(c) $SO_2Cl_2 + HI \rightarrow H_2S + H_2O + HCl + I_2$

(d) $NO + CH_4 \rightarrow HCN + H_2O + H_2$

10.17 Balance the following chemical equations.

(a) $CH_4 + O_2 \rightarrow CO_2 + H_2O$

(b) $C_6H_6 + O_2 \rightarrow CO_2 + H_2O$

(c) $C_4H_8O_2 + O_2 \rightarrow CO_2 + H_2O$

(d) $C_5H_{10}O + O_2 \rightarrow CO_2 + H_2O$

10.18 Balance the following chemical equations.

(a) $C_2H_4 + O_2 \rightarrow CO_2 + H_2O$

(b) $C_6H_{12} + O_2 \rightarrow CO_2 + H_2O$

(c) $C_3H_6O + O_2 \rightarrow CO_2 + H_2O$

(d) $C_5H_{10}O_2 + O_2 \rightarrow CO_2 + H_2O$

10.19 Balance the following chemical equations.

(a) $Ca(OH)_2 + HNO_3 \rightarrow Ca(NO_3)_2 + H_2O$

(b) $BaCl_2 + (NH_4)_2SO_4 \rightarrow BaSO_4 + NH_4Cl$

(c) $Fe(OH)_3 + H_2SO_4 \rightarrow Fe_2(SO_4)_3 + H_2O$

(d) $Na_3PO_4 + AgNO_3 \rightarrow NaNO_3 + Ag_3PO_4$

10.20 Balance the following chemical equations.

(a) $Al + Sn(NO_3)_2 \rightarrow Al(NO_3)_3 + Sn$

(b) $Na_2CO_3 + Mg(NO_3)_2 \rightarrow MgCO_3 + NaNO_3$

(c) $Al(NO_3)_3 + H_2SO_4 \rightarrow Al_2(SO_4)_3 + HNO_3$

(d) $Ba(C_2H_3O_2)_2 + (NH_4)_3PO_4 \rightarrow Ba_3(PO_4)_2 + NH_4C_2H_3O_2$

10.21 Each of the following *mathematically balanced* chemical equations is in a *nonconventional* form. Through coefficient adjustment, change each of these equations to conventional form without unbalancing them.

(a) $3 AgNO_3 + 3 KCl \rightarrow 3 KNO_3 + 3 AgCl$

(b) $2 CS_2 + 6 O_2 \rightarrow 2 CO_2 + 4 SO_2$

(c) $H_2 + \frac{1}{2}O_2 \rightarrow H_2O$

(d) $Ag_2CO_3 \rightarrow 2\,Ag + CO_2 + \frac{1}{2}\,O_2$

10.22 Each of the following *mathematically balanced* chemical equations is in a *nonconventional* form. Through coefficient adjustment, change each of these equations to conventional form without unbalancing them.

(a) $2\,Cu(NO_3)_2 + 2\,Fe \rightarrow 2\,Cu + 2\,Fe(NO_3)_2$

(b) $2\,N_2H_4 + 4\,H_2O_2 \rightarrow 2\,N_2 + 8\,H_2O$

(c) $Li_3N \rightarrow 3\,Li + \frac{1}{2}\,N_2$

(d) $2\,HNO_3 \rightarrow 2\,NO_2 + H_2O + \frac{1}{2}\,O_2$

Classes of Chemical Reactions (Sec. 10.5)

10.23 Classify each of the following chemical reactions as synthesis, decomposition, single-replacement, or double-replacement.

(a) $SO_3 + H_2O \rightarrow H_2SO_4$

(b) $2\,H_2 + O_2 \rightarrow 2\,H_2O$

(c) $Na_2CO_3 + Ca(OH)_2 \rightarrow CaCO_3 + 2\,NaOH$

(d) $Cu(NO_3)_2 + Fe \rightarrow Cu + Fe(NO_3)_2$

10.24 Classify each of the following chemical reactions as synthesis, decomposition, single-replacement, or double-replacement.

(a) $3\,CuSO_4 + 2\,Al \rightarrow Al_2(SO_4)_3 + 3\,Cu$

(b) $K_2CO_3 \rightarrow K_2O + CO_2$

(c) $2\,AgNO_3 + K_2SO_4 \rightarrow Ag_2SO_4 + 2\,KNO_3$

(d) $2\,SO_2 + O_2 \rightarrow 2\,SO_3$

10.25 Identify the products of, and then write a balanced chemical equation for, each of the following chemical reactions.

(a) $Zn + Cu(NO_3)_2 \rightarrow ? + ?$ (single-replacement reaction)

(b) $Ca + O_2 \rightarrow ?$ (synthesis reaction)

(c) $K_2SO_4 + Ba(NO_3)_2 \rightarrow ? + ?$ (double-replacement reaction)

(d) $Ag_2O \rightarrow ? + ?$ (decomposition reaction)

10.26 Identify the products of, and then write a balanced chemical equation for, each of the following chemical reactions.

(a) $AlCl_3 \rightarrow ? + ?$ (decomposition reaction)

(b) $Cu(NO_3)_2 + Na_2CO_3 \rightarrow ? + ?$ (double-replacement reaction)

(c) $Al + Ni(NO_3)_2 \rightarrow ? + ?$ (single-replacement reaction)

(d) $Be + N_2 \rightarrow ?$ (synthesis reaction)

10.27 Write a balanced chemical equation for the thermal decomposition of each of the following metal carbonates to its metal oxide and carbon dioxide.

(a) K_2CO_3 (b) $CaCO_3$

(c) $NiCO_3$ (d) $Fe_2(CO_3)_3$

10.28 Write a balanced chemical equation for the thermal decomposition of each of the following metal carbonates to its metal oxide and carbon dioxide.

(a) $BeCO_3$ (b) Li_2CO_3

(c) $ZnCO_3$ (d) Cs_2CO_3

10.29 Write a balanced chemical equation for the combustion of each of the following hydrocarbons in air.

(a) CH_4 (b) C_6H_6

(c) C_6H_{12} (d) C_3H_4

10.30 Write a balanced chemical equation for the combustion of each of the following hydrocarbons in air.

(a) C_5H_{12} (b) C_4H_6

(c) C_7H_8 (d) C_8H_{18}

10.31 Write a balanced chemical equation for the combustion of each of the following carbon–hydrogen–oxygen compounds in air.

(a) CH_2O (b) C_3H_6O

(c) CH_2O_2 (d) $C_4H_8O_2$

10.32 Write a balanced chemical equation for the combustion of each of the following carbon–hydrogen–oxygen compounds in air.

(a) C_2H_4O (b) $C_5H_{10}O$

(c) $C_2H_4O_2$ (d) $C_3H_6O_2$

10.33 Write a balanced chemical equation for the combustion in air of each of the following compounds.

(a) C_2H_7N, where NO_2 is one of the products

(b) CH_4S, where SO_2 is one of the products

10.34 Write a balanced chemical equation for the combustion in air of each of the following compounds.

(a) C_2H_6S, where SO_2 is one of the products

(b) CH_5N, where NO_2 is one of the products

10.35 Indicate to which of the following types of chemical reactions each of the statements listed applies: *combination, decomposition, single-replacement, double-replacement*, and *combustion*.

(a) An element may be a reactant.

(b) An element may be a product.

(c) A compound may be a reactant.

(d) A compound may be a product.

10.36 Indicate to which of the following types of chemical reactions each of the statements listed applies: *combination, decomposition, single-replacement, double-replacement*, and *combustion*.

(a) Two reactants are required.

(b) Only one reactant is present.

(c) Two products are required.

(d) Only one product is present.

Chemical Equations and the Mole Concept (Sec. 10.6)

10.37 Consider the general chemical equation

$$3\,A + 2\,B \rightarrow C + 3\,D$$

(a) How many molecules of B will react with three molecules of A?

(b) How many molecules of A will react with six molecules of B?

(c) How many molecules of D are produced when four molecules of B react?

(d) How many moles of B will react with three moles of A?

10.38 Consider the general chemical equation

$$A + 3\,B \rightarrow C + 2\,D$$

(a) How many molecules of A will react with three molecules of B?

(b) How many molecules of B will react with three molecules of A?

(c) How many molecules of C are produced when two molecules of A react?

(d) How many moles of B will react with two moles of A?

10.39 Write the 12 mole-to-mole conversion factors that can be derived from the balanced equation

$$4\,NH_3 + 3\,O_2 \rightarrow 2\,N_2 + 6\,H_2O$$

10.40 Write the 12 mole-to-mole conversion factors that can be derived from the balanced equation

$$CS_2 + 3\,O_2 \rightarrow CO_2 + 2\,SO_2$$

10.41 Using each of the following chemical equations, calculate the number of moles of the first-listed reactant that are needed to produce 3.00 moles of N_2.

(a) $2\,NaN_3 \rightarrow 2\,Na + 3\,N_2$

(b) $3\,CO + 2\,NaCN \rightarrow Na_2CO_3 + 4\,C + N_2$

(c) $2\,NH_2Cl + N_2H_4 \rightarrow 2\,NH_4Cl + N_2$

(d) $4\,C_3H_5O_9N_3 \rightarrow 12\,CO_2 + 6\,N_2 + O_2 + 10\,H_2O$

10.42 Using each of the following chemical equations, calculate the number of moles of the first-listed reactant that are needed to produce 4.00 moles of N_2.

(a) $4\,NH_3 + 3\,O_2 \rightarrow 2\,N_2 + 6\,H_2O$

(b) $(NH_4)_2Cr_2O_7 \rightarrow Cr_2O_3 + N_2 + 4\,H_2O$

(c) $N_2H_4 + 2\,H_2O_2 \rightarrow N_2 + 4\,H_2O$

(d) $2\,Li_3N \rightarrow 6\,Li + N_2$

10.43 How many moles of the first-listed reactant in each of the following chemical equations will completely react with 1.42 moles of the second-listed reactant?

(a) $C_7H_{16} + 11\,O_2 \rightarrow 7\,CO_2 + 8\,H_2O$

(b) $2\,HCl + CaCO_3 \rightarrow CaCl_2 + CO_2 + H_2O$

(c) $Na_2SO_4 + 2\,C \rightarrow Na_2S + 2\,CO_2$

(d) $4\,Na_2CO_3 + Fe_3Br_8 \rightarrow 8\,NaBr + 4\,CO_2 + Fe_3O_4$

10.44 How many moles of the first-listed reactant in each of the following chemical equations will completely react with 2.03 moles of the second-listed reactant?

(a) $3\,O_2 + CS_2 \rightarrow CO_2 + 2\,SO_2$

(b) $FeO + CO \rightarrow Fe + CO_2$

(c) $2\,C_8H_{18} + 25\,O_2 \rightarrow 16\,CO_2 + 18\,H_2O$

(d) $Fe_3O_4 + CO \rightarrow 3\,FeO + CO_2$

10.45 Using each of the following equations, calculate the total number of moles of products that can be obtained from the decomposition of 1.75 moles of the reactant.

(a) $2\,NH_4NO_3 \rightarrow 2\,N_2 + O_2 + 4\,H_2O$

(b) $2\,NaClO_3 \rightarrow 2\,NaCl + 3\,O_2$

(c) $2\,KNO_3 \rightarrow 2\,KNO_2 + O_2$

(d) $4\,I_4O_9 \rightarrow 6\,I_2O_5 + 2\,I_2 + 3\,O_2$

10.46 Using each of the following equations, calculate the total number of moles of products that can be obtained from the decomposition of 2.25 moles of the reactant.

(a) $2\,Ag_2CO_3 \rightarrow 4\,Ag + 2\,CO_2 + O_2$

(b) $2\,KClO_3 \rightarrow 2\,KCl + 3\,O_2$

(c) $4\,HNO_3 \rightarrow 4\,NO_2 + 2\,H_2O + O_2$

(d) $2\,H_2O_2 \rightarrow 2\,H_2O + O_2$

10.47 For the chemical reaction

$$CH_4(g) + 4\,Cl_2(g) \rightarrow CCl_4(l) + 4\,HCl(g)$$

(a) how many moles of Cl_2 are needed to produce 4.75 moles of CCl_4?

(b) how many moles of HCl will be produced from 0.083 mole of CH_4?

(c) how many moles of CH_4 are needed to react with 2.30 moles of Cl_2?

(d) how many moles of CCl_4 are produced at the same time that 1.23 moles of HCl are produced?

10.48 For the chemical reaction

$$4\,FeS_2(s) + 11\,O_2(g) \rightarrow 2\,Fe_2O_3(s) + 8\,SO_2(g)$$

(a) how many moles of O_2 are needed to produce 3.50 moles of SO_2?

(b) how many moles of Fe_2O_3 will be produced from 1.02 moles of FeS_2?

(c) how many moles of FeS_2 are needed to react with 5.40 moles of O_2?

(d) how many moles of Fe_2O_3 are produced at the same time that 0.908 mole of SO_2 is produced?

Chemical Equations and the Law of Conservation of Mass (Sec. 10.7)

10.49 Consider the general chemical reaction

$$2\,A + 3\,B \longrightarrow C$$

(a) If 5.00 g of A reacts with 3.20 g of B, what is the mass, in grams, of C that is formed?

(b) If 2.70 g of A reacts to produce 7.50 g of C, what is the mass, in grams, of B that reacts?

10.50 Consider the general chemical reaction

$$3\,A + B \longrightarrow 2\,C$$

(a) If 6.00 g of A reacts with 7.50 g of B, what is the mass, in grams of C that is formed?

(b) If 3.00 g of B reacts to produce 8.00 g of C, what is the mass, in grams, of A that reacts?

10.51 Verify the law of conservation of mass using the molar masses of reactants and products for each substance in the following balanced equations.

(a) $SiO_2 + 3\,C \rightarrow 2\,CO + SiC$

(b) $CH_4 + 4\,Cl_2 \rightarrow 4\,HCl + CCl_4$

(c) $H_2O_2 + H_2S \rightarrow 2\,H_2O + S$

(d) $Mg + 2\,HCl \rightarrow MgCl_2 + H_2$

10.52 Verify the law of conservation of mass using the molar masses of reactants and products for each substance in the following balanced equations.

(a) $5\,O_2 + C_3H_8 \rightarrow 3\,CO_2 + 4\,H_2O$

(b) $4\,NH_3 + 3\,O_2 \rightarrow 2\,N_2 + 6\,H_2O$

(c) $3\,NO_2 + H_2O \rightarrow 2\,HNO_3 + NO$

(d) $6\,HCl + 2\,Al \rightarrow 3\,H_2 + 2\,AlCl_3$

Stoichiometry (Sec. 10.8)

10.53 How many grams of oxygen can be obtained by the decomposition of 7.00 moles of reactant in each of the following chemical reactions?

(a) $2\,KClO_3 \rightarrow 2\,KCl + 3\,O_2$

(b) $2\,CuO \rightarrow 2\,Cu + O_2$

(c) $2\,NaNO_3 \rightarrow 2\,NaNO_2 + O_2$

(d) $4\,HNO_3 \rightarrow 4\,NO_2 + 2\,H_2O + O_2$

10.54 How many grams of oxygen can be obtained by the decomposition of 2.50 moles of reactant in each of the following chemical reactions?

(a) $2\,KClO_4 \rightarrow 2\,KCl + 4\,O_2$

(b) $2\,HgO \rightarrow 2\,Hg + O_2$

(c) $2\,H_2O \rightarrow 2\,H_2 + O_2$

(d) $2\,Ag_2CO_3 \rightarrow 4\,Ag + 2\,CO_2 + O_2$

10.55 How much nitric acid (HNO_3), in grams, is needed to produce 1.00 mole of water in each of the following chemical reactions?

(a) $Cu + 4\,HNO_3 \rightarrow Cu(NO_3)_2 + 2\,NO_2 + 2\,H_2O$

(b) $Al_2O_3 + 6\,HNO_3 \rightarrow 2\,Al(NO_3)_3 + 3\,H_2O$

(c) $Au + HNO_3 + 4\,HCl \rightarrow HAuCl_4 + NO + 2\,H_2O$

(d) $4\,Zn + 10\,HNO_3 \rightarrow 4\,Zn(NO_3)_2 + NH_4NO_3 + 3\,H_2O$

10.56 How much nitric acid (HNO_3), in grams, is needed to produce 2.00 moles of water in each of the following chemical reactions?

(a) $Fe_2O_3 + 6\,HNO_3 \rightarrow 2\,Fe(NO_3)_3 + 3\,H_2O$

(b) $4\,HNO_3 \rightarrow 4\,NO_2 + O_2 + 2\,H_2O$

(c) $3\,Cu + 8\,HNO_3 \rightarrow 3\,Cu(NO_3)_2 + 2\,NO + 4\,H_2O$

(d) $8\,Al + 30\,HNO_3 \rightarrow 8\,Al(NO_3)_3 + 3\,NH_4NO_3 + 9\,H_2O$

10.57 How many grams of the second-listed reactant in each of the following reactions are needed to react completely with 1.772 g of the first-listed reactant?

(a) $SiO_2 + 3\,C \rightarrow 2\,CO + SiC$

(b) $5\,O_2 + C_3H_8 \rightarrow 3\,CO_2 + 4\,H_2O$

(c) $CH_4 + 4\,Cl_2 \rightarrow 4\,HCl + CCl_4$

(d) $3\,NO_2 + H_2O \rightarrow 2\,HNO_3 + NO$

10.58 How many grams of the second-listed reactant in each of the following reactions are needed to react completely with 12.56 g of the first-listed reactant?

(a) $H_2O_2 + H_2S \rightarrow 2\,H_2O + S$

(b) $4\,NH_3 + 3\,O_2 \rightarrow 2\,N_2 + 6\,H_2O$

(c) $Mg + 2\,HCl \rightarrow MgCl_2 + H_2$

(d) $6\,HCl + 2\,Al \rightarrow 3\,H_2 + 2\,AlCl_3$

10.59 Silicon carbide, SiC, used as an abrasive on sandpaper, is prepared using the following chemical reaction

$$SiO_2(s) + 3\,C(s) \rightarrow SiC(s) + 2\,CO(g)$$

(a) How many grams of SiO_2 are needed to react with 1.50 moles of C?

(b) How many grams of CO are produced when 1.37 moles of SiO_2 react?

(c) How many grams of SiC are produced at the same time that 3.33 moles of CO are produced?

(d) How many grams of C must react in order to produce 0.575 mole of SiC?

10.60 In the atmosphere, the air pollutant nitrogen dioxide (NO_2) reacts with water to produce nitric acid (HNO_3). The reaction for the formation of nitric acid is

$$3\,NO_2(g) + H_2O(l) \rightarrow 2\,HNO_3(aq) + NO(g)$$

(a) How many grams of NO_2 are needed to react with 2.30 moles of H_2O?

(b) How many grams of NO are produced when 2.04 moles of H_2O react?

(c) How many grams of HNO_3 are produced at the same time that 0.500 mole of NO are produced?

(d) How many grams of NO_2 must react in order to produce 1.23 moles of HNO_3?

10.61 One way to remove gaseous carbon dioxide (CO_2) from the air in a spacecraft is to let canisters of solid lithium hydroxide (LiOH) absorb it according to the reaction

$$2\,LiOH(s) + CO_2(g) \rightarrow Li_2CO_3(s) + H_2O(l)$$

Based on this equation, how many grams of LiOH must be used to achieve the following?

(a) absorb 4.50 moles of CO_2

(b) absorb 3.00×10^{24} molecules of CO_2

(c) produce 10.0 g of H_2O

(d) produce 10.0 g of Li_2CO_3

10.62 Tungsten (W) metal, used to make incandescent light bulb filaments, is produced by the reaction

$$WO_3(s) + 3\,H_2(g) \rightarrow W(s) + 3\,H_2O(l)$$

Based on this equation, how many grams of WO_3 are needed to produce each of the following?

(a) 10.00 g of W

(b) 1 billion (1.00×10^9) molecules of H_2O

(c) 2.53 moles of H_2O

(d) 250,000 atoms of W

10.63 Hydrofluoric acid, HF, cannot be stored in glass bottles because it attacks silicate compounds present in the glass. For example, sodium silicate, Na_2SiO_3, reacts with HF in the following way:

$$Na_2SiO_3 + 8\,HF \rightarrow H_2SiF_6 + 2\,NaF + 3\,H_2O$$

(a) How many moles of Na_2SiO_3 must react to produce 25.00 g of NaF?

(b) How many grams of HF must react to produce 27.00 g of H_2O?

(c) How many molecules of H_2SiF_6 are produced from the reaction of 2.000 g of Na_2SiO_3?

(d) How many grams of HF are needed to react with 50.00 g of Na_2SiO_3?

10.64 Potassium thiosulfate, $K_2S_2O_3$, is used to remove any excess chlorine from fibers and fabrics that have been bleached with that gas.

$$K_2S_2O_3 + 4\,Cl_2 + 5\,H_2O \rightarrow 2\,KHSO_4 + 8\,HCl$$

(a) How many moles of $K_2S_2O_3$ must react to produce 2.500 g of HCl?

(b) How many grams of Cl_2 must react to produce 20.00 g of $KHSO_4$?

(c) How many molecules of HCl are produced at the same time that 2.000 g of $KHSO_4$ is produced?

(d) How many grams of H_2O are consumed as 12.50 g of Cl_2 reacts?

10.65 How many grams of sodium (Na) are needed to react completely with 16.5 g of sulfur (S) in the synthesis of Na_2S?

10.66 How many grams of beryllium (Be) are needed to react completely with 45.0 g of nitrogen (N_2) in the synthesis of Be_3N_2?

10.67 When chromium metal reacts with chlorine gas, a violet solid with the formula $CrCl_3$ is formed.

$$2\,Cr + 3\,Cl_2 \rightarrow 2\,CrCl_3$$

How many grams of Cr and how many grams of Cl_2 are needed to produce 200.0 g of $CrCl_3$?

10.68 Black silver sulfide can be produced from the reaction of silver metal with sulfur.

$$2\,Ag + S \rightarrow Ag_2S$$

How many grams of Ag and how many grams of S are needed to produce 150.0 g of Ag_2S?

Limiting Reactant Calculations (Sec. 10.9)

10.69 What will be the limiting reactant in the production of "three-nut, four-bolt" combinations from a collection of 216 nuts and 284 bolts?

10.70 What will be the limiting reactant in the production of "five-nut, four-bolt" combinations from a collection of 785 nuts and 660 bolts?

10.71 A model airplane kit is designed to contain two wings, one fuselage, four engines, and six wheels. How many model airplane kits can a manufacturer produce from a parts inventory of 426 wings, 224 fuselages, 860 engines, and 1578 wheels?

10.72 A model car kit is designed to contain one body, four wheels, two bumpers, and one steering wheel. How many model car kits can a manufacturer produce from a parts inventory of 137 bodies, 532 wheels, 246 bumpers, and 139 steering wheels?

10.73 At high temperatures and pressures, nitrogen will react with hydrogen to produce ammonia as shown by the equation

$$N_2 + 3\,H_2 \rightarrow 2\,NH_3$$

For each of the following combinations of reactants, decide which is the limiting reactant.

(a) 1.25 moles of N_2 and 3.65 moles of H_2

(b) 2.60 moles of N_2 and 8.00 moles of H_2

(c) 327 molecules of N_2 and 975 molecules of H_2

(d) 44.0 g of N_2 and 3.00 moles of H_2

10.74 Aluminum oxide can be prepared by the direct reaction of the elements as shown by the equation

$$4\,Al + 3\,O_2 \rightarrow 2\,Al_2O_3$$

For each of the following combinations of reactants, decide which is the limiting reactant.

(a) 3.00 moles of Al and 4.00 moles of O_2

(b) 7.00 moles of Al and 5.40 moles of O_2

(c) 975 molecules of Al and 833 molecules of O_2

(d) 16.2 g of Al and 0.40 mole of O_2

10.75 Magnesium nitride can be prepared by the direct reaction of the elements as shown by the equation

$$3\,Mg + N_2 \rightarrow Mg_3N_2$$

How many grams of magnesium nitride can be produced from the following amounts of reactants?

(a) 10.0 g of Mg and 10.0 g of N_2

(b) 20.0 g of Mg and 10.0 g of N_2

(c) 30.0 g of Mg and 10.0 g of N_2

(d) 40.0 g of Mg and 10.0 g of N_2

10.76 Under appropriate conditions water can be produced from the reaction of the elements hydrogen and oxygen as shown by the equation

$$2\,H_2 + O_2 \rightarrow 2\,H_2O$$

How many grams of water can be produced from the following amounts of reactants?

(a) 10.0 g of H_2 and 40.0 g of O_2

(b) 10.0 g of H_2 and 60.0 g of O_2

(c) 10.0 g of H_2 and 80.0 g of O_2

(d) 10.0 g of H_2 and 100.0 g of O_2

10.77 Determine how many $CoCl_3$ formula units can be produced from a reaction mixture containing 525 cobalt atoms and 525 HCl molecules according to the following reaction.

$$2\,Co\ +\ 6\,HCl \rightarrow 2\,CoCl_3\ +\ 3\,H_2$$

10.78 Determine how many $NiCl_2$ formula units can be produced from a reaction mixture containing 782 nickel atoms and 782 HCl molecules according to the following reaction.

$$Ni\ +\ 2\,HCl \rightarrow NiCl_2\ +\ H_2$$

10.79 If 70.0 g of Fe_3O_4 and 12.0 g of O_2 are present in a reaction mixture, determine how many grams of each reactant will be left unreacted upon completion of the following reaction.

$$4\,Fe_3O_4\ +\ O_2 \rightarrow 6\,Fe_2O_3$$

10.80 If 70.0 g of $TiCl_4$ and 16.0 g of Ti are present in a reaction mixture, determine how many grams of each reactant will be left unreacted upon completion of the following reaction.

$$3\,TiCl_4\ +\ Ti \rightarrow 4\,TiCl_3$$

10.81 Determine the number of grams of each of the products that can be made from 8.00 g of SCl_2 and 4.00 g of NaF by the following reaction.

$$3\,SCl_2\ +\ 4\,NaF \rightarrow SF_4\ +\ S_2Cl_2\ +\ 4\,NaCl$$

10.82 Determine the number of grams of each of the products that can be made from 100.0 g of Na_2CO_3 and 300.0 g of Fe_3Br_8 by the following reaction.

$$4\,Na_2CO_3\ +\ Fe_3Br_8 \rightarrow 8\,NaBr\ +\ 4\,CO_2\ +\ Fe_3O_4$$

Theoretical Yield and Percent Yield (Sec. 10.10)

10.83 Because of "sloppiness" in his procedures, a student was able to isolate only 16.0 g of a desired product from a chemical reaction rather than the 52.0 g that were theoretically possible. What was the percent yield of product that the student obtained?

10.84 The theoretical yield of product for a particular reaction is 25.31 g. A very "meticulous" student isolates 24.79 g of product when the reaction is run. What is the percent yield that this student obtained?

10.85 Aluminum and sulfur react to form aluminum sulfide by the equation

$$2\,Al\ +\ 3\,S \rightarrow Al_2S_3$$

In a certain experiment, 125 g of Al_2S_3 are produced from 75.0 g of Al and 300.0 g of S.

(a) What is the theoretical yield of Al_2S_3?

(b) What is the percent yield of Al_2S_3?

10.86 Aluminum and oxygen react to form aluminum oxide by the equation

$$4\,Al\ +\ 3\,O_2 \rightarrow 2\,Al_2O_3$$

In a certain experiment, 125 g of Al_2O_3 are produced from 75.0 g of Al and 200.0 g of O_2.

(a) What is the theoretical yield of Al_2O_3?

(b) What is the percent yield of Al_2O_3?

10.87 If 74.30 g of HCl were produced from 2.13 g of H_2 and an excess of Cl_2 according to the reaction

$$H_2\ +\ Cl_2 \rightarrow 2\,HCl$$

what was the percent yield of HCl?

10.88 If 115.7 g of Ca_3N_2 were produced from 28.2 g of N_2 and an excess of Ca according to the reaction

$$3\,Ca\ +\ N_2 \rightarrow Ca_3N_2$$

what was the percent yield of Ca_3N_2?

10.89 Under appropriate reaction conditions Al and S produce Al_2S_3 according to the equation

$$2\,Al\ +\ 3\,S \rightarrow Al_2S_3$$

In a certain experiment with 55.0 g of Al and an excess of S, a percent yield of 85.6% was obtained. What was the actual yield of Al_2S_3, in grams, for this experiment?

10.90 Under appropriate reaction conditions Ag and S produce Ag_2S according to the equation

$$2\,Ag\ +\ S \rightarrow Ag_2S$$

In a certain experiment with 75.0 g of Ag and an excess of S, a percent yield of 72.9% was obtained. What was the actual yield of Ag_2S, in grams, for this experiment?

10.91 If the percent yield for the reaction

$$2\,CO\ +\ O_2 \rightarrow 2\,CO_2$$

were 57.8%, what mass of product, in grams, could be produced from a reactant mixture containing 35.0 g of each reactant?

10.92 If the percent yield for the reaction

$$2\,C\ +\ O_2 \rightarrow 2\,CO$$

were 34.3%, what mass of product, in grams, could be produced from a reactant mixture containing 2.25 g of each reactant?

Simultaneous Chemical Reactions (Sec. 10.11)

10.93 A mixture of composition 60.0% ZnS and 40.0% CuS is heated in air until the sulfides are completely converted to oxides as shown by the following equations.

$$2\,ZnS\ +\ 3\,O_2 \rightarrow 2\,ZnO\ +\ 2\,SO_2$$
$$2\,CuS\ +\ 3\,O_2 \rightarrow 2\,CuO\ +\ 2\,SO_2$$

How many grams of SO_2 are produced from the reaction of 82.5 g of the sulfide mixture?

10.94 A mixture of composition 50.0% H_2S and 50.0% CH_4 is reacted with oxygen, producing SO_2, CO_2, and H_2O. The equations for the reactions are

$$2\,H_2S + 3\,O_2 \rightarrow 2\,SO_2 + 2\,H_2O$$
$$CH_4 + 2\,O_2 \rightarrow CO_2 + 2\,H_2O$$

How many grams of H_2O are produced from the reaction of 65.0 g of mixture?

10.95 A mixture of composition 70.0% methane (CH_4) and 30.0% ethane (C_2H_6) by mass is burned in oxygen to produce CO_2 and H_2O. The reactions that occur are

$$CH_4 + 2\,O_2 \rightarrow CO_2 + 2\,H_2O$$
$$2\,C_2H_6 + 7\,O_2 \rightarrow 4\,CO_2 + 6\,H_2O$$

How many grams of O_2 are needed to react completely with 75.0 g of mixture?

10.96 A mixture of composition 64.0% Zn and 36.0% Sn by mass is dissolved in hydrochloric acid (HCl) to produce the metal chlorides and H_2. The reactions that occur are

$$Zn + 2\,HCl \rightarrow ZnCl_2 + H_2$$
$$Sn + 2\,HCl \rightarrow SnCl_2 + H_2$$

How many grams of HCl are needed to react completely with 50.0 g of mixture?

Sequential Chemical Reactions (Sec. 10.11)

10.97 Consider the following two-step reaction sequence.

$$2\,NaClO_3 \rightarrow 2\,NaCl + 3\,O_2$$
$$S + O_2 \rightarrow SO_2$$

Assuming that all of the oxygen generated in the first step is consumed in the second step,

(a) how many moles of sodium chlorate $(NaClO_3)$ is needed in the first step to produce 6.00 moles of sulfur dioxide (SO_2) in the second step?

(b) how many grams of sodium chlorate $(NaClO_3)$ is needed in the first step to produce 20.0 grams of sulfur dioxide (SO_2) in the second step?

10.98 Consider the following two-step reaction sequence.

$$2\,NaCl + 2\,H_2O \rightarrow 2\,NaOH + H_2 + Cl_2$$
$$Ni + Cl_2 \rightarrow NiCl_2$$

Assuming that all of the chlorine generated in the first step is consumed in the second step,

(a) how many moles of sodium chloride (NaCl) is needed in the first step to produce 8.00 moles of nickel(II) chloride $(NiCl_2)$ in the second step?

(b) how many grams of sodium chloride (NaCl) is needed in the first step to produce 50.0 grams of nickel(II) chloride $(NiCl_2)$ in the second step?

10.99 Acid rain contains both nitric acid and sulfuric acid. The nitric acid is formed from atmospheric N_2 in a three-step process.

$$N_2 + O_2 \rightarrow 2\,NO$$
$$2\,NO + O_2 \rightarrow 2\,NO_2$$
$$4\,NO_2 + 2\,H_2O + O_2 \rightarrow 4\,HNO_3$$

(a) How many moles of HNO_3 result from the reaction of 2.00 moles of N_2 in the first step?

(b) How many grams of HNO_3 result from the reaction of 2.00 grams of N_2 in the first step?

10.100 Acid rain contains both nitric acid and sulfuric acid. The sulfuric acid is formed from S (primarily in coal) in a three-step process.

$$S + O_2 \rightarrow SO_2$$
$$2\,SO_2 + O_2 \rightarrow 2\,SO_3$$
$$SO_3 + H_2O \rightarrow H_2SO_4$$

(a) How many moles of H_2SO_4 result from the reaction of 5.00 moles of S in the first step?

(b) How many grams of H_2SO_4 result from the reaction of 5.00 grams of S in the first step?

10.101 The following process has been used for obtaining iodine from oilfield brines.

$$NaI + AgNO_3 \rightarrow AgI + NaNO_3$$
$$2\,AgI + Fe \rightarrow FeI_2 + 2\,Ag$$
$$2\,FeI_2 + 3\,Cl_2 \rightarrow 2\,FeCl_3 + 2\,I_2$$

How much $AgNO_3$, in grams, is required in the first step for every 5.00 g of I_2 produced in the third step?

10.102 Sodium bicarbonate, $NaHCO_3$, can be prepared from sodium sulfate, Na_2SO_4, using the following three-step process.

$$Na_2SO_4 + 4\,C \rightarrow Na_2S + 4\,CO$$
$$Na_2S + CaCO_3 \rightarrow CaS + Na_2CO_3$$
$$Na_2CO_3 + H_2O + CO_2 \rightarrow 2\,NaHCO_3$$

How much carbon, C, in grams, is required in the first step for every 10.00 g of $NaHCO_3$ produced in the third step?

Additional Problems

10.103 For each of the following *unbalanced* chemical equations, determine how many moles of the second-listed reactant would be required to react exactly with 3.00 moles of the first-listed reactant.

(a) $CH_4 + O_2 \longrightarrow CO_2 + H_2O$

(b) $PCl_3 + H_2O \longrightarrow H_3PO_3 + HCl$

(c) $NaOH + H_3PO_4 \longrightarrow Na_3PO_4 + H_2O$

(d) $NaClO_2 + Cl_2 \longrightarrow ClO_2 + NaCl$

10.104 For each of the following *unbalanced* chemical equations, determine how many moles of the second-listed reactant would be required to react exactly with 3.00 moles of the first-listed reactant.

(a) $CH_4 + H_2O \longrightarrow H_2 + CO$

(b) $C_3H_8 + O_2 \longrightarrow CO_2 + H_2O$

(c) $LiOH + H_2SO_4 \longrightarrow Li_2SO_4 + H_2O$

(d) $SCl_4 + H_2O \longrightarrow SO_2 + HCl$

10.105 Ammonium dichromate decomposes according to the following reaction.

$$(NH_4)_2Cr_2O_7 \rightarrow N_2 + 4\,H_2O + Cr_2O_3$$

How many grams of each of the products can be formed from the decomposition of 75.0 g of $(NH_4)_2Cr_2O_7$?

10.106 Nitrous acid decomposes according to the following reaction.

$$3\,HNO_2 \rightarrow 2\,NO + HNO_3 + H_2O$$

How many grams of each of the products can be formed from the decomposition of 63.5 g of HNO_2?

10.107 Hydrogen sulfide burns in oxygen to form sulfur dioxide and water.

$$2\,H_2S + 3\,O_2 \rightarrow 2\,SO_2 + 2\,H_2O$$

How many grams of hydrogen sulfide must react in order to produce a total of 100.0 g of products?

10.108 Carbon disulfide burns in oxygen to form carbon dioxide and sulfur dioxide.

$$CS_2 + 3\,O_2 \rightarrow CO_2 + 2\,SO_2$$

How many grams of carbon disulfide must react in order to produce a total of 50.0 g of products?

10.109 From each of the following pairs of ratios, determine whether reactant A or reactant B is the limiting reactant.

(a) required ratio: 3 moles A to 1 mole B; ratio actually present: 2 moles A to 0.5 mole B

(b) required ratio: 3 moles A to 2 moles B; ratio actually present: 4.8 moles A to 3.6 moles B

(c) required ratio: 1 mole A to 2 moles B; ratio actually present: 3.5 moles A to 6.9 moles B

(d) required ratio: 4 moles A to 3 moles B; ratio actually present: 1.9 moles A to 1.6 moles B

10.110 From each of the following pairs of ratios, determine whether reactant A or reactant B is the limiting reactant.

(a) required ratio: 2 moles A to 1 mole B; ratio actually present: 0.40 mole A to 0.30 mole B

(b) required ratio: 2 moles A to 3 moles B; ratio actually present: 2.4 moles A to 2.7 moles B

(c) required ratio: 1 mole A to 1 mole B; ratio actually present: 1.3 moles A to 1.5 moles B

(d) required ratio: 3 moles A to 4 moles B; ratio actually present: 3.6 moles A to 4.4 moles B

10.111 Consider the balanced chemical equation

$$3\,A + 2\,B \longrightarrow C + 2\,D$$

When 6.0 g of A reacts completely with 8.0 g of B, 9.0 g of C and 5.0 g of D are produced.

(a) Which has a greater molar mass, A or B?

(b) Which has a greater molar mass, A or C?

(c) Which has a greater molar mass, A or D?

(d) If the molar mass of A is 30.0 grams, what are the molar masses of B, C, and D?

10.112 Consider the balanced chemical equation

$$2\,A + B \longrightarrow C + 2\,D$$

When 8.0 g of A reacts completely with 6.0 g of B, 9.0 g of C and 5.0 g of D are produced.

(a) Which has a greater molar mass, A or B?

(b) Which has a greater molar mass, A or C?

(c) Which has a greater molar mass, A or D?

(d) If the molar mass of A is 40.0 grams, what are the molar masses of B, C, and D?

10.113 Pure, dry NO gas can be made by the following reaction.

$$3\,KNO_2 + KNO_3 + Cr_2O_3 \rightarrow 4\,NO + 2\,K_2CrO_4$$

How many grams of NO can be produced from a reaction mixture containing 2.00 moles each of KNO_2, KNO_3, and Cr_2O_3?

10.114 Sodium cyanide, NaCN, can be made by the following reaction.

$$Na_2CO_3 + 4\,C + N_2 \rightarrow 2\,NaCN + 3\,CO$$

How many grams of CO can be produced from a reaction mixture containing 3.00 moles each of Na_2CO_3, C, and N_2?

10.115 An *impure* sample of $CuSO_4$ weighing 7.53 g was dissolved in water. The dissolved $CuSO_4$, but not the impurities, then reacted with excess zinc.

$$CuSO_4 + Zn \rightarrow ZnSO_4 + Cu$$

What was the mass percent $CuSO_4$ in the sample if 1.33 g of Cu were produced?

10.116 An *impure* sample of $Hg(NO_3)_2$ weighing 64.5 g was dissolved in water. The dissolved $Hg(NO_3)_2$, but not the impurities, then reacted with excess Mg metal.

$$Hg(NO_3)_2 + Mg \rightarrow Mg(NO_3)_2 + Hg$$

What was the mass percent $Hg(NO_3)_2$ in the sample if 23.6 g of Hg were produced?

10.117 Silver oxide (Ag_2O) decomposes completely at high temperatures to produce metallic silver and oxygen gas. A 1.80-g sample of *impure* silver oxide yielded 0.115 g of O_2. Assuming that Ag_2O was the only source of O_2, what was the mass percent of Ag_2O in the sample?

10.118 Gold(III) oxide (Au_2O_3) decomposes completely at high temperatures to produce metallic gold and oxygen gas. A 2.21-g sample of *impure* gold(III) oxide yielded 0.233 g of O_2. Assuming that Au_2O_3 was the only source of O_2, what was the mass percent of Au_2O_3 in the sample?

10.119 The reaction between 113.4 g of I_2O_5 and 132.2 g of BrF_3 was found to produce 97.0 g of IF_5. The equation for the reaction is

$$6\ I_2O_5 + 20\ BrF_3 \rightarrow 12\ IF_5 + 15\ O_2 + 10\ Br_2$$

What is the percent yield of IF_5?

10.120 The reaction between 20.0 g of NH_3 and 20.0 g of CH_4 with an excess of oxygen was found to produce 15.0 g of HCN. The equation for the reaction is

$$2\ NH_3 + 3\ O_2 + 2\ CH_4 \rightarrow 2\ HCN + 6\ H_2O$$

What is the percent yield of HCN?

10.121 Copper metal can be recovered from an ore containing $CuCO_3$ by the decomposition reaction

$$2\ CuCO_3 \rightarrow 2\ Cu + 2\ CO_2 + O_2$$

What mass of copper ore, in tons, is needed to produce 500.0 lb of Cu if the ore is 13.22% by mass $CuCO_3$? Assume complete decomposition of the $CuCO_3$.

10.122 Silver metal can be recovered from an ore containing Ag_2CO_3 by the decomposition reaction

$$2\ Ag_2CO_3 \rightarrow 4\ Ag + 2\ CO_2 + O_2$$

What mass of silver ore, in tons, is needed to produce 500.0 lb of Ag if the ore is 24.21% by mass Ag_2CO_3?

10.123 A 13.20-g sample of a mixture of $CaCO_3$ and $NaHCO_3$ was heated, and the compounds decomposed as follows.

$$CaCO_3 \rightarrow CaO + CO_2$$
$$2\ NaHCO_3 \rightarrow Na_2CO_3 + CO_2 + H_2O$$

The decomposition of the sample yields 4.35 g of CO_2 and 0.873 g of H_2O. What percentage, by mass, of the original sample was $CaCO_3$?

10.124 A 4.00-g sample of a mixture of H_2S and CS_2 was burned in oxygen. The equations for the reactions are

$$2\ H_2S + 3\ O_2 \rightarrow 2\ H_2O + 2\ SO_2$$
$$CS_2 + 3\ O_2 \rightarrow CO_2 + 2\ SO_2$$

If 7.32 g of SO_2 and 0.577 g of CO_2 were produced along with some H_2O, what percentage, by mass, of the original sample was H_2S?

Cumulative Problems

10.125 Write a balanced equation for each of the following chemical reactions.

(a) zinc + silver nitrate $\rightarrow$ zinc nitrate + silver

(b) hydrochloric acid + sodium hydroxide $\rightarrow$ sodium chloride + water

(c) phosphorus trichloride + chloride $\rightarrow$ phosphorus pentachloride

(d) copper + oxygen $\rightarrow$ copper(II) oxide

10.126 Write a balanced equation for each of the following chemical reactions.

(a) sodium oxide + sulfur trioxide $\rightarrow$ sodium sulfate

(b) barium carbonate $\rightarrow$ barium oxide + carbon dioxide

(c) aluminum + iron(II) oxide $\rightarrow$ iron + aluminum oxide

(d) ammonia + phosphoric acid $\rightarrow$ ammonium phosphate

10.127 After the following equation was balanced, the name of one of the reactants was substituted for its formula.

$$2\ \text{cyclopropane} + 9\ O_2 \rightarrow 6\ CO_2 + 6\ H_2O$$

Using only the information found within this equation, determine the molecular formula of cyclopropane.

10.128 After the following equation was balanced, the name of one of the reactants was substituted for its formula.

$$2\ \text{butyne} + 11\ O_2 \rightarrow 8\ CO_2 + 6\ H_2O$$

Using only the information found within this equation, determine the molecular formula of butyne.

10.129 Write a balanced chemical equation for the reaction in which Cu_2S and O_2 are reactants and SO_2 and a copper oxide containing 88.82% copper by mass are products. The molecular and empirical formulas of the copper oxide are the same.

10.130 Write a balanced chemical equation for the reaction in which NH_3 and O_2 are reactants and H_2O and a nitrogen oxide containing 46.68% nitrogen by mass are products. The molecular and empirical formulas of the nitrogen oxide are the same.

10.131 Write a balanced chemical equation for the reaction in which CO_2 and H_2O are the products, O_2 is one of the reactants, and a compound with an empirical formula of CH and a formula mass of 78.12 amu is the other reactant.

10.132 Write a balanced chemical equation for the reaction in which CO_2 and H_2O are the products, O_2 is one of the reactants, and a compound with an empirical formula of C_3H_5 and a formula mass of 82.16 amu is the other reactant.

10.133 Fifty (50.000) grams of Be are reacted with an excess of F_2 to produce BeF_2. The equation for the reaction is

$$Be + F_2 \rightarrow BeF_2$$

Calculate the mass of BeF_2 produced using each of the following specifications.

(a) mass, in grams, to three significant figures

(b) mass, in kilograms, to five significant figures

(c) mass, in micrograms, to four significant figures

(d) mass, in pounds, to four significant figures

10.134 Eighty (80.000) grams of Na are reacted with an excess of P to produce Na_3P. The equation for the reaction is

$$3\ Na + P \rightarrow Na_3P$$

Calculate the mass of Na_3P produced using each of the following specifications.

(a) mass, in grams, to four significant figures

(b) mass, in milligrams, to three significant figures

(c) mass, in pounds, to five significant figures

(d) mass, in ounces, to four significant figures

10.135 If 100.0 g of $KClO_3$ and 200.0 g of HCl are allowed to react according to the equation

$$2 KClO_3 + 4 HCl \rightarrow 2 KCl + 2 ClO_2 + Cl_2 + 2 H_2O$$

what is the combined total number of moles of chlorine-containing products produced?

10.136 If 50.0 g of SO_2Cl_2 and 200.0 g of HI are allowed to react according to the equation

$$SO_2Cl_2 + 8 HI \rightarrow H_2S + 2 H_2O + 2 HCl + 4 I_2$$

what is the combined total number of moles of hydrogen-containing products produced?

10.137 The reusable booster rockets of the U.S. space shuttle employ a mixture of aluminum and ammonium perchlorate for fuel. The chemical reaction that occurs is

$$3 Al + 3 NH_4ClO_4 \rightarrow Al_2O_3 + AlCl_3 + 3 NO + 6 H_2O$$

How many moles of electrons are present in the $AlCl_3$ produced when 10.0 g of Al react?

10.138 The fluoride in many toothpastes is tin(II) fluoride produced by the reaction of Sn metal with gaseous HF.

$$Sn + 2 HF \rightarrow SnF_2 + H_2$$

How many moles of electrons are present in the HF consumed when 25.0 g of H_2 are produced?

10.139 If the products produced by the reaction of 500.0 g of $CaCl_2$ with excess Na_2CO_3 according to the equation

$$CaCl_2 + Na_2CO_3 \rightarrow 2 NaCl + CaCO_3$$

were broken up into ions, how many positive ions would result?

10.140 If the products produced by the reaction of 220.0 g of $AgNO_3$ with excess K_3PO_4 according to the equation

$$K_3PO_4 + 3 AgNO_3 \rightarrow Ag_3PO_4 + 3 KNO_3$$

were broken up into ions, how many negative ions would result?

10.141 The concentration of an aqueous NaBr solution, whose density is 1.046 g/mL, is 6.00% by mass. Determine the volume, in milliliters, of NaBr solution needed to prepare 10.0 g of AgBr by the following reaction.

$$NaBr + AgNO_3 \rightarrow AgBr + NaNO_3$$

10.142 The concentration of an aqueous NH_3 solution, whose density is 0.979 g/mL, is 4.50% by mass. Determine the volume, in milliliters, of NH_3 solution needed to prepare 10.0 g of NH_4NO_3 by the following reaction.

$$NH_3 + HNO_3 \rightarrow NH_4NO_3$$

10.143 A three-step process for producing nitric acid, HNO_3, from gaseous ammonia, NH_3, is

$$4 NH_3 + 5 O_2 \rightarrow 4 NO + 6 H_2O$$
$$2 NO + O_2 \rightarrow 2 NO_2$$
$$3 NO_2 + H_2O \rightarrow 2 HNO_3 + NO$$

Assuming yields, respectively, of 85.2%, 82.7%, and 87.0% for the nitrogen-containing products in the three steps, how many grams of nitric acid can be produced from 75.0 mL of ammonia with a density of 0.695 g/L?

10.144 A three-step process for producing sulfuric acid, H_2SO_4, from gaseous sulfur dioxide, SO_2, is

$$2 SO_2 + O_2 \rightarrow 2 SO_3$$
$$SO_3 + H_2SO_4 \rightarrow H_2S_2O_7$$
$$H_2S_2O_7 + H_2O \rightarrow 2 H_2SO_4$$

Assuming yields, respectively, of 63.1%, 87.5%, and 73.8% for the three steps, how many grams of sulfuric acid can be produced from 125 mL of sulfur dioxide with a density of 0.773 g/L?

10.145 A particular coal contains 4.3% sulfur by mass as an impurity. When the coal is burned the S is converted to gaseous SO_2. The SO_2 enters the exhaust gases where it is removed by reaction with powdered CaO to produce solid $CaSO_3$. How much by-product $CaSO_3$, in tons, is produced by the burning of 1.0 ton of coal?

10.146 A particular coal contains 3.5% sulfur by mass as an impurity. When the coal is burned the S is converted to gaseous SO_2. The SO_2 enters the exhaust gases where it is removed by reaction with powdered CaO to produce solid $CaSO_3$. How much coal would have to be burned, in tons, to produce 2.0 tons of by-product $CaSO_3$?

Answers to Practice Exercises

10.1 $2 Fe_2O_3 + 3 C \rightarrow 4 Fe + 3 CO_2$

10.2 $2 C_2H_6 + 7 O_2 \rightarrow 4 CO_2 + 6 H_2O$

10.3 **(a)** synthesis **(b)** decomposition
(c) single replacement **(d)** double replacement

10.4 $C_2H_6O + 3 O_2 \rightarrow 2 CO_2 + 3 H_2O$

10.5 **(a)** 2.46 moles H_2O **(b)** 7.22 moles O_2

10.6 212.08 g (reactants) = 212.08 g (products)

10.7 0.800 g $Mg(OH)_2$

10.8 18.2 g O_2

10.9 0.104 mole $KClO_3$

10.10 2.66×10^{-12} g O_2

10.11 Nuts are the limiting reactant

10.12 Pb is the limiting reactant

10.13 55.5 g H_2O

10.14 **(a)** 23.1 g Cr **(b)** 93.9%

10.15 4.54 moles CO_2

10.16 602 g HNO_3

11

States of Matter

11.1 Factors that Determine Physical State

The physical state of a substance is determined by (1) what it is, that is, its chemical identity; (2) its temperature; and (3) the pressure to which it is subjected.

At room temperature and pressure some substances are solids (gold, sodium chloride, etc.), others are liquids (water, mercury, etc.), and still others are gases (oxygen, carbon dioxide, etc.). Thus chemical identity must be a determining factor for physical state, since all three states are observed at room temperature and pressure. On the other hand, when the physical state of a single substance is considered, temperature and pressure are determining variables. Liquid water can be changed to a solid by lowering the temperature or to a gas by raising the temperature.

We tend to characterize a substance almost exclusively in terms of its most common physical state, that is, the state in which it is found at room temperature and pressure. Oxygen is almost always thought of as a gas, its most common state; gold is almost always thought of as a solid, its most common state. A major reason for such "single-state" characterization is the narrow range of temperatures encountered on this planet. Most substances are never encountered in more than one state under "natural" conditions. We must be careful not to fall into the error of assuming that the commonly observed state of a substance is the *only* state in which it can exist. Under laboratory conditions, states other than the "natural" one can be obtained for almost all substances. Figure 11.1 shows the temperature ranges for the solid, liquid, and gaseous states of a few elements and compounds. As can be seen in this figure, the size and location of the physical-state

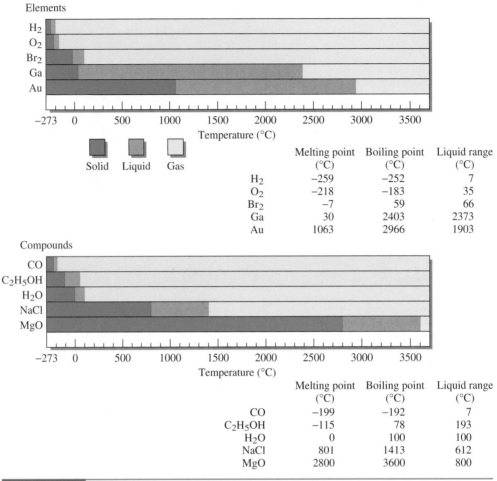

Elements

	Melting point (°C)	Boiling point (°C)	Liquid range (°C)
H_2	−259	−252	7
O_2	−218	−183	35
Br_2	−7	59	66
Ga	30	2403	2373
Au	1063	2966	1903

Compounds

	Melting point (°C)	Boiling point (°C)	Liquid range (°C)
CO	−199	−192	7
C_2H_5OH	−115	78	193
H_2O	0	100	100
NaCl	801	1413	612
MgO	2800	3600	800

Figure 11.1

Solid, liquid, and gaseous state temperature ranges for selected elements and compounds.

temperature ranges vary widely among chemical substances. Extremely high temperatures are required to obtain some substances in the gaseous state; other substances are gases at temperatures below room temperature. The size of a given physical-state range also varies dramatically. For example, the elements H_2 and O_2 are liquids over a very narrow temperature range, whereas the elements Ga and Au remain liquids over ranges of hundreds of degrees.

A large majority of the naturally occurring elements (75 out of 88) are solids at room temperature and pressure. Of the remaining thirteen naturally occurring elements, eleven are gases and two (bromine and mercury) are liquids. The abbreviated periodic table in Figure 11.2 identifies the "nonsolid" elements. There are two naturally occurring elements, cesium of group IA and gallium of group IIIA, that have melting points between 25 and 30°C. Temperatures in this range are reached on hot summer days. Thus, at such times, there could be four elements rather than two that are liquids.

Explanations for the experimentally observed physical-state variations among substances can be derived from the concepts of atomic structure and bonding we have considered in previous chapters. Such explanations are found in later sections of this chapter.

> Water is the only substance that is commonly encountered in all three physical states at temperatures normally found on Earth.

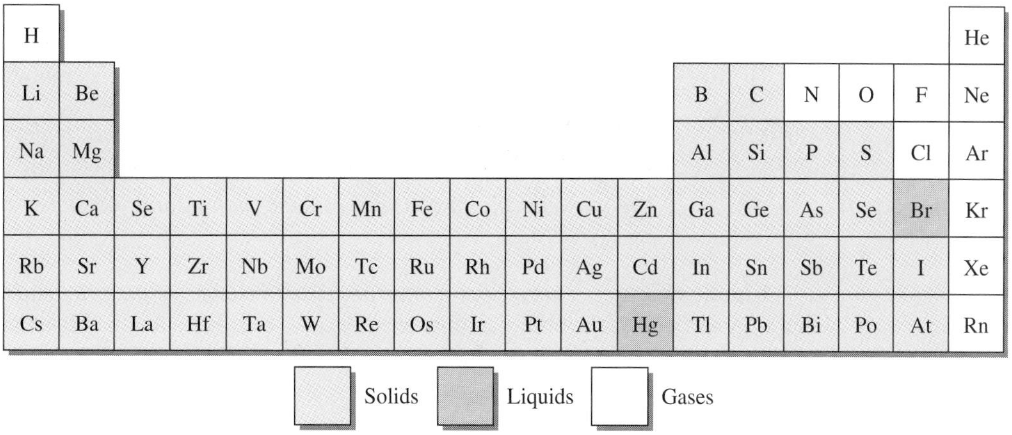

H																	He
Li	Be											B	C	N	O	F	Ne
Na	Mg											Al	Si	P	S	Cl	Ar
K	Ca	Se	Ti	V	Cr	Mn	Fe	Co	Ni	Cu	Zn	Ga	Ge	As	Se	Br	Kr
Rb	Sr	Y	Zr	Nb	Mo	Tc	Ru	Rh	Pd	Ag	Cd	In	Sn	Sb	Te	I	Xe
Cs	Ba	La	Hf	Ta	W	Re	Os	Ir	Pt	Au	Hg	Tl	Pb	Bi	Po	At	Rn

☐ Solids   ▨ Liquids   ☐ Gases

Figure 11.2

Physical states of the elements at room temperature and pressure.

11.2   Property Differences among Physical States

The differences among solids, liquids, and gases are so great that only a few gross distinguishing features need to be mentioned to differentiate them clearly. Certain obvious differences among the three states of matter are apparent to even the most casual observer— differences related to (1) volume and shape, (2) density, (3) compressibility, and (4) thermal expansion. These distinguishing properties are compared in Table 11.1 for the three states of matter. The properties of volume and density have been discussed in detail previously (Secs. 3.4 and 3.8, respectively). **Compressibility** *is a measure of the volume change resulting from a pressure change.* **Thermal expansion** *is a measure of the volume change resulting from a temperature change.*

The contents of Table 11.1 should be studied in detail; the data given will serve as the starting point for further discussions about the states of matter.

11.3   The Kinetic Molecular Theory of Matter

The **kinetic molecular theory of matter** *is a set of five statements used to explain the physical behavior of the three states of matter (solids, liquids, and gases).* The basic idea of this

Table 11.1   Distinguishing Properties of Solids, Liquids, and Gases

Property	Solid State	Liquid State	Gaseous State
Volume and shape	definite volume and definite shape	definite volume and indefinite shape; takes the shape of container to the extent it is filled	indefinite volume and indefinite shape; takes the volume and shape of container that it fills
Density	high	high, but usually lower than corresponding solid	low
Compressibility	small	small, but usually greater than corresponding solid	large
Thermal expansion	very small: about 0.01% per °C	small: about 0.1% per °C	moderate: about 0.3% per °C

The word *kinetic* comes from the Greek *kinesis*, which means "movement." The kinetic molecular theory deals with the movement of particles.

theory is that the particles (atoms, molecules, or ions) present in a substance, independent of the physical state of the substance, are always in motion.

The five statements of the kinetic molecular theory of matter are as follows:

STATEMENT 1 *Matter is ultimately composed of tiny particles (atoms, molecules, or ions) that have definite and characteristic sizes that do not change.*

STATEMENT 2 *The particles are in constant random motion and therefore possess kinetic energy.*

Kinetic energy *is energy that matter possesses because of particle motion.* An object that is in motion has the ability to transfer its kinetic energy to another object upon collision with that object (see Fig. 11.3).

STATEMENT 3 *The particles interact with each other through attractions and repulsions and therefore possess potential energy.*

Water behind a dam represents potential energy because of *position*. Uncontrolled water release can cause massive destruction (a flood) as the potential energy is converted to other forms of energy. A compressed spring can spontaneously expand and do work as the result of potential energy associated with *condition*. Potential energy associated with *composition* is released when gasoline is burned. Electrostatic interactions are also potential energy associated with composition of a substance.

Potential energy *is stored energy that matter possesses as a result of its position, condition, and/or composition.* The potential energy of greatest importance when considering the three states of matter is that which originates from *electrostatic forces* among charged particles. An **electrostatic force** *is an attractive force or repulsive force that occurs between charged particles.* Particles of opposite charge (one positive and one negative) attract one another, and particles of like charge (both positive or both negative) repel one another.

STATEMENT 4 *The kinetic energy (velocity) of the particles increases as the temperature is increased.*

The *average* kinetic energy (velocity) of all particles in a system depends on temperature; the higher the temperature the higher the average kinetic energy of the system.

STATEMENT 5 *The particles in a system transfer energy to each other through elastic collisions.*

In an *elastic* collision, the total kinetic energy remains constant; no kinetic energy is lost. The difference between an *elastic* and an *inelastic* collision is illustrated by comparing the collision of two hard steel spheres with the collision of two masses of putty. The collision of spheres approximates an elastic collision (the spheres bounce off one another and continue moving); the putty collision has none of these characteristics (the masses "glob" together with no resulting movement).

Two consequences of the elasticity of particle collisions are that (1) the energy of any given particle is continually changing and (2) particle energies in a system are not all the same; a range of particle energies is always encountered.

Figure 11.3

The kinetic energy of water falling through the penstocks of a large dam is used to turn huge turbines and generate electricity. (Ian Murphy/ Tony Stone Images)

The relative influence of kinetic energy and potential energy in a chemical system is the major consideration in using kinetic molecular theory to explain the general properties of the solid, liquid, and gaseous states of matter. The important question is whether the kinetic energy or the potential energy dominates the energetics of the chemical system under study.

Kinetic energy may be considered a *disruptive force* within the chemical system, tending to make the particles of the system increasingly independent of each other. As the result of energy of motion, the particles will tend to move away from each other. Potential energy may be considered a *cohesive force* tending to cause order and stability among the particles of the system.

The role that temperature plays in determining the state of a system is related to kinetic energy magnitude. Kinetic energy increases as temperature increases (statement 4 of the kinetic molecular theory). Thus, the higher the temperature the greater the magnitude of disruptive influences within a chemical system. The magnitude of potential energy is essentially independent of temperature change. Neither charge nor separation distance, the two factors on which the magnitude of potential energy depends, is affected significantly by temperature change.

Sections 11.4, 11.5, and 11.6 deal, respectively, with kinetic molecular theory explanations for the general properties of the solid, liquid, and gaseous states.

11.4 The Solid State

A **solid** *is the physical state characterized by a dominance of potential energy (cohesive forces) over kinetic energy (disruptive forces).* The particles in a solid are drawn close together in a regular pattern by the strong cohesive forces present. Each particle occupies a fixed position about which it vibrates. An explanation of the characteristic properties of solids is obtained from this model.

1. *Definite volume and definite shape.* The strong cohesive forces hold the particles in essentially fixed positions, resulting in definite volume and definite shape.
2. *High density.* The constituent particles of solids are located as close together as possible. Therefore, large numbers of particles are contained in a unit volume, resulting in a high density.
3. *Small compressibility.* Since there is very little space between particles, increased pressure cannot push them any closer together and therefore has little effect on the solid's volume.
4. *Very small thermal expansion.* An increase in temperature increases the kinetic energy (disruptive forces), thereby causing more vibrational motion of the particles. Each particle "occupies" a slightly larger volume. The result is a slight expansion of the solid. The strong cohesive forces prevent this effect from becoming very large.

> **Some solids, such as foam rubber, are compressible because they are full of gas (air).**

11.5 The Liquid State

The liquid state consists of particles randomly packed relatively close to each other. The molecules are in constant random motion, freely sliding over one another but without sufficient energy to separate from each other. A **liquid** *is the physical state characterized by potential energy (cohesive forces) and kinetic energy (disruptive forces) of about the same magnitude.* The fact that the particles freely slide over each other indicates the influence of disruptive forces, but the fact that the particles do not separate indicates a fairly strong influence from cohesive forces. The characteristic properties of liquids are explained by this model.

1. *Definite volume and indefinite shape.* Attractive forces are strong enough to restrict particles to movement within a definite volume. They are not strong enough, however, to prevent the particles from moving over each other in a random manner, limited only by the container walls. Thus liquids have no definite shape, with the exception that they maintain a horizontal upper surface in containers that are not completely filled.

2. *High density.* The particles in a liquid are not widely separated; they essentially touch each other. Therefore, there will be a large number of particles per unit volume and a resultant high density.

3. *Small compressibility.* Since the particles in a liquid essentially touch each other, there is very little empty space. Therefore, a pressure increase cannot squeeze the particles much closer together.

4. *Small thermal expansion.* Most of the particle movement in a liquid involves particles sliding over each other. The increased particle velocity that accompanies a temperature increase results in a small increase in such motion. The net effect is an increase in the effective volume a particle "occupies," which causes a slight volume increase in the liquid.

> The measurement of temperature (thermometers) is based on the thermal expansion of liquid mercury in a narrow tube. As the temperature increases, the increased mercury volume raises the height of the mercury column in the tube.

11.6 The Gaseous State

A **gas** *is the physical state characterized by a complete dominance of kinetic energy (disruptive forces) over potential energy (cohesive forces).* As a result, the particles of a gas are essentially independent of one another and move in a totally random manner. Under ordinary pressure, the particles are relatively far apart except, of course, when they collide with each other. In between collisions with each other or with the container walls, gas particles travel in straight lines. The particle velocities and resultant collision frequencies are extremely high; at room temperature and pressure the collisions experienced by one molecule are of the order of 10^{10} collisions per second.

The kinetic theory explanation of gaseous state properties follows the same pattern we saw earlier for solids and liquids.

1. *Indefinite volume and indefinite shape.* The attractive (cohesive) forces between particles have been overcome by kinetic energy, and the particles are free to travel in all directions. Therefore, the particles completely fill the container the gas is in and assume its shape.

2. *Low density.* The particles of a gas are widely separated. There are relatively few of them in a given volume, which means little mass per unit volume.

3. *Large compressibility.* Particles in a gas are widely separated; a gas is mostly empty space. When pressure is applied, the particles are easily pushed closer together, decreasing the amount of empty space and the volume of the gas (see Fig. 11.4).

Figure 11.4

The compression of a gas involves decreasing the amount of empty space in the container. Particles present do not change in size.

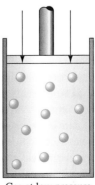

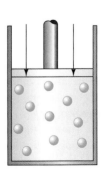

Gas at low pressure   Gas at higher pressure

4. *Moderate thermal expansion.* An increase in temperature means an increase in particle velocity. The increased kinetic energy of the particles enables them to push back whatever barrier is confining them into a given volume. Hence, the volume increases.

It must be understood that the size of the particles is not changed during expansion or compression of gases, solids, or liquids. The particles merely move farther apart or closer together; the space between them is what changes.

11.7 A Comparison of Solids, Liquids, and Gases

Two obvious conclusions about the similarities and differences between the various states of matter may be drawn from a comparison of the descriptive materials in Sections 11.4 through 11.6.

1. One of the states of matter, the gaseous state, is markedly different from the other two states.

2. Two of the states of matter, the solid and the liquid states, have many similar characteristics.

These two conclusions are illustrated diagrammatically in Figure 11.5.

The average distance between particles is only slightly different in the solid and liquid states but markedly different in the gaseous state. Roughly speaking, at ordinary temperatures and pressures, particles in a liquid are about 10% and particles in a gas about 1000% farther apart than those in the solid state. The distance ratio between particles in the three states (solid to liquid to gas) is thus 1 to 1.1 to 10.

> **The difference in distance between molecules in gases and liquids explains why it is easier to walk through air than through water.**

11.8 Endothermic and Exothermic Changes of State

A **change of state** *is a process in which a substance is transformed from one physical state to another physical state.* Changes of state were previously considered in Section 4.4. The terminology associated with various changes of state—evaporation, condensation, sublimation, and so on—was introduced at that time (Fig. 4.4).

Changes of state are usually accomplished through heating or cooling a substance. (Pressure change is also a factor in some systems.) Changes of state may be classified according to whether heat (thermal energy) is absorbed or released. An **endothermic change**

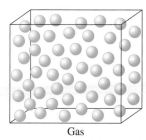
Gas
Molecules far apart and disordered
Negligible interactions between molecules

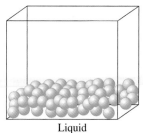
Liquid
Intermediate situation

Solid
Molecules close together and ordered
Strong interactions between molecules

Figure 11.5

Similarities and differences, at the molecular level, among the three states of matter.

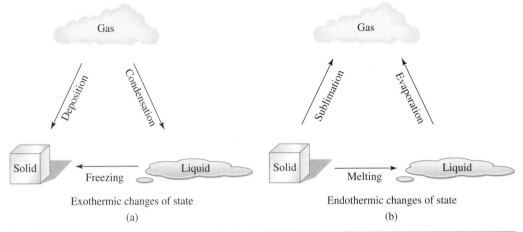

Figure 11.6

Exothermic (a) and endothermic (b) changes of state.

Although the processes of sublimation and deposition are not common in everyday life, they are still encountered. Dry ice sublimes, as do mothballs placed in a clothing storage area. It is because of sublimation that ice cubes left in a freezer get smaller as time passes. Ice or snow forming in clouds (from water vapor) during the winter season is an example of deposition.

of state is a change of state that requires the input (absorption) of heat energy. The endothermic changes of state are melting, sublimation, and evaporation. An **exothermic change of state** *is a change of state that requires heat energy to be given up (released).* Exothermic changes of state are the reverse of endothermic changes of state and include deposition, condensation, and freezing. Figure 11.6 summarizes the classification of changes of state as exothermic or endothermic.

11.9   Heat Energy and Specific Heat

Energy can exist in any of several forms. Common forms include radiant (light) energy, chemical energy, heat (thermal) energy, electrical energy, and mechanical energy. These forms of energy are interconvertible. The heating of a home is a process that illustrates energy interconversion. As the result of burning natural gas or some other fuel, chemical energy is converted into heat energy. In large conventional power plants that are used to produce electricity, the heat energy obtained from burning coal is used to change water into steam, which can then turn a turbine (mechanical energy) to produce electricity (electrical energy).

The form of energy most often encountered when considering physical and chemical changes is *heat energy.* For this reason, we will consider further particulars about this form of energy.

Heat Energy Units

In addition to the various forms *of energy, there are two* types *of energy: potential and kinetic (Sec. 11.3). The basis for determining energy types depends on whether the energy is available but not being used (potential) or is actually in use (kinetic).*

The two most commonly used units for expressing the amount of heat energy released or absorbed in a process are the *joule* and the *calorie.* Today, most heat energy calculations are carried out using the joule unit because this unit is more compatible with other metric system units than is the calorie.

A **joule (J)** *is the base unit for energy in the metric system.* The joule unit, abbreviated as J, is suitable for measuring all types of energy, not just heat energy. The name *joule* (pronounced *jool*, rhymes with pool) comes from the name of the English physicist James Prescott Joule (1818–1889), an early researcher in the area of energy studies.

It is convenient to think of the joule in terms of the amount of heat energy required to raise the temperature of one gram of water from 14.5°C to 15.5°C (one degree Celsius), which is 4.184 J. In this context, the joule unit and the calorie unit are easily related. The calorie was originally defined as the amount of heat energy necessary to raise the temperature of one gram of water from 14.5°C to 15.5°C. Thus,

$$4.184 \text{ joules} = 1 \text{ calorie}$$

Both the joule and the calorie involve relatively small amounts of energy, so the kilojoule (kJ) and the kilocalorie (kcal) are often used instead. It follows from the relationship between joules and calories that

$$4.184 \text{ kJ} = 1 \text{ kcal}$$

In discussions involving nutrition, the energy content of foods, or dietary tables, the unit *Calories* (spelled with a capital C) is used. The dietetic Calorie is actually 1 kilocalorie (1000 cal). The statement that an oatmeal raisin cookie contains 60 Calories means that 60 kcal (60,000 cal) of energy is released when the cookie is metabolized (undergoes chemical change) within the body.

> The Btu unit used to rate air conditioners and heaters is an English system unit for heat. "Btu" stands for British thermal unit and is the amount of heat energy it takes to raise the temperature of one pound of water one degree Fahrenheit. One Btu is the equivalent of 0.818 kcal.

EXAMPLE 11.1

Interrelationships Among Heat Energy Units

One gram of a combustible substance burns to produce 55.2 kJ of heat energy. Express this amount of heat energy, 55.2 kJ, in the following units.

(a) joules   **(b)** kilocalories   **(c)** calories

SOLUTION

(a) The relationship between joules and kilojoules is

$$10^3 \text{ J} = 1 \text{ kJ}$$

Therefore, using dimensional analysis, we have

$$55.2 \text{ kJ} \times \frac{10^3 \text{ J}}{1 \text{ kJ}} = 55,200 \text{ J} \quad \text{(calculator and correct answer)}$$

(b) The relationship between kilocalories and kilojoules is

$$1 \text{ kcal} = 4.184 \text{ kJ}$$

The one-step dimensional analysis setup for this problem is

$$55.2 \text{ kJ} \times \frac{1 \text{ kcal}}{4.184 \text{ kJ}} = 13.193116 \text{ kcal} \quad \text{(calculator answer)}$$

$$= 13.2 \text{ kcal} \qquad \text{(correct answer)}$$

(c) There will be two conversion factors in this part, derived, respectively, from the relationship 1 kcal = 4.184 kJ and 10^3 cal = 1 kcal. The pathway will be

$$\text{kJ} \rightarrow \text{kcal} \rightarrow \text{cal}$$

$$55.2 \text{ kJ} \times \frac{1 \text{ kcal}}{4.184 \text{ kJ}} \times \frac{10^3 \text{ cal}}{1 \text{ kcal}} = 13193.116 \text{ cal} \quad \text{(calculator answer)}$$

$$= 13,200 \text{ cal} \qquad \text{(correct answer)}$$

There can be only three significant figures in the correct answer.

Practice Exercise 11.1

One gram of acetylene gas, a gas used in welding torches, undergoes combustion (burns) to produce 11,900 calories of heat energy. Express the energy obtained from this reaction in

(a) kilocalories   **(b)** kilojoules   **(c)** joules

> Answers to practice exercises are located at the end of the chapter.

Table 11.2	Specific Heats of Selected Pure Substances	
Substance	**Physical State**	**Specific Heat [J/(g · °C)]**
Aluminum	solid	0.908
Copper	solid	0.382
Ethyl alcohol	liquid	2.42
Gold	solid	0.13
Iron	solid	0.444
Nitrogen	gas	1.0
Oxygen	gas	0.92
Silver	solid	0.24
Sodium chloride	solid	0.88
Water (ice)	solid	2.09
Water	liquid	4.18
Water (steam)	gas	2.03

Specific Heat

For every pure substance in a given state (solid, liquid, or gas), we can measure a physical property called the *specific heat* of that substance. **Specific heat** *is the amount of heat energy needed to raise the temperature of one gram of a substance in a specific physical state by 1°C.* The units most commonly used for specific heat, in scientific work, are joules per gram per degree Celsius [J/(g · °C)]. Specific heats for a number of substances in various states are given in Table 11.2. Note the three different entries for water in this table—one for each of the physical states. As these entries point out, the magnitude of the specific heat for a substance changes when the physical state of the substance changes. (Water is one of the few substances that we routinely encounter in all three physical states, hence the three specific-heat values in the table for this substance.)

The lower the specific heat of a substance, the more its temperature will change when it absorbs a given amount of heat. Metals generally have low specific heats. This means they heat up quickly and also cool quickly.

Liquid water has one of the highest specific heats known (see Table 11.2). This high specific heat makes water a very effective coolant. The moderate climates of geographical areas where large amounts of water are present—for example, the Hawaiian Islands—are related to water's ability to absorb large amounts of heat without undergoing drastic temperature changes. Desert areas, areas that lack water, are the areas where the extremes of high temperature are encountered on Earth. The temperature of a living organism remains relatively constant because of the large amounts of water present in it.

The amount of heat energy needed to cause a fixed amount of a substance to undergo a specific temperature change (within a range that causes no change of state) can easily be calculated if the substance's specific heat is known. The specific heat [in J/(g · °C)] is multiplied by the mass (in grams) and by the temperature change (in degrees Celsius) to eliminate the units of g and °C and obtain the unit joules.

> A walk on the beach on a sunny day will demonstrate differences in specific heat. The sand, which has a relatively low specific heat, heats up rapidly in the sun. The water, with its higher specific heat, stays cool.

$$\text{heat absorbed} = \text{specific heat} \times \text{mass} \times \text{temperature change}$$

$$= \frac{J}{\cancel{g} \cdot \cancel{°C}} \times \cancel{g} \times \cancel{°C}$$

$$= J$$

The temperature change, denoted as ΔT, is always calculated as a positive number; the lower temperature is always subtracted from the higher temperature.

EXAMPLE 11.2

Calculating the Amount of Heat Energy Absorbed by a Substance Undergoing a Specific Temperature Increase

The element beryllium is a relatively unreactive steel-gray metal with a specific heat of $1.8 \text{ J/g} \cdot \text{°C}$. How many joules of heat energy must 24.0 g of beryllium metal absorb for its temperature to increase from 25°C to 55°C?

SOLUTION

We will use the equation

$$\text{heat absorbed} = \text{specific heat} \times \text{mass} \times \text{temperature change}$$

in solving the problem. All the quantities on the right-hand side of the equation are known. Both specific heat and mass are directly given in the problem statement. The temperature change can be calculated using the two temperatures given in the problem statement.

Substituting the known quantities into the equation gives

$$\text{heat absorbed} = \frac{1.8 \text{ J}}{\text{g} \cdot \text{°C}} \times 24.0 \text{ g} \times (55 - 25)\text{°C}$$

$$= 1296 \text{ J} \quad (\text{calculator answer})$$

$$= 1300 \text{ J} \quad (\textbf{correct answer})$$

CHEMICAL EXTENSION

Beryllium, element 4, is a *light* metal, with a density one-third that of aluminum. Its largest use is in the manufacture of beryllium–copper alloys. Addition of 2% beryllium to copper produces an alloy with a strength six times that of pure copper. Such alloys find use as moving parts in aircraft engines, as components of precision instruments, and in nonsparking tools needed in industries that deal with flammable materials.

In general, beryllium compounds are quite toxic, and tough safety standards must be followed in fabricating beryllium-containing products. Inhalation of the dust of beryllium compounds can result in chronic lung problems.

Practice Exercise 11.2

Calculate the number of joules of heat energy needed to increase the temperature of 50.0 g of copper metal from 21.0°C to 80.0°C. Copper's specific heat is found in Table 11.2.

EXAMPLE 11.3

Calculating the Temperature Change Caused by Addition of a Specific Amount of Heat Energy

One cup of dry roasted, salted, shelled pistachio nuts has a caloric value of 78.7 Cal (78,700 cal). What would be the temperature change in a quart of water (944 g) at 10°C if the same amount of energy were added to it?

SOLUTION

The equation

$$\text{heat absorbed} = \text{specific heat} \times \text{mass} \times \text{temperature change}$$

is rearranged to isolate temperature change on the left side of the equation.

$$\text{temperature change } (°C) = \frac{\text{heat absorbed (J)}}{\text{mass (g)} \times \text{specific heat [J/(g} \cdot °C)]}}$$

In this equation the heat absorbed will be the heat energy (caloric value) supplied by the pistachio nuts. Since this is given in calories, we will need to change it to joule units.

$$78,700 \text{ cal} \times \frac{4.184 \text{ J}}{1 \text{ cal}} = 329,280.8 \text{ J} \quad \text{(calculator answer)}$$

$$= 329,000 \text{ J} \quad \textbf{(correct answer)}$$

Substituting the values 329,000 J (heat absorbed), 944 g (mass of water), and 4.18 J/g · °C (specific heat of water; see Table 11.2) in the previously derived equation for change in temperature (°C) gives

$$\text{temperature change } (°C) = \frac{329,000 \text{ J}}{944 \text{ g} \times 4.18 \text{ J/(g} \cdot °C)}$$

$$= 83.37726°C \quad \text{(calculator answer)}$$

$$= 83.4°C \quad \textbf{(correct answer)}$$

The temperature of the water will increase from 10°C to 93°C, an increase of 83°.

Practice Exercise 11.3

A one-half cup serving of zucchini has a caloric value of approximately 15 Cal (15,000 cal). What would be the temperature change in a cup of water (237 g) at 20°C if the same amount of energy were added to it?

A quantity closely related to specific heat is that of *heat capacity*. **Heat capacity** *is the amount of heat energy needed to raise the temperature of a given quantity of a substance in a specific physical state by 1°C.* The relationship between heat capacity and specific heat is

$$\text{heat capacity} = \text{grams} \times \text{specific heat}$$

Common units for heat capacity are J/°C. Heat capacity refers to a property of a whole object (its entire mass), while specific heat refers to the heat capacity per unit mass (1 g). If either heat capacity or specific heat is known, the other quantity can always be calculated from the known quantity.

EXAMPLE 11.4

Using Heat Capacity Data to Calculate Specific Heat

A 80.0-g sample of the metal gold has a heat capacity of 10.5 J/°C. What is the specific heat of gold?

SOLUTION

The relationship between heat capacity and specific heat is

$$\text{heat capacity} = \text{mass} \times \text{specific heat}$$

Rearranging this equation to isolate specific heat on one side gives

$$\text{specific heat} = \frac{\text{heat capacity}}{\text{mass}}$$

$$= \frac{10.5 \text{ J/°C}}{80.0 \text{ g}}$$

$$= 0.13125 \text{ J/(g} \cdot {}^\circ\text{C)} \quad \text{(calculator answer)}$$

$$= 0.131 \text{ J/(g} \cdot {}^\circ\text{C)} \quad \text{(correct answer)}$$

CHEMICAL EXTENSION

Gold is the most malleable and ductile of all metals. Use of gold in jewelry accounts for about three-fourths of the world's annual production of this element. Industrial applications, especially electronics, consumes another 10–15% of production, with the remainder divided among medical and dental uses, coinage, and bar stock for governmental and private holdings.

Pure gold is too soft for normal use in jewelry, so it is alloyed with copper, silver, and other metals, with a resulting increase in hardness. The amount of gold in such alloys is specified in karats: 14-karat gold is an alloy with 14 out of 24 parts (by mass) of gold, and 18-karat gold contains 18 out of 24 parts gold. "White" gold contains 14 parts Au, 4 parts Cu, 4 parts Ni, and 2 parts Zn; "yellow" gold contains 14 parts Au, 6 parts Cu, and 4 parts Ag.

Practice Exercise 11.4

A 35.0-g sample of the metal iron has a heat capacity of 15.5 J/°C. What is the specific heat, in J/g · °C, of iron?

Another common type of heat energy calculation involves the transfer of heat from one substance to another. The following two generalizations always apply to such situations.

1. Heat always flows from the warmer body to the colder body.
2. The heat lost by the warmer body is equal to the heat gained by the colder body.

EXAMPLE 11.5

Calculation Involving "Heat lost is equal to heat gained."

A 125-g piece of a rock of unknown specific heat is heated to 93°C and then the rock is dropped into 100.0 g of water at 19°C. The temperature of the water rises to 31°C. What is the specific heat, in J/g · °C, of the rock?

SOLUTION

The heat lost by the rock (the hotter body) is equal to the heat gained by the water (the colder body):

$$\text{heat lost (rock)} = \text{heat gained (water)}$$

Using the general equation for heat lost or heat gained, which is

$$\text{heat lost or gained} = \text{specific heat} \times \text{mass} \times \text{temperature change}$$

we have for our present situation

$$\underbrace{\text{specific heat} \times 125 \text{ g} \times 62{}^\circ\text{C}}_{\text{heat lost by rock}} = \underbrace{4.18 \frac{\text{J}}{\text{g} \cdot {}^\circ\text{C}} \times 100.0 \text{ g} \times 12{}^\circ\text{C}}_{\text{heat gained by water}}$$

Solving this equation for the specific heat of the rock gives

$$\text{specific heat of rock} = \frac{4.18\dfrac{J}{g \cdot °C} \times 100.0 \text{ g} \times 12°C}{125 \text{ g} \times 62°C}$$

$$= 0.6472258 \text{ J/g} \cdot °C \quad \text{(calculator answer)}$$

$$= 0.65 \text{ J/g} \cdot °C \qquad \textbf{(correct answer)}$$

Practice Exercise 11.5

How many grams of copper can be heated from 20°C to 30°C by the heat energy released when 245 g of aluminum cools from 80°C to 50°C? Needed specific heats are found in Table 11.2.

11.10 Temperature Changes as a Substance Is Heated

For a given pure substance, molecules in the gaseous state contain more energy than molecules in the liquid state, which in turn contain more energy than molecules in the solid state. This fact is obvious: We know that it takes energy (heat) to melt a solid and still more energy (heat) to change the resulting liquid to a gas. Additional information concerning the relationship of energy to the states of matter can be obtained by a closer examination of what happens, step by step, to a solid (which is below its melting point) as heat is continuously supplied, causing it to melt and ultimately to change to a gas.

The heating curve shown in Figure 11.7 gives the steps involved in changing a solid to a gas, using water as the example. As the ice (solid) is heated, its temperature rises until the melting point for ice is reached. The temperature increase indicates that the added heat causes an increase in the kinetic energy of the water molecules (recall Sec. 11.3, kinetic molecular theory). Once the melting point is reached, the temperature remains constant while the ice melts. The constant temperature during the melting process indicates that the added heat has increased the potential energy of the water molecules without increasing their kinetic energy—the intermolecular attractions are being weakened by the increase in potential energy. The net result is that the ice becomes liquid water. The addition of more heat to the liquid water increases its temperature until the boiling point is reached. During this stage the water molecules are again gaining kinetic energy. At the boiling point another state

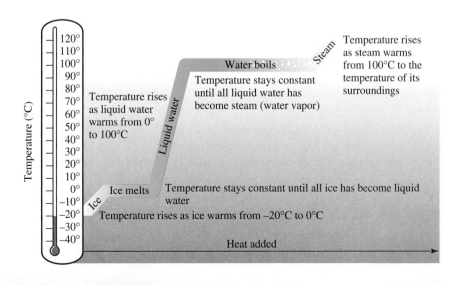

Figure 11.7

A heating curve depicting the addition of heat (at a constant rate) to a solid (ice) at a temperature below its melting point until it becomes a gas (steam) at a temperature above its boiling point. Note that when heat is added during a transition between physical states, the temperature does not change.

change occurs (liquid to gas) as heat is added with the temperature remaining constant. The constant temperature again indicates an increase in potential energy. Once the system is completely changed to steam (water vapor), the temperature again increases with further heating.

The actual amount of energy that must be added to a system to cause it to undergo the series of changes just described depends on values of three properties of the substance. These properties are (1) specific heat, (2) heat of fusion, and (3) heat of vaporization. Specific heat was discussed in Section 11.9. Heats of fusion and vaporization are the subject of Section 11.11.

11.11 Energy and Changes of State

When heat energy is added to a solid, its temperature rises until the melting point is reached, with the amount of heat added governed by the specific heat (Sec. 11.9) of the solid. Once the melting point is reached, the temperature then remains constant while the solid changes to a liquid (Sec. 11.10). The "energetics" of the system during this transition from the solid state to the liquid state depend on the value of the substance's *heat of fusion*. (The term *fusion* means melting.)

The **heat of fusion** *is the amount of heat energy absorbed in the conversion of one gram of a solid to a liquid at the solid's melting point.* Units for heat of fusion are joules per gram (J/g). Note that these units do not involve temperature (degrees) as was the case for specific heat [J/(g · °C)]. No temperature units are needed because the temperature remains constant during a change of state.

The reverse of the fusion process is solidification (or freezing). The **heat of solidification** *is the amount of heat energy evolved in the conversion of one gram of a liquid to a solid at the liquid's freezing point.* The heat of solidification always has the *same numerical value* as the heat of fusion. The only difference between these two entities is in the *direction* of heat flow (in or out). Heat of solidification is associated with an exothermic process, and heat of fusion with an endothermic process. The amount of heat required to melt 50.0 g of ice at its melting point is the same as the amount of heat that must be removed to freeze 50.0 g of water at its freezing point. The heats of fusion (or solidification) for selected substances are given in Table 11.3 in the units joules per gram.

Table 11.3 also gives a second heat of fusion (solidification) value for each substance, the *molar* heat of fusion (solidification), which has the units kJ/mole. In some types of calculations kJ/mole are more convenient units to use.

The magnitude of the heat of fusion for a given solid depends on the intermolecular forces of attraction in the solid state. The strength of such forces is the subject of Section 11.16.

Ice cubes cool drinks not only by being cold, but also by absorbing heat as they melt. Each gram of ice absorbs 334 joules from the liquid as it melts. The amount of heat it takes to melt 1 gram of ice can cool 1 gram of water from 80°C to 0°C.

Table 11.3	Heats of Fusion (or Solidification) for Various Substances at Their Melting (or Freezing) Points		
		Heat of Fusion (or Solidification)	
Solid	**Melting (or Freezing) Point (°C)**	**J/g**	**kJ/mole**
Methane	−182	59	0.94
Ethyl alcohol	−117	109	5.01
Carbon tetrachloride	−23	16.3	2.51
Water	0	334	6.01
Benzene	6	126	9.87
Aluminum	658	393	10.6
Copper	1083	205	13.0

The general equation for calculating the amount of heat absorbed as a substance changes from a solid to a liquid is

$$\text{heat absorbed (J)} = \text{heat of fusion (J/g)} \times \text{mass(g)}$$

Similarly, for the amount of heat released as a liquid freezes to a solid we have

$$\text{heat released (J)} = \text{heat of solidification (J/g)} \times \text{mass (g)}$$

Example 11.6 illustrates the use of the first of these two equations.

EXAMPLE 11.6

Calculating the Heat Energy Absorbed as a Substance Melts

How much heat energy, in joules, is required to melt 25.1 g of aluminum at its melting point of 658°C?

SOLUTION

The heat of fusion for aluminum, from Table 11.3, is 393 J/g. Therefore, the total amount of heat energy required is

$$\text{Heat absorbed} = 25.1 \text{ g} \times \frac{393 \text{ J}}{\text{g}} = 9864.3 \text{ J} \quad \text{(calculator answer)}$$

$$= 9860 \text{ J} \quad \textbf{(correct answer)}$$

Practice Exercise 11.6

How much heat energy, in joules, is required to melt 35.2 g of ice at its melting point of 0°C?

Principles similar to those just considered for solid–liquid or liquid–solid changes apply to changes between the liquid and gaseous states. Here the specific energy quantities involved are *heats of vaporization* and *heats of condensation*. The **heat of vaporization** *is the amount of heat energy absorbed in the conversion of one gram of a liquid to a gas at the liquid's boiling point.* Heats of vaporization are usually measured at the normal boiling point (Sec. 11.15) of the liquid.

The reverse of the vaporization (evaporation) process is condensation (liquefaction). The **heat of condensation** *is the amount of heat energy evolved in the conversion of one gram of a gas to a liquid at the liquid's boiling point.* The heat of vaporization and the heat of condensation will always have the same numerical value because these two quantities characterize processes that are the "reverse" of each other. The only difference is direction of heat flow; one process is endothermic (vaporization), and the other is exothermic (condensation). Table 11.4 gives heats of vaporization (or condensation) for selected substances.

The general equation for calculating the amount of heat absorbed as a substance changes from a liquid to a gas is

$$\text{heat absorbed (J)} = \text{heat of vaporization (J/g)} \times \text{mass (g)}$$

Similarly, for the amount of heat released as a gas condenses to a liquid, we have

$$\text{heat released (J)} = \text{heat of condensation (J/g)} \times \text{mass (g)}$$

Steam at 100°C is much more dangerous than water at 100°C. If steam contacts the skin, it is not only hot, it also releases its heat of condensation as it condenses on the skin. This exothermic release of energy can cause severe burns.

Table 11.4	Heats of Vaporization (or Condensation) for Various Substances at Their Boiling Point		

Liquid	Boiling Point (°C)	Heat of Vaporization (or Condensation)	
		J/g	kJ/mole
Methane	−161	65.0	10.4
Ammonia	−33	1380	23.4
Diethyl ether	34.6	375	27.8
Carbon tetrachloride	77	195	30.0
Ethyl alcohol	78.3	837	38.6
Benzene	80.1	394	30.8
Water	100	2260	40.7

EXAMPLE 11.7

Calculating the Heat of Vaporization from Given Thermochemical Data

5744 joules of heat energy are required to vaporize 15.0 g of an unknown liquid at its boiling point. What is the heat of vaporization, in joules per gram, of this unknown liquid?

SOLUTION

Rearranging the equation

$$\text{heat absorbed} = \text{heat of vaporization} \times \text{mass}$$

to isolate "heat of vaporization," which is our desired quantity, gives

$$\text{heat of vaporization} = \frac{\text{heat absorbed}}{\text{mass}}$$

Substituting the given quantities into this equation gives

$$\text{heat of vaporization} = \frac{5744 \text{ J}}{15.0 \text{ g}} = 382.93333 \text{ J/g} \quad \text{(calculator answer)}$$

$$= 383 \text{ J/g} \quad \text{(correct answer)}$$

Practice Exercise 11.7

The vaporization of 20.0 g of ethyl alcohol at its boiling point requires 16,740 J of heat energy. Using this information, calculate the heat of vaporization, in joules per gram, of ethyl alcohol.

> Refrigerators work by letting a fluid (usually a Freon) evaporate in coils inside the refrigerator. The endothermic evaporation absorbs heat from inside the refrigerator. The compressor recondenses the gas back to a fluid outside of the cooling coils. (This is where most of the electricity is consumed.) The heat released in the condensation is radiated out of the coils in the rear. That is why it is always warm at the back of a refrigerator.

11.12 Heat Energy Calculations

In doing "bookkeeping" on heat energy added to or removed from a chemical system, the following two generalizations always apply.

1. The amount of energy added to or removed from a chemical system that undergoes a temperature change *but no change in state* is governed by the substance's specific heat.

The way in which specific heat is used in calculations involving this situation was considered in Section 11.9 (Examples 11.2 and 11.3).

2. The amount of energy added to or removed from a chemical system that undergoes a change of state *at constant temperature* is governed by heat of fusion or solidification (solid–liquid changes) and heat of vaporization or condensation (liquid–gas changes). The way in which these entities are used in change-of-state situations was considered in Section 11.11 (Examples 11.6 and 11.7).

Figure 11.8 is a useful summary diagram of common "energy-change-of-temperature" and "energy-change-of-state" situations that occur. Also, at the bottom of this diagram are given the key equations needed for doing heat–energy calculations. These equations are presented with some new symbolism (abbreviations). The new symbolism is as follows:

$$Q = \text{heat absorbed or heat released}$$

$$C = \text{specific heat of the substance in a given physical state}$$

$$m = \text{mass of the substance}$$

$$\Delta T = \text{change in temperature of the substance}$$

$$\Delta H_{\text{fus}} = \text{heat of fusion of the substance}$$

$$\Delta H_{\text{sol}} = \text{heat of solidification of the substance}$$

$$\Delta H_{\text{vap}} = \text{heat of vaporization of the substance}$$

$$\Delta H_{\text{con}} = \text{heat of condensation of the substance}$$

In problem-solving situations where a pure substance experiences both a temperature change and one or more changes of state (a common situation), the calculational equations

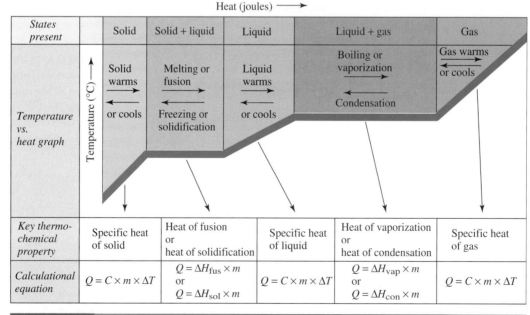

States present		Solid	Solid + liquid	Liquid	Liquid + gas	Gas
Temperature vs. heat graph	Temperature (°C)	Solid warms → ← or cools	Melting or fusion → ← Freezing or solidification	Liquid warms → ← or cools	Boiling or vaporization → ← Condensation	Gas warms → ← or cools
Key thermo-chemical property		Specific heat of solid	Heat of fusion or heat of solidification	Specific heat of liquid	Heat of vaporization or heat of condensation	Specific heat of gas
Calculational equation		$Q = C \times m \times \Delta T$	$Q = \Delta H_{\text{fus}} \times m$ or $Q = \Delta H_{\text{sol}} \times m$	$Q = C \times m \times \Delta T$	$Q = \Delta H_{\text{vap}} \times m$ or $Q = \Delta H_{\text{con}} \times m$	$Q = C \times m \times \Delta T$

Heat (joules) →

Figure 11.8

A temperature–heat–energy graph showing what happens as a solid, below its melting point, is heated to higher and higher temperatures.

of Figure 11.8 are used in an additive manner. Examples 11.8 and 11.9 illustrate problem solving in such "combination" situations.

EXAMPLE 11.8

Thermochemical Calculation Involving Both Change of State and Change of Temperature

Calculate the heat released, in joules, when 324 g of steam at 100°C are condensed and the resulting liquid water is cooled to 45°C.

SOLUTION

We may consider this process to occur in two steps.

1. Condensing the steam to liquid water at 100°C.
2. Cooling the liquid water from 100°C to 45°C.

We will calculate the amount of heat released in each of the steps, then add these amounts together to get our final answer.

STEP 1 $\text{gas}_{100°C} \rightarrow \text{liquid}_{100°C}$

This step involves a change of state, from the gaseous to the liquid state (the fourth column of the heating curve of Figure 11.7). The heat energy equation for this change is

$$\text{heat released} = \text{heat of condensation} \times \text{mass}$$

The heat of condensation of H_2O, from Table 11.4, is 2260 J/g. Therefore, we have

$$\text{heat released} = \frac{2260 \text{ J}}{g} \times 324 \text{ } g = 732{,}240 \text{ J} \quad \text{(calculator answer)}$$

$$= 732{,}000 \text{ J} \quad \textbf{(correct answer)}$$

STEP 2 $\text{liquid}_{100°C} \rightarrow \text{liquid}_{45°C}$

From the middle column of Figure 11.4 we note that the heat energy equation for the change in temperature of a liquid substance is

$$\text{heat released} = \text{specific heat} \times \text{mass} \times \text{temperature change}$$

From the values given in Table 11.2 we find that the specific heat of water (as a liquid) is 4.18 J/g · °C. The temperature change is 55°C, and the mass is given as 324 g. Plugging these values into our equation gives

$$\text{heat released} = \frac{4.18 \text{ J}}{g \cdot °C} \times 324 \text{ } g \times 55°C = 74{,}487.6 \text{ J} \quad \text{(calculator answer)}$$

$$= 74{,}000 \text{ J} \quad \textbf{(correct answer)}$$

The temperature change limits the correct answer to two significant figures. The total heat released is the sum of that released in each step.

From step 1:   732,000 J
From step 2:   74,000 J
              806,000 J (calculator and **correct answer**)

CHEMICAL EXTENSION

Water's high specific heat and high heats of fusion and vaporization, together with its abundance, make it an ideal coolant. Large bodies of water exert a temperature-moderating effect on their surroundings because of water's ability to absorb large amounts of heat energy when evaporating (heat of vaporization) and to release large amounts of heat energy when freezing (heat of solidification).

In the heat of a summer day, extensive water evaporation occurs, and in the process energy is absorbed from the surroundings. The net effect of the evaporation is a lowering of the temperature of the surroundings. In the cool of the evening, some of the water vapor condenses back to the liquid state, releasing heat that raises the temperature of the surroundings. In this manner the temperature variation between night and day is reduced. In the winter a similar process occurs: water freezes on cold days and releases heat to the surroundings.

The hottest and coldest regions on Earth are all inland regions, those distant from the moderating effects of large bodies of water.

Practice Exercise 11.8

Calculate the heat released when 40.0 g of steam at 115°C condenses to water at 100°C in a radiator of a steam heating system.

EXAMPLE 11.9

Thermochemical Calculation Involving All Three States of Matter

Calculate the amount of heat, in joules, needed to convert 23.6 g of ice at −27°C to steam at 121°C.

SOLUTION

We may consider the change process to occur in five steps.

1. Heating the ice from −27 to 0°C, its melting point. (This change corresponds to the first column of the heating curve of Figure 11.7.)

2. Melting the ice, while the temperature remains at 0°C. (This change corresponds to the second column of the heating curve of Figure 11.7.)

3. Heating the liquid water from 0 to 100°C, its boiling point. (This change corresponds to the third column of the heating curve of Figure 11.7.)

4. Evaporating the water, while the temperature remains at 100°C. (This change corresponds to the fourth column of the heating curve of Figure 11.7.)

5. Heating the steam from 100 to 121°C. (This change corresponds to the last column of the heating curve of Figure 11.7.)

We will calculate the amount of heat required in each of the steps and then add these amounts together to get our desired answer. In doing this we will need the following heat energy values associated with water.

STEP 1   specific heat of ice, 2.09 J/g · °C (Table 11.2)

STEP 2   heat of fusion, 334 J/g (Table 11.3)

STEP 3   specific heat of water, 4.18 J/g · °C (Table 11.2)

STEP 4   heat of vaporization, 2260 J/g (Table 11.4)

STEP 5   specific heat of steam, 2.03 J/g · °C (Table 11.2)

STEP 1 solid$_{-27°C}$ → solid$_{0°C}$

heat required = specific heat × mass × temperature change

$$= \frac{2.09 \text{ J}}{\text{g} \cdot °C} \times 23.6 \text{ g} \times 27°C = 1331.748 \text{ J} \quad \text{(calculator answer)}$$

$$= 1300 \text{ J} \qquad \textbf{(correct answer)}$$

STEP 2 solid$_{0°C}$ → liquid$_{0°C}$

heat required = heat of fusion × mass

$$= \frac{334 \text{ J}}{\text{g}} \times 23.6 \text{ g} = 7882.4 \text{ J} \quad \text{(calculator answer)}$$

$$= 7880 \text{ J} \qquad \textbf{(correct answer)}$$

STEP 3 liquid$_{0°C}$ → liquid$_{100°C}$

heat required = specific heat × mass × temperature change

$$= \frac{4.18 \text{ J}}{\text{g} \cdot °C} \times 23.6 \text{ g} \times 100°C = 9864.8 \text{ J} \quad \text{(calculator answer)}$$

$$= 9860 \text{ J} \qquad \textbf{(correct answer)}$$

STEP 4 liquid$_{100°C}$ → gas$_{100°C}$

heat required = heat of vaporization × mass

$$= \frac{2260 \text{ J}}{\text{g}} \times 23.6 \text{ g} = 53,336 \text{ J} \quad \text{(calculator answer)}$$

$$= 53,300 \text{ J} \quad \textbf{(correct answer)}$$

STEP 5 gas$_{100°C}$ → gas$_{121°C}$

heat required = specific heat × mass × temperature change

$$= \frac{2.03 \text{ J}}{\text{g} \cdot °C} \times 23.6 \text{ g} \times 21°C = 1006.068 \text{ J} \quad \text{(calculator answer)}$$

$$= 1\overline{0}00 \text{ J} \qquad \textbf{(correct answer)}$$

The total heat required is the sum of the results of the five steps.

$$(1300 + 7880 + 9860 + 53,300 + 1\overline{0}00) \text{ J} = 73,340 \text{ J} \quad \text{(calculator answer)}$$

$$= 73,300 \text{ J} \quad \textbf{(correct answer)}$$

(The answer must be rounded to the hundreds place. In adding, the numbers 1300, 53,300, and $1\overline{0}00$ are limiting, with uncertainty in the hundreds place.)

CHEMICAL EXTENSION

All water as it occurs in nature is impure in a chemical sense. Impurities present include suspended matter, microorganisms, dissolved gases, and dissolved minerals. Minerals dissolved in water produce ions.

The major distinction between *fresh water* and *salt water* (seawater) is the number of ions present. Seawater has a concentration of dissolved ions about 500 times greater than that of fresh water. The dominant ions in fresh water and seawater differ. In seawater, Na^+ is the dominant cation, and Cl^- is the dominant anion. This contrasts with fresh water, where Ca^{2+} and Mg^{2+} are the most abundant cations and HCO_3^- is the most abundant anion.

Fresh water purified for drinking purposes still contains its ions; suspended particles, disease-causing agents, and objectionable odors have been removed. The ions present are not harmful to health and give water its "taste"; water without any ions present would taste flat to most people.

Hard water contains the ions Ca^{2+}, Mg^{2+}, and Fe^{2+}. The presence of these ions does not affect the drinkability of water but does affect other uses for water such as washing. *Soft water* is water in which the offending hard-water ions have been exchanged for Na^+ and K^+ ions. These ions are compatible with the nondrinking uses for water. People with high blood pressure or kidney problems are often advised to avoid drinking soft water because of its high sodium content.

Practice Exercise 11.9

Calculate the amount of heat, in joules, needed to convert 12.8 g of ice at $-18°C$ to steam at $121°C$.

11.13 Evaporation of Liquids

> For a liquid to evaporate, its molecules must gain enough kinetic energy to overcome the attractive forces between them.

Evaporation *is the process by which molecules escape from a liquid phase to the gas phase.* It is a familiar process. We are all aware that water left in an open container at room temperature will slowly disappear by evaporation.

The phenomenon of evaporation can readily be explained using kinetic molecular theory. Statement 5 of this theory (Sec. 11.3) indicates that not all the molecules in a liquid (or solid or gas) possess the same kinetic energy. At any given instant some molecules will have above-average kinetic energies and others below-average kinetic energies as a result of collisions between molecules. A given molecule's energy constantly changes as a result of collisions with neighboring molecules. Molecules considerably above average in kinetic energy can overcome the attractive forces (potential energy) that are holding them in the liquid and escape if they are at the liquid surface and are moving in a favorable direction relative to the surface.

Note that evaporation is a surface phenomenon. Molecules within the interior of a liquid are surrounded on all sides by other molecules, making escape very improbable. Surface molecules are subject to fewer attractive forces since they are not completely surrounded by other molecules; escape is much more probable (see Fig. 11.9). Liquid

Figure 11.9

Molecules at the surface of a liquid experience fewer intermolecular attractions than do molecules within the interior of a liquid. Molecules within the interior of a liquid experience intermolecular attraction to neighboring molecules in all directions. Molecules on the surface of a liquid are attracted by other surface molecules and by molecules below the surface.

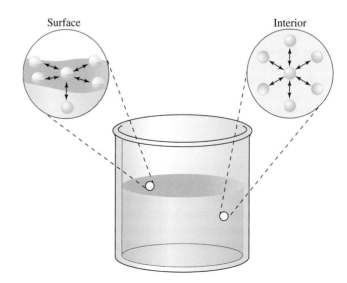

Surface   Interior

surface area is an important factor in determining the rate at which evaporation occurs. Increased surface area results in an increased evaporation rate; a greater fraction of molecules occupy "surface" locations.

Rate of Evaporation and Temperature

Water evaporates faster from a glass of hot water than from a glass of cold water. Why is this so? A certain minimum kinetic energy is required for molecules to overcome the attractions of neighboring molecules. As the temperature of a liquid increases, a larger fraction of the molecules present possess this needed minimum kinetic energy. Consequently, the rate of evaporation always increases as liquid temperature increases. Figure 11.10 contrasts the fraction of molecules possessing the needed minimum kinetic energy for escape at two temperatures. Note that at both the lower and higher temperatures a broad distribution of kinetic energies is present and that at each temperature some molecules possess the needed minimum kinetic energy. However, at the higher temperature a larger fraction of molecules present have the requisite kinetic energy so the rate of evaporation increases.

The escape of high-energy molecules from a liquid during evaporation affects the liquid in two ways: the amount of liquid decreases, and the liquid temperature is lowered. The temperature lowering reflects the fact that the average kinetic energy of the remaining molecules is lower than the preevaporation value due to the loss of the most energetic molecules. (Analogously, if all the tall people are removed from a classroom of students, the average height of the remaining students decreases.) A lower average kinetic energy corresponds to a lower temperature (statement 4 of kinetic molecular theory); hence a cooling effect is produced.

Evaporative cooling is important in many processes. Our own bodies use evaporation to maintain a constant temperature. In hot summer weather and also during strenuous exercise, the human body produces perspiration. Perspiration production is an effective method for body temperature regulation because evaporation of the perspiration cools our skin. The cooling effect of evaporation is quite noticeable when someone first comes out of a swimming pool on a hot day (especially if a breeze is blowing). A canvas water bag keeps water cool because some of the water seeps through the canvas and evaporates, a process that removes heat from the remaining water. In medicine, the local skin anesthetic ethyl chloride (C_2H_5Cl) exerts its effect through evaporative cooling (freezing). The minimum kinetic energy molecules of this substance need to acquire to escape (evaporate) is very low, resulting in a very rapid evaporation rate. Evaporation is so fast that the cooling "freezes" tissue near the surface of the skin, with temporary loss of feeling in the region of application. Certain alcohols also evaporate quite rapidly. An alcohol "rub" is sometimes used to reduce body temperature when a high fever is present.

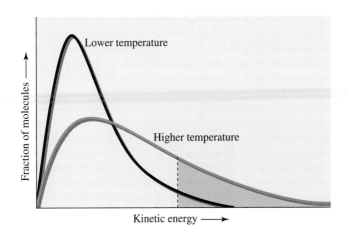

Figure 11.10

Kinetic energy distributions of molecules of a liquid at two different temperatures. The dashed line represents the minimum kinetic energy required for molecules of the liquid to overcome attractive forces and escape into the gas phase. Molecules in the shaded area have the necessary energy to overcome attractions.

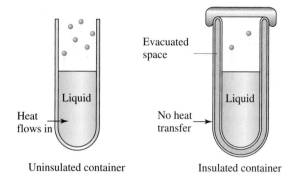

Figure 11.11

For an evaporative cooling effect to be measured, a container that minimizes heat flow into the container from its surroundings must be used.

Uninsulated container

Insulated container

For a liquid in a container, the decrease in temperature that occurs as a result of evaporation can be actually measured only if the container is an *insulated* one. When a liquid evaporates from a noninsulated container, there is sufficient heat flow from the surroundings into the container to counterbalance the loss of energy in the escaping molecules and thus prevent any cooling effect. The Thermos bottle is an insulated container that minimizes heat transfer, making it useful for maintaining liquids at a cool temperature for a short period of time. Figure 11.11 contrasts the evaporation process occurring in an uninsulated container with that which occurs within an insulated one. The temperature of liquid and surroundings is the same in an uninsulated container; in an insulated one, liquid temperature drops below that of the surroundings because of evaporative cooling.

Collectively, the molecules that escape from an evaporating liquid are often referred to as *vapor* rather than gas. A **vapor** *is the gaseous state of a substance at a temperature and pressure at which the substance is normally a liquid or solid.* For example, at room temperature and atmospheric pressure the normal state for water is the liquid state. Molecules that escape (evaporate) from liquid water at these conditions are called water vapor.

Evaporation and Equilibrium

The evaporative behavior of a liquid in a *closed* container is quite different from that in an *open* container. In a closed container we observe that some liquid evaporation occurs, as indicated by a drop in liquid level. However, unlike the open container system, the liquid level, with time, ceases to drop (becomes constant), an indication that not all of the liquid will evaporate.

Kinetic molecular theory explains these observations in the following way. The molecules that do evaporate are unable to move completely away from the liquid as they did in the open container. They find themselves confined in a fixed space immediately above the liquid (see Fig. 11.12a). These "trapped" vapor molecules undergo many random collisions with the container walls, other vapor molecules, and the liquid surface. Molecules colliding with the liquid surface may be recaptured by the liquid. Thus, two processes—evaporation (escape) and condensation (recapture)—take place in the closed container.

> Keeping a lid on a pot of water allows the pot of water to boil faster. Without a lid, the heat of vaporization absorbed when water turns to steam escapes into the air. With a lid on, the water recondenses inside the lid and releases its heat of condensation back to the pot.

Figure 11.12

Evaporation of a liquid in a closed container.

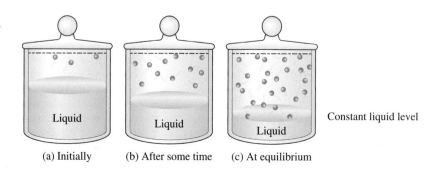

(a) Initially   (b) After some time   (c) At equilibrium

Constant liquid level

In a closed container, for a short time, the rate of evaporation exceeds the rate of condensation and the liquid level drops. However, as more and more of the liquid evaporates, the number of vapor molecules increases and the chance of their recapture through striking the liquid surface also increases. Eventually the rate of condensation becomes equal to the rate of evaporation and the liquid level stops dropping (see Fig. 11.12c). At this point the number of molecules that escape in a given time is the same as the number recaptured; a steady-state condition has been reached. The amounts of liquid and vapor in the container are not changing, even though both evaporation and condensation are still occurring.

This steady-state situation, which will continue as long as the temperature of the system remains constant, is an example of an *equilibrium state*. An **equilibrium state** *is a situation in which two opposite processes take place at equal rates*. For systems in a state of equilibrium, no net macroscopic changes can be detected. However, the system is dynamic; both forward and reverse processes are still occurring but in a manner such that they balance each other.

11.14 Vapor Pressure of Liquids

For a liquid–vapor system in equilibrium in a closed container, the vapor in the fixed space immediately above the liquid exerts a constant pressure on the liquid surface and the walls of the container. This pressure is called the liquid's *vapor pressure*. **Vapor pressure** *is the pressure exerted by a vapor above a liquid when the liquid and vapor are in equilibrium*.

The magnitude of a vapor pressure depends on the nature and temperature of the liquid. Liquids with strong attractive forces between molecules will have lower vapor pressures than liquids in which only weak attractive forces exist between particles. Substances with high vapor pressures evaporate readily; that is, they are *volatile*. A **volatile substance** *is a substance that readily evaporates at room temperature because of a high vapor pressure*.

The vapor pressures of all liquids increase with temperature. Why? An increase in temperature results in more molecules having the minimum energy required for evaporation. Hence, at equilibrium the pressure of the vapor is greater. Table 11.5 shows the variation in vapor pressure with increasing temperature for water.

When the temperature of a liquid–vapor system in equilibrium is changed, the equilibrium is upset. The system immediately begins the process of establishing a new equilibrium. Let us consider a case where the temperature is increased. The higher temperature signifies that energy has been added to the system. More molecules will have the minimum energy needed to escape. Thus, immediately after the temperature is increased, molecules begin to escape at a rate greater than that at which they are recaptured. With time, however, the rates of escape and recapture will again become equal. The new rates will, however, be different

Table 11.5 Vapor Pressure of Water at Various Temperatures

Temperature (°C)	Vapor Pressure (mm Hg)*	Temperature (°C)	Vapor Pressure (mm Hg)*
0	4.6	60	149.4
10	9.2	70	233.7
20	17.5	80	355.1
30	31.8	90	525.8
40	55.3	100	760.0
50	92.5		

*The units used to specify vapor pressure in this table will be discussed in detail in Section 12.2.

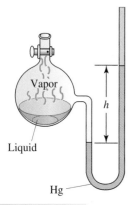

Apparatus for determining the vapor pressure of a liquid.

from those for the previous equilibrium. Since energy has been added to the system, the rates will be higher, resulting in a higher vapor pressure.

The size (volume) of the space that the vapor occupies does not affect the magnitude of the vapor pressure. A larger fixed space will enable more molecules to be present in the vapor at equilibrium. However, the larger number of molecules spread over a larger volume results in the same pressure as a small number of molecules in a small volume.

Vapor pressures for liquids are commonly measured in an apparatus similar to that pictured in Figure 11.13. At the moment liquid is added to the vessel, the space above the liquid is filled only with air and the levels of mercury in the U-tube are equal. With time, liquid evaporates and equilibrium is established. The vapor pressure of the liquid is proportional to the difference between the heights of the mercury columns. The larger the vapor pressure, the greater the extent to which the mercury column is "pushed up."

11.15 | Boiling and Boiling Points

Usually, for a molecule to escape from the liquid state, it must be on the surface of the liquid. **Boiling** *is a special form of evaporation in which conversion from the liquid to the vapor state occurs within the body of a liquid through bubble formation.* This phenomenon begins to occur when the vapor pressure of a liquid, which is steadily increasing as a liquid is heated, reaches a value equal to that of the prevailing external pressure on the liquid; for liquids in open containers this value is atmospheric pressure. When these two pressures become equal, bubbles of vapor form around any speck of dust or any irregularity associated with the surface of the container. Being less dense than the liquid itself, these vapor bubbles quickly rise to the surface and escape. The quick ascent of the bubbles causes the agitation associated with a boiling liquid.

Let us consider this "bubble phenomenon" in more detail. As the heating of a liquid begins, the first small bubbles that form on the bottom and sides of the container are bubbles of dissolved air (oxygen and nitrogen), which have been driven out of solution by the rising liquid temperature. (The solubilities of oxygen and nitrogen in liquids decrease with increasing temperature.)

As the liquid is heated further, larger bubbles form and begin to rise. These bubbles, usually also formed on the bottom of the container (where the heat is being applied and the liquid is hottest), are vapor bubbles rather than air bubbles. Initially, these vapor bubbles "disappear" as they rise, never reaching the liquid surface. Their disappearance is related to vapor pressure. In the hotter lower portions of the liquid, the liquid's vapor pressure is high enough to sustain bubble formation. [For a bubble to exist, the pressure within it (vapor pressure) must equal external pressure (atmospheric pressure).] In the cooler, higher portions of the liquid, where the vapor pressure is lower, the bubbles are collapsed by external pressure. Finally, with further heating, the temperature throughout the liquid becomes high enough to sustain bubble formation. Then bubbles rise all the way to the surface and escape (see Figure 11.14). At this point we say the liquid is boiling.

Like evaporation, boiling is actually a cooling process. When heat is taken away from a boiling liquid, boiling ceases almost immediately. It is the highest-energy molecules that are escaping. Quickly the temperature of the remaining molecules drops below the *boiling point* of the liquid.

> **Anyone who has let a pan of water boil dry knows boiling is an endothermic process. As long as the water is boiling, the pan stays at a "cool" 100°C. Once all the water boils away, the pan turns red hot.**

Bubble formation associated with a liquid that is boiling.

Factors that Affect Boiling Point

A **boiling point** *is the temperature of a liquid at which the vapor pressure of the liquid becomes equal to the external (atmospheric) pressure exerted on the liquid.* Since atmospheric

Table 11.6	Variation of the Boiling Point of Water with Elevation	
Location	**Elevation (ft above sea level)**	**Boiling Point of Water (°C)**
San Francisco, CA	0	100.0
Salt Lake City, UT	4,390	95.6
Denver, CO	5,280	95.0
La Paz, Bolivia	12,795	91.4
Mount Everest	28,028	76.5

pressure fluctuates from day to day, so does the boiling point of a liquid. To compare the boiling points of different liquids the external pressure must be the same. The boiling point of a liquid most often used for comparison and tabulation purposes (in reference books, for example) is the *normal boiling point*. A **normal boiling point** is *the temperature of a liquid at which the liquid boils under a pressure of 760 mm Hg*.

At any given location the changes in the boiling point of liquids due to *natural variation* in atmospheric pressure seldom exceed a few degrees; in the case of water the maximum is about 2°C. However, variations in boiling points *between* locations at different elevations can be quite striking, as shown by the data in Table 11.6.

The boiling point of a liquid can be increased by increasing the external pressure. Use is made of this principle in the operation of a pressure cooker. Foods cook faster in pressure cookers because the elevated pressure causes water to boil above 100°C. An increase in temperature of only 10°C will cause food to cook in approximately half the normal time. (Cooking involves chemical reactions, and the rate of a chemical reaction generally doubles with every 10°C increase in temperature.) Table 11.7 gives the boiling temperatures reached by water in normal household pressure cookers. Hospitals use the same principle in sterilizing instruments and laundry in autoclaves; sufficiently high temperatures are reached to destroy bacteria.

Liquids that have high normal boiling points or that undergo undesirable chemical reactions at boiling temperatures can be made to boil at low temperatures by reducing the external pressure. This principle is used in the preparation of numerous food products including frozen fruit juice concentrates. At a reduced pressure some of the water in a fruit juice is boiled away, concentrating the juice without having to heat it to a high temperature. Heating to a high temperature would cause changes that would spoil the taste of the juice and/or reduce its nutritional value.

The converse of the pressure cooker "phenomenon" is that food cooks slower at reduced pressures. The pressure reduction associated with higher altitudes means that food cooked over a campfire in the mountains requires longer cooking times.

Table 11.7	Boiling Point of Water in a Pressure Cooker	
Pressure Above Atmospheric		
lb/in.²	**mm Hg**	**Boiling Point of Water (°C)**
5	259	108
10	517	116
15	776	121

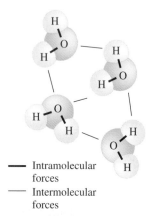

—— Intramolecular
forces
—— Intermolecular
forces

Figure 11.15

Intermolecular forces
exist *between* molecules.
Intramolecular forces exist
within molecules and hold
the molecules together; they
are the chemical bonds
within the molecule. Intermolecular forces are much
weaker than intramolecular
forces.

11.16 Intermolecular Forces in Liquids

For a liquid in an open container to boil, its vapor pressure must reach atmospheric pressure. For some substances this occurs at temperatures well below zero; for example, oxygen has a boiling point of −183°C. Other substances do not boil until the temperature is much higher. Mercury, for example, has a boiling point of 357°C, which is 540°C higher than that of oxygen. An explanation for this variation involves a consideration of the nature of the *intermolecular forces* that must be overcome in order for molecules (or atoms) to escape from the liquid state to the vapor state. An **intermolecular force** *is an attractive force that acts between a molecule and another molecule*. [Strictly speaking, the term *intermolecular force* refers only to the attraction between *molecules*, but in practice it is used to refer to interactions in liquids that involve all types of particles (molecules, ions, and atoms).]

Intermolecular forces are similar in one way to the previously discussed *intra*molecular forces (*within* molecules) involved in covalent bonding (Sec. 7.10). They are electrostatic in origin; that is, they involve positive-negative interactions (Sec. 11.3). A major difference between inter- and intramolecular forces is their magnitude; the former are much weaker. However, intermolecular forces, despite their relative weakness, are sufficiently strong to influence the behavior of liquids, often in a very dramatic way. Figure 11.15 pictorially contrasts *intermolecular* forces and *intramolecular* forces for water molecules.

Five types of intermolecular forces that can be present in a liquid are considered in this section. They are (1) dipole–dipole interactions, (2) hydrogen bonds, (3) London forces, (4) ion–dipole interactions, and (5) ion–ion interactions.

Dipole–Dipole Interactions

A **dipole–dipole interaction** *is an intermolecular attractive force that occurs between polar molecules*. Polar molecules, it should be recalled, are electrically unsymmetrical (Sec. 7.20), that is, they have a positive end and a negative end. This uneven charge distribution in a polar molecule is called a *dipole;* the molecule has two "poles" or ends, one pole being more negative than the other. Hence, the terminology *dipole–dipole interactions* for interactions between polar molecules.

When polar molecules approach each other, they tend to line up so that the relatively positive end of one molecule is directed toward the relatively negative end of the other molecule. As a result, there is an electrostatic attraction between the molecules. The greater the polarity of the molecules, the greater the strength of the dipole–dipole interaction. Figure 11.16 shows the many dipole–dipole interactions possible for a random arrangement of polar ClF molecules.

Hydrogen Bonds

Unusually strong dipole–dipole interactions are found among hydrogen-containing molecules in which hydrogen is covalently bonded to a highly electronegative element of small

Figure 11.16

Dipole–dipole interactions between
randomly arranged ClF molecules.

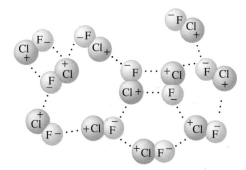

atomic size (fluorine, oxygen, and nitrogen). Two factors account for the extra strength of these dipole–dipole interactions.

1. The highly electronegative element to which hydrogen is covalently bonded dominates the electron-sharing process to such a degree that the hydrogen atom is left with significant partial positive charge (Sec. 7.19).

$$\overset{\delta^+}{H}\!-\!\overset{\delta^-}{F} \qquad \overset{\delta^+}{H}\!-\!\overset{\delta^-}{O} \qquad \overset{\delta^+}{H}\!-\!\overset{\delta^-}{N}$$

The hydrogen atom is essentially a "bare" nucleus, because it has no electrons besides the one attracted to the electronegative element—a unique property of hydrogen.

2. The small size of the hydrogen atom allows the "bare" nucleus to approach closely and be strongly attracted to an unshared pair of electrons on the electronegative atom of another molecule.

Dipole–dipole interactions of the type we are now considering are given a special name, *hydrogen bonds*. A **hydrogen bond** *is an extra strong dipole–dipole interaction involving a hydrogen atom covalently bonded to a small, very electronegative atom (F, O, or N) and an unshared pair of electrons on another small, very electronegative atom (F, O, or N).*

Water is the most commonly encountered substance wherein hydrogen bonding is significant. Figure 11.17 depicts the process of hydrogen bonding among water molecules. Note that each oxygen atom in water can participate in two hydrogen bonds—one involving each of its nonbonding electron pairs.

The two molecules that participate in a hydrogen bond need not be identical. Hydrogen bond formation is possible whenever two molecules, the same or different, have the following characteristics.

1. One molecule has a hydrogen atom attached by a covalent bond to an atom of nitrogen, oxygen, or fluorine.

2. The other molecule has a nitrogen, oxygen, or fluorine atom present that possesses one or more nonbonding electron pairs.

Figure 11.18 shows additional examples of hydrogen bonding between simple molecules.

> The three elements that have significant hydrogen bonding ability are fluorine, oxygen, and nitrogen. They are all very electronegative elements of small atomic size. Chlorine has the same electronegativity as nitrogen, but its larger atomic size causes it to have little hydrogen bonding ability.

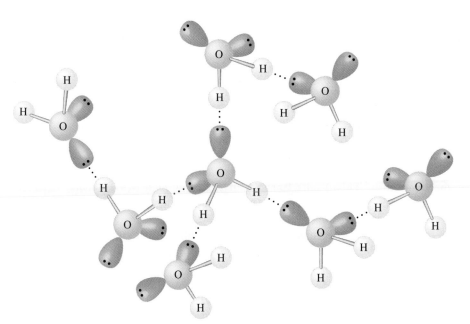

Figure 11.17

Depiction of hydrogen bonding among water molecules. The dotted lines are the hydrogen bonds.

It is not an overstatement to say that hydrogen bonding makes life possible. Were it not for hydrogen bonding, water, the most abundant compound in the human body, would be a gas at room temperature. Life as we know it could not exist under such a condition. Also, many other molecules of biological importance, such as DNA and proteins, contain O—H and N—H bonds, and hydrogen bonding plays a role in the behavior of these substances. Certain bonds in such compounds must be capable of breaking and reforming with relative ease. Only hydrogen bonds have just the right strengths to permit this.

EXAMPLE 11.10

Predicting Whether Molecules Can Participate in Hydrogen Bonding

Indicate whether hydrogen bonding is possible between two molecules of each of the following substances.

(a) Nitrogen trifluoride **(b)** Ethyl alcohol **(c)** Formaldehyde

SOLUTION

(a) Hydrogen bonding cannot occur. No hydrogen atoms are present.

(b) Hydrogen bonding should occur, because we have an O—H bond and an oxygen atom with unshared electron pairs.

(c) Hydrogen bonding should not occur. We have an oxygen atom with unshared electron pairs, but no N—H, O—H, or F—H bond is present.

Practice Exercise 11.10

Indicate whether hydrogen bonding is possible between two molecules of each of the following substances.

(a) Ethyl amine **(b)** Methyl alcohol **(c)** Diethyl ether

The force generated from the expansion of water upon freezing can break the water's container. Antifreeze is added to car radiators in the winter to prevent the water present from freezing and cracking the engine block; sufficient force is generated from water expansion upon freezing to burst even iron and copper parts. Also, the weathering of rocks and concrete and the formation of potholes in the streets are hastened by the expansion of water upon freezing.

Consequences of Hydrogen Bonding

The vapor pressures (Sec. 11.14) of liquids that have significant hydrogen bonding are much lower than those of similar liquids wherein little or no hydrogen bonding occurs. This is because the presence of hydrogen bonds makes it more difficult for molecules to escape from the condensed state; additional energy is needed to overcome the hydrogen bonds. For this reason, boiling points are much higher for liquids in which hydrogen bonding occurs. The effect that hydrogen bonding has on boiling point can be seen by comparing water's boiling point with those of other hydrogen compounds of group VIA elements—H_2S, H_2Se, and H_2Te (see Fig. 11.19). Water is the only compound in the series where significant hydrogen bonding occurs.

Water is one of the few substances known in which the solid form (ice) is less dense than the liquid form. This results from water expanding as it freezes. This "expansion phenomenon" is directly related to hydrogen bonding. In the solid state, hydrogen bonds, which must have specific orientations, that is, have directional nature, cause the water molecules to be farther apart than they are in the liquid state. (In the liquid state hydrogen bonds are continually being formed and broken; in the solid state, hydrogen bonds are "permanent.")

Because ice is less dense than liquid water, lakes freeze from top to bottom and not vice versa. If this "density inversion" for water did not exist (no hydrogen bonding), surface ice would sink to the bottom of a lake as it was formed, and water bodies in colder climates

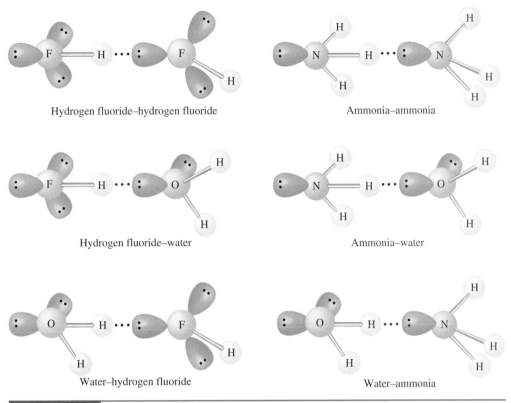

Hydrogen fluoride–hydrogen fluoride

Ammonia–ammonia

Hydrogen fluoride–water

Ammonia–water

Water–hydrogen fluoride

Water–ammonia

Figure 11.18

Depiction of hydrogen bonding between various simple molecules. The dotted lines are the hydrogen bonds.

would freeze solid in winter. This would be a tremendous problem for fish; they would have to evolve ice picks.

London Forces

Named after the German physicist Fritz London (1900–1954), who first postulated their existence, *London forces* are the weakest type of intermolecular force. A **London force** *is a weak temporary dipole–dipole interaction that occurs between an atom or molecule (polar or nonpolar) and another atom or molecule (polar or nonpolar).* The origin of London forces is more difficult to visualize than that of dipole–dipole interactions.

London forces result from momentary (temporary) uneven electron distributions in molecules. Most of the time electrons can be visualized as being distributed in a molecule in a definite pattern determined by their energies and the electronegativities of the atoms. However, there is a small statistical chance (probability) that the electrons will deviate from their normal pattern. For example, in the case of a nonpolar diatomic molecule, more electron density may temporarily be located on one side of the molecule than on the other. This condition causes the molecule to become polar for an instant. The negative side of this instantaneously polar molecule will tend to repel electrons of adjoining molecules and cause these molecules to also become polar (an *induced polarity*). The original (statistical) polar molecule and all the molecules with induced polarity are then attracted to each other. This happens many, many times per second throughout the liquid, resulting in a net attractive force. Figure 11.20 depicts the situation present when London forces exist.

When plant (or human) cells freeze, the ice crystals formed cause cell membranes to rupture, killing the cells. The slower the freezing process, the larger the ice crystals that form. Frozen-food manufacturers freeze foods so rapidly that the ice crystals formed are very small; this minimizes the damage to the cellular structure of the food. Food processed in this manner is said to have been "flash-frozen."

Figure 11.19

Boiling points of the hydrogen compounds of group VIA elements. Water is the only one of the four compounds in which significant hydrogen bonding occurs.

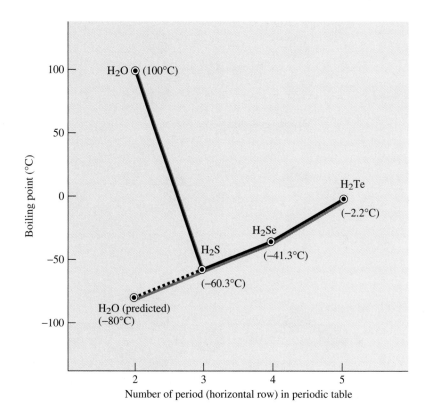

Figure 11.20

A London force. The instantaneous (temporary) polarity present in atom A results in induced polarity in atom B.

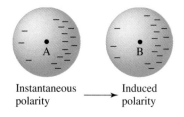

Instantaneous polarity $\longrightarrow$ Induced polarity

As an analogy for London forces, consider what happens when a bucket filled with water is moved. The water will "slosh" from side to side. This is similar to the movement of electrons. The "sloshing" from side to side is instantaneous; a given "slosh" quickly disappears. Uneven electron distribution is likewise an instantaneous situation.

The strength of London forces depends on the ease with which an electron distribution in a molecule can be distorted (polarized) by the polarity present in another molecule. In large-diameter molecules the outermost electrons are necessarily located farther from the nucleus than are the outermost electrons in small molecules. The farther electrons are located from the nucleus the weaker the attractive forces that act on them, the more freedom they have, and the more susceptible they are to polarization. This leads to the observation that for *related* molecules boiling points increase with molecular mass, which usually parallels size. This trend is reflected in the boiling points given in Table 11.8 for two series of related substances: the noble gases and the elements of group VIIA.

If two molecules have approximately the same molecular mass and polarity but are of different diameter, the smaller one will have the lower boiling point, since its electrons are less susceptible to polarization. The nonpolar SF_6 and $C_{10}H_{22}$ molecules (molecular masses 146 and 142 amu, respectively) reflect this trend with boiling points of -64 and $174°C$, respectively.

Of the three types of intermolecular forces discussed so far (dipole–dipole interactions, hydrogen bonds, and London forces), the London forces are the most common and, in the majority of cases, the most prevalent. They are the only attractive forces present between *nonpolar* molecules, and it is estimated that London forces contribute 85% of the total intermolecular force in the *polar* HCl molecule. Only where hydrogen bonding is involved do

Table 11.8	Boiling Point Trends for Related Series of Nonpolar Molecules	
Substance	Molecular Mass (amu)	Boiling Point (°C)
Noble Gases		
He	4.0	−269
Ne	20.2	−246
Ar	39.9	−186
Kr	83.8	−153
Xe	131.3	−107
Rn	222	−62
Group VIIA Elements		
F_2	39.0	−187
Cl_2	70.9	−35
Br_2	159.8	+59
I_2	253.8	+184

London forces play a minor role. In water, for example, about 80% of the intermolecular attraction is attributed to hydrogen bonding and only 20% to London forces.

Ion–Dipole Interactions

An **ion–dipole interaction** *is an intermolecular attractive force between an ion and a polar molecule.* Polar molecules are dipoles; they have a positive end and a negative end. A positive ion (cation) can attract the negative end of dipoles and a negative ion (anion) can attract the positive end of dipoles, as shown in Figure 11.21. Ion–dipole interactions are an important consideration in aqueous solutions in which dissolved ionic compounds are present, a topic to be considered in Chapter 13.

Ion–Ion Interactions

Our discussion of intermolecular forces to this point has assumed that the particles making up a liquid are primarily molecules or atoms. This is a valid assumption for liquids encountered at normal temperatures. However, at extremely high temperatures, liquids exist in which no molecules are present. Liquids obtained by melting ionic compounds, which always requires an extremely high temperature, contain only ions. The attractive forces in such liquids are those that result from *ion–ion interactions* among positive and negative ions.

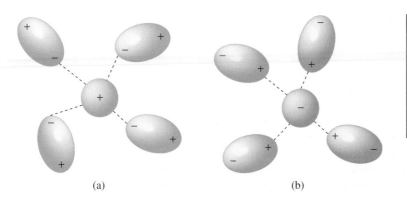

(a) (b)

Figure 11.21

Ion-dipole interactions. (a) The negative ends of polar molecules are attracted by a positive ion, and (b) the positive ends of polar molecules are attracted by a negative ion. Such attractions are important when ionic compounds dissolve in water since water is a polar molecule.

An **ion–ion interaction** *is an intermolecular attractive force between oppositely charged ions present in liquid state (molten) ionic compounds.*

11.17   Types of Solids

Solids may be classified into two categories: *crystalline solids* and *amorphous solids*. Most solids are crystalline. A **crystalline solid** *is a solid that is characterized by a regular three-dimensional arrangement of the atoms, ions, or molecules present.* This regular arrangement of particles is manifested in the outward appearance of the solid; crystalline features are discernible. These features are often very obvious; at other times, however, the crystals may be so tiny as to be visible only with the aid of a microscope. Many times the crystals are imperfect, making it difficult to discern them, but they are there.

An **amorphous solid** *is a solid that is characterized by a random, nonrepetitive three-dimensional arrangement of the atoms, ions, or molecules present.* The word *amorphous* comes from the Greek *amorphos* meaning "without form." Examples of amorphous solids include glass, tar, rubber, and many plastics. In some textbooks amorphous solids are referred to as *supercooled liquids,* since they have the molecular disorder associated with liquids. However, they differ from liquids and resemble solids in being rigid and immobile.

Most amorphous solids consist of very long chainlike molecules that are randomly intertwined. To form a crystalline material from such a solid, the intertwined molecules would have to become regularly arranged in the melted state. When such a liquid is cooled to a solid, molecular motion decreases to the point where "solidification" takes place before the untwining can occur, resulting in a disordered structure.

The remainder of this section deals with the more commonly encountered crystalline solid. Indeed, in this text the word *solid* without a modifier will always mean a *crystalline solid.*

The highly ordered pattern of particles found in a crystalline solid is called a *crystal lattice.* The positions occupied by the particles in the lattice are called *crystal lattice sites.* Crystalline solids may be classified into five groups based on the type of particles at the crystal lattice sites and the forces that hold the particles together. A consideration of these classifications provides some insight into the reasons some solids have very low and others very high melting points and some solids are more volatile than others. The five classes of crystalline solids are ionic, polar molecular, nonpolar molecular, covalent network, and metallic.

An **ionic solid** *is a solid that consists of positive and negative ions arranged in such a way that each ion is surrounded by nearest neighbors of the opposite charge.* The structure of such solids was previously discussed in Section 7.8. Any given ion is bonded by electrostatic attractions to all the ions of opposite charge immediately surrounding it. Since these interionic attractions are relatively strong, ionic solids have high melting points and negligible vapor pressures. Figure 11.22 shows a two-dimensional cross section and also a three-dimensional view of the arrangement of ions for the compound NaCl.

A **polar molecular solid** *is a solid that has polar molecules at the crystal lattice sites.* The molecules are oriented so that the positive end of each molecule is near the negative end of an adjacent molecule in the lattice. Dipole–dipole interactions and London forces (Sec. 11.16) hold the molecules in the lattice positions. Since these forces are relatively weak compared to other types of electrostatic forces (ionic and covalent bonds), the resulting solids have moderate melting points and moderate vapor pressures.

A **nonpolar molecular solid** *is a solid that has nonpolar molecules (or atoms in the case of the noble gases) at the crystal lattice sites.* The lattice site particles are held together

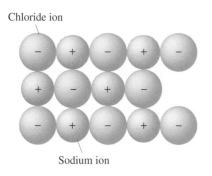

Chloride ion

Sodium ion

Chloride ion

Sodium ion

Figure 11.22

Two-dimensional cross section and three-dimensional view of the ion arrangement in the ionic compound sodium chloride (NaCl).

solely by London forces, the weakest of electrostatic forces. Consequently, these solids have low melting points and are volatile.

A **covalent network solid** *is a solid that has atoms at the crystal lattice sites that are linked together by covalent bonds into a vast three-dimensional array.* A crystal of such a solid can be visualized as a gigantic molecule. Such a molecule, for the form of carbon called diamond, is shown in Figure 11.23. Each lattice site in diamond is occupied by a carbon atom. Covalent network solids are not common, but they are noteworthy because of their extremely high melting points and their low volatility. Covalent bonds must be broken to cause physical change in such solids. The atoms at the lattice sites in a covalent network solid need not all be the same kind, as was the case for diamond. Examples of covalent network solids where different kinds of atoms are found at the lattice sites include quartz (SiO_2, the main component of sand) and silicon carbide (SiC, an abrasive used in sandpaper and other similar materials).

A **metallic solid** *has metal atoms occupying the crystal lattice sites.* The nature of bonding in such solids is not completely understood. However, the solids are often considered to have a lattice of metal ions, with the outer electrons of each metal atom free to move about in the lattice. The bonding results from the interaction of the electrons with the various nuclei. The movement of the free electrons also accounts for the electrical conductivity of metals. The melting points of metallic solids show a wide range. Their volatility is generally low.

Table 11.9 gives a summary of the various types of crystalline solids and their distinguishing characteristics along with examples of each type.

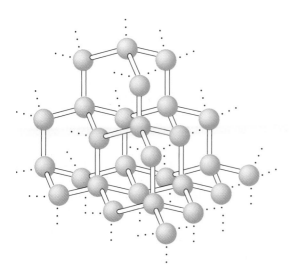

Figure 11.23

The diamond crystal lattice, which is a covalent network solid. Each carbon atom is tetrahedrally bonded to four other carbon atoms.

Table 11.9	Characteristics and Examples of the Various Types of Crystalline Solids			
Type	**Particles Occupying Lattice Sites**	**Forces Between Particles**	**Properties of Solids**	**Examples**
Ionic	positive and negative ions	electrostatic attraction between oppositely charged ions	high melting points; nonvolatile	$NaCl$ $MgSO_4$ KBr
Polar molecular	polar molecules	dipole–dipole attractions and London forces	moderate melting points; moderate volatility	H_2O NH_3 SO_2
Nonpolar molecular	nonpolar molecules (or atoms)	London forces	low melting points; volatile	CO_2 CH_4 I_2 O_2 Ar
Covalent network	atoms	covalent bonds between atoms	extremely high melting points; nonvolatile	C (diamond) SiO_2 (sand) SiC
Metallic	metal atoms	attraction between outer electrons and positive atomic centers	variable melting points; low volatility	Cu Ag Au Fe Al

Summary

1. **Physical State Differences** The physical state of a substance is determined by (1) its chemical identity, (2) its temperature, and (3) the pressure to which it is subjected. Distinguishing properties of the physical states include (1) definiteness of volume and shape, (2) relative density, (3) relative compressibility, and (4) relative thermal expansion.

2. **Kinetic Molecular Theory of Matter** The kinetic molecular theory of matter is a set of five statements that explains the physical behavior of the three states of matter. Two premises of this theory are that the particles (atoms, molecules, or ions) present in a substance are (1) in constant motion and therefore possess kinetic energy and (2) interact with each other through attractions and repulsions and therefore possess potential energy.

3. **Kinetic and Potential Energy** Kinetic energy is energy that matter possesses because of the movement (motion) of its particles. Potential energy is energy that matter possesses as a result of its position, condition, and/or composition. The solid state is characterized by a dominance of potential energy over kinetic energy. The liquid state is characterized by potential energy and kinetic energy of about the same magnitude. The gaseous state is characterized by a complete dominance of kinetic energy over potential energy.

4. **Endothermic and Exothermic Changes of State** Endothermic changes of state, which require the input of heat energy, are melting, sublimation, and evaporation. Exothermic changes of state, which require the release of heat energy, are deposition, condensation, and freezing.

5. **Specific Heat and Heat Capacity** The most commonly used unit of measurement for heat energy is the joule. The specific heat of a substance is the amount of heat energy that is necessary to raise the temperature of 1 gram of the substance in a specific physical state by 1°C. Heat capacity is the amount of heat energy needed to raise the temperature of a given quantity of a substance in a specific physical state by 1°C.

6. **Temperature Change as a Substance Is Heated** As a solid is heated, its temperature rises until the melting point is reached, and then it remains constant during the change of state from solid to liquid. Heating the resultant liquid causes the temperature to rise again until the boiling point is reached. The temperature then remains constant during

the change of state from liquid to gas. Heating the resultant gas again causes the temperature to rise.

7. **Heat Energy Calculations** Important physical quantities when "bookkeeping" on heat energy added or removed from a chemical system are (1) heat of fusion/solidification, (2) heat of vaporization/condensation, and (3) specific heat for each of the three physical states.

8. **Evaporation** The evaporation rate for a liquid depends on the liquid's temperature and on its surface area. The evaporative behavior of a liquid in a closed container is dependent on both the rate of evaporation and rate of condensation of the liquid. With time, a steady-state situation (an equilibrium state) is reached in which the rates of these two opposite processes become equal.

9. **Vapor Pressure** The pressure exerted by a vapor in equilibrium with its liquid is the vapor pressure of the liquid. Vapor pressure increases as liquid temperature increases.

10. **Boiling and Boiling Points** Boiling is a special form of evaporation in which bubbles of vapor form within the liquid and rise to the surface. The boiling point of a liquid is the temperature at which the vapor pressure of the liquid becomes equal to the external (atmospheric) pressure exerted on the liquid. The boiling point of a liquid increases or decreases as the prevailing atmospheric pressure increases or decreases.

11. **Intermolecular Forces** Intermolecular forces are the attractive interactions between the particles (molecules, ions, atoms) present in a chemical system. In the liquid state, five different types of intermolecular forces can be present, depending on the composition of the liquid: (1) dipole–dipole interactions, (2) hydrogen bonds, (3) London forces, (4) ion–dipole interactions, and (5) ion–ion interactions.

12. **Types of Solids** Solids are classified into two types: crystalline and amorphous. Crystalline solids have a regular three-dimensional arrangement of the atoms, ions, or molecules present. Amorphous solids have a random, nonrepetitive three-dimensional arrangement of the particles present. Crystalline solids are further classified, based on the type of particle present at the lattice sites of the solid, as (1) ionic, (2) polar molecular, (3) nonpolar molecular, (4) covalent network, and (5) metallic.

Key Terms

The new terms defined in this chapter are

amorphous solid   *Sec. 11.17*

boiling   *Sec. 11.15*

boiling point   *Sec. 11.15*

change of state   *Sec. 11.8*

compressibility   *Sec. 11.2*

covalent network solid   *Sec. 11.17*

crystalline solid   *Sec. 11.17*

dipole–dipole interaction   *Sec. 11.16*

electrostatic force   *Sec. 11.3*

endothermic change of state   *Sec. 11.8*

equilibrium state   *Sec. 11.13*

evaporation   *Sec. 11.13*

exothermic change of state   *Sec. 11.8*

gas   *Sec. 11.6*

heat capacity   *Sec. 11.9*

heat of condensation   *Sec. 11.11*

heat of fusion   *Sec. 11.11*

heat of solidification   *Sec. 11.11*

heat of vaporization   *Sec. 11.11*

hydrogen bond   *Sec. 11.16*

intermolecular force   *Sec. 11.16*

ion–dipole interaction   *Sec. 11.16*

ion–ion interaction   *Sec. 11.16*

ionic solid   *Sec. 11.17*

joule   *Sec. 11.9*

kinetic energy   *Sec. 11.3*

kinetic molecular theory of matter   *Sec. 11.3*

liquid   *Sec. 11.5*

London force   *Sec. 11.16*

metallic solid   *Sec. 11.17*

nonpolar molecular solid   *Sec. 11.17*

normal boiling point   *Sec. 11.15*

polar molecular solid   *Sec. 11.17*

potential energy   *Sec. 11.3*

solid   *Sec. 11.4*

specific heat   *Sec. 11.9*

thermal expansion   *Sec. 11.2*

vapor   *Sec. 11.13*

vapor pressure   *Sec. 11.14*

volatile substance   *Sec. 11.14*

Intermolecular Forces in Liquids (Sec. 11.16)

11.75 Describe the molecular conditions necessary for the existence of a dipole–dipole interaction.

11.76 Describe the molecular conditions necessary for the existence of a London force.

11.77 In liquids, what is the relationship between boiling point and the strength of intermolecular forces?

11.78 In liquids, what is the relationship between vapor pressure magnitude and the strength of intermolecular forces?

11.79 For liquid state samples of the following diatomic substances, classify the dominant intermolecular forces present as London forces, dipole–dipole interactions, or hydrogen bonds.

(a) H_2 **(b)** HF   **(c)** CO   **(d)** F_2

11.80 For liquid state samples of the following diatomic substances, classify the dominant intermolecular forces present as London forces, dipole–dipole interactions, or hydrogen bonds.

(a) O_2 **(b)** HCl   **(c)** Cl_2 **(d)** BrCl

11.81 In which of the following substances, in the pure liquid state, would hydrogen bonding occur?

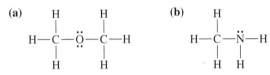

11.82 In which of the following substances, in the pure liquid state, would hydrogen bonding occur?

(a)

H H
| |
H—C—C—H
| |
H H

(b)

:C̈l—N̈—H
|
H

(c) H—N̈—N̈—H
| |
H H

(d)

H Ö:
| ||
H—C—C—H
|
H

11.83 In each of the following pairs of molecules, predict which member of the pair would be expected to have the higher boiling point and indicate why.

(a) F_2 and Cl_2 **(b)** HF and HBr
(c) N_2 and NO   **(d)** C_2H_6 and O_2

11.84 In each of the following pairs of molecules, predict which member of the pair would be expected to have the higher boiling point and indicate why.

(a) Cl_2 and Br_2 **(b)** H_2O and H_2S
(c) O_2 and CO   **(d)** C_3H_8 and CO_2

11.85 In each of the following pairs of molecules, predict which member of the pair would be expected to have the greatest polarizability.

(a) CH_4 and SiH_4 **(b)** SiH_4 and $SiCl_4$
(c) $SiCl_4$ and $GeBr_4$ **(d)** N_2 and C_2H_4

11.86 In each of the following pairs of molecules, predict which member of the pair would be expected to have the greatest polarizability.

(a) SiH_4 and GeH_4 **(b)** $GeCl_4$ and $GeBr_4$
(c) F_2 and Cl_2 **(d)** CO_2 and C_3H_8

Types of Solids (Sec. 11.17)

11.87 Indicate whether each of the following statements about various types of crystalline solids is *true* or *false*.

(a) In covalent network solids, covalent bonds link the particles present at the lattice sites.

(b) Nonpolar molecular solids have moderately high melting points.

(c) Both ionic and covalent network solids are nonvolatile substances.

(d) Polar molecular, nonpolar molecular, and covalent network solids all have molecules present at the lattice sites.

11.88 Indicate whether each of the following statements about various types of crystalline solids is *true* or *false*.

(a) Metallic solids have metal atoms occupying the lattice sites.

(b) Nonpolar molecular solids usually have high volatility.

(c) Ionic molecular solids generally have low melting points because of the presence of ions at the lattice sites.

(d) In polar molecular solids, dipole–dipole interactions are important forces between lattice site particles.

11.89 For each of the following pairs of substances, predict which member of the pair will have the higher melting point and indicate why.

(a) NaCl and Cl_2 **(b)** SiO_2 and CO_2
(c) Cu and CO   **(d)** NaF and MgO

11.90 For each of the following pairs of substances, predict which member of the pair will have the higher melting point and indicate why.

(a) HCl and HBr   **(b)** Ar and NO
(c) Na_2O and SiC   **(d)** CaO and AlN

Additional Problems

11.91 The vapor pressure of $SnCl_4$ reaches 400 mm Hg at 92°C. The vapor pressure of SnI_4 reaches 400 mm Hg at 315°C.

(a) At 100°C, which substance should evaporate at the faster rate?

(b) Which substance should have the higher normal boiling point?

(c) Which substance should have the stronger intermolecular forces?

(d) At 80°C, which substance should have the lower vapor pressure?

11.92 The vapor pressure of PBr_3 reaches 400 mm Hg at 150°C. The vapor pressure of PCl_3 reaches 400 mm Hg at 57°C.

(a) At 100°C, which substance should evaporate at the faster rate?

(b) Which substance should have the lower normal boiling point?

(c) Which substance should have the weaker intermolecular forces?

(d) At 50°C, which substance should have the higher vapor pressure?

11.93 If heat is supplied at an identical, constant rate in each case, which would take longer, heating 50.0 g of water from 0.0 to 80.0°C or vaporizing 50.0 g of water at 100.0°C?

11.94 If heat is supplied at an identical, constant rate in each case, which would take longer, heating 80.0 g of water from 20.0 to 100.0°C or melting 80.0 g of ice at 0.0°C?

11.95 A quantity of ice at 0.0°C was added to 40.0 g of water at 19.0°C in an insulated container. All of the ice melted, and the water temperature decreased to 0.0°C. How many grams of ice were added?

11.96 A quantity of ice at 0.0°C was added to 55.0 g of water at 25.3°C in an insulated container. All of the ice melted, and the water temperature decreased to 0.0°C. How many grams of ice were added?

11.97 If 10.0 g of ice at −10.0°C and 30.0 g of liquid water at 80.0°C are mixed in an insulated container, what will the final temperature of the liquid be?

11.98 If 10.0 g of ice at −20.0°C and 60.0 g of liquid water at 70.0°C are mixed in an insulated container, what will the final temperature of the liquid be?

11.99 A 500.0-g piece of metal at 50.0°C is placed in 100.0 g of water at 10.0°C in an insulated container. The metal and water come to the same temperature of 23.4°C. What is the specific heat, in joules per gram per degree Celsius, of the metal?

11.100 A 100.0-g piece of metal at 80.0°C is placed in 200.0 g of water at 20.0°C in an insulated container. The metal and water come to the same temperature of 33.6°C. What is the specific heat, in joules per gram per degree Celsius, of the metal?

11.101 Heat is added to a 25.3-g sample of a metal at the constant rate of 136 J/sec. After the sample reaches the metal's melting point temperature, the temperature remains constant for 4.3 min. Calculate the heat of fusion for this metal in joules per gram.

11.102 Heat is added to a 35.2-g sample of a metal at the constant rate of 189 J/sec. After the sample reaches the metal's melting point temperature, the temperature remains constant for 5.2 min. Calculate the heat of fusion for this metal in joules per gram.

Cumulative Problems

11.103 The specific heat of the metal silver is 0.24 J/(g·°C). How many joules of energy are needed to raise the temperature of a block of silver containing 2.50 moles of silver atoms by 10.0°C?

11.104 The specific heat of the metal gold is 0.13 J/(g·°C). How many joules of energy are needed to raise the temperature of a block of gold containing 1.40 moles of gold atoms by 15.0°C?

11.105 The human body contains approximately 5.7 L of blood. Assuming that the density of blood is 1.06 g/mL and that the specific heats of blood and water are the same, how many kilojoules of energy are required to raise the temperature of this amount of blood by 1.0°C?

11.106 Assuming that the density of blood is 1.06 g/mL and that the specific heats of blood and water are the same, how many liters of blood are present in a human body if 36.0 kJ of energy raises the temperature of the blood present by 1.5°C?

11.107 The specific heat of substance A is 0.88 J/(g·°C) and that of substance B is 2.1 J/(g·°C). You are given an unknown that could be pure substance A, pure substance B, or a homogeneous mixture of A and B. In the laboratory you determine that it requires 59.5 J of heat energy to raise the temperature of a 35.0-g sample of the unknown by 1.0°C. What conclusions can you make about the identity of your unknown from these data?

11.108 The specific heat of substance C is 0.93 J/(g·°C) and that of substance D is 1.8 J/(g·°C). You are given an unknown that could be pure substance C, pure substance D, or a homogeneous mixture of C and D. In the laboratory you determine that it requires 23.3 J of heat energy to raise the temperature of a 25.0-g sample of the unknown by 1.0°C. What conclusions can you make about the identity of your unknown from these data?

11.109 To solidify a 10.0-g sample of a liquid at its freezing point, 1485 J of heat must be removed. To solidify a 10.0-mL sample of this same liquid at its freezing point, 1151 J of heat must be removed. What is the density, in grams per milliliter, of this liquid?

11.110 To solidify a 10.0-mL sample of a liquid at its freezing point, 1063 J of heat must be removed. To solidify a 15.0-g sample of this same liquid at its freezing point, 1276 J of heat must be removed. What is the density, in grams per milliliter, of this liquid?

11.111 The heat of vaporization for a substance is 18.1 kJ/mole. Determine the molecular mass of this substance given that 1.00 g, at its boiling point, releases 602 J of heat as it condenses.

11.112 The heat of vaporization for a substance is 36.6 kJ/mole. Determine the molecular mass of this substance given that 1.00 g, at its boiling point, releases 314 J of heat as it condenses.

11.113 How many kilojoules of heat energy are needed to melt completely at its melting point a pure sample of copper that was obtained from the complete decomposition of 52.0 g of Cu_2O?

11.114 How many kilojoules of heat energy are needed to melt completely at its melting point a pure sample of aluminum that was obtained from the complete decomposition of 75.0 g of Al_2O_3?

11.115 How many joules of heat energy are needed to heat a sample of liquid benzene (C_6H_6) from a temperature 10°C below its boiling point to its boiling point, given that the sample contains 6.32×10^{24} hydrogen atoms and that the specific heat of liquid benzene is 1.74 J/(g·°C)?

11.116 How many joules of heat energy must be removed from a sample of liquid benzene (C_6H_6) to lower its temperature from 10°C above its melting point to its melting point, given that the sample contains 4.03×10^{23} carbon atoms and the specific heat of liquid benzene is 1.74 J/(g · °C)?

11.117 A nuclear power plant produces 3.3×10^6 kJ of waste heat energy every second. If this heat is to be dissipated by evaporating 25°C water (in cooling towers), how many gallons of water per day would be needed? The density of water is 0.997 g/mL at 25°C. Atmospheric pressure is 1.00 atm. The heat of vaporization of water at 25°C is 2420 J/g.

11.118 A nuclear power plant produces 3.3×10^6 kJ of waste heat energy every second. If this waste heat is dumped into a river that is 42 ft wide and has an average depth of 5.1 ft and a current of 3.0 mi/hr, what is the temperature difference between the water upstream and the water downstream? Assume that the density of water is 1.0 g/mL.

Answers to Practice Exercises

11.1 (a) 11.9 kcal   **(b)** 49.8 kJ   **(c)** 49,800 J

11.2 1130 J

11.3 64°C

11.4 0.443 J/g · °C

11.5 1700 g

11.6 11,800 J

11.7 837 J/g

11.8 91,600 J

11.9 39,600 J

11.10 (a) yes   **(b)** yes   **(c)** no

12

Gas Laws

12.1 Properties of Some Common Gases

The word *gas* is used to refer to a substance that is normally in the gaseous state at ordinary temperatures and pressures. The word *vapor* (Sec. 11.13) describes the gaseous form of any substance that is a liquid or solid at normal temperatures and pressures. Thus, we speak of oxygen gas and water vapor.

The normal state for 11 of the elements is that of a gas (Sec. 11.1). A listing of these elements along with some of their properties is given in Table 12.1. This table also lists some common compounds that are gases at room temperature and pressure.

The three most commonly encountered elemental gases—hydrogen, oxygen, and nitrogen—are colorless and odorless. So also are all of the noble gases (Sec. 6.11). From these observations one should not conclude that all gases, or even all elemental gases, are colorless and odorless. Note from Table 12.1 the pale yellow and greenish yellow colors associated with elemental fluorine and chlorine and the irritating odors of both of them. Many gaseous compounds have pungent odors.

Table 12.1 also brings to your attention again (recall Sec. 10.2) that all of the elemental gases, except for the noble gases, exist in the form of diatomic molecules (H_2, O_2, N_2, F_2, Cl_2).

Many of the gaseous compounds listed in Table 12.1 are colorless, have odors, and are toxic. Note that odor and toxicity do not have to go together. Carbon monoxide, a deadly poison, is odorless but toxic.

Table 12.1	Color, Odor, and Toxicity of Elements and Common Compounds That Are Gases at Ordinary Temperatures and Pressures	

Element		Properties
H_2	hydrogen	colorless, odorless
O_2	oxygen	colorless, odorless
N_2	nitrogen	colorless, odorless
Cl_2	chlorine	greenish yellow, choking odor, toxic
F_2	fluorine	pale yellow, pungent-odor, toxic
He	helium	colorless, odorless
Ne	neon	colorless, odorless
Ar	argon	colorless, odorless
Kr	krypton	colorless, odorless
Xe	xenon	colorless, odorless
Rn	radon	colorless, odorless

Compound		Properties
CO_2	carbon dioxide	colorless, faintly pungent odor
CO	carbon monoxide	colorless, odorless, toxic
NH_3	ammonia	colorless, pungent odor, toxic
CH_4	methane	colorless, odorless
SO_2	sulfur dioxide	colorless, pungent choking odor, toxic
H_2S	hydrogen sulfide	colorless, rotten egg odor, toxic
HCl	hydrogen chloride	colorless, choking odor, toxic
NO_2	nitrogen dioxide	reddish brown, irritating odor, toxic

12.2   Gas Law Variables

The behavior of a gas can be described reasonably well by *simple* quantitative relationships called *gas laws*. **Gas laws** *are generalizations that describe in mathematical terms the relationships among the amount, pressure, temperature, and volume of a gas.*

It is only the gaseous state that is describable by *simple* mathematical laws. Laws describing liquid and solid state behavior are mathematically more complex. Consequently, quantitative treatments of these latter behaviors will not be given in this text.

Before we discuss the mathematical form of the various gas laws, some comments concerning the major variables involved in gas law calculations—amount, volume, temperature, and pressure—are in order. Three of these four variables, amount, volume, and temperature, have been discussed previously (Secs. 9.5, 3.4, and 3.11, respectively). Amount is usually specified in terms of *moles* of gas present. The units of *liter* or *milliliter* are usually used in specifying gas volume. Only one of the three temperature scales discussed in Section 3.11, the *Kelvin* scale, can be used in gas law calculations if the results are to be valid. Therefore, you should be thoroughly familiar with the conversion of Celsius and Fahrenheit scale readings to kelvins (Sec. 3.11). We have not yet discussed *pressure*, the fourth variable. Comments concerning pressure will occupy the remainder of this section.

Pressure *is the force applied per unit area, that is, the total force on a surface divided by the area of that surface.*

A sharp knife cuts better than a dull knife. Because the pushing force acts on a smaller area in a sharp knife, the pressure is greater and the knife cuts better with less effort.

$$P \text{ (pressure)} = \frac{F \text{ (force)}}{A \text{ (area)}}$$

Note that pressure and force are not the same. Identical forces give rise to different pressures if they are acting on areas of different size. For areas of the same size, the larger the force, the greater the pressure.

Barometers, manometers, and gauges are the instruments commonly used by chemists to measure gas pressure. Barometers and manometers measure pressure in terms of the height of a column of mercury, whereas gauges are usually calibrated in terms of force per area, for example, in pounds per square inch.

The air that surrounds the Earth exerts a pressure on all objects with which it has contact. A **barometer** *is a device used for measuring atmospheric pressure.* It was invented by the Italian physicist Evangelista Torricelli (1608–1647) in 1643. The essential components of a barometer are shown in Figure 12.1. A barometer can be constructed by filling a long glass tube, sealed at one end, all the way to the top with mercury and then inverting the tube (without letting any air in) into a dish of mercury. The mercury in the tube falls until the pressure from the mass of the mercury in the tube is just balanced by the pressure of the atmosphere on the mercury in the dish. The pressure of the atmosphere is then expressed in terms of the height of the supported column of mercury.

Mercury is the liquid of choice in a barometer for two reasons: (1) it is a very dense liquid, and therefore only a short glass tube is needed; and (2) it has a very low vapor pressure, so the pressure reading does not have to be corrected for vapor pressure.

The height of the mercury column is most often expressed in millimeters or inches. Millimeters of mercury (mm Hg) are used in laboratory work. The most common use of inches of mercury (in. Hg) is in weather reporting.

The pressure of the atmosphere varies with altitude, decreasing at the rate of approximately 25 mm Hg per 1000 ft increase in altitude. It also fluctuates with weather conditions. Recall the terminology used in a weather report: high-pressure front, low-pressure front, and so on. At sea level the height of a column of mercury in a barometer fluctuates with weather conditions between about 740 and 770 mm Hg and averages about 760 mm Hg. Another pressure unit, the atmosphere (atm), is defined in terms of this average sea-level pressure.

$$1 \text{ atm} = 760 \text{ mm Hg}$$

Because of its size, 760 times larger than 1 mm Hg, the atmosphere unit is frequently used to express the high pressures encountered in many industrial processes and in some experimental work.

Blood pressure is measured with the aid of an apparatus known as a sphygmomanometer, which is essentially a barometer tube connected to an inflatable cuff by a hollow tube. A "normal" blood pressure is 120:80; this ratio means a systolic pressure of 120 mm Hg above atmospheric pressure and a diastolic pressure of 80 mm Hg above atmospheric pressure.

The popping feeling in the ears encountered driving up a steep mountainous road is caused by the decrease in pressure with altitude.

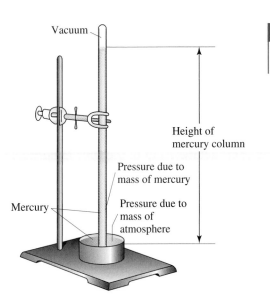

Figure 12.1

The essential components of a mercury barometer.

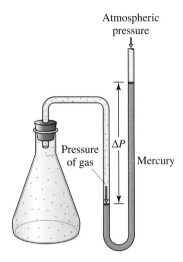

Figure 12.2

Pressure measurement by means of a manometer.

It is impractical to measure the pressure of gases other than air with a barometer. One cannot usually introduce a mercury barometer directly into a container of a gas. A **manometer** *is a device used to measure gas pressures in a laboratory.* It is a U-tube filled with mercury. One side of the U-tube is connected to the container in which the pressure is to be measured, and the other side is connected to a region of known pressure. Such an arrangement is depicted in Figure 12.2. The gas in the container exerts a pressure on the Hg on that side of the tube, while atmospheric pressure pushes on the open end. A difference in the heights of the Hg columns in the two arms of the manometer indicates a pressure difference between the gas and the atmosphere. In Figure 12.2 the gas pressure in the container exceeds atmospheric pressure by an amount, ΔP, that is equal to the difference in the heights of the Hg columns in the two arms.

$$P_{\text{container}} = P_{\text{atm}} + \Delta P$$

Pressures of contained gas samples can also be measured by special gas pressure gauges attached to their containers. Such gauges are commonly found on tanks (cylinders) of gas purchased commercially. These gauges are most often calibrated in terms of pounds per square inch (lb/in.^{2} or psi). The relationship between psi and atmospheres is

$$14.68 \text{ psi} = 1 \text{ atm}$$

Table 12.2 summarizes the relationships among the various pressure units we have discussed and also lists one additional pressure unit, the *pascal*. The pascal is the SI unit (Sec. 3.1) of pressure.

The values 1 atm and 760 mm Hg are exact numbers since they arise from a definition. Consequently, these two values have an infinite number of significant figures. All other

> Automobile tire pressure is measured in psi units. The pressure inside the tire is the pressure shown by the gauge plus atmospheric pressure.

Table 12.2	**Units of Pressure and Their Relationship to the Unit Atmosphere**	
Unit	**Relationship to Atmosphere**	**Area of Use**
Atmosphere	—	gas law calculations
Millimeters of mercury	760 mm Hg = 1 atm	gas law calculations
Inches of mercury	29.92 in. Hg = 1 atm	weather reports
Pounds per square inch	14.68 psi = 1 atm	stored or bottled gases
Pascal	1.013×10^5 Pa = 1 atm	calculations requiring SI units

values in Table 12.2 are not exact numbers; they are given to four significant figures in the table.

The equalities of Table 12.2 can be used to construct conversion factors for problem solving in the usual way (Sec. 3.6). Example 12.1 illustrates the use of such conversion factors in the context of change in pressure unit problems.

EXAMPLE 12.1

Interconverting between Various Pressure Units

At a certain altitude, a weather balloon recorded a barometric pressure of 367 mm Hg. Express this barometric pressure in

(a) atmospheres **(b)** inches Hg **(c)** pounds per square inch

SOLUTION

(a) The given quantity is 367 mm Hg, and the unit of the desired quantity is atmospheres.

$$367 \text{ mm Hg} = ? \text{ atm}$$

The conversion factor relating these two units is

$$\frac{1 \text{ atm}}{760 \text{ mm Hg}}$$

The result of using this conversion factor is

$$367 \cancel{\text{ mm Hg}} \times \frac{1 \text{ atm}}{760 \cancel{\text{ mm Hg}}} = 0.48289473 \text{ atm} \quad \text{(calculator answer)}$$

$$= 0.483 \text{ atm} \qquad \textbf{(correct answer)}$$

(b) This time, the unit for the desired quantity is in. Hg.

$$367 \text{ mm Hg} = ? \text{ in. Hg}$$

A direct relationship between mm Hg and in. Hg is not given in Table 12.1. However, the table does give the relationships of both inches and millimeters of mercury to atmospheres. Hence, we can use atmospheres as an intermediate step in the unit conversion sequence.

$$\text{mm Hg} \longrightarrow \text{atm} \longrightarrow \text{in. Hg}$$

The dimensional analysis setup is

$$367 \cancel{\text{ mm Hg}} \times \frac{1 \cancel{\text{ atm}}}{760 \cancel{\text{ mm Hg}}} \times \frac{29.92 \text{ in. Hg}}{1 \cancel{\text{ atm}}}$$

$$= 14.44821 \text{ in. Hg} \quad \text{(calculator answer)}$$

$$= 14.4 \text{ in. Hg} \qquad \textbf{(correct answer)}$$

(c) The problem to be solved here is

$$367 \text{ mm Hg} = ? \text{ psi}$$

The unit conversion sequence will be

$$\text{mm Hg} \longrightarrow \text{atm} \longrightarrow \text{psi}$$

The dimensional analysis sequence of conversion factors, for this pathway, is

$$367 \; \cancel{mm \; Hg} \times \frac{1 \; \cancel{atm}}{760 \; \cancel{mm \; Hg}} \times \frac{14.68 \; psi}{1 \; \cancel{atm}} = 7.0888947 \; psi \quad \text{(calculator answer)}$$

$$= 7.09 \; psi \qquad \textbf{(correct answer)}$$

Practice Exercise 12.1

A weather reporter gives the day's barometric pressure as 30.12 in. of mercury. Express this pressure in the units of

(a) atmospheres **(b)** millimeters of mercury **(c)** pounds per square inch

> **Answers to practice exercises are located at the end of the chapter.**

Pressure Readings and Significant Figures

Standard procedure in obtaining pressures that are based on the height of a column of mercury (barometric and manometric readings) is to estimate the column height to the closest millimeter. Thus, such pressure readings have an uncertainty in the "ones place," that is, to the closest millimeter of mercury.

> **A similar "common sense" approach to significant figures for temperature readings was considered in Section 3.11.**

When considering significant figures for pressure readings, the preceding operational procedure must be taken into account. Millimeter of mercury pressure readings of 750, 730, 720, 650, etc., are considered to have three significant figures even though no decimal point is explicitly shown after the zero (Sec. 2.5). A pressure reading of 700 mm Hg or 600 mm Hg is considered to possess three significant figures.

Note that the previous paragraph applies to "ordinary" pressures obtained using mercury column heights. A very high pressure, such as 19,000 mm Hg (25 atmospheres), would be assumed to be a two-significant-figure pressure reading rather than a five-significant-figure reading, without further information being given. Obviously, this pressure was not determined by reading the height of a column of mercury.

12.3 Boyle's Law: A Pressure–Volume Relationship

Of the several relationships that exist between gas law variables, the first to be discovered was the one that relates gas pressure to gas volume. It was formulated over 300 years ago, in 1662, by the British chemist and physicist Robert Boyle (1627–1691; see "The Human Side of Chemistry 12") and is known as *Boyle's law*. **Boyle's law** *states that the volume of a fixed mass of gas is* inversely proportional *to the pressure applied to the gas if the temperature is kept constant.* This means that if the pressure on the gas increases, the volume decreases proportionally; and conversely, if the pressure is decreased, the volume will increase. Doubling the pressure cuts the volume in half, tripling the pressure cuts the volume to one-third its original value, quadrupling the pressure cuts the volume to one-fourth, and so on. Any time two quantities are *inversely proportional*, as pressure and volume are (Boyle's law), one increases as the other decreases. Data illustrating Boyle's law are given in Figure 12.3.

> **The phrase "fixed mass of gas" in the statement of Boyle's law means that the number of moles of gas present is a constant.**

Boyle's law can be illustrated physically quite simply with the J-tube apparatus shown in Figure 12.4. The pressure on the trapped gas is increased by adding mercury to the J-tube. The volume of the trapped gas decreases as the pressure is raised.

Boyle's law can be stated mathematically as

$$P \times V = \text{constant}$$

The Human Side of Chemistry 12

Robert Boyle (1627–1691)

Robert Boyle, one of the most prominent of seventeenth-century scientists, spent most of his life in England although he was born in Ireland. Born to wealth and nobility, the 14th of 15 children of the Earl of Cork, he was an infant prodigy. At the age of 14 he was in Italy studying the works of the recently deceased Galileo.

Like most men of the seventeenth century who devoted themselves to science, Boyle was self-taught. Primarily known as a chemist and natural philosopher, he was best known for the gas law that bears his name. However, he made many other significant contributions in chemistry and physics.

During his lifetime his name was known throughout the learned world as a leading advocate of the "experimental philosophy." Through his efforts the true value of experimental investigation was first realized.

Historically, before chemistry could become a flourishing science, a period of "housecleaning" was needed to get rid of accumulated false concepts of previous centuries. Perhaps Boyle's greatest work in chemistry was in this area. He was very instrumental in demolishing the false dogma of only four elements—earth, air, water, and fire.

His book *The Sceptical Chymist* (1661), which attacked many of the false concepts of previous years, influenced generations of chemists. In this book he developed the concept of "primary particles," a forerunner of our present concept of atoms and elements. He attempted to explain different kinds of matter in terms of the organization and motion of these primary particles.

Boyle was a prolific writer on not only scientific but also philosophical and religious topics. In later years he became increasingly interested in religion. Upon his death, Boyle left a sum of money to fund the Boyle lectures, which were to be not on science, but on the defense of Christianity against unbelievers.

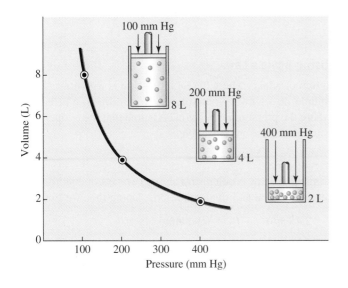

Figure 12.3

Data illustrating the "inverse proportionality" associated with Boyle's law.

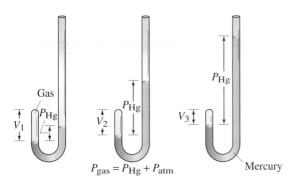

Figure 12.4

An apparatus for demonstrating Boyle's law. The volume of the trapped gas in the closed end of the tube (V_1, V_2, and V_3) decreases as mercury is added through the open end of the tube.

In this expression, V is the volume of the gas at a given temperature and P is the pressure. This expression thus indicates that at constant temperature the product of the pressure times the volume is always the same (or constant). (Note that Boyle's law is valid only if the temperature of the gas does not change.)

An alternative, and more useful, mathematical form of Boyle's law can be derived by considering the following situation. Suppose a gas is at an initial pressure P_1 and has a volume V_1. (We will use the subscript 1 to indicate the initial conditions.) Now imagine that the pressure is changed to some final pressure P_2. The volume will also change, and we will call the final volume V_2. (We will use the subscript 2 to indicate the final conditions.) According to Boyle's law,

$$P_1 \times V_1 = \text{constant}$$

After the change in pressure and volume, we have

$$P_2 \times V_2 = \text{constant}$$

The constant is the same in both cases; we are dealing with the same sample of gas. Thus, we can combine the two PV products to give the equation

$$P_1 \times V_1 = P_2 \times V_2$$

When we know any three of the four quantities in this equation, we can calculate the fourth, which will usually be the final pressure, P_2, or the final volume, V_2, as illustrated in Examples 12.2 and 12.3.

EXAMPLE 12.2

Calculating Final Volume Using Boyle's Law

An air bubble forms at the bottom of a lake, where the total pressure is 2.45 atm. At this pressure, the bubble has a volume of 3.1 mL. What volume, in milliliters, will it have when it rises to the surface, where the pressure is 0.93 atm? Assume that the temperature remains constant, as does the amount of gas in the bubble.

SOLUTION

A suggested first step in working all gas law problems involving two sets of conditions is to analyze the given data in terms of initial and final conditions. Doing this, we find that

$$P_1 = 2.45 \text{ atm} \qquad P_2 = 0.93 \text{ atm}$$
$$V_1 = 3.1 \text{ mL} \qquad V_2 = ? \text{ mL}$$

Next, we rearrange Boyle's law to isolate V_2 (the quantity to be calculated) on one side of the equation. This is accomplished by dividing both sides of the Boyle's law equation by P_2.

$$P_1V_1 = P_2V_2 \quad (\textbf{Boyle's law})$$

$$\frac{P_1V_1}{P_2} = \frac{P_2V_2}{P_2} \quad (\text{division of each side of the equation by } P_2)$$

$$V_2 = V_1 \times \frac{P_1}{P_2}$$

Substituting the given data in the rearranged equation and doing the arithmetic gives

$$V_2 = 3.1 \text{ mL} \times \frac{2.45 \text{ atm}}{0.93 \text{ atm}} = 8.1666666 \text{ mL} \quad (\text{calculator answer})$$

$$= 8.2 \text{ mL} \qquad (\textbf{correct answer})$$

In gas law problems, when possible, it is always a good idea to check qualitatively the reasonableness of your final answer. This is a double check against two types of errors frequently made: improper substitution of variables into the equation and improper rearrangement of the equation. Decreasing the pressure on a fixed amount of gas at constant temperature, which is the case in this problem, should result in a volume increase for the gas. Our answer is consistent with this conclusion.

Practice Exercise 12.2

An inflated air mattress has a volume of 114 L at a pressure of 0.925 atm. The mattress, still inflated, is taken to a new location where the pressure is 0.975 atm. What will be the air-mattress volume, in liters, at this higher pressure, assuming that temperature remains constant?

EXAMPLE 12.3

Calculating Final Volume Using Boyle's Law with Pressure Values Given in Different Units

On an airliner, an inflated toy has a volume of 0.310 L at 25°C and 745 mm Hg. During flight, the cabin pressure unexpectedly drops to 0.810 atm with the temperature remaining the same. What is the volume, in liters, of the toy at the new conditions?

SOLUTION

Analyzing the given data in terms of initial and final conditions gives

$$P_1 = 745 \text{ mm Hg} \quad P_2 = 0.810 \text{ atm}$$

$$V_1 = 0.310 \text{ L} \qquad V_2 = ? \text{ L}$$

The temperature is constant; it will not enter into the calculation.

A slight complication exists with the pressure values; they are not given in the same units. We must make the units the same before proceeding with the calculation. It does not matter whether they are both millimeters of mercury or both atmospheres. The same answer is obtained with either unit. Let us arbitrarily decide to change atmospheres to mm Hg.

$$0.810 \text{ atm} \times \frac{760 \text{ mm Hg}}{1 \text{ atm}} = 615.6 \text{ mm Hg} \quad \text{(calculator answer)}$$
$$= 616 \text{ mm Hg} \quad \textbf{(correct answer)}$$

Our given conditions are now

$$P_1 = 745 \text{ mm Hg} \qquad P_2 = 616 \text{ mm Hg}$$

$$V_1 = 0.310 \text{ L} \qquad V_2 = ? \text{ L}$$

Boyle's law with V_2 isolated on the left side has the form

$$V_2 = V_1 \times \frac{P_1}{P_2}$$

Plugging the given quantities into this equation and doing the arithmetic gives

$$V_2 = 0.310 \text{ L} \times \frac{745 \text{ mm Hg}}{616 \text{ mm Hg}} = 0.37491883 \text{ L} \quad \text{(calculator answer)}$$
$$= 0.375 \text{ L} \qquad \textbf{(correct answer)}$$

The answer is reasonable. Decreased pressure, at constant temperature, should produce a volume increase.

In solving this problem we arbitrarily chose to use mm Hg pressure units. If we had used atmosphere pressure units instead, the answer obtained would still be the same, as we now illustrate.

$$745 \ \cancel{\text{mm Hg}} \times \frac{1 \ \text{atm}}{760 \ \cancel{\text{mm Hg}}} = 0.98026315 \ \text{atm} \quad \text{(calculator answer)}$$

$$= 0.980 \ \text{atm} \quad \textbf{(correct answer)}$$

Using $P_1 = 0.980$ atm and $P_2 = 0.810$ atm, we get

$$V_2 = 0.310 \ \text{L} \times \frac{0.980 \ \cancel{\text{atm}}}{0.810 \ \cancel{\text{atm}}} = 0.37506172 \ \text{L} \quad \text{(calculator answer)}$$

$$= 0.375 \ \text{L} \quad \textbf{(correct answer)}$$

Practice Exercise 12.3

A sample of cyclopropane (C_3H_6), a colorless gas with a pleasant odor that finds use as a general anesthetic, occupies a volume of 2.00 L at 27°C and 1.00 atm pressure. What volume will this sample occupy at the same temperature but at a pressure of 655 mm Hg?

Boyle's law is consistent with kinetic molecular theory (Sec. 11.3). The theory states that the pressure a gas exerts results from collisions of the gas molecules with the sides of the container. The pressure of the gas at a given temperature is proportional to the number of collisions within a given area on the container wall in a given time. If the volume of a container holding a specific number of gas molecules is increased, the total wall area of the container will also increase and the number of collisions in a given area (the pressure) will decrease due to the greater wall area. Conversely, if the volume of the container is decreased, the wall area will be smaller and there will be more collisions in a given wall area. This means an increase in pressure. Figure 12.5 illustrates this idea.

The phenomenon described by Boyle's law has practical importance. Helium-filled research balloons, used to study the upper atmosphere, are only half-filled with helium when launched. As the balloon ascends, it encounters lower and lower pressures. As the pressure decreases, the balloon expands until it reaches full inflation. A balloon launched at full inflation would burst in the upper atmosphere because of the reduced external pressure.

> Filling a medical syringe with a liquid demonstrates Boyle's law. As the plunger is drawn out of the syringe, the increase in volume inside the syringe chamber results in decreased pressure there. The liquid, which is at atmospheric pressure, flows into this reduced-pressure area. This liquid is then expelled from the chamber by pushing the plunger back in. This ejection of liquid does not involve Boyle's law; a liquid is incompressible and mechanical force pushes it out.

2-L flask
1-L flask
The volume is decreased by one-half.
(a)

A given molecule hits container walls twice as often.
(b)

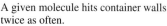

Figure 12.5

The pressure exerted by a gas is doubled when the volume of the gas, at constant temperature, is cut by one-half.

Breathing is an example of Boyle's law in action, as is the operation of a respirator, a machine designed to help patients with respiration difficulties to breathe. A respirator contains a movable diaphragm that works in opposition to the patient's lungs. When the diaphragm is moved out so that the volume inside the respirator increases, the lower pressure in the respirator allows air to expand out of the patient's lungs. When the diaphragm is moved in the opposite direction, the higher pressure inside the respirator compresses the air into the lungs and causes them to increase in volume.

12.4 Charles's Law: A Temperature–Volume Relationship

The relationship between the temperature and the volume of a fixed amount of gas at constant pressure is called *Charles's law*, after the French scientist Jacques Charles (1746–1823; see "The Human Side of Chemistry 13"). In 1787, over 100 years after the discovery of Boyle's law, Charles showed that a simple mathematical relationship exists between the volume and temperature of a fixed amount of gas, at constant pressure, *provided the temperature is expressed in kelvins*. **Charles's law** *states that the volume of a fixed mass of gas is* directly proportional *to its Kelvin temperature if the pressure is kept constant*. Contained within the wording of this law is the phrase *directly proportional*; this contrasts with Boyle's law, which contains the phrase *inversely proportional*. Any time a *direct* proportion exists between two quantities, one increases when the other increases, and one decreases when the other decreases. Thus a direct proportion and an inverse proportion portray "opposite" behaviors. The direct proportion relationship of Charles's law means that if the temperature increases the volume will also increase and if the temperature decreases the volume will also decrease. Data illustrating Charles's law are given in Figure 12.6.

Charles's law can be qualitatively illustrated by a balloon filled with air. If the balloon is placed near a heat source, such as a light bulb that has been on for some time, the heat will cause the balloon to increase in size (volume). The change in volume is usually apparent. Cooling the air in the balloon will cause the balloon to shrink (see Fig. 12.7).

The Human Side of Chemistry 13

Jacques Alexandre César Charles (1746–1823)

Jacques Alexandre César Charles, a French physicist, had an early career completely removed from science. He worked at a routine job in the Bureau of Finances in Paris. All was well until, during a period of government austerity, he was discharged from his job. Unemployed, he decided to study physics.

During this time of transition in Charles's life, hot air ballooning, both unmanned and manned, was in its first stages of development. Charles was best known to his colleagues for his contributions to the developing science of

ballooning. He was the first to design a hydrogen-filled, rather than air-filled, balloon for manned flight. In constructing the balloon Charles showed much innovation in the design of scientific apparatus.

The first hot-air balloon flight with passengers was made November 17, 1783. Burning straw was the source of heat for such hot-air balloons. A few days later Charles made an ascent in his hydrogen-filled balloon. His landing put him in a rural area. The peasants in the area, terrified at their first exposure to "manned flight," attacked and destroyed the balloon using pitchforks.

In the process of working with balloons, about 1787, Charles observed that several different gases expanded

the same way when heated. From further observations he was able to estimate the degree of thermal expansion for gases as a function of temperature. This led to what is now known as Charles's law.

Charles never published his work. He did, however, tell another scientist, Joseph Louis Gay-Lussac ("The Human Side of Chemistry 14"), about it, and it was Gay-Lussac who made it public. In an article on the expansion of gases by heat, published in 1802, Gay-Lussac described, criticized, and considerably improved upon Charles's experimental procedures.

Charles's contributions to science were limited to those associated with his ballooning.

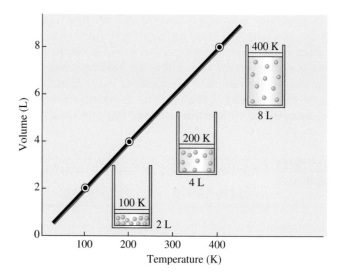

Figure 12.6

Data illustrating the "direct proportionality" associated with Charles's law.

Figure 12.7

As liquid nitrogen ($-196°C$) is poured over a balloon, the gas in the balloon is cooled and the volume decreases. (Richard Megna/ Fundamental Photographs)

Charles's law can be stated mathematically as

$$\frac{V}{T} = \text{constant}$$

In this expression V is the volume of a gas at a given pressure and T is the temperature, expressed on the Kelvin temperature scale. A consideration of two sets of temperature–volume conditions for a gas, in a manner similar to that done for Boyle's law, leads to the following useful form of the law.

$$\frac{V_1}{T_1} = \frac{V_2}{T_2}$$

When you use the mathematical form of Charles's law, the temperatures used *must always be* Kelvin scale temperatures.

Again, note that in any of the mathematical expressions of Charles's laws, or any other gas law, the symbol T is understood to mean the Kelvin temperature.

EXAMPLE 12.4

Calculating Final Volume Using Charles's Law

A helium-filled birthday balloon has a volume of 223 mL at a temperature of 24°C (room temperature). The balloon is placed in a car overnight during the winter where the temperature drops to $-13°C$. What is the balloon's volume, in milliliters, when it is removed from the car in the morning (assuming no helium has escaped)?

SOLUTION

Writing all the given data in the form of initial and final conditions, we have

$$V_1 = 223 \text{ mL} \qquad V_2 = ? \text{ mL}$$

$$T_1 = 24°C = 297 \text{ K} \quad T_2 = -13°C = 260 \text{ K}$$

Note that both of the given temperatures have been converted to Kelvin scale readings by adding 273 to them.

Rearranging Charles's law to isolate V_2, the desired quantity, on one side of the equation is accomplished by multiplying each side of the equation by T_2.

$$\frac{V_1}{T_1} = \frac{V_2}{T_2} \qquad (\textbf{Charles's law})$$

$$\frac{V_1 T_2}{T_1} = \frac{V_2 T_2}{T_2} \qquad (\text{multiplication of each side by } T_2)$$

$$V_2 = V_1 \times \frac{T_2}{T_1}$$

Substituting the given data into the equation and doing the arithmetic gives

$$V_2 = 223 \text{ mL} \times \frac{260 \text{ K}}{297 \text{ K}}$$

$$= 195.21885 \text{ mL} \quad (\text{calculator answer})$$

$$= 195 \text{ mL} \qquad (\textbf{correct answer})$$

Our answer is consistent with what reasoning says it should be. Decreasing the temperature of a gas, at constant pressure, should result in a volume decrease.

Practice Exercise 12.4

At constant pressure, a 675-mL sample of N_2 gas is warmed from 23°C (a comfortable room temperature) to 38°C (the temperature outdoors on a hot summer day). What is the new volume of the N_2 gas?

Figure 12.8 is a graph showing four sets of volume–temperature data for a gas, each data set being at a different constant pressure. The plot of each data set gives a straight line.

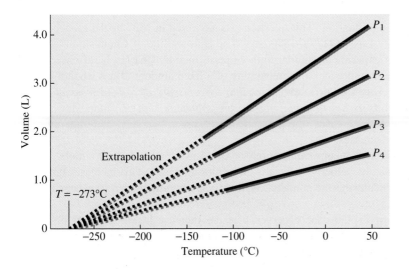

Figure 12.8

A graph showing the variation of the volume of a gas with temperature at a constant pressure. Each line in the graph represents a different fixed pressure for the gas. The pressures increase from P_1 to P_4. The solid portion of each line represents the temperature region before the gas condenses to a liquid. The extrapolated (dashed) portion of each line represents the hypothetical volume of the gas if it had not condensed. All extrapolated lines intersect at the same point, a point corresponding to zero volume and a temperature of $-273°C$.

These four straight lines can be used to show the basis for the use of the Kelvin temperature scale in gas law calculations. If each of these lines is extrapolated (extended) to lower temperatures (the dashed portions of the lines in Fig. 12.8), we find that they all intersect at a common point on the temperature axis. This point of intersection, corresponding to a temperature value of $-273°C$, is the point at which the volume of the gas "would become" zero. Notice that the words "would become" in the last sentence are in quotation marks. In reality, the volume of a gas never reaches zero; all gases would liquefy before they reach a temperature of $-273°C$, and Charles's law would no longer apply. Thus, the extrapolated portions of the four straight lines portray a hypothetical situation. It is not, however, a situation without significance.

The Scottish mathematician and physicist William Thomson (1824–1907), better known as Lord Kelvin, was the first to recognize the importance of the "zero-volume" temperature value of $-273°C$. It is the lowest temperature that is theoretically attainable, a temperature now referred to as *absolute zero*. It is this temperature that is now used as the starting point for the temperature scale called the Kelvin temperature scale. When temperatures are specified in kelvins, Charles's law assumes the simple mathematical form in which it was previously given in this section.

Charles's law is readily understood in terms of kinetic molecular theory. The theory states that when the temperature of a gas increases, the velocity of the gas particles increases. The speedier particles hit the container walls harder and more often. For the pressure of the gas to remain constant, it is necessary for the container volume to increase. Moving in a larger volume, the particles will hit the container walls less often, and thus the pressure can remain the same. A similar argument applies if the temperature of the gas is lowered, only this time the velocity of the molecules decreases and the wall area (volume) must decrease to increase the number of collisions in a given area in a given time.

Charles's law is the principle used in the operation of a convection heater. When air comes in contact with the heating element it expands (its density becomes less). The hot, less dense air rises, causing continuous circulation of warm air. This same principle has ramifications in closed rooms in which there is not effective air circulation. The warmer and less dense air stays near the top of the room. This is desirable in the summer but not in the winter.

> Thunder can be related to Charles's law. A typical lightning bolt heats the air around it to 30,000°C in a few microseconds. The heated air expands rapidly and pushes against the cooler surrounding air. This creates the shock wave of sound known as thunder.

12.5 Gay-Lussac's Law: A Temperature–Pressure Relationship

If a gas is placed in a rigid container, one that cannot expand, the volume of the gas must remain constant. What is the relationship between the pressure and temperature of a fixed amount of gas in such a situation? This question was answered in 1802 by the French scientist Joseph Louis Gay-Lussac (1778–1850; see "The Human Side of Chemistry 14"). Performing a number of experiments similar to those performed by Charles, he discovered the relationship between pressure and the temperature of a fixed amount of gas when it is held at constant volume. **Gay-Lussac's law** *states that the pressure of a fixed mass of gas is* directly proportional *to its Kelvin temperature if the volume is kept constant.* Thus, as the temperature of the gas increases, the pressure increases; conversely, as temperature is decreased, pressure decreases.

The kinetic molecular theory explanation for Gay-Lussac's law is very simple. As the temperature increases, the velocity of the gas molecules increases. This increases the number of times the molecules hit the container walls in a given time, which translates into an increase in pressure.

Gay-Lussac's law explains the observed pressure increase inside an automobile tire after a car has been driven for a period of time. It also explains why aerosol cans should not be disposed of in a fire. The aerosol can keeps the gas at a constant volume. The elevated

The Human Side of Chemistry 14

Joseph Louis Gay-Lussac (1778–1850)

Joseph Louis Gay-Lussac, the eldest of five children, is considered one of the greatest scholars France has produced. Because of the French Revolution and the uncertainties created by it, Gay-Lussac's formal education had to be delayed until he was 16. He quickly distinguished himself as a student.

His grandfather was a physician and his father a lawyer. The family surname was actually Gay, the surname by which he was baptized. His father added the appendage Lussac (which has geographical connotations) to the family name to distinguish his family from many others in the area with the same surname.

Gay-Lussac devoted nearly all of his life to pure and applied science. Much of it, from 1809 on, was spent as a professor at the Ecole Polytechnique in Paris. He did have a brief political career (1831–1838) as an elected French legislator.

Gay-Lussac's first major research (1801–1802) involved the thermal expansion of gases. Here he followed up on work performed earlier by Charles. His greatest achievement, which came in 1808, was work involving the ratio, by volume, in which gases combine during chemical reactions. From this work came the law of combining volumes that now bears his name (Sec. 12.8).

He was the first to isolate the elements potassium and boron. With a coworker, he made the first thorough studies of iodine and cyanogen. He was an early pioneer in the area of volumetric analysis. The "Gay-Lussac tower," an essential part of sulfuric acid plants for many years, is evidence of his interest in chemical technology.

temperatures encountered in the fire increase the pressure of the confined gas to the point that the can will probably explode.

Mathematical forms of Gay-Lussac's law are

$$\frac{P}{T} = \text{constant} \quad \text{and} \quad \frac{P_1}{T_1} = \frac{P_2}{T_2}$$

Again, note that the symbol T is understood to mean the Kelvin temperature. Any pressure unit is acceptable; P_1 and P_2 must, however, be in the same units.

EXAMPLE 12.5

Calculating Final Temperature Using Gay-Lussac's Law

A pressurized can of shaving cream contains a gas under an internal pressure of 1.065 atm at 24°C (room temperature). The can itself is able to withstand an internal pressure of 3.00 atm before it explodes. What temperature, in degrees Celsius, is needed to elevate the internal pressure within the can to its maximum value?

SOLUTION

Writing all the given data in the form of initial and final conditions, we have

$$P_1 = 1.065 \text{ atm} \qquad P_2 = 3.00 \text{ atm}$$
$$T_1 = 24°C = 297 \text{ K} \quad T_2 = ? \text{ K}$$

Rearrangement of Gay-Lussac's law to isolate T_2, the quantity we desire, on the left side of the equation gives

$$T_2 = T_1 \times \frac{P_2}{P_1}$$

Substituting the given data into the equation and doing the arithmetic gives

$$T_2 = 297 \text{ K} \times \frac{3.00 \text{ atm}}{1.065 \text{ atm}} = 836.61971 \text{ K (calculator answer)}$$
$$= 837 \text{ K} \qquad \textbf{(correct answer)}$$

Our answer is in kelvins. We were asked for the temperature in degrees Celsius. Subtracting 273 from the Kelvin temperature will give us the Celsius temperature.

$$T(°C) = 837\text{ K} - 273 = 564°C$$

The final answer is reasonable. Increased pressure, at constant volume, should produce a temperature increase.

Practice Exercise 12.5

An aerosol spray can contains gas under a pressure of 2.00 atm at 27°C. The can itself is only able to withstand a pressure of 3.00 atm. To what temperature, in degrees Celsius, could the can and its contents be heated before the can explodes?

12.6 The Combined Gas Law

The **combined gas law** *is the expression obtained from mathematically combining Boyle's, Charles's, and Gay-Lussac's laws*. Its mathematical form is

$$\frac{P_1V_1}{T_1} = \frac{P_2V_2}{T_2}$$

This combined gas law is a much more versatile equation than the individual gas laws. With it, for a fixed amount of gas, a change in any one of the other three gas law variables brought about by changes in *both* of the remaining two variables can be calculated. Each of the individual gas laws requires that an additional variable besides amount of gas be held constant.

The combined gas law reduces (simplifies) to each of the equations for the individual gas laws when the appropriate variable is held constant. These reduction relationships are given in Table 12.3. Because of the ease with which they can be derived from the combined gas law, you need not memorize the mathematical forms for the individual gas laws if you know the mathematical form for the combined gas law.

Table 12.3 **Relationship of the Individual Gas Laws to the Combined Gas Law**

Law	Constancy Requirement (for a fixed mass of gas)	Mathematical Form of the Law
Combined gas law	none	$\dfrac{P_1V_1}{T_1} = \dfrac{P_2V_2}{T_2}$
Boyle's law	$T_1 = T_2$	Since T_1 and T_2 are equal, substitute T_1 for T_2 in the combined gas law and cancel. $\dfrac{P_1V_1}{T_1} = \dfrac{P_2V_2}{T_1}$ or $P_1V_1 = P_2V_2$
Charles's law	$P_1 = P_2$	Since P_1 and P_2 are equal, substitute P_1 for P_2 in the combined gas law and cancel. $\dfrac{P_1V_1}{T_1} = \dfrac{P_1V_2}{T_2}$ or $\dfrac{V_1}{T_1} = \dfrac{V_2}{T_2}$
Gay-Lussac's law	$V_1 = V_2$	Since V_1 and V_2 are equal, substitute V_1 for V_2 in the combined gas law and cancel. $\dfrac{P_1V_1}{T_1} = \dfrac{P_2V_1}{T_2}$ or $\dfrac{P_1}{T_1} = \dfrac{P_2}{T_2}$

The three most used forms of the combined gas law are those that isolate V_2, P_2, and T_2 on the left side of the equation.

$$V_2 = V_1 \times \frac{P_1}{P_2} \times \frac{T_2}{T_1}$$

$$P_2 = P_1 \times \frac{V_1}{V_2} \times \frac{T_2}{T_1}$$

$$T_2 = T_1 \times \frac{P_2}{P_1} \times \frac{V_2}{V_1}$$

Students frequently have questions about the algebra involved in accomplishing these rearrangements. Example 12.6 should clear up any such questions. Examples 12.7 and 12.8 illustrate the use of the combined gas law equation.

EXAMPLE 12.6

Mathematical Manipulation of the Combined Gas Law

Rearrange the standard form of the combined gas law equation such that the variable P_2 is by itself on the left side of the equation.

SOLUTION

In rearranging the standard form of the combined gas law into various formats, the following rule from algebra is useful.

If two fractions are equal,

$$\frac{a}{b} = \frac{c}{d}$$

then the numerator of the first fraction (a) times the denominator of the second fraction (d) is equal to the numerator of the second fraction (c) times the denominator of the first fraction (b).

$$\text{If } \frac{a}{b} = \frac{c}{d}, \quad \text{then} \quad a \times d = c \times b$$

Applying this rule to the standard form of the combined gas law gives

$$\frac{P_1 V_1}{T_1} = \frac{P_2 V_2}{T_2} \longrightarrow P_1 V_1 T_2 = P_2 V_2 T_1$$

With the combined gas law in the form $P_1 V_1 T_2 = P_2 V_2 T_1$, any of the six variables can be isolated by a simple division. To isolate P_2, we divide both sides of the equation by $V_2 T_1$ (the other quantities on the same side of the equation as P_2).

$$\frac{P_1 V_1 T_2}{V_2 T_1} = \frac{P_2 \cancel{V_2} \cancel{T_1}}{\cancel{V_2} \cancel{T_1}}$$

$$P_2 = \frac{P_1 V_1 T_2}{V_2 T_1} \quad \text{or} \quad P_2 = P_1 \times \frac{V_1}{V_2} \times \frac{T_2}{T_1}$$

Practice Exercise 12.6

Rearrange the standard form of the combined gas law equation such that the variable V_2 is by itself on the left side of the equation.

EXAMPLE 12.7

Calculating Final Volume Using the Combined Gas Law

Hydrogen gas (H_2) is the least dense of all gases. A sample of H_2 gas is found to occupy a volume of 1.23 L at 755 mm Hg and 0°C. What volume, in liters, will this same gas sample occupy at 735 mm Hg pressure and a temperature of 50°C?

SOLUTION

Writing all the given data in the form of initial and final conditions, we have

$$P_1 = 755 \text{ mm Hg} \qquad P_2 = 735 \text{ mm Hg}$$
$$V_1 = 1.23 \text{ L} \qquad V_2 = ? \text{ L}$$
$$T_1 = 0°C = 273 \text{ K} \qquad T_2 = 50°C = 323 \text{ K}$$

Rearrangement of the combined gas law expression to isolate V_2 on the left side gives

$$V_2 = V_1 \times \frac{P_1}{P_2} \times \frac{T_2}{T_1}$$

Substituting numerical values into this equation and doing the arithmetic gives

$$V_2 = 1.23 \text{ L} \times \frac{755 \text{ mm Hg}}{735 \text{ mm Hg}} \times \frac{323 \text{ K}}{273 \text{ K}}$$

$$= 1.494874 \text{ L} \quad \text{(calculator answer)}$$

$$= 1.49 \text{ L} \qquad \text{(correct answer)}$$

CHEMICAL EXTENSION

The Earth's atmosphere contains only trace amounts of H_2 gas even though hydrogen is the most abundant element in outer space (Sec. 4.8). The reason for this is the high velocities at which H_2 molecules travel. The smaller the mass of a molecule, the greater its velocity at any given temperature. With the smallest mass of any molecule, gaseous H_2 molecules obtain molecular speeds sufficient for the H_2 molecules to overcome the gravitational attraction of the Earth and escape to outer space.

Practice Exercise 12.7

On a winter day, a person takes in a breath of 425 mL of cold air $(-12°C)$. The atmospheric pressure is 682 mm Hg. What is the volume of this air, in milliliters, in the lungs where the temperature is 37°C and the pressure is 684 mm Hg?

EXAMPLE 12.8

Calculating Final Temperature Using the Combined Gas Law

A sample of helium gas [the gas used in lighter-than-air airships (blimps)] occupies a volume of 180.0 mL at a pressure of 0.800 atm and a temperature of 29°C. What will be the temperature of the gas, in degrees Celsius, if the volume of the container is decreased to 90.0 mL and the pressure is increased to 1.60 atm?

SOLUTION

Writing all the given data in the form of initial and final conditions, we have

$$P_1 = 0.800 \text{ atm} \qquad P_2 = 1.60 \text{ atm}$$
$$V_1 = 180.0 \text{ mL} \qquad V_2 = 90.0 \text{ mL}$$
$$T_1 = 29°C = 302 \text{ K} \qquad T_2 = ?°C$$

Rearrangement of the combined gas law to isolate T_2 on the left side of the equation gives

$$T_2 = T_1 \times \frac{P_2}{P_1} \times \frac{V_2}{V_1}$$

Substituting the given data into this equation and doing the arithmetic gives

$$T_2 = 302 \text{ K} \times \frac{1.60 \text{ atm}}{0.800 \text{ atm}} \times \frac{90.0 \text{ mL}}{180.0 \text{ mL}}$$

$$= 302 \text{ K} \quad (\text{calculator and } \textbf{correct answer} \text{ in kelvins})$$

Converting the temperature to the Celsius scale by subtracting 273 gives 29°C as the final answer.

$$T(°C) = 302 \text{ K} - 273 = 29°C$$

The temperature did not change! The pressure correction factor, considered by itself, would cause the temperature to increase by a factor of 2; the pressure was doubled. The volume correction factor, considered by itself, would cause the temperature to decrease by a factor of 2; the volume was halved. Considered together, the effects of the two factors cancel each other and result in the temperature not changing.

CHEMICAL EXTENSION

Helium is the only element discovered extraterrestrially before being found on Earth. In 1868, spectroscopic studies of the Sun's solar flares, during an eclipse, revealed the existence of a bright yellow line in the solar flare light that could not be ascribed to any known element. The name helium (Greek *helios* means "sun") was given to this hypothetical element.

In 1895 terrestrial evidence for helium's existence came when chemical treatment of several minerals found in Kansas produced small amounts of an unreactive gas whose spectrum matched that previously observed in the solar flares.

It is now known that helium is produced continuously in the Earth's rocks by the radioactive decay of uranium and thorium present in small amounts in the rocks. Some of this helium, through the process of diffusion through rock formations, becomes incorporated into natural gas deposits, and it can be isolated from the natural gas during processing of the natural gas.

Helium-containing natural gases serve as the only economical source of helium at present. Because of dwindling natural gas supplies, the U.S. government (in the early 1960s) began stockpiling helium to assure its availability in the future. The helium storage program involves placing recovered helium into partially depleted natural gas wells that serve as storage containers.

Practice Exercise 12.8

A sample of argon gas (the gas used in electric light bulbs) occupies a volume of 80.0 mL at a pressure of 1.10 atm and a temperature of 29°C. What will be the temperature of the gas, in degrees Celsius, if the volume of the container is decreased to 40.0 mL and the pressure is increased to 2.20 atm?

12.7 Standard Conditions for Temperature and Pressure

The volumes of liquids and solids change only slightly with temperature and pressure changes (Sec. 11.2). This is not the case for volumes of gases. As the gas law discussions of previous sections pointedly show, the volume of a fixed amount of gas can change markedly with changes in temperature and pressure.

The selection of 0°C and 1 atm as the reference points called standard temperature and standard pressure for gases is an arbitrary decision in the same way that the selection of ^{12}C as the reference point for the amu scale (Sec. 5.9) was an arbitrary decision.

Gas volumes can be compared only if the gases are at the same temperature and pressure. It is convenient to specify a particular temperature and pressure as standards for comparison purposes. **Standard temperature for gases** *is 0°C (273.15 K).* **Standard pressure for gases** *is 1 atm (760 mm Hg).* **STP conditions for gases** *are those of standard temperature and standard pressure.*

In terms of significant figures, we note that in the definitions of standard temperature and standard pressure the values 0°C, 1 atm, and 760 mm Hg are defined (exact) numbers and the value 273.15 is an experimental value. In problem solving, standard temperature on the Kelvin scale is usually specified to three significant figures (273 K) rather than five significant figures (273.15 K). This "rule of thumb" is based on the general practice of specifying laboratory-measured temperatures to the closest whole number of degrees. The value 273.15 K is available, however, for use when temperatures involve tenths or hundredths of a degree.

Example 12.9 is a problem involving STP conditions.

EXAMPLE 12.9

Using STP Conditions in a Combined Gas Law Calculation

A sample of neon gas [the gas employed in luminescent lighting (neon signs)] occupies a volume of 23.4 L at STP. What volume, in liters, will it occupy at a pressure of 0.750 atm and a temperature of −20°C?

SOLUTION

STP conditions denote a temperature of 0°C and a pressure of 1 atm. Writing all the given data in the form of initial and final conditions, we have

$$
\begin{aligned}
P_1 &= 1 \text{ atm} & P_2 &= 0.750 \text{ atm} \\
V_1 &= 23.4 \text{ L} & V_2 &= ? \text{ L} \\
T_1 &= 0°\text{C} = 273 \text{ K} & T_2 &= -20°\text{C} = 253 \text{ K}
\end{aligned}
$$

Rearrangement of the combined gas law expression to isolate V_2 on the left side gives

$$ V_2 = V_1 \times \frac{P_1}{P_2} \times \frac{T_2}{T_1} $$

Substituting numerical values into the equation gives

$$ V_2 = 23.4 \text{ L} \times \frac{1 \text{ atm}}{0.750 \text{ atm}} \times \frac{253 \text{ K}}{273 \text{ K}} $$

$$ = 28.914285 \text{ L} \quad \text{(calculator answer)} $$

$$ = 28.9 \text{ L} \qquad \text{(correct answer)} $$

In this problem the pressure correction factor (P_1/P_2) and the temperature correction factor (T_1/T_2) work in opposite directions. The decreased pressure should cause the volume to increase. The decreased temperature should cause the volume to decrease. The fact that, overall, the volume increases indicates that the pressure change is the larger change.

CHEMICAL EXTENSION

Neon, along with helium, argon, krypton, and xenon, constitute the noble gas family of elements (Sec. 6.11). All members of this family are colorless, odorless, tasteless, unreactive gases with extremely low melting points and boiling points.

Neon particularly, but the other noble gases also, is used in luminescent lighting ("neon signs"). Each of the noble gases produces a particular color when an electric current flows through it. Neon is orange-red, helium a yellowish flesh tone, argon is lavender, krypton silver-white, and xenon blue. Colors can be modified by mixing gases, adding mercury vapor, using colored glass, and glass coated with fluorescent powders.

All the noble gases are also used in lasers. A helium–neon laser was the first continuously operating gas laser. It emits a red light.

Practice Exercise 12.9

A sample of H_2 gas occupies a volume of 1.37 L at STP. What volume, in liters, will it occupy at a pressure of 4.00 atm and a temperature of 340°C?

12.8 Gay-Lussac's Law of Combining Volumes

Gases are involved as reactants or products in many chemical reactions. In such reactions it is usually easier to determine the volumes rather than the masses of the gases involved.

A very simple relationship exists between the volumes of different gases consumed or produced in chemical reactions, provided the volumes are all determined at the same temperature and pressure. This relationship, known as *Gay-Lussac's law of combining volumes*, was first formulated in 1808 by Joseph Louis Gay-Lussac (the same Gay-Lussac discussed in Sec. 12.5). **Gay-Lussac's law of combining volumes** *states that the volumes of different gases that participate in a chemical reaction, measured at the same temperature and pressure, are in the same ratio as the coefficients for these gases in the balanced equation for the reaction.* For example, 1 volume of nitrogen and 3 volumes of hydrogen react to give 2 volumes of ammonia (NH_3).

$$N_2(g) + 3\,H_2(g) \longrightarrow 2\,NH_3(g)$$

1 volume **3 volumes** **2 volumes**

"Volume" is used here in the general sense of relative volume in any units. It could, for example, be 1 L of N_2, 3 L of H_2, and 2 L of NH_3 or 0.1 mL of N_2, 0.3 mL of H_2, and 0.2 mL of NH_3. Note that in making volume comparisons, the units must always be the same; if the volume of N_2 is measured in milliliters, the volumes of H_2 and NH_3 must also be in milliliters.

It is important to remember that these volume relationships apply only to gases, and then only when all gaseous volumes are measured at the same temperature and pressure. The volumes of solids and liquids involved in reactions cannot be treated this way.

The previously given statement of the law of combining volumes is a modern version of the law. At the time the law was first formulated by Gay-Lussac, chemists were still struggling with the difference between atoms and molecules. Equations as we now write them were unknown; formulas for substances were still to be determined. The original statement of the law simply noted that the ratio of volumes was always a ratio of small whole numbers. Gay-Lussac's work was based solely on volume measurements and had nothing to do with chemical equations. The explanation for the small whole numbers and the linkup with equations came later.

This volume interpretation for the coefficients in a balanced chemical equation involving gaseous substances is the third interpretation we have encountered for equation coefficients. In Section 10.3 we learned that equation coefficients can be interpreted in terms of molecules and in Section 10.6 a molar interpretation for equation coefficients was considered. Thus, for the equation

$$4\,NH_3(g) + 3\,O_2(g) \longrightarrow 2\,N_2(g) + 6\,H_2O(g)$$

it is correct to say

$$4 \; molecules \; NH_3(g) + 3 \; molecules \; O_2(g) \longrightarrow 2 \; molecules \; N_2(g) + 6 \; molecules \; H_2O(g)$$

and

$$4 \; moles \; NH_3(g) + 3 \; moles \; O_2(g) \longrightarrow 2 \; moles \; N_2(g) + 6 \; moles \; H_2O(g)$$

and, as a result of Gay-Lussac's law of combining volumes,

$$4 \; volumes \; NH_3(g) + 3 \; volumes \; O_2(g) \longrightarrow 2 \; volumes \; N_2(g) + 6 \; volumes \; H_2O(g)$$

Again, all the volumes must be measured at the same temperature and pressure for these volume relationships to be valid.

Table 12.4 summarizes the three ways of interpreting equation coefficients in chemical equations where gaseous substances are involved.

In calculations involving chemical reactions, where two or more gases are participants, this volume interpretation of coefficients can be used to generate conversion factors useful in problem solving. Consider the balanced equation

$$2 \, CO(g) + O_2(g) \rightarrow 2 \, CO_2(g)$$

At constant temperature and pressure, three volume–volume relationships are obtainable from this equation.

2 volumes CO	produce	2 volumes CO_2
1 volume O_2	produces	2 volumes CO_2
2 volumes CO	react with	1 volume O_2

From these three relationships, six conversion factors can be written.
From the first relationship:

$$\frac{2 \text{ volumes CO}}{2 \text{ volumes CO}_2} \quad \text{and} \quad \frac{2 \text{ volumes CO}_2}{2 \text{ volumes CO}}$$

From the second relationship:

$$\frac{1 \text{ volume O}_2}{2 \text{ volumes CO}_2} \quad \text{and} \quad \frac{2 \text{ volumes CO}_2}{1 \text{ volume O}_2}$$

From the third relationship:

$$\frac{2 \text{ volumes CO}}{1 \text{ volume O}_2} \quad \text{and} \quad \frac{1 \text{ volume O}_2}{2 \text{ volumes CO}}$$

The more gaseous reactants and products there are in a chemical reaction, the greater the number of volume–volume conversion factors obtainable from the equation for the chemical reaction. The use of volume–volume conversion factors is illustrated in Example 12.10.

Table 12.4 **Ways in Which Equation Coefficients May Be Interpreted**

For the General Equation	2A (g)	+	3B (g)	$\longrightarrow$	C (g)	+	2D (g)
The ratio of molecules is	2	:	3	:	1	:	2
The ratio of moles is	2	:	3	:	1	:	2
The ratio of volumes of gas (at the same temperature and pressure) is	2	:	3	:	1	:	2

EXAMPLE 12.10

Using Gay-Lussac's Law of Combining Volumes to Determine Reactant–Product Volume Relationships

Nitrogen (N_2) reacts with hydrogen (H_2) to produce ammonia (NH_3) as shown by the equation

$$N_2(g) + 3\,H_2(g) \longrightarrow 2\,NH_3(g)$$

What volume of H_2, in liters, at 750 mm Hg and 25°C is required to produce 1.75 L of NH_3 at the same temperature and pressure?

SOLUTION

STEP 1 The given quantity is 1.75 L of NH_3 and the desired quantity is liters of H_2.

$$1.75\ \text{L}\ NH_3 = ?\ \text{L}\ H_2$$

STEP 2 This is a volume-to-volume problem. The conversion factor needed for this one-step problem is derived from the coefficients of H_2 and NH_3 in the equation for the chemical reaction. The equation tells us that at constant temperature and pressure three volumes of H_2 are needed to prepare two volumes of NH_3. From this relationship the conversion factor

$$\frac{3\ \text{L}\ H_2}{2\ \text{L}\ NH_3}$$

is obtained.

> Note that when *Gay-Lussac's law of combining volumes* does apply, you do not need to know the exact temperature and pressure values for the gases since they are not used in the calculation. All you need to know is that the temperature and pressure, whatever they may be, are the same for all of the gases.

STEP 3 The dimensional analysis setup for this problem is

$$1.75\ \cancel{\text{L}\ NH_3} \times \frac{3\ \text{L}\ H_2}{2\ \cancel{\text{L}\ NH_3}}$$

Note that it makes no difference what the temperature and pressure are, as long as they are the same for the two gases involved in the calculation.

STEP 4 Doing the arithmetic, after cancellation of units, gives

$$\frac{1.75 \times 3}{2}\ \text{L}\ H_2 = 2.625\ \text{L}\ H_2\ \text{(calculator answer)}$$

$$= 2.62\ \text{L}\ H_2 \quad \textbf{(correct answer)}$$

> If gases are not at identical temperatures and pressures, *Gay-Lussac's law of combining volumes* does not apply. In such cases, the best approach is to convert volume information to a mole basis using other gas laws. Then use the mole interpretation for the coefficients in the chemical equation. Section 12.14 considers this approach in detail.

CHEMICAL EXTENSION

Gaseous N_2 and H_2 do not react to form NH_3 at room temperature and pressure. For a successful reaction to occur, high pressures (100–1000 atm), high temperatures (400–550°C), and a catalyst are required. The needed conditions for successful reaction were first worked out by the German chemist Fritz Haber (1868–1934), and commercially this reaction is known as the Haber process for ammonia production. The ammonia so produced is then used as fertilizer.

The raw materials for the Haber process are air and natural gas. Fractional distillation of liquid air supplies the needed N_2. Natural gas is the source of H_2.

You might expect that H_2O would be the H_2 source. However, the expense of obtaining H_2 from H_2O prohibits this; electrical energy (electrolysis) is required to break down the water. Currently, methane (CH_4), the main component of natural gas, is the source of H_2. Steam reforming of methane, a process in which CH_4 reacts with water vapor at high temperatures, produces the H_2.

$$CH_4(g) + 2\,H_2O(g) \xrightarrow[\text{high T}]{\text{catalyst}} CO_2(g) + 4\,H_2(g)$$

Practice Exercise 12.10

Hydrogen, H_2, reacts with acetylene, C_2H_2, according to the equation

$$2\,H_2(g) + C_2H_2(g) \longrightarrow C_2H_6(g)$$

What volume of H_2, in liters, is required to react with 13.7 L of C_2H_2? Assume that both gases are at the same conditions of temperature and pressure.

12.9   Volumes of Gases and the Limiting Reactant Concept

In Section 10.9, we considered the concept of a limiting reactant and did calculations that determined limiting reactant identity. These calculations were based on moles of product formed from given masses of various reactants. Limiting reactant calculations can also be made using volumes of gases and the volume interpretation of equation coefficients. Example 12.11 illustrates how volume-based limiting reactant calculations are carried out.

EXAMPLE 12.11

Calculating a Limiting Reactant Using Volumes of Gases

Many high-temperature welding torches use an acetylene–oxygen mixture as a fuel. The equation for the reaction between these two substances is

$$2\,C_2H_2(g) + 5\,O_2(g) \longrightarrow 4\,CO_2(g) + 2\,H_2O(g)$$

If a fuel mixture contains 30.0 L of C_2H_2 and 80.0 L of O_2, which mixture component is the limiting reactant?

SOLUTION

To determine a limiting reactant using volumes, we calculate how many volumes of gaseous product can be formed from given volumes of reactants, assuming constant conditions of temperature and pressure. In this particular problem there are two gaseous products: CO_2 and H_2O. It is sufficient to calculate the volume of just one of the products. The decision as to which product is used for the calculation is arbitrary; we will choose CO_2. We go through the calculation twice, once for each reactant. For C_2H_2:

$$30.0 \; \cancel{\text{L } C_2H_2} \times \frac{4 \text{ L } CO_2}{2 \; \cancel{\text{L } C_2H_2}} = 60 \text{ L } CO_2 \quad \text{(calculator answer)}$$

$$= 60.0 \text{ L } CO_2 \; \textbf{(correct answer)}$$

The conversion factor used was obtained from the coefficients of the balanced equation. For O_2:

$$80.0 \text{ L } \cancel{O_2} \times \frac{4 \text{ L } CO_2}{5 \text{ L } \cancel{O_2}} = 64 \text{ L } CO_2 \quad \text{(calculator answer)}$$

$$= 64.0 \text{ L } CO_2 \quad \textbf{(correct answer)}$$

The limiting reactant is the reactant that will produce the fewest number of liters of CO_2, which in this case is C_2H_2.

CHEMICAL EXTENSION

When acetylene (C_2H_2), a very reactive, colorless gas, is burned with oxygen in an oxyacetylene welding torch, a flame temperature of about 3000°C is reached. A flame of such high temperature can be used to cut through and to weld metals. Only about 10% of acetylene production, however, finds this end use. The majority of acetylene production is converted to chemical intermediates needed to make various types of plastics, fibers, and resins.

Early models of cars had headlights that used acetylene produced by the action of slowly dripping water on calcium carbide (CaC_2). The same type of lamp, which was also used by miners, is still often used by spelunkers (cave explorers).

Practice Exercise 12.11

Hydrogen sulfide (H_2S) reacts with oxygen, O_2, according to the equation

$$2 H_2S(g) + 3 O_2(g) \rightarrow 2 SO_2(g) + 2 H_2O(g)$$

If 33 L of H_2S and 45 L of O_2 at the same temperature and pressure are allowed to react, which gas is the limiting reactant?

12.10 Avogadro's Law: A Volume–Quantity Relationship

In the year 1811, Amedeo Avogadro, the Avogadro involved in Avogadro's number (Sec. 9.5), published work in which he proposed an explanation for Gay-Lussac's observations (Sec. 12.8). His proposal, a hypothesis at the time it was first published, is now known as *Avogadro's law*. Its validity has been demonstrated in a number of ways.

Avogadro's law *states that equal volumes of different gases, measured at the same temperature and pressure, contain equal numbers of molecules.* Although Avogadro's law seems very simple in terms of today's scientific knowledge, it was a very astute conclusion at the time it was first proposed. At that time scientists were still struggling with the differences between atoms and molecules, and Avogadro's reasoning eventually led to the realization that many of the common gaseous elements, such as hydrogen, oxygen, nitrogen, and chlorine, occur naturally as diatomic molecules (H_2, O_2, N_2, Cl_2) rather than single atoms.

Equal volumes of gases at the same temperature and pressure, because they contain equal numbers of molecules, must also contain equal numbers of moles of molecules. Thus, Avogadro's law can also be stated in terms of volumes and moles rather than volumes and molecules.

Avogadro's law can also be stated in another alternative form, which uses terminology similar to that used in Boyle's, Charles's, and Gay-Lussac's laws. This alternative form deals

Boyle's law, Charles's law, Gay-Lussac's law, and the combined gas law all share a common feature— in each case a fixed amount of gas is present. What happens if the amount of gas present in a sample is changed? Avogadro's law deals with this situation.

with two samples of the same gas rather than two samples of different gases. **Avogadro's law** *states that the volume of a gas is directly proportional to the number of moles of gas present if the temperature and pressure are kept constant.* When the original number of moles of gas present is doubled, the volume of the gas increases twofold. Halving the number of moles present halves the volume.

Experimentally, it is easy to show the direct relationship between volume and amount of gas present at constant temperature and pressure. All that is needed is an inflated balloon. Adding more gas to the balloon increases the volume of the balloon. The more gas added to the balloon, the greater the volume increase. Conversely, letting gas out of the balloon decreases its volume. It should be noted that this balloon demonstration is an approximate rather than exact demonstration for the law, since the tension of the rubber material from which the balloon is made exerts an effect on the proportionality between volume and moles.

Mathematical statements of Avogadro's law, where n represents the number of moles, are

$$\frac{V}{n} = \text{constant} \quad \text{and} \quad \frac{V_1}{n_1} = \frac{V_2}{n_2}$$

Note the similarity in mathematical form between this law and the laws of Charles and Gay-Lussac. The similarity results from all three laws being direct proportionality relationships between two variables.

Avogadro's law and the combined gas law can be combined to give the expression

$$\frac{P_1 V_1}{n_1 T_1} = \frac{P_2 V_2}{n_2 T_2}$$

This equation covers the situation where none of the four variables, P, T, V, and n, is constant. With it, a change in any one of the four variables brought about by changes in the other three variables can be calculated.

EXAMPLE 12.12

Calculating Final Volume Using Avogadro's Law

A balloon containing 2.00 moles of He has a volume of 0.500 L at a given temperature and pressure. If 0.48 mole of He gas is removed from the balloon without changing temperature and pressure conditions, what will be the new volume, in liters, of the balloon?

SOLUTION

We will use the equation

$$\frac{V_1}{n_1} = \frac{V_2}{n_2}$$

in solving the problem.

Writing the given data in terms of initial and final conditions, we have

$$V_1 = 0.500 \text{ L} \qquad V_2 = ? \text{ L}$$

$$n_1 = 2.00 \text{ moles} \qquad n_2 = (2.00 - 0.48)\text{moles} = 1.52 \text{ moles}$$

Rearrangement of the Avogadro's law expression to isolate V_2 on the left side gives

$$V_2 = V_1 \times \frac{n_2}{n_1}$$

Substituting numerical values into the equation gives

$$V_2 = 0.500 \text{ L} \times \frac{1.52 \text{ moles}}{2.00 \text{ moles}} = 0.38 \text{ L} \quad \text{(calculator answer)}$$

$$= 0.380 \text{ L} \quad \textbf{(correct answer)}$$

Our answer is consistent with reasoning. Decreasing the number of moles of gas present in the balloon at constant temperature and pressure should decrease the volume of the balloon.

CHEMICAL EXTENSION

Helium gas is twice as dense as hydrogen under the same conditions. However, helium, because it is nonflammable, is used instead of hydrogen to provide buoyancy in airships. Almost everyone has heard of the "*Hindenburg* disaster," the disastrous fire in the German hydrogen-filled airship *Hindenburg* as it landed in Lakehurst, New Jersey, in 1937. The *Hindenburg* was actually designed to use helium. Hydrogen had to be used, however, because the United States refused to supply Germany with helium, fearing the gas would be used for military purposes.

There are numerous applications for helium-filled airships today, besides the Goodyear blimp and other advertising blimps. Other applications include use at radar posts by the U.S. Coast Guard to identify illegal drug-carrying flights and in instruments to study the upper canopy of the rainforest in the Amazon.

Practice Exercise 12.12

A balloon containing 1.83 moles of He has a volume of 0.673 L at a given temperature and pressure. If an additional 0.50 mole of He gas is introduced into the balloon without changing temperature and pressure conditions, what will be the new volume, in liters, of the balloon?

12.11 Molar Volume of a Gas

The **molar volume of a gas** *is the volume of one mole of gas, which is 22.414 liters at STP conditions*. It follows from Avogadro's law (Sec. 12.10) that all gases will have the same molar volume. To visualize a volume this size, 22.414 L, think of standard-sized basketballs. The volume occupied by three standard-sized basketballs is very close to 22.414 L.

The fact that all gases have the same molar volume at STP (or any other temperature–pressure combination) is a property unique to the gaseous state. Similar statements cannot be made about the liquid and solid states. The reason for the difference can be understood by considering the relationship between the volume the molecules occupy in the given physical state and the volume of the molecules themselves. In the gaseous state, because the molecules are so far apart—a gas is mostly empty space—the volume of the gas molecules themselves is negligible compared to the total volume. This is not the case in the liquid and solid states, where the molecules are in close contact with each other.

It is informative to contrast the properties of molar volume and molar mass, the latter having been discussed in Section 9.5. The *molar mass* is a constant for a given substance, and is independent of temperature and pressure. Each substance has its own unique molar mass. The *molar volume* of a gas has a temperature and pressure specification (STP conditions). All gases have the same molar volume at the same temperature and pressure (STP conditions).

The volume of a basketball is 7.5 L, a soccer ball, 6.0 L, and a football, 4.4 L.

22.414 L is a little over half the volume of a 10-gallon aquarium.

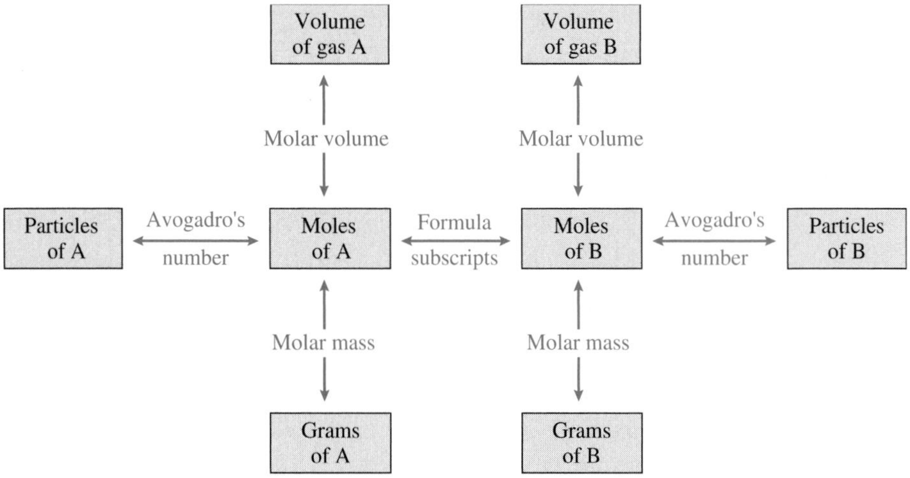

Figure 12.9

Quantitative relationships needed for solving mass-to-volume and volume-to-mass chemical-formula-based problems.

The molar volume concept is very useful in a variety of types of calculations. When used in calculations, the concept "translates" into a conversion factor having either of two forms:

$$\frac{1 \text{ mole gas}}{22.41 \text{ L gas}} \quad \text{or} \quad \frac{22.41 \text{ L gas}}{1 \text{ mole gas}}$$

In conversion factors, the four-significant-figure value of 22.41 for molar volume is usually adequate, just as the four-significant-figure value of 6.022×10^{23} is usually adequate for Avogadro's number (Sec. 9.7). A five-significant-figure value of 22.414 for molar volume is available, however, when needed.

A most common type of problem involving these conversion factors is one where the volume of a gas at STP is known and you are asked to calculate from it the moles, grams, or particles of gas present, or vice versa. Figure 12.9 summarizes the relationships needed in performing calculations of this general type.

Perhaps you recognize Figure 12.9 as being very similar to a diagram you have already encountered many times in problem solving, Figure 9.5. This "new diagram" differs from Figure 9.5 in only one way; volume boxes have been added at the top of the diagram. The diagram in Figure 12.9 is used in the same way as the earlier one. The given and desired quantities are determined, and the arrows of the diagram are used to "map out" the pathway to be used in going from the given quantity to the desired quantity.

Examples 12.13 and 12.14 illustrate the usefulness of Figure 12.9. Example 12.13 is a simple "volume of gas A" to "grams of gas A" problem. Example 12.14 illustrates the important point that equal volumes of different gases at STP do not have equal masses since their molecules will be different.

EXAMPLE 12.13

Calculating the Volume of a Gaseous Sample at STP Given Its Mass

Fluorine gas, F_2, is one of the most reactive of all gases. What would be the volume, in liters, of a 15.0-g sample of this gas at STP conditions?

SOLUTION

STEP 1 The given quantity is 15.0 g of F_2, and the unit of the desired quantity is liters of F_2.

$$15.0 \text{ g } F_2 = ? \text{ L } F_2$$

STEP 2 In terms of Figure 12.9, this problem is a "grams of A" to "volume of gas A" problem. The pathway is

Grams of A $\xrightarrow[\text{mass}]{\text{Molar}}$ Moles of A $\xrightarrow[\text{volume}]{\text{Molar}}$ Volume of gas A

Using dimensional analysis, we set up this sequence of conversion factors as

$$15.0 \text{ g } F_2 \times \frac{1 \text{ mole } F_2}{38.00 \text{ g } F_2} \times \frac{22.41 \text{ L } F_2}{1 \text{ mole } F_2}$$

grams A $\longrightarrow$ moles A $\longrightarrow$ volume A

Note how the molar volume relationship that 1 mole of gas is equal to 22.41 L is used as the last conversion factor.

STEP 3 The solution, obtained by doing the arithmetic, is

$$\frac{15.0 \times 1 \times 22.41}{38.00 \times 1} \text{ L } F_2 = 8.8460526 \text{ L } F_2 \quad \text{(calculator answer)}$$

$$= 8.85 \text{ L } F_2 \quad \textbf{(correct answer)}$$

CHEMICAL EXTENSION

The element fluorine, in the form of F_2 gas, is the most reactive of all the elements. It reacts with every other element in the periodic table except for helium, neon, and argon. (There are no known compounds of these latter three elements; that is, they will not react with anything.) Asbestos, silicon, and water burst into flame in the presence of fluorine gas.

Because of its extreme reactivity, it is difficult to find a container to hold gaseous F_2. It attacks both glass and quartz and causes most metals to burst into flame. Containers made of certain copper–nickel alloys can be used to handle fluorine. The fluorine still reacts with these alloys; the result, however, is a coating of metal fluoride that protects the metal from further reaction with the fluorine.

Practice Exercise 12.13

Hydrogen sulfide, H_2S, is a colorless gas with a strong foul odor. This gas is responsible for the smell of rotten eggs. What would be the volume, in liters, of a 25.0 g sample of this gas at STP conditions?

EXAMPLE 12.14

Calculating the Mass of a Gaseous Sample at STP Given Its Volume

What is the mass, in grams, of 2.50 L of each of the following gases at STP conditions?

(a) carbon monoxide (CO)   **(b)** carbon dioxide (CO_2)

SOLUTION

(a)

STEP 1 The given quantity is 2.50 L of CO at STP, and the desired quantity is grams of CO.

$$2.50 \text{ L CO} = \text{ g CO}$$

STEP 2 This is a "volume of gas A" to "grams of A" problem. The pathway appropriate for solving this problem, as indicated in Figure 12.9, is

$$\boxed{\text{Volume of A}} \xrightarrow[\text{volume}]{\text{Molar}} \boxed{\text{Moles of A}} \xrightarrow[\text{mass}]{\text{Molar}} \boxed{\text{Grams of gas A}}$$

The setup, from dimensional analysis, is

$$2.50 \text{ L CO} \times \frac{1 \text{ mole CO}}{22.41 \text{ L CO}} \times \frac{28.01 \text{g CO}}{1 \text{ mole CO}}$$

STEP 3 Collecting the numerical terms after cancellation of units and doing the arithmetic gives

$$\frac{2.50 \times 1 \times 28.01}{22.41 \times 1} \text{g CO} = 3.1247211 \text{ g CO} \quad \text{(calculator answer)}$$
$$= 3.12 \text{ g CO} \quad \text{(correct answer)}$$

(b) The analysis and setup for this part are identical to those in (a) except that CO_2 and its molecular mass replace CO and its molecular mass. Thus, the setup is

$$2.50 \text{ L CO}_2 \times \frac{1 \text{ mole CO}_2}{22.41 \text{ L CO}_2} \times \frac{44.01 \text{ g CO}_2}{1 \text{ mole CO}_2} = 4.9096385 \text{ g CO}_2 \quad \text{(calculator answer)}$$
$$= 4.91 \text{ g CO}_2 \quad \text{(correct answer)}$$

Equal volumes of different gases at STP do not have equal masses; 2.50 L of CO has a mass of 3.12 g, and 2.50 L of CO_2 has a mass of 4.91 g.

Practice Exercise 12.14

What is the mass, in grams, of 4.00 L of each of the following gases at STP conditions?

(a) sulfur dioxide (SO_2) **(b)** sulfur trioxide (SO_3)

The density of a gas at STP conditions can be calculated by dividing the molar mass (grams/mole) of the substance by its molar volume (liters/mole).

$$\text{density (STP)} = \frac{\text{molar mass}}{\text{molar volume(STP)}}$$

$$\text{density (STP)} = \frac{\dfrac{\text{grams}}{\text{mole}}}{\dfrac{\text{liters}}{\text{mole}}} = \frac{\text{grams} \times \text{mole}}{\text{mole} \times \text{liters}} = \frac{\text{grams}}{\text{liters}}$$

The units of density, g/L, are, indeed, obtained from this division.

Since molar volume is the same for all gases, the density of a gas at a given temperature and pressure is a function of (depends only on) the molar mass.

Rearrangement of the density equation enables one to calculate molar mass:

$$\text{molar mass} = \text{density (STP)} \times \text{molar volume (STP)}$$

Calculating the Density at STP Using Molar Mass and Molar Volume

Gaseous hydrogen cyanide, HCN, is highly poisonous and has been used to execute prisoners convicted of murder. Calculate the density of HCN gas at STP conditions, in grams per liter.

SOLUTION

The molar mass and molar volume relationships needed to solve this problem are

$$\text{molar mass} = \frac{27.03 \text{ g HCN}}{1 \text{ mole HCN}}, \quad \text{molar volume} = \frac{22.41 \text{ L HCN}}{1 \text{ mole HCN}}$$

Substituting into the density equation gives

$$\text{density} = \frac{\text{molar mass}}{\text{molar volume}} = \frac{\dfrac{27.03 \text{ g HCN}}{1 \text{ mole HCN}}}{\dfrac{22.41 \text{ L HCN}}{1 \text{ mole HCN}}}$$

$$= \frac{27.03 \text{ g HCN} \times 1 \text{ mole HCN}}{1 \text{ mole HCN} \times 22.41 \text{ L HCN}}$$

$$= \frac{27.03 \times 1}{1 \times 22.41} \frac{\text{g}}{\text{L}} \text{HCN}$$

$$= 1.2061579 \frac{\text{g}}{\text{L}} \text{HCN} \quad \text{(calculator answer)}$$

$$= 1.206 \frac{\text{g}}{\text{L}} \text{HCN} \quad \text{(correct answer)}$$

CHEMICAL EXTENSION

The death of people in house fires is often caused by toxic gases rather than actual physical burns. It is common knowledge that carbon monoxide is formed in smoldering fires. It is not so common knowledge that hydrogen cyanide is often formed when plastics burn. Hydrogen cyanide may be as equally important as carbon monoxide in some house fire deaths.

The standard antidote for hydrogen cyanide poisoning, which must be administered quickly because of the rapid action of HCN, is sodium thiosulfate ($Na_2S_2O_3$) solution. A sulfur atom is transferred from the thiosulfate ion to the cyanide ion, converting it to the nonpoisonous thiocyanate ion (SCN^-).

$$HCN + Na_2S_2O_3 \longrightarrow HSCN + Na_2SO_3$$

One medical recommendation is that fire department crews carry hypodermic syringes filled with sodium thiosulfate solution for emergency treatment of smoke inhalation victims.

Practice Exercise 12.15

Gaseous hydrogen, H_2, is the lightest of all gases. Calculate the density of H_2 gas at STP conditions, in grams per liter.

EXAMPLE 12.16

Calculating the Volume at STP of a Component of a Compound

If the Xe atoms present in 25.0 g of solid XeO_3 are converted to Xe gas, what volume, in liters, will the Xe gas occupy at STP conditions?

SOLUTION

STEP 1 The given quantity is 25.0 g of XeO_3, and the desired quantity is liters of Xe at STP. In the jargon of Figure 12.9, this is a "grams of A" to "volume of gas B" problem.

STEP 2 The sequence of conversion factors needed for this problem, using Figure 12.9 as a guide, follows the pathway

$$\boxed{\text{Grams of A}} \xrightarrow[\text{mass}]{\text{Molar}} \boxed{\text{Moles of A}} \xrightarrow[\text{subscript}]{\text{Formula}} \boxed{\text{Moles of B}} \xrightarrow[\text{volume}]{\text{Molar}} \boxed{\text{Volume of gas B}}$$

Translating this pathway into a dimensional analysis setup gives

$$25.0 \text{ g XeO}_3 \times \frac{1 \text{ mole XeO}_3}{179.29 \text{ g XeO}_3} \times \frac{1 \text{ mole Xe}}{1 \text{ mole XeO}_3} \times \frac{22.41 \text{ L Xe}}{1 \text{ mole Xe}}$$

STEP 3 Collecting numerical terms after cancellation of units and doing the arithmetic gives

$$\frac{25.0 \times 1 \times 1 \times 22.41}{179.29 \times 1 \times 1} \text{ L Xe} = 3.1248257 \text{ L Xe} \quad \text{(calculator answer)}$$

$$= 3.12 \text{ L Xe} \quad \textbf{(correct answer)}$$

CHEMICAL EXTENSION

Xenon, with a density approximately five times that of air, is the heaviest of the stable (nonradioactive) noble gases. It is the rarest (least abundant) of all the stable naturally occurring elements. Traces of xenon (produced from radioactive materials) are found in minerals and meteorites. Air, which is the commercial source for xenon, contains 8 parts per billion of xenon.

When subjected to an electric discharge, xenon gas produces a brilliant white light. In photography, it is used in strobe lights, which supply extremely short bursts of light in a continuous manner. With such lights, cameras can "freeze" on photographic film a rapidly moving object.

Practice Exercise 12.16

If the Xe atoms present in 14.5 g of gaseous XeF_2 are converted to Xe gas, what volume, in liters, will the Xe gas occupy at STP conditions?

12.12 The Ideal Gas Law

The **ideal gas law** *describes the relationships among the four variables temperature* (T), *pressure* (P), *volume* (V), *and moles of gas* (n) *for gaseous substances under one set of conditions.*

In previous sections we have discussed three independent relationships dealing with the volume of a gas. They are

$$\text{Boyle's law:} \quad V = k \times \frac{1}{P} \quad (n \text{ and } T \text{ constant})$$

$$\text{Charles's law:} \quad V = kT \quad (n \text{ and } P \text{ constant})$$

$$\text{Avogadro's law:} \quad V = kn \quad (P \text{ and } T \text{ constant})$$

We can combine these three equations into a single expression

$$V = \frac{kTn}{P}$$

since we know (from mathematics) that if a quantity is independently proportional to two or more quantities, it is also proportional to their product. This combined equation is a mathematical statement of the ideal gas law. Any gas that obeys the individual laws of Boyle, Charles, and Avogadro will also obey the ideal gas law.

An alternate statement of the ideal gas law is

$$PV = nRT$$

This form of the ideal gas law, besides being rearranged, differs from the previous form in that the proportionality constant k has been given the symbol R. The constant R is called the *ideal gas constant*. To use the ideal gas equation we must know the value of R. This can be determined by substituting STP values for P, V, T, and n into the ideal gas equation and solving for R. For 1 mole of gas at STP, $P = 1$ atm, $V = 22.414$ L, $T = 273.15$ K, and $n = 1$ mole. Substituting these values into the ideal gas equation arranged to have R isolated on the left side gives

$$R = \frac{PV}{nT} = \frac{(1 \text{ atm})(22.414 \text{ L})}{(1 \text{ mole})(273.15 \text{ K})} = 0.082057477 \frac{\text{atm} \cdot \text{L}}{\text{mole} \cdot \text{K}} \quad \text{(calculator answer)}$$

$$= 0.082057 \frac{\text{atm} \cdot \text{L}}{\text{mole} \cdot \text{K}} \quad \textbf{(correct answer)}$$

Notice the complex units associated with R—the four variables temperature, pressure, volume, and moles—are all involved. This will always be the case.

The value of R is dependent on the units used to express pressure and volume. If we use 760 mm Hg instead of 1 atm as standard pressure, then R would have the value 62.364 and the units $(\text{mm Hg} \cdot \text{L})/(\text{mole} \cdot \text{K})$.

$$R = \frac{PV}{nT} = \frac{(760 \text{ mm Hg})(22.414 \text{ L})}{(1 \text{ mole})(273.15 \text{ K})} = 62.363682 \frac{\text{mm Hg} \cdot \text{L}}{\text{mole} \cdot \text{K}} \quad \text{(calculator answer)}$$

$$= 62.364 \frac{\text{mm Hg} \cdot \text{L}}{\text{mole} \cdot \text{K}} \quad \textbf{(correct answer)}$$

These two values of R, to four significant figures, should be memorized, since it is in these forms that they are usually used in problem solving.

$$0.08206 \frac{\text{atm} \cdot \text{L}}{\text{mole} \cdot \text{K}} \qquad 62.36 \frac{\text{mm Hg} \cdot \text{L}}{\text{mole} \cdot \text{K}}$$

When pressure units other than millimeters of mercury or atmosphere and volume units other than liters are encountered, convert them to one of these units and then use the appropriate known R value.

If three of the four variables in the ideal gas equation are known, then the fourth can be calculated by the equation. The ideal gas equation is used in calculations when *one* set of conditions is given with one missing variable. The combined gas law (Sec. 12.6) is used when *two* sets of conditions are given with one missing variable. Examples 12.17 and 12.18 illustrate the use of the ideal gas equation.

EXAMPLE 12.17

Calculating the Volume of a Gas Using the Ideal Gas Law

The colorless, odorless, tasteless gas carbon monoxide, CO, is a by-product of incomplete combustion of any material that contains the element carbon. Calculate the volume, in liters, occupied by 1.52 moles of this gas at 0.992 atm pressure and a temperature of 65°C.

SOLUTION

This problem deals with only one set of conditions, a situation where the ideal gas equation is applicable. Three of the four variables in the ideal gas equation (P, n, and T) are given, and the fourth (V) is to be calculated.

$$P = 0.992 \text{ atm} \quad n = 1.52 \text{ moles}$$
$$V = ? \text{ L} \quad T = 65°C = 338 \text{K}$$

Rearranging the ideal gas equation to isolate V on the left side of the equation gives

$$V = \frac{nRT}{P}$$

Since the pressure is given in atmospheres and the volume unit is liters, the appropriate R value is

$$R = 0.08206 \frac{\text{atm} \cdot \text{L}}{\text{mole} \cdot \text{K}}$$

Substituting the given numerical values into the equation and canceling units gives

$$V = \frac{(1.52 \text{ moles})\left(0.08206 \dfrac{\text{atm} \cdot \text{L}}{\text{mole} \cdot \text{K}}\right)(338 \text{ K})}{0.992 \text{ atm}}$$

Note how all of the parts of the ideal gas constant unit cancel except for one, the volume part. Doing the arithmetic, we get as an answer 42.5 L CO.

$$V = \frac{1.52 \times 0.08206 \times 338}{0.992} \text{L CO}$$

$$= 42.499138 \text{ L CO} \quad \text{(calculator answer)}$$

$$= 42.5 \text{ L CO} \quad \text{(correct answer)}$$

CHEMICAL EXTENSION

Almost everyone is aware that carbon monoxide is toxic to human beings. It acts by reducing the oxygen-carrying capacity of the blood. Normally, hemoglobin present in blood picks up oxygen in the lungs and distributes it to cells throughout the body. Carbon monoxide bonds to hemoglobin at the location where oxygen normally bonds, thus preventing the oxygen from attaching itself to the hemoglobin. Someone who dies from carbon monoxide poisoning actually dies from lack of oxygen.

There is no way for a person to tell that carbon monoxide is present (without using a CO detector or chemical tests) because CO is a nonirritating, colorless, odorless, tasteless gas. Drowsiness is usually the first symptom. There are many other common gases, including air pollutants, that are more toxic than CO. However, they have properties, such as odor or irritant, that give warning of their presence.

Practice Exercise 12.17

Ozone, a form of oxygen with the formula O_3, is naturally found in the upper atmosphere where it screens out 95–99% of the ultraviolet solar radiation headed for Earth. Calculate the volume, in liters, occupied by 1.52 moles of ozone at 0.882 atm pressure and a temperature of 75°C.

EXAMPLE 12.18

Calculating the Temperature of a Gas Using the Ideal Gas Law

Dinitrogen monoxide (nitrous oxide), N_2O, is a naturally present constituent, in trace amounts, of the atmosphere. Its source is decomposition reactions occurring in soils. What would be the temperature, in degrees Celsius, of a 67.4-g sample of N_2O gas under a pressure of 5.00 atm in a 7.00-L container?

SOLUTION

The amount of N_2O present is given in grams rather than moles. The grams need to be changed to moles prior to using the ideal gas law.

$$67.4 \text{ g N}_2\text{O} \times \frac{1 \text{ mole N}_2\text{O}}{44.02 \text{ g N}_2\text{O}} = 1.5311222 \text{ moles N}_2\text{O} \quad \text{(calculator answer)}$$

$$= 1.53 \text{ moles N}_2\text{O} \quad \textbf{(correct answer)}$$

Three of the four variables in the ideal gas equation (P, V, and n) are now known, and the fourth (T) is to be calculated.

$$P = 5.00 \text{ atm} \quad n = 1.53 \text{ moles}$$
$$V = 7.00 \text{ L} \quad T = ? \text{ K}$$

Rearranging the ideal gas equation to isolate T on the left side gives

$$T = \frac{PV}{nR}$$

Since the pressure is given in atmospheres and the volume in liters, the value of R to be used is

$$R = 0.08206 \frac{\text{atm} \cdot \text{L}}{\text{mole} \cdot \text{K}}$$

Substituting numerical values into the equation gives

$$T = \frac{(5.00 \text{ atm})(7.00 \text{ L})}{(1.53 \text{ moles})\left(0.08206 \dfrac{\text{atm} \cdot \text{L}}{\text{mole} \cdot \text{K}}\right)}$$

Notice again how the gas constant units, except for K, cancel. After cancellation, the expression $1/(1/K)$ remains. This expression is equivalent to K. That this is the case can be easily shown. All we need to do is multiply both the numerator and denominator of the fraction by K.

$$\frac{1 \times K}{\frac{1}{K} \times K} = K$$

Doing the arithmetic, we get as an answer 279 K for the temperature of the O_2 gas.

$$T = \frac{(5.00)(7.00)}{(1.53)(0.08206)} K = 278.7694 \text{ K} \quad \text{(calculator answer)}$$

$$= 279 \text{ K} \qquad \textbf{(correct answer)}$$

The calculated temperature is in kelvins. To convert to degrees Celsius, the unit specified in the problem statement, we subtract 273 from the Kelvin temperature.

$$T(°C) = 279 \text{ K} - 273 = 6°C$$

CHEMICAL EXTENSION

Dinitrogen monoxide, a sweet-smelling, colorless gas, is also known by the names *nitrous oxide* and *laughing gas*. As early as 1800, it was noted that when N_2O is inhaled in relatively small amounts, it produces a state often accompanied by either convulsive laughter or crying. When taken in larger doses, it produces fast and efficient relief from pain.

Being fat-soluble, nitrous oxide is used as a propellant gas in whipped cream dispensers. It is the only common gas other than oxygen to support combustion. When N_2O is decomposed, it produces a gaseous mixture that is one-third oxygen:

$$2\,N_2O(g) \longrightarrow 2\,N_2(g) + O_2(g)$$

Since the oxygen content (mass percent) of N_2O is 36% and that of air is 21%, a candle glows brighter in nitrous oxide than in air. N_2O, itself, is not flammable in air.

Practice Exercise 12.18

Decomposition processes in wet locations where oxygen is not available, such as swamps and natural wetlands, produce the gas methane, CH_4. Calculate the temperature, in degrees Celsius, of a 24.56-gram sample of CH_4 gas under a pressure of 6.00 atm in an 8.00-L container.

A word about the phrase "ideal gas" in the name *ideal gas law* is in order. An **ideal gas** *is a gas that obeys exactly all of the statements of kinetic molecular theory (Sec. 11.3) and obeys exactly the ideal gas law.* Real gases are not ideal gases; that is, real gases do not obey exactly the ideal gas equation. Nonetheless, for real gases under ordinary conditions of temperature and pressure, deviations from ideal gas behavior are small, and the ideal gas law (as well as the other laws discussed in this chapter) gives accurate information about gas behavior. It is only at low temperatures and/or high pressures (near liquefaction conditions) that ideal gas behavior breaks down. At low temperatures, because of the slower motion of the molecules, attractive forces between molecules begin to become important. At high pressures, where the molecules are forced closer together, attractive forces also cause deviation from behavior predicted by the ideal gas law.

12.13 Molar Mass, Density, and the Ideal Gas Law

The ideal gas law can be used to determine other properties of gases. In this section, we will use this law to determine the molar mass of a gas and the density of a gas.

The Molar Mass of a Gas

For compounds that are gases at convenient temperatures and pressures, the ideal gas law provides a basis for determining molar mass. Required information is the identity of the gas, its mass, and the conditions it is under. There are two approaches to molar mass determination using the ideal gas law, both of which are illustrated in Example 12.19.

EXAMPLE 12.19

Calculating the Molar Mass of a Gas Using the Ideal Gas Law

A 3.30-g sample of chlorine gas (Cl_2) occupies a volume of 1.20 L at 741 mm Hg and 33°C. Based on this data, calculate the molar mass of chlorine gas.

SOLUTION METHOD A

We will first use the ideal gas equation, $PV = nRT$, to find the number of moles of gas. Then, knowing the mass of that number of moles, the mass of one mole (the molar mass) can be calculated.

Rearranging the ideal gas equation so that the number of moles (n) is isolated on one side of the equation gives

$$n = \frac{PV}{RT}$$

All the quantities on the right side of the equation are known.

$$P = 741 \text{ mm Hg} \quad T = 33°C + 273 = 306 \text{ K}$$

$$V = 1.20 \text{ L} \quad R = 62.36 \frac{\text{mm Hg} \cdot \text{L}}{\text{mole} \cdot \text{K}}$$

Substitution of these values into the equation gives

$$n = \frac{(741 \text{ mm Hg})(1.20 \text{ L})}{\left(62.36 \dfrac{\text{mm Hg} \cdot \text{L}}{\text{mole} \cdot \text{K}}\right)(306 \text{ K})}$$

Doing the arithmetic, we obtain a value of 0.0466 mole of Cl_2.

$$n = \frac{741 \times 1.20}{62.36 \times 306} \text{ mole } Cl_2 = 0.046598498 \text{ mole } O_2 \quad \text{(calculator answer)}$$

$$= 0.0466 \text{ mole } O_2 \quad \text{(correct answer)}$$

The 0.0466 mole of Cl_2 gas has a mass of 3.30 grams (given in the problem statement).

$$0.0466 \text{ mole } Cl_2 = 3.30 \text{ grams } Cl_2$$

This equality can be used as a conversion factor to determine the mass, in grams, of 1 mole of Cl_2 gas, which is the molar mass of Cl_2.

$$1 \text{ mole } Cl_2 \times \frac{3.30 \text{ g } Cl_2}{0.0466 \text{ mole } Cl_2} = 70.81545 \text{ g } Cl_2 \quad \text{(calculator answer)}$$

$$= 70.8 \text{ g } Cl_2 \quad \text{(correct answer)}$$

Because molecules of chlorine gas are diatomic (Cl_2), its molar mass, calculated using a table of atomic masses (periodic table), is 70.90 g. Thus, the experimental molar mass value just calculated and the molar mass obtained from a table of atomic masses are in close agreement.

SOLUTION METHOD B

An alternate method for calculating molar mass of a gas involves incorporating the quantity molar mass (*MM*) directly into the ideal gas equation before using it and then solving directly for molar mass.

Dividing the mass of a substance (grams) by its molar mass (grams/mole) gives the moles of the substance.

$$\frac{\text{mass}(m)}{\text{molar mass}(MM)} = \frac{\cancel{g}}{\cancel{g}/\text{mole}} = \text{mole}$$

Thus,

$$n = \frac{m}{MM}$$

Substituting this equality in the ideal gas equation for moles (*n*) gives

$$PV = \left(\frac{m}{MM}\right)RT$$

Rearranging this equation by solving for molar mass (*MM*), we have

$$MM = \frac{mRT}{PV}$$

This equation can be used to directly calculate molar mass since all quantities on the right side of the equation are known.

$$m = 3.30 \text{ g}, \quad R = 62.36\frac{\text{mm Hg} \cdot \text{L}}{\text{mole} \cdot \text{K}}, \quad T = 33°\text{C} + 273 = 306 \text{ K}$$

$$P = 741 \text{ mm Hg}, \quad V = 1.20 \text{ L}$$

Substitution of these known values in the equation gives

$$MM = \frac{(3.30 \text{ g})\left(62.36\dfrac{\cancel{\text{mm Hg}} \cdot \cancel{\text{L}}}{\text{mole} \cdot \cancel{\text{K}}}\right)(306 \cancel{\text{K}})}{(741 \cancel{\text{mm Hg}})(1.20 \cancel{\text{L}})}$$

All units cancel except for grams per mole, the units of molar mass.

Doing the arithmetic, we obtain a value of 70.8 for the molar mass of Cl_2, the same value obtained using method A for solving this problem.

$$MM = \frac{3.30 \times 62.36 \times 306}{741 \times 1.20}\frac{\text{g}}{\text{mole}} = 70.817732\frac{\text{g}}{\text{mole}} \quad \text{(calculator answer)}$$

$$= 70.8\frac{\text{g}}{\text{mole}} \quad \text{(correct answer)}$$

CHEMICAL EXTENSION

Chlorine is a yellow-green, dense, sharp-smelling, toxic gas. Inhaling even a small amount of Cl_2 gas can cause extensive lung damage. Interestingly, the color of Cl_2 diminishes with temperature, and it is almost colorless at $-195°$C.

The toxicity of low concentrations of chlorine toward microorganisms leads to chlorine's use in disinfecting wastewater, swimming pools, and drinking water supplies. Waterborne disease-causing agents have been virtually eliminated from drinking supplies in developed countries because of chlorination.

The bleaching action of aqueous chlorine solutions is one of the well-known properties of chlorine. Before the bleaching effects of aqueous chlorine solution were known, linen had to be left in the sun for weeks at a time to make it white. The linen cloth was placed over wooden frames with the wooden frames arranged in large arrays in what were known as bleach fields or bleachers. This is the origin of the name bleachers for the bench seats at athletic fields that have a resemblance in structure. (Today most linen is bleached with hydrogen peroxide solution rather than chlorine.)

Practice Exercise 12.19

A 4.25-g sample of the refrigerant Freon-12 is vaporized and found to occupy a volume of 1.36 L at a temperature of 50°C and a pressure of 521 mm Hg. What is the molar mass of Freon-12?

The Density of a Gas

The ideal gas law can be rearranged to calculate density. First, we isolate n/V on the left side of the ideal gas law equation.

$$\frac{n}{V} = \frac{P}{RT}$$

The quantity n/V has the units moles/liter, units close to those for density, which are grams/liter. To change from moles/liter units to grams/liter units, we multiply both sides of the equation by molar mass.

$$\frac{n}{V}(MM) = \frac{P}{RT}(MM)$$

The quantity on the left side of this equation, $n(MM)/V$, has the units grams per liter and therefore represents density (d).

$$\frac{n}{V}(MM) = \frac{\cancel{\text{moles}}}{\text{liter}}\left(\frac{\text{grams}}{\cancel{\text{mole}}}\right) = \frac{\text{grams}}{\text{liter}} = d$$

Thus, the density of a gas can be calculated using the equation

$$d = \frac{P(MM)}{RT}$$

Before doing a calculation using this "density equation," let us consider "messages" about the density of a gas conveyed by this equation.

1. The density of a gas is directly proportional (Sec. 12.4) to pressure. The higher the pressure, the more dense the gas.
2. The density of a gas is directly proportional to its molar mass. The higher the molar mass, the more dense the gas. At a given temperature and pressure, CO_2 gas (with a molar mass of 44.01 g/mole) will have a higher density than O_2 gas (with a molar mass of 32.00 g/mole).
3. The density of a gas is inversely proportional (Sec. 12.3) to temperature. The higher the temperature, the less dense the gas.

EXAMPLE 12.20

Calculating the Density of a Gas Using the Ideal Gas Law in Modified Form

In gas law calculations, air is often considered to be a single gas with a molar mass of 29 g/mole. On this basis, calculate the density of air, in grams per liter, on a hot summer day (41°C) when the atmospheric pressure is 0.91 atm.

SOLUTION

The ideal gas equation in the modified form

$$d = \frac{P(MM)}{RT}$$

is used to calculate the density of a gas.

All the quantities on the right side of this equation are known.

$$P = 0.91 \text{ atm}, \qquad MM = 29 \text{ g/mole}$$

$$R = 0.08206 \frac{\text{atm} \cdot \text{L}}{\text{mole} \cdot \text{K}}, \quad T = 41°C = 314 \text{ K}$$

Substitution of these values into the equation gives

$$d = \frac{(0.91 \text{ atm})\left(29 \frac{\text{g}}{\text{mole}}\right)}{\left(0.08206 \frac{\text{atm} \cdot \text{L}}{\text{mole} \cdot \text{K}}\right)(314 \text{ K})}$$

All units cancel except for the desired ones, grams per liter.

Doing the arithmetic, we obtain a value of 1.0 g/L for the density of air at the specified temperature and pressure.

$$d = \frac{0.91 \times 29}{0.08206 \times 314} \frac{\text{g}}{\text{L}} = 1.0241845 \frac{\text{g}}{\text{L}} \quad \text{(calculator answer)}$$

$$= 1.0 \frac{\text{g}}{\text{L}} \qquad \textbf{(correct answer)}$$

Practice Exercise 12.20

Calculate the density of carbon dioxide gas, CO_2, in grams per liter, at 1.21 atm pressure and a temperature of 35°C.

12.14 Gas Laws and Chemical Equations

In Section 10.8 we learned to calculate the mass of any component (reactant or product) of a chemical reaction, given the balanced equation for the reaction and mass of any other component—a mass-to-mass (or gram-to-gram) problem. A simple extension of the procedures involved in solving such problems will enable us to do mass-to-volume and volume-to-mass calculations for reactions where at least one gas is involved.

If you are given the number of liters of any gaseous component of a reaction and are asked to calculate the mass of any other component (gaseous or otherwise)—a *volume-to-mass problem*—first convert the given volume of gas to moles of gas, using the principles in this chapter. If the reaction is at STP conditions, use molar volume as the conversion factor to go from volume to moles. At other temperatures and pressures, use the ideal gas law (or the combined gas law and molar volume) to make the conversion. Once you obtain moles of gas the rest of the calculation is a standard mole-to-mass conversion (Sec. 10.8).

If you are given the number of grams of any component (gaseous or otherwise) in a chemical reaction and are asked to calculate the volume of any gaseous component (reactant or product)—a *mass-to-volume problem*—use standard procedures to calculate the moles of gas and then the new procedures of this chapter to go from moles to volume.

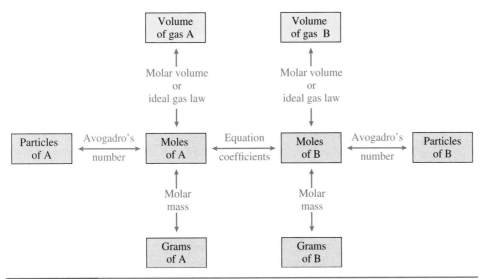

Figure 12.10

Quantitative relationships needed for solving mass-to-volume and volume-to-mass chemical-equation-based problems.

Figure 12.10 summarizes the relationships between the conversion factors needed to solve mass-to-volume and volume-to-mass types of chemical-equation-based problems. This diagram should seem very familiar to you. It differs from Figure 10.4 in only one way: volume boxes have been added to the top of the diagram. It is almost identical to Figure 12.9, differing in some of the conversion factors listed on the arrows. For example, Figure 12.9 has only molar volume listed on the arrows between volume and moles boxes, while Figure 12.10 has both molar volume and the ideal gas law listed on these arrows.

EXAMPLE 12.21

Calculating the Volume of a Gas that Reacts in a Chemical Reaction

Lithium metal is one of the few substances with which nitrogen gas (N_2) will react.

$$6\,Li(s) + N_2(g) \longrightarrow 2\,Li_3N(s)$$

What volume, in liters, of N_2 at STP will completely react with 75.0 g of Li metal?

SOLUTION

STEP 1 The given quantity is 75.0 g of Li, and the desired quantity is liters of N_2 at STP conditions.

$$75.0\text{ g Li} = ?\text{ L }N_2$$

STEP 2 This is a "grams of A" to "volume of gas B" problem. The pathway used in solving it, in terms of Figure 12.10, is

$$\boxed{\begin{array}{c}\text{Grams}\\\text{of A}\end{array}} \xrightarrow[\text{mass}]{\text{Molar}} \boxed{\begin{array}{c}\text{Moles}\\\text{of A}\end{array}} \xrightarrow[\text{coefficients}]{\text{Equation}} \boxed{\begin{array}{c}\text{Moles}\\\text{of B}\end{array}} \xrightarrow[\text{volume}]{\text{Molar}} \boxed{\begin{array}{c}\text{Volume}\\\text{of gas B}\end{array}}$$

The dimensional analysis setup for the calculation is

$$75.0 \text{ g Li} \times \frac{1 \text{ mole Li}}{6.94 \text{ g Li}} \times \frac{1 \text{ mole } N_2}{6 \text{ moles Li}} \times \frac{22.41 \text{ L } N_2}{1 \text{ mole } N_2}$$

STEP 3 The solution, obtained from combining all the numerical factors, is

$$\frac{75.0 \times 1 \times 1 \times 22.41}{6.94 \times 6 \times 1} \text{L O}_2 = 40.363832 \text{ L N}_2 \quad \text{(calculator answer)}$$

$$= 40.4 \text{ L N}_2 \qquad \text{(correct answer)}$$

CHEMICAL EXTENSION

Nitrogen gas (N_2) is a colorless, odorless, tasteless gas that is not very reactive at room temperature. Its lack of reactivity relates to the presence of a triple bond in the N_2 molecule.

Industrially, N_2 gas is obtained from air by liquefaction and then fractional distillation. The 13°C difference in boiling points between N_2 and O_2, air's major components, is sufficient to separate the two gases. The chief use of N_2 is as a blanketing agent (inert atmosphere) where the presence of air would involve danger of fire or explosion (preparation of highly reactive compounds) or result in the oxidation of the substance.

Nitrogen is relatively insoluble in water and blood at atmospheric pressure. The increased solubility of N_2 in blood at higher pressures presents problems for deep-sea divers. The diver's blood becomes saturated with air (N_2 and O_2) at high pressure. Oxygen is used by the cells, but nitrogen accumulates. If a diver is brought to the surface too quickly, the rapid release of N_2 forms bubbles that block blood capillaries and cause the "bends," so named because the condition is painful enough to cause the diver to double up in pain. To avoid this problem, which can be fatal, oxygen–helium mixtures are now used for deep diving, because helium has a much lower blood solubility than N_2.

Practice Exercise 12.21

Aluminum and oxygen react to produce aluminum oxide as shown by the equation

$$4 \text{ Al(s)} + 3 \text{ O}_2(\text{g}) \longrightarrow 2 \text{ Al}_2\text{O}_3(\text{s})$$

What volume, in liters, of O_2 at STP conditions will completely react with 75.0 g of Al?

EXAMPLE 12.22

Calculating the Mass of a Reactant from the Volume of a Gaseous Product

Oxygen gas can be generated by heating $KClO_3$ to a high temperature.

$$2 \text{ KClO}_3(\text{s}) \longrightarrow 2 \text{ KCl(s)} + 3 \text{ O}_2(\text{g})$$

How much $KClO_3$, in grams, is needed to generate 7.50 L of O_2 at a pressure of 1.00 atm and a temperature of 37°C?

SOLUTION

STEP 1 The given quantity is 7.50 L of O_2, and the desired quantity is grams of $KClO_3$.

$$7.50 \text{ L O}_2 = ? \text{ g KClO}_3$$

STEP 2 This is a "volume of gas A" to "grams of B" problem. The pathway, in terms of Figure 12.10, is

Volume of gas A	Ideal gas law	Moles of A	Equation coefficients	Moles of B	Molar mass	Grams of B

The first conversion step, from volume of O_2 to moles of O_2, can be made by applying the ideal gas law.

$$n = \frac{PV}{RT} = \frac{(1.00 \text{ atm}(7.50 \text{ L})}{\left(0.08206 \dfrac{\text{atm} \cdot \text{L})}{\text{mole} \cdot \text{K}}\right)(310 \text{ K})}$$

$$= 0.29482754 \text{ mole } O_2 \quad \text{(calculator answer)}$$

$$= 0.295 \text{ mole } O_2 \qquad \textbf{(correct answer)}$$

The dimensional analysis setup for the remaining conversion steps is

$$0.295 \text{ mole } O_2 \times \frac{2 \text{ moles KClO}_3}{3 \text{ moles } O_2} \times \frac{122.55 \text{ g KClO}_3}{1 \text{ mole KClO}_3}$$

STEP 3 The solution, obtained from combining all the numerical factors, is

$$\frac{0.295 \times 2 \times 122.55}{3 \times 1} \text{ g KClO}_3 = 24.1015 \text{ g KClO}_3 \quad \text{(calculator answer)}$$

$$= 24.1 \text{ g KClO}_3 \qquad \textbf{(correct answer)}$$

An alternative method can be used for converting the liters of O_2 to moles of O_2. The combined gas law can be used to calculate the STP volume of oxygen, and then the molar volume relationship can be used to calculate the moles of oxygen.

The combined gas law, rearranged to isolate V_2 on the left side, is

$$V_2 = V_1 \times \frac{P_1}{P_2} \times \frac{T_2}{T_1}$$

The given values for the variables are

$$
\begin{array}{ll}
P_1 = 1.00 \text{ atm} & P_2 = 1 \text{ atm (STP)} \\
V_1 = 7.50 \text{ L} & V_2 = ? \text{ L} \\
T_1 = 37°C = 310 \text{ K} & T_2 = 0°C = 273 \text{ K (STP)}
\end{array}
$$

Substituting these values into the combined gas law equation, we get

$$V_2 = 7.50 \text{ L} \times \frac{1.00 \text{ atm}}{1 \text{ atm}} \times \frac{273 \text{ K}}{310 \text{ K}}$$

$$= 6.6048387 \text{ L } O_2 \quad \text{(calculator answer)}$$

$$= 6.60 \text{ L } O_2 \qquad \textbf{(correct answer)}$$

Now, changing the STP volume to moles by the molar volume relationship, we get

$$6.60 \text{ L } O_2 \times \frac{1 \text{ mole } O_2}{22.41 \text{ L } O_2} = 0.29451137 \text{ mole } O_2 \quad \text{(calculator answer)}$$

$$= 0.295 \text{ mole } O_2 \qquad \textbf{(correct answer)}$$

This is the same value for moles of O_2 we previously obtained with the ideal gas law.

CHEMICAL EXTENSION

The solubility of O_2 in water is much lower than most gases. Solubilities in milliliters at STP per 1000 mL of H_2O for the gases N_2, O_2, CO_2, HCl, and NH_3 are, respectively, 23.5, 48.9, 1713, 79,800, and 1,130,000. Nevertheless, the concentration of O_2 in natural waters is sufficient to support aquatic life under "normal" conditions.

The solubility of O_2, as well as that of most other gases, decreases with increasing temperature. Thus, cold waters can support larger fish populations than warm waters.

On hot summer days the temperatures of *shallow* waters sometimes reach a point where the amount of dissolved oxygen is not sufficient to support life; under these conditions, suffocated fish are often found on the surface. Such a situation is often called "thermal pollution." Thermal pollution actually has a "double-barreled" effect on fish. In the hotter water, fish require more oxygen because of an increased respiration rate and the available oxygen in such water has been decreased.

Practice Exercise 12.22

Lead(IV) chloride ($PbCl_4$) can be produced from its constituent elements as shown by the equation

$$Pb(s) + 2\,Cl_2(g) \longrightarrow PbCl_4(s)$$

How many grams of $PbCl_4$ can be produced from the reaction of 5.00 L of Cl_2 at 1.33 atm pressure and a temperature of 7°C with an excess of Pb metal?

12.15 Mixtures of Gases

Many of the original experiments from which the gas laws we have considered in this chapter were "discovered" were based on the behavior of samples of air. Air is a mixture of gases. The simple gas laws and the ideal gas law apply not only to individual gases but also to *mixtures* of gases that do not react with each other.

Often the gas laws can be applied to gaseous mixtures by using, for the value of n (moles of gas), the sum of the moles of the various components present in a gaseous mixture. This approach to using gas laws for mixtures is illustrated in Example 12.23.

EXAMPLE 12.23

Applying the Ideal Gas Law to a Mixture of Gases

A gaseous mixture consists of 3.00 g of N_2 and 7.00 g of Ne. What is the volume, in liters, occupied by this mixture if it is under a pressure of 2.00 atm and at a temperature of 27°C?

SOLUTION

We first calculate the number of moles of each of the gases present and then add these quantities together to obtain the total moles of gas present.

$$n_{N_2}: \quad 3.00 \text{ g N}_2 \times \frac{1 \text{ mole N}_2}{28.02 \text{ g N}_2} = 0.10706638 \text{ mole N}_2 \quad \text{(calculator answer)}$$

$$= 0.107 \text{ mole N}_2 \qquad \text{(correct answer)}$$

$$n_{Ne}: \quad 7.00 \text{ g Ne} \times \frac{1 \text{ mole Ne}}{20.18 \text{ g Ne}} = 0.34687809 \text{ mole Ne} \quad \text{(calculator answer)}$$

$$= 0.347 \text{ mole Ne} \qquad \text{(correct answer)}$$

$$n_{total} = n_{N_2} + n_{Ne} = (0.107 + 0.347) \text{ mole} = 0.454 \text{ mole}$$

(calculator and **correct answer**)

Rearranging the ideal gas law to isolate V on the left side of the equation gives

$$V = \frac{nRT}{p}$$

All of the quantities on the right side of the equation are known.

$$n_{total} = 0.454 \text{ mole}, \quad R = 0.08206 \frac{\text{atm} \cdot \text{L}}{\text{mole} \cdot \text{K}} \quad T = 300 \text{ K}, \quad P = 2.00 \text{ atm}$$

Substituting in our equation, canceling units, and doing the arithmetic gives

$$V = \frac{(0.454 \text{ mole})\left(0.08206 \frac{\text{atm} \cdot \text{L}}{\text{mole} \cdot \text{K}}\right)(300 \text{ K})}{2.00 \text{ atm}}$$

$$= 5.588286 \quad \text{(calculator answer)}$$

$$= 5.59 \quad \text{(\textbf{correct answer})}$$

Practice Exercise 12.23

A gaseous mixture consists of 5.00 g of CO_2 and 8.00 g of He. What is the volume, in liters, occupied by this mixture if it is under a pressure of 1.50 atm and at a temperature of 27°C?

12.16 Dalton's Law of Partial Pressures

In a mixture of gases that do not react with each other, each type of molecule moves about in the container as if the other kinds were not there. This type of behavior is possible because attractions between molecules in the gaseous state are negligible at most temperatures and pressures and because a gas is mostly empty space (Sec. 11.6). Each gas in the mixture occupies the entire volume of the container; that is, it distributes itself uniformly throughout the container. The molecules of each type strike the walls of the container as frequently and with the same energy as though they were the only gas in the mixture. Consequently, the pressure exerted by a gas in a mixture is the same as it would be if the gas were alone in the same container under the same conditions.

John Dalton—the same John Dalton discussed in Section 5.1—was the first to notice this independent behavior of gases in mixtures. In 1803 he published a summary statement concerning such behavior, which is now known as *Dalton's law of partial pressures*. **Dalton's law of partial pressures** *states that the total pressure exerted by a mixture of gases is the sum of the partial pressures of the individual gases.* A new term, *partial pressure*, is used in stating Dalton's law. A **partial pressure** *is the pressure that a gas in mixture would exert if it were the only gas present under the same conditions.*

Expressed mathematically, Dalton's law states that

$$P_{total} = P_1 + P_2 + P_3 + \cdots$$

where P_{Total} is the total pressure of a gaseous mixture and P_1, P_2, P_3, and so on are the partial pressures of the individual gaseous components of the mixture. (When the identity of a gas is known, its molecular formula is used as a subscript in the partial-pressure notation; for example, P_{CO_2} is the partial pressure of carbon dioxide in a mixture.)

Figure 12.11

An illustration of Dalton's law of partial pressures.

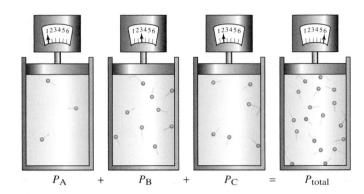

$$P_A \quad + \quad P_B \quad + \quad P_C \quad = \quad P_{total}$$

To illustrate Dalton's law, consider the four identical gas containers shown in Figure 12.11. Suppose we place amounts of three different gases (represented by A, B, and C) into three of the containers and measure the pressure exerted by each sample. We then place all three samples in the fourth container and measure the pressure exerted by this mixture of gases. It is found that

$$P_{total} = P_A + P_B + P_C$$

Using the pressures given in Figure 12.11, we see that

$$P_{total} = 1 + 3 + 2 = 6$$

EXAMPLE 12.24

Using Dalton's Law to Calculate a Partial Pressure

An unknown quantity of the noble gas xenon (Xe) is added to a cylinder already containing a mixture of the noble gases helium (He) and argon (Ar) at partial pressures, respectively, of 3.00 atm and 1.00 atm. After the Xe addition, the total pressure in the cylinder is 5.80 atm. What is the partial pressure, in atmospheres, of the Xe gas?

SOLUTION

The partial pressures of the He and Ar in the mixture will not be affected by the addition of the Xe (Dalton's law). Thus, the partial pressures of the He and Ar remain at 3.00 atm and 1.00 atm, respectively.

The sum of the partial pressures of He and Ar is 4.00 atm.

$$3.00 \text{ atm} + 1.00 \text{ atm} = 4.00 \text{ atm}$$

The difference between this pressure sum and the total pressure in the cylinder is caused by the Xe present. Thus, the partial pressure of the Xe is

$$P_{total} - (P_{He} + P_{Ar}) = P_{Xe}$$

$$5.80 \text{ atm} - 4.00 \text{ atm} = 1.8 \text{ atm} \quad \text{(calculator answer)}$$

$$= 1.80 \text{ atm} \quad \textbf{(correct answer)}$$

Practice Exercise 12.24

A gaseous mixture contains the three noble gases helium, argon, and krypton. The total pressure exerted by the mixture is 1.57 atm, and the partial pressures of the helium and argon are 0.33 and 0.39 atm, respectively. What is the partial pressure of the krypton present in the mixture?

The validity of Dalton's law of partial pressures is easily demonstrated by using the ideal gas law (Sec. 12.12). For a mixture of three gases (A, B, and C) the total pressure is given by the expression

$$P_{total} = n_{total}\frac{RT}{V}$$

The total number of moles present is the sum of the moles of A, B, and C; that is,

$$n_{total} = n_A + n_B + n_C$$

Substituting this equation into the previous one gives

$$P_{total} = (n_A + n_B + n_C)\frac{RT}{V}$$

Expanding the right side of this equation results in the expression

$$P_{total} = n_A\frac{RT}{V} + n_B\frac{RT}{V} + n_C\frac{RT}{V}$$

The three individual terms on the right side of this equation are, respectively, the partial pressure of A, B, and C.

$$P_A = n_A\frac{RT}{V}, \ P_B = n_B\frac{RT}{V}, \ P_C = n_C\frac{RT}{V}$$

Substitution of this information into the previous equation gives

$$P_{total} = P_A + P_B + P_C$$

which is a statement of Dalton's law of partial pressures for a mixture of three gases.

In the preceding derivation, two of the expressions that are encountered are

$$P_A = n_A\frac{RT}{V} \quad \text{and} \quad P_{total} = n_{total}\frac{RT}{V}$$

If we divide the first of these expressions by the second, we obtain

$$\frac{P_A}{P_{total}} = \frac{n_A\frac{\cancel{RT}}{\cancel{V}}}{n_{total}\frac{\cancel{RT}}{\cancel{V}}} = \frac{n_A}{n_{total}} \quad \text{or} \quad \frac{P_A}{P_{total}} = \frac{n_A}{n_{total}}$$

Rearrangement of this equation gives

$$P_A = P_{total} \times \frac{n_A}{n_{total}}$$

The fraction n_A/n_{total} in this equation is called the *mole fraction* of A in the mixture. It is the fraction of the total moles that is accounted for by gas A. A **mole fraction** *is a dimensionless quantity that gives the ratio of the number of moles of a component in a mixture to the number of moles of all components present.* The symbol X is used to denote a mole fraction.

$$X_A = \frac{n_A}{n_{total}}$$

Using mole fractions and total pressure, we can calculate the partial pressures of the individual components of a gaseous mixture. The partial pressure of a gas in a mixture is equal to its mole fraction multiplied by the total pressure.

$$P_A = X_A \times P_{total}$$

Example 12.25 shows how mole fractions are calculated, and Example 12.26 shows how mole fractions are used in obtaining partial pressures.

EXAMPLE 12.25

Calculating the Mole Fraction for a Gas in a Gaseous Mixture

A gaseous mixture contains 10.0 g each of the gases N_2, O_2, and Ar. What is the mole fraction of each gas in the mixture?

SOLUTION

We first calculate the number of moles of each gas present.

$$N_2: \quad 10.0 \text{ g } N_2 \times \frac{1 \text{ mole } N_2}{28.02 \text{ g } N_2} = 0.35688793 \text{ mole } N_2 \quad \text{(calculator answer)}$$

$$= 0.357 \text{ mole } N_2 \quad \textbf{(correct answer)}$$

$$O_2: \quad 10.0 \text{ g } O_2 \times \frac{1 \text{ mole } O_2}{32.00 \text{ g } O_2} = 0.3125 \text{ mole } O_2 \quad \text{(calculator answer)}$$

$$= 0.312 \text{ mole } O_2 \quad \textbf{(correct answer)}$$

$$Ar: \quad 10.0 \text{ g Ar} \times \frac{1 \text{ mole Ar}}{39.95 \text{ g Ar}} = 0.25031289 \text{ mole Ar} \quad \text{(calculator answer)}$$

$$= 0.250 \text{ mole Ar} \quad \textbf{(correct answer)}$$

The total number of moles of gas present is

$$n_{total} = n_{N_2} + n_{O_2} + n_{Ar}$$
$$= (0.357 + 0.312 + 0.250) \text{ mole}$$
$$= 0.919 \text{ mole} \quad \text{(calculator and } \textbf{correct answer)}$$

Mole fractions are calculated as ratios of individual component moles to total moles.

$$X_{N_2} = \frac{0.357 \text{ mole}}{0.919 \text{ mole}} = 0.38846572 \quad \text{(calculator answer)}$$

$$= 0.388 \quad \textbf{(correct answer)}$$

$$X_{O_2} = \frac{0.312 \text{ mole}}{0.919 \text{ mole}} = 0.33949945 \quad \text{(calculator answer)}$$

$$= 0.339 \quad \textbf{(correct answer)}$$

$$X_{Ar} = \frac{0.250 \text{ mole}}{0.919 \text{ mole}} = 0.27203482 \quad \text{(calculator answer)}$$

$$= 0.272 \quad \textbf{(correct answer)}$$

The sum of all mole fractions should always add to one.

$$0.388 + 0.339 + 0.272 = 0.999 \quad \text{(calculator and } \textbf{correct answer)}$$

Rounding errors cause the sum to be 0.999 rather than 1.000.

Practice Exercise 12.24

A gaseous mixture contains 5.00 g each of the gases Ar, Kr, and Xe. What is the mole fraction of each gas in the mixture?

EXAMPLE 12.26

Obtaining Partial Pressures Using Mole Fractions

A mixture of gases contains 4.23 moles of neon (Ne), 0.93 mole of argon (Ar), and 7.65 moles of hydrogen (H_2). Calculate the partial pressures of the gases if the total pressure is 5.00 atm at a certain temperature.

SOLUTION

We first calculate the mole fraction of each gas.

$$X_{Ne} = \frac{4.23 \text{ moles}}{(4.23 + 0.93 + 7.65) \text{moles}} = 0.33021077 \quad \text{(calculator answer)}$$
$$= 0.330 \qquad \textbf{(correct answer)}$$

$$X_{Ar} = \frac{0.93 \text{ mole}}{(4.23 + 0.93 + 7.65) \text{moles}} = 0.072599531 \quad \text{(calculator answer)}$$
$$= 0.073 \qquad \textbf{(correct answer)}$$

$$X_{H_2} = \frac{7.65 \text{ moles}}{(4.23 + 0.93 + 7.65) \text{moles}} = 0.59718969 \quad \text{(calculator answer)}$$
$$= 0.597 \qquad \textbf{(correct answer)}$$

To calculate partial pressures, we rearrange the equation

$$\frac{P_A}{P_{total}} = X_A$$

to isolate the partial pressure on a side by itself.

$$P_A = X_A \times P_{total}$$

Substituting known quantities into this equation gives the partial pressures.

$$P_{Ne} = 0.330 \times 5.00 \text{ atm} = 1.65 \text{ atm} \quad \text{(calculator and \textbf{correct answer})}$$

$$P_{Ar} = 0.073 \times 5.00 \text{ atm} = 0.365 \text{ atm} \quad \text{(calculator answer)}$$
$$= 0.36 \text{ atm} \quad \textbf{(correct answer)}$$

$$P_{H_2} = 0.597 \times 5.00 \text{ atm} = 2.985 \text{ atm} \quad \text{(calculator answer)}$$
$$= 2.98 \text{ atm} \quad \textbf{(correct answer)}$$

Practice Exercise 12.26

A sample of natural gas contains 6.20 moles of methane (CH_4), 0.317 mole of ethane (C_2H_6), and 0.0872 mole of propane (C_3H_8). If the total pressure of the gases is 2.43 atm, what are the partial pressures, in atmospheres, of the gases?

EXAMPLE 12.27

Calculating the Partial Pressure of a Gas

Calculate the partial pressure of O_2, in atm, in a gaseous mixture with a volume of 2.50 L at 20°C, given that the mixture composition is 0.50 mole O_2 and 0.75 mole N_2.

SOLUTION

This problem differs from Example 12.26 in that both the temperature and volume of the gaseous mixture are given. This added information will enable us to calculate the partial pressure of O_2 using two different methods: (1) mole fractions, and (2) the ideal gas law.

Using the mole fraction method, we calculate first the total pressure of the gaseous mixture based on the presence of 1.25 total moles of gas (0.50 mole + 0.75 mole):

$$P_{total} = \frac{n_{total}\, RT}{V} = \frac{(1.25 \text{ moles})\left(0.08206\dfrac{\text{atm} \cdot \text{L}}{\text{mole} \cdot \text{K}}\right)(293 \text{ K})}{2.50 \text{ L}}$$

$$= 12.02179 \text{ atm} \quad (\text{calculator answer})$$

$$= 12.0 \text{ atm} \quad\quad (\textbf{correct answer})$$

Next we calculate the mole fraction of O_2:

$$X_{O_2} = \frac{n_{O_2}}{n_{total}} = \frac{0.50 \text{ mole}}{(0.50 + 0.75 \text{ mole})} = 0.4 \quad (\text{calculator answer})$$

$$= 0.40 \quad (\textbf{correct answer})$$

The partial pressure of the O_2 is obtained using the mole fraction O_2 and the total pressure:

$$P_{O_2} = X_{O_2}\, P_{total} = 0.40 \times 12.0 \text{ atm} = 4.8 \text{ atm} \quad (\text{calculator and correct answer})$$

This problem can also be solved without the use of mole fractions, using the concept that the partial pressure exerted by the O_2 is determined by the number of moles of O_2 present. With this approach, we substitute directly into the ideal gas law the quantities given in the problem statement.

$$P_{O_2} = \frac{n_{O_2}\, RT}{V} = \frac{(0.50 \text{ mole})\left(0.08206\dfrac{\text{atm} \cdot \text{L}}{\text{mole} \cdot \text{K}}\right)(293 \text{ K})}{2.50 \text{ L}}$$

$$= 4.808716 \text{ atm} \quad (\text{calculator answer})$$

$$= 4.8 \text{ atm} \quad\quad (\textbf{correct answer})$$

Practice Exercise 12.27

Calculate the partial pressure of N_2, in atm, in a gaseous mixture with a volume of 2.00 L at 30°C, given that the mixture composition is 0.10 mole N_2 and 0.90 mole He.

The air we breathe is a most important mixture of gases. The composition of clean air from which all water vapor has been removed (dry air) is found to be virtually constant over the entire Earth. Table 12.5 gives the composition of clean, dry air in terms of mole fractions. All components that have a mole fraction of at least 1×10^{-5} (0.001%) are listed.

Atmospheric pressure is the sum of the partial pressures of the gaseous components present in air. Table 12.5 also gives the partial pressure of each component of air in a situation where total atmospheric pressure is 760 mm Hg.

Table 12.5	The Major Components of Clean, Dry Air		
Gaseous Component	**Formula**	**Mole Fraction**	**Partial Pressure (mm Hg) When Total Pressure Is 760.0 mm Hg**
Nitrogen	N_2	0.78084	593.4
Oxygen	O_2	0.20948	159.2
Argon	Ar	9.34×10^{-3}	7.1
Carbon dioxide	CO_2	3.1×10^{-4}	0.2
Neon	Ne	2×10^{-5}	0.02
Helium	He	1×10^{-5}	0.01

The composition of air is not absolutely constant. The variability in composition is caused predominantly by the presence of water vapor, a substance not listed in Table 12.5 because those statistics are for *dry* air. The amount of water vapor in air varies between almost zero and a mole fraction of 0.05–0.06, depending on weather and temperature.

A common application of Dalton's law of partial pressures is encountered in the laboratory preparation of gases. Such gases are often collected by displacement of water. Figure 12.12 shows O_2, prepared from the decomposition of $KClO_3$, being collected by water displacement. A gas collected by water displacement is never pure. It always contains some water vapor. The total pressure exerted by the gaseous mixture is the sum of the partial pressures of the gas being collected and the water vapor.

$$P_{total} = P_{gas} + P_{H_2O}$$

The pressure exerted by the water vapor in the mixture will be constant at any given temperature if sufficient time has been allowed to establish equilibrium conditions.

For gases collected by water displacement, the partial pressure of the water vapor can be obtained from a table showing the variation of water vapor pressure with temperature (see Table 12.6). Thus the partial pressure of the collected gas is easily determined.

$$P_{gas} = P_{atm} - P_{H_2O}$$

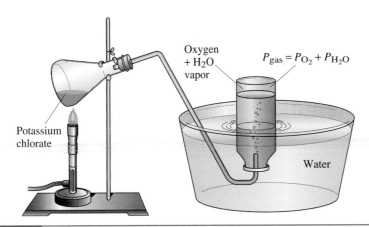

Oxygen + H_2O vapor

$P_{gas} = P_{O_2} + P_{H_2O}$

Potassium chlorate

Water

Figure 12.12

Collection of oxygen gas by water displacement. Potassium chlorate ($KClO_3$) decomposes to form oxygen (O_2), which is collected over water.

Table 12.6	Vapor Pressure of Water at Various Temperatures				
T(°C)	Vapor Pressure (mm Hg)	T(°C)	Vapor Pressure (mm Hg)	T(°C)	Vapor Pressure (mm Hg)
15	12.8	22	19.8	29	30.0
16	13.6	23	21.1	30	31.8
17	14.5	24	22.4	31	33.7
18	15.5	25	23.8	32	35.7
19	16.5	26	25.2	33	37.7
20	17.5	27	26.7	34	39.9
21	18.7	28	28.3	35	42.2

EXAMPLE 12.28

Calculating the Partial Pressure of a Gas Collected over Water

What is the partial pressure of oxygen collected over water at 17°C on a day when the barometric pressure is 743 mm Hg?

SOLUTION

From Table 12.6 we determine that water has a vapor pressure of 14.5 mm Hg at 17°C. Using the equation

$$P_{O_2} = P_{atm} - P_{H_2O}$$

we find the partial pressure of the oxygen to be 728 mm Hg.

$$P_{O_2} = (743 - 14.5) \text{ mm Hg} = 728.5 \text{ mm Hg} \quad \text{(calculator answer)}$$
$$= 728 \text{ mm Hg} \quad \textbf{(correct answer)}$$

Practice Exercise 12.28

What is the partial pressure of nitrogen collected over water at 33°C on a day when the barometric pressure is 657 mm Hg?

When considering the components of a gaseous mixture, a useful set of equalities for any given component (A) of the gaseous mixture, at a given temperature and pressure, is

$$\text{mole percent A} = \text{pressure percent A} = \text{volume percent A}$$

where

$$\text{mole \% A} = \frac{n_A}{n_{total}} \times 100$$

$$\text{pressure \% A} = \frac{P_A}{P_{total}} \times 100$$

$$\text{volume \% A} = \frac{V_A}{V_{total}} \times 100$$

Example 12.29 is a calculation showing the equality of these three quantities.

EXAMPLE 12.29

Expressing Gaseous Mixture Composition in Mole Percent, Pressure Percent, and Volume Percent

A 17.92-L flask, at STP, contains 0.200 mole of O_2, 0.300 mole of N_2, and 0.300 mole of Ar. For this gaseous mixture, calculate the

(a) mole percent of O_2 present
(b) pressure percent of O_2 present
(c) volume percent of O_2 present

SOLUTION

(a) The total number of moles of gas present is 0.800 mole (0.200 mole + 0.300 mole + 0.300 mole). The mole percent O_2 is

$$\text{mole \% } O_2 = \frac{n_{O_2}}{n_{total}} \times 100 = \frac{0.200 \text{ mole}}{0.800 \text{ mole}} \times 100 = 25\% \quad \text{(calculator answer)}$$

$$= 25.0\% \text{ (correct answer)}$$

(b) Since conditions are specified as STP, the total pressure in the flask is 1.00 atm. The partial pressure of the O_2 is

$$P_{O_2} = X_{O_2} \times P_{total}$$

$$= \frac{0.200 \text{ mole}}{0.800 \text{ mole}} \times 1.00 \text{ atm}$$

$$= 0.25 \text{ atm} \quad \text{(calculator answer)}$$

$$= 0.250 \text{ atm (correct answer)}$$

The pressure percent O_2 is

$$\text{pressure \% } O_2 = \frac{P_{O_2}}{P_{total}} \times 100 = \frac{0.250 \text{ atm}}{1.00 \text{ atm}} \times 100$$

$$= 25\% \quad \text{(calculator answer)}$$

$$= 25.0\% \text{ (correct answer)}$$

(c) The total volume of the flask is given as 17.92 L. The volume of the oxygen present, if it were alone at STP conditions, in a different container, is

$$0.200 \text{ mole } O_2 \times \frac{22.41 \text{ L } O_2}{1 \text{ mole } O_2} = 4.482 \text{ L } O_2 \quad \text{(calculator answer)}$$

$$= 4.48 \text{ L } O_2 \quad \text{(correct answer)}$$

The volume percent O_2 is

$$\text{volume \% } O_2 = \frac{V_{O_2}}{V_{total}} \times 100 = \frac{4.48 \text{ L}}{17.92 \text{ L}} \times 100$$

$$= 25\% \quad \text{(calculator answer)}$$

$$= 25.0\% \quad \text{(correct answer)}$$

Note, from the answers to parts (a), (b), and (c), that for O_2 the

$$\text{mole percent} = \text{pressure percent} = \text{volume percent}$$

Practice Exercise 12.29

A 20.16-L flask, at STP, contains 0.200 mole of O_2, 0.300 mole of N_2, and 0.400 mole of Ar. For this gaseous mixture, calculate the

(a) mole percent of N_2 present
(b) pressure percent of N_2 present
(c) volume percent of N_2 present

Summary

1. **Gas Laws** Gas laws are generalizations that describe in mathematical terms the relationships among the amount, pressure, temperature, and volume of a gas. When gas laws are used, it is necessary to express the temperature of the gas on the Kelvin scale. Pressure is usually expressed in atmospheres or millimeters of mercury.

2. **Boyle's Law** Boyle's law, the pressure–volume law, states that the volume of a fixed quantity of a gas is *inversely proportional* to the pressure applied to the gas if the temperature is kept constant. This means that when the pressure on the gas increases, the volume decreases proportionally; conversely, when the volume decreases, the pressure increases.

3. **Charles's Law** Charles's law, the temperature–volume law, states that the volume of a fixed quantity of gas is *directly proportional* to its Kelvin temperature if the pressure is kept constant. This means that when the temperature increases, the volume also increases and that when the temperature decreases, the volume also decreases.

4. **Gay-Lussac's Law** Gay-Lussac's law, the temperature–pressure law, states that the pressure of a fixed quantity of gas is *directly proportional* to its Kelvin temperature if the volume is kept constant. As the temperature of the gas increases, the pressure also increases; conversely, as the temperature is decreased, pressure decreases.

5. **Combined Gas Law** The combined gas law is an expression obtained by mathematically combining Boyle's, Charles's, and Gay-Lussac's laws. A change in pressure, temperature, or volume of a fixed quantity of gas that is brought about by changes in the other two variables can be calculated by using this law.

6. **STP Conditions** STP conditions are those of standard temperature (0°C) and standard pressure (1 atm).

7. **Gay-Lussac's Law of Combining Volumes** The law of combining volumes states that the volumes of different gases involved in a reaction, measured at the same temperature and pressure, are in the same ratio as the coefficients for these gases in the balanced equation for the reaction.

8. **Avogadro's Law** Avogadro's law, the volume–quantity law, states that equal volumes of different gases, measured at the same temperature and pressure, contain equal numbers of molecules. An alternative statement of the law is that the volume of a gas, at constant temperature and pressure, is *directly proportional* to the number of moles of gas present.

9. **Molar Volume of a Gas** The molar volume of a gas, experimentally determined to be 22.414 L, is the volume occupied by one mole of a gas at STP conditions. The molar volume and molar mass of a gas can be used to calculate the density of the gas at STP conditions.

10. **Ideal Gas Law** The ideal gas law describes the relationships among the four gas law variables temperature, pressure, volume, and moles for a gas under one set of conditions. This law is used to calculate any one of the gas law variables (P, V, T, n) given the other three. Equations derived from the ideal gas law by substitution of variables can be used to calculate the density, and molar mass of a gas.

11. **Gas Laws and Chemical Equations** A simple extension of the procedures used in mass-to-mass stoichiometric calculations enables mass-to-volume and volume-to-mass calculations to be carried out for reactions where at least one gas is involved. The extension is based on the relationship between mass (moles) and volume as given by the ideal gas law.

12. **Dalton's Law of Partial Pressures** Dalton's law of partial pressures states that the total pressure exerted by a mixture of gases that do not react with each other is the sum of the partial pressures of the individual gases present. A partial pressure is the pressure that a gas in a mixture would exert if it were present alone under the same conditions. Dalton's law of partial pressures, in modified form, can also be used to specify gas mixture concentrations in terms of mole percent, volume percent, and pressure percent.

Key Terms

The new terms defined in this chapter are

Avogadro's law   *Sec. 12.10*
barometer   *Sec. 12.2*
Boyle's law   *Sec. 12.3*
Charles's law   *Sec. 12.4*
combined gas law   *Sec. 12.6*
Dalton's law of partial pressures   *Sec. 12.16*
gas laws   *Sec. 12.2*
Gay-Lussac's law   *Sec. 12.5*
Gay-Lussac's law of combining volumes   *Sec. 12.8*

ideal gas   *Sec. 12.12*
ideal gas law   *Sec. 12.12*
manometer   *Sec. 12.2*
molar volume of a gas   *Sec. 12.11*
mole fraction   *Sec. 12.16*
partial pressure   *Sec. 12.16*
pressure   *Sec. 12.2*
standard pressure   *Sec. 12.7*
standard temperature   *Sec. 12.7*
STP conditions   *Sec. 12.7*

Practice Problems

Measurement of Pressure (Sec. 12.2)

12.1 If the helium gas in a steel cylinder is at a pressure of 6.20 atm, what is the pressure in each of the following pressure units?

(a) millimeters of mercury   **(b)** inches of mercury
(c) pounds per square inch   **(d)** centimeters of mercury

12.2 If the oxygen gas in a steel cylinder is at a pressure of 9570 millimeters of mercury, what is the pressure in each of the following pressure units?

(a) inches of mercury   **(b)** pounds per square inch
(c) atmospheres   **(d)** centimeters of mercury

12.3 For each of the following pairs of pressure measurements, indicate whether the first listed measurement is larger than, equal to, or smaller than the second listed measurement.

(a) 457 mm Hg and 1.07 atm
(b) 14.68 lb/in.^{2} and 760 mm Hg
(c) 29.9 in. Hg and 585 mm Hg
(d) 4.639 atm and 68.10 psi

12.4 For each of the following pairs of pressure measurements, indicate whether the first listed measurement is larger than, equal to, or smaller than the second listed measurement.

(a) 0.998 atm and 762 mm Hg
(b) 29.92 psi and 29.92 in. Hg
(c) 17.8 psi and 545 mm Hg
(d) 1.34 atm and 40.1 in. Hg

12.5 The mercury level in the arm of a manometer (see Fig. 12.2) that is open to the atmosphere is found to be 237 mm higher than the mercury level in the arm of the manometer connected to the container of gas. Measured barometric pressure is 762 mm Hg. What is the pressure, in millimeters of mercury, of the gas in the container?

12.6 The mercury level in the arm of a manometer (see Fig. 12.2) that is open to the atmosphere is found to be 35 mm lower than the mercury level in the arm of the manometer connected to the

container of gas. Measured barometric pressure is 743 mm Hg. What is the pressure, in millimeters of mercury, of the gas in the container?

Boyle's Law (Sec. 12.3)

12.7 A sample of gas occupies a volume of 3.00 L at 27°C and a pressure of 1.00 atm. Without actually doing a calculation, predict whether the volume will increase or decrease when the pressure is changed, at constant temperature, to the following values.

(a) 2.00 atm   **(b)** 0.952 atm
(c) 1.20 atm   **(d)** 775 mm Hg

12.8 A sample of gas occupies a volume of 2.00 L at 33°C and a pressure of 0.50 atm. Without actually doing a calculation, predict whether the volume will increase or decrease when the pressure is changed, at constant temperature, to the following values.

(a) 0.75 atm   **(b)** 0.333 atm
(c) 1.25 atm   **(d)** 200 mm Hg

12.9 A sample of O_2 gas occupies a volume of 2.00 L at 27°C and 2.00 atm pressure. What volume, in liters, will this O_2 sample occupy at the same temperature but at each of the following pressures?

(a) 3.13 atm   **(b)** 0.723 atm
(c) 762 mm Hg   **(d)** 37.2 mm Hg

12.10 A sample of N_2 gas occupies a volume of 3.00 L at 37°C and 3.00 atm pressure. What volume, in liters, will this N_2 sample occupy at the same temperature but at each of the following pressures?

(a) 2.98 atm   **(b)** 10.5 atm
(c) 453 mm Hg   **(d)** 54.2 mm Hg

12.11 At constant temperature, the pressure on a sample of H_2 gas is decreased from 4.0 atm to 2.5 atm. What was the original volume, in milliliters, of the gas sample if this action increases the sample volume to each of the following amounts?

(a) 425 mL   **(b)** 25.4 mL   **(c)** 1.08 L   **(d)** 4.68 L

12.12 At constant temperature, the pressure on a sample of H_2 gas is increased from 2.5 atm to 4.0 atm. What was the original volume, in milliliters, of the gas sample if this action decreases the sample volume to each of the following amounts?

(a) 322 mL   **(b)** 15.0 mL   **(c)** 2.24 L   **(d)** 0.88 L

12.13 A sample of Cl_2 gas at a pressure of 645 mm Hg is transferred to a new container having a volume one-third that of the original container. What pressure, in millimeters of mercury, does the Cl_2 exert in the new container? Assume that the temperature does not change.

12.14 A sample of F_2 gas at a pressure of 1.03 atm is transferred to a new container having a volume 2.50 times that of the original container. What pressure, in atmospheres, does the F_2 exert in the new container? Assume that the temperature does not change.

Charles's Law (Sec. 12.4)

12.15 A sample of gas occupies a volume of 3.00 L at 27°C and a pressure of 1.00 atm. Without actually doing a calculation, predict whether the volume will increase or decrease when the temperature is changed, at constant pressure, to the following values.

(a) 37°C   **(b)** 17°C   **(c)** 375°C   **(d)** 295 K

12.16 A sample of gas occupies a volume of 2.00 L at 33°C and a pressure of 0.50 atm. Without actually doing a calculation, predict whether the volume will increase or decrease when the temperature is changed, at constant pressure, to the following values.

(a) 23°C   **(b)** 63°C   **(c)** −23°C   **(d)** 303 K

12.17 A sample of carbon dioxide gas, CO_2, has a volume of 5.00 L at 35°C. What volume, in liters, will this CO_2 gas occupy at each of the following temperatures if the pressure is held constant?

(a) 125°C   **(b)** 5°C   **(c)** −5°C   **(d)** 985°C

12.18 A sample of nitrogen dioxide gas, NO_2, has a volume of 1.50 L at 23°C. What volume, in liters, will this NO_2 gas occupy at each of the following temperatures if the pressure is held constant?

(a) 123°C   **(b)** 223°C   **(c)** −25°C   **(d)** 883°C

12.19 At constant pressure, the temperature of a sample of He gas is decreased from 73°C to 0°C. What was the original volume of the gas sample, in milliliters, if this action decreases the sample volume to each of the following amounts?

(a) 15.2 mL   **(b)** 879 mL   **(c)** 1.20 L   **(d)** 10.7 L

12.20 At constant pressure, the temperature of a sample of Ne gas is increased from 0°C to 73°C. What was the original volume of the gas sample, in milliliters, if this action increases the sample volume to each of the following amounts?

(a) 17.5 mL   **(b)** 742 mL   **(c)** 3.42 L   **(d)** 0.90 L

12.21 At constant pressure, at what temperature, in degrees Celsius, will a gas have exactly one-half the volume that it has at room temperature (24°C)?

12.22 At constant pressure, at what temperature, in degrees Celsius, will a gas have exactly double the volume that it has at room temperature (24°C)?

Gay-Lussac's Law (Sec. 12.5)

12.23 A sample of gas occupies a volume of 3.00 L at 27°C and a pressure of 1.00 atm. Without actually doing a calculation, predict whether the pressure will increase or decrease when the temperature is changed, at constant volume, to the following values.

(a) 37°C   **(b)** 17°C   **(c)** 375°C   **(d)** 295 K

12.24 A sample of gas occupies a volume of 2.00 L at 33°C and a pressure of 0.50 atm. Without actually doing a calculation, predict whether the pressure will increase or decrease when the temperature is changed, at constant value, to the following values.

(a) 23°C   **(b)** 63°C   **(c)** −23°C   **(d)** 303 K

12.25 A sample of air exerts a pressure of 1.00 atm at 22°C. If the sample volume remains constant, what is the new pressure, in atmospheres, exerted by the air when it is heated to each of the following temperatures?

(a) 122°C   **(b)** 222°C   **(c)** 422°C   **(d)** 722°C

12.26 A sample of air exerts a pressure of 2.00 atm at 37°C. If the sample volume remains constant, what is the new pressure, in atmospheres, exerted by the air when it is cooled to each of the following temperatures?

(a) 31°C   **(b)** 6°C   **(c)** −37°C   **(d)** −137°C

12.27 At constant volume, the temperature of a sample of sulfur dioxide gas, SO_2, is decreased from 97°C to 27°C. What was the original pressure of the gas, in atmospheres, if this action decreases the sample pressure to the following?

(a) 3.00 atm   **(b)** 1.00 atm

(c) 1394 mm Hg   **(d)** 375 mm Hg

12.28 At constant volume, the temperature of a sample of carbon monoxide gas, CO, is decreased from 122°C to 22°. What was the original pressure of the gas, in millimeters of mercury, if this action decreases the sample pressure to the following?

(a) 762 mm Hg   **(b)** 662 mm Hg

(c) 1.05 atm   **(d)** 25.0 atm

12.29 A spray can is empty except for the propellant gas, which exerts a pressure of 1.2 atm at 24°C. If the can is thrown into a fire (485°C), what will be the pressure, in atmospheres, inside the hot can?

12.30 A spray can is empty except for the propellant gas, which exerts a pressure of 1.2 atm at 24°C. If the can is placed in a refrigerator (3°C), what will be the pressure, in atmospheres, inside the cold can?

The Combined Gas Law (Sec. 12.6)

12.31 Rearrange the standard form of the combined gas law to result in the following.

(a) The variable T_2 is isolated on the left side of the equation.

(b) The quantity V_2/P_1 is isolated on the left side of the equation.

12.32 Rearrange the standard form of the combined gas law to result in the following.

(a) The variable T_1 is isolated on the left side of the equation.

(b) The quantity P_2/V_1 is isolated on the left side of the equation.

12.33 What is the new volume, in milliliters, of a 3.00-mL sample of air at 0.980 atm and 230°C that is compressed and cooled to each of the following sets of conditions?

(a) 185°C and 1.50 atm

(b) 35°C and 2.00 atm

(c) −35°C and 4.00 atm

(d) −125°C and 5.67 atm

12.34 What is the new volume, in liters, of a 25.0-L sample of air at 1.11 atm and 152°C that is compressed and cooled to each of the following sets of conditions?

(a) 25°C and 2.00 atm

(b) −25°C and 3.00 atm

(c) −75°C and 5.00 atm

(d) −125°C and 7.75 atm

12.35 A sample of CO_2 gas has a volume of 15.2 L at a pressure of 1.35 atm and a temperature of 33°C. Determine the following for this gas sample.

(a) volume, in liters, at $T = 35$°C and $P = 3.50$ atm

(b) volume, in milliliters, at $T = 97$°C and $P = 6.70$ atm

(c) pressure, in atmospheres, at $T = 42$°C and $V = 10.0$ L

(d) temperature, in degrees Celsius, at $P = 7.00$ atm and $V = 0.973$ L

12.36 A sample of NO_2 gas has a volume of 37.3 mL at a pressure of 621 mm Hg and a temperature of 52°C. Determine the following for this gas sample.

(a) volume, in milliliters, at $T = 35$°C and $P = 650$ mm Hg

(b) volume, in liters, at $T = 43$°C and $P = 1.11$ atm

(c) pressure, in millimeters of mercury, at $T = 125$°C and $V = 52.4$ mL

(d) temperature, in degrees Celsius, at $P = 775$ mm Hg and $V = 23.0$ mL

12.37 A sample of ammonia gas, NH_3, in a 375-mL container at a pressure of 1.03 atm and a temperature of 27°C, is transferred to a container with a volume of 1.25 L.

(a) What is the new pressure, in millimeters of mercury, if no change in temperature occurs?

(b) What is the new temperature, in degrees Celsius, if no change in pressure occurs?

12.38 A sample of nitrous oxide gas, N_2O, in a 475-mL container at a pressure of 676 mm Hg and a temperature of 22°C, is transferred to a container with a volume of 5.00 L.

(a) What is the new pressure, in atmospheres, if no change in temperature occurs?

(b) What is the new temperature, in degrees Celsius, if no change in pressure occurs?

12.39 A sample of nitrous oxide gas, N_2O, in a nonrigid container at a temperature of 33°C, occupies a certain volume at a certain pressure. What will be its temperature, in degrees Celsius, in each of the following situations?

(a) Both pressure and volume are tripled.

(b) Both pressure and volume are cut in half.

(c) The pressure is tripled and the volume is cut in half.

(d) The pressure is cut in half and the volume is doubled.

12.40 A sample of nitric oxide gas, NO, in a nonrigid container, occupies a volume of 2.50 L at a certain temperature and pressure. What will be its volume, in liters, in each of the following situations?

(a) Both pressure and Kelvin temperature are doubled.

(b) Both pressure and Kelvin temperature are cut by one-third.

(c) The pressure is doubled, and the Kelvin temperature is cut by one-third.

(d) The pressure is cut in half, and the Kelvin temperature is tripled.

STP Conditions (Sec. 12.7)

12.41 What is the volume, in liters, at STP of 3.50 L of methane gas, CH_4, at each of the following initial conditions?

(a) 30°C and 1.00 atm

(b) 0°C and 3.00 atm

(c) 55°C and 1.25 atm

(d) 135°C and 852 mm Hg

12.42 What is the volume, in liters, at STP of 4.25 L of acetylene gas, C_2H_2, at each of the following initial conditions?

(a) 52°C and 760 mm Hg

(b) 0°C and 452 mm Hg

(c) −8°C and 683 mm Hg

(d) 125°C and 1.25 atm

12.43 A quantity of air has a volume of 1.00 L at STP. What volume, in liters, will the air occupy if the pressure and temperature are changed to the following values?

(a) 1.08 atm and 8°C

(b) 6.20 atm and 875°C

(c) 0.500 atm and −15°C

(d) 680 mm Hg and −30°C

12.44 A quantity of air has a volume of 856 mL at STP. What volume, in milliliters, will the air occupy if the pressure and temperature are changed to the following values?

(a) 589 mm Hg and 11°C

(b) 11 mm Hg and 589°C

(c) 1575 mm Hg and −25°C

(d) 1.01 atm and −1°C

Gay-Lussac's Law of Combining Volumes (Sec. 12.8)

12.45 Ammonia reacts with oxygen to form nitrogen (N_2) and water:

$$4 NH_3(g) + 3 O_2(g) \longrightarrow 2 N_2(g) + 6 H_2O(g)$$

It can also react with oxygen to form nitric oxide (NO) and water:

$$4 NH_3(g) + 5 O_2(g) \longrightarrow 4 NO(g) + 6 H_2O(g)$$

Which of these two reactions occurred if 1.60 liters of NH_3 are found by experiment to react with 2.00 liters of O_2? Assume that both gas volumes are measured at the same temperature and pressure.

12.46 Methane, CH_4, reacts with steam to form hydrogen and carbon monoxide:

$$CH_4(g) + H_2O(g) \longrightarrow 3 H_2(g) + CO(g)$$

It can also react with steam to form hydrogen and carbon dioxide:

$$CH_4(g) + 2 H_2O(g) \longrightarrow 4 H_2(g) + CO_2(g)$$

Which of these two reactions occurred if 2.33 L of methane are found by experiment to react with 2.33 L of steam? Assume that both gas volumes are measured at the same temperature and pressure.

12.47 The equation for the combustion of the fuel propane, C_3H_8, is

$$C_3H_8(g) + 5 O_2(g) \longrightarrow 3 CO_2(g) + 4 H_2O(g)$$

(a) How many liters of C_3H_8 must be burned to produce 1.30 L of CO_2 if both volumes are measured at STP?

(b) How many liters of C_3H_8 must be burned to produce 1.30 L of H_2O if both volumes are measured at 2.00 atm and 56°C?

12.48 The equation for the combustion of the fuel butane, C_4H_{10}, is

$$2 C_4H_{10}(g) + 13 O_2(g) \longrightarrow 8 CO_2(g) + 10 H_2O(g)$$

(a) How many liters of C_4H_{10} must be burned to produce 2.60 L of CO_2 if both volumes are measured at STP?

(b) How many liters of C_4H_{10} must be burned to produce 2.60 L of H_2O if both volumes are measured at 1.75 atm and 43°C?

12.49 For the reaction in Problem 12.47, how many liters of C_3H_8 must be burned to produce a combined total of 0.75 liter of gaseous products if all volumes are measured at the same temperature and pressure?

12.50 For the reaction in Problem 12.48, how many liters of C_4H_{10} must be burned to produce a combined total of 1.75 liters of gaseous products if all volumes are measured at the same temperature and pressure?

Gas Volumes and Limiting Reactants (Sec. 12.9)

12.51 At high temperatures and pressures, nitrogen will react with hydrogen to produce ammonia as shown by the equation

$$N_2(g) + 3 H_2(g) \longrightarrow 2 NH_3(g)$$

For each of the following combinations of volumes of gases, decide which is the limiting reactant. (Assume all gases are at the same conditions.)

(a) 1.00 L of N_2 and 1.50 L of H_2

(b) 2.00 L of N_2 and 5.50 L of H_2

(c) 1.00 L of N_2 and 4.00 L of H_2

(d) 3.00 L of N_2 and 1.00 L of H_2

12.52 At normal atmospheric conditions, nitric oxide will react with oxygen (from the air) to produce nitrogen dioxide as shown by the equation

$$2 NO(g) + O_2(g) \longrightarrow 2 NO_2(g)$$

For each of the following combinations of volumes of gases, decide which is the limiting reactant. (Assume all gases are at the same conditions.)

(a) 2.00 L of NO and 1.50 L of O_2

(b) 2.00 L of NO and 5.50 L of O_2

(c) 5.00 L of NO and 2.00 L of O_2

(d) 3.00 L of NO and 1.00 L of O_2

12.53 Under appropriate conditions, the reaction between methane (CH_4) gas and steam proceeds as shown by the equation

$$CH_4(g) + 2 H_2O(g) \longrightarrow CO_2(g) + 4 H_2(g)$$

How many liters of each of the products can be produced from the following volumes of reactants? Assume all gases are at the same conditions.

(a) 45.0 L of CH_4 and 45.0 L of H_2O

(b) 45.0 L of CH_4 and 66.0 L of H_2O

(c) 64.0 L of CH_4 and 134 L of H_2O

(d) 16.0 L of CH_4 and 32.0 L of H_2O

12.54 Under appropriate conditions, the reaction between methane (CH_4) gas and oxygen gas proceeds as shown by the equation

$$CH_4(g) + 2 O_2(g) \longrightarrow CO_2(g) + 2 H_2O(g)$$

How many liters of each of the products can be produced from the following volumes of reactants? Assume all gases are at the same conditions.

(a) 45.0 L of CH_4 and 45.0 L of O_2

(b) 45.0 L of CH_4 and 65.0 L of O_2

(c) 64.0 L of CH_4 and 142 L of O_2

(d) 56.0 L of CH_4 and 112 L of O_2

Avogadro's Law (Sec. 12.10)

12.55 Consider two samples of Ar gas at the same temperature and pressure. The first sample contains 0.573 mole Ar and has a volume of 37.2 mL. How many moles of Ar are present in the second sample if it has a volume of 50.7 mL?

12.56 Consider two samples of Xe gas at the same temperature and pressure. The first sample contains 1.00 mole of Xe and has a volume of 123 mL. How many moles of Xe are present in the second sample if it has a volume of 57 mL?

12.57 At a certain temperature and pressure, 3.75 g of CO gas occupies a volume of 1.25 L. What would be the volume, in liters, of 0.300 mole of CO gas at the same temperature and pressure?

12.58 At a certain temperature and pressure, 6.00 g of SO_2 gas occupies a volume of 3.50 L. What would be the volume, in liters, of 2.00 moles of SO_2 gas at the same temperature and pressure?

12.59 A balloon containing 1.83 moles of He has a volume of 0.673 L at a certain temperature and pressure. How many *grams* of He would have to be added to the balloon in order for the volume to increase to 0.811 L at the same temperature and pressure?

12.60 A balloon containing 1.83 moles of He has a volume of 0.673 L at a certain temperature and pressure. How many *grams* of He would have to be removed from the balloon in order for the volume to decrease to 0.455 L at the same temperature and pressure?

12.61 A 0.625-mole sample of N_2 gas at 1.50 atm and 45°C occupies a volume of 1.33 L. What volume, in liters, would a 0.625-mole sample of N_2O gas occupy at the same temperature and pressure?

12.62 A 1.34-mole sample of Cl_2 gas at 2.00 atm and 32°C occupies a volume of 85.0 mL. What volume, in liters, would a 1.34-mole sample of Cl_2O gas occupy at the same temperature and pressure?

12.63 In each of the following pairs of gas samples, select the pair member that would have the larger volume at 27°C and 1.00 atm.

(a) 0.400 mole N_2 and 0.450 mole O_2
(b) 2.32 moles CH_4 and 2.00 moles C_2H_6
(c) 100.0 g NO_2 and 100.0 g N_2O
(d) 100.0 g CO and 100.0 g CO_2

12.64 In each of the following pairs of gas samples, select the pair member that would have the smaller volume at 33°C and 2.00 atm.

(a) 0.300 mole He and 0.200 mole H_2
(b) 1.05 moles SO_2 and 2.10 moles SO_3
(c) 50.0 g F_2 and 50.0 g OF_2
(d) 50.0 g NH_3 and 50.0 g PH_3

12.65 A gaseous nitrogen–fluorine compound decomposes to give nitrogen and fluorine gas. Using Avogadro's law, determine the molecular formula of this compound given that 2.88 L of the compound decompose to form 1.44 L of N_2 and 4.32 L of F_2, all volumes being measured at the same temperature and pressure.

12.66 A gaseous nitrogen–oxygen compound decomposes to give nitrogen and oxygen gas. Using Avogadro's law, determine the molecular formula of this compound given that 1.24 L of the compound decompose to form 1.24 L of N_2 and 2.48 L of O_2, all volumes being measured at the same temperature and pressure.

12.67 A 0.500-mole sample of a gas has a volume of 2.0 L at a pressure of 1.5 atm and a temperature of 27°C. What will be the volume of the gas, in liters, if the following changes are made to the system: the amount of gas is decreased by 0.100 mole, the pressure is decreased by 0.50 atm, and the temperature is increased by 20°C?

12.68 A 2.00-mole sample of a gas has a volume of 4.0 L at a pressure of 0.50 atm and a temperature of 35°C. What will be the volume of the gas, in liters, if the following changes are made to the system: the amount of gas is increased by 0.100 mole, the pressure is increased by 0.50 atm, and the temperature is decreased by 20°C?

Molar Volume (Sec. 12.11)

12.69 What is the volume, in liters, at STP occupied by 1.25 moles of each of the following gases?

(a) N_2 **(b)** NH_3 **(c)** CO_2 **(d)** Cl_2

12.70 What is the volume, in liters, at STP occupied by 1.75 moles of each of the following gases?

(a) O_3 **(b)** O_2 **(c)** NO_2 **(d)** C_2H_6

12.71 In each of the following pairs of gas samples, select the pair member that occupies the larger volume at STP.

(a) 24.5 g N_2 and 24.5 g NH_3
(b) 30.0 g O_2 and 30.0 g O_3
(c) 10.0 g SO_2 and 20.0 g NO_2
(d) 15.0 g N_2O and 20.0 g NO

12.72 In each of the following pairs of gas samples, select the pair member that occupies the larger volume at STP.

(a) 10.0 g H_2 and 10.0 g He
(b) 25.0 g F_2 and 25.0 g Cl_2
(c) 15.0 g CH_4 and 5.00 g PH_3
(d) 100.0 g HCN and 2.00 g UF_6

12.73 Under STP conditions, what is the mass, in grams, of 23.7-L samples of each of the following gases?

(a) Ar   **(b)** N_2O **(c)** SO_3 **(d)** PH_3

12.74 Under STP conditions, what is the mass, in grams, of 35.2-L samples of each of the following gases?

(a) Kr   **(b)** S_2O **(c)** O_3 **(d)** NH_3

12.75 Calculate the density at STP conditions, in grams per liter, of each of the following gases.

(a) N_2O **(b)** N_2O_4 **(c)** NO_2 **(d)** NO

12.76 Calculate the density at STP conditions, in grams per liter, of each of the following gases.

(a) CH_4 **(b)** C_2H_2 **(c)** C_2H_4 **(d)** C_2H_6

12.77 Which gas in each of the following pairs of gases will have the greater density at STP?

(a) O_2 and O_3 **(b)** NH_3 and PH_3
(c) CO and CO_2 **(d)** F_2 and CH_4

12.78 Which gas in each of the following pairs of gases will have the greater density at STP?

(a) SO_2 and S_2O (b) F_2 and OF_2

(c) NO and CO   (d) SF_6 and UF_6

12.79 Calculate molar masses for gases with the following densities at STP.

(a) 1.97 g/L   (b) 1.25 g/L

(c) 0.714 g/L   (d) 3.17 g/L

12.80 Calculate molar masses for gases with the following densities at STP.

(a) 1.70 g/L   (b) 0.897 g/L

(c) 1.16 g/L   (d) 0.759 g/L

12.81 If the Xe atoms present in 22.5 g of XeO_2F_2 are converted to gaseous Xe, what volume, in liters, will they occupy at STP conditions?

12.82 If the Xe atoms present in 235 g of $XeOF_4$ are converted to gaseous Xe, what volume, in liters, will they occupy at STP conditions?

The Ideal Gas Law (Sec. 12.12)

12.83 Using the ideal gas law, calculate the volume, in liters, of 1.20 moles of Cl_2 gas at each of the following sets of conditions.

(a) STP

(b) 73°C and 1.54 atm

(c) 525°C and 15.0 atm

(d) −23°C and 765 mm Hg

12.84 Using the ideal gas law, calculate the volume, in liters, of 1.15 moles of F_2 gas at each of the following sets of conditions.

(a) STP

(b) 315°C and 456 mm Hg

(c) −45°C and 4.50 atm

(d) 23°C and 762 mm Hg

12.85 How many moles of H_2 gas does it take to fill a 6.00-L container at a pressure of 3.67 atm at 25°C?

12.86 How many moles of He gas does it take to fill a 5.00-L container at a pressure of 2.50 atm at 35°C?

12.87 If 0.332 mole of He gas has a volume of 1275 mL and a pressure of 5.78 atm, what is its temperature in degrees Celsius?

12.88 If 0.504 mole of Ar gas has a volume of 1975 mL and a pressure of 4.60 atm, what is its temperature in degrees Celsius?

12.89 What is the pressure, in atmospheres, inside a 6.00-L container that contains the following amounts of N_2 gas at 25°C?

(a) 0.30 mole   (b) 1.20 moles

(c) 0.30 g   (d) 1.20 g

12.90 What is the pressure, in millimeters of mercury, inside a 4.00-L container that contains the following amounts of O_2 gas at 40°C?

(a) 0.72 mole   (b) 4.5 moles

(c) 0.72 g   (d) 4.5 g

12.91 1.14 moles of the noble gas argon occupy a volume of 3.00 L at a temperature of 152°C and a pressure of 13.3 atm. Use this information to calculate the value of the ideal gas constant R in the units of atm · L/mole · K.

12.92 0.317 mole of the noble gas helium occupies a volume of 8.67 L at a temperature of 25°C and a pressure of 679 mm Hg. Use this information to calculate the value of the ideal gas constant R in the units of mm Hg · L/mole · K.

12.93 A 1.00-mole sample of liquid water is placed in a flexible sealed container and allowed to evaporate. After complete evaporation, what will be the container volume, in liters, at 127°C and 0.908 atm pressure?

12.94 A 1.00-mole sample of dry ice (solid CO_2) is placed in a flexible sealed container and allowed to sublime. After all the CO_2 has changed from solid to gas, what will be the container volume, in liters, at 23°C and 0.983 atm pressure?

12.95 Calculate the mass, in grams, of each of the following quantities of gas.

(a) 25.0 L of C_2H_6 at 0.972 atm and 29°C

(b) 2.22 L of HCl at 854 mm Hg and 75°C

(c) 5.50 L of SO_2 at STP

(d) 783 mL of N_2O at 359 mm Hg and 273°C

12.96 Calculate the mass, in grams, of each of the following quantities of gas.

(a) 3.00 L of NO at 1.23 atm and 35°C

(b) 1.37 L of CO_2 at 498 mm Hg and 285°C

(c) 3.50 L of HF at STP

(d) 1780 mL of C_2H_2 at 3.00 atm and 585°C

12.97 A 30.0-L cylinder contains 100.0 g of Cl_2 at 23°C. How many grams of Cl_2 must be added to the container to increase the pressure in the cylinder to 1.65 atm? Assume that the temperature remains constant.

12.98 A 50.0-L cylinder contains 100.0 g of F_2 at 30°C. How many grams of F_2 must be removed from the container to decrease the pressure in the cylinder to 1.00 atm? Assume that the temperature remains constant.

Molar Mass, Density, and the Ideal Gas Law (Sec. 12.13)

12.99 A 1.305-g sample of carbon monoxide gas (CO) occupies a volume of 1.20 L at 741 mm Hg pressure and 33°C. Based on these data, calculate the molar mass of carbon monoxide.

12.100 A 1.305-g sample of nitrogen gas (N_2) occupies a volume of 1.20 L at 0.975 atm pressure and 33°C. Based on these data, calculate the molar mass of nitrogen gas.

12.101 A 125-mL flask contains 0.450 g of a gaseous compound at 75°C and 1.00 atm pressure. What is the molar mass of the compound?

12.102 A 250-mL flask contains 0.350 g of a gaseous compound at 85°C and 1.00 atm pressure. What is the molar mass of the compound?

12.103 Determine the molar mass of a liquid whose vapor filled a 125-mL flask at 90°C and 1.20 atm pressure. The mass of the vapor is 0.4537 g.

12.104 Determine the molar mass of a liquid whose vapor filled a 125-mL flask at 75°C and 1.00 atm pressure. The mass of the vapor is 0.5545 g.

12.105 A gas is either CO or CO_2. Based on the following information about a sample of the gas, what is its identity? A 0.902-L sample of the gas exerts a pressure of 1.65 atm, has a mass of 1.733 g, and is at a temperature of 20°C.

12.106 A gas is either CH_4 or HF. Based on the following information about a sample of the gas, what is its identity? A 1.98-L sample of the gas exerts a pressure of 2.00 atm, has a mass of 3.20 g, and is at a temperature of 30°C.

12.107 Calculate the density of H_2S gas, in grams per liter, at each of the following temperature–pressure conditions.

(a) 24°C and 675 mm Hg

(b) 24°C and 1.20 atm

(c) 370°C and 1.30 atm

(d) −25°C and 452 mm Hg

12.108 Calculate the density of SCl_2 gas, in grams per liter, at each of the following temperature–pressure conditions.

(a) 27°C and 1.20 atm

(b) 37°C and 5.00 atm

(c) 227°C and 4.00 atm

(d) 550°C and 1.00 atm

12.109 What pressure, in atmospheres, is required to cause each of the following gases to have a density of 1.00 g/L at 47°C?

(a) N_2 (b) Xe   (c) ClF   (d) N_2O

12.110 What pressure, in atmospheres, is required to cause each of the following gases to have a density of 1.00 g/L at 53°C?

(a) O_2 (b) O_3 (c) SO_2 (d) CH_4

12.111 A gas is either NO or CO. Based on the following information about a sample of the gas, what is its identity? A sample of the gas at a temperature of 27°C and a pressure of 2.00 atm has a density of 2.27 g/L.

12.112 A gas is either H_2S or HF. Based on the following information about a sample of the gas, what is its identity? A sample of the gas at a temperature of 127°C and a pressure of 3.00 atm has a density of 3.12 g/L.

Gas Laws and Chemical Equations (Sec. 12.14)

12.113 A mixture of 25.0 g of NO and an excess of O_2 reacts according to the balanced equation

$$2\,NO(g) + O_2(g) \longrightarrow 2\,NO_2(g)$$

How many liters of NO_2, at STP, are produced?

12.114 A mixture of 25.0 g of H_2 and an excess of N_2 reacts according to the balanced equation

$$3\,H_2(g) + N_2(g) \longrightarrow 2\,NH_3(g)$$

How many liters of NH_3, at STP, are produced?

12.115 A sample of O_2 with a volume of 25.0 L at 27°C and 1.00 atm is reacted with excess N_2 to produce NO. The equation for the reaction is

$$O_2(g) + N_2(g) \longrightarrow 2\,NO(g)$$

How many grams of NO are produced?

12.116 A sample of H_2 with a volume of 35.0 L at 35°C and 1.35 atm is reacted with excess O_2 to produce H_2O. The equation for the reaction is

$$2\,H_2(g) + O_2(g) \longrightarrow 2\,H_2O(g)$$

How many grams of H_2O are produced?

12.117 Hydrogen gas can be produced in the laboratory through reaction of magnesium metal with hydrochloric acid:

$$Mg(s) + 2\,HCl(aq) \longrightarrow MgCl_2(aq) + H_2(g)$$

What volume, in liters, of H_2 at 23°C and 0.980 atm pressure can be produced from the reaction of 12.0 g of Mg with an excess of HCl?

12.118 A common laboratory preparation for O_2 gas involves the thermal decomposition of potassium nitrate:

$$2\,KNO_3(s) \longrightarrow 2\,KNO_2(s) + O_2(g)$$

What volume, in liters, of O_2 at 35°C and 1.31 atm pressure can be produced from the decomposition of 35.0 g of KNO_3?

12.119 Ammonium nitrate, NH_4NO_3, can decompose explosively when heated to a high temperature according to the equation

$$2\,NH_4NO_3(s) \longrightarrow 2\,N_2(g) + 4\,H_2O(g) + O_2(g)$$

If a 100.0-g sample of ammonium nitrate decomposes at 450°C, how many liters of gaseous products would be formed? Assume that atmospheric pressure is 1.00 atm.

12.120 The industrial explosive nitroglycerin, $C_3H_5N_3O_9$, detonates according to the equation

$$4\,C_3H_5N_3O_9(s) \longrightarrow 6\,N_2(g) + O_2(g) + 12\,CO_2(g)$$
$$+ 10\,H_2O(g)$$

At a detonation temperature of 1950°C, how many liters of gaseous products would be formed from 100.0 g of nitroglycerin? Assume that atmospheric pressure is 1.00 atm.

12.121 How many liters of NO_2 gas at 21°C and 2.31 atm must be consumed in producing 75.0 L of NO gas at 38°C and 645 mm Hg according to the following reaction?

$$3\,NO_2(g) + H_2O(l) \longrightarrow 2\,HNO_3(aq) + NO(g)$$

12.122 How many liters of Cl_2 gas at 25°C and 1.50 atm are needed to react completely with 3.42 L of NH_3 gas at 50°C and 2.50 atm according to the following reaction?

$$2\,NH_3(g) + 3\,Cl_2(g) \longrightarrow N_2(g) + 6\,HCl(g)$$

Mixtures of Gas (Sec. 12.15)

12.123 A gaseous mixture consists of 3.00 moles of N_2 and 3.00 moles of O_2. What is the volume, in liters, occupied by this mixture if it is under a pressure of 20.00 atm and at a temperature of 27°C?

12.124 A gaseous mixture consists of 4.00 moles of N_2 and 5.00 moles of O_2. What is the volume, in liters, occupied by this mixture if it is under a pressure of 20.00 atm and at a temperature of 27°C?

12.125 What would be the volume, in liters, of a gaseous mixture at 1.00 atm pressure and 27°C if its composition is each of the following?

(a) 3.00 g Ne and 3.00 g Ar

(b) 4.00 g Ne and 2.00 g Ar

(c) 3.00 moles Ne and 5.00 g Ar

(d) 4.00 moles Ne and 4.00 moles Ar

12.126 What would be the volume, in liters, of a gaseous mixture at 1.00 atm pressure and 27°C if its composition is each of the following?

(a) 4.00 g He and 8.00 g H_2

(b) 4.00 g He and 8.00 g N_2

(c) 3.00 moles He and 3.00 moles N_2

(d) 3.00 moles He and 3.00 moles H_2

12.127 What would be the pressure, in atm, of a gaseous mixture that occupies a volume of 27.0 L at 20°C if 3.00 moles each of Ar, Ne, and He are present?

12.128 What would be the pressure, in atm, of a gaseous mixture that occupies a volume of 2.00 L at 40°C if 3.00 grams each of Ar, Ne, and He are present?

Dalton's Law of Partial Pressures (Sec. 12.16)

12.129 Helium gas is added to an empty gas cylinder until the pressure reaches 9.0 atm. Neon gas is then added to the cylinder until the total pressure is 14.0 atm. Argon gas is then added to the cylinder until the total cylinder pressure reaches 29.0 atm. What is the partial pressure of each gas in the cylinder?

12.130 Nitrogen gas (N_2) is added to an empty gas cylinder until the pressure reaches 8.0 atm. Helium gas is then added to the cylinder until the total pressure is 10.0 atm. Carbon monoxide gas (CO) is then added to the cylinder until the total cylinder pressure reaches 20.0 atm. What is the partial pressure of each gas in the cylinder?

12.131 A mixture of H_2, N_2, and Ar gases is present in a steel cylinder. The total pressure within the cylinder is 675 mm Hg and the partial pressures of N_2 and Ar are, respectively, 354 mm Hg and 235 mm Hg. If CO_2 gas is added to the mixture, at constant temperature, until the total pressure reaches 842 mm Hg, what is the partial pressure, in millimeters of Hg, of the following?

(a) CO_2 **(b)** N_2 **(c)** Ar   **(d)** H_2

12.132 A mixture of O_2, He, and Ne gases is present in a steel cylinder. The total pressure within the cylinder is 652 mm Hg and the partial pressures of He and Ne are, respectively, 251 mm Hg and 152 mm Hg. If CO_2 gas is added to the mixture, at constant temperature, until the total pressure reaches 704 mm Hg, what is the partial pressure, in millimeters of Hg, of the following?

(a) CO_2 **(b)** He   **(c)** Ne   **(d)** O_2

12.133 A gaseous mixture contains 25.0 g of each of the gases CO, CO_2, and H_2S. Assuming that the gases do not react with each other, calculate the following.

(a) the mole fraction of each gas present in the mixture

(b) the partial pressure of each gas present given that the total pressure exerted by the gaseous mixture is 1.72 atm

12.134 A gaseous mixture contains 15.0 g of each of the gases HCl, H_2S, and Xe. Assuming that the gases do not react with each other, calculate the following.

(a) the mole fraction of each gas present in the mixture

(b) the partial pressure of each gas present given that the total pressure exerted by the gaseous mixture is 2.24 atm

12.135 Calculate the partial pressure of O_2, in atmospheres, in a gaseous mixture with a volume of 2.50 L at 20°C given that the mixture composition is

(a) 0.50 mole O_2 and 0.50 mole N_2

(b) 0.50 mole O_2 and 0.75 mole N_2

(c) 0.50 mole O_2, 0.75 mole N_2, and 0.75 mole Ar

(d) 0.50 g O_2 and 0.75 g N_2

12.136 Calculate the partial pressure of Xe, in atmospheres, in a gaseous mixture with a volume of 1.20 L at 32°C given that the mixture composition is

(a) 0.40 mole Xe and 0.40 mole Ne

(b) 0.40 mole Xe and 0.60 mole Ne

(c) 0.40 mole Xe, 0.60 mole Ne, and 1.25 moles O_2

(d) 0.40 g Xe and 0.60 g Ne

12.137 What is the partial pressure of O_2 in a gaseous mixture whose total pressure is 1.20 atm given the following mixture compositions?

(a) 0.40 mole O_2 and 0.40 mole Ne

(b) 0.40 mole O_2 and 0.80 mole Ne

(c) an equal number of moles of O_2, N_2, and H_2

(d) an equal number of molecules of O_2, N_2, and H_2

12.138 What is the partial pressure of Xe in a gaseous mixture whose total pressure is 1.55 atm given the following mixture compositions?

(a) 0.50 mole Xe and 0.50 mole Ne

(b) 0.50 mole Xe and 1.00 mole Ne

(c) an equal number of moles of Xe, Ne, and He

(d) an equal number of atoms of Xe, Ne, and He

12.139 What is the total pressure in a flask that contains 4.0 moles He, 2.0 moles Ne, and 0.50 mole Ar, and in which the partial pressure of Ar is 0.40 atm?

12.140 What is the total pressure in a flask that contains 2.0 moles H_2, 6.0 moles O_2, and 0.50 mole N_2, and in which the partial pressure of H_2 is 0.80 atm?

12.141 Calculate the partial pressure of O_2, in atm, in a gaseous mixture of O_2 and N_2 given that

(a) the total pressure is 6.00 atm and the mole fraction of N_2 is 0.150.

(b) the partial pressure of N_2 is 2.26 atm and the mole fraction N_2 is 0.180.

12.142 Calculate the partial pressure of O_2, in atm, in a gaseous mixture of O_2 and N_2 given that

(a) the total pressure is 5.00 atm and the mole fraction of N_2 is 0.800.

(b) the partial pressure of N_2 is 3.00 atm and the mole fraction N_2 is 0.220.

12.143 What is the mole fraction of each gas in a mixture having the partial pressures of 0.500 atm of He, 0.250 atm of Ar, and 0.350 atm of Xe?

12.144 What is the mole fraction of each gas in a mixture having the particle pressures of 0.350 of Ne, 0.550 atm of Kr, and 0.750 atm of He?

12.145 A sample of N_2 gas of mass 20.0 g is present in a vessel at 0°C and 1.00 atm. What will be the final pressure, in atm, in the vessel after 8.00 g of Ar is pumped into the vessel at constant temperature?

12.146 A sample of O_2 gas of mass 25.0 g is present in a vessel at 35°C and 2.00 atm. What will be the final pressure, in atm, in the vessel after 20.00 g of He is pumped into the vessel at constant temperature?

12.147 A sample of ammonia, NH_3, is *completely* decomposed to its constituent elements.

$$2\,NH_3(g) \longrightarrow N_2(g) + 3\,H_2(g)$$

If the total pressure of the N_2 and H_2 produced is 852 mm Hg, calculate the partial pressures, in millimeters of mercury, of N_2 and H_2.

12.148 A sample of steam, H_2O, is *completely* decomposed to its constituent elements.

$$2\,H_2O(g) \longrightarrow 2\,H_2(g) + O_2(g)$$

If the total pressure of the H_2 and O_2 produced is 1.35 atm, calculate the partial pressures, in atmospheres, of H_2 and O_2.

12.149 What would be the partial pressure, in millimeters of mercury, of O_2 collected over water at the following conditions of temperature and atmospheric pressure?

(a) 19°C and 743 mm Hg

(b) 28°C and 645 mm Hg

(c) 34°C and 762 mm Hg

(d) 21°C and 0.933 atm

12.150 What would be the partial pressure, in millimeters of mercury, of O_2 collected over water at the following conditions of temperature and atmospheric pressure?

(a) 15°C and 632 mm Hg

(b) 35°C and 749 mm Hg

(c) 31°C and 682 mm Hg

(d) 26°C and 0.975 atm

12.151 A 24.64-L flask, at STP, contains 0.100 mole of He, 0.200 mole of Ne, and 0.800 mole of Ar. For this gaseous mixture, calculate the

(a) mole fraction He

(b) mole percent Ar

(c) pressure percent Ne

(d) volume percent He

12.152 A 15.68-L flask, at STP, contains 0.200 mole of Ar, 0.400 mole of Kr, and 0.100 mole of Xe. For this gaseous mixture, calculate the

(a) mole fraction Ar

(b) mole percent Kr

(c) pressure percent Xe

(d) volume percent Ar

12.153 Three containers of gases are combined into a single large container; that is, 2.0 L of O_2 at STP, 3.0 L of Ar at STP, and 3.0 L of Ne at STP are put into an 8.0-L container at STP. Calculate the following items pertaining to the gaseous mixture.

(a) volume percent O_2

(b) mole percent Ar

(c) pressure percent Ne

(d) partial pressure O_2

12.154 Three containers of gases are combined into a single large container; that is, 1.0 L of N_2 at STP, 4.0 L of He at STP, and 1.0 L of Xe at STP are put into a 6.0-L container at STP. Calculate the following items pertaining to the gaseous mixture.

(a) volume percent He

(b) mole percent Xe

(c) pressure percent He

(d) partial pressure N_2

Additional Problems

12.155 How many molecules of carbon dioxide (CO_2) gas are contained in 1.00 L of CO_2 at STP?

12.156 How many molecules of hydrogen sulfide (H_2S) gas are contained in 2.00 L of H_2S at STP?

12.157 A near-vacuum pressure of 0.0010 mm Hg is readily obtained in a laboratory by means of a vacuum pump. Calculate the number of molecules in 1.00 mL of O_2 gas at this pressure and 23°C.

12.158 A near-vacuum pressure of 0.0010 mm Hg is readily obtained in a laboratory by means of a vacuum pump. Calculate the number of atoms in 1.00 mL of Xe gas at this pressure and 27°C.

12.159 At a particular temperature and pressure, 8.00 g of N_2 gas occupy 6.00 L. What would be the volume, in liters, occupied by 1.00×10^{23} molecules of NH_3 at the same temperature and pressure?

12.160 At a particular temperature and pressure, 2.00×10^{23} molecules of N_2 gas occupy 5.00 L. What would be the volume, in liters, occupied by 25.7 g of SO_2 at the same temperature and pressure?

12.161 A piece of Al is placed in a 1.00-L container with pure O_2. The O_2 is at a pressure of 1.00 atm and a temperature of 25°C. One hour later the pressure has dropped to 0.880 atm and the temperature has dropped to 22°C. Calculate the number of grams of O_2 that reacted with the Al.

12.162 A piece of Ca is placed in a 1.00-L container with pure N_2. The N_2 is at a pressure of 1.12 atm and a temperature of 26°C. One hour later the pressure has dropped to 0.924 atm and the temperature has dropped to 24°C. Calculate the number of grams of N_2 that reacted with the Ca.

12.163 At constant temperature, the pressure on a sample of H_2 gas is increased from 1.50 atm to 3.50 atm. This action decreases the sample volume by 25.0 mL. What was the original volume, in milliliters, of the gas sample?

12.164 At constant temperature, the pressure on a sample of HCl gas is decreased from 3.50 atm to 1.50 atm. This action increases the sample volume by 75.0 mL. What was the original volume, in milliliters, of the gas sample?

12.165 A large flask (of unknown volume) is filled with air until the pressure reaches 3.6 atm. The flask is then attached to a second evacuated flask of known volume, and the air from the first flask is allowed to expand into the second flask. The final pressure of the air (in both flasks) is 2.6 atm, and the volume of the second flask is 5.21 L. Calculate the volume, in liters, of the first flask.

12.166 A large 4.2-L flask is filled with air to a pressure that is unknown. This flask is then attached to a second evacuated flask of known volume, and the air from the first flask is allowed to expand into the second flask. The final pressure of the air (in both flasks) is 2.6 atm, and the volume of the second flask is 5.21 L. Calculate the original pressure, in atmospheres, in the first flask.

12.167 The volume of a fixed quantity of gas, at constant temperature, is decreased by 20.0%. What is the resulting percentage increase in the pressure of the gas?

12.168 The volume of a fixed quantity of gas, at constant temperature, is increased by 30.0%. What is the resulting percentage decrease in the pressure of the gas?

12.169 The volume of a fixed quantity of gas, at constant pressure, is increased by 50.0%. What is the resulting percentage increase in the Kelvin temperature of the gas?

12.170 The volume of a fixed quantity of gas, at constant pressure, is decreased by 10.0%. What is the resulting percentage decrease in the Kelvin temperature of the gas?

12.171 Calculate the ratio of the densities of O_2 and N_2 at

(a) STP conditions   **(b)** 1.25 atm and 25°C

12.172 Calculate the ratio of the densities of He and Ne at

(a) STP conditions

(b) 2.30 atm and 57°C

12.173 At constant temperature, a 2.000-L mixture of gases is produced by combining 1.000 L of N_2 at 350.0 mm Hg, 6.000 L of O_2 at 300.0 mm Hg, and 1.000 L of H_2 at 250.0 mm Hg. What is the pressure, in millimeters of mercury, of the mixture? Assume that no chemical reactions occur.

12.174 At constant temperature, a 2.000-L mixture of gases is produced by combining 2.000 L of N_2 at 250.0 mm Hg, 4.000 L of O_2 at 250.0 mm Hg, and 1.000 L of H_2 at 600.0 mm Hg. What is the pressure, in millimeters of mercury, of the mixture? Assume that no chemical reactions occur.

12.175 Suppose 532 mL of Ne gas at 20°C and 1.04 atm and 376 mL of SF_6 gas at 20°C and 0.97 atm are put into a 275-mL flask. Assuming that the temperature remains constant, calculate the partial pressure, in atmospheres, of the SF_6 gas in the mixture.

12.176 Suppose 734 mL of Ar gas at 18°C and 1.25 atm and 252 mL of HF gas at 18°C and 2.25 atm are put into a 545-mL flask. Assuming that the temperature remains constant, calculate the partial pressure, in atmospheres, of the HF gas in the mixture.

12.177 A mixture of 15.0 g of Ar and 15.0 g of CH_4 occupies a 4.0-L container at 8.80 atm and 54°C. What is the partial pressure, in atmospheres, of Ar in the mixture?

12.178 A mixture of 30.0 g of Ar and 15.0 g of CH_4 occupies a 4.0-L container at 11.9 atm and 27°C. What is the partial pressure, in atmospheres, of CH_4 in the mixture?

12.179 Suppose 30.0 mL of N_2 gas at 27°C and 645 mm Hg pressure are added to a 40.0-mL container that already contains He at 37°C and 765 mm Hg. If the resulting mixture is brought to 32°C, what is the total pressure, in millimeters of mercury, of the mixture?

12.180 Suppose 50.0 mL of Xe gas at 45°C and 0.998 atm pressure are added to a 100.0-mL container that already contains He at 37°C and 765 mm Hg. If the resulting mixture is warmed to 75°C, what is the total pressure, in millimeters of mercury, of the mixture?

Cumulative Problems

12.181 Which sample contains more molecules: 20.0 L of steam at 135°C and 1.02 atm pressure or 10.5 mL of ice with a density of 0.917 g/mL at 0°C and 1.02 atm pressure?

12.182 Which sample contains more molecules: 30.0 L of steam at 135°C and 0.879 atm pressure or 10.5 mL of liquid water with a density of 0.998 g/mL at 20°C and 0.879 atm pressure?

12.183 A 2.24-L sample of a gaseous compound has a mass of 7.5 g at STP. What is the molecular mass of a molecule of this compound in atomic mass units?

12.184 A 4.48-L sample of a gaseous compound has a mass of 20.0 g at STP. What is the molecular mass of a molecule of this compound in atomic mass units?

12.185 What is the value of x in the formula PH_x if the density of the PH_x gas is 1.517 g/L at 0°C and 1.00 atm?

12.186 What is the value of x in the formula P_2H_x if the density of P_2H_x gas is 2.944 g/L at 0°C and 1.00 atm?

12.187 The composition of a gaseous mixture in terms of mass percent is 75.0% HCl, 5.00% H_2, and 20.0% He. For this mixture, calculate the following.

(a) the mole fraction of each component

(b) the partial pressure of each component, given that the total pressure is 1.20 atm

(c) a weighted average molar mass for the mixture as a whole

(d) the density of the mixture, at STP, based on the weighted average molar mass

12.188 The composition of a gaseous mixture in terms of mass percent is 43.0% Ar, 15.0% He, and 42.0% H_2S. For this mixture, calculate the following.

(a) the mole fraction of each component

(b) the partial pressure of each component, given that the total pressure is 3.50 atm

(c) a weighted average molar mass for the mixture as a whole

(d) the density of the mixture, at STP, based on the weighted average molar mass

12.189 A 0.581-g sample of a gaseous compound containing only carbon and hydrogen contains 0.480 g of carbon and 0.101 g of hydrogen. At STP, 33.6 mL of the gas have a mass of 0.0869 g. What is the molecular formula for the compound?

12.190 A 6.01-g sample of a gaseous compound containing only carbon and hydrogen contains 4.80 g of carbon and 1.21 g of hydrogen. At STP, 762 mL of the gas have a mass of 1.02 g. What is the molecular formula for the compound?

12.191 The elemental analysis of a certain compound is 24.3% C, 4.1% H, and 71.6% Cl by mass. If 0.132 g of compound vapor occupies 41.4 mL at 741 mm Hg pressure and 96°C, what are the molar mass and molecular formula of the compound?

12.192 The elemental analysis of a certain compound is 88.82% C and 11.18% H by mass. A 20.87-mg sample of compound vapor occupies 11.63 mL at 1.016 atm pressure and 100°C. What are the molar mass and the molecular formula of the compound?

12.193 Ammonia, NH_3, burns in oxygen to form nitric oxide, NO, and water.

$$4 NH_3(g) + 5 O_2(g) \longrightarrow 4 NO(g) + 6 H_2O(g)$$

Calculate the total volume, in liters, of products formed when 60.0 L of NH_3 burn in the presence of 60.0 L of O_2. Assume that all volume measurements are made at the same temperature and pressure.

12.194 Methane, CH_4, burns in oxygen to form carbon dioxide, CO_2, and water.

$$CH_4(g) + 2 O_2(g) \longrightarrow CO_2(g) + 2 H_2O(g)$$

Calculate the total volume, in liters, of products formed when 30.0 L of CH_4 burn in the presence of 50.0 L of O_2. Assume that all volume measurements are made at the same temperature and pressure.

12.195 A 1.75-L sample of H_2S, measured at 25.0°C and 625 mm Hg, is mixed with 5.75 L of O_2, measured at 10.0°C and 715 mm Hg, and the mixture is allowed to react.

$$2 H_2S(g) + 3 O_2(g) \longrightarrow 2 SO_2(g) + 2 H_2O(g)$$

How much H_2O, in grams, is produced?

12.196 A 2.00-L sample of N_2, measured at 30.0°C and 1.08 atm, is mixed with 4.00 L of O_2, measured at 25.0°C and 0.118 atm, and the mixture is allowed to react.

$$N_2(g) + O_2(g) \longrightarrow 2 NO(g)$$

How much NO, in grams, is produced?

12.197 A 22.0-g sample of NH_3 reacts with an excess of Cl_2 gas according to the equation

$$2 NH_3(g) + 3 Cl_2(g) \longrightarrow N_2(g) + 6 HCl(g)$$

What volume, in cubic meters at STP, of HCl gas is produced?

12.198 A 43.0-g sample of NO reacts with an excess of O_2 gas according to the equation

$$2 NO(g) + O_2(g) \longrightarrow 2 NO_2(g)$$

What volume, in cubic meters at STP, of NO_2 gas is produced?

12.199 Ammonia gas reacts with hydrogen chloride gas according to the equation

$$NH_3(g) + HCl(g) \longrightarrow NH_4Cl(s)$$

If 7.00 g of NH_3 are reacted with 12.0 g of HCl in a 1.00-L container at 25°C, what will be the final pressure, in atmospheres, in the reaction container?

12.200 At elevated temperatures phosphorus reacts with oxygen according to the equation

$$4 P(g) + 5 O_2(g) \longrightarrow P_4O_{10}(s)$$

If 50.0 g of P are reacted with 25.0 g of O_2 in an 8.00-L container at 175°C, what will be the final pressure, in atmospheres, in the reaction container?

12.201 An impure 60.2-g sample of KNO_3 was heated until all of the KNO_3 had decomposed.

$$2 KNO_3(s) \longrightarrow 2 KNO_2(s) + O_2(g)$$

The oxygen produced occupied 4.22 L at STP. What was the percent purity of the KNO_3 sample? Assume that KNO_3 is the only source for the O_2.

12.202 An impure 50.0-g sample of $CaCO_3$ was heated until all of the $CaCO_3$ had decomposed.

$$CaCO_3(s) \longrightarrow CaO(s) + CO_2(g)$$

The carbon dioxide produced occupied 6.63 L at STP. What was the percent purity of the $CaCO_3$ sample? Assume that $CaCO_3$ is the only source of the CO_2.

12.203 If the price of nitrogen gas (N_2) is \$5.20 per thousand cubic feet at 0°C and 1.00 atm pressure, what is the price of the gas per gram of nitrogen?

12.204 If the price of oxygen gas (O_2) is \$5.40 per thousand cubic feet at 0°C and 1.00 atm pressure, what is the price of the gas per gram of oxygen?

Answers to Practice Exercises

12.1 **(a)** 1.007 atm   **(b)** 765.1 mm Hg   **(c)** 14.78 psi

12.2 108 L

12.3 2.32 L

12.4 709 mL

12.5 177°C

12.6 $V_2 = V_1 \times \dfrac{P_1}{P_2} \times \dfrac{T_2}{T_1}$

12.7 503 mL

12.8 29°C

12.9 0.769 L

12.10 27.4 L H_2

12.11 O_2 is the limiting reactant

12.12 0.857 L He

12.13 16.4 L H_2S

12.14 **(a)** 11.4 g SO_2 **(b)** 14.3 g SO_3

12.15 0.0901 g/L H_2

12.16 1.92 L Xe

12.17 49.2 L O_3

12.18 109°C

12.19 121 g/mole

12.20 2.11 g/L

12.21 46.7 L O_2

12.22 50.4 g $PbCl_4$

12.23 34.6 L

12.24 0.85 atm

12.25 $X_{Ar} = 0.561$, $X_{Kr} = 0.268$, $X_{Xe} = 0.171$

12.26 $P_{CH_4} = 2.28$ atm, $P_{C_2H_6} = 0.117$ atm, $P_{C_3H_8} = 0.0321$ atm

12.27 1.2 atm

12.28 619 mm Hg

12.29 **(a)** 33.3%   **(b)** 33.3%   **(c)** 33.3%

13

Solutions

13.1 Characteristics of Solutions

A **solution** *is a homogeneous mixture of two or more substances, with each substance retaining its chemical identity.* To achieve a homogeneous mixture, the intermingling of components must be on the molecular level; that is, the particles present must be of atomic or molecular size.

In discussing solutions it is often convenient to categorize the components of a solution as *solvent* and *solute(s)*. A **solvent** *is the component of the solution present in the greatest amount.* The solvent may be thought of as the medium in which the other substances present are *dissolved.* A **solute** *is a solution component present in a small amount relative to that of solvent.* More than one solute may be present in the same solution. For example, both sugar and salt (two solutes) may be dissolved in water (solvent).

In most situations we will encounter, the solutes present in a solution will be of more interest to us than the solvent. The solutes are the "active ingredients" in the solution. They are the substances that may react when solutions are mixed.

Solutions used in the laboratory are usually liquids, and the solvent is almost always water. However, as we shall see shortly, gaseous solutions and solid solutions of several types do exist.

A solution, since it is homogeneous, will have the same properties throughout. No matter from where we take a sample in a solution it will have the same composition as that of any other sample from the solution. The composition of a solution can be varied, usually within certain limits, by changing the relative amounts of solvent and solute present. (If the composition limits are violated, a heterogeneous mixture is formed.)

501

Table 13.1	Examples of Various Types of Solutions

Solution Type (solute listed first)	Example
Gaseous Solutions	
Gas dissolved in gas	Dry air (oxygen and other gases dissolved in nitrogen)
Liquid dissolved in gas*	Wet air (water vapor in air)
Solid dissolved in gas*	Moth repellent (or moth balls) sublimed into air
Liquid Solutions	
Gas dissolved in liquid	Carbonated beverage (carbon dioxide in water)
Liquid dissolved in liquid	Vinegar (acetic acid dissolved in water)
Solid dissolved in liquid	Salt water
Solid Solutions	
Gas dissolved in solid	Hydrogen in platinum
Liquid dissolved in solid	Dental filling (mercury dissolved in silver)
Solid dissolved in solid	Sterling silver (copper dissolved in silver)

An alternative viewpoint is that liquid-in-gas and solid-in-gas solutions do not actually exist as true solutions. From this viewpoint water vapor or moth repellent in air is considered to be a gas-in-gas solution since the water or moth repellent must evaporate or sublime first in order to enter the air.

> "All solutions are mixtures" is a valid statement. However, the reverse statement, "All mixtures are solutions," is not valid. Only those mixtures that are *homogeneous* are solutions.

Two-component solutions can be classified into nine types according to the physical states of the solvent and solute before mixing. These types, along with an example of each, are listed in Table 13.1. Solutions in which the final state of the solution components is liquid are the most common and are the type that are emphasized in this book.

The physical state of the solute becomes that of the solvent when a solution is formed. For example, solid naphthalene (moth repellent) must be sublimed (Sec. 4.4) for it to dissolve in air. Pulverizing a solid to a fine powder and dispersing it in air does not produce a solution. (Dust particles in air would be an example of this.) The particles of the solid must be subdivided to the molecular level; the solid must sublime. Similarly, fog is a suspension of water droplets in air; the droplets are large enough to reflect light, a fact that becomes evident when we drive an automobile on a foggy night. Thus, fog is not a solution. Water vapor, however, is present in solution form in air. When hydrogen gas dissolves in platinum metal (a gas-in-solid solution), the gas molecules take up fixed positions in the metal lattice. The gas is "solidified" as a result.

13.2 Solubility

In addition to *solvent* and *solute*, several other terms are useful in describing characteristics of solutions. **Solubility** *is the maximum amount of solute that will dissolve in a given amount of solvent.* Numerous factors affect the numerical value of a solute's solubility in a given solvent, including the nature of the solvent itself, the temperature, and in some cases the pressure and the presence of other solutes. Common units for expressing solubility are grams of solute per 100 g of solvent.

Effect of Temperature on Solubility

Most solids become more soluble in water with increasing temperature. The data in Table 13.2 illustrate this temperature-solubility pattern. Here, the solubilities of selected ionic solids in water are given at three different temperatures. Note from the values in this table that a temperature-caused solubility increase can be dramatic ($AgNO_3$, 680% increase by going

Table 13.2	Solubilities of Various Compounds in Water at 0°C, 50°C, and 100°C		
	Solubility (g solute/100 g H₂0)		
Solute	0°C	50°C	100°C
Lead(II) bromide ($PbBr_2$)	0.455	1.94	4.75
Silver sulfate (Ag_2SO_4)	0.573	1.08	1.41
Copper(II) sulfate ($CuSO_4$)	14.3	33.3	75.4
Sodium chloride (NaCl)	35.7	37.0	39.8
Silver nitrate ($AgNO_3$)	122	455	952
Cesium chloride (CsCl)	161.4	218.5	270.5

from 0°C to 100°C) to slight (NaCl, 11% increase in going from 0°C to 100°C). Figure 13.1 shows the solubility change with temperature, at 10°C intervals, for the molecular substance sucrose ($C_{12}H_{22}O_{11}$, table sugar) in water.

The use of specific units for solubility, as in Table 13.2, allows us to compare solubilities quite precisely. Such precision is often unnecessary, and instead *qualitative* statements about solubilities are made by using terms such as *very soluble, slightly soluble*, and so forth. The guidelines for the use of such terms are given in Table 13.3.

In contrast to the solubilities of solids, gas solubilities in water decrease with increasing temperature. For example, both N_2 and O_2, the major components of air, are less soluble in hot water than in cold water.

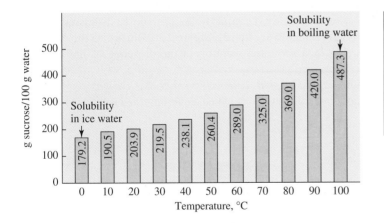

Figure 13.1

Solubility in water of the molecular compound sucrose ($C_{12}H_{22}O_{11}$, table sugar) as a function of temperature in the range 0°C to 100°C.

Table 13.3	Qualitative Solubility Terms
Solute Solubility (g solute/100 g solvent)	Qualitative Solubility Description
Less than 0.1	insoluble
0.1–1	slightly soluble
1–10	soluble
Greater than 10	very soluble

Effect of Pressure on Solubility

Pressure has little effect on the solubility of solids and liquids in water. However, it has a major effect on the solubility of gases in water. The amount of gas that will dissolve in a liquid at a given temperature is directly proportional to the pressure of the gas above the liquid. In other words, as the pressure of a gas above a liquid increases, the solubility of the gas increases; conversely, as the pressure of the gas decreases, its solubility decreases.

Respiratory therapy takes advantage of the fact that increased pressure increases the solubility of a gas. Patients with lung problems who are unable to get sufficient oxygen from air are given an oxygen-enriched mixture of gases to breathe. The larger oxygen partial pressure in the enriched mixture translates into increased oxygen uptake in the patient's lungs.

Amount of Solute in a Solution

> When the amount of dissolved solute in a solution corresponds to the solute's solubility in the solvent, the solution is a saturated solution.

A **saturated solution** *is a solution that contains the maximum amount of solute that can be dissolved under the conditions at which the solution exists.* A saturated solution together with excess undissolved solute is an equilibrium situation where the rate of dissolution of undissolved solute is equal to the rate of crystallization of dissolved solute. Consider the process of adding table sugar (sucrose) to a container of water. Initially the sugar dissolves as the solution is stirred. Finally, as we add more sugar, a point is reached where no amount of stirring will cause the added sugar to dissolve. Sugar remains as a solid on the bottom of the container; the solution is *saturated*. Although it appears to the eye that nothing is happening once the saturation point is reached, on the molecular level this is not the case. Solid sugar from the bottom of the container is continuously dissolving in the water, and an equal amount of sugar is coming out of solution. Accordingly, the net number of sugar molecules in the liquid remains the same, and outwardly it appears that the dissolution process has stopped. This equilibrium situation in the saturated solution is somewhat similar to the previously discussed evaporation of a liquid in a closed container (Sec. 11.12). Figure 13.2 illustrates the dynamic equilibrium process occurring in a saturated solution in the presence of undissolved excess solute.

An **unsaturated solution** *is a solution that contains less solute than the maximum amount that could dissolve.* The majority of solutions encountered are *unsaturated* solutions.

Figure 13.2

The dynamic equilibrium process occurring in a saturated solution that contains undissolved solute.

Saturated solution

Undissolved solute

Sometimes it is possible to exceed the maximum solubility of a compound, producing a *supersaturated solution*. A **supersaturated solution** *is a solution that contains more dissolved solute than that needed for a saturated solution.* An indirect rather than a direct procedure is needed for preparing a supersaturated solution; it involves the slow cooling, without agitation of any kind, of a high-temperature saturated solution in which no excess solid solute is present. Even though solute solubility decreases as the temperature is reduced, the excess solute remains in solution. A supersaturated solution is an unstable situation; with time, excess solute will crystallize out, and the solution will revert to a saturated solution. A supersaturated solution will produce crystals rapidly, often in a dramatic manner, if it is slightly disturbed or if it is "seeded" with a tiny crystal of solute.

The terms *dilute* and *concentrated* are also used to convey qualitative information about the degree of saturation of a solution. A **dilute solution** *is a solution that contains a small amount of solute in solution relative to the amount that could dissolve.* On the other hand, a **concentrated solution** *is a solution that contains a large amount of solute relative to the amount that could dissolve.* A concentrated solution need not be a saturated solution.

Aqueous and Nonaqueous Solutions

A set of solution terms that relates to the identity of the solvent present is *aqueous* and *nonaqueous*. An **aqueous solution** *is a solution in which water is the solvent.* The presence of water is not a prerequisite for a solution, however. A **nonaqueous solution** *is a solution in which a substance other than water is the solvent.* Alcohol-based solutions are often encountered in a medical setting.

> When the term *solution* is used, it is generally assumed that "aqueous solution" is meant, unless the context makes it clear that the solvent is not water.

13.3 Solution Formation

In a solution, solute particles are uniformly dispersed throughout the solvent. Considering what happens at the molecular level during the solution process will help us to understand how this is achieved. Let us consider in detail the process of dissolving sodium chloride, a typical ionic solid, in water.

Figure 13.3 shows what is thought to happen when sodium chloride is placed in water. The polar water molecules become oriented so that the negative oxygen portion points toward positive sodium ions and the positive hydrogen portions point toward negative chloride ions. As the polar water molecules begin to surround ions on the crystal surface, they exert sufficient attraction to cause these ions to break away from the crystal surface. After leaving the crystal, an ion retains its surrounding group of water molecules; it has become a *hydrated ion*. As each hydrated ion leaves the surface, other ions are exposed to the water, and the crystal is picked apart ion by ion. Once in solution, the hydrated ions are uniformly distributed by stirring or by random collisions with other molecules or ions.

The random motion of solute ions in solution causes them to collide with each other, with solvent molecules, and occasionally with the surface of the undissolved solute. Ions undergoing this last type of collision occasionally stick to the solid surface and thus leave the solution. When the number of ions in solution is low, the chances for collision with the undissolved solute are low. However, as the number of ions in solution increases, so do the chances for such collisions, and more ions are recaptured by the undissolved solute. Eventually, the number of ions in solution reaches such a level that ions return to the undissolved solute at the same rate as other ions leave. At this point the solution is saturated, and the equilibrium process discussed in the last section is in operation.

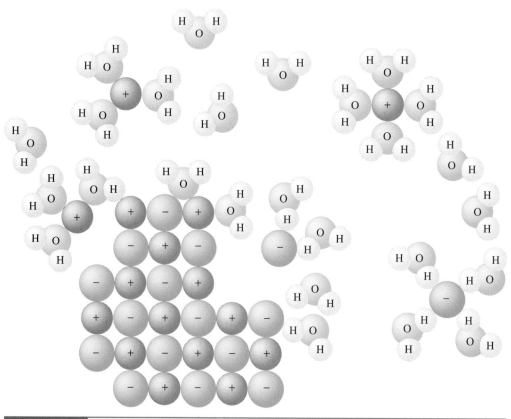

Figure 13.3

The solution process for an ionic solid in water.

Factors Affecting the Rate of Solution Formation

The rate at which a solution forms is governed by how rapidly the solute particles are distributed throughout the solvent. Three factors that affect the rate of solution formation are:

1. *The state of subdivision of the solute.* A crushed aspirin tablet will dissolve in water more rapidly than a whole aspirin tablet. The more compact whole aspirin tablet has less surface area, and thus fewer solvent molecules can interact with it at a given time.

2. *The degree of agitation during solution preparation.* Stirring solution components disperses the solute particles more rapidly, increasing the possibilities for solute–solvent interactions. Hence, the rate of solution formation is increased.

3. *The temperature of the solution components.* Solution formation occurs more rapidly as the temperature is increased. At a higher temperature, both solute and solvent molecules move more rapidly (Sec. 11.3) so more interactions between them occur within a given time period.

13.4 Solubility Rules

In this section we present some rules for qualitatively predicting solubilities. These rules summarize in a concise form the results of thousands of experimental solute–solvent solubility determinations.

A very useful generalization that relates polarity to solubility is *substances of like polarity tend to be more soluble in each other than substances that differ in polarity*. This conclusion is often expressed as the simple phrase "*like dissolves like*." Polar substances, in general, are good solvents for other polar substances but not for nonpolar substances. Similarly, nonpolar substances exhibit greater solubility in nonpolar solvents than they do in polar solvents.

A consideration of the intermolecular forces present in a solution reveals the basis for the "like dissolves like" rule. The same types of intermolecular forces present in pure liquids (Sec. 11.16) also operate in solutions. However, the situation is more complex for solutions because there are three types of interactions present: solute–solute interactions, solute–solvent interactions, and solvent–solvent interactions.

Solutions form when these three types of interactions are similar in nature and in magnitude (see Fig. 13.4). Cholesterol ($C_{27}H_{46}O$), a nonpolar substance, is soluble in fat (also a nonpolar material) because of the similar magnitude of the London forces (Sec. 11.16) present. Cholesterol has limited solubility in water because of the dissimilarity of the intermolecular forces, which are London forces for cholesterol and hydrogen bonding for water (Sec. 11.16). Sodium chloride is soluble in water because the strong Na^+ ion-water and Cl^- ion-water attractions (ion–dipole attractions; Sec. 11.16) are similar in nature to the strong attractions between Na^+ and Cl^- ions (ion–ion attractions; Sec. 11.16) and the strong hydrogen bonds between water molecules.

The generalization "like dissolves like" is a useful tool for predicting solubility behavior in many, but not all, solute–solvent situations. Results that agree with the generalization are almost always obtained in the cases of gas-in-liquid and liquid-in-liquid solutions and for solid-in-liquid solutions in which the solute is not an ionic compound. For example, NH_3 gas (a polar gas) is much more soluble in H_2O (a polar liquid) than is O_2 gas (a nonpolar gas). (The actual solubilities of NH_3 and O_2 in water at 20°C are, respectively, 51.8 g/100 g H_2O and 0.0043 g/100 g H_2O.)

For those solid-in-liquid solutions in which the solute is an ionic compound—a very common situation—the rule "like dissolves like" is not adequate. One would predict that since all ionic compounds are polar, they all would dissolve in polar solvents such as water. This is not the case. The failure of the generalization here is related to the complexity of the factors involved in determining the magnitude of the solute–solute (ion–ion) and solute–solvent (ion–polar solvent molecule) interactions. Among other things, both the charge on and size of the ions in the solute must be considered. Changes in these factors affect both types of interactions, but not to the same extent.

Some guidelines concerning the solubility of ionic compounds in water, which should be used in place of "like dissolves like," are given in Table 13.4.

There is no such thing as an absolutely insoluble ionic compound; all dissolve to a slight extent. Thus, the insoluble classification in Table 13.4 really means ionic compounds that have a very limited solubility in water.

You should become thoroughly familiar with the rules in Table 13.4. They find extensive use in chemical discussions. We will next encounter them in Section 14.7 when the topic of net ionic equations is presented.

The principle of "like dissolves like" is used in offset lithography printing. Most books, including this one, are printed with this process.

1. The printing plate is treated with a nonpolar, greasy lacquer in places where the image is to be printed on the paper.
2. The plate is moistened. The lacquer does not adsorb the polar water. The rest of the plate does.
3. The plate is inked. The ink is nonpolar and dissolves in the lacquer but not in the moistened areas.
4. Paper is pressed against the plate.
5. The ink transfers from the lacquer to the paper.

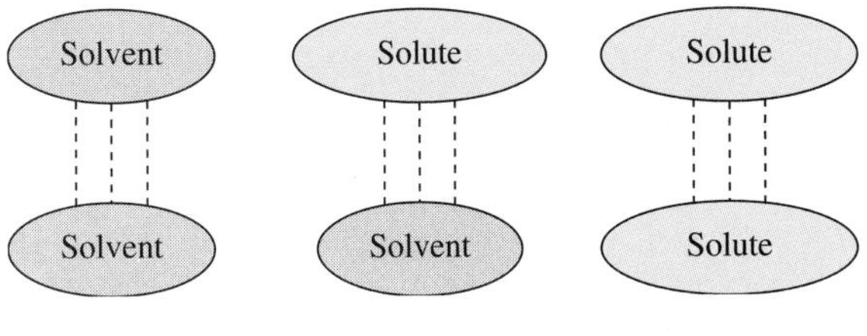

Figure 13.4

Solution formation occurs when these three kinds of intermolecular forces are similar in nature and in magnitude.

| Table 13.4 | Solubility Guidelines for Ionic Compounds in Water |

Soluble Compounds	Important Exceptions
Compounds containing the following ions are soluble with exceptions as noted.	
Group IA (Li^+, Na^+, K^+, etc.)	none
Ammonium NH_4^+	none
Acetate ($C_2H_3O_2^-$)	none
Nitrate (NO_3^-)	none
Chloride (Cl^-), bromide (Br^-), and iodide (I^-)	Ag^+, Pb^{2+}, Hg_2^{2+}
Sulfate (SO_4^{2-})	Ca^{2+}, Sr^{2+}, Ba^{2+}, Pb^{2+}

Insoluble Compounds	Important Exceptions
Compounds containing the following ions are insoluble with exceptions as noted.	
Carbonate (CO_3^{2-})	group IA and NH_4^+
Phosphate (PO_4^{3-})	group IA and NH_4^+
Sulfide (S^{2-})	groups IA and IIA and NH_4^+
Hydroxide (OH^-)	group IA, Ba^{2+}, Sr^{2+}, Ca^{2+}

EXAMPLE 13.1

Using Solubility Guidelines to Predict Compound Solubilities

Predict the solubility of each of the following solutes in the solvent indicated.

(a) acetone (a polar liquid) in water
(b) ammonia (a polar gas) in benzene (a nonpolar liquid)
(c) NaBr (an ionic solid) in water
(d) $CaCO_3$ (an ionic solid) in water
(e) AgCl (an ionic solid) in water

SOLUTION

(a) Soluble. Acetone is polar, as is water. Like dissolves like.
(b) Insoluble. Since the two substances are of unlike polarity, they should be relatively insoluble in each other.
(c) Soluble. Table 13.4 indicates that all compounds containing Na^+ are soluble.
(d) Insoluble. Table 13.4 indicates that all carbonates are insoluble except those of group IA ions and NH_4^+ ions.
(e) Insoluble. Table 13.4 indicates that silver is an exception to the rule that all chlorides are soluble.

Practice Exercise 13.1

Predict the solubility of each of the following solutes in the solvent indicated.

(a) SO_2 (a polar gas) in water
(b) paraffin wax (a nonpolar solid) in CCl_4 (a nonpolar liquid)
(c) $AgNO_3$ (an ionic solid) in water
(d) $BaSO_4$ (an ionic solid) in water
(e) NH_4Cl (an ionic solid) in water

Answers to practice exercises are located at the end of the chapter.

13.5 Solution Concentrations

In Section 13.2 we learned that, in general, there is a limit to the amount of solute that can be dissolved in a specified amount of solvent and also that a solution is said to be saturated when this maximum amount of solute has been dissolved. The amount of dissolved solute in a saturated solution is given by the solute's solubility.

Most solutions chemists deal with are *unsaturated* rather than saturated solutions. The amount of solute present in an unsaturated solution is specified by stating the *concentration* of the solution. A **concentration** *is the amount of solute present in a specified amount of solvent or a specified amount of solution.* Thus, concentration is a ratio of two quantities, being either the ratio

$$\frac{\text{Amount of solute}}{\text{Amount of solvent}} \quad \text{or} \quad \frac{\text{Amount of solute}}{\text{Amount of solution}}$$

In specifying a concentration, what are the units used to indicate the amounts of solute and solvent or solution present? In practice, a number of different unit combinations are used, with the choice of units depending on the use to be made of the concentration units. In each of the next four sections we shall discuss a commonly encountered set of units used to express solution concentration. The concentration expressions to be discussed are (1) percentage of solute (Sec. 13.6), (2) parts per million and parts per billion (Sec. 13.7), (3) molarity (Sec. 13.8), and (4) molality (Sec. 13.9).

13.6 Concentration: Percentage of Solute

The concentration of a solution is often specified in terms of the percentage of solute in the total amount of solution. Since the amounts of solute and solution present can be stated in terms of either mass or volume, different types of percent units exist. The three most common are

1. Percent by mass (or mass–mass percent)
2. Percent by volume (or volume–volume percent)
3. Mass–volume percent

The percent unit most frequently used by chemists is *percent by mass* (or mass–mass percent). **Percent by mass** *is the mass of solute divided by the total mass of solution multiplied by 100 (to put the value in terms of percentage).* (Percentage is always part of the whole divided by the whole times 100; see Sec. 3.10.)

$$\text{percent by mass} = \frac{\text{mass of solute}}{\text{mass of solution}} \times 100$$

The solute and solution masses must be in the same units but any units are allowed. The mass of solution is equal to the mass of *solute* plus the mass of *solvent.*

$$\text{percent by mass} = \frac{\text{mass of solute}}{\text{mass of solute} + \text{mass of solvent}} \times 100$$

A solution of 5.0% by mass concentration would contain 5.0 g of solute in 100.0 g of solution (5.0 g of solute and 95.0 g of solvent). Thus, percent by mass gives directly the number of grams of solute in 100 g of solution. The abbreviation for percent by mass is % (m/m).

The concentration of butterfat in milk is expressed in terms of mass percent. When you buy 2% milk, you are buying milk that contains 2 grams of butterfat per 100 grams of milk.

EXAMPLE 13.2

Calculating Mass Percent from Mass of Solute and Mass of Solvent

What is the percent by mass, % (m/m), concentration of sucrose (table sugar) in a solution made by dissolving 5.4 g of sucrose in 75.0 g of water?

SOLUTION

To calculate percent by mass, we need both mass of solute and mass of solution.

$$\text{percent by mass} = \frac{\text{mass of solute}}{\text{mass of solution}} \times 100$$

The mass of solute is given (5.4 g) and the mass of solution is calculated by adding together mass of solute and mass of solvent.

mass of solution = 5.4 g + 75.0 g = 80.4 g   (calculator and **correct answer**)

Substituting known values into the defining equation for percent by mass gives

$$\text{percent by mass} = \frac{5.4 \text{ g}}{80.4 \text{ g}} \times 100 = 6.7164179\% \quad \text{(calculator answer)}$$

$$= 6.7\% \qquad \text{(\textbf{correct answer})}$$

CHEMICAL EXTENSION

Sucrose, whose chemical formula is $C_{12}H_{22}O_{11}$, is the carbohydrate most people refer to as simply "sugar"; it is our common table sugar. It occurs in many fruits, in the nectar of flowers, and in the juices of many plants. Sugarcane (20% sucrose by mass) and sugar beets (17% sucrose by mass) are the commercial sources for this substance.

Sucrose is the most-used food additive in the United States. Annual per capita consumption of sucrose is approximately 100 pounds, two-thirds of which is "food additive sugar" and one-third of which is "sugar naturally present" in foods. Percent by mass sucrose content is 10% for a cola-type drink, 12% for fruit juice, 11% for Koolaid, and 44% for milk chocolate candy.

Practice Exercise 13.2

A solution of hydrogen sulfide, H_2S, in water is prepared by bubbling H_2S gas into water. Calculate the solution concentration, as percent by mass, given that 0.290 g of H_2S dissolves in 75.00 g of water.

EXAMPLE 13.3

Calculating the Mass of Solute Needed to Produce a Solution of a Given Mass Percent Concentration

How many grams of iodine must be added to 25.0 g of ethyl alcohol to prepare a 5.00% (m/m) ethyl alcohol solution of iodine?

SOLUTION

Often, when a solution concentration is given as part of a problem statement, the concentration information is used in the form of a conversion factor in solving the problem. That will be the case in this problem.

The given quantity is 25.0 g of ethyl alcohol (grams of solvent), and the desired quantity is grams of iodine (grams of solute).

$$25.0 \text{ g ethyl alcohol} = ? \text{ g iodine}$$

The conversion factor relating these two quantities (solvent and solute) is obtained from the given concentration. In this 5.00% (m/m) iodine solution, there are 5.00 g of iodine for every 95.00 g of ethyl alcohol.

$$100.00 \text{ g solution} - 5.00 \text{ g iodine} = 95.00 \text{ g ethyl alcohol}$$

This relationship between grams of solute and grams of solvent (5.00 to 95.00) gives us the needed conversion factor

$$\frac{5.00 \text{ g iodine}}{95.00 \text{ g ethyl alcohol}}$$

Dimensional analysis gives the problem setup, which is solved in the following manner.

$$25.0 \text{ g ethyl alcohol} \times \frac{5.00 \text{ g iodine}}{95.00 \text{ g ethyl alcohol}} = 1.3157894 \text{ g iodine} \quad \text{(calculator answer)}$$

$$= 1.32 \text{ g iodine} \quad \textbf{(correct answer)}$$

CHEMICAL EXTENSION

Elemental iodine is a violet-black solid that readily sublimes at room temperature to produce violet I_2 vapor. Iodine occurs in seawater and seaweed, from which it can be commercially extracted. An iodine-in-ethyl-alcohol solution, tincture of iodine, has use as an antiseptic for treating minor injuries to the skin.

Iodine is necessary for the proper functioning of the human body; for this reason table salt is "iodized." The hormone thyroxine, produced by the thyroid gland, has molecules in which four iodine atoms are present $(C_{15}H_{11}I_4NO_4)$. The more thyroxine produced, the higher the basal metabolic rate of a person. The clinical manifestation of iodine deficiency in the body is *goiter*, characterized by an enlarged thyroid gland.

Practice Exercise 13.3

How many grams of lithium nitrate, $LiNO_3$, must be added to 15.0 g of water to prepare a 7.50% (m/m) solution of lithium nitrate?

The proof system used for alcoholic beverages is twice the volume–volume percent. Forty proof is 20% (v/v) alcohol; 100 proof is 50% (v/v) alcohol.

Percent by volume (or volume–volume percent) finds use as a concentration unit when both the solute and solvent are liquids or gases. In such cases it is often more convenient to measure volumes than masses. **Percent by volume** *is the volume of solute divided by the total volume of solution multiplied by 100.*

$$\text{percent by volume} = \frac{\text{volume of solute}}{\text{volume of solution}} \times 100$$

Solute and solution volumes must always be expressed in the same units when this expression is used. The abbreviation for percent by volume is % (v/v).

The numerical value of a concentration expressed as a percent by volume gives directly the number of milliliters of solute in 100 mL of solution. Thus, a 100-mL sample of a 5.0% alcohol-in-water solution contains 5.0 mL of alcohol dissolved in enough water to give 100 mL of solution. Note that such a 5.0%-by-volume solution could not be made by adding 5 mL of alcohol to 95 mL of water, since volumes of liquids are not usually additive. Differences in the way molecules are packed as well as in the distances between molecules almost always result in the volume of a solution being less than the sum of the volumes of solute and solvent. For example, the final volume resulting from the addition of 50.0 mL of ethyl alcohol to 50.0 mL of water is 96.5 mL of solution.

When volumes of two different liquids are combined, the volumes are not additive. This process is somewhat analogous to pouring marbles and golf balls together. The marbles can fill in the spaces between the golf balls. This results in the "mixed" volume being less than the sum of the "premixed" volumes.

EXAMPLE 13.4

Calculating the Percent by Volume Concentration of a Solution

A windshield washer solution is made by mixing 37.8 mL of methyl alcohol (CH_4O) with 56.2 mL of water to produce 80.0 mL of solution. What is the concentration of methyl alcohol in the solution expressed as percent by volume methyl alcohol?

SOLUTION

To calculate this percent by volume, the volumes of methyl alcohol and solution are needed. Both are given in this problem.

$$\text{methyl alcohol volume} = 37.8 \text{ mL}$$

$$\text{solution volume} = 80.0 \text{ mL}$$

Note that the solution volume is not the sum of the solute and solvent volumes. As previously mentioned, liquid volumes of different substances are generally not additive.

Substituting the given values into the equation

$$\text{percent by volume} = \frac{\text{volume of methyl alcohol}}{\text{volume of solution}} \times 100$$

gives

$$\text{percent by volume} = \frac{37.8 \text{ mL}}{80.0 \text{ mL}} \times 100 = 47.25\% \quad \text{(calculator answer)}$$

$$= 47.2\% \quad \textbf{(correct answer)}$$

CHEMICAL EXTENSION

Methyl alcohol, with only one carbon atom, is the simplest of all alcohols. This colorless liquid is a good fuel for internal combustion engines. Since 1965 all racing cars at the Indianapolis Speedway have been fueled with methyl alcohol. (Methyl alcohol fires are easier to put out than gasoline fires, because water mixes with and dilutes methyl alcohol.) Methyl alcohol also has excellent solvent properties, and it is the solvent of choice for paints, shellacs, and varnishes.

Appearancewise and odorwise, methyl alcohol and ethyl alcohol (drinking alcohol) cannot be distinguished from each other. Mistaking methyl alcohol for drinking alcohol has tragic consequences. Methyl alcohol is extremely toxic. It is toxic to the optic nerve and can cause blindness; ingesting as little as 1 oz (30 mL) of methyl alcohol can cause death.

Practice Exercise 13.4

The final volume of a solution resulting from the addition of 50.0 mL of ethyl alcohol (C_2H_6O) to 50.0 mL of water is 96.5 mL. What is the volume percent of ethyl alcohol in the solution?

The third type of percentage unit in common use is *mass–volume* percent. This unit, which is often encountered in hospital and industrial settings, is particularly convenient to use when working with a solid solute (which is easily weighed) and a liquid solvent. Concentrations are specified using this unit when dealing with physiological fluids such as blood and urine. **Mass–volume percent** *is the mass of solute (in grams) divided by the total volume of solution (in milliliters) multiplied by 100.*

"Saline solution" used in hypodermic solutions is 0.92% NaCl (m/v).

$$\text{mass-volume percent} = \frac{\text{mass of solute (g)}}{\text{volume of solution (mL)}} \times 100$$

Note that specific mass and volume units are given in the definition of mass–volume percent. This is necessary because the units do not cancel as was the case with mass percent and volume percent. The abbreviation for mass–volume percent is % (m/v).

EXAMPLE 13.5

Calculating the Mass of Solute Present in a Solution of a Given Mass–Volume Percent Concentration

Vinegar is a 5.0% (m/v) aqueous solution of acetic acid $(HC_2H_3O_2)$. How much acetic acid, in grams, is present in one teaspoon (5.0 mL) of vinegar?

SOLUTION

The given quantity is 5.0 mL of vinegar, and the desired quantity is grams of acetic acid.

$$5.0 \text{ mL vinegar} = ? \text{ g acetic acid}$$

The given concentration of 5.0% (m/v), which means 5.0 g acetic acid per 100 mL vinegar, can be used as a conversion factor to go from milliliters of vinegar to grams of acetic acid. The setup for the conversion is

$$5.0 \text{ mL vinegar} \times \frac{5.0 \text{ g acetic acid}}{100 \text{ mL vinegar}}$$

Doing the arithmetic, after cancellation of units, gives

$$\frac{5.0 \times 5.0}{100} \text{g acetic acid} = 0.25 \text{ g acetic acid} \quad \text{(calculator and \textbf{correct answer})}$$

CHEMICAL EXTENSION

It is acetic acid, a weak acid, that gives vinegar its tartness (sour taste); vinegar is a 4–8% (m/v) acetic acid solution with additional flavoring agents also present.

Vinegar is produced by fermentation of grapes, apples, and other fruits in the presence of ample oxygen. Ethyl alcohol is the first fermentation product; further fermentation changes the ethyl alcohol to acetic acid.

Acetic acid solutions with concentrations exceeding 50% (m/v) are corrosive and can damage human tissue. Concentrated acetic acid spilled on the skin causes no immediate pain; however, painful blisters begin to form in approximately 30 minutes.

Pure acetic acid is known as *glacial* acetic acid because it freezes on a moderately cold day (f.p. = 17°C), producing icy-looking crystals.

Practice Exercise 13.5

Saline solution, a 0.92% (m/v) sodium chloride (NaCl) solution, is often administered intravenously to hospital patients. How many grams of sodium chloride are required to prepare 345 mL of saline solution?

When a percent concentration is given without specifying which of the three types of percent concentration it is (not a desirable situation), it is assumed to mean percent by mass. Thus, a 5% NaCl solution is assumed to be a 5% (m/m) NaCl solution.

For dilute aqueous solutions, % (m/m) and % (m/v) concentrations are almost the same, because mass in grams of the solution equals the volume in milliliters when the density is close to 1.00 g/mL, as it is for pure water and for dilute aqueous solutions.

13.7 Concentration: Parts per Million and Parts per Billion

The concentration units parts per million (ppm) and parts per billion (ppb) find use when dealing with extremely dilute solutions. Environmental chemists frequently use such units in specifying the concentrations of the minute amounts of trace pollutants or toxic chemicals in air and water samples.

Parts per million and *parts per billion* units are closely related to percentage concentration units. Not only are the defining equations very similar, but also various forms of the units exist. Because amounts of solute and solution present may be stated in terms of either mass or volume, there are three different forms for each unit: mass–mass (m/m), volume–volume (v/v), and mass–volume (m/v).

A **part per million** *(ppm) is one part of solute per million parts of solution.* In terms of defining equations, we can write

$$\text{ppm (m/m)} = \frac{\text{mass of solute}}{\text{mass of solution}} \times 10^6$$

$$\text{ppm (v/v)} = \frac{\text{volume of solute}}{\text{volume of solution}} \times 10^6$$

$$\text{ppm (m/v)} = \frac{\text{mass of solute (g)}}{\text{volume of solution (mL)}} \times 10^6$$

> A ppm is the equivalent of one second in 11 days and 12 hours.

Note that the units of grams and milliliters are specified in the last of the three defining equations, but that no units are given in the first two equations. For the first two equations, the only unit restriction is that the units be the same for both numerator and denominator.

A **part per billion** *(ppb) is one part of solute per billion parts of solution.* The mathematical defining equations for the three types of part-per-billion units are identical to those just shown for parts per million except that a multiplicative factor of 10^9 instead of 10^6 is used.

> A ppb is the equivalent of one second in 31 years and 8 months.

The use of parts per million and parts per billion in specifying concentration often avoids the very small numbers that result when other concentration units are used. For example, a pollutant in water might be present at a level of 0.0013 g per 100 mL of solution. In terms of mass–volume percent, this concentration is 0.0013%. In parts per million, however, the concentration is 13.

$$\text{ppm (m/v)} = \frac{0.0013 \text{ g}}{100 \text{ mL}} \times 10^6 = 13$$

The only difference in the ways in which percent concentrations and parts per million or billion are calculated is in the multiplicative factor used. For percentages it is 10^2, for parts per million 10^6, and for parts per billion 10^9. An alternative name for percentage concentration units would be *parts per hundred*.

For dilute aqueous solutions, because water has a density of 1.00 g/mL, the following relationships are valid.

$$1 \text{ ppm (m/v)} = 1 \text{ milligram/liter (mg/L)}$$

$$1 \text{ ppb (m/v)} = 1 \text{ microgram/liter(µg/L)}$$

EXAMPLE 13.6

Expressing Concentrations in Parts per Million and Parts per Billion

The concentration of sodium fluoride, NaF, in a town's fluoridated tap water is found to be 32.3 mg of NaF per 20.0 kg of tap water. Express this NaF concentration in (a) ppm (m/m) and (b) ppb (m/m).

SOLUTION

(a) The defining equation for ppm (m/m) is

$$\text{ppm (m/m)} = \frac{\text{mass of solute}}{\text{mass of solution}} \times 10^6$$

The two masses in this equation must be in the same units. Let us use grams as our mass unit. Expressing the given quantities in terms of grams, we have

$$32.3 \text{ mg NaF} = 3.23 \times 10^{-2} \text{ g NaF}$$

$$20.0 \text{ kg tap water} = 2.00 \times 10^4 \text{ g tap water}$$

Substituting these gram quantities into the defining equation for ppm (m/m) gives

$$\text{ppm (m/m)} = \frac{3.23 \times 10^{-2} \text{ g}}{2.00 \times 10^4 \text{ g}} \times 10^6$$

$$= 1.615 \quad \text{(calculator answer)}$$

$$= 1.62 \quad \textbf{(correct answer)}$$

(b) For parts per billion we have

$$\text{ppb (m/m)} = \frac{3.23 \times 10^{-2} \text{ g}}{2.00 \times 10^4 \text{ g}} \times 10^9$$

$$= 1615 \quad \text{(calculator answer)}$$

$$= 1620 \quad \textbf{(correct answer)}$$

Any time a concentration is expressed in both parts per million and parts per billion, the parts per billion value will be 1000 times larger than the parts per million value.

CHEMICAL EXTENSION

Fluoridation is the process in which the F^- ion is purposely added to drinking water. Many communities fluoridate their drinking water by adding approximately 1 ppm (m/v) fluoride ion. The purpose of fluoridation is to reduce dental caries.

Tooth enamel is a complex calcium phosphate called hydroxyapatite, $Ca_{10}(PO_4)_6(OH)_2$. Fluoride ions readily replace the hydroxide ions, OH^-, present in hydroxyapatite, producing fluorapatite, $Ca_{10}(PO_4)_6F_2$. Fluorapatite, which is a harder mineral, is less soluble in acid and less reactive than its predecessor. This decreased solubility and decreased reactivity afford greater protection against cavities. In 1971, only 29% of 9-year-olds in the United States were cavity free; in 1986, cavity-free 9-year-olds had increased to 65%, evidence that fluoridated water is helping with the tooth-decay problem.

There is some opposition to fluoridation of water because many fluorine-containing compounds (at concentrations much greater than in fluoridated drinking water) exert toxic effects.

Practice Exercise 13.6

A 500.0-mg aspirin tablet is found to contain 19 μg of a nontoxic contaminant. What is the concentration of the contaminant in (a) ppm (m/m) and (b) ppb (m/m)?

EXAMPLE 13.7

Calculating the Volume of Solute Present Given a Parts per Million Concentration

The carbon monoxide, CO, content of the tobacco smoke that reaches a smoker's lungs can be as high as 375 ppm (v/v). At this concentration, how much CO, in milliliters, would be present in a sample of air the size of a standard-sized basketball (7.5 L)?

SOLUTION

The defining equation for ppm (v/v) is

$$\text{ppm (v/v)} = \frac{\text{volume solute}}{\text{volume solution}} \times 10^6$$

The solute is CO. Rearranging the equation to isolate volume of solute on the left gives

$$\text{volume solute (mL)} = \frac{\text{ppm (v/v)} \times \text{volume solution (mL)}}{10^6}$$

Substituting the known values into the equation, remembering to change 7.5 L to 7500 mL so that the answer will have the unit of milliliters, gives

$$\text{volume CO} = \frac{375 \times 7500 \text{ mL}}{10^6}$$
$$= 2.8125 \text{ mL} \quad \text{(calculator answer)}$$
$$= 2.8 \text{ mL} \quad \text{(correct answer)}$$

CHEMICAL EXTENSION

The extent of the adverse effects of CO on the oxygen-transport system of the human body is measured in terms of the amount of the blood's hemoglobin that has been converted to carboxyhemoglobin. Carboxyhemoglobin is hemoglobin that CO has "deactivated."

Symptoms of CO poisoning such as headache, fatigue, and dizziness are present when 10% of a person's hemoglobin is tied up as carboxyhemoglobin. When the percent increases to 20%, death can result unless the victim is removed from the CO-polluted atmosphere. Studies show that the concentration of carboxyhemoglobin in a smoker's blood is two to five times that of a nonsmoker. In smokers, carboxyhemoglobin levels are around 5%.

The background concentration of CO in relatively clean, unpolluted air rarely exceeds 1.0 ppm (v/v). By comparison, in city traffic, concentrations of 50 ppm (v/v) are often reached, and CO concentrations may go as high as 140 ppm (v/v) in a traffic jam. The inhaled smoke from cigarettes contains about 400 ppm (v/v) CO.

Practice Exercise 13.7

The ozone, O_3, content of smog is approximately 0.5 ppm (v/v). At this concentration, how much O_3, in milliliters, would be present in a sample of smog the size of a 2-L soft drink container?

13.8 Concentration: Molarity

Molarity *is a solution concentration unit that gives the number of moles of solute per liter of solution.*

$$\text{molarity (M)} = \frac{\text{moles of solute}}{\text{liters of solution}}$$

A solution containing 1 mole of KBr in 1 L of solution has a molarity of 1 and is said to be a 1-M (1 *molar*) solution. Note that the abbreviation for molarity is a capital M.

When a solution is to be used for a chemical reaction, concentration is almost always expressed in units of molarity. A major reason for this is the fact that the amount of solute is expressed in moles, a most convenient unit for dealing with stoichiometry in chemical reactions. Because chemical reactions occur between molecules and atoms, a unit that counts particles, as the mole does, is desirable.

To find the molarity of a solution we need to know the number of moles of solute present and the solution volume in liters and then take the ratio of the two quantities. An alternative to knowing the number of moles of solute is knowledge about the grams of solute present and the solute's molar mass.

> In preparing 100 mL of a solution of a specific molarity, enough solvent is added to a weighed amount of solute to give a *final* volume of 100 mL. The weighed solute is not added to a *starting* volume of 100 mL; this would produce a final volume greater than 100 mL, because the solute increases the total volume.

EXAMPLE 13.8

Calculating the Molarity of a Solution from Mass/Amount and Volume Data

Determine the molarities of the following solutions.

(a) 1.45 moles of KCl dissolved in enough water to give 875 mL of solution

(b) 57.2 g of NH_4Br dissolved in enough water to give 2.15 L of solution

SOLUTION

(a) The number of moles of solute is given in the problem statement.

$$\text{moles of solute} = 1.45 \text{ moles KCl}$$

The volume of the solution is also given in the problem statement, but not in the right units. Molarity requires liters for volume units. Making the unit change gives

$$875 \text{ mL} \times \frac{10^{-3} \text{ L}}{1 \text{ mL}} = 0.875 \text{ L} \quad \text{(calculator and \textbf{correct answer})}$$

The molarity of the solution is obtained by substituting the known quantities into the equation

$$\text{molarity (M)} = \frac{\text{moles of solute}}{\text{L of solution}}$$

which gives

$$M = \frac{1.45 \text{ moles of KCl}}{0.875 \text{ L solution}} = 1.6571428 \frac{\text{moles KCl}}{\text{L solution}} \quad \text{(calculator answer)}$$

$$= 1.66 \frac{\text{moles KCl}}{\text{L solution}} \quad \text{(\textbf{correct answer})}$$

Note that the units of molarity are always moles per liter.

(b) This time the volume of solution is given in the right units, liters.

$$\text{volume of solution} = 2.15 \text{ L}$$

The moles of solute must be calculated from the grams of solute (given) and the solute's formula mass, which is 97.95 amu (calculated from a table of atomic masses).

$$57.2 \text{ g } \cancel{\text{NH}_4\text{Br}} \times \frac{1 \text{ mole NH}_4\text{Br}}{97.95 \text{ g } \cancel{\text{NH}_4\text{Br}}} = 0.58397141 \text{ mole NH}_4\text{Br} \quad \text{(calculator answer)}$$

$$= 0.584 \text{ mole NH}_4\text{Br} \qquad \textbf{(correct answer)}$$

Substituting the known quantities into the defining equation for molarity gives

$$M = \frac{0.584 \text{ mole NH}_4\text{Br}}{2.15 \text{ L solution}} = 0.2716279 \frac{\text{mole NH}_4\text{Br}}{\text{L solution}} \quad \text{(calculator answer)}$$

$$= 0.272 \frac{\text{mole NH}_4\text{Br}}{\text{L solution}} \qquad \textbf{(correct answer)}$$

Practice Exercise 13.8

What is the molarity of a solution prepared by dissolving 25.0 g of NaOH in enough water to give 2.50 L of solution?

When you perform molarity concentration calculations, you must know the *identity* of the solute. You cannot calculate moles of solute without knowing the chemical formula of the solute. When you perform percent concentration calculations, the *identity* of the solute is not used in the calculation; all you need is the *mass* or *volume* of the solute.

As the previous example indicates, when you perform a molarity calculation the chemical formula of the solute is always needed. You cannot calculate moles of solute without knowing the chemical formula of the solute. In contrast, when you perform percent concentration calculations (or parts per million or billion)—Sections 13.6 and 13.7—the chemical formula of the solute is not used in the calculation.

The number of moles of solute present in a known volume of solution is an easily calculated quantity if the molarity of the solution is known. In doing such a calculation, molarity serves as a conversion factor relating liters of solution to moles of solute.

EXAMPLE 13.9

Calculating the Amount of Solute Present in a Given Amount of Solution Using Molarity as a Conversion Factor

Citric acid, $H_3C_6H_5O_7$, is the substance that gives lemon juice and other citrus fruit juices a sour taste. How many grams of citric acid are present in 125 mL of a 0.400 M citric acid solution?

SOLUTION

The given quantity is 125 mL of solution, and the desired quantity is grams of $H_3C_6H_5O_7$.

$$125 \text{ mL solution} = ? \text{ g } H_3C_6H_5O_7$$

The pathway to be used in solving this problem is

$$\text{mL solution} \longrightarrow \text{L solution} \longrightarrow \text{moles } H_3C_6H_5O_7 \longrightarrow \text{g } H_3C_6H_5O_7$$

The given molarity (0.400 M) will serve as the conversion factor for the second unit change; the molecular mass of citric acid (which must be calculated because it is not given) is used in accomplishing the third unit change.

The dimensional analysis setup for this pathway is

$$125 \text{ mL solution} \times \frac{10^{-3} \text{ L solution}}{1 \text{ mL solution}} \times \frac{0.400 \text{ mole H}_3\text{C}_6\text{H}_5\text{O}_7}{1 \text{ L solution}}$$
$$\times \frac{192.14 \text{ g H}_3\text{C}_6\text{H}_5\text{O}_7}{1 \text{ mole H}_3\text{C}_6\text{H}_5\text{O}_7}$$

Canceling units and doing the arithmetic gives

$$\frac{125 \times 10^{-3} \times 0.400 \times 192.14}{1 \times 1 \times 1} \text{ g H}_3\text{C}_6\text{H}_5\text{O}_7$$
$$= 9.607 \text{ g H}_3\text{C}_6\text{H}_5\text{O}_7 \quad \text{(calculator answer)}$$
$$= 9.61 \text{ g H}_3\text{C}_6\text{H}_5\text{O}_7 \quad \textbf{(correct answer)}$$

CHEMICAL EXTENSION

In the pure state, citric acid is a white crystalline solid that readily absorbs moisture from the air. The presence of citric acid in citrus fruits is a well-known fact. Its natural occurrence is, however, not limited to citrus fruits, as is shown by the following citric acid mass percent concentration values: lemons (4.0–8.0), grapefruit (1.2–2.1), tangerines (0.9–1.2), oranges (0.6–1.0), raspberries (1.0–1.3), strawberries (0.6–0.8), tomatoes (0.3), and potatoes (0.3–0.5).

Citric acid's flavoring and acidity-controlling properties make it ideal for use in beverages. The citric acid concentration of fruit-flavored carbonated beverages falls in the range of 0.10–0.25 % (m/m). Citric acid is used in jams, jellies, and preserves for tartness and to create the right acidity for optimum gelation. In fresh salads, citric acid prevents enzymatic browning reactions, and in frozen fruits it is used to inhibit color and flavor deterioration. Addition of citric acid to seafood retards microbial growth by increasing acidity.

Practice Exercise 13.9

How many grams of ascorbic acid (vitamin C), $C_6H_8O_6$, are present in 125 mL of a 0.400-M vitamin C solution?

EXAMPLE 13.10

Calculating the Amount of Solution Needed to Supply a Given Amount of Solute

A typical dose of iron(II) sulfate ($FeSO_4$) used in the treatment of iron-deficiency anemia is 0.35 g. How many milliliters of a 0.10 M iron(II) sulfate solution would be needed to supply this dose?

SOLUTION

The given quantity is 0.35 g of $FeSO_4$, and the desired quantity is milliliters of $FeSO_4$ solution.

$$0.35 \text{ g FeSO}_4 = ? \text{ mL FeSO}_4 \text{ solution}$$

The pathway to be used to solve this problem is

$$\text{g FeSO}_4 \longrightarrow \text{moles FeSO}_4 \longrightarrow \text{L FeSO}_4 \text{ solution} \longrightarrow \text{mL FeSO}_4 \text{ solution}$$

We accomplish the first unit conversion by using the formula mass of $FeSO_4$ (which must be calculated) as a conversion factor. The second unit conversion involves the use of the given molarity as a conversion factor.

$$0.35 \text{ g FeSO}_4 \times \left(\frac{1 \text{ mole FeSO}_4}{151.90 \text{ g FeSO}_4}\right) \times \left(\frac{1 \text{ L solution}}{0.10 \text{ mole FeSO}_4}\right) \times \left(\frac{1 \text{ mL solution}}{10^{-3} \text{ L solution}}\right)$$

Canceling units and doing the arithmetic, we find that

$$\left(\frac{0.35 \times 1 \times 1 \times 1}{151.90 \times 0.10 \times 10^{-3}}\right) \text{ mL solution}$$

$$= 23.041474 \text{ mL solution} \quad \text{(calculator answer)}$$

$$= 23 \text{ mL solution} \qquad \text{(correct answer)}$$

CHEMICAL EXTENSION

Iron(II) ions (Fe^{2+}) play vital roles in a number of body processes. About 70% of the iron in the body occurs in hemoglobin and myoglobin, where it functions in the transport of oxygen. Iron-poor blood has less oxygen-carrying capacity and hence results in a general weakening of the body (anemia). In addition to oxygen transport, iron is an essential part of many of the body's enzymes.

Iron is absorbed by the body as Fe^{2+} ion; however, most iron in foods is found as Fe^{3+} ion. Vitamin C aids in changing Fe^{3+} ion to Fe^{2+} ion. Dietary iron supplements frequently contain vitamin C.

Cooking utensils are often a good source of dietary iron. The iron content of 100 g of spaghetti sauce simmered in a glass pan has been measured at 3 mg, but the same amount of the same type of sauce simmered in an iron skillet contained 87 mg of iron. The iron content of scrambled eggs is tripled when they are cooked in an iron skillet.

Practice Exercise 13.10

How many liters of 0.100 M aqueous sodium hydroxide (NaOH) solution can be prepared from 10.0 g of sodium hydroxide?

Molarity and mass percent are probably the two most commonly used concentration units. The need to convert from one to the other often arises. Such a conversion can easily be done provided the density of the solution is known. Figure 13.5 shows schematically the steps involved in converting one of these concentration units to the other.

Figure 13.5

A "road-map" diagram showing the steps involved in converting from percent by mass to molarity or vice versa. (For the reverse process, reverse the direction of the arrows in the diagram.)

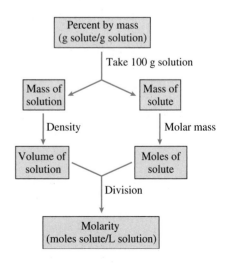

EXAMPLE 13.11

Using Mass Percent Concentration and Density to Calculate Molarity

The skin irritation that accompanies insect bites is often caused by formic acid $(HCHO_2)$. A 40.00%-by-mass aqueous solution of formic acid has a density of 1.098 g/mL. What is the molarity of this solution?

SOLUTION

Calculate the moles of solute and liters of solution present in a sample of this solution. Since solution concentration is independent of sample size, any size sample can be the basis for the calculation. To simplify the math, take a 100.0-g sample of solution.

STEP 1 *Moles of solute.* The given quantity is 100.0 g of solution, and the desired quantity is moles of $HCHO_2$.

$$100.0 \text{ g solution} = ? \text{ moles } HCHO_2$$

The known mass percent concentration will be the basis for the conversion factor that takes us from grams of solution to grams of solute. The pathway to be used in solving this problem is

$$\text{g solution} \longrightarrow \text{g solute} \longrightarrow \text{moles solute}$$

The setup is

$$100.0 \text{ g solution} \times \frac{40.00 \text{ g } HCHO_2}{100 \text{ g solution}} \times \frac{1 \text{ mole } HCHO_2}{46.03 \text{ g } HCHO_2}$$

$$= 0.86899847 \text{ mole } HCHO_2 \quad \text{(calculator answer)}$$

$$= 0.8690 \text{ mole } HCHO_2 \quad \textbf{(correct answer)}$$

STEP 2 *Liters of solution.* The density of the solution is used as a conversion factor in obtaining the volume of solution. The pathway for the calculation is

$$\text{g solution} \longrightarrow \text{mL solution} \longrightarrow \text{L solution}$$

The setup is

$$100.0 \text{ g solution} \times \frac{1 \text{ mL solution}}{1.098 \text{ g solution}} \times \frac{10^{-3} \text{ L solution}}{1 \text{ mL solution}}$$

$$= 0.091074681 \text{ L solution} \quad \text{(calculator answer)}$$

$$= 0.09107 \text{ L solution} \quad \textbf{(correct answer)}$$

STEP 3 *Molarity.* With both moles of solute and liters of solution known, the molarity is obtained by substitution into the defining equation for molarity:

$$M = \frac{\text{moles } HCHO_2}{\text{L solution}} = \frac{0.8690 \text{ mole } HCHO_2}{0.09107 \text{ L solution}}$$

$$= 9.5421104 \frac{\text{moles } HCHO_2}{\text{L solution}} \quad \text{(calculator answer)}$$

$$= 9.542 \frac{\text{moles } HCHO_2}{\text{L solution}} \quad \textbf{(correct answer)}$$

CHEMICAL EXTENSION

The "formic" in formic acid's name reflects its historical source. "Formica" is Latin for "ant," and formica acid used to be obtained by boiling ants. The bite or sting of an ant smarts because the ant injects formic acid under the skin as it bites. In the Amazon, fire ant colonies can become so large that nearby streams become polluted by their formic acid. (The stings of wasps and bees also contain formic acid as well as other irritants.)

In the pure state formic acid, also called methanoic acid, is a colorless liquid with a pungent odor. Potential symptoms of overexposure to formic acid include eye irritation, lacrimation (crying), and throat irritation.

Industrially, formic acid is used as an acidulant in the dyeing of natural and synthetic fibers and in leather tanning. It is also used in the manufacture of numerous other chemicals.

Practice Exercise 13.11

A 15.00% by mass aqueous solution of silver nitrate $(AgNO_3)$ has a density of 1.141 g/mL. What is the molarity of this solution?

Molar concentrations do not give information about the amount of *solvent* present. All that is known is that enough solvent is present to give a specific volume of *solution*. The amount of solvent present in a solution of a known molarity can be calculated if the density of the solution is known. Without the density it cannot be calculated. However, in dilute solutions the volume of solvent is approximately the same as the volume of solution.

EXAMPLE 13.12

Using Molarity and Density to Calculate the Amount of Solvent Present in a Solution

Large amounts of sulfuric acid (H_2SO_4) are used in the production of phosphate fertilizers. A 2.324-M H_2SO_4 solution has a density of 1.142 g/mL. How many grams of solvent (water) are present in 25.0 mL of this solution?

SOLUTION

To find the grams of solvent present we must first find the grams of solute (H_2SO_4) and the grams of solution. The grams of solvent present is then obtained by subtraction.

$$\text{g solvent} = \text{g solution} - \text{g solute}$$

STEP 1 *Grams of solution.* The volume of solution is given. Density, used as a conversion factor, will enable us to convert this volume to grams of solution.

$$25.0 \text{ mL solution} \times \frac{1.142 \text{ g solution}}{1 \text{ mL solution}} = 28.55 \text{ g solution} \quad \text{(calculator answer)}$$
$$= 28.6 \text{ g solution} \quad \textbf{(correct answer)}$$

STEP 2 *Grams of solute.* We will use the molarity of the solution as a conversion factor in obtaining the grams of solute. The setup for this calculation is similar to that in Example 13.9.

$$25.0 \text{ mL solution} \times \frac{10^{-3} \text{ L solution}}{1 \text{ mL solution}} \times \frac{2.324 \text{ moles } H_2SO_4}{1 \text{ L solution}} \times \frac{98.08 \text{ g } H_2SO_4}{1 \text{ mole } H_2SO_4}$$
$$= 5.698448 \text{ g } H_2SO_4 \quad \text{(calculator answer)}$$
$$= 5.70 \text{ g } H_2SO_4 \quad \textbf{(correct answer)}$$

STEP 3 *Grams of solvent.* The grams of solvent will be the difference in mass between the grams of solution and the grams of solute.

28.6 g solution − 5.70 g solute = 22.9 g solvent (calculator and **correct answer**)

CHEMICAL EXTENSION

Sulfuric acid is one of three extremely important industrial acids; the other two are nitric acid and hydrochloric acid. Advantages of sulfuric acid use over the other two acids are that it is the least expensive to make, and it is the only one of the three that can be prepared and shipped in an almost pure form.

Pure sulfuric acid is a thick and syrupy liquid. At room temperature, it is stable and nonvolatile. All the reactions of concentrated sulfuric acid solutions are very exothermic; that is, much heat is generated. Even diluting a sulfuric acid solution with water produces significant heat.

A remarkable property of sulfuric acid is its great affinity for water, that is, its dehydrating ability. For example, when this acid is mixed with sugar, it dehydrates the sugar (extracts hydrogen and oxygen atoms to form water) and leaves behind a large porous mass of black carbon. Often, sulfuric acid is used as a dehydrating agent in industrial processes.

Practice Exercise 13.12

A 0.900-M acetic acid $(HC_2H_3O_2)$ solution has a density of 1.01 g/mL. How many grams of solvent are present in 125 mL of this solution?

13.9 Concentration: Molality

Molality is a concentration unit based on a fixed amount of *solvent* and is used in areas where this is a concern. Despite this unit having a name very similar to molarity, molality differs distinctly from molarity; molarity is a unit based on a fixed amount of *solution* rather than a fixed amount of *solvent*. **Molality** *is a solution concentration unit that gives the number of moles of solute per kilogram of solvent.*

$$\text{molality } (m) = \frac{\text{moles of solute}}{\text{kilograms of solvent}}$$

Note that the abbreviation for molality is an italic lowercase *m*.

Molality also finds use, in preference to molarity, in experimental situations where changes in temperature are of concern. Molality is a temperature-independent concentration unit; molarity is not. To be temperature independent, a concentration unit cannot involve a volume measurement. Volumes of solutions change (expand or contract) with changes in temperature. A change in temperature thus means a change in concentration, even though the amount of solute remains constant, if a concentration unit has a volume dependency. Volume changes caused by temperature change are usually very, very small; consequently, temperature independence or dependence is a factor in only the most precise experimental measurements.

Careful note should be taken of the fact that the same letter of the alphabet is used as an abbreviation for both molality and molarity—a lowercase, italic *m* for molality (*m*) and a capitalized M for molarity (M).

In dilute aqueous solutions molarity and molality are practically identical in numerical value. This results from dilute aqueous solution having a density of 1.0 g/mL. Molarity and molality have significantly different values when the solvent has a density that is not equal to unity or when the solution is concentrated.

EXAMPLE 13.13

Calculating the Molality of a Solution

What would be the molality of a solution made by dissolving 3.50 g of sodium chloride (NaCl) in 225 g of H_2O?

SOLUTION

To calculate molality, the number of moles of solute and the solvent mass in kilograms must be known.

In this problem the solvent mass is given, but in grams rather than kilograms. We thus need to change the grams unit to kilograms.

$$225 \text{ g } H_2O \times \frac{1 \text{ kg } H_2O}{10^3 \text{ g } H_2O} = 0.225 \text{ kg } H_2O \quad \text{(calculator and correct answer)}$$

Information about the solute is given in terms of grams. We can calculate moles of solute from the given information by using molar mass (which is not given and must be calculated) as a conversion factor.

$$3.50 \text{ g NaCl} \times \frac{1 \text{ mole NaCl}}{58.44 \text{ g NaCl}} = 0.059890485 \text{ mole NaCl} \quad \text{(calculator answer)}$$

$$= 0.0599 \text{ mole NaCl} \quad \text{(correct answer)}$$

Substituting moles of solute and kilograms of solvent into the defining equation for molality gives

$$m = \frac{\text{moles solute}}{\text{kg solvent}} = \frac{0.0599 \text{ mole NaCl}}{0.225 \text{ kg } H_2O}$$

$$= 0.26622222 \frac{\text{mole NaCl}}{\text{kg } H_2O} \quad \text{(calculator answer)}$$

$$= 0.266 \frac{\text{mole NaCl}}{\text{kg } H_2O}$$

$$= 0.266 \, m \quad \text{(correct answer)}$$

CHEMICAL EXTENSION

Sodium chloride, common table salt, is mentioned through recorded history. It was one of the earliest commodities to be traded. Roman soldiers were partially paid in salt (sal) from whence comes the word *salary*. During the Middle Ages, the Catholic Church controlled European salt mines—a source of wealth as well as power. The French Revolution was caused in part by unhappiness with "salt taxes."

The phrases "you are the salt of the earth" (a positive comment) and "you are not worth your salt" (a negative comment) reflect the important role of "salt" in history.

In today's world, salt (NaCl) is still a very important chemical. More NaCl is used in chemical manufacturing than any other mineral; annual world consumption exceeds 150 million tons. Important chemicals obtained from salt include sodium hydroxide (NaOH), chlorine gas (Cl_2), and sodium carbonate (Na_2CO_3).

Practice Exercise 13.13

Calculate the molality of a solution made by dissolving 25.0 g of potassium carbonate (K_2CO_3) in 725 g of H_2O.

EXAMPLE 13.14

Calculating the Amount of Solute Needed to Prepare a Solution of Specified Molality

Calculate the number of grams of isopropyl alcohol, C_3H_8O, which must be added to 275 g of water to prepare a 2.00-*m* solution of isopropyl alcohol.

SOLUTION

The given quantity is 275 g of H_2O, and the desired quantity is grams of C_3H_8O.

$$275 \text{ g } H_2O = ? \text{ g } C_3H_8O$$

The pathway to be used in solving this problem is

$$\text{g solvent} \longrightarrow \text{kg solvent} \longrightarrow \text{moles solute} \longrightarrow \text{g solute}$$

The molality of the solution, which is given, will serve as a conversion factor to effect the change from kilograms of solvent to moles of solute.

The dimensional analysis setup for the problem is

$$275 \text{ g } H_2O \times \frac{1 \text{ kg } H_2O}{10^3 \text{ g } H_2O} \times \frac{2.00 \text{ moles } C_3H_8O}{1 \text{ kg } H_2O} \times \frac{60.11 \text{ g } C_3H_8O}{1 \text{ mole } C_3H_8O}$$

The second conversion factor involves the numerical value of the molality, and the third conversion factor is based on the molar mass of C_3H_8O.

Canceling units and then doing the arithmetic gives

$$\frac{275 \times 1 \times 2.00 \times 60.11}{10^3 \times 1 \times 1} \text{g } C_3H_8O = 33.0605 \text{ g } C_3H_8O \quad \text{(calculator answer)}$$

$$= 33.1 \text{ g } C_3H_8O \quad \textbf{(correct answer)}$$

CHEMICAL EXTENSION

Isopropyl alcohol is a colorless liquid, with a "medicinal" odor that we commonly associate with doctors' offices. A 70% isopropyl alcohol–30% water solution (by volume) is marketed as *rubbing alcohol.* Isopropyl alcohol's rapid evaporation rate creates a dramatic cooling effect when it is applied to the skin, hence its use for alcohol rubs to combat high body temperature.

It is used as a skin cleaner and disinfectant before drawing blood or giving injections. Interestingly, *pure* isopropyl alcohol is less effective than the commonly used 70% (v/v) solution as a disinfectant. The pure alcohol coagulates bacterial protein surfaces so fast that a barrier to further penetration by the alcohol is formed. The 70% (v/v) solution acts more slowly and complete penetration is achieved before coagulation; complete penetration is what kills the bacteria.

Practice Exercise 13.14

Calculate the number of grams of potassium hydroxide, KOH, that must be added to 25.0 g of water to prepare a 0.0100-*m* solution.

Interconversion between molarity and molality concentration units requires knowledge of solution density. Example 13.15 is a sample molality-to-molarity interconversion, and Example 13.16 is the opposite process, a sample molarity-to-molality interconversion.

EXAMPLE 13.15

Calculating Molarity from Molality and Density

Calculate the molarity of an 8.92-*m* (molal) ethyl alcohol (C_2H_6O) solution whose density is 0.927 g/mL.

SOLUTION

The defining equation for molarity involves moles of solute (numerator) and liters of solution (denominator). The defining equation for molality involves moles of solute (numerator) and kilograms of solvent (denominator). The numerators of the two defining equations are the same and the denominators are different. The essence of this problem is, thus, the conversion of kilograms of solvent to liters of solution.

$$\text{kg solvent} \longrightarrow \text{L solution}$$

To convert kilograms of solvent to liters of solution requires three steps, with each step involving a different type of calculation.

STEP 1 kilograms solvent $\longrightarrow$ mass of solute

STEP 2 mass of solute $\longrightarrow$ mass of solution

STEP 3 mass of solution $\longrightarrow$ liters of solution

The given molality (8.92 *m*), which has 1 kilogram of solvent as its basis, is the starting point for the calculation. Information associated with this molality is

$$1 \text{ kilogram } H_2O = 1000 \text{ grams } H_2O = 8.92 \text{ moles } C_2H_6O$$

STEP 1 *Calculation of mass of solute*
The mass of solute needed is that present in 1 kilogram of solvent. It is calculated using the given molality and the molar mass of the solute.

$$1 \text{ kg } H_2O \times \frac{8.92 \text{ moles } C_2H_6O}{1 \text{ kg } H_2O} \times \frac{46.08 \text{ g } C_2H_6O}{1 \text{ mole } C_2H_6O}$$

$$= 411.0336 \text{ g } C_2H_6O \quad \text{(calculator answer)}$$

$$= 411 \text{ g } C_2H_6O \quad \textbf{(correct answer)}$$

STEP 2 *Calculation of the mass of solution*
The mass of solution is obtained from the mass of solute (calculated in step 1) and the mass of solvent (our basis amount). This is a simple addition problem.

$$\underset{\text{solvent}}{1000 \text{ g}} + \underset{\text{solute}}{411 \text{ g}} = \underset{\text{solution}}{1411 \text{ g}} \quad \text{(calculator and correct answer)}$$

STEP 3 *Calculation of liters of solution*
This calculation requires use of the mass of solution (calculated in step 2) and the density of the solution (given in the problem statement).

$$1411 \text{ g solution} \times \frac{1 \text{ mL solution}}{0.927 \text{ g solution}} \times \frac{10^{-3} \text{ L solution}}{1 \text{ mL solution}}$$

$$= 1.5221143 \text{ L solution} \quad \text{(calculator answer)}$$

$$= 1.52 \text{ L solution} \quad \textbf{(correct answer)}$$

We are now ready to calculate the molarity of the solution.

$$\text{molarity} = \frac{\text{moles of solute (given in the basis statement)}}{\text{liters of solution (calculated in step 3)}}$$

$$= \frac{8.92 \text{ moles}}{1.52 \text{ L}} = 5.868421 \frac{\text{moles solute}}{\text{L solution}} \quad \text{(calculator answer)}$$

$$= 5.87 \text{ M} \qquad \textbf{(correct answer)}$$

Molarity and molality are approximately equal when a solution is dilute and the solution density is about 1 g/mL. Neither of these conditions applies in this problem; hence, molarity (5.87) and molality (8.92) are quite different.

CHEMICAL EXTENSION

Ethyl alcohol is the compound that is simply called "alcohol" by the layperson. Besides being the active ingredient in alcoholic beverages (*drinking alcohol*), it is also used in the pharmaceutical industry as a solvent (*tinctures* are ethyl alcohol solutions), as a medicinal ingredient (cough syrup often contains as much as 20% by volume ethyl alcohol), and as an industrial solvent.

All alcohol destined for human consumption (alcoholic beverages) must be produced by carefully controlled fermentation of sugars obtained from various grains. This is the basis for calling ethyl alcohol *grain alcohol*. Ethyl alcohol produced by fermentation always contains about 5% by volume water. *Absolute alcohol* is ethyl alcohol from which all traces of water have been removed. In certain chemical laboratory settings pure ethyl alcohol is needed.

Ethyl alcohol used for industrial purposes, usually as a solvent, is rendered unfit for human consumption by the addition of small amounts of toxic substances. Ethyl alcohol treated in this manner is called *denatured alcohol*.

Practice Exercise 13.15

Calculate the molarity of a 2.73-*m* (molal) methyl alcohol (CH_4O) solution whose density is 0.976 g/mL.

EXAMPLE 13.16

Calculating Molality from Molarity and Density

Calculate the molality of an 1.08-M lactose ($C_{12}H_{22}O_{11}$) solution whose density is 1.102 g/mL.

SOLUTION

The defining equation for molarity involves moles of solute (numerator) and liters of solution (denominator). The defining equation for molality involves moles of solute (numerator) and kilograms of solvent (denominator). The numerators of the two defining equations are the same and the denominators are different. The essence of this problem is, thus, the conversion of liters of solution to kilograms of solvent.

$$\text{L solution} \longrightarrow \text{kg solvent}$$

To convert liters of solution to kilograms of solvent requires three steps, with each step involving a different type of calculation.

STEP 1 liters of solution $\longrightarrow$ mass of solute

STEP 2 mass of solute $\longrightarrow$ mass of solution

STEP 3 mass of solution $\longrightarrow$ kilograms of solvent

The given molarity (1.08 M), which has 1 liter of solution as its basis, is the starting point for the calculation. Information associated with this molarity is

$$1 \text{ L solution} = 1000 \text{ mL solution} = 1.08 \text{ moles } C_{12}H_{22}O_{11}$$

STEP 1 *Calculation of mass of solute*

The mass of solute needed is that present in 1 L of solution. It is calculated using the given molarity and the molar mass of the solute.

$$1 \text{ L solution} \times \frac{1.08 \text{ moles } C_{12}H_{22}O_{11}}{1 \text{ L solution}} \times \frac{342.34 \text{ g } C_{12}H_{22}O_{11}}{1 \text{ mole } C_{12}H_{22}O_{11}}$$

$$= 369.7272 \text{ g } C_{12}H_{22}O_{11} \quad \text{(calculator answer)}$$

$$= 37\overline{0} \text{ g } C_{12}H_{22}O_{11} \quad \textbf{(correct answer)}$$

STEP 2 *Calculation of the mass of solution*

The mass of solution is obtained using the given density for the solution.

$$1 \text{ L solution} \times \frac{1 \text{ mL solution}}{10^{-3} \text{ L solution}} \times \frac{1.102 \text{ g solution}}{1 \text{ mL solution}}$$

$$= 1102 \text{ g solution} \quad \text{(calculator and } \textbf{correct answer)}$$

STEP 3 *Calculation of kilograms of solvent*

This calculation requires the mass of solute (step 1) and the mass of solution (step 2). The former is subtracted from the latter.

$$1102 \text{ g} - 37\overline{0} \text{ g} = 732 \text{ g} \quad \text{(calculator and } \textbf{correct answer)}$$

solution   solute   solvent

$$732 \text{ g solvent} \times \frac{1 \text{ kg solvent}}{10^3 \text{ g solvent}} = 0.732 \text{ kg solvent}$$

$$\text{(calculator and } \textbf{correct answer)}$$

We are now ready to calculate the molality of the solution.

$$\text{molality} = \frac{\text{moles solute (given in the basis statement)}}{\text{kilograms solvent (calculated in step 3)}}$$

$$= \frac{1.08 \text{ moles}}{0.732 \text{ kg}} = 1.4754098 \frac{\text{moles solute}}{\text{kg}} \quad \text{(calculator answer)}$$

$$= 1.48 \text{ m} \quad \textbf{(correct answer)}$$

CHEMICAL EXTENSION

Lactose, also known as *milk sugar*, is the major carbohydrate found in milk. Its sweetness is 16% that of sucrose (table sugar). The lactose content of mother's milk obtained by nursing infants [7–8%(m/m)] is almost twice that in cow's milk [4–5%(m/m)]. Lactose supplies about 40% of the energy in human milk.

Many adults lack the enzyme *lactase*, necessary for the digestion of lactose; this causes *lactose intolerance*. When lactose molecules remain in the intestine undigested, they attract water to themselves, causing fullness, discomfort, cramping, nausea, and diarrhea. The level of *lactase* in humans varies with age. Most people have sufficient lactase during childhood years when milk is a needed calcium source. In adulthood, lactase levels decrease and lactose intolerance develops. This explains changes in milk-drinking habits of many adults. About one-third of Americans suffer from lactose intolerance.

The level of the enzyme lactase in humans varies widely among ethnic groups, indicating that the trait is genetically determined (inherited). The occurrence of lactose intolerance is lowest among Scandinavians and other northern Europeans and highest among native North Americans, Southeast Asians, Africans, and Greeks.

Practice Exercise 13.16

Calculate the molality of a 4.53-M aqueous ammonia (NH_3) solution whose density is 0.9651 g/mL.

Examples 13.15 and 13.16 illustrate "opposite" processes: "molality to molarity" versus "molarity to molality." The similarities and differences in these opposite processes are contrasted in Figure 13.6. In both cases, the second and third steps involve calculation of mass of solute and mass of solution, respectively. Note, however, that calculation of these two quantities is carried out in a different manner for the two processes.

13.10 Dilution

A common problem encountered when working with solutions in the laboratory is that of *diluting* a solution of known concentration (usually called a stock solution) to a lower concentration. **Dilution** *is the process in which more solvent is added to a specific volume of solution to lower its concentration.* Dilution *always* lowers the concentration of a solution. The same amount of solute is present, but it is now distributed in a larger amount of solvent (the original solvent plus the added solvent).

Since laboratory solutions are almost always liquids, dilution is normally a volumetric procedure. Most often, a solution of a specific molarity is prepared by adding solvent to a solution until a specific total volume is reached.

With molar concentration units, a very simple mathematical relationship exists between the volumes and molarities of the diluted and stock solutions. This relationship is derived from the fact that the same amount of solute is present in both solutions; only solvent is added in a dilution procedure.

$$\text{moles solute}_{\text{stock solution}} = \text{moles solute}_{\text{diluted solution}}$$

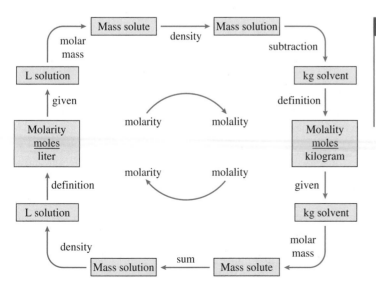

Figure 13.6

The steps involved in converting from molality to molarity concentration units and vice versa are similar but not identical. Density is needed in both cases, as are the mass of solute and mass of solvent. How the latter two quantities are calculated and used differs in the two cases.

The number of moles of solute in both solutions is given by the expression

$$\text{moles solute} = \text{molarity } (M) \times \text{liters of solution } (V)$$

(This equation is just a rearrangement of the defining equation for molarity to isolate moles of solute on the left side.) Substitution of this second expression into the first one gives the equation

$$M_s \times V_s = M_d \times V_d$$

In this equation M_s and V_s are the molarity and volume of the stock solution (the solution to be diluted) and M_d and V_d the molarity and volume of the solution resulting from the dilution. Because volume appears on both sides of the equation, any volume unit, not just liters, may be used as long as it is the same on both sides of the equation. Again, the validity of this equation is based on there being no change in the amount of solute present.

EXAMPLE 13.17

Calculating the Molarity of a Solution After It Has Been Diluted

What is the molarity of the solution prepared by diluting 65 mL of 0.95-M nitric acid (HNO_3) solution to a final volume of 135 mL through addition of solvent?

SOLUTION

Three of the four variables in the equation

$$M_s \times V_s = M_d \times V_d$$

are known.

$$M_s = 0.95 \text{ M} \quad M_d = ? \text{ M}$$
$$V_s = 65 \text{ mL} \quad V_d = 135 \text{ mL}$$

Rearranging the equation to isolate M_d on the left side and substituting the known variables into it gives

$$M_d = M_s \times \frac{V_s}{V_d}$$

$$= 0.95 \text{ M} \times \frac{65 \text{ mL}}{135 \text{ mL}} = 0.4574074 \text{ M} \quad \text{(calculator answer)}$$

$$= 0.46 \text{ M} \quad \text{(correct answer)}$$

Thus, the diluted solution's concentration is 0.46 M.

Practice Exercise 13.17

What is the molarity of the solution prepared by diluting 75 mL of 1.50-M silver nitrate ($AgNO_3$) solution to a final volume of 225 mL?

EXAMPLE 13.18

Calculating the Amount of Solvent That Must Be Added to a Solution to Dilute It to a Specified Concentration

How much solvent, in milliliters, must be added to 200.0 mL of a 1.25-M sodium chloride ($NaCl$) solution to decrease its concentration to 0.770 M?

SOLUTION

The volume of solvent added is equal to the difference between the final and initial volumes. The initial volume is known. The final volume can be calculated using the equation

$$M_s \times V_s = M_d \times V_d$$

Once the final volume is known, the difference between the two volumes can be obtained.

Substituting the known quantities into the dilution equation, rearranged to isolate V_d on the left side, gives

$$V_d = V_s \times \frac{M_s}{M_d}$$

$$= 200.0 \text{ mL} \times \frac{1.25 \text{ M}}{0.770 \text{ M}} = 324.67532 \text{ mL} \quad \text{(calculator answer)}$$

$$= 325 \text{ mL} \qquad \text{(correct answer)}$$

The solvent added is

$$V_d - V_s = (325 - 200.0) \text{ mL} = 125 \text{ mL} \quad \text{(calculator and correct answer)}$$

CHEMICAL EXTENSION

"Pass the salt" is a commonly heard phrase at the dinner table. Sodium chloride, in the form of "table salt," is the most used food additive for food after its preparation. The table salt in common use is not pure sodium chloride. Commercially produced table salt always contains its own additives.

Most table salt is *iodized* salt. Iodized salt contains 0.02% by mass potassium iodide (KI). Iodide ion (I^-) has long been known to be essential to thyroid gland function. It is essential for the formation of thyroxin, a hormone secreted by the thyroid gland that is responsible in part for the growth, development, and maintenance of body tissues. Lack of iodine in the diet results in an enlarged thyroid gland, a condition called *goiter.*

Table salt contains anticaking agents. Anticaking agents are added to salt to prevent its caking in humid weather. The presence of anticaking agents in table salt is the basis for the long-used Morton's advertising slogan "When it rains it pours." A commonly used anticaking agent in table salt is magnesium silicate, Mg_2SiO_4.

Practice Exercise 13.18

How much solvent, in milliliters, must be added to 50.0 mL of 2.20-M potassium chloride (KCl) solution to decrease its concentration to 0.0113 M?

When two "like" solutions—that is, solutions that contain the same solute and the same solvent—of differing known molarities and volumes are mixed together, the molarity of the newly formed solution can be calculated by using the same principles that apply in a simple dilution problem.

Again, the key concept involves the amount of solute present; it is constant. The sum of the amounts of solute present in the individual solutions prior to mixing is the same as the total amount of solute present in the solution after mixing. No solute is lost or gained in the mixing process. Thus, we can write

moles of solute$_{\text{first solution}}$ + moles of solute$_{\text{second solution}}$ = moles of solute$_{\text{combined solution}}$

Substituting the expression $(M \times V)$ for moles of solute in this equation gives

$$(M_1 \times V_1) + (M_2 \times V_2) = M_3 \times V_3$$

where the subscripts 1 and 2 denote the solutions to be mixed and the subscript 3 is the solution resulting from the mixing. Again, this expression is valid only when the solutions that are mixed are "like" solutions.

EXAMPLE 13.19

Calculating Molarity When Two "Like" Solutions of Differing Concentration Are Mixed

What is the molarity of the solution obtained by mixing 50.0 mL of 2.25-M hydrochloric acid (HCl) solution with 160.0 mL of 1.25-M hydrochloric acid solution?

SOLUTION

Five of the six variables in the equation

$$(M_1 \times V_1) + (M_2 \times V_2) = M_3 \times V_3$$

are known:

$$M_1 = 2.25 \text{ M} \quad V_1 = 50.0 \text{ mL}$$
$$M_2 = 1.25 \text{ M} \quad V_2 = 160.0 \text{ mL}$$
$$M_3 = ? \text{ M} \quad V_3 = 210.0 \text{ mL}$$

Note that in the mixing process we consider the volumes of the solution to be additive; that is,

$$V_3 = V_1 + V_2$$

This is a valid assumption for "like" solutions.

Solving our equation for M_3 and then substituting the known quantities into it gives

$$M_3 = \frac{(M_1 \times V_1) + (M_2 \times V_2)}{V_3} = \frac{(2.25 \text{ M} \times 50.0 \text{ mL}) + (1.25 \text{ M} \times 160.0 \text{ mL})}{(210.0 \text{ mL})}$$

$$= 1.4880952 \text{ M} \quad \text{(calculator answer)}$$

$$= 1.49 \text{ M} \quad \text{(correct answer)}$$

CHEMICAL EXTENSION

Hydrochloric acid (HCl) is one of the "big three" industrial acids, the other two being sulfuric acid (H_2SO_4) and nitric acid (HNO_3).

Since hydrochloric acid readily attacks most metals, it is stored in glass or plastic containers. The largest single use for hydrochloric acid involves metal "pickling." The pickling process removes oxide coatings from the surface of metals. For example, steel pickling removes iron oxide (rust). Metals are pickled in preparation for plating or reuse as scrap metal.

Hydrochloric acid readily dissolves limestone ($CaCO_3$). It is often injected into old oil wells to increase production by dissolving some of the subterranean limestone. Carbon dioxide gas produced as the limestone dissolves helps repressurize the wells (see Fig. 13.7). In the home, hydrochloric acid is found as the active ingredient in toilet bowl cleaners. The buildup that forms in toilet bowls is mainly calcium carbonate deposited from hard water, along with discolorations due mainly to iron compounds. Hydrochloric acid, sold under the name *muriatic acid* in home improvement stores, is used to clean bathtubs and also to clean mortar from brick and stone surfaces.

Figure 13.7

Bubble formation (CO_2 generation) occurs on this piece of limestone rock as hydrochloric acid is dropped upon it. (Carey Van Loon)

Practice Exercise 13.19

What is the molarity of the solution obtained by mixing 50.0 mL of 1.25-M ammonium chloride (NH_4Cl) solution with 175 mL of 0.125-M ammonium chloride solution?

In the solution of Example 13.19 the given liquid volumes were considered additive. In Section 13.6, when discussing volume percent, it was stressed that volumes were not additive. Why the difference? Volumes of different liquids (Sec. 13.6) are not additive; volumes of the same liquid (Example 13.19) are additive.

13.11 Molarity and Chemical Equations

Section 10.8 introduced a general problem-solving procedure for setting up problems that involve chemical equations. With this procedure, if information is given about one reactant or product in a chemical reaction (number of grams, moles, or particles), similar information can easily be obtained for any other reactant or product.

In Section 12.13 this procedure was refined to allow us to do mass-to-volume or volume-to-mass calculations for reactions when at least one reactant or product is a gas.

This section further refines our problem-solving procedure to deal efficiently with reactions that occur in aqueous solution. Of primary importance in this new area of problem solving will be *solution volume*. In most situations, solution volume is more conveniently determined than solution mass.

When solution concentrations are expressed in terms of molarity, a direct relationship exists between solution volume (in liters) and moles of solute present. The definition of molarity itself gives the relationship; molarity is the ratio of moles of solute to volume (in liters) of solution. Thus, molarity is the connection that links volume of solution to the other common problem-solving parameters, such as moles and grams. Figure 13.8 shows diagrammatically the place that volume of solution occupies, relative to other parameters, in the overall scheme of chemical-equation-based problem solving. This diagram is a simple extension of Figure 12.10; "volume of solution" boxes have been added. It is used in the same way as Figure 12.10 was.

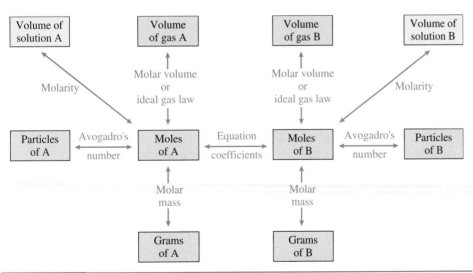

Figure 13.8

Conversion factor relationships needed to solve problems involving chemical reactions that occur in aqueous solution.

EXAMPLE 13.20

Calculating the Volume of a Reactant Given the Volume and Concentration of Another Reactant

The fizz produced when an Alka-Seltzer tablet is dissolved in water is due to the reaction between sodium bicarbonate, $NaHCO_3$, and citric acid, $C_6H_8O_7$.

$$3\ NaHCO_3(aq) + C_6H_8O_7(aq) \longrightarrow 3\ CO_2(g) + 3\ H_2O(l) + Na_3C_6H_5O_7(aq)$$

If this reaction were run in a laboratory, what volume, in liters, of 2.50-M $NaHCO_3$ solution is needed to react completely with 0.025 L of 3.50-M $C_6H_8O_7$ solution?

SOLUTION

STEP 1 The given quantity is 0.025 L of $C_6H_8O_7$ solution, and the desired quantity is liters of $NaHCO_3$ solution.

$$0.025\ L\ C_6H_8O_7 = ?\ L\ NaHCO_3$$

STEP 2 This problem is a "volume of solution A" to "volume of solution B" problem. The pathway used in solving it, in terms of Figure 13.8, is

$$\boxed{\begin{array}{c}\text{Volume of}\\\text{solution A}\end{array}} \xrightarrow{\text{Molarity}} \boxed{\begin{array}{c}\text{Moles}\\\text{of A}\end{array}} \xrightarrow[\text{coefficients}]{\text{Equation}} \boxed{\begin{array}{c}\text{Moles}\\\text{of B}\end{array}} \xrightarrow{\text{Molarity}} \boxed{\begin{array}{c}\text{Volume of}\\\text{solution B}\end{array}}$$

STEP 3 The dimensional analysis setup for the calculation is

$$0.025\ L\ \cancel{C_6H_8O_7} \times \frac{3.50\ \text{moles}\ \cancel{C_6H_8O_7}}{1\ L\ \cancel{C_6H_8O_7}} \times \frac{3\ \text{moles}\ \cancel{NaHCO_3}}{1\ \text{mole}\ \cancel{C_6H_8O_7}} \times \frac{1\ L\ NaHCO_3}{2.50\ \text{moles}\ \cancel{NaHCO_3}}$$

STEP 4 Combining all the numerical factors gives

$$\frac{0.025 \times 3.50 \times 3 \times 1}{1 \times 1 \times 2.50}\ L\ NaHCO_3 = 0.105\ L\ NaHCO_3\ \text{(calculator answer)}$$

$$= 0.10\ L\ NaHCO_3\ \text{(correct answer)}$$

Practice Exercise 13.20

What volume, in liters, of a 3.40-M potassium hydroxide (KOH) solution is needed to react completely with 0.100 L of a 6.72-M sulfuric acid (H_2SO_4) solution according to the following equation?

$$2\ KOH(aq) + H_2SO_4(aq) \longrightarrow K_2SO_4(aq) + 2\ H_2O(l)$$

EXAMPLE 13.21

Calculating the Mass of Product Produced from a Reactant of Known Volume and Concentration

How many grams of lead(II) chloride can be produced from the reaction of 1.05 L of 0.470-M potassium chloride (KCl) solution with an excess of 4.00-M lead(II) nitrate $\left[(Pb(NO_3)_2)\right]$ solution according to the following equation?

$$2\ KCl(aq) + Pb(NO_3)_2(aq) \longrightarrow PbCl_2(s) + 2\ KNO_3(aq)$$

SOLUTION

STEP 1 The given quantity is 1.05 L of KCl solution, and the desired quantity is grams of $PbCl_2$.

$$1.05 \text{ L KCl} = ? \text{ g } PbCl_2$$

STEP 2 This is a "volume of solution A" to "grams of B" problem. The pathway, in terms of Figure 13.8, is

$$\boxed{\begin{array}{c} \text{Volume of} \\ \text{solution A} \end{array}} \xrightarrow{\text{Molarity}} \boxed{\begin{array}{c} \text{Moles} \\ \text{of A} \end{array}} \xrightarrow[\text{coefficients}]{\text{Equation}} \boxed{\begin{array}{c} \text{Moles} \\ \text{of B} \end{array}} \xrightarrow[\text{mass}]{\text{Molar}} \boxed{\begin{array}{c} \text{Grams} \\ \text{of B} \end{array}}$$

STEP 3 The dimensional analysis setup for the calculation is

$$1.05 \text{ L KCl} \times \frac{0.470 \text{ mole KCl}}{1 \text{ L KCl}} \times \frac{1 \text{ mole } PbCl_2}{2 \text{ moles KCl}} \times \frac{278.1 \text{ g } PbCl_2}{1 \text{ mole } PbCl_2}$$

STEP 4 The answer, obtained from combining all of the numerical factors, is

$$\frac{1.05 \times 0.470 \times 1 \times 278.1}{1 \times 2 \times 1} \text{ g } PbCl_2 = 68.621175 \text{ g } PbCl_2 \quad \text{(calculator answer)}$$

$$= 68.6 \text{ g } PbCl_2 \qquad \textbf{(correct answer)}$$

Note that the concentration of $Pb(NO_3)_2$ solution, given as 4.00 M in the problem statement, did not enter into the calculation. This is because the $Pb(NO_3)_2$ solution is present in excess; we know that we have enough of it. If a specific volume of $Pb(NO_3)_2$ solution had been given in the problem statement, we would have had to determine the limiting reactant [$Pb(NO_3)_2$ or KCl] as the first step in working the problem. The concept of a limiting reactant was discussed in Section 10.9.

Practice Exercise 13.21

How many grams of $BaCrO_4$ can be produced from the reaction of 1.05 L of 0.470-M $BaCl_2$ solution with an excess of 2.00-M K_2CrO_4 solution according to the following equation?

$$BaCl_2(aq) + K_2CrO_4(aq) \longrightarrow BaCrO_4(s) + 2 \text{ KCl}(aq)$$

EXAMPLE 13.22

Calculating the Gaseous Volume of a Product from the Solution Volume of a Reactant

What volume, in liters of nitrogen monoxide gas, NO, measured at STP, can be produced from 1.75 L of 0.550-M nitric acid, HNO_3, and an excess of 0.650-M hydrosulfuric acid, H_2S, according to the following reaction?

$$2 \text{ } HNO_3(aq) + 3 \text{ } H_2S(aq) \longrightarrow 2 \text{ NO}(g) + 3 \text{ S}(s) + 4 \text{ } H_2O(l)$$

SOLUTION

STEP 1 The given quantity is 1.75 L of HNO_3 solution, and the desired quantity is liters of NO gas at STP.

$$1.75 \text{ L } HNO_3 = ? \text{ L NO (at STP)}$$

STEP 2 This is a "volume of solution A" to "volume of gas B" problem. From Figure 13.8 the pathway for solving the problem is

$$\boxed{\text{Volume of solution A}} \xrightarrow{\text{Molarity}} \boxed{\text{Moles of A}} \xrightarrow[\text{coefficients}]{\text{Equation}} \boxed{\text{Moles of B}} \xrightarrow[\text{volume}]{\text{Molar}} \boxed{\text{Volume of gas B}}$$

STEP 3 The dimensional analysis setup for the calculation is

$$1.75 \text{ L HNO}_3 \times \frac{0.550 \text{ mole HNO}_3}{1 \text{ L HNO}_3} \times \frac{2 \text{ moles NO}}{2 \text{ moles HNO}_3} \times \frac{22.41 \text{ L NO}}{1 \text{ mole NO}}$$

The last conversion factor is derived from the fact that one mole of any gas occupies 22.41 L at STP conditions (Sec. 12.11).

STEP 4 The result, obtained by combining all the numerical factors, is

$$\frac{1.75 \times 0.550 \times 2 \times 22.41}{1 \times 2 \times 1} \text{L NO} = 21.569625 \text{ L NO} \quad \text{(calculator answer)}$$

$$= 21.6 \text{ L NO} \qquad \textbf{(correct answer)}$$

Practice Exercise 13.22

What volume, in liters, of H_2 gas measured at STP can be produced from 0.575 L of 1.22 M HBr solution and an excess of Zn according to the following equation?

$$2 \text{ HBr(aq)} + \text{Zn(s)} \longrightarrow \text{ZnBr}_2\text{(aq)} + \text{H}_2\text{(g)}$$

13.12 Calculations Involving Volume: A Summary

Many calculations in this chapter have involved the quantity called *volume*. Such was the case also in the previous chapter, where calculations involving the gas laws were considered. Even earlier, in Chapter 3 volume was a central concept when calculations involving density were first considered. Figure 13.9 summarizes, in diagrammatic form, the various ways we have encountered volume to this point in a calculational setting.

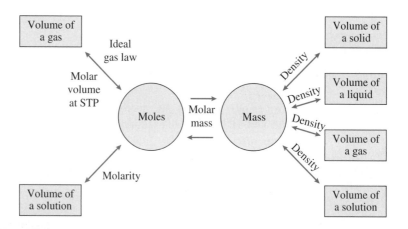

Figure 13.9

A summary of the various ways in which the quantity *volume* enters into chemical calculations.

Summary

1. **Solution Characteristics and Components** A solution is a homogeneous (uniform) mixture. The component of a solution that is present in the greatest amount is the *solvent*. A *solute* is a solution component that is present in a small amount relative to the solvent. The composition and properties of a solution are dependent on the ratio of solutes(s) to solvent. Solutes are present as individual particles (molecules, atoms, or ions).

2. **Solubility** The solubility of a solute is the maximum amount of solute that will dissolve in a given amount of solvent. The extent to which a solute dissolves in a solvent depends on the polarities of solute(s) and solvent, the temperature, and the pressure. A saturated solution contains the maximum amount of solute that can be dissolved under the conditions at which the solution exists. A supersaturated solution, an unstable situation, contains more dissolved solute than that needed for a saturated solution.

3. **Solubility Rules** Substances of like polarity tend to be more soluble in each other than substances that differ in polarity. This conclusion is often expressed as the simple phrase "like dissolves like." This generalization is not adequate for predicting the solubility of ionic compounds in water. More detailed guidelines are needed for this situation.

4. **Solution Concentration** The concentration of a solution is the amount of solute present in a specified amount of solvent or a specified amount of solution. A number of different concentration units are used, with the choice of units depending on the use to be made of the concentration units.

5. **Percent Concentration Units** The concentration of a solution in terms of percentage solute can be expressed in three different ways: (1) percent by mass (mass–mass percent), (2) percent by volume (volume–volume percent), and (3) mass–volume percent. All three types of percentages are in common use.

6. **Parts per Million and Parts per Billion** The concentration units parts per million and parts per billion find use when dealing with extremely dilute solutions. These units are closely related to percentage concentration units, differing only in the multiplicative factors used.

7. **Molarity** The molarity concentration unit, abbreviated M, is a ratio giving the number of moles of solute per liter of solution. It is the most used concentration unit in a chemical laboratory. Molarity finds use as a conversion factor in mass-to-volume calculations that involve solution volume.

8. **Molality** The molality concentration unit, abbreviated *m*, is a ratio giving the number of moles of solute per kilogram of solvent. Molality finds use, in preference to molarity, in experimental situations where changes in temperature are of concern since solution volume is dependent on temperature. In dilute solution molality and molarity are practically identical in numerical value.

9. **Dilution** Dilution is the process in which more solvent is added to a solution to lower its concentration. Most dilutions are carried out by adding a predetermined volume of solvent to a specific volume of stock solution of known concentration.

Key Terms

The new terms defined in this chapter are

aqueous solution   *Sec. 13.2*

concentrated solution   *Sec. 13.2*

concentration   *Sec. 13.5*

dilute solution   *Sec. 13.2*

dilution   *Sec. 13.10*

mass–volume percent   *Sec. 13.6*

molality   *Sec. 13.9*

molarity   *Sec. 13.8*

nonaqueous solution   *Sec. 13.2*

parts per billion   *Sec. 13.7*

parts per million   *Sec. 13.7*

percent by mass   *Sec. 13.6*

percent by volume   *Sec. 13.6*

saturated solution   *Sec. 13.2*

solubility   *Sec. 13.2*

solute   *Sec. 13.1*

solution   *Sec. 13.1*

solvent   *Sec. 13.1*

supersaturated solution   *Sec. 13.2*

unsaturated solution   *Sec. 13.2*

Practice Problems

Characteristics of Solutions (Sec. 13.1)

13.1 Indicate whether each of the following statements about the general properties of solutions is *true* or *false*.

(a) A solution may contain more than one solute.

(b) All solutions are homogeneous mixtures.

(c) Every part of a solution has exactly the same properties as every other part.

(d) The solutes present in a solution will "settle out" with time if the solution is left undisturbed.

13.2 Indicate whether each of the following statements about the general properties of solutions is *true* or *false*.

(a) All solutions have a variable composition.

(b) For a solution to form, the solute and solvent must chemically react with each other.

(c) Solutes are present as individual particles (molecules, atoms, ions) in a solution.

(d) A general characteristic of all solutions is the liquid state.

13.3 Identify the *solute* and the *solvent* in solutions with the following compositions.

(a) 5.00 g of sodium chloride and 50.0 g of water

(b) 4.00 g of sucrose and 1000 g of water

(c) 2.00 mL of water and 20.0 mL of ethyl alcohol

(d) 60.0 mL of methyl alcohol and 20.0 mL of ethyl alcohol

13.4 Identify the *solute* and the *solvent* in solutions with the following compositions.

(a) 5.00 g of sodium bromide and 200.0 g of water

(b) 50.0 g of silver nitrate and 1000 g of water

(c) 50.0 mL of water and 100.0 mL of methyl alcohol

(d) 50.0 mL of isopropyl alcohol and 20.0 mL of ethyl alcohol

Solubility (Sec. 13.2)

13.5 Use Table 13.2 to determine whether each of the following silver nitrate ($AgNO_3$) solutions is *unsaturated*, *saturated*, or *supersaturated*.

(a) 455 g of $AgNO_3$ in 100 g of H_2O at 100°C

(b) 465 g of $AgNO_3$ in 100 g of H_2O at 50°C

(c) 910 g of $AgNO_3$ in 200 g of H_2O at 50°C

(d) 55 g of $AgNO_3$ in 50 g of H_2O at 0°C

13.6 Use Table 13.2 to determine whether each of the following silver sulfate (Ag_2SO_4) solutions is *unsaturated*, *saturated*, or *supersaturated*.

(a) 1.08 g of Ag_2SO_4 in 100 g of H_2O at 100°C

(b) 0.54 g of Ag_2SO_4 in 50 g of H_2O at 50°C

(c) 1.08 g of Ag_2SO_4 in 200 g of H_2O at 50°C

(d) 0.25 g of Ag_2SO_4 in 50 g of H_2O at 0°C

13.7 Using the solubilities given in Table 13.2, characterize each of the following solids as *insoluble*, *slightly soluble*, *soluble*, or *very soluble* in water at the indicated temperature.

(a) lead(II) bromide at 50°C

(b) cesium chloride at 0°C

(c) silver nitrate at 100°C

(d) silver sulfate at 0°C

13.8 Using the solubilities given in Table 13.2, characterize each of the following solids as *insoluble*, *slightly soluble*, *soluble*, or *very soluble* in water at the indicated temperature.

(a) copper(II) sulfate at 100°C

(b) lead(II) bromide at 0°C

(c) silver nitrate at 50°C

(d) sodium chloride at 0°C

13.9 Based on the solubilities in Table 13.2, characterize each of the following sodium chloride (NaCl) solutions as *dilute* or *concentrated*.

(a) 35.0 g of NaCl in 100 g of H_2O at 100°C

(b) 35.0 g of NaCl in 100 g of H_2O at 50°C

(c) 3.50 g of NaCl in 10 g of H_2O at 50°C

(d) 3.50 g of NaCl in 100 g of H_2O at 50°C

13.10 Based on the solubilities in Table 13.2, characterize each of the following copper(II) sulfate solutions as *dilute* or *concentrated*.

(a) 14.3 g of $CuSO_4$ in 100 g of H_2O at 100°C

(b) 14.3 g of $CuSO_4$ in 100 g of H_2O at 0°C

(c) 14.3 g of $CuSO_4$ in 50 g of H_2O at 50°C

(d) 1.43 g of $CuSO_4$ in 10 g of H_2O at 0°C

13.11 For each of the following pairs of solutions, select the solution for which solute solubility is greatest.

(a) ammonia gas in water with P(pressure) = 1 atm and T(temperature) = 50°C
ammonia gas in water with P = 1 atm and T = 90°C

(b) carbon dioxide gas in water with P = 2 atm and T = 50°C
carbon dioxide gas in water with P = 1 atm and T = 50°C

(c) table salt in water with P = 1 atm and T = 60°C
table salt in water with P = 1 atm and T = 50°C

(d) table sugar in water with P = 2 atm and T = 40°C
table sugar in water with P = 1 atm and T = 70°C

13.12 For each of the following pairs of solutions, select the solution for which solute solubility is greatest.

(a) oxygen gas in water with P = 1 atm and T = 10°C
oxygen gas in water with P = 1 atm and T = 20°C

(b) nitrogen gas in water with P = 2 atm and T = 50°C
nitrogen gas in water with P = 1 atm and T = 70°C

(c) table salt in water with P = 1 atm and T = 40°C

table salt in water with P = 1 atm and T = 70°C

(d) table sugar in water with P = 3 atm and T = 30°C

table sugar in water with P = 1 atm and T = 80°C

13.13 The solubility of $Pb(NO_3)_2$ at 70°C is 110 g per 100 g of water. At 40°C the solubility drops to 78 g per 100 g of water. A 200-g quantity of $Pb(NO_3)_2$ is stirred into 200 mL of water.

(a) At 70°C how many grams, if any, of crystals of $Pb(NO_3)_2$ will settle out of solution?

(b) After cooling the solution to 40°C, how many grams, if any, of crystals of $Pb(NO_3)_2$ will settle out of solution?

13.14 The solubility of KNO_3 at 60°C is 94 g per 100 g of water. At 20°C the solubility drops to 56 g per 100 g of water. A 150-g quantity of KNO_3 is stirred into 200 mL of water.

(a) At 60°C how many grams, if any, of crystals of KNO_3 will settle out of solution?

(b) After cooling the solution to 20°C, how many grams, if any, of crystals of KNO_3 will settle out of solution?

Solution Formation (Sec. 13.3)

13.15 Match each of the following statements about the dissolving of the ionic solid NaCl in water with the term *hydrated ion, hydrogen atom*, or *oxygen atom*.

(a) A Na^+ ion surrounded with water molecules

(b) A Cl^- ion surrounded with water molecules

(c) The portion of a water molecule that is attracted to a Na^+ ion

(d) The portion of a water molecule that is attracted to a Cl^- ion

13.16 Match each of the following statements about the dissolving of the ionic solid KBr in water with the term *hydrated ion, hydrogen atom*, or *oxygen atom*.

(a) A K^+ ion surrounded with water molecules

(b) A Br^- ion surrounded with water molecules

(c) The portion of a water molecule that is attracted to a K^+ ion

(d) The portion of a water molecule that is attracted to a Br^- ion

13.17 Indicate whether each of the following actions will *increase* or *decrease* the rate of dissolving of a sugar cube in water.

(a) Cooling the sugar cube–water mixture

(b) Stirring the sugar cube–water mixture

(c) Breaking the sugar cube up into smaller chunks

(d) Crushing the sugar cube to give a granulated form of sugar

13.18 Indicate whether each of the following actions will *increase* or *decrease* the rate of dissolving of table salt in water.

(a) Heating the table salt–water mixture

(b) Agitating the table salt–water mixture

(c) Heating the table salt prior to adding it to the water

(d) Heating the water prior to adding the table salt to it

Solubility Rules (Sec. 13.4)

13.19 Predict whether the following solutes are *very soluble* or *slightly soluble* in water.

(a) NO_2 (a polar gas)

(b) CCl_4 (a nonpolar liquid)

(c) CH_4 (a nonpolar gas)

(d) $C_6H_{12}O_6$ (a polar nonionic solid)

13.20 Predict whether the following solutes are *very soluble* or *slightly soluble* in water.

(a) F_2 (a nonpolar gas)

(b) CO_2 (a nonpolar gas)

(c) $C_{12}H_{22}O_{11}$ (a polar nonionic solid)

(d) C_2H_6O (a polar liquid)

13.21 Indicate whether each of the following anions forms compounds that are *generally soluble* or *generally insoluble* in water.

(a) NO_3^- **(b)** Cl^- **(c)** S^{2-} **(d)** SO_4^{2-}

13.22 Indicate whether each of the following anions forms compounds that are *generally soluble* or *generally insoluble* in water.

(a) $C_2H_3O_2^-$ **(b)** CO_3^{2-} **(c)** PO_4^{3-} **(d)** Br^-

13.23 On the basis of the general solubility rules for ionic compounds given in Table 13.4, indicate which of the rules covers each of the following substances' solubility situation.

(a) silver carbonate

(b) magnesium sulfide

(c) ammonium cyanide

(d) calcium sulfate

13.24 On the basis of the general solubility rules for ionic compounds given in Table 13.4, indicate which of the rules covers each of the following substances' solubility situation.

(a) copper(II) carbonate

(b) potassium cyanide

(c) aluminum nitrate

(d) silver phosphate

13.25 Classify each of the following types of ionic compounds into the solubility categories *soluble, soluble with exceptions, insoluble*, or *insoluble with exceptions*.

(a) chlorides and sulfates

(b) nitrates and ammonium-ion-containing

(c) carbonates and phosphates

(d) sodium-ion-containing and potassium-ion-containing

13.26 Classify each of the following types of ionic compounds into the solubility categories *soluble, soluble with exceptions, insoluble*, or *insoluble with exceptions*.

(a) nitrates and sodium-ion-containing

(b) chlorides and bromides

(c) hydroxides and phosphates

(d) sulfates and iodides

13.27 In which of the following pairs of ionic compounds do both members of the pair have like solubility (both soluble or both insoluble) in water?

(a) $NaNO_3$ and $Pb(NO_3)_2$

(b) KCl and AgCl

(c) $BeSO_4$ and $MgSO_4$

(d) $FePO_4$ and $Ca_3(PO_4)_2$

13.28 In which of the following pairs of ionic compounds do both members of the pair have like solubility (both soluble or both insoluble) in water?

(a) $NaC_2H_3O_2$ and $Mg(C_2H_3O_2)_2$

(b) NH_4F and NH_4Cl

(c) KOH and $Ba(OH)_2$

(d) Al_2S_3 and CuS

13.29 Which of the following ions would react with both Mg^{2+} and Ca^{2+} ions to form water-insoluble compounds?

(a) sulfate   (b) phosphate

(c) sulfide   (d) nitrate

13.30 Which of the following ions would react with both Cu^{2+} and Ba^{2+} ions to form water-insoluble compounds?

(a) chloride   (b) carbonate

(c) hydroxide   (d) acetate

Mass Percent (Sec. 13.6)

13.31 Calculate the mass percent of sodium iodide (NaI) in each of the following solutions.

(a) 6.43 g of NaI dissolved in 85.0 g of H_2O

(b) 3.23 g of NaI dissolved in 175.00 g of H_2O

(c) 10.3 g of NaI dissolved in 53.0 g of solution

(d) 0.030 mole of NaI dissolved in 100.0 g of H_2O

13.32 Calculate the mass percent of potassium hydroxide (KOH) in each of the following solutions.

(a) 10.2 g of KOH dissolved in 135 g of H_2O

(b) 3.14 g of KOH dissolved in 53.14 g of H_2O

(c) 1.33 g of KOH dissolved in 23.50 g of solution

(d) 0.500 mole of KOH dissolved in 1375 g of H_2O

13.33 How many grams of solute are dissolved in the following amounts of solution?

(a) 35.0 g of 2.00% (m/m) NaCl solution

(b) 125 g of 3.50% (m/m) $AgNO_3$ solution

(c) 1355 g of 10.00% (m/m) K_2SO_4 solution

(d) 43.3 g of 8.25% (m/m) HCl solution

13.34 How many grams of solute are dissolved in the following amounts of solution?

(a) 134 g of 3.00% (m/m) KNO_3 solution

(b) 75.02 g of 9.735% (m/m) NaOH solution

(c) 1576 g of 0.800% (m/m) HI solution

(d) 1.25 g of 12.0% (m/m) NH_4Cl solution

13.35 What mass of water, in grams, is needed to prepare each of the following calcium chloride ($CaCl_2$) solutions?

(a) 5.75 g of 10.00% (m/m) $CaCl_2$ solution

(b) 57.5 g of 10.00% (m/m) $CaCl_2$ solution

(c) 57.5 g of 1.00% (m/m) $CaCl_2$ solution

(d) 2.3 g of 0.80% (m/m) $CaCl_2$ solution

13.36 What mass of water, in grams, is needed to prepare each of the following lithium nitrate ($LiNO_3$) solutions?

(a) 34.7 g of 5.00% (m/m) $LiNO_3$ solution

(b) 3.47 g of 5.00% (m/m) $LiNO_3$ solution

(c) 235 g of 12.75% (m/m) $LiNO_3$ solution

(d) 1352 g of 0.0032% (m/m) $LiNO_3$ solution

13.37 How many grams of water must be added to 50.0 g of each of the following solutes to prepare a 5.00% (m/m) solution?

(a) NaCl   (b) KCl

(c) Na_2SO_4 (d) $LiNO_3$

13.38 How many grams of water must be added to 20.0 g of each of the following solutes to prepare a 2.00% (m/m) solution?

(a) NaOH   (b) LiBr

(c) Li_2SO_4 (d) $Ca(NO_3)_2$

Volume Percent (Sec. 13.6)

13.39 What is the volume percent ethyl alcohol in a solution containing 257 mL of ethyl alcohol and enough water to give the following amounts of solution?

(a) 325 mL   (b) 675 mL

(c) 1.23 L   (d) 5.000 L

13.40 What is the volume percent water in a solution containing 35.0 mL of water and enough ethyl alcohol to give the following amounts of solution?

(a) 45.0 mL   (b) 675 mL

(c) 1.08 L   (d) 4.500 L

13.41 The final volume of a solution made by adding 360.6 mL of methyl alcohol to 667.2 mL of water is 1000.0 mL. Determine the volume percent of the following.

(a) methyl alcohol in the solution

(b) water in the solution

13.42 The final volume of a solution made by adding 678.2 mL of methyl alcohol to 358.4 mL of water is 1000.0 mL. Determine the volume percent of the following.

(a) methyl alcohol in the solution

(b) water in the solution

13.43 How much hydrogen peroxide (H_2O_2), in milliliters, is needed to prepare 375 mL of a 3.00% (v/v) solution of hydrogen peroxide in water?

13.44 How much hydrogen peroxide (H_2O_2), in milliliters, is needed to prepare 525 mL of a 2.00% (v/v) solution of hydrogen peroxide in water?

13.45 What volume of water, in gallons, is contained in 4.00 gal of a 35.0% (v/v) solution of water in acetone?

13.46 What volume of water, in quarts, is contained in 3.50 qt of a 2.00% (v/v) solution of water in acetone?

Mass–Volume Percent (Sec. 13.6)

13.47 Calculate the concentration, as mass–volume percent, for potassium iodide (KI) solutions with the following characteristics.

(a) 2.00 g solute, 75.0 mL solution

(b) 15.00 g solute, 1.25 L solution

(c) 2.00 moles solute, 10.00 L solution

(d) 0.0020 mole solute, 5.00 mL solution

13.48 Calculate the concentration, as mass–volume percent, for magnesium fluoride (MgF_2) solutions with the following characteristics.

(a) 5.00 g solute, 125 mL solution

(b) 25.0 g solute, 2.20 L solution

(c) 0.150 mole solute, 105 mL solution

(d) 0.32 mole solute, 0.100 L solution

13.49 How many milliliters of a 6.0% (m/v) sodium nitrate ($NaNO_3$) solution would fulfill each of the following requirements?

(a) contain 45.0 g of $NaNO_3$

(b) supply 2.00 g of $NaNO_3$

13.50 How many milliliters of a 3.5% (m/v) potassium acetate ($KC_2H_3O_2$) solution would fulfill each of the following requirements?

(a) contain 6.60 g of $KC_2H_3O_2$

(b) supply 25.00 g of $KC_2H_3O_2$

13.51 Determine how many grams of sodium phosphate (Na_3PO_4) would

(a) be needed to prepare 455 mL of a 2.50% (m/v) Na_3PO_4 solution.

(b) be present in 50.0 L of a 7.50% (m/v) Na_3PO_4 solution.

13.52 Determine how many grams of potassium carbonate (K_2CO_3) would

(a) be needed to prepare 4.55 mL of a 15.00% (m/v) K_2CO_3 solution.

(b) be present in 1.06 L of a 0.800% (m/v) K_2CO_3 solution.

13.53 Calculate the concentration, as mass–volume percent cesium chloride (CsCl), for a solution prepared by adding 5.0 g of CsCl to 20.0 g of H_2O to give a solution with a density of 1.18 g/mL.

13.54 Calculate the concentration, as mass–volume percent ammonium sulfate [$(NH_4)_2SO_4$], for a solution prepared by adding 3.0 g of $(NH_4)_2SO_4$ to 17.0 g of H_2O to give a solution with a density of 1.09 g/mL.

Parts per Million and Parts per Billion (Sec. 13.7)

13.55 What is the concentration of sodium chloride (NaCl), in ppm (m/m), in each of the following NaCl solutions?

(a) 37.5 mg of NaCl in 21.0 kg of water

(b) 2.12 cg of NaCl in 125 g of water

(c) 1.00 μg of NaCl in 32.0 dg of water

(d) 35.7 mg of NaCl in 15.7 g of water

13.56 What is the concentration of sodium bromide (NaBr), in ppm (m/m), in each of the following NaBr solutions?

(a) 37.5 cg of NaBr in 33.0 kg of water

(b) 2.12 mg of NaBr in 375 dg of water

(c) 3.00 μg of NaBr in 45.0 g of water

(d) 125 dg of NaBr in 255 kg of water

13.57 What is the concentration of each of the solutions in Problem 13.55 in parts per billion (m/m)?

13.58 What is the concentration of each of the solutions in Problem 13.56 in parts per billion (m/m)?

13.59 Fish generally need an oxygen concentration in water of at least 5 ppm (m/v) for survival. Will river water that contains 7 mg of O_2 per liter contain sufficient O_2 to sustain fish life?

13.60 A carbon dioxide concentration in water of 200 ppm (m/v) or higher is lethal to fish. Will river water that contains 0.62 g of dissolved CO_2 per 2.0 L be toxic to fish?

13.61 A typical concentration of the air pollutant sulfur dioxide (SO_2) in urban atmospheres is 0.087 ppm (v/v). At this concentration, how many milliliters of SO_2 are present in 5.000 L of air?

13.62 A typical concentration of the air pollutant nitrogen dioxide (NO_2) in urban atmospheres is 0.30 ppm (v/v). At this concentration, how many liters of air would be needed to extract 1.00 mL of NO_2?

13.63 Determine how much ammonia (NH_3), in grams, must be present in a 725-mL sample of air to give the following NH_3 concentrations.

(a) 3.6 ppm (m/v)   **(b)** 7.5 ppb (m/v)   **(c)** 1.2% (m/v)

13.64 Determine how much hydrogen sulfide (H_2S), in grams, must be present in a 475-mL sample of air to give the following H_2S concentrations.

(a) 2.0 ppm (m/v)   **(b)** 9.7 ppb (m/v)   **(c)** 5.2% (m/v)

Molarity (Sec. 13.8)

13.65 Calculate the molarity of each of the following aqueous sodium hydroxide (NaOH) solutions.

(a) 2.0 moles NaOH in 0.50 L of solution

(b) 13.7 g NaOH in 90.0 mL of solution

(c) 53.0 g NaOH in 1.255 L of solution

(d) 0.0020 mole NaOH in 5.00 mL of solution

13.66 Calculate the molarity of each of the following aqueous potassium chloride (KCl) solutions.

(a) 1.45 moles KCl in 2.50 L of solution

(b) 12.5 g KCl in 85.0 mL of solution

(c) 27.0 g KCl in 1.055 L of solution

(d) 0.0500 mole KCl in 12.0 mL of solution

13.67 Calculate the number of grams of solute in each of the following nitric acid (HNO_3) solutions.

(a) 35.0 mL of a 6.00-M solution

(b) 10.0 mL of a 0.600-M solution

(c) 375 L of a 1.00-M solution

(d) 375 g of a 7.91-M solution with a density of 1.25 g/mL

13.68 Calculate the number of grams of solute in each of the following sulfuric acid (H_2SO_4) solutions.

(a) 27.0 mL of a 3.00-M solution

(b) 20.0 mL of a 6.00-M solution

(c) 125 L of a 0.100-M solution

(d) 125 g of a 7.50-M solution with a density of 1.42 g/mL

13.69 Calculate the volume, in milliliters, of the following sodium thiosulfate $(Na_2S_2O_3)$ solutions needed to provide the indicated amounts of solute.

(a) 2.50 g of $Na_2S_2O_3$ from a 0.468-M solution

(b) 125 g of $Na_2S_2O_3$ from a 3.50-M solution

(c) 4.50 moles of $Na_2S_2O_3$ from a 2.50-M solution

(d) 0.0015 mole of $Na_2S_2O_3$ from a 0.990-M solution

13.70 Calculate the volume, in milliliters, of the following sodium sulfate (Na_2SO_4) solutions needed to provide the indicated amounts of solute.

(a) 2.50 g of Na_2SO_4 from a 0.468-M solution

(b) 125 g of Na_2SO_4 from a 3.50-M solution

(c) 4.50 moles of Na_2SO_4 from a 2.50-M solution

(d) 0.0015 mole of Na_2SO_4 from a 0.990-M solution

13.71 How many liters of 0.775-M solution can be prepared from 55.0 g of each of the following solutes?

(a) HBr   **(b)** NH_4Br **(c)** $MgBr_2$ **(d)** CsBr

13.72 How many liters of 1.30-M solution can be prepared from 34.3 g of each of the following solutes?

(a) HI   **(b)** NH_4I **(c)** BaI_2 **(d)** LiI

13.73 The density of an 88.00% (m/m) methyl alcohol (CH_4O) solution is 0.8274 g/mL. What is the molarity of the solution?

13.74 The density of a 60.00% (m/m) ethyl alcohol (C_2H_6O) solution is 0.8937 g/mL. What is the molarity of the solution?

13.75 The density of a 2.019-M sodium bromide (NaBr) solution is 1.157 g/mL. What is the concentration of this solution expressed as % (m/m) NaBr?

13.76 The density of a 2.687-M sodium acetate $(NaC_2H_3O_2)$ solution is 1.104 g/mL. What is the concentration of this solution expressed as % (m/m) $NaC_2H_3O_2$?

13.77 What is the molarity of a 15.0% (m/v) sodium hydroxide (NaOH) solution?

13.78 What is the molarity of a 15.0% (m/v) nitric acid (HNO_3) solution?

Molality (Sec. 13.9)

13.79 Calculate the molality of each of the following sucrose $(C_{12}H_{22}O_{11})$ solutions.

(a) 16.5 g of sucrose in 1.35 kg of water

(b) 3.15 moles of sucrose in 455 g of water

(c) 0.0356 g of sucrose in 13.0 g of water

(d) 45.0 g of sucrose in enough water to give 318 mL of solution with a density of 1.06 g/mL

13.80 Calculate the molality of each of the following glucose $(C_6H_{12}O_6)$ solutions.

(a) 23.0 g of glucose in 2.40 kg of water

(b) 2.00 moles of glucose in 975 g of water

(c) 0.230 g of glucose in 22.0 g of water

(d) 30.0 g of glucose in enough water to give 312 mL of solution with a density of 1.04 g/mL

13.81 Calculate the number of grams of each solute that must be added to 125 g of water to prepare a 0.400-*m* solution of

(a) $Al(NO_3)_3$ **(b)** $MgCl_2$

(c) Na_3PO_4 **(d)** K_2SO_4

13.82 Calculate the number of grams of each solute that must be added to 235 g of water to prepare a 0.600-*m* solution of

(a) $Be(NO_3)_2$ **(b)** $CaBr_2$

(c) Na_2CO_3 **(d)** $Al_2(SO_4)_3$

13.83 How many grams of water must be added to 80.0 g of sodium chloride (NaCl) to prepare the following molal solutions?

(a) 0.050 *m*   **(b)** 0.23 *m*

(c) 1.345 *m*   **(d)** 2.8 *m*

13.84 How many grams of water must be added to 70.0 g of potassium chloride (KCl) to prepare the following molal solutions?

(a) 0.010 *m*   **(b)** 0.45 *m*

(c) 1.23 *m*   **(d)** 2.45 *m*

13.85 An aqueous solution of oxalic acid $(H_2C_2O_4)$ is 0.568 M and has a density of 1.022 g/mL. What is the molality of the solution?

13.86 An aqueous solution of citric acid $(H_3C_6H_5O_7)$ is 0.655 M and has a density of 1.049 g/mL. What is the molality of the solution?

13.87 An aqueous solution of acetic acid $(HC_2H_3O_2)$ is 0.796 *m* and has a density of 1.004 g/mL. What is the molarity of the solution?

13.88 An aqueous solution of tartaric acid $(H_2C_4H_4O_6)$ is 0.278 *m* and has a density of 1.006 g/mL. What is the molarity of the solution?

13.89 Calculate the molality of a 14.0%-by-mass nitric acid (HNO_3) solution.

13.90 Calculate the molality of a 23.0%-by-mass acetic acid $(HC_2H_3O_2)$ solution.

Dilution (Sec. 13.10)

13.91 What is the molarity of a solution prepared by diluting 25.0 mL of 0.400-M potassium hydroxide (KOH) to each of the following volumes?

(a) 50.0 mL   (b) 83.0 mL   (c) 375 mL   (d) 2.67 L

13.92 What is the molarity of a solution prepared by diluting 50.0 mL of 0.300-M sodium nitrate $(NaNO_3)$ to each of the following volumes?

(a) 60.0 mL   (b) 97.0 mL   (c) 452 mL   (d) 8.75 L

13.93 What is the molarity of the solution prepared by concentrating, by evaporation of solvent, 1353 mL of 0.500-M ammonium chloride (NH_4Cl) solution to each of the following final volumes?

(a) 1125 mL   (b) 1.06 L   (c) 975 mL   (d) 297.5 mL

13.94 What is the molarity of the solution prepared by concentrating, by evaporation of solvent, 2212 mL of 0.400-M potassium sulfate (K_2SO_4) solution to each of the following final volumes?

(a) 1875 mL   (b) 1.25 L   (c) 853 mL   (d) 553 mL

13.95 How many milliliters of 6.0-M potassium hydroxide (KOH) solution are required to produce, using dilution, the following KOH solutions?

(a) 30.0 mL of 5.0-M solution
(b) 6.5 L of 1.0-M solution
(c) 275 mL of 5.9-M solution
(d) 3.0 mL of 0.10-M solution

13.96 How many milliliters of 3.0-M potassium chloride (KCl) solution are required to produce, using dilution, the following KCl solutions?

(a) 20.0 mL of 2.5-M solution
(b) 3.5 L of 1.0-M solution
(c) 352 mL of 2.9-M solution
(d) 4.2 mL of 0.25-M solution

13.97 In each of the following silver nitrate $(AgNO_3)$ solutions, how many milliliters of water should be added to obtain a solution that has a concentration of 0.100 M?

(a) 20.0 mL of a 2.00-M solution
(b) 20.0 mL of a 0.250-M solution
(c) 358 mL of a 0.950-M solution
(d) 2.3 L of a 6.00-M solution

13.98 In each of the following sodium nitrate $(NaNO_3)$ solutions, how many milliliters of water should be added to obtain a solution that has a concentration of 0.200 M?

(a) 30.0 mL of a 4.00-M solution
(b) 30.0 mL of a 0.400-M solution
(c) 785 mL of a 0.230-M solution
(d) 1.25 L of a 1.50-M solution

13.99 What will be the final concentration of each of the following solutions if the volume of the solution is increased by 20.0 mL by adding water?

(a) 25.0 mL of 6.0-M Na_2SO_4
(b) 100.0 mL of 3.0-M K_2SO_4
(c) 0.155 L of 10.0-M CsCl
(d) 2.00 mL of 0.100-M $MgCl_2$

13.100 What will be the final concentration of each of the following solutions if the volume of the solution is increased by 20.0 mL by adding water?

(a) 50.0 mL of 2.0-M KNO_3
(b) 50.0 mL of 3.0-M $AgNO_3$
(c) 1.0000 L of 1.2131-M $NaNO_3$
(d) 1.0000 mL of 1.000-M $LiNO_3$

13.101 What would be the molarity of a solution obtained when 275 mL of 6.00-M sodium hydroxide (NaOH) solution is mixed with each of the following?

(a) 3.254 L of H_2O
(b) 125 mL of 6.00-M NaOH solution
(c) 125 mL of 2.00-M NaOH solution
(d) 27 mL of 5.80-M NaOH solution

13.102 What would be the molarity of a solution obtained when 352 mL of 4.00-M sodium bromide (NaBr) solution is mixed with each of the following?

(a) 425 mL of water
(b) 225 mL of 4.00-M NaBr solution
(c) 225 mL of 2.00-M NaBr solution
(d) 15 mL of 4.20-M NaBr solution

Molarity and Chemical Equations (Sec. 13.11)

13.103 What volume, in liters, of 1.00-M $Pb(NO_3)_2$ is needed to react completely with 0.500 L of 4.00-M NaCl according to the following equation?

$$Pb(NO_3)_2(aq) + 2\,NaCl(aq) \longrightarrow PbCl_2(s) + 2\,NaNO_3(aq)$$

13.104 What volume, in milliliters, of 0.300-M $CaCl_2$ is needed to react completely with 40.0 mL of 0.200-M H_3PO_4 according to the following equation?

$$3\,CaCl_2(aq) + 2\,H_3PO_4(aq) \longrightarrow Ca_3(PO_4)_2(s) + 6\,HCl(aq)$$

13.105 How many grams of S can be produced from the reaction of 30.0 mL of 12.0-M HNO_3 with an excess of 0.035-M H_2S solution according to the following equation?

$$2\,HNO_3(aq) + 3\,H_2S(aq) \longrightarrow 2\,NO(g)$$
$$+ 3\,S(s) + 4\,H_2O(l)$$

13.106 How many grams of Ag_3PO_4 can be produced from the reaction of 2.50 L of 0.200-M $AgNO_3$ with an excess of 0.750-M K_3PO_4 solution according to the following equation?

$$3\,AgNO_3(aq) + K_3PO_4(aq) \longrightarrow Ag_3PO_4(s) + 3\,KNO_3(aq)$$

13.107 What volume, in milliliters, of 0.50-M H_2SO_4 is required to react with 18.0 g of nickel according to the following equation?

$$Ni(s) + H_2SO_4(aq) \longrightarrow NiSO_4(aq) + H_2(g)$$

13.108 What volume, in milliliters, of 1.50-M HNO_3 is required to react with 100.0 g of tin according to the following equation?

$$\underline{\quad} Sn(s) + 2\,HNO_3(aq) \longrightarrow 2\,Sn(NO_3)_2(aq) + H_2(g)$$

13.109 What is the molarity of a 37.5-mL sample of HNO_3 solution that will completely react with 23.7 mL of 0.100-M NaOH according to the following equation?

$$HNO_3(aq) + NaOH(aq) \longrightarrow NaNO_3(aq) + H_2O(l)$$

13.110 What is the molarity of a 50.0-mL sample of H_2SO_4 solution that will completely react with 40.0 mL of 0.200-M $Mg(OH)_2$ according to the following equation?

$$\underline{\quad} H_2SO_4(aq) + Mg(OH)_2(aq) \longrightarrow MgSO_4(aq) + 2\,H_2O(l)$$

13.111 What volume, in liters, of NO gas measured at STP can be produced from 50.0 mL of 6.0-M HNO_3 solution and an excess of Cu metal according to the following reaction?

$$8\,HNO_3(aq) + 3\,Cu(s) \longrightarrow 3\,Cu(NO_3)_2(aq) + 2\,NO(g)$$
$$+ 4\,H_2O(l)$$

13.112 What volume, in liters, of H_2 gas measured at STP can be produced from 50.0 mL of 3.0-M HBr solution and an excess of Zn metal according to the following reaction?

$$\underline{\quad} 2\,HBr(aq) + Zn(s) \longrightarrow ZnBr_2(aq) + H_2(g)$$

13.113 What is the molarity of a 1.75-L $Ca(OH)_2$ solution that would completely react with 2.00 L of CO_2 gas measured at STP according to the following reaction?

$$CO_2(g) + Ca(OH)_2(aq) \longrightarrow CaCO_3(s) + H_2O(l)$$

13.114 What is the molarity of a 5.00-L NaOH solution that would completely react with 4.00 L of CO_2 gas measured at STP according to the following reaction?

$$CO_2(g) + 2\,NaOH(aq) \longrightarrow Na_2CO_3(aq) + H_2O(l)$$

Additional Problems

13.115 In each of the following sets of ionic compounds, identify the members of the set that are soluble in water.

(a) $Be_3(PO_4)_2$, $AlPO_4$, $FePO_4$, $(NH_4)_3PO_4$

(b) $Cu(OH)_2$, $Be(OH)_2$, $Ca(OH)_2$, $Zn(OH)_2$

(c) Ag_3PO_4, $AgNO_3$, $AgCl$, $AgBr$

(d) CaS, $Ca(NO_3)_2$, $CaSO_4$, $Ca(C_2H_3O_2)_2$

13.116 In each of the following sets of ionic compounds, identify the members of the set that are soluble in water.

(a) K_2CO_3, $MgCO_3$, $NiCO_3$, $Al_2(CO_3)_3$

(b) MgS, Rb_2S, Al_2S_3, CaS

(c) $Pb(OH)_2$, $PbCl_2$, $Pb(NO_3)_2$, $PbSO_4$

(d) BaS, $BaCl_2$, $BaSO_4$, $Ba(OH)_2$

13.117 The solubility of $CuSO_4$ in water at 50°C is 33.3 g/100 g H_2O. If 400.0 g of a 75% saturated $CuSO_4$ solution at 50°C is heated to evaporate the water completely, how much solid $CuSO_4$ should be recovered?

13.118 The solubility of NaCl in water at 50°C is 37.0 g/100 g H_2O. If 300.0 g of a 85% saturated NaCl solution at 50°C is heated to evaporate the water completely, how much solid NaCl should be recovered?

13.119 After all the water is evaporated from 254 mL of a $AgNO_3$ solution, 45.2 g of $AgNO_3$ remain. Express the original concentration of the $AgNO_3$ solution in each of the following units.

(a) mass–volume percent

(b) molarity

13.120 After all the water is evaporated from 10.0 mL of a CsCl solution, 3.75 g of CsCl remains. Express the original concentration of CsCl solution in each of the following units.

(a) mass–volume percent

(b) molarity

13.121 What mass, in grams, of Na_2SO_4 would be required to prepare 425 mL of a 1.55% (m/m) Na_2SO_4 solution whose density is 1.02 g/mL?

13.122 What mass, in grams, of NaCl would be required to prepare 275 mL of a 30.0% (m/m) NaCl solution whose density is 1.18 g/mL?

13.123 A 3.000-M $NaNO_3$ solution has a density of 1.161 g/mL at 20°C. How many grams of solvent are present in 1.375 L of this solution?

13.124 A 0.157-M NaCl solution has a density of 1.09 g/mL at 20°C. How many grams of solvent are present in 80.0 mL of this solution?

13.125 Calculate the volume, in milliliters, of 0.125-M Na_2SO_4 solution needed to provide each of the following.

(a) 10.0 g of Na_2SO_4

(b) 2.5 g of Na^+ ion

(c) 0.567 mole of Na_2SO_4

(d) 0.112 mole of SO_4^{2-} ion

13.126 Calculate the volume, in milliliters, of 1.25-M $Mg(NO_3)_2$ solution needed to provide each of the following.

(a) 15.7 g of $Mg(NO_3)_2$

(b) 3.57 g of Mg^{2+} ion

(c) 1.2 moles of $Mg(NO_3)_2$

(d) 0.57 mole of NO_3^- ion

13.127 A solution is made by diluting 225 mL of a 0.245-M aluminum nitrate [$Al(NO_3)_3$] solution with water to a final volume of 0.750 L. Calculate the following.

(a) the molarities of Al^{3+} ion and NO_3^- ion in the original solution

(b) the molarities of $Al(NO_3)_3$, Al^{3+} ion, and NO_3^- ion in the diluted solution

13.128 A solution is made by diluting 315 mL of a 0.115-M potassium phosphate (K_3PO_4) solution with water to a final volume of 0.650 L. Calculate the following.

(a) the molarities of K^+ ion and PO_4^{3-} ion in the original solution
(b) the molarities of K_3PO_4, K^+ ion, and PO_4^{3-} ion in the diluted solution

13.129 A solution is made by mixing 175 mL of 0.100-M K_3PO_4 with 27 mL of 0.200-M KCl. Assuming that the volumes are additive, what are the molar concentrations of the following ions in the new solution?

(a) K^+ ion   (b) Cl^- ion   (c) PO_4^{3-} ion

13.130 A solution is made by mixing 50.0 mL of 0.300-M Na_2SO_4 with 30.0 mL of 0.900-M K_2SO_4. Assuming that the volumes are additive, what are the molar concentrations of the following ions in the new solution?

(a) Na^+ ion   (b) K^+ ion   (c) SO_4^{2-} ion

13.131 A solute concentration is 3.74 ppm (m/m). What would this concentration be in the units of milligram of solute per kilogram of solution?

13.132 A solute concentration is 5.14 ppm (m/m). What would this concentration be in the units of microgram of solute per milligram of solution?

13.133 A solution is prepared by dissolving 1.00 g of NaCl in enough water to make 10.00 mL of solution. A 1.00-mL portion of this solution is then diluted to a final volume of 10.00 mL. What is the molarity of the final NaCl solution?

13.134 A solution is prepared by dissolving 30.0 g of Na_2SO_4 in enough water to make 750.0 mL of solution. A 10.00-mL portion of this solution is then diluted to a final volume of 100.0 mL. What is the molarity of the final Na_2SO_4 solution?

13.135 How many milliliters of 38.0% (m/m) HCl (density of 1.19 g/mL) are needed to make, using a dilution procedure, 1.00 L of 0.100-M HCl?

13.136 How many milliliters of 20.0% (m/m) NaCl (density of 1.15 g/mL) are needed to make, using a dilution procedure, 3.50 L of 0.150-M NaCl?

13.137 How many grams of water should you add to a 1.23-m NaCl solution containing 1.50 kg H_2O to reduce the molality to 1.00 m?

13.138 How many grams of water should you add to a 0.0883-m NaCl solution containing 0.650 kg H_2O to reduce the molality to 0.0100 m?

13.139 Calculate the total mass, in grams, and the total volume, in milliliters, of a 2.16-m H_3PO_4 solution containing 52.0 g of solute. The density of the solution is 1.12 g/mL.

13.140 Calculate the total mass, in grams, and the total volume, in milliliters, of a 0.710-m $H_3C_6H_5O_7$ (citric acid) solution containing 23.0 g of solute. The density of the solution is 1.05 g/mL.

13.141 An aqueous solution having a density of 0.980 g/mL is prepared by dissolving 11.3 mL of CH_4O (density of 0.793 g/mL) in

enough water to produce 75.0 mL of solution. Express the percent CH_3OH in this solution as

(a) % (m/v)   (b) % (m/m)   (c) % (v/v)

13.142 An aqueous solution having a density of 0.993 g/mL is prepared by dissolving 20.0 mL of C_2H_6O (density of 0.789 g/mL) in enough water to produce 85.0 mL of solution. Express the percent C_2H_6O in this solution as

(a) % (m/v)   (b) % (m/m)   (c) % (v/v)

13.143 The concentration of a KCl solution is 0.273 molal and 0.271 molar. What is the density of the solution, in grams per milliliter?

13.144 The concentration of a $Pb(NO_3)_2$ solution is 0.953 molal and 0.907 molar. What is the density of the solution, in grams per milliliter?

Cumulative Problems

13.145 Identify the insoluble substance(s) formed when each of the following pairs of soluble substances react in aqueous solution through a double-replacement reaction.

(a) NaCl and $AgNO_3$
(b) $Ba(C_2H_3O_2)_2$ and K_3PO_4
(c) $Pb(NO_3)_2$ and Ag_2SO_4
(d) $CuSO_4$ and BaS

13.146 Identify the insoluble substance(s) formed when each of the following pairs of soluble substances react in aqueous solution through a double-replacement reaction.

(a) $MgCl_2$ and $Ba(OH)_2$
(b) NH_4Cl and $Pb(NO_3)_2$
(c) MgS and Na_2CO_3
(d) $SrCl_2$ and Ag_2SO_4

13.147 How many liters of NH_3 gas at 25°C and 1.46 atm pressure are required to prepare 2.00 L of a 3.50-M solution of NH_3?

13.148 How many liters of HCl gas at 35°C and 1.05 atm pressure are required to prepare 4.00 L of a 0.500-M solution of HCl?

13.149 Calculate the theoretical yield, in grams, of AgCl formed from the reaction of 6.41 g of $ZnCl_2$ with 40.0 mL of a 0.404-M $AgNO_3$ solution according to the reaction

$$ZnCl_2(s) + 2\,AgNO_3(aq) \longrightarrow Zn(NO_3)_2(aq) + 2\,AgCl(s)$$

13.150 Calculate the theoretical yield, in grams, of AgCl formed from the reaction of 1.00 g of KCl with 100.0 mL of a 0.0250-M $AgC_2H_3O_2$ solution according to the reaction

$$KCl(s) + AgC_2H_3O_2(aq) \longrightarrow KC_2H_3O_2(aq) + AgCl(s)$$

13.151 What mass, in grams, of $BaCrO_4$ would be produced by mixing 0.350 L of a 3.25-M $BaCl_2$ solution with 0.450 L of a 4.50-M K_2CrO_4 solution? The two solutions react according to the equation

$$BaCl_2(aq) + K_2CrO_4(aq) \longrightarrow BaCrO_4(s) + 2\,KCl(aq)$$

13.152 What mass, in grams, of $BaSO_4$ would be produced by mixing 1.53 L of a 4.50-M Na_2SO_4 solution with 3.20 L of a 2.50-M $Ba(NO_3)_2$ solution? The two solutions react according to the equation

$$Na_2SO_4(aq) + Ba(NO_3)_2(aq) \longrightarrow$$
$$2\,NaNO_3(aq) + BaSO_4(s)$$

13.153 A 1.25-g sample of *impure* Na_2CO_3 is found to react completely with 70.0 mL of 0.125-M HCl. The equation for the reaction is

$$Na_2CO_3(s) + 2\,HCl(aq)$$
$$\longrightarrow 2\,NaCl(aq) + CO_2(g) + H_2O(l)$$

What is the mass percent Na_2CO_3 in the impure sample?

13.154 A 5.00-g sample of *impure* $CaCO_3$ is found to react completely with 100.0 mL of 0.100-M H_2SO_4. The equation for the reaction is

$$CaCO_3(s) + H_2SO_4(aq) \longrightarrow$$
$$CaSO_4(s) + CO_2(g) + H_2O(l)$$

What is the mass percent $CaCO_3$ in the impure sample?

13.155 Magnesium, calcium, and zinc all react with hydrochloric acid as follows (where M represents any of these metals).

$$M(s) + 2\,HCl(aq) \longrightarrow MCl_2(aq) + H_2(g)$$

A sample of one of these metals reacts completely with the acid in 27.9 mL of 2.48-M HCl, and the resulting solution is evaporated to dryness. The residue MCl_2 has a mass of 4.72 g. What is the identity of the metal used?

13.156 Iron, nickel, and tin all react with hydrochloric acid as follows (where M represents any of these metals).

$$M(s) + 2\,HCl(aq) \longrightarrow MCl_2(aq) + H_2(g)$$

A sample of one of these metals reacts completely with the acid in 34.2 mL of 4.00-M HCl, and the resulting solution is evaporated to dryness. The residue MCl_2 has a mass of 8.87 g. What is the identity of the metal used?

13.157 A quantity of sodium peroxide (Na_2O_2) is added to water, and the following reaction occurs.

$$2\,Na_2O_2(s) + 2\,H_2O(l) \longrightarrow 4\,NaOH(aq) + O_2(g)$$

If 70.0 mL of O_2 gas (at STP) and 150.0 mL of NaOH solution are produced, what is the molarity of the NaOH solution?

13.158 A quantity of lithium nitride (Li_3N) is added to water, and the following reaction occurs.

$$Li_3N(s) + 3\,H_2O(l) \longrightarrow 3\,LiOH(aq) + NH_3(g)$$

If 100.0 mL of NH_3 gas (at STP) and 255 mL of LiOH solution are produced, what is the molarity of the LiOH solution?

Answers to Practice Exercises

13.1 **(a)** soluble   **(b)** soluble   **(c)** soluble
       (d) insoluble   **(e)** soluble

13.2 0.385% (m/m)

13.3 1.22 g $LiNO_3$

13.4 51.8% (v/v)

13.5 3.2 g NaCl

13.6 **(a)** 38 ppm (m/m)   **(b)** 38,000 ppb (m/m)

13.7 0.001 mL O_3

13.8 0.250 M NaOH

13.9 8.81 g $C_6H_8O_6$

13.10 2.50 L NaOH

13.11 1.008 M $AgNO_3$

13.12 119 g solvent

13.13 0.250 *m* K_2CO_3

13.14 0.0140 g KOH

13.15 2.46 M CH_4O

13.16 5.10 *m* $C_{12}H_{22}O_{11}$

13.17 0.50 M $AgNO_3$

13.18 9680 mL solvent

13.19 0.375 M NH_4Cl

13.20 0.395 L KOH

13.21 125 g $BaCrO_4$

13.22 7.86 L H_2

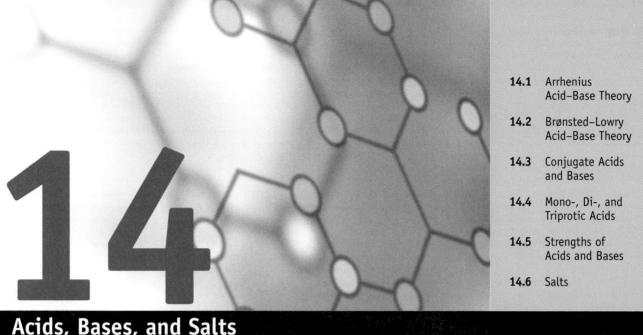

14

Acids, Bases, and Salts

14.1 Arrhenius Acid–Base Theory

Acids and bases are among the most common and important compounds known. Aqueous solutions of acids and bases are key materials in both biological systems and chemical industrial processes.

Historically, as early as the seventeenth century, acids and bases were recognized as important groups of compounds. Such early recognition was based on what the substances did rather than on their chemical composition.

Early known facts about acids include

1. Acids, when dissolved in water, have a sour taste. (The name acid comes from the Latin word *acidus*, which means "sour.") Note that although taste was once an acceptable criterion for identifying a chemical, it is not anymore. It is not a wise idea to taste chemicals when in a chemical laboratory.
2. Acids cause the dye litmus to change from blue to red. (Litmus is a naturally occurring vegetable dye obtained from lichens.)
3. When certain metals, such as zinc and iron, are placed in acids, they dissolve, liberating hydrogen gas.

Early known characteristics of bases include

1. Bases, when dissolved in water, have a bitter taste.
2. Bases cause the dye litmus to change from red to blue.

3. When fats are placed in base solutions, they dissolve.

4. Base solutions feel slippery or soapy to the touch. (The bases themselves are not slippery but react with fats in the skin to form new slippery or soapy compounds.)

It was not until 1884 that acids and bases were defined in terms of chemical composition. In that year, the Swedish chemist Svante August Arrhenius (1859–1927; see "The Human Side of Chemistry 15") proposed that acids and bases be defined in terms of the species they form upon dissolution in water. His definitions are the simplest and most commonly used today. An **Arrhenius acid** *is a hydrogen-containing compound that, in water, produces hydrogen ions* (H^+). The acidic species in Arrhenius theory is, thus, the hydrogen ion. An **Arrhenius base** *is a hydroxide-containing compound that, in water, produces hydroxide ions* (OH^-). The basic species in Arrhenius theory is, thus, the hydroxide ion.

Two common examples of acids, according to the Arrhenius definition, are the substances HNO_3 and HCl.

$$HNO_3(l) \xrightarrow{H_2O} H^+(aq) + NO_3^-(aq)$$

$$HCl(g) \xrightarrow{H_2O} H^+(aq) + Cl^-(aq)$$

Arrhenius acids in the pure state (not in solution) are covalent compounds; that is, they do not contain H^+ ions. The H^+ ions are produced when the Arrhenius acid interacts with the water, a process called *ionization*. **Ionization** *is the process whereby ions are produced from a molecular compound when it is dissolved in a solvent.*

Two common examples of Arrhenius bases are NaOH and KOH.

$$NaOH(s) \xrightarrow{H_2O} Na^+(aq) + OH^-(aq)$$

$$KOH(s) \xrightarrow{H_2O} K^+(aq) + OH^-(aq)$$

The Human Side of Chemistry 15

Svante August Arrhenius (1859–1927)

Svante August Arrhenius, born in 1859 near Uppsala, Sweden, is considered one of the founders of modern physical chemistry.

His roots are those of a Swedish farming family. An infant prodigy, on his own accord and against his parents' wishes, he taught himself to read at age 3.

In 1884, he proposed his definitions for acids and bases. Simultaneously he shook the world of chemistry by presenting his theory of ionic dissociation, which was that ionic substances, when dissolved in water, dissociate into ions.

This theory, which came directly from his university doctoral work in Stockholm, was given a hostile reception by many other chemists, including his mentors at the University of Stockholm. He was awarded his doctoral degree with reluctance and given the lowest possible passing grade. It was the opinion of his teachers that his theory was "too farfetched."

Completely rebuffed by Swedish scientists, he decided to approach the scientific world elsewhere. He did find acceptance in some places and, building on this acceptance, he further developed and refined his theory.

Acceptance in his homeland eventually came. In 1903, Arrhenius was awarded the Nobel Prize in chemistry for his "farfetched" ideas concerning ions in

solution. In 1905, the King of Sweden founded the Nobel Institute for Physical Research at Stockholm and installed Arrhenius as director. This appointment came as a counteroffer to that of a major professorship in Berlin. Arrhenius remained at the Nobel Institute until shortly before his death.

Arrhenius made other major contributions to chemistry besides those dealing with ions, acids, and bases. In 1889, while studying how rates of reactions increased with temperature, he worked out the concept of energy of activation (to be discussed in Chapter 16). In later years he became interested in such diverse things as serum chemistry and astronomy. He spent considerable time speculating on the origin of life on Earth.

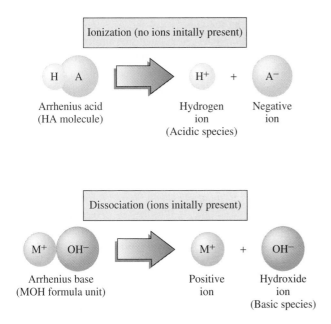

Figure 14.1

The difference between the aqueous solution processes of ionization (Arrhenius acids) and dissociation (Arrhenius bases). Ionization is the production of ions from a *molecular* compound that has been dissolved in a solvent. Dissociation is the production of ions from an *ionic* compound that has been dissolved in a solvent.

Arrhenius bases are ionic compounds in the pure state, in direct contrast to acids. When such compounds dissolve in water, the already existent OH^- ions are released, a process called *dissociation*. **Dissociation** *is the process whereby the already existent ions in an ionic compound separate when the ionic compound is dissolved in a solvent.* Figure 14.1 contrasts the processes of ionization (Arrhenius acids) and dissociation (Arrhenius bases).

14.2 Brønsted–Lowry Acid–Base Theory

Although widely used, Arrhenius acid–base theory has some shortcomings. Two disadvantages are that it is restricted to aqueous solution and it does not explain why compounds like ammonia (NH_3), which do not contain hydroxide ion, produce a basic water solution.

In 1923, Johannes Nicolaus Brønsted (1879–1947), a Danish chemist, and Thomas Martin Lowry (1874–1936), a British chemist, independently and almost simultaneously proposed broadened definitions for acids and bases—definitions that applied in both aqueous and nonaqueous solutions and that also explained how some nonhydroxide-containing substances, when added to water, produce basic solutions.

A **Brønsted–Lowry acid** *is any substance that can donate a proton* (H^+) *to some other substance.* A **Brønsted–Lowry base** *is any substance that can accept a proton* (H^+) *from some other substance.* In simpler terms, a Brønsted–Lowry *acid* is a *proton donor* (or hydrogen ion donor) and a Brønsted–Lowry *base* is a *proton acceptor* (or hydrogen ion acceptor).

Three important additional concepts associated with Brønsted–Lowry acid–base theory are

1. Any chemical reaction involving a Brønsted–Lowry acid must also involve a Brønsted–Lowry base. You cannot have one without the other. Proton donation (from an acid) cannot occur unless an acceptor (a base) is present.
2. All the acids and bases included in the Arrhenius theory (Sec. 14.1) are also acids and bases according to the Brønsted–Lowry theory. However, the converse is not true; some substances not considered Arrhenius bases are Brønsted–Lowry bases.

The terms *hydrogen ion* and *proton* are used synonymously in acid–base discussions. Why? The predominant hydrogen isotope, ^1_1H, is unique in that no neutrons are present; it consists of a proton and an electron. Thus, the ion $^1_1H^+$, a hydrogen atom that has lost its only electron, is simply a proton.

3. The identity of the acidic species in *aqueous solution* is not the Arrhenius H^+ ion but rather the H_3O^+ ion. Hydrogen ions in solution react with water. The attraction between a hydrogen ion and a water molecule is sufficiently strong to bond the hydrogen ion to the water molecule to form a *hydronium ion* (H_3O^+). The bond between them is a coordinate covalent bond (Sec. 7.14) because both electrons are furnished by the oxygen atom.

$$H^+ + :\ddot{O}-H \longrightarrow \left[H:\overset{..}{\underset{..}{O}}-H\right]^+$$

Coordinate covalent bond

Hydronium ion

The Brønsted–Lowry acid–base definitions can best be illustrated by example. Consider the formation reaction for hydrochloric acid, which involves the dissolving of hydrogen chloride gas in water.

$$\overset{\frown}{H}:\ddot{\underset{..}{Cl}}: + :\overset{..}{\underset{H}{O}}:H \longrightarrow \left[H:\overset{..}{\underset{H}{O}}:H\right]^+ + \left[:\overset{..}{\underset{..}{Cl}}:\right]^-$$

Coordinate covalent bond

Acid   Base

The HCl behaves as a Brønsted–Lowry acid by donating a proton to a water molecule. Note that a hydronium ion is formed as a result. The base in this reaction is water because it has accepted a proton; no hydroxide ions are involved. The Brønsted–Lowry definition of a base includes all species that accept a proton; hydroxide ions can do this, but so can many other substances.

It is not necessary that a water molecule be one of the reactants in a Brønsted–Lowry acid–base reaction or that the reaction take place in the liquid state. An important application of Brønsted–Lowry acid–base theory is to gas-phase reactions. The white solid haze that often covers glassware in a chemistry laboratory results from the gas-phase reaction between HCl and NH_3.

$$\overset{\frown}{H}:\ddot{\underset{..}{Cl}}: + :\overset{H}{\underset{H}{N}}:H \longrightarrow \left[H:\overset{H}{\underset{H}{N}}:H\right]^+ + \left[:\overset{..}{\underset{..}{Cl}}:\right]^-$$

> A Brønsted–Lowry base—a proton acceptor—must contain an atom that possesses a pair of unshared electrons that can be used in forming a coordinate covalent bond to an incoming proton (from a Brønsted–Lowry acid).

This is a Brønsted–Lowry acid–base reaction, because the HCl molecules donate protons to the NH_3, forming NH_4^+ and Cl^- ions. These ions instantaneously combine to form the white solid NH_4Cl.

Another example of a Brønsted–Lowry acid–base reaction involves the dissolving of ammonia (a nonhydroxide base) in water. In the following equation, note how a hydroxide ion is produced as the result of the transfer of a proton from water (written as HOH) to the ammonia.

$$NH_3(g) + \overset{\frown}{H}OH(l) \longrightarrow NH_4^+(aq) + OH^-(aq)$$

Base   Acid

As an ionic solid dissolves in water to produce an aqueous solution, the solid ionic lattice breaks up, producing individual ions that are free to move about in the solution (Sec. 13.3). Ions so formed can function as Brønsted–Lowry acids or bases, as is illustrated in the following two equations.

$$\overset{\frown}{\text{H}}\text{CO}_3{}^-(\text{aq}) + \text{H}_2\text{O}(\text{l}) \longrightarrow \text{CO}_3{}^{2-}(\text{aq}) + \text{H}_3\text{O}^+(\text{aq})$$

Acid   Base

$$\overset{\frown}{\text{H}}_2\text{PO}_4{}^-(\text{aq}) + \text{OH}^-(\text{aq}) \longrightarrow \text{HPO}_4{}^{2-}(\text{aq}) + \text{H}_2\text{O}(\text{l})$$

Acid   Base

14.3 Conjugate Acids and Bases

For most Brønsted–Lowry acid–base reactions, 100% proton transfer does not occur. Instead, a state of equilibrium (Sec. 11.13) is reached in which a forward reaction and a reverse reaction are occurring at an equal rate.

The equilibrium mixture of a Brønsted–Lowry acid–base reaction always has *two* acids and *two* bases present. To illustrate this, consider the acid–base reaction involving hydrogen fluoride and water.

$$\text{HF}(\text{aq}) + \text{H}_2\text{O}(\text{l}) \rightleftharpoons \text{H}_3\text{O}^+(\text{aq}) + \text{F}^-(\text{aq})$$

(The double arrows in this equation indicate a state of equilibrium—both a forward and a reverse reaction are occurring.) For the forward reaction, the HF molecules donate protons to water molecules. Thus, the HF is functioning as an acid and the H_2O is functioning as a base.

$$\underset{\text{acid}}{\text{HF}(\text{aq})} + \underset{\text{base}}{\text{H}_2\text{O}(\text{l})} \longrightarrow \text{H}_3\text{O}^+(\text{aq}) + \text{F}^-(\text{aq})$$

For the reverse reaction, the one going from right to left, a different picture emerges. Here, H_3O^+ is functioning as an acid (by donating a proton), and F^- behaves as a base (by accepting the proton).

$$\underset{\text{acid}}{\text{H}_3\text{O}^+(\text{aq})} + \underset{\text{base}}{\text{F}^-(\text{aq})} \longrightarrow \text{HF}(\text{aq}) + \text{H}_2\text{O}(\text{l})$$

The two acids and two bases involved in a Brønsted–Lowry equilibrium situation can be grouped into two *conjugate acid–base pairs*. A **conjugate acid–base pair** *is two species, one an acid and one a base, that differ from each other through the loss or gain of a proton* (H^+ *ion*). The two conjugate acid–base pairs in our example are (HF and F^-) and (H_3O^+ and H_2O).

Conjugate pair

$$\text{HF}(\text{aq}) + \text{H}_2\text{O}(\text{l}) \rightleftharpoons \text{H}_3\text{O}^+(\text{aq}) + \text{F}^-(\text{aq})$$

Acid   Base   Acid   Base

Conjugate pair

Abbreviated notation for specifying a conjugate acid–base pair is "acid/base." Using this notation, the two conjugate acid–base pairs in the preceding example are HF/F^- and H_3O^+/H_2O. The acid is always written first in such notation.

For any given conjugate acid–base pair,

1. The acid in the acid–base pair always has one *more* H atom and one *fewer* negative charge than the base. Note this relationship for the HF/F^- conjugate acid–base pair.

2. The base in the acid–base pair always has one *fewer* H atom and one *more* negative charge than the acid. Note this relationship for the HF/F^- conjugate acid–base pair.

The acid in a conjugate acid–base pair is called the *conjugate acid* of the base, and the base in the conjugate acid–base pair is called the *conjugate base* of the acid. A **conjugate acid** *is the species formed when a proton* (H^+ *ion*) *is added to a Brønsted–Lowry base*. The H_3O^+ ion is the conjugate acid of a H_2O molecule. A **conjugate base** *is the species that remains when a proton* (H^+ *ion*) *is removed from a Brønsted–Lowry acid*. The H_2O molecule is the conjugate base of the H_3O^+ ion. Every acid has a conjugate base, and every base has a conjugate acid. In general terms, these relationships can be diagrammed as follows.

Conjugate **means "coupled" or "joined together" (as in a pair).**

$$HA + B \rightleftharpoons HB^+ + A^-$$

Acid   Base   Conjugate   Conjugate
               acid   base

EXAMPLE 14.1

Determining the Members of a Conjugate Acid–Base Pair

Identify the conjugate acid–base pairs in the following reaction.

$$HBr(aq) + H_2O(l) \longrightarrow H_3O^+(aq) + Br^-(aq)$$

SOLUTION

To determine the conjugate acid–base pairs, we look for formulas that differ only by one H^+ ion. For this reaction, one pair must be HBr and Br^-, and the other pair must be H_2O and H_3O^+. In each pair, the acid is the substance with one more hydrogen atom, so the two *acids* are HBr and H_3O^+, and the two *bases* are Br^- and H_2O.

Conjugate pair

$$HBr(aq) + H_2O(l) \rightleftharpoons H_3O^+(aq) + Br^-(aq)$$

Conjugate pair

Practice Exercise 14.1

Answers to practice exercises are located at the end of the chapter.

Identify the conjugate acid–base pairs in the following reaction:

$$HCN(aq) + H_2O(l) \rightleftharpoons H_3O^+(aq) + CN^-(aq)$$

EXAMPLE 14.2

Determining the Formula of One Member of a Conjugate Acid–Base Pair When Given the Other Member

Write formulas for the following.

(a) the conjugate base of HCO_3^- **(b)** the conjugate acid of PO_4^{3-}

SOLUTION

(a) A conjugate base can always be found by removing one H^+ ion from a given acid. Removing one H^+ (both the atom and the charge) from HCO_3^- leaves CO_3^{2-}. Thus, CO_3^{2-} is the *conjugate base* of HCO_3^-.

(b) A conjugate acid can always be found by adding one H^+ ion to a given base. Adding one H^+ (both the atom and the charge) to PO_4^{3-} produces HPO_4^{2-}. Thus, HPO_4^{2-} is the *conjugate acid* of PO_4^{3-}.

Practice Exercise 14.2

Write formulas for the following.

(a) the conjugate base of HSO_4^-

(b) the conjugate acid of HPO_4^{2-}

Some molecules or ions are able to function as either an acid or a base, depending on the kind of substance with which they react. Such molecules are said to be *amphoteric*. An **amphoteric substance** *is a substance that can either lose or accept a proton* (H^+ *ion*) *and thus can function as either an acid or a base.*

Water is the most common example of an amphoteric substance. In the first of the following two reactions, water functions as a base and in the second it functions as an acid.

$$HNO_3(l) + H_2O(l) \rightleftharpoons H_3O^+(aq) + NO_3^-(aq)$$
$$\text{acid} \qquad \text{base}$$

$$NH_3(g) + H_2O(l) \rightleftharpoons NH_4^+(aq) + OH^-(aq)$$
$$\text{base} \qquad \text{acid}$$

Another example of an amphoteric substance is the hydrogen carbonate ion.

$$HCO_3^-(aq) + OH^-(aq) \rightleftharpoons CO_3^{2-}(aq) + H_2O(l)$$
$$\text{acid} \qquad \text{base}$$

$$HCO_3^-(aq) + H_3O^+(aq) \rightleftharpoons H_2CO_3(aq) + H_2O(l)$$
$$\text{base} \qquad \text{acid}$$

> The term *amphoteric* comes from the Greek *amphoteres*, which means "partly one and partly the other." Just as an amphibian is an animal that lives partly on land and partly in the water, an amphoteric substance is sometimes an acid and sometimes a base.

14.4 Mono-, Di-, and Triprotic Acids

Acids can be classified according to the number of hydrogen ions (protons) they can transfer per molecule during an acid–base reaction. A **monoprotic acid** *is an acid that can transfer only one* H^+ *ion (proton) per molecule during an acid–base reaction.* Hydrochloric acid (HCl) and nitric acid (HNO_3) are both monoprotic acids.

A **diprotic acid** *is an acid that can transfer two* H^+ *ions (two protons) per molecule during an acid–base reaction.* Sulfuric acid (H_2SO_4) and carbonic acid (H_2CO_3) are examples of diprotic acids. The transfer of protons for a diprotic acid always occurs in steps. For H_2SO_4, the two steps are as follows.

$$H_2SO_4(aq) + H_2O(l) \longrightarrow H_3O^+(aq) + HSO_4^-(aq)$$

$$HSO_4^-(aq) + H_2O(l) \longrightarrow H_3O^+(aq) + SO_4^{2-}(aq)$$

A few *triprotic acids* exist. A **triprotic acid** *is an acid that can transfer three* H^+ *ions (three protons) per molecule during an acid–base reaction.* Phosphoric acid, H_3PO_4, is the most common triprotic acid. The three proton-transfer steps for this acid are as follows.

$$H_3PO_4(aq) + H_2O(l) \longrightarrow H_3O^+(aq) + H_2PO_4^-(aq)$$

$$H_2PO_4^-(aq) + H_2O(l) \longrightarrow H_3O^+(aq) + HPO_4^{2-}(aq)$$

$$HPO_4^{2-}(aq) + H_2O(l) \longrightarrow H_3O^+(aq) + PO_4^{3-}(aq)$$

A **polyprotic acid** *is an acid that can transfer two or more H^+ ions (protons) per molecule during an acid–base reaction.* Both diprotic acids and triprotic acids are examples of polyprotic acids.

The number of hydrogen atoms present in one molecule of an acid cannot always be used to classify the acid as mono-, di-, or triprotic. For example, a molecule of acetic acid contains four hydrogen atoms, and yet it is a monoprotic acid. Only one of the hydrogen atoms in acetic acid is acidic. An **acidic hydrogen atom** *is a hydrogen atom in an acid molecule that can be transferred to a base during an acid–base reaction.*

Whether or not a hydrogen atom is acidic is related to its location in a molecule, that is, to which other atom it is bonded. Let us consider our previously mentioned acetic acid example in more detail by looking at the structure of this acid. A *structural* equation for the acidic behavior of acetic acid is

Note the structure of the acetic acid molecule (reactant side of the equation): one hydrogen atom is bonded to an oxygen atom, and the other three hydrogen atoms are each bonded to a carbon atom. It is only the hydrogen atom bonded to the oxygen atom that is acidic. The hydrogen atoms bonded to the carbon atom are too tightly held to be removed by reaction with water molecules. Water has very little effect on a carbon–hydrogen bond because it is essentially nonpolar (Sec. 7.19). On the other hand, the hydrogen bonded to oxygen is involved in a very polar bond because of oxygen's large electronegativity (Sec. 7.18). Water, which is a polar molecule, readily attacks polar bonds but has very little effect on nonpolar bonds.

We now see why the formula for acetic acid is usually written as $HC_2H_3O_2$ rather than as $C_2H_4O_2$. In the situation where some hydrogens are easily removed (acidic) and others are not (nonacidic), it is accepted procedure to write the acidic hydrogens first, separated from the other hydrogens in the formula. Citric acid, the principal acid in citrus fruits, is another example of an acid that contains both acidic and nonacidic hydrogens. Its formula, $H_3C_6H_5O_7$, indicates that three of the eight hydrogen atoms present in a molecule are acidic. Table 14.1 gives the formulas, classifications, and common occurrences of selected mono-, di-, and triprotic acids, many of which contain nonacidic hydrogen atoms.

We have focused our attention on acids in the preceding discussion. It should be noted that similar concepts can be applied to bases. From an Arrhenius standpoint, bases can release more than one hydroxide ion; for example, $Ca(OH)_2$ is a base that produces two OH^- ions per molecule. From a Brønsted–Lowry viewpoint, bases exist that can accept more than one proton, in a stepwise manner; for example, the PO_4^{3-} ion is a Brønsted–Lowry base that can ultimately accept three protons through reaction with three H_3O^+ ions:

$$PO_4^{3-} \xrightarrow{H^+} HPO_4^{2-} \xrightarrow{H^+} H_2PO_4^- \xrightarrow{H^+} H_3PO_4$$

14.5 Strengths of Acids and Bases

Brønsted–Lowry acids vary in their ability to transfer protons and produce hydronium ions in aqueous solution. Such acids are classified as strong or weak on the basis of the extent that proton transfer occurs in aqueous solution. A **strong acid** *is an acid that, in aqueous solution, transfers 100%, or very nearly 100%, of its acidic hydrogen atoms to water.* Thus, if an acid is strong, almost all of the acid molecules present give up protons to water. This extensive transfer of protons produces many hydronium ions (the acidic species) within the solution. A **weak**

Table 14.1	Selected Common Mono-, Di-, and Triprotic Acids			
Name	Formula	Classification	Number of Nonacidic Hydrogen Atoms	Common Occurrence
Acetic acid	$HC_2H_3O_2$	monoprotic	three	vinegar
Lactic acid	$HC_3H_5O_3$	monoprotic	five	sour milk, cheese; produced during muscle contraction
Salicylic acid	$HC_7H_5O_3$	monoprotic	five	present in chemically combined form in aspirin
Hydrochloric acid	HCl	monoprotic	zero	constituent of gastric juice; industrial cleaning agent
Nitric acid	HNO_3	monoprotic	zero	used in urinalysis test for protein; used in manufacture of dyes and explosives
Tartaric acid	$H_2C_4H_4O_6$	diprotic	four	grapes
Carbonic acid	H_2CO_3	diprotic	zero	carbonated beverages; produced in the body from carbon dioxide
Sulfuric acid	H_2SO_4	diprotic	zero	storage batteries; manufacture of fertilizer
Citric acid	$H_3C_6H_5O_7$	triprotic	five	citrus fruits
Phosphoric acid	H_3PO_4	triprotic	zero	found in dissociated form (HPO_4^{2-}, $H_2PO_4^-$) in intracellular fluid; component of DNA

acid *is an acid that, in aqueous solution, transfers only a small percentage of its acidic hydrogen atoms to water.* The extent of proton transfer for weak acids is usually less than 5%. The actual percentage of molecules involved in proton transfer to water depends on the molecular structure of the acid; molecular polarity and the strength and polarity of individual bonds are important factors in determining whether an acid is strong or weak.

A graphical representation of the differences between strong and weak acids, in terms of species present in solution, is given in Figure 14.2. The formula HA represents the acid, and H_3O^+ and A^- are the products from the proton transfer to H_2O.

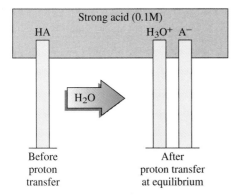

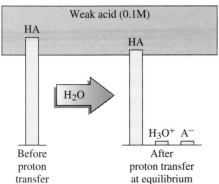

Figure 14.2

A comparison of the number of H_3O^+ ions (the acidic species) present in strong acid and weak acid solutions of the same concentration.

It is important not to confuse the terms *strong* and *weak* with the terms *concentrated* and *dilute*. *Strong* and *weak* apply to the *extent of proton transfer,* not to the concentration of acid or base. *Concentrated* and *dilute* are relative concentration terms. Stomach acid (gastric juice) is a dilute (not weak) solution of a strong acid (HCl); it is 5% by mass hydrochloric acid.

A 0.1-M solution of nitric acid (HNO_3) or sulfuric acid (H_2SO_4), when spilled on your clothes and not immediately washed off, will "eat" holes in your clothing. If 0.1-M solutions of either acetic acid ($HC_2H_3O_2$) or carbonic acid (H_2CO_3) were spilled on your clothes, the previously noted corrosive effects would not be observed. Why? All four acid solutions are of equal concentration; all are 0.1-M solutions. The difference in behavior relates to the *strength* of the acids; nitric and sulfuric acids are *strong* acids, whereas acetic and carbonic acids are *weak* acids. The number of H_3O^+ ions (the active species) present in the strong acid solutions is many times greater than for the weak acid solutions even though all the solutions had the same number of acid molecules present (before reaction with water).

There are very few strong acids; the formulas and structures of the seven most commonly encountered strong acids are given in Table 14.2. You should know the identity of these seven strong acids; you will need such knowledge to write net ionic equations, the topic of Section 14.7.

The vast majority of acids that exist are weak acids. Familiar weak acids include acetic acid ($HC_2H_3O_2$), the acidic component of vinegar, and carbonic acid (H_2CO_3), found in carbonated beverages. Weak acids are not all equally weak; proton transfer occurs to a greater extent for some weak acids than for others. Table 14.3 gives percent proton-transfer values for selected weak acids. The calculational techniques needed to determine percent proton-transfer values, such as those in Table 14.3, will not be considered in this text.

For polyprotic acids the stepwise proton-transfer sequence that occurs (Sec. 14.4) can be used to determine relative acid strengths for the related acidic species. Consider the two-step proton-transfer process for carbonic acid.

$$H_2CO_3(aq) + H_2O(l) \longrightarrow H_3O^+(aq) + HCO_3^-(aq)$$
$$HCO_3^-(aq) + H_2O(l) \longrightarrow H_3O^+(aq) + CO_3^{2-}(aq)$$

Table 14.2	Commonly Encountered Strong Acids	
Name[*]	**Molecular Formula**	**Molecular Structure**
Nitric acid	HNO_3	H—O—N—O (with =O below N)
Sulfuric acid	H_2SO_4	H—O—S—O—H (with O above and O below S)
Perchloric acid	$HClO_4$	H—O—Cl—O (with O above and O below Cl)
Chloric acid	$HClO_3$	H—O—Cl—O (with O below Cl)
Hydrochloric acid	HCl	H—Cl
Hydrobromic acid	HBr	H—Br
Hydroiodic acid	HI	H—I

[*]*Nomenclature for acids was discussed in Section 8.6.*

Table 14.3	Percent Proton-Transfer Values for 1.0 M Solutions (at 25°C) of Selected Weak Acids	
Name of Acid	**Molecular Formula**	**Percent Proton Transfer**
Phosphoric acid	H_3PO_4	8.3
Nitrous acid	HNO_2	2.7
Hydrofluoric acid	HF	2.5
Acetic acid	$HC_2H_3O_2$	0.42
Carbonic acid	H_2CO_3	0.065
Dihydrogen phosphate ion	$H_2PO_4^-$	0.025
Hydrocyanic acid	HCN	0.0020
Hydrogen carbonate ion	HCO_3^-	0.00075
Hydrogen phosphate ion	HPO_4^{2-}	0.000047

The second proton is not as easily removed as the first because it must be pulled away from a negatively charged particle, HCO_3^-. Accordingly, HCO_3^- is a weaker acid than H_2CO_3. In general, each successive step in a stepwise proton-transfer process occurs to a lesser extent than the previous step. Thus, for triprotic H_3PO_4, the parent H_3PO_4 species (first-step reactant) is a stronger acid than $H_2PO_4^-$ (second-step reactant), which in turn is a stronger acid than HPO_4^{2-} (third-step reactant). This ordering of the phosphoric acid–derived species is reflected in the values given in Table 14.3.

Just as there are strong acids and weak acids, there are also strong bases and weak bases. As with acids, there are only a few strong bases. Strong bases are limited to the hydroxides of groups IA and IIA of the periodic table and are listed in Table 14.4. Of the strong bases, only NaOH and KOH are commonly used in the chemical laboratory. The low solubility of the group IIA hydroxides in water limits their use. However, despite this low solubility, these hydroxides are still considered to be strong bases because whatever dissolves dissociates into ions 100%.

Only one of the many weak bases that exist is fairly common—aqueous ammonia. In this solution of ammonia gas (NH_3) in water, small amounts of OH^- ions are produced through the reaction of NH_3 molecules with water.

$$NH_3(g) + H_2O(l) \longrightarrow NH_4^+(aq) + OH^-(aq)$$

A solution of ammonia in water is most properly called *aqueous ammonia,* although it is also commonly called ammonium hydroxide, its commercial name. Aqueous ammonia is the preferred designation, since most of the NH_3 present is in molecular form. Only a very few NH_3 molecules have reacted with the water to give ammonium (NH_4^+) and hydroxide (OH^-) ions.

Table 14.4	Common Strong Bases
Group IA Hydroxides	**Group IIA Hydroxides**
LiOH	
NaOH	
KOH	$Ca(OH)_2$
RbOH	$Sr(OH)_2$
CsOH	$Ba(OH)_2$

14.6 Salts

The title of this chapter is "Acids, Bases, and Salts." In preceding sections, we have discussed acids and bases, but not salts. What is a salt? To a nonscientist the word *salt* connotes a white granular substance used as a seasoning for food. To the chemist it has a much broader meaning. Sodium chloride, or table salt, is only one of thousands of salts known to a chemist. "Pass the salt" is a very ambiguous request to a chemist.

From a chemical viewpoint, a **salt** *is an ionic compound containing a metal ion or polyatomic ion as the positive ion and a nonmetal ion or polyatomic ion (except hydroxide) as the negative ion.* (Ionic compounds containing hydroxide ion are bases rather than salts.)

Many salts occur in nature, and numerous others have been prepared in the laboratory. The wide variety of uses found for salts can be seen from Table 14.5, a listing of selected salts and their uses.

Much information concerning salts has been presented in previous chapters, although the term *salt* was not explicitly used in these discussions. Formula writing and nomenclature for binary ionic compounds (salts) were covered in Sections 7.7 and 8.3. Many salts, as shown in Table 14.5, contain polyatomic ions such as nitrate and sulfate. Such ions were discussed in Sections 7.9 and 8.4. The solubility of ionic compounds (salts) in water was the topic of Section 13.4.

In solution all common salts are dissociated into ions (Sec. 13.3). Even if a salt is only slightly soluble, the small amount that does dissolve completely dissociates. Thus, the terms weak and strong, used to denote qualitatively the percent dissociation of acids and bases, are not applicable to common salts. We do not use the terms *strong salt* and *weak salt.*

Acids, bases, and salts are related in that a salt is one of the products resulting from the reaction of an acid with a hydroxide base. This particular type of reaction, called neutralization, will be discussed in Section 14.8.

14.7 Ionic and Net Ionic Equations

Soluble strong acids, soluble strong bases, and soluble salts all dissociate 100% (or nearly 100%) in aqueous solution to produce ions (Secs. 14.5 and 14.6). It is extremely useful to discuss the reactions of such acids, bases, and salts in aqueous solution in terms of the ions present. This is most easily done using a new type of equation—a net ionic equation.

Table 14.5 Some Common Salts and Their Uses

Name	Formula	Uses
Ammonium nitrate	NH_4NO_3	fertilizer; explosives
Barium sulfate	$BaSO_4$	enhancer for X-rays of gastrointestinal tract
Calcium carbonate	$CaCO_3$	chalk; limestone
Calcium chloride	$CaCl_2$	drying agent for removal of small amounts of water
Iron(II) sulfate	$FeSO_4$	treatment for anemia
Potassium chloride	KCl	"salt" substitute for low-sodium diets
Sodium chloride	$NaCl$	table salt; used as a de-icer (to melt ice)
Sodium bicarbonate	$NaHCO_3$	ingredient in baking powder
Sodium hypochlorite	$NaClO$	bleaching agent
Silver bromide	$AgBr$	light-sensitive material in photographic film
Tin(II) fluoride	SnF_2	toothpaste additive

Up to this point in the text, most equations we have used have been *molecular equations,* equations where the complete formulas of all reactants and products are shown. Molecular equations are the starting point for deriving *ionic equations*, which in turn lead to *net ionic equations*. An **ionic equation** *is an equation in which the formulas of the predominant form of each compound in aqueous solution are used; dissociated/ionized compounds are written as ions, and undissociated compounds are written in molecular form.* Net ionic equations are derived from ionic equations. A **net ionic equation** *is an ionic equation from which nonparticipating (spectator) species have been eliminated.*

The differences between molecular, ionic, and net ionic equations can best be illustrated by examples. Let us consider the chemical reaction that results when a solution of potassium chloride (KCl) is mixed with a solution of silver nitrate $(AgNO_3)$. An insoluble salt, silver chloride (AgCl), is produced as a result of the mixing. The *molecular equation* for this reaction is

$$AgNO_3(aq) + KCl(aq) \longrightarrow KNO_3(aq) + AgCl(s)$$

Three of the four substances involved in this reaction—$AgNO_3$, KCl, and KNO_3—are soluble salts and thus exist in solution in ionic form. This is shown by writing the *ionic equation* for the reaction.

$$\underbrace{Ag^+(aq) + NO_3^-(aq)}_{AgNO_3 \text{ in ionic form}} + \underbrace{K^+(aq) + Cl^-(aq)}_{KCl \text{ in ionic form}} \longrightarrow \underbrace{K^+(aq) + NO_3^-(aq)}_{KNO_3 \text{ in ionic form}} + AgCl(s)$$

In this equation each of the three soluble salts is shown in dissociated (ionic) form rather than in undissociated form. A close look at this ionic equation shows that the potassium ions (K^+) and nitrate ions (NO_3^-) appear on both sides of the equation, indicating that they did not undergo any chemical change. In other words, they are *spectator ions*; they did not participate in the reaction. The *net ionic equation* for this reaction is written by dropping (canceling) all spectator ions from the ionic equation. (The spectator ions dropped must occur in equal numbers on both sides of the equation.) In our case, the net ionic equation becomes

$$Ag^+(aq) + Cl^-(aq) \longrightarrow AgCl(s)$$

This net ionic equation indicates that the product AgCl was formed by the reaction of silver ions (Ag^+) with chloride ions (Cl^-). It totally ignores the presence of those ions that are not taking part in the reaction. Thus, a net ionic equation focuses on only those species in a solution actually involved in a chemical reaction. It does not give all species present in the solution.

If you can write equations in molecular form, you will find it a straightforward process to convert these equations to net ionic form. Follow these three steps.

1. Check the given molecular equation to make sure that it is balanced.
2. Expand the molecular equation into an ionic equation.
3. Convert the ionic equation into a net ionic equation by eliminating spectator ions.

In expanding a molecular equation into an ionic equation (step 2), you must decide whether to write each reactant and product in dissociated (ionic) form or undissociated (molecular) form. The following rules serve as guidelines in making such decisions.

1. Soluble compounds that completely dissociate/ionize in aqueous solution are written in ionic form. They include
   (a) all soluble salts (see Sec. 13.4 for solubility rules)
   (b) all strong acids (see Table 14.2)
   (c) all strong bases (see Table 14.4)

2. Soluble weak acids and weak bases are written in molecular form, since they are incompletely dissociated/ionized in solution and thus exist predominantly in the undissociated/un-ionized form. The following are considered to be weak acids or weak bases.

   (a) All acids not listed in Table 14.2 as strong acids. Common examples are HNO_2, HF, H_2S, $HC_2H_3O_2$, H_2CO_3, and H_3PO_4.

   (b) All bases not listed in Table 14.4 as strong bases. Aqueous ammonia (NH_3) is the most common weak base.

3. All insoluble substances (solids, liquids, and gases), whether ionic or covalent, exist as molecules or neutral ionic units and are written as such.

4. All soluble covalent substances, for example, carbon dioxide (CO_2) or sucrose ($C_{12}H_{22}O_{11}$), are written in molecular form.

5. If water, the solvent, appears in the equation, it is written in molecular form.

Now, we apply these guidelines by writing some net ionic equations.

EXAMPLE 14.3

Converting a Molecular Equation into a Net Ionic Equation

Write the net ionic equation for the following aqueous solution reaction.

$$MgCl_2 + AgNO_3 \longrightarrow Mg(NO_3)_2 + AgCl \quad \text{(unbalanced equation)}$$

SOLUTION

STEP 1 To balance the given molecular equation, the coefficient 2 must be placed in front of both $AgNO_3$ and $AgCl$.

$$MgCl_2 + 2\,AgNO_3 \longrightarrow Mg(NO_3)_2 + 2\,AgCl$$

STEP 2 A decision must be made whether to write each reactant and product in ionic or molecular form. Let us consider them one by one.

$MgCl_2$: This compound is a salt. The solubility rules indicate that it is soluble. Thus, $MgCl_2$ will be written in ionic form: $Mg^{2+} + 2\,Cl^-$. Note that three ions (one Mg^{2+} ion and two Cl^- ions) are produced from the dissociation of one $MgCl_2$ unit.

$AgNO_3$: This compound is also a soluble salt; it will be written in ionic form. Each $AgNO_3$ unit (there are two) produces one Ag^+ ion and one NO_3^- ion.

$Mg(NO_3)_2$: All nitrate salts are soluble. Thus, $Mg(NO_3)_2$ will be written in ionic form. Three ions are produced upon dissociation of one $Mg(NO_3)_2$ unit: one Mg^{2+} and two NO_3^-.

$AgCl$: The solubility rules indicate that this compound is an insoluble salt. Thus, it will be written in molecular form in the ionic equation.

The ionic equation will have the form

$$Mg^{2+} + 2\,Cl^- + 2\,Ag^+ + 2\,NO_3^- \longrightarrow Mg^{2+} + 2\,NO_3^- + 2\,AgCl$$

Note how the coefficient 2 in front of $AgNO_3$ and $AgCl$ in the molecular equation affects the ionic equation. The dissociation of two $AgNO_3$ units produces two Ag^+ ions and two NO_3^- ions. Similarly, two $AgCl$ units are present.

STEP 3 Inspection of the ionic equation shows that Mg^{2+} ion and the two NO_3^- ions are spectator ions. Cancellation of these ions from the equation will give the net ionic equation.

$$Mg^{2+} + 2\,Cl^- + 2\,Ag^+ + 2\,\overline{NO_3^-} \longrightarrow \overline{Mg^{2+}} + 2\,\overline{NO_3^-} + 2\,AgCl$$

$$2\,Ag^+ + 2\,Cl^- \longrightarrow 2\,AgCl$$

The coefficients in the net ionic equation should be the smallest set of numbers that correctly balance the equation. In this case, all the coefficients are divisible by 2. Dividing by 2, we get

$$Ag^+ + Cl^- \longrightarrow AgCl$$

Practice Exercise 14.3

Write the net ionic equation for the following aqueous solution reaction.

$$KCl + AgC_2H_3O_2 \longrightarrow KC_2H_3O_2 + AgCl$$

EXAMPLE 14.4

Converting a Molecular Equation into a Net Ionic Equation

Write the net ionic equation for the following aqueous solution reaction.

$$H_2S + AlI_3 \longrightarrow Al_2S_3 + HI \quad \text{(unbalanced equation)}$$

SOLUTION

STEP 1 Balancing the molecular equation, we get

$$3\,H_2S + 2\,AlI_3 \longrightarrow Al_2S_3 + 6\,HI$$

STEP 2 The expansion of the molecular equation into an ionic equation is accomplished by the following analysis.

H_2S: This is a weak acid. All weak acids are written in molecular form in ionic equations.

AlI_3: This is a soluble salt. All iodide salts are soluble, with three exceptions; this is not one of the exceptions. Soluble salts are written in ionic form in ionic equations.

Al_2S_3: This is an insoluble salt. All sulfides are insoluble except for groups IA and IIA and NH_4^+. Thus, Al_2S_3 will remain in molecular form in the ionic equation.

HI: This compound is an acid. It is one of the seven strong acids listed in Table 14.2. Strong acids are written in ionic form.

The ionic equation for the reaction is

$$3\,H_2S + 2\,Al^{3+} + 6\,I^- \longrightarrow Al_2S_3 + 6\,H^+ + 6\,I^-$$

Note again that the coefficients present in the balanced molecular equation must be taken into consideration when determining the total number of ions produced from dissociation. On dissociation an AlI_3 unit produces four ions: one Al^{3+} ion and three I^- ions. This number must be doubled for the ionic equation because AlI_3 carries the coefficient 2 in the balanced molecular equation. Similar considerations apply to HI in this equation.

STEP 3　Inspection of the ionic equation shows that only I^- ions (six of them) are spectator ions. Cancellation of these ions from the equation will give the net ionic equation.

$$3\,H_2S + 2\,Al^{3+} + \cancel{6\,I^-} \longrightarrow Al_2S_3 + 6\,H^+ + \cancel{6\,I^-}$$

$$3\,H_2S + 2\,Al^{3+} \longrightarrow Al_2S_3 + 6\,H^+$$

Practice Exercise 14.4

Write the net ionic equation for the following aqueous solution reaction.

$$H_2CO_3 + MgCl_2 \longrightarrow MgCO_3 + HCl \quad \text{(unbalanced equation)}$$

EXAMPLE 14.5

Converting a Molecular Equation into a Net Ionic Equation

Write the net ionic equation for the following aqueous solution reaction.

$$HNO_3 + LiOH \longrightarrow LiNO_3 + H_2O$$

SOLUTION

STEP 1　All coefficients in this equation are 1; the equation is balanced as written.

$$HNO_3 + LiOH \longrightarrow LiNO_3 + H_2O$$

STEP 2　The expansion of the molecular equation into an ionic equation is based on the following analysis.

HNO_3:　This compound is an acid. It is one of the seven strong acids listed in Table 14.2. Strong acids are written in ionic form in ionic equations.

$LiOH$:　This compound is a base. It is one of the strong bases listed in Table 14.4. Strong bases are written in ionic form in ionic equations.

$LiNO_3$:　This compound is a soluble salt. All nitrate salts are soluble. Thus, $LiNO_3$ is written in ionic form.

H_2O:　This compound is a covalent compound; two nonmetals are present. Covalent compounds are always written in molecular form.

The ionic equation for the reaction, using the above information, is

$$H^+ + NO_3^- + Li^+ + OH^- \longrightarrow Li^+ + NO_3^- + H_2O$$

STEP 3　Inspection of the ionic equation shows that NO_3^- ions and Li^+ ions are spectator ions. Cancellation of these ions from the equation gives the net ionic equation.

$$H^+ + \cancel{NO_3^-} + \cancel{Li^+} + OH^- \longrightarrow \cancel{Li^+} + \cancel{NO_3^-} + H_2O$$

$$H^+ + OH^- \longrightarrow H_2O$$

Practice Exercise 14.5

Write the net ionic equation for the following aqueous solution reaction.

$$HNO_3 + Ba(OH)_2 \longrightarrow Ba(NO_3)_2 + H_2O \quad \text{(unbalanced equation)}$$

14.8 Reactions of Acids

All acids have some unique properties that adapt them for use in specific situations. In addition, all acids have certain chemical properties in common, properties related to the presence of H_3O^+ ions in aqueous solution. In this section we consider three types of chemical reactions that acids characteristically undergo.

1. Acids react with active metals to produce hydrogen gas and a salt.
2. Acids react with hydroxide bases to produce a salt and water.
3. Acids react with carbonates and bicarbonates to produce carbon dioxide, a salt, and water.

Reaction with Metals

Acids react with many, but not all, metals. When they do react, the metal dissolves and hydrogen gas (H_2) is liberated. In the reaction the metal atoms lose electrons and become metal ions. The lost electrons are taken up by the hydrogen ions (protons) of the acid; the hydrogen ions become electrically neutral, combine into molecules, and emerge from the reaction mixture as hydrogen gas. Illustrative of the reaction of an acid and a metal is the reaction between zinc and sulfuric acid.

$$\text{molecular equation:} \quad Zn + H_2SO_4 \longrightarrow ZnSO_4 + H_2$$
$$\text{net ionic equation:} \quad Zn + 2\,H^+ \longrightarrow Zn^{2+} + H_2$$

In terms of the reaction types discussed in Section 10.5, the reaction of an acid with a metal to produce hydrogen gas is a *single-replacement reaction;* the metal replaces the hydrogen from the acid. Recall, from Section 10.5, that a single-replacement reaction has the general form

$$X + YZ \longrightarrow Y + XZ$$

Metals can be arranged in a reactivity order based on their ability to react with acids. Such an ordering for the more common metals is given in Table 14.6. Any metal above hydrogen in the activity series will dissolve in an acid solution and form H_2. The closer a metal is to the top of the series, the more rapid the reaction. Those metals below hydrogen in the series do not dissolve in an acid to form H_2.

As noted in Table 14.6, the most active metals (those nearest the top in the activity series) also react with water. Again, hydrogen gas is produced. In the cases of potassium and sodium, the reaction is sometimes violent enough to cause explosions as the result of H_2 ignition. The equation for the reaction of potassium with water, which is also a single-replacement reaction, is

$$\text{molecular equation:} \quad 2\,K + 2\,H_2O \longrightarrow 2\,KOH + H_2$$
$$\text{net ionic equation:} \quad 2\,K + 2\,H_2O \longrightarrow 2\,K^+ + 2\,OH^- + H_2$$

Note that the resulting solution is basic when a metal reacts with water; hydroxide ions are produced.

Reaction with Bases

When Arrhenius acids and bases are mixed, they react with each other; their acidic and basic properties disappear, and we say that they have *neutralized* each other. **Neutralization** *is the reaction between an acid and a hydroxide base to form a salt and water.* The hydrogen ions from the acid combine with the hydroxide ions from the base to form water. The salt formed

Because they do not react with the acidic components of skin secretions (sweat, etc.), gold, platinum, and silver are good metals for jewelry. Jewelry made with these metals will not tarnish like jewelry made from cheaper metals.

Hydrochloric acid (HCl), which is necessary for proper digestion of food, is present in the gastric juices of the human stomach. Overeating and emotional factors can cause the stomach to produce too much hydrochloric acid, a condition called acid indigestion or heartburn. Substances known as *antacids* provide symptomatic relief from this condition. Over-the-counter antacids such as Maalox, Tums, and Alka-Seltzer contain one or more *basic* substances [often $Mg(OH)_2$] that are capable of neutralizing the excess hydrochloric acid present. The neutralization reaction that occurs when stomach acid reacts with a typical antacid is

$$2\,HCl + Mg(OH)_2 \longrightarrow MgCl_2 + 2\,H_2O$$

Table 14.6 Activity Series for Common Metals

Increasing tendency to react ↑		Metal	Symbol	Remarks
	React with H^+ ions to liberate hydrogen gas	Potassium	K	react violently with cold water
		Sodium	Na	
		Calcium	Ca	reacts slowly with cold water
		Magnesium	Mg	react slowly with hot water (steam)
		Aluminum	Al	
		Zinc	Zn	
		Chromium	Cr	
		Iron	Fe	
		Nickel	Ni	
		Tin	Sn	
		Lead	Pb	
		Hydrogen	H	
	Do not react with H^+ ions	Copper	Cu	
		Mercury	Hg	
		Silver	Ag	
		Platinum	Pt	
		Gold	Au	

contains the negative ion from the acid and the positive ion from the base. Neutralization is a *double-replacement* reaction (Sec. 10.5).

$$AX + BY \longrightarrow AY + BX$$
$$\underset{\text{acid}}{HCl} + \underset{\text{base}}{KOH} \longrightarrow \underset{\text{water}}{HOH} + \underset{\text{salt}}{KCl}$$

Any time an acid is completely reacted with a base, neutralization occurs. It does not matter whether the acid and base are strong or weak. Sodium hydroxide (a strong base) and nitric acid (a strong acid) react as follows.

$$\text{molecular equation:} \quad HNO_3 + NaOH \longrightarrow NaNO_3 + H_2O$$
$$\text{net ionic equation:} \quad H^+ + OH^- \longrightarrow H_2O$$

The equations for the reaction of potassium hydroxide (a strong base) with hydrocyanic acid (a weak acid) are

$$\text{molecular equation:} \quad HCN + KOH \longrightarrow KCN + H_2O$$
$$\text{net ionic equation:} \quad HCN + OH^- \longrightarrow CN^- + H_2O$$

Note that in each case the products are a salt ($NaNO_3$ in the first reaction, KCN in the second) and water. Note also that the net ionic equations for the two neutralization reactions are different. In the second set of equations the acid must remain written in molecular form because it is a weak acid.

In any acid–base neutralization reaction, the amounts of H^+ ion and OH^- ion that react are equal. These two ions always react in a one-to-one ratio to form water.

$$H^+ + OH^- \longrightarrow H_2O\ (HOH)$$

This constant reaction ratio between the two ions enables us to balance chemical equations for neutralization reactions quickly.

Let us consider the neutralization reaction between H_3PO_4 and KOH to give a salt and water.

$$H_3PO_4 + KOH \longrightarrow salt + H_2O$$

Because H_3PO_4 is triprotic and the base KOH contains only one OH^- ion, we will need three times as many base molecules as acid molecules. Thus we place the coefficient 3 in front of the formula for KOH in the chemical equations; this gives three H^+ reacting with three OH^- to produce three H_2O molecules.

$$H_3PO_4 + 3 KOH \longrightarrow salt + 3 H_2O$$

The salt formed is K_3PO_4; there are three K^+ ions and one $PO_4{}^{3-}$ ion on the left side of the equation, which combine to give the salt. The balanced equation for the neutralization is thus

$$H_3PO_4 + 3 KOH \longrightarrow K_3PO_4 + 3 H_2O$$

Reaction with Carbonates and Bicarbonates

Carbon dioxide gas (CO_2), water, and a salt are always the products of the reaction of acids with carbonates or bicarbonates, as illustrated by the following equations.

molecular equation: $\quad 2 HCl + Na_2CO_3 \longrightarrow 2 NaCl + CO_2 + H_2O$

molecular equation: $\quad\; HCl + NaHCO_3 \longrightarrow NaCl + CO_2 + H_2O$

Baking powder is a mixture of a bicarbonate and an acid-forming solid. The addition of water to this mixture generates the acid that then reacts with the bicarbonate to release carbon dioxide into the batter. It is the generated carbon dioxide that causes the batter to rise. Baking soda is pure $NaHCO_3$. To cause it to release carbon dioxide, an acid-containing substance, such as buttermilk, sour milk, or fruit juice, must be added to it.

> **The reaction equation for the "volcanoes" that children enjoy making by mixing vinegar [5% acetic acid (v/v)] and baking soda is**
> $HC_2H_3O_2 + NaHCO_3 \longrightarrow$
> $CO_2 + H_2O + NaC_2H_3O_2.$
> **The carbon dioxide gas makes the foamy "volcano."**

14.9 Reactions of Bases

The most important characteristic reaction of bases is their reaction with acids (neutralization), discussed in the preceding section. Another characteristic reaction, that of bases with certain salts, is discussed in Section 14.10.

Bases react with fats and oils and convert them into smaller, soluble molecules. For this reason most household cleaning products contain basic substances. Lye (impure NaOH) is an active ingredient in numerous drain cleaners. Also, many advertisements for liquid household cleaners emphasize the fact that aqueous ammonia (a weak base) is present in the product.

In Section 14.1 we noted that one of the general properties of bases is a "slippery" or "soapy" feeling to the touch. The bases themselves are not slippery; the slipperiness results as the bases react with fats and oils in the skin to form "slippery" or "soapy" compounds.

14.10 Reactions of Salts

Dissolved salts will react with metals, acids, bases, and other salts under specific conditions.

1. Salts react with some metals to convert the metallic ion of the salt to free metal and the free metal to its salt.

2. Salts react with some acid solutions to form other acids and salts.

3. Salts react with some base solutions to form other bases and salts.
4. Salts react with some solutions of other salts to form new salts.

The tendency for salts to react with metals is related to the relative positions of the two involved metals in the activity series (Table 14.6). For salts to react with acids, bases, or other salts, one of the reaction products must be (1) an insoluble salt, (2) a gas that is evolved from the solution, or (3) an undissociated soluble species, such as a weak acid or a weak base. The formation of any of these products serves as the driving force to cause the reaction to occur.

Reaction with Metals

If an iron nail is placed in a solution of copper sulfate $(CuSO_4)$, metallic copper will be deposited on the nail and some of the iron will dissolve.

$$\text{molecular equation:} \quad Fe(s) + CuSO_4(aq) \longrightarrow Cu(s) + FeSO_4(aq)$$

$$\text{net ionic equation:} \quad Fe + Cu^{2+} \longrightarrow Cu + Fe^{2+}$$

One metal has replaced the other: a single-replacement reaction (Sec. 10.5) has occurred. This type of reaction will occur only if the metal going into solution is above the replaced metal in the activity series. Iron is above copper in the activity series and can replace it. If a strip of copper were placed in a solution of $FeSO_4$—just the opposite situation to what we have been discussing—no reaction would occur because copper is below iron in the activity series.

Reaction with Acids

For a salt to react with an acid, a new weaker acid, a new insoluble salt, or a gaseous compound must be one of the products.

An example of a reaction in which the formation of an *insoluble salt* is the driving force for the reaction to occur is

$$\text{molecular equation:} \quad AgNO_3(aq) + HCl(aq) \longrightarrow AgCl(s) + HNO_3(aq)$$

$$\text{net ionic equation:} \quad Ag^+ + Cl^- \longrightarrow AgCl$$

This is a double-replacement reaction (Sec. 10.5); the silver and hydrogen have traded partners.

The conclusion that this reaction will occur comes from a consideration of the possible recombinations of the reacting species. In a solution made by mixing silver nitrate $(AgNO_3)$ and hydrochloric acid (HCl), four kinds of ions are present initially (before any reaction occurs): Ag^+ and NO_3^- (since $AgNO_3$ is a soluble salt) and H^+ and Cl^- (since HCl is a strong acid). The question is whether these ions can get together in new appropriate combinations. The possible new combinations of oppositely charged ions are $H^+NO_3^-$ and Ag^+Cl^-. The first of these combinations would result in the formation of the strong acid HNO_3. Strong acids in solution exist in dissociated form; therefore, these ions will not combine. The second combination does occur because AgCl is an insoluble salt. Thus, the overall reaction takes place as a result of the formation of this insoluble salt. The net result of the reaction is that the original ions exchange partners.

The double-replacement reaction of sodium fluoride (a soluble salt) with hydrochloric acid (a strong acid) illustrates the case where formation of a *new weaker acid* is the driving force for the reaction.

$$\text{molecular equation:} \quad HCl(aq) + NaF(aq) \longrightarrow NaCl(aq) + HF(aq)$$

$$\text{net ionic equation:} \quad H^+ + F^- \longrightarrow HF$$

Using an analysis pattern similar to that in the previous example, we find that four types of ions are present initially: Na^+ and F^- (from the soluble salt) and H^+ and Cl^- (from the strong acid). Possible new combinations are Na^+Cl^- and H^+F^-. Sodium chloride, the result of the first combination, will not form because this salt is soluble. The combination of H^+ ion with F^- ion does occur because it yields the weak acid HF. In solution weak acids exist predominantly in molecular form. In all reactions of this general type, the acid formed in the reaction must be weaker than the reactant acid. If the reactant acid is strong, as in this example, such a determination is obvious. If both the reactant and product acids are weak, information such as that given in Table 14.3 would be needed to predict which of the two acids is the weaker.

The most common type of reaction in which the driving force is the *evolution of a gas* involves a carbonate or bicarbonate. This type of reaction was discussed in Section 14.8.

Note that a reaction does not always occur when acid and salt solutions are mixed. Consider the possible reaction of NaCl and HNO_3 solutions. Initially, four types of ions are present: Na^+ and Cl^- (from the soluble salt) and H^+ and NO_3^- (from the strong acid). The new combinations, if a reaction did occur, would be $Na^+NO_3^-$ (a soluble salt) and H^+Cl^- (a strong acid). Since both the products would exist in dissociated form in solution, no recombination of ions occurs; hence, no reaction occurs.

Reaction with Bases

The criteria for the reaction of bases with salts are similar to those for acid–salt reactions, except that weaker base formation replaces weaker acid formation as one of the three driving forces. An example of a base–salt reaction involving the formation of an *insoluble salt* is

molecular equation: $\quad Ba(OH)_2(aq) + Na_2SO_4(aq) \longrightarrow BaSO_4(s) + 2\,NaOH(aq)$

net ionic equation: $\quad\quad\quad Ba^{2+} + SO_4^{2-} \longrightarrow BaSO_4$

The most common situation in which *gas evolution* is the driving force for base–salt reactions is where ammonium salts are involved. In such cases, ammonia gas is given off, as illustrated by the reaction of NH_4Cl and KOH.

molecular equation: $\quad NH_4Cl(aq) + KOH(aq) \longrightarrow KCl(aq) + NH_3(g) + H_2O(l)$

net ionic equation: $\quad\quad\quad NH_4^+ + OH^- \longrightarrow NH_3 + H_2O$

Reaction of Salts with Each Other

Two different salt solutions will react when mixed, in a double-replacement reaction, only if an *insoluble salt* can be formed. Consider the following possible reactions.

molecular equation: $\quad AgNO_3(aq) + NaCl(aq) \longrightarrow AgCl(s) + NaNO_3(aq)$

molecular equation: $\quad KNO_3(aq) + NaCl(aq) \longrightarrow KCl(aq) + NaNO_3(aq)$

The first reaction occurs because AgCl is an insoluble salt. The second reaction does not occur since both of the possible products are soluble salts, which means there is no driving force for the reaction.

EXAMPLE 14.6

Predicting Whether a Reaction Will Occur and Writing a Net Ionic Equation for the Reaction If It Does Occur

Write molecular, ionic, and net ionic equations for the reaction that occurs, if any, when 0.1-M solutions of the following substances are mixed.

(a) $Fe(NO_3)_2$ and K_2S

(b) $CaCl_2$ and H_2SO_4

(c) HNO_3 and $NaC_2H_3O_2$

SOLUTION

(a) Both the reactants are soluble salts. Two different salt solutions will react when mixed only if an insoluble salt can be formed.

In a solution made by mixing $Fe(NO_3)_2$ and K_2S, four kinds of ions are present initially (before any reaction occurs): Fe^{2+} and NO_3^- [from the $Fe(NO_3)_2$] and K^+ and S^{2-} (from the K_2S). The possible new combinations of oppositely charged ions are Fe^{2+} with S^{2-} and K^+ with NO_3^-.

Original ion combinations   *Possible new combinations*

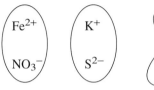 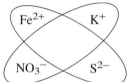

The first one of these new combinations, the formation of FeS, is the one that will be the driving force for the reaction to occur. FeS is an insoluble salt. The second new combination, the formation of KNO_3, does not occur because KNO_3 is a soluble salt and soluble salts exist in dissociated form in solution. The equations for the reaction are

Molecular: $$Fe(NO_3)_2 + K_2S \longrightarrow FeS + 2\,KNO_3$$

Ionic: $$Fe^{2+} + 2\,\cancel{NO_3^-} + 2\,\cancel{K^+} + S^{2-} \longrightarrow FeS + 2\,\cancel{K^+} + 2\,\cancel{NO_3^-}$$

Net ionic: $$Fe^{2+} + S^{2-} \longrightarrow FeS$$

(b) One of the reactants, $CaCl_2$, is a soluble salt, and the other reactant, H_2SO_4, is a strong acid. Both reactants exist in solution in dissociated form; thus, four types of ions are present in the mixed solution (before any reaction occurs): Ca^{2+}, Cl^-, H^+, and SO_4^{2-}. The conclusion that a reaction will occur comes from a consideration of the possible new combinations of the reacting species.

Original ion combinations   *Possible new combinations*

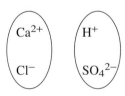 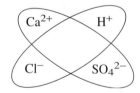

The first one of these new combinations, the formation of $CaSO_4$, is the driving force for the reaction to occur; $CaSO_4$ is an insoluble salt. The other new combination, the formation of HCl, does not occur because HCl is a strong acid and will exist in solution in dissociated form. The equations for the reaction are

Molecular: $$CaCl_2 + H_2SO_4 \longrightarrow CaSO_4 + 2\,HCl$$

Ionic: $$Ca^{2+} + 2\,\cancel{Cl^-} + 2\,\cancel{H^+} + SO_4^{2-} \longrightarrow CaSO_4 + 2\,\cancel{H^+} + 2\,\cancel{Cl^-}$$

Net ionic: $$Ca^{2+} + SO_4^{2-} \longrightarrow CaSO_4$$

(c) The reactants are a strong acid (HNO_3) and a soluble salt $(NaC_2H_3O_2)$. Both are dissociated in solution; hence, H^+, NO_3^-, Na^+, and $C_2H_3O_2^-$ ions are present in the reaction mixture (before any reaction occurs). Possible new combinations of the reacting species are

Original ion combinations   *Possible new combinations*

H^+ Na^+ H^+ Na^+

NO_3^- $C_2H_3O_2^-$ NO_3^- $C_2H_3O_2^-$

Weak acid formation is the driving force for the reaction; acetic acid $(HC_2H_3O_2)$ forms from the combination of H^+ and $C_2H_3O_2^-$ ions. The Na^+ and NO_3^- will not combine because a soluble salt would be the product. The equations for the reaction are

Molecular: $HNO_3 + NaC_2H_3O_2 \longrightarrow HC_2H_3O_2 + NaNO_3$

Ionic: $H^+ + \cancel{NO_3^-} + \cancel{Na^+} + C_2H_3O_2^- \longrightarrow HC_2H_3O_2 + \cancel{Na^+} + \cancel{NO_3^-}$

Net ionic: $H^+ + C_2H_3O_2^- \longrightarrow HC_2H_3O_2$

Practice Exercise 14.6

Write molecular, ionic, and net ionic equations for the reaction that occurs, if any, when 0.1-M solutions of the following substances are mixed.

(a) $Fe(NO_3)_3$ and Na_3PO_4

(b) $CaCl_2$ and HNO_3

(c) HCl and Na_2S

14.11   Self-Ionization of Water

Although we usually think of water as a molecular (covalent) substance, experiments show that a *small* percentage of water molecules in pure water interact with one another to form ions, a process that is called *self-ionization*. This interaction can be thought of as a Brønsted–Lowry acid–base reaction (Sec. 14.2) involving the transfer of a proton from one water molecule to another (see Figure 14.3):

$$H_2O + H_2O \longrightarrow H_3O^+ + OH^-$$

or simply as the formation of ions from a single water molecule (Arrhenius theory; Sec. 14.1):

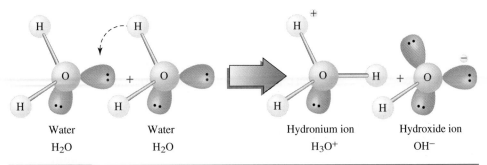

| Water | Water | Hydronium ion | Hydroxide ion |
| H_2O | H_2O | H_3O^+ | OH^- |

Figure 14.3

Brønsted–Lowry acid–base reaction between two water molecules to produce hydronium ion and hydroxide ion.

$$H_2O \, (HOH) \longrightarrow H^+ + OH^-$$

From either viewpoint, the net result is the formation of *equal amounts* of hydronium (hydrogen) ion and hydroxide ion.

The dissociation of water molecules is part of an equilibrium situation. Individual water molecules are continually dissociating. This process is balanced by hydroxide and hydronium ions recombining to form water at the same rate. At equilibrium, at 25°C, the H_3O^+ and OH^- ion concentrations are each 1.00×10^{-7} M (0.000000100 M). This very, very small concentration is equivalent to there being one H_3O^+ and one OH^- ion present for every 550,000,000 undissociated water molecules. Even though the H_3O^+ and OH^- ion concentrations are very minute, they are important, as we shall shortly see.

Ion Product Constant for Water

Experimentally it is found that, at any given temperature, the product of the concentrations of H_3O^+ ion and OH^- ion in water is a constant. We can calculate the value of this constant at 24°C, since we know that the concentration of each ion is 1.00×10^{-7} M at this temperature. The brackets [ ] specifically denote ion concentration in moles per liter.

$$[H_3O^+] \times [OH^-] = \text{constant}$$
$$(1.00 \times 10^{-7}) \times (1.00 \times 10^{-7}) = 1.00 \times 10^{-14}$$

Ion product constant for water *is the numerical value* (1.00×10^{-14}) *associated with the product of the* H_3O^+ *ion and* OH^- *ion molar concentrations in water.* Note that the ion concentrations must be expressed in moles per liter (M) in order to obtain the value 1.00×10^{-14} for the ion product for water. The general expression for the ion product constant for water is

$$[H_3O^+] \times [OH^-] = \text{ion product for water} = 1.00 \times 10^{-14}$$

The ion product constant expression for water is valid not only in pure water but also when solutes are present in the water. At all times, the product of the hydronium and hydroxide ion molarities in an aqueous solution, at 24°C, must equal 1.00×10^{-14}. Thus if the $[H_3O^+]$ is increased by the addition of an acidic solute, the $[OH^-]$ must decrease until the expression

$$[H_3O^+] \times [OH^-] = 1.00 \times 10^{-14}$$

is satisfied. Similarly, if OH^- ions are added to the water, the $[H_3O^+]$ must correspondingly decrease. The extent of the decrease in $[H_3O^+]$ or $[OH^-]$, as the result of the addition of a quantity of the other ion, is easily calculated by the ion product expression.

If we know the concentration of the other ion, we can calculate the concentration of either H_3O^+ or OH^- present in an aqueous solution by simply rearranging the ion product expression.

$$[H_3O^+] = \frac{1.00 \times 10^{-14}}{[OH^-]} \quad \text{or} \quad [OH^-] = \frac{1.00 \times 10^{-14}}{[H_3O^+]}$$

EXAMPLE 14.7

Calculating the Hydroxide Ion Concentration of a Solution with a Known Hydronium Ion Concentration

Sufficient acidic solute is added to a quantity of water to produce $[H_3O^+] = 7.50 \times 10^{-5}$. What is the $[OH^-]$ in this solution?

SOLUTION

The [OH⁻] can be calculated using the ion product constant expression for water. Solving this expression for [OH⁻] gives

$$[OH^-] = \frac{1.00 \times 10^{-14}}{[H_3O^+]}$$

Substituting into this expression the known [H_3O^+] and doing the arithmetic gives

$$[OH^-] = \frac{1.00 \times 10^{-14}}{7.50 \times 10^{-5}} = 1.3333333 \times 10^{-10} \quad \text{(calculator answer)}$$

$$= 1.33 \times 10^{-10} \quad \textbf{(correct answer)}$$

Practice Exercise 14.7

Sufficient acidic solute is added to a quantity of water to produce [H_3O^+] = 4.50×10^{-2}. What is the [OH⁻] in this solution?

The relationship between [H_3O^+] and [OH⁻] is that of an inverse proportion; when one increases, the other decreases. If [H_3O^+] increases by a factor of 10^2, the [OH⁻] decreases by the same factor, 10^2. A graphical portrayal of this increase–decrease relationship for [H_3O^+] and [OH⁻] is given in Figure 14.4.

> **Neither [H_3O^+] nor [OH⁻] is ever zero in an aqueous solution.**

Acidic, Basic, and Neutral Solutions

Small amounts of both H_3O^+ ion and OH^- ion are present in all aqueous solutions. What, then, determines whether a given solution is acidic or basic? It is the relative amounts of

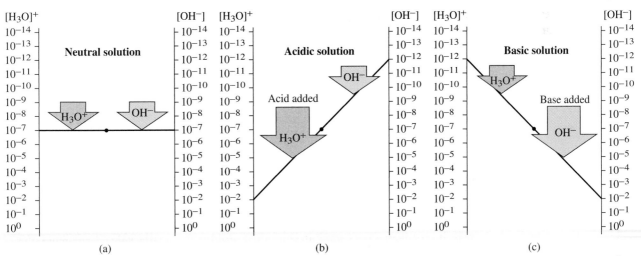

(a) In pure water the concentration of hydronium ions, [H_3O^+], and that of hydroxide ions, [OH⁻], are equal. Both are 1.00×10^{-7} M at 24°C.

(b) If [H_3O^+] is increased by a factor of 10^5 (from 10^{-7} M to 10^{-2} M), then [OH⁻] is decreased by a factor of 10^5 (from 10^{-7} M to 10^{-12} M).

(c) If [OH⁻] is increased by a factor of 10^5 (from 10^{-7} M to 10^{-2} M), then [H_3O^+] is decreased by a factor of 10^5 (from 10^{-7} M to 10^{-12} M).

Figure 14.4

The relationship between [H_3O^+] and [OH⁻] in aqueous solution is an inverse proportion; when [H_3O^+] is increased, [OH⁻] decreases, and vice versa.

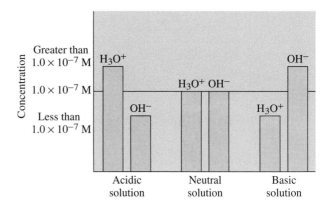

Figure 14.5

Relative molar concentrations of H_3O^+ and OH^- ions at 24°C in acidic, neutral, and basic solutions.

these two ions present. An **acidic solution** *is a solution in which the concentration of H_3O^+ ion is greater than that of OH^- ion.* Acids are substances that, when added to water, increase the concentration of H_3O^+ ion. A **basic solution** *is a solution in which the concentration of OH^- ion is greater than that of H_3O^+ ion.* Bases are substances that, when added to water, increase the concentration of OH^- ion. A **neutral solution** *is a solution in which the concentrations of H_3O^+ and OH^- ions are equal.* Figure 14.5 summarizes the relationships between $[H_3O^+]$ and $[OH^-]$ that we have just considered.

> A basic solution is also often referred to as an *alkaline solution.*

14.12 The pH Scale

Hydronium ion concentrations in aqueous solution range from relatively high values (10 M) to extremely small ones (10^{-14} M). It is inconvenient to work with numbers that extend over such a wide range; a hydronium ion concentration of 10 M is 1000 trillion times larger than a hydronium ion concentration of 10^{-14} M. The *pH scale*, proposed by the Danish chemist Sören Peter Lauritz Sörensen (1868–1939) in 1909, is a more practical way to handle such a wide range of numbers. The **pH scale** *is a scale of small numbers that is used to specify molar hydronium ion concentration in an aqueous solution.*

The calculation of pH scale values involves the use of logarithms. The **pH** *of a solution is the negative logarithm of the solution's molar hydronium ion concentration.* Expressed mathematically, the definition of pH is

$$pH = -\log [H_3O^+]$$

Logarithms are simply exponents. The *common logarithm,* abbreviated *log,* which is the type of logarithm used in the definition of pH, is based on powers of 10. *For a number expressed in scientific notation that has a coefficient of 1, the log of that number is the value of the exponent.* For instance, the log of 1×10^{-8} is -8.0, and the log of 1×10^6 is 6.0. Table 14.7 gives more examples of the relationship between powers of 10 and logarithmic values for numbers in scientific notation whose coefficients are 1. Note from this table that log values may be either positive or negative depending on the sign of the exponent. (How we determine significant figures in logarithmic calculations is discussed later in this section.)

Integral pH Values

It is easy to calculate the pH value for a solution when the molar hydronium ion concentration is an exact power of 10, for example, 1×10^{-4}. In this situation the pH is given directly by the negative of the exponent value on the power of 10.

Table 14.7	Logarithm Values for Selected Numbers	
Number	**Number Expressed as a Power of 10**	**Common Logarithm**
10,000	1×10^4	4.0
1,000	1×10^3	3.0
100	1×10^2	2.0
10	1×10^1	1.0
1	1×10^0	0.0
0.1	1×10^{-1}	-1.0
0.01	1×10^{-2}	-2.0
0.001	1×10^{-3}	-3.0
0.0001	1×10^{-4}	-4.0

$$[H_3O^+] = 1 \times 10^{-x}$$
$$pH = x$$

Thus, if the hydronium ion concentration is 1×10^{-9}, the pH will be 9.0.

This simple relationship between pH and power of 10 is obtained from the formal definition of pH as follows.

$$pH = -\log[H_3O^+]$$
$$= -\log[1 \times 10^{-x}]$$
$$= -(-x)$$
$$= x$$

Again, it should be noted that this simple relationship is valid only when the coefficient in the exponential expression for the hydronium ion concentration is 1. How the pH is calculated when the coefficient is not 1 will be covered later in this section.

> The *p* in pH comes from the German word *potenz*, which means "power," as in "power of 10."

EXAMPLE 14.8

Calculating the pH of a Solution When Given Its Hydronium Ion or Hydroxide Ion Concentration

Calculate the pH for each of the following solutions.

(a) $[H_3O^+] = 1 \times 10^{-3}$ **(b)** $[H_3O^+] = 1 \times 10^{-9}$

(c) $[OH^-] = 1 \times 10^{-4}$

SOLUTION

(a) Let us use the formal definition of pH in obtaining this first pH value.

$$pH = -\log[H_3O^+]$$

This expression indicates that to obtain a pH we must first take the logarithm of the molar hydronium ion concentration and then change the sign of that logarithm.

The logarithm of 1×10^{-3} is -3.0. Thus, we have

$$pH = -\log(1 \times 10^{-3})$$
$$= -(-3.0)$$
$$= 3.0$$

A formal discussion of significant figure rules as they apply to logarithms is presented later in this section.

(b) Let us use the shorter, more direct way for obtaining pH this time, the method based on the relationship

$$[H_3O^+] = 1 \times 10^{-x}$$

$$pH = x$$

Since the power of 10 is -9 in this case, the pH will be 9.0.

(c) The given quantity involves hydroxide ion rather than hydronium ion. Thus, we must first calculate the hydronium ion concentration, and then the pH.

$$[H_3O^+] = \frac{1.00 \times 10^{-14}}{[OH^-]} = \frac{1.00 \times 10^{-14}}{1 \times 10^{-4}}$$

$$= 1 \times 10^{-10} \quad \text{(calculator and \textbf{correct answer})}$$

A solution with a hydronium ion concentration of 1×10^{-10} M will have a pH of 10.0.

Practice Exercise 14.8

Calculate the pH for each of the following solutions.

(a) $[H_3O^+] = 1 \times 10^{-6}$ **(b)** $[H_3O^+] = 1 \times 10^{-12}$ **(c)** $[OH^-] = 1 \times 10^{-3}$

Since pH is simply another way of expressing hydronium ion concentration, acidic, neutral, and basic solutions can be identified by their pH values. A neutral solution ($[H_3O^+] = 1.0 \times 10^{-7}$) has a pH of 7.00. Values of pH less than 7.00 correspond to acidic solutions. The lower the pH value, the greater the acidity. Values of pH greater than 7.00 represent basic solutions. The higher the pH value, the greater the basicity. The relationships between $[H_3O^+]$, $[OH^-]$, and pH are summarized in Table 14.8. Note that a change of *one* unit in pH corresponds to a *tenfold* increase or decrease in $[H_3O^+]$. Also note that *lowering* the pH always corresponds to *increasing* the H_3O^+ ion concentration.

Table 14.8	The pH Scale		
pH	$[H_3O^+]$	$[OH^-]$	
0.0	1	10^{-14}	
1.0	10^{-1}	10^{-13}	
2.0	10^{-2}	10^{-12}	
3.0	10^{-3}	10^{-11}	Acidic
4.0	10^{-4}	10^{-10}	
5.0	10^{-5}	10^{-9}	
6.0	10^{-6}	10^{-8}	
7.0	10^{-7}	10^{-7}	Neutral
8.0	10^{-8}	10^{-6}	
9.0	10^{-9}	10^{-5}	
10.0	10^{-10}	10^{-4}	
11.0	10^{-11}	10^{-3}	Basic
12.0	10^{-12}	10^{-2}	
13.0	10^{-13}	10^{-1}	
14.0	10^{-14}	1	

Table 14.9	Approximate pH Values of Some Common Substances

	pH	
	0	1 M HCl (0.0)
	1	0.1 M HCl (1.0), gastric juice (1.6–1.8), lime juice (1.8–2.0)
	2	soft drinks (2.0–4.0), vinegar (2.4–3.4)
Acidic	3	grapefruit (3.0–3.3), peaches (3.4–3.6)
	4	tomatoes (4.0–4.4), human urine (4.8–8.4)
	5	carrots (4.9–5.3), peas (5.8–6.4)
	6	human saliva (6.2–7.4), cow's milk (6.3–6.6), drinking water (6.5–8.0)
Neutral	7	pure water (7.0), human blood (7.35–7.45), fresh eggs (7.6–8.0)
	8	seawater (8.3), soaps, shampoos (8.0–9.0)
	9	detergents (9.0–10.0)
	10	milk of magnesia (9.9–10.1)
Basic	11	household ammonia (11.5–12.0)
	12	liquid bleach (12.0)
	13	0.1 M NaOH (13.0)
	14	1 M NaOH (14.0)

Table 14.9 lists the pH values of a number of common substances. Except for gastric juice, most human body fluids have pH values within a couple of units of neutrality. Almost all foods are acidic. Tart taste is associated with food of low pH.

Nonintegral pH Values

Note that some of the pH values in Table 14.9 are nonintegral, that is, not whole numbers. Nonintegral pH values result from molar hydronium ion concentrations where the coefficient in the exponential expression for concentration has a value other than 1. For example, consider the following matchups between hydronium ion concentration and pH.

$$[H_3O^+] = 6.3 \times 10^{-5} \qquad pH = 4.20$$

$$[H_3O^+] = 4.0 \times 10^{-5} \qquad pH = 4.40$$

$$[H_3O^+] = 2.0 \times 10^{-5} \qquad pH = 4.70$$

Obtaining nonintegral pH values like these from hydronium ion concentrations requires an electronic calculator that allows for the input of exponential numbers and has a base 10 logarithm key (log).

In using an electronic calculator, depending on the model you have, you can obtain logarithm values simply by pressing the log key after having entered the number whose log value is desired or vice versa. For pH, you must remember that after obtaining the log value you must change its sign because of the negative sign in the defining equation for pH.

Significant figure considerations for log values involve a concept not previously encountered. It can best be illustrated by considering some actual log values. Consider the following five related numbers and their log values.

Number	Logarithm
2.43×10^0	0.38560627
2.43×10^2	2.38560627
2.43×10^4	4.38560627
2.43×10^7	7.38560627
2.43×10^{11}	11.38560627

This tabulation shows that

1. The number to the left of the decimal point in each logarithm (called the *characteristic*) is related only to the exponent of 10 in the number whose logarithm was taken.
2. The number to the right of the decimal point in each logarithm (called the *mantissa*) is related only to the coefficient in the exponential notation form of the number. Since the coefficient is 2.43 in each case, the mantissas are all the same (0.38560627).

Combining generalizations (1) and (2) gives us the significant figure rule for logarithms. *In a logarithm the digits to the left of the decimal point are not counted as significant figures.* These digits relate to the placement of the decimal point in the number. From our previous significant figure work (Sec. 2.5) they are somewhat analogous to the leading zeros (which are not significant) in a number such as 0.0000243.

Thus, the *coefficient* of the number whose logarithm has been taken and the *mantissa* of the logarithm must have the same number of digits. Rewriting the previous tabulation of logarithms to the correct number of significant figures gives

$$
\begin{array}{cc}
2.43 \times 10^{0} & 0.386 \\
2.43 \times 10^{2} & 2.386 \\
2.43 \times 10^{4} & 4.386 \\
2.43 \times 10^{7} & 7.386 \\
\underbrace{2.43 \times 10^{11}}_{\text{3 digits}} & \underbrace{11.386}_{\text{3 digits}}
\end{array}
$$

Now we can understand the following number–logarithm relationships that were given, without explanation, earlier in this section.

$$\log 1 \times 10^{-4} = -4.0$$

$$\log 1.0 \times 10^{-9} = -9.00$$

The first exponential number has one significant figure, and the second one has two significant figures.

EXAMPLE 14.9

Calculating the pH of a Solution Given Its Hydronium Ion Concentration

Calculate the pH of a solution with $[H_3O^+] = 3.9 \times 10^{-5}$.

SOLUTION

With an electronic calculator, we first enter the number 3.9×10^{-5} into the calculator. We then use the log key to obtain the logarithm value, -4.4089353. (With some calculators, the log key is pressed before entering the number.)

$$
\begin{aligned}
pH &= \log (3.9 \times 10^{-5}) \\
&= -(-4.4089353) \\
&= 4.4089353 \quad \text{(calculator answer)} \\
&= 4.41 \quad \quad \text{(correct answer)}
\end{aligned}
$$

The given hydronium ion concentration has two significant figures. Therefore, the logarithm should have two significant figures, as 4.41 does. In a logarithm only the digits to the right of the decimal place are considered significant.

Practice Exercise 14.9

Calculate the pH of a solution with $[H_3O^+] = 7.9 \times 10^{-11}$.

It is frequently necessary to calculate the hydronium ion concentration for a solution from its pH value. This type of calculation, which is the reverse of that just illustrated, is shown in Example 14.10.

EXAMPLE 14.10

Calculating the Molar Hydronium Ion Concentration of a Solution from the Solution's pH

The pH of a solution is 5.70. What is the molar hydronium ion concentration for this solution?

SOLUTION

Because the pH is between 5 and 6, we know immediately that $[H_3O^+]$ will be between 10^{-5} and 10^{-6} M. From the defining equation for pH, we have

$$[pH] = -\log [H_3O^+] = 5.70$$

$$\log [H_3O^+] = -5.70$$

To find $[H_3O^+]$ we need to determine the *antilog* of -5.70.

How an antilog is obtained using a calculator depends on the type of calculator you have. Many calculators have an antilog function (sometimes labed INV log) that performs this operation. If this key is present, then

1. Enter the number -5.70. Note that it is the *negative* of the pH that is entered into the calculator.
2. Press the INV log key (or an inverse key and then a log key). The result is the desired hydronium ion concentration.

$$\log [H_3O^+] = -5.70$$

$$\text{antilog } [H_3O^+] = 1.9952623 \times 10^{-6} \quad \text{(calculator answer)}$$

$$[H_3O^+] = 2.0 \times 10^{-6} \quad \text{(correct answer)}$$

(With some calculators, step 2 and step 1 are reversed.) Remember that the original pH value was a two-significant-figure pH.

Some calculators use a 10^x key to perform the antilog operation. Use of this key is based on the mathematical identity

$$\text{antilog } X = 10^x$$

For our case this means

$$\text{antilog } -5.70 = 10^{-5.70}$$

If the 10^x key is present, then

1. Enter the number -5.70 (the negative of the pH).
2. Press the function key 10^x. The result is the desired hydronium ion concentration.

$$[H_3O^+] = 10^{-5.70} = 1.9952623 \times 10^{-6} \quad \text{(calculator answer)}$$
$$= 2.0 \times 10^{-6} \quad \text{(correct answer)}$$

Practice Exercise 14.10

The pH of a solution is 8.40. What is the molar hydronium ion concentration for this solution?

14.13 Hydrolysis of Salts

The addition of an acid to water produces an acidic solution. The addition of a base to water produces a basic solution. What type of solution is produced when a salt is added to water? Since salts are the products of acid–base neutralizations, a logical supposition would be that salts dissolve in water to produce neutral (pH = 7) solutions. Such is the case for a *few* salts. Aqueous solutions of *most* salts, however, are either acidic or basic rather than neutral. Let us consider why this is so.

When a salt is dissolved in water it completely dissociates, that is, it completely breaks up into the ions of which it is composed (Sec. 13.3). For many salts, one or more of the ions so produced is reactive toward water. The ensuing reaction, which is called *hydrolysis*, causes the solution to have a nonneutral pH. **Hydrolysis** *is the reaction of a substance with water to produce hydronium ion or hydroxide ion or both.*

Types of Salt Hydrolysis

Not all salts hydrolyze. Which ones do and which ones do not? Of those salts that do hydrolyze, which ones produce acidic solutions and which ones produce basic solutions? The following guidelines, based on the neutralization "parentage" of a salt, that is, on the acid and base that will produce the salt through neutralization, can be used to answer these questions.

1. The salt of a *strong acid* and a *strong base* does not hydrolyze and therefore its aqueous solution is neutral.
2. The salt of a *strong acid* and a *weak base* hydrolyzes to produce an acidic solution.
3. The salt of a *weak acid* and a *strong base* hydrolyzes to produce a basic solution.
4. The salt of a *weak acid* and a *weak base* hydrolyzes to produce a slightly acidic, neutral, or slightly basic solution depending on the relative weaknesses of the acid and base.

The first prerequisite for using these guidelines is the ability to classify a salt into one of the four categories mentioned in the guidelines. This classification is accomplished by writing the neutralization equation (Sec. 14.8) that produces the salt and then specifying the strength (strong or weak) of the involved acid and base. The "parent" acid and base for the salt are identified by pairing the negative ion of the salt with H^+ (to form the acid) and pairing the positive ion of the salt with OH^- (to form the base). The following two equations illustrate the overall procedure.

$$\text{Na}\,\text{OH} + \text{H}\,\text{Cl} \longrightarrow \text{H}_2\text{O} + \text{NaCl}$$

Strong base   Strong acid   Strong acid–strong base salt

$$\text{K}\,\text{OH} + \text{H}\,\text{CN} \longrightarrow \text{H}_2\text{O} + \text{KCN}$$

Strong base   Weak acid   Weak acid–strong base salt

Table 14.10	Neutralization "Parentage" of Salts and the Nature of the Aqueous Solutions They Form	

Type of Salt	Nature of Aqueous Solution	Examples
Weak base–strong acid	Acidic	NH_4Cl, NH_4NO_3
Strong base–weak acid	Basic	$NaC_2H_3O_2$, K_2CO_3
Weak base–weak acid	Depends on the salt	$NH_4C_2H_3O_2$, NH_4NO_2
Strong base–strong acid	Neutral	$NaCl$, KBr

Note that knowledge of which acids and bases are strong and which are weak (Sec. 14.5) is a necessary part of the classification process. Once the salt has been classified, the guideline that is appropriate for the situation is easily selected.

Table 14.10 summarizes the concepts of this section to this point in our discussion.

EXAMPLE 14.11

Predicting Whether a Salt's Aqueous Solution Will Be Acidic, Basic, or Neutral

Determine the acid–base "parentage" of each of the following salts and then use this information to predict whether each salt's aqueous solution is acidic, basic, or neutral.

(a) Sodium acetate, $NaC_2H_3O_2$

(b) Ammonium chloride, NH_4Cl

(c) Potassium chloride, KCl

(d) Ammonium fluoride, NH_4F

SOLUTION

(a) The ions present are Na^+ and $C_2H_3O_2^-$. The "parent" base of Na^+ is $NaOH$, a strong base. The "parent" acid of $C_2H_3O_2^-$ is $HC_2H_3O_2$, a weak acid. Thus, the acid–base neutralization that produces this salt is

$$NaOH + HC_2H_3O_2 \longrightarrow H_2O + NaC_2H_3O$$

strong base   weak acid   weak acid–strong base salt

The solution of a weak acid–strong base salt (guideline 3) produces a basic solution.

(b) The ions present are NH_4^+ and Cl^-. The "parent" base of NH_4^+ is NH_3, a weak base. The "parent" acid of Cl^- is HCl, a strong acid. This "parentage" will produce a strong acid–weak base salt through neutralization. Such a salt gives an acidic solution upon hydrolysis (guideline 2).

(c) The ions present are K^+ and Cl^-. The "parent" base is KOH (a strong base) and the "parent" acid is HCl (a strong acid). The salt produced from neutralization involving this acid–base pair will be a strong acid–strong base salt. Such salts do not hydrolyze. The aqueous solution is neutral (guideline 1).

(d) The ions present are NH_4^+ and F^-. Both ions are of weak "parentage"; NH_3 is a weak base and HF is a weak acid. Thus, NH_4F is a weak acid–weak base salt. This is a guideline 4 situation. In this situation you cannot predict the effect of hydrolysis unless you know the relative strengths of the weak acid and weak base, that is, which is the weaker of the two. Guidelines to determine such information are not given in this text and thus we cannot predict the final acidity of the solution.

Practice Exercise 14.11

Predict whether an aqueous solution of each of the following salts will be acidic, basic, or neutral.

(a) sodium bromide, NaBr

(b) potassium cyanide, KCN

(c) ammonium iodide, NH_4I

(d) barium chloride, $BaCl_2$

Chemical Equations for Salt Hydrolysis Reactions

Salt hydrolysis reactions are Brønsted–Lowry acid–base (proton transfer) reactions (Sec. 14.2). Such reactions are of the following two general types.

1. *Basic hydrolysis*: The reaction of the *negative ion* from a salt with water to produce the ion's conjugate acid and hydroxide ion. Examples of such reactions are:

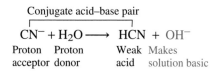

The only negative ions that undergo hydrolysis are those of "weak acid parentage." The driving force for the reaction is the formation of the weak acid.

2. *Acidic hydrolysis*: The reaction of the *positive ion* from a salt with water to produce the ion's conjugate base and hydronium ion. The most common ion to undergo this type of reaction is the NH_4^+ ion.

$$\text{NH}_4^+ + \text{H}_2\text{O} \longrightarrow \text{NH}_3 + \text{H}_3\text{O}^+$$

The only positive ions that undergo hydrolysis are those of "weak base parentage." The driving force for the reaction is the formation of the weak base.

EXAMPLE 14.12

Writing Net Ionic Equations for Hydrolysis Reactions

For each of the following salts, identify the ion or ions present that will hydrolyze and then write net ionic equations for hydrolysis reactions that occur.

(a) sodium fluoride, NaF

(b) potassium bromide, KBr

(c) ammonium nitrate, NH_4NO_3

(d) ammonium cyanide, NH_4CN

SOLUTION

(a) The ions produced when NaF dissolves are Na^+ and F^-. The Na^+ ion will not hydrolyze since its "parent" base, NaOH, is *strong*. The F^- ion will hydrolyze since its "parent" acid, HF, is *weak*. The equation for the hydrolysis reaction is

$$F^- + H_2O \longrightarrow HF + OH^-$$

The product OH^- causes the solution to be basic.

(b) Dissolution of KBr in water produces K^+ and Br^- ions. Neither of these ions will hydrolyze. The "parent" base of K^+ is KOH (strong) and the "parent" acid of Br^- is HBr (strong).

(c) This salt ionizes to produce NH_4^+ and NO_3^- ions. The NH_4^+ ion is associated with the *weak* base NH_3 and the NO_3^- ion is associated with the *strong* acid HNO_3. The former will hydrolyze, the latter will not. The hydrolysis reaction, which produces an acidic solution (H_3O^+), is

$$NH_4^+ + H_2O \longrightarrow NH_3 + H_3O^+$$

(d) Both the NH_4^+ ion (from the weak base NH_3) and the CN^- ion (from the weak acid HCN) will hydrolyze.

$$NH_4^+ + H_2O \longrightarrow NH_3 + H_3O^+$$
$$CN^- + H_2O \longrightarrow HCN + OH^-$$

The pH of the solution will be determined by the reaction that occurs to the greater extent. If the first reaction occurs to the greater extent the solution will be acidic; conversely, if the second reaction is dominant a basic solution results. (In this course you are not expected to be able to make such a determination, which involves comparison of the relative acid and base strengths of HCN and NH_3. For the record, the CN^- hydrolysis dominates and the solution is basic.)

Practice Exercise 14.12

For each of the following salts, identify the ion or ions present that will hydrolyze and then write net ionic equations for the hydrolysis reactions that occur.

(a) potassium chloride, KCl

(b) ammonium chloride, NH_4Cl

The pH change that accompanies salt hydrolysis can be significant—differing from neutrality by 2 to 4 units. Table 14.11 shows the range of pH values encountered for selected 0.1 M aqueous salt solutions after hydrolysis has occurred.

14.14 Buffers

A **buffer** *is a solution that resists major changes in pH when small amounts of acid or base are added to it.* Buffers are used in a laboratory setting to maintain optimum pH conditions for chemical reactions. Many commercial products contain buffers. Examples include buffered aspirin (Bufferin) and pH-controlled hair shampoos. Naturally occurring biochemical buffers are a necessary aspect of life processes. Most human body fluids are buffer solutions. For example, a buffer system maintains blood's pH at a value close to 7.4, an optimum pH for oxygen transport through the blood.

Table 14.11	pH Values of Selected 0.1 M Aqueous Salt Solutions at 24°C		
Name of salt	**Formula of salt**	**pH**	**Category of salt**
Ammonium nitrate	NH_4NO_3	5.1	strong acid–weak base
Ammonium nitrite	NH_4NO_2	6.3	weak acid–weak base
Ammonium acetate	$NH_4C_2H_3O_2$	7.0	weak acid–weak base
Sodium chloride	NaCl	7.0	strong acid–strong base
Sodium fluoride	NaF	8.1	weak acid–strong base
Sodium acetate	$NaC_2H_3O_2$	8.9	weak acid–strong base
Ammonium cyanide	NH_4CN	9.3	weak acid–weak base
Sodium cyanide	NaCN	11.1	weak acid–strong base

> A less common type of buffer involves a weak base and its conjugate acid. We will not consider this type of buffer here.

Buffers contain two chemical species: (1) a substance to react with and remove added base, and (2) a substance to react with and remove added acid. Typically, a buffer system is composed of a weak acid *and* its conjugate base—that is, a conjugate acid–base pair (Section 14.3). Conjugate acid–base pairs commonly used as buffers include $HC_3H_3O_2/C_2H_3O_2^-$, $H_2PO_4^-/HPO_4^{2-}$, and H_2CO_3/HCO_3^-.

EXAMPLE 14.13

Recognizing Pairs of Chemical Substances That Can Function as a Buffer in Aqueous Solution

Predict whether each of the following pairs of substances could function as a buffer system in aqueous solution.

(a) HCl and NaCl   (b) HCN and KCN
(c) HCl and HCN   (d) NaCN and KCN

SOLUTION

Buffer solutions contain either a weak acid and a salt of that weak acid or a weak base and a salt of that weak base. The salt supplies the conjugate base of the acid or the conjugate acid of the base.

(a) No. We have an acid and a salt of that acid. However, the acid is a strong acid rather than a weak acid.

(b) Yes. HCN is a weak acid, and KCN is a salt of that weak acid. The conjugate acid–base pair HCN/CN^- is present.

(c) No. Both HCl and HCN are acids. No salt is present.

(d) No. Both NaCN and KCN are salts. No weak acid is present.

Practice Exercise 14.13

Predict whether each of the following pairs of substances could function as a buffer system in aqueous solution.

(a) HCl and NaOH   (b) $HC_2H_3O_2$ and $KC_2H_3O_2$
(c) NaCl and NaCN   (d) HCN and $HC_2H_3O_2$

Chemical Equations for Buffer Action

As an illustration of buffer action, consider a buffer solution containing approximately equal concentrations of acetic acid (a weak acid) and sodium acetate (a salt of this weak acid). This buffer solution resists pH change by the following mechanisms.

1. When a small amount of a strong acid such as HCl is added to this buffer solution, the newly added H_3O^+ ions react with the acetate ions from the sodium acetate to give acetic acid.

$$H_3O^+ + C_2H_3O_2^- \longrightarrow HC_2H_3O_2 + H_2O$$

   Most of the added H_3O^+ ions are incorporated into acetic acid molecules, and the pH changes very little.

2. When a small amount of a strong base such as NaOH is added to this buffer solution, the newly added OH^- ions react with the acetic acid (neutralization) to give acetate ions and water.

$$OH^- + HC_2H_3O_2 \longrightarrow C_2H_3O_2^- + H_2O$$

   Most of the added OH^- ions are converted to water, and the pH changes only slightly.

The reactions that are responsible for the buffering action in the acetic acid/acetate ion system can be summarized as follows:

$$C_2H_3O_2^- \underset{OH^-}{\overset{H_3O^+}{\rightleftharpoons}} HC_2H_3O_2$$

Note that one member of the buffer pair (acetate ion) removes excess H_3O^+ ion and that the other (acetic acid) removes excess OH^- ion. The buffering action always results in the active species being converted to its partner species.

EXAMPLE 14.14

Writing Equations for Reactions That Occur in a Buffered Solution

Write an equation for each of the following buffering actions.

(a) The response of $H_2PO_4^-/HPO_4^{2-}$ buffer to the addition of H_3O^+ ions

(b) The response of HCN/CN^- buffer to the addition of OH^- ions

SOLUTION

(a) The base in a conjugate acid–base pair is the species that responds to the addition of acid. (Recall, from Section 14.3, that the base in a conjugate acid–base pair always has one less hydrogen than the acid.) The base for this reaction is HPO_4^{2-}. The equation for the buffering action is

$$H_3O^+ + HPO_4^{2-} \longrightarrow H_2PO_4^- + H_2O$$

In the buffering response, the base is always converted into its conjugate acid.

(b) The acid in a conjugate acid–base pair is the species that responds to the addition of base. The acid for this reaction is HCN. The equation for the buffering action is

$$OH^- + HCN \longrightarrow CN^- + H_2O$$

Water will always be one of the products of the buffering action in an aqueous solution.

Practice Exercise 14.14

Write an equation for each of the following buffering actions.

(a) The response of H_2CO_3/HCO_3^- buffer to the addition of H_3O^+ ions

(b) The response of $H_2PO_4^-/HPO_4^{2-}$ buffer to the addition of OH^- ions

A false notion about buffers is that they will hold the pH of a solution *absolutely* constant. The addition of even small amounts of a strong acid or a strong base to any solution, buffered or not, will lead to a change in pH. The important concept is that the change in pH will be much less when a buffer is present than in an unbuffered solution (see Table 14.12).

Buffer systems have their limits. If large amounts of H_3O^+ or OH^- ions are added to a buffer, the buffer capacity can be exceeded; then the buffer system is overwhelmed and the pH changes. For example, if large amounts of H_3O^+ were added to the acetate/acetic acid buffer previously discussed, the H_3O^+ ion would react with acetate ion until the acetate ion was depleted. Then the pH would begin to decrease rapidly as free H_3O^+ ions accumulated in solution.

A common misconception about buffers is that the "starting" pH of a buffered solution is always 7.0 (a neutral solution). This is false. One can buffer a solution at any desired pH. A pH 7.4 buffer will hold the pH of the solution near pH 7.4, whereas a pH 9.3 buffer will tend to hold the pH of a solution near pH 9.3. The starting pH of a buffer is determined by the degree of weakness of the weak acid used and by the concentration of the acid and its conjugate base.

14.15 Acid–Base Titrations

Determining the concentration of acid or base in a solution is a regular activity in many laboratories. The concentration of an acid or base in a solution and the solution's pH are two different entities. The pH of a solution gives information about the concentration of hydrogen (hydronium) ions in solution. Only dissociated molecules influence the pH value. The concentration of an acid or base solution gives information about the *total number* of acid or base molecules present; both dissociated and undissociated molecules are counted.

The procedure most frequently used to determine the concentration of an acidic or basic solution is an *acid–base titration*. An **acid–base titration** *is a procedure in which a measured volume of acid or base of known concentration is exactly reacted with a measured volume of a base or an acid of unknown concentration.*

Table 14.12	A Comparison of pH Changes in Buffered and Unbuffered Solutions
Unbuffered solution	
1 liter water	pH = 7.0
1 liter water + 0.01 mole strong base (NaOH)	pH = 12.0
1 liter water + 0.01 mole strong acid (HCl)	pH = 2.0
Buffered solution	
1 liter buffer*	pH = 7.2
1 liter buffer* + 0.01 mole strong base (NaOH)	pH = 7.3
1 liter buffer* + 0.01 mole mole strong acid (HCl)	pH = 7.1

*Buffer = 0.1 M HPO_4^{2-} + 0.1 M $H_2PO_4^-$

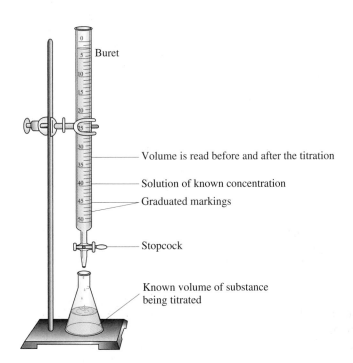

Figure 14.6

Use of a buret in a titration procedure.

Buret

Volume is read before and after the titration

Solution of known concentration

Graduated markings

Stopcock

Known volume of substance being titrated

Suppose we want to determine the concentration of an acid solution by titration. We would first measure out a *known volume* of the acid solution into a flask. We would then slowly add a solution of base of *known concentration* to the flask by means of a buret (see Fig. 14.6). Base addition continues until all the acid has completely reacted with added base. The *volume of base* needed to reach this point is obtained from the buret readings. Knowing the original volume of acid, the concentration of the base, and the volume of added base, we can calculate the concentration of the acid (Sec. 14.16).

To complete a titration successfully, we must be able to detect when the reaction between acid and base is complete. One way to do this is to add an *indicator* to the solution being titrated. An **indicator** *is a compound that exhibits different colors depending on the pH of its surroundings.* Typically, an indicator is a weak acid or weak base whose conjugate base or acid is a different color. An indicator is selected that will change color at a pH corresponding as nearly as possible to the pH of the solution when the titration is complete. This pH can be calculated ahead of time based on the identities of the acid and base involved in the titration.

Example 14.15 shows how titration data are used to calculate the molarity of an acid solution of unknown concentration.

EXAMPLE 14.15

Calculating an Unknown Molarity Using Acid–Base Titration Data

In an acid–base titration, 32.7 mL of 0.100 M KOH is required to neutralize completely 50.0 mL of H_3PO_4. Calculate the molarity of the H_3PO_4 solution.

SOLUTION

The first thing we must do is write the balanced equation for the neutralization reaction. Since the acid is triprotic and the base is monobasic, it will take three moles of base to neutralize one mole of acid. The neutralization equation is

$$H_3PO_4 + 3\ KOH \longrightarrow K_3PO_4 + 3\ H_2O$$

Next, we calculate the number of moles of H_3PO_4 that reacted with the KOH. This is a "volume of solution A" to "moles of B" problem (Fig. 13.8).

$$32.7 \text{ mL KOH} = ? \text{ moles } H_3PO_4$$

The pathway for the calculation, using dimensional analysis, is

$$\text{mL KOH} \longrightarrow \text{L KOH} \longrightarrow \text{moles KOH} \longrightarrow \text{moles } H_3PO_4$$

The sequence of conversion factors that effect this series of unit changes is:

$$32.7 \cancel{\text{ mL KOH}} \times \frac{10^{-3} \cancel{\text{ L KOH}}}{1 \cancel{\text{ mL KOH}}} \times \frac{0.100 \text{ mole } \cancel{\text{KOH}}}{1 \cancel{\text{ L KOH}}} \times \frac{1 \text{ mole } H_3PO_4}{3 \text{ moles } \cancel{\text{KOH}}}$$

The first conversion factor comes from the definition of a milliliter, the second conversion factor derives from the definition of molarity, and the third conversion factor uses the coefficients in the balanced chemical equation for the titration.

The number of moles of H_3PO_4 that react is obtained by combining the numbers in the dimensional analysis setup as indicated.

$$\frac{32.7 \times 10^{-3} \times 0.100 \times 1}{1 \times 1 \times 3} \text{ mole } H_3PO_4 = 0.00109 \text{ mole } H_3PO_4$$

(calculator and **correct answer**)

Now that we know how many moles of H_3PO_4 reacted, we calculate the molarity of the H_3PO_4 solution using the definition for molarity and the volume of acid given in the problem statement.

$$\text{Molarity } H_3PO_4 = \frac{\text{moles } H_3PO_4}{\text{L solution}} = \frac{0.00109 \text{ mole } H_3PO_4}{0.0500 \text{ L solution}}$$

$$= 0.0218 \text{ M } H_3PO_4$$

(calculator and **correct answer**)

Note that the units in the denominator of the molarity equation must be liters (0.0500) rather than milliliters (50.0).

Practice Exercise 14.15

In an acid–base titration, 50.2 mL of 0.252 M NaOH is required to neutralize 32.7 mL of H_2SO_4. Calculate the molarity of the H_2SO_4 solution.

14.16 Acid and Base Stock Solutions

Acids and bases are used so often in most laboratories that stock solutions (Sec. 13.10) of the most common ones are made readily available at each work space. The concentrations of such solutions are traditionally the same from laboratory to laboratory and are given in Table 14.13. Ordinarily the concentrations of the stock solutions are not given on their containers. Only the name of the acid or base and the term *dilute (dil)* or *concentrated (conc)* are found. It is assumed that students or researchers know what is implied by the designations *dil* and *conc* in each specific case; that is, that they know the information found in Table 14.13. Note from Table 14.13 that the designations *dil* and *conc* do not have a constant meaning in terms of molarity. For example, concentrated solutions of sulfuric acid, nitric acid, and hydrochloric acid have molarities of 18, 16, and 12, respectively. There is not as much variation in the meaning of the term dilute; except for sulfuric acid, all of the listed dilute solutions are 6 M.

| Table 14.13 | Concentrations of Common Laboratory Stock Solutions of Acids and Bases |

Label Description	Chemical Formula	Concentration Molarity
Acids		
Dilute hydrochloric acid	HCl	6
Concentrated hydrochloric acid	HCl	12
Dilute nitric acid	HNO_3	6
Concentrated nitric acid	HNO_3	16
Dilute sulfuric acid	H_2SO_4	3
Concentrated sulfuric acid	H_2SO_4	18
Dilute acetic acid	$HC_2H_3O_2$	6
Concentrated acetic acid[*]	$HC_2H_3O_2$	18
Bases		
Dilute aqueous ammonia[†]	$NH_3(aq)$	6
Concentrated aqueous ammonia[†]	$NH_3(aq)$	15
Dilute sodium hydroxide[‡]	NaOH	6

[*]*Often labeled "glacial acetic acid."*
[†]*Often labeled "ammonium hydroxide."*
[‡]*Often labeled simply "sodium hydroxide."*

Summary

1. **Arrhenius Acid–Base Theory** An Arrhenius acid is a hydrogen-containing compound that, in water, ionizes to produce hydrogen (H^+) ions. An Arrhenius base is a hydroxide-containing compound that, in water, dissociates to produce hydroxide (OH^-) ions.

2. **Brønsted–Lowry Acid–Base Theory** A Brønsted–Lowry acid is any substance that can donate a proton (H^+) to another substance. A Brønsted–Lowry base is any substance that can accept a proton (H^+) from another substance. Proton donation (from an acid) does not occur unless an acceptor (a base) is present.

3. **Conjugate Acids and Conjugate Bases** The conjugate base of an acid is the species that remains when the acid loses a proton. The conjugate acid of a base is the species formed when the base accepts a proton. A conjugate acid–base pair is two species that differ by one proton.

4. **Acids Classification** Acids can be classified, according to the number of hydrogen ions (protons) they can transfer per molecule during an acid–base reaction, as *monoprotic, diprotic,* and *triprotic. Polyprotic* acids are acids that can transfer two or more hydrogen ions during an acid–base reaction. Acids can be classified as *strong* or *weak* based on the extent to which proton transfer occurs in aqueous solution. A strong acid completely

transfers its protons to water. A weak acid transfers only a small percentage of its protons to water.

5. **Salts** A salt is an ionic compound containing a metal ion or polyatomic ion as the positive ion and a nonmetal ion or polyatomic ion (except hydroxide) as the negative ion. Ionic compounds containing hydroxide ion are bases rather than salts.

6. **Ionic and Net Ionic Equations** An ionic equation is an equation in which the formulas of the predominant form of each compound in aqueous solution are used; dissociated and ionized compounds are written as ions, and undissociated and un-ionized compounds are written in molecular form. A net ionic equation is an ionic equation from which nonparticipating (spectator) species have been eliminated.

7. **Reactions of Acids** Acids react with active metals to produce hydrogen gas and a salt. They react with hydroxide bases to produce a salt and water (neutralization). They react with carbonates and bicarbonates to produce carbon dioxide, a salt, and water.

8. **Reactions of Bases** The most characteristic reaction of bases is their reaction with acids (neutralization). They also react with fats and oils and convert them into smaller, soluble molecules.

Summary (*Continued*)

9. **Reactions of Salts** Salts react with some metals to convert the metallic ion of the salt to free metal and the free metal to its salt. They react with some acid solutions to form other acids and salts and with some base solutions to form other bases and salts. Salts react with some solutions of other salts to form new salts.

10. **Self-Ionization of Water** In pure water, a small number of water molecules [1.0×10^{-7} M] donate protons to other water molecules to produce small concentrations [1.0×10^{-7} M] of hydronium and hydroxide ions. Ion product constant for water is the name given to the numerical value 1.00×10^{-14} associated with the product of the hydronium and hydroxide ion molar concentrations in pure water.

11. **The pH Scale** The pH scale is a scale of small numbers that is used to specify molar hydronium ion concentration in an aqueous solution. The calculation of pH scale values involves the use of logarithms. The pH of a solution is the negative logarithm of the solution's molar hydronium ion concentration.

12. **Acidic and Basic Solutions** An acidic solution has a higher hydronium ion concentration than hydroxide ion concentration. Conversely, a basic solution has a higher hydroxide ion concentration than hydronium ion concentration. An acidic solution has a pH of less than 7.00. A basic solution has a pH greater than 7.00. A neutral solution has a pH of 7.00.

13. **Hydrolysis of Salts** Salt hydrolysis is a reaction in which a salt interacts with water to produce an acidic or a basic solution. Only salts that contain the conjugate base of a weak acid and/or the conjugate acid of a weak base hydrolyze.

14. **Buffers** A buffer is a solution that resists major changes in pH when small amounts of acid or base are added to the solution. The resistance to pH change in most buffers is caused by the presence of a weak acid and a salt of its conjugate base.

15. **Acid–Base Titrations** An acid–base titration is a procedure in which an acid–base neutralization reaction is used to determine the unknown concentration of an acid or base. A measured volume of an acid or a base of known concentration is exactly reacted with a measured volume of a base or an acid of unknown concentration. An indicator is used to detect when the neutralization process is complete.

16. **Acid and Base Stock Solutions** Acids and bases are used so often in most laboratories that stock solutions of the most common ones are made available. The concentrations of these stock solutions are traditionally the same from laboratory to laboratory.

Key Terms

The new terms defined in this chapter are

acid–base titration   *Sec 14.15*

acidic hydrogen atom   *Sec. 14.4*

acidic solution   *Sec. 14.11*

amphoteric substance   *Sec. 14.3*

Arrhenius acid   *Sec. 14.1*

Arrhenius base   *Sec. 14.1*

basic solution   *Sec. 14.11*

Brønsted–Lowry acid   *Sec. 14.2*

Brønsted–Lowry base   *Sec. 14.2*

Buffer   *Sec 14.14*

conjugate acid   *Sec. 14.3*

conjugate acid–base pair   *Sec. 14.3*

conjugate base   *Sec. 14.3*

diprotic acid   *Sec. 14.4*

dissociation   *Sec. 14.1*

hydrolysis   *Sec. 14.13*

indicator   *Sec. 14.15*

ion product constant for water   *Sec. 14.11*

ionic equation   *Sec. 14.7*

ionization   *Sec. 14.1*

monoprotic acid   *Sec. 14.4*

net ionic equation   *Sec. 14.7*

neutralization   *Sec. 14.8*

neutral solution   *Sec. 14.11*

pH   *Sec. 14.12*

pH scale   *Sec. 14.12*

polyprotic acid   *Sec. 14.4*

salt   *Sec. 14.6*

strong acid   *Sec. 14.5*

triprotic acid   *Sec. 14.4*

weak acid   *Sec. 14.5*

Practice Problems

Acid–Base Definitions (Secs. 14.1 and 14.2)

14.1 In Arrhenius acid–base theory:

(a) What ion is responsible for the properties of acidic solutions?

(b) What term is used to describe the formation of ions, in aqueous solution, from an ionic compound?

14.2 In Arrhenius acid–base theory:

(a) What ion is responsible for the properties of basic solutions?

(b) What term is used to describe the formation of ions, in aqueous solution, from a molecular compound?

14.3 Classify each of the following properties as that of an Arrhenius acid or that of an Arrhenius base.

(a) has a sour taste

(b) changes the color of blue litmus paper to red

14.4 Classify each of the following properties as that of an Arrhenius acid or that of an Arrhenius base.

(a) has a bitter taste

(b) changes the color of red litmus paper to blue

14.5 Write equations for the ionization/dissociation of the following Arrhenius acids and bases in water.

(a) HBr (hydrobromic acid)

(b) $HClO_2$ (chlorous acid)

(c) LiOH (lithium hydroxide)

(d) $Ba(OH)_2$ (barium hydroxide)

14.6 Write equations for the ionization/dissociation of the following Arrhenius acids and bases in water.

(a) $HClO_3$ (chloric acid)

(b) HI (hydroiodic acid)

(c) H_2SO_4 (sulfuric acid)

(d) CsOH (cesium hydroxide)

14.7 In each of the following reactions, decide whether the first listed species is a Brønsted–Lowry acid or base.

(a) $NH_4^+ + OH^- \longrightarrow H_2O + NH_3$

(b) $HS^- + H_2O \longrightarrow H_2S + OH^-$

(c) $HClO_4 + NO_2^- \longrightarrow HNO_2 + ClO_4^-$

(d) $H_2O + HC_2O_4^- \longrightarrow H_3O^+ + C_2O_4^{2-}$

14.8 In each of the following reactions, decide whether the first listed species is a Brønsted–Lowry acid or base.

(a) $HClO_4 + H_2O \longrightarrow H_3O^+ + ClO_4^-$

(b) $HClO + NH_3 \longrightarrow NH_4^+ + ClO^-$

(c) $H_2O + CN^- \longrightarrow HCN + OH^-$

(d) $NH_3 + H_3O^+ \longrightarrow NH_4^+ + H_2O$

14.9 Write equations to illustrate the acid–base reactions that can take place between the following Brønsted–Lowry acids and bases.

(a) acid, HBr; base, H_2O

(b) acid, H_2O; base, N_3^-

(c) acid, H_2S; base, H_2O

(d) acid, $HClO_4$; base, NO_2^-

14.10 Write equations to illustrate the acid–base reactions that can take place between the following Brønsted–Lowry acids and bases.

(a) acid, H_3PO_4; base, NH_3

(b) acid, H_3O^+; base, OH^-

(c) acid, HSO_4^-; base, H_2O

(d) acid, H_2O; base, S^{2-}

Conjugate Acids and Bases (Sec. 14.3)

14.11 Write the formula of each of the following conjugate acids or bases.

(a) conjugate base of H_2SO_3

(b) conjugate acid of CN^-

(c) conjugate acid of S^{2-}

(d) conjugate base of HClO

14.12 Write the formula of each of the following conjugate acids or bases.

(a) conjugate acid of HCO_3^-

(b) conjugate base of NH_3

(c) conjugate acid of $H_2PO_4^-$

(d) conjugate base of H_3O^+

14.13 Identify the two conjugate acid–base pairs involved in each of the following reactions.

(a) $H_2C_2O_4 + ClO^- \rightleftharpoons HC_2O_4^- + HClO$

(b) $HSO_4^- + H_2O \rightleftharpoons H_3O^+ + SO_4^{2-}$

(c) $HPO_4^{2-} + NH_4^+ \rightleftharpoons NH_3 + H_2PO_4^-$

(d) $HCO_3^- + H_2O \rightleftharpoons OH^- + H_2CO_3$

14.14 Identify the two conjugate acid–base pairs involved in each of the following reactions.

(a) $SO_4^{2-} + H_2O \rightleftharpoons HSO_4^- + OH^-$

(b) $CN^- + H_2O \rightleftharpoons HCN + OH^-$

(c) $HSO_4^- + HCO_3^- \rightleftharpoons SO_4^{2-} + H_2CO_3$

(d) $H_3PO_4 + PO_4^{3-} \rightleftharpoons H_2PO_4^- + HPO_4^{2-}$

14.15 In which of the following pairs of substances do the two members of the pair constitute a conjugate acid–base pair?

(a) HCN and CN^-

(b) H_3PO_4 and PO_4^{3-}

(c) HCO_3^- and HSO_4^-

(d) NH_4^+ and NH_3

14.16 In which of the following pairs of substances do the two members of the pair constitute a conjugate acid–base pair?

(a) HN_3 and N_3^-
(b) H_2SO_4 and SO_4^{2-}
(c) H_2CO_3 and HSO_4^-
(d) NH_3 and NH_2^-

14.17 For each of the following amphoteric substances, write the two equations needed to describe its behavior in aqueous solution.

(a) HS^- (b) HPO_4^{2-} (c) HCO_3^- (d) $H_2PO_3^-$

14.18 For each of the following amphoteric substances, write the two equations needed to describe its behavior in aqueous solution.

(a) $H_2PO_4^-$ (b) HSO_3^- (c) $HC_2O_4^-$ (d) PH_3

Polyprotic Acids (Sec. 14.4)

14.19 Classify each of the following acids as monoprotic, diprotic, or triprotic.

(a) $HClO_3$ (chloric acid)
(b) $H_2C_2O_4$ (oxalic acid)
(c) $HC_3H_3O_3$ (pyruvic acid)
(d) $H_2C_3H_2O_4$ (malonic acid)

14.20 Classify each of the following acids as monoprotic, diprotic, or triprotic.

(a) HNO_3 (nitric acid)
(b) H_2SeO_4 (selenic acid)
(c) $HC_3H_5O_3$ (lactic acid)
(d) $H_2C_4H_4O_4$ (succinic acid)

14.21 How many acidic hydrogen atoms and how many nonacidic hydrogen atoms are present in each of the following molecules?

(a) HNO_3 (nitric acid)
(b) $H_2CH_4O_4$ (succinic acid)
(c) $HC_4H_7O_2$ (butyric acid)
(d) CH_4 (methane)

14.22 How many acidic hydrogen atoms and how many nonacidic hydrogen atoms are present in each of the following molecules?

(a) H_2CO_3 (carbonic acid)
(b) $H_2C_3H_2O_4$ (malonic acid)
(c) NH_3 (ammonia)
(d) $HC_3H_5O_2$ (propanoic acid)

14.23 Write equations for the stepwise proton-transfer process that occurs in aqueous solution for each of the following acids.

(a) $H_2C_4H_4O_4$ (succinic acid)
(b) H_2SO_3 (sulfurous acid)

14.24 Write equations for the stepwise proton-transfer process that occurs in aqueous solution for each of the following acids.

(a) $H_2C_3H_2O_4$ (malonic acid)
(b) $H_3C_6H_5O_7$ (citric acid)

14.25 The formula for salicylic acid is preferably written as $HC_7H_5O_3$ rather than $C_7H_6O_3$. Explain why this is so.

14.26 The formula for tartaric acid is preferably written as $H_2C_4H_4O_6$ rather than $C_4H_6O_6$. Explain why this is so.

14.27 Pyruvic acid, which is produced in metabolic reactions within the human body, has the following structure.

Would you predict this acid to be mono-, di-, tri-, or tetraprotic? Give your reasoning for your answer.

14.28 Succinic acid, a biologically important substance, has the following structure.

How many acidic hydrogen atoms are present in the structure? Give your reasoning for your answer.

Strength of Acids and Bases (Sec. 14.5)

14.29 Classify each of the acids in Problem 14.19 as a *strong acid* or a *weak acid*.

14.30 Classify each of the acids in Problem 14.20 as a *strong acid* or a *weak acid*.

14.31 For which of the following pairs of acids are both members of the pair of "like strength," that is, both strong or both weak?

(a) H_2SO_4 and H_2SO_3 (b) $HClO_4$ and HCN
(c) HF and HI (d) HNO_2 and $HClO_2$

14.32 For which of the following pairs of acids are both members of the pair of "like strength," that is, both strong or both weak?

(a) HNO_3 and HNO_2 (b) HCl and HBr
(c) H_3PO_4 and HCN (d) H_2CO_3 and $H_2C_2O_4$

14.33 In which of the following acid and base combinations are both a strong acid and a strong base present?

(a) H_2SO_4 and KOH (b) HNO_3 and $NaOH$
(c) H_3PO_4 and $LiOH$ (d) HF and $Ba(OH)_2$

14.34 In which of the following acid and base combinations are both a strong acid and a strong base present?

(a) HCl and $NaOH$ (b) H_2CO_3 and $Ca(OH)_2$
(c) $HC_2H_3O_2$ and KOH (d) $HClO_4$ and $Sr(OH)_2$

14.35 With the help of Table 14.3 when necessary, indicate which acid in each of the following pairs of acids is the stronger.

(a) HNO_3 and HNO_2 (b) $HClO_4$ and $HC_2H_3O_2$

(c) H_3PO_4 and HCN (d) H_2CO_3 and HF

14.36 With the help of Table 14.3 when necessary, indicate which acid in each of the following pairs of acids is the stronger.

(a) H_2SO_4 and H_2SO_3 (b) HCl and HF

(c) HCN and HNO_2 (d) $HC_2H_3O_2$ and H_3PO_4

Salts (Sec. 14.6)

14.37 Identify each of the following substances as an acid, a base, or a salt.

(a) NH_4NO_3 (b) KOH (c) Na_2SO_4 (d) HCN

14.38 Identify each of the following substances as an acid, a base, or a salt.

(a) $CaCO_3$ (b) HF (c) KF (d) $LiOH$

14.39 Give the formula and name of the positive and negative ions present in each of the following salts.

(a) Na_3PO_4 (b) $LiNO_3$ (c) NH_4Cl (d) KCN

14.40 Give the formula and name of the positive and negative ions present in each of the following salts.

(a) K_2SO_4 (b) $MgCO_3$ (c) K_2S (d) NH_4NO_3

14.41 Indicate whether each of the salts in Problem 14.39 is soluble or insoluble in water (Sec. 13.4).

14.42 Indicate whether each of the salts in Problem 14.40 is soluble or insoluble in water (Sec. 13.4).

14.43 Write a balanced equation for the dissociation in water of each of the following soluble salts into ions.

(a) NaI (b) BaS (c) Li_2SO_4 (d) $Al(NO_3)_3$

14.44 Write a balanced equation for the dissociation in water of each of the following soluble salts into ions.

(a) $NaC_2H_3O_2$ (b) $Be(NO_3)_2$

(c) $MgCl_2$ (d) $(NH_4)_2CO_3$

Ionic and Net Ionic Equations (Sec. 14.7)

14.45 Classify the following equations for reactions occurring in aqueous solution as *molecular*, *ionic*, or *net ionic*.

(a) $MgCO_3 + 2 HBr \longrightarrow MgBr_2 + H_2O + CO_2$

(b) $2 OH^- + H_2CO_3 \longrightarrow CO_3^{2-} + 2 H_2O$

(c) $K^+ + OH^- + H^+ + I^- \longrightarrow K^+ + I^- + H_2O$

(d) $Ag^+ + Cl^- \longrightarrow AgCl$

14.46 Classify the following equations for reactions occurring in aqueous solution as *molecular*, *ionic*, or *net ionic*.

(a) $CaCO_3 + 2 H^+ + 2 NO_3^- \longrightarrow$
$$Ca^{2+} + 2 NO_3^- + H_2O + CO_2$$

(b) $Ni + Cu^{2+} \longrightarrow Cu + Ni^{2+}$

(c) $NaCl + AgNO_3 \longrightarrow NaNO_3 + AgCl$

(d) $Ca^{2+} + 2 OH^- \longrightarrow Ca(OH)_2$

14.47 Write a balanced net ionic equation for each of the following reactions, each of which occurs in aqueous solution.

(a) $2 NaBr + Pb(NO_3)_2 \longrightarrow 2 NaNO_3 + PbBr_2$

(b) $FeCl_3 + 3 NaOH \longrightarrow Fe(OH)_3 + 3 NaCl$

(c) $Zn + 2 HCl \longrightarrow ZnCl_2 + H_2$

(d) $H_2S + 2 KOH \longrightarrow K_2S + 2 H_2O$

14.48 Write a balanced net ionic equation for each of the following reactions, each of which occurs in aqueous solution.

(a) $CaCl_2 + CuSO_4 \longrightarrow CaSO_4 + CuCl_2$

(b) $Ca(NO_3)_2 + K_2CO_3 \longrightarrow CaCO_3 + 2 KNO_3$

(c) $Mg + 2 HBr \longrightarrow MgBr_2 + H_2$

(d) $H_3PO_4 + 3 NaOH \longrightarrow Na_3PO_4 + 3 H_2O$

14.49 Write a balanced net ionic equation for each of the following reactions, each of which occurs in aqueous solution.

(a) $Pb + 2 AgNO_3 \longrightarrow 2 Ag + Pb(NO_3)_2$

(b) $Cl_2 + 2 NaBr \longrightarrow 2 NaCl + Br_2$

(c) $2 Al(NO_3)_3 + 3 Na_2S \longrightarrow Al_2S_3 + 6 NaNO_3$

(d) $NaC_2H_3O_2 + NH_4Cl \longrightarrow NH_4C_2H_3O_2 + NaCl$

14.50 Write a balanced net ionic equation for each of the following reactions, each of which occurs in aqueous solution.

(a) $Ni + Cu(NO_3)_2 \longrightarrow Ni(NO_3)_2 + Cu$

(b) $Br_2 + 2 NaI \longrightarrow 2 NaBr + I_2$

(c) $Hg(NO_3)_2 + K_2S \longrightarrow 2 KNO_3 + HgS$

(d) $(NH_4)_2SO_4 + 2 NaBr \longrightarrow 2 NH_4Br + Na_2SO_4$

Reactions of Acids and Bases (Secs. 14.8 and 14.9)

14.51 On the basis of the activity series (Table 14.6) predict whether a reaction takes place when

(a) iron metal is added to hydrochloric acid.

(b) potassium metal is added to cold water.

(c) gold metal is added to hydrochloric acid.

(d) aluminum metal is added to hot water (steam).

14.52 On the basis of the activity series (Table 14.6) predict whether a reaction takes place when

(a) chromium metal is added to hydrochloric acid.

(b) silver metal is added to hot water (steam).

(c) sodium metal is added to cold water.

(d) copper metal is added to hydrochloric acid.

14.53 Write the balanced molecular equation for each of the following hydrogen-gas-producing reactions.

(a) Nickel metal is added to hydrochloric acid, giving $NiCl_2$.

(b) Calcium metal is added to cold water.

(c) Magnesium metal is added to hydrochloric acid.

(d) Zinc metal is added to hot water (steam).

14.54 Write the balanced molecular equation for each of the following hydrogen-gas-producing reactions.

(a) Magnesium metal is added to hot water (steam).

(b) Calcium metal is added to hydrochloric acid.

(c) Tin metal is added to hydrochloric acid, giving $SnCl_2$.

(d) Lead metal is added to hydrochloric acid, giving $PbCl_2$.

14.55 Indicate whether each of the following reactions is an acid–base neutralization reaction.

(a) $NaCl + AgNO_3 \longrightarrow AgCl + NaNO_3$

(b) $HNO_3 + KOH \longrightarrow NaNO_3 + H_2O$

(c) $HBr + KOH \longrightarrow KBr + H_2O$

(d) $H_2SO_4 + Pb(NO_3)_2 \longrightarrow PbSO_4 + 2\,HNO_3$

14.56 Indicate whether each of the following reactions is an acid–base neutralization reaction.

(a) $H_2S + CuSO_4 \longrightarrow H_2SO_4 + CuS$

(b) $HCN + LiOH \longrightarrow LiCN + H_2O$

(c) $H_2SO_4 + Ba(OH)_2 \longrightarrow BaSO_4 + 2\,H_2O$

(d) $Ni + 2\,HCl \longrightarrow NiCl_2 + H_2$

14.57 Without writing a chemical equation, specify the molecular ratio in which each of the following acids and bases will react in a neutralization reaction.

(a) HNO_3 and $NaOH$ **(b)** H_2SO_4 and $NaOH$

(c) H_2SO_4 and $Sr(OH)_2$ **(d)** HNO_3 and $Ba(OH)_2$

14.58 Without writing a chemical equation, specify the molecular ratio in which each of the following acids and bases will react in a neutralization reaction.

(a) HCl and KOH **(b)** H_2CO_3 and KOH

(c) HCl and $Ca(OH)_2$ **(d)** H_2CO_3 and $Ca(OH)_2$

14.59 Write a balanced molecular equation to represent each of the following acid–base neutralization reactions.

(a) HBr and $Sr(OH)_2$ **(b)** $HC_2H_3O_2$ and $LiOH$

(c) H_2SO_4 and $Mg(OH)_2$ **(d)** H_3PO_4 and KOH

14.60 Write a balanced molecular equation to represent each of the following acid–base neutralization reactions.

(a) HCl and $NaOH$ **(b)** HNO_3 and KOH

(c) H_2S and $Ba(OH)_2$ **(d)** $HClO_4$ and $LiOH$

14.61 Write a net ionic equation to represent each of the acid–base neutralization reactions in Problem 14.59.

14.62 Write a net ionic equation to represent each of the acid–base neutralization reactions in Problem 14.60.

14.63 Give the formulas of the acid and base needed to prepare the following salts by neutralization reactions.

(a) Na_3PO_4 **(b)** KCN

(c) $BeCl_2$ **(d)** $Ca(C_2H_3O_2)_2$

14.64 Give the formulas of the acid and base needed to prepare the following salts by neutralization reactions.

(a) $LiNO_3$ **(b)** K_2S **(c)** $Al_2(SO_4)_3$ **(d)** $NaBr$

14.65 Write a molecular equation for the action of HCl on each of the following. (Zn metal forms a +2 ion in solution.)

(a) Zn **(b)** $NaOH$ **(c)** Na_2CO_3 **(d)** $NaHCO_3$

14.66 Write a molecular equation for the action of HNO_3 on each of the following. (Ni metal forms a +2 ion in solution.)

(a) Ni **(b)** KOH **(c)** Li_2CO_3 **(d)** $LiHCO_3$

Reactions of Salts (Sec. 14.10)

14.67 On the basis of the activity series (Table 14.6) predict whether or not a reaction occurs when each of the following metals is added to a nickel(II) nitrate solution.

(a) Cu **(b)** Zn **(c)** Au **(d)** Fe

14.68 On the basis of the activity series (Table 14.6) predict whether or not a reaction occurs when each of the following metals is added to an iron(II) nitrate solution.

(a) Ag **(b)** Ni **(c)** Cr **(d)** Pb

14.69 Write a net ionic equation for each of the following metal-replacement reactions. Assume that metals going into solution form +2 ions.

(a) Iron is added to $CuSO_4$ solution.

(b) Tin is added to $AgNO_3$ solution.

(c) Zinc is added to $NiCl_2$ solution.

(d) Chromium is added to $Pb(C_2H_3O_2)_2$ solution.

14.70 Write a net ionic equation for each of the following metal-replacement reactions. Assume that metals going into solution form +2 ions.

(a) Lead is added to Cu_2SO_4 solution.

(b) Mercury is added to $Au(NO_3)_3$ solution.

(c) Chromium is added to $FeCl_2$ solution.

(d) Iron is added to $Ni(C_2H_3O_3)_2$ solution.

14.71 Indicate the driving force (condition) that causes each of the following acid–salt reactions to occur.

(a) $Ba(NO_3)_2 + H_2SO_4 \longrightarrow BaSO_4 + 2\,HNO_3$

(b) $3\,CaCl_2 + 2\,H_3PO_4 \longrightarrow Ca_3(PO_4)_2 + 6\,HCl$

(c) $AgC_2H_3O_2 + HCl \longrightarrow AgCl + HC_2H_3O_2$

(d) $K_2CO_3 + 2\,HNO_3 \longrightarrow 2\,KNO_3 + CO_2 + H_2O$

14.72 Indicate the driving force (condition) that causes each of the following acid–salt reactions to occur.

(a) $NaCN + HCl \longrightarrow NaCl + HCN$

(b) $3\,MgSO_4 + 2\,H_3PO_4 \longrightarrow Mg_3(PO_4)_2 + 3\,H_2SO_4$

(c) $Li_2CO_3 + 2\,HBr \longrightarrow 2\,LiBr + CO_2 + H_2O$

(d) $Na_3PO_4 + 3\,HCl \longrightarrow 3\,NaCl + H_3PO_4$

14.73 Complete and balance a net ionic equation for the reaction, if any, between each of the following pairs of aqueous solutions. If no reaction occurs, write "no reaction."

(a) $Al(NO_3)_3$ and $(NH_4)_2S$ (b) HCl and $Ba(OH)_2$
(c) Na_2SO_4 and HNO_3 (d) KNO_3 and $HC_2H_3O_2$

14.74 Complete and balance a net ionic equation for the reaction, if any, between each of the following pairs of aqueous solutions. If no reaction occurs, write "no reaction."

(a) H_2CO_3 and KOH (b) K_3PO_4 and HCl
(c) $CaCl_2$ and HNO_3 (d) $Fe(NO_3)_2$ and $MgSO_4$

Hydronium Ion and Hydroxide Ion Concentrations (Sec. 14.11)

14.75 What is the molar H_3O^+ ion concentration in solutions with the following OH^- ion concentrations?

(a) 2.0×10^{-4} M (b) 7.3×10^{-7} M
(c) 3.0×10^{-10} M (d) 2.5×10^{-8} M

14.76 What is the molar H_3O^+ ion concentration in solutions with the following OH^- ion concentrations?

(a) 9.7×10^{-6} M (b) 3.8×10^{-2} M
(c) 3.3×10^{-7} M (d) 9.3×10^{-11} M

14.77 Indicate whether each of the solutions in Problem 14.75 is acidic, basic, or neutral.

14.78 Indicate whether each of the solutions in Problem 14.76 is acidic, basic, or neutral.

14.79 What is the molar OH^- ion concentration in solutions with the following H_3O^+ ion concentrations?

(a) 2.7×10^{-4} M (b) 7.5×10^{-9} M
(c) 1.0×10^{-7} M (d) 5.0×10^{-8} M

14.80 What is the molar OH^- ion concentration in solutions with the following H_3O^+ ion concentrations?

(a) 3.3×10^{-2} M (b) 6.9×10^{-12} M
(c) 1.5×10^{-6} M (d) 4.7×10^{-8} M

14.81 Indicate whether each of the solutions in Problem 14.79 is acidic, basic, or neutral.

14.82 Indicate whether each of the solutions in Problem 14.80 is acidic, basic, or neutral.

The pH Scale (Sec. 14.12)

14.83 Calculate the pH of solutions with the following hydronium ion concentrations.

(a) 1×10^{-4} M (b) 1×10^{-9} M
(c) 0.00001 M (d) 0.000000001 M

14.84 Calculate the pH of solutions with the following hydronium ion concentrations.

(a) 1×10^{-3} M (b) 1×10^{-11} M
(c) 0.0001 M (d) 0.0000001 M

14.85 Calculate the pH of solutions with the following hydronium ion concentrations.

(a) 4×10^{-2} M (b) 7×10^{-4} M
(c) 8×10^{-10} M (d) 5×10^{-7} M

14.86 Calculate the pH of solutions with the following hydronium ion concentrations.

(a) 6×10^{-4} M (b) 6×10^{-8} M
(c) 9×10^{-3} M (d) 7×10^{-6} M

14.87 Calculate the pH of solutions with the following hydronium ion concentrations.

(a) 3×10^{-3} M (b) 3.0×10^{-3} M
(c) 3.00×10^{-3} M (d) 3.000×10^{-3} M

14.88 Calculate the pH of solutions with the following hydronium ion concentrations.

(a) 7×10^{-6} M (b) 7.0×10^{-6} M
(c) 7.00×10^{-3} M (d) 7.000×10^{-3} M

14.89 In which of the following pairs of pH values do both values represent acidic solution conditions?

(a) 5.31 and 6.31 (b) 6.31 and 7.31
(c) 7.31 and 8.31 (d) 6.90 and 7.00

14.90 In which of the following pairs of pH values do both values represent basic solution conditions?

(a) 5.92 and 6.92 (b) 6.92 and 7.92
(c) 7.92 and 8.92 (d) 7.01 and 7.10

14.91 What is the molar hydronium ion concentration associated with each of the following pH values?

(a) 3.0 (b) 5.0 (c) 5.7 (d) 6.3

14.92 What is the molar hydronium ion concentration associated with each of the following pH values?

(a) 4.0 (b) 8.0 (c) 8.2 (d) 10.1

14.93 What is the molar hydronium ion concentration associated with each of the following pH values?

(a) 2.43 (b) 3.43 (c) 7.43 (d) 7.45

14.94 What is the molar hydronium ion concentration associated with each of the following pH values?

(a) 4.05 (b) 5.05 (c) 8.05 (d) 8.15

14.95 Solution A has $[OH^-] = 4.3 \times 10^{-4}$. Solution B has $[H_3O^+] = 7.3 \times 10^{-10}$.

(a) Which solution is more basic?
(b) Which solution has the lower pH?

14.96 Solution A has $[H_3O^+] = 2.7 \times 10^{-6}$. Solution B has $[OH^-] = 4.5 \times 10^{-8}$.

(a) Which solution is more acidic?
(b) Which solution has the higher pH?

14.97 A solution has a pH of 4.500. What will be the pH of this solution if the hydronium ion concentration is

(a) doubled?
(b) quadrupled?
(c) increased by a factor of 10?
(d) increased by a factor of 1000?

14.98 A solution has a pH of 3.699. What will be the pH of this solution if the hydronium ion concentration is

(a) tripled?

(b) cut in half?

(c) increased by a factor of 100?

(d) decreased by a factor of 10^4?

14.99 Calculate the pH of each of the following solutions of strong acids or strong bases.

(a) 6.3×10^{-3} M HNO_3 (b) 0.20 M HCl

(c) 0.000021 M H_2SO_4 (d) 2.3×10^{-4} M NaOH

14.100 Calculate the pH of each of the following solutions of strong acids or strong bases.

(a) 4.02×10^{-5} M HCl   (b) 4.02×10^{-5} M H_2SO_4

(c) 0.00035 M HNO_3 (d) 5.7×10^{-3} M KOH

Hydrolysis of Salts (Sec. 14.13)

14.101 Identify the ion (or ions) present, if any, that will undergo hydrolysis in aqueous solution in each of the following salts.

(a) Na_3PO_4 (b) NaCN

(c) NH_4Cl (d) LiCl

14.102 Identify the ion (or ions) present, if any, that will undergo hydrolysis in aqueous solution in each of the following salts.

(a) $KC_2H_3O_2$ (b) NH_4F

(c) $Ca(CN)_2$ (d) NaBr

14.103 Predict whether each of the following aqueous salt solutions will be acidic, basic, or neutral.

(a) Na_2SO_4 (b) LiCN

(c) NH_4Br (d) KI

14.104 Predict whether each of the following aqueous salt solutions will be acidic, basic, or neutral.

(a) KNO_3 (b) NaCN

(c) $LiC_2H_3O_2$ (d) $(NH_4)_2SO_4$

14.105 Write a net ionic equation for the hydrolysis of each of the following salts in aqueous solution.

(a) NH_4Cl (b) $NaC_2H_3O_2$

(c) KF   (d) LiCN

14.106 Write a net ionic equation for the hydrolysis of each of the following salts in aqueous solution.

(a) NaF   (b) KCN

(c) $LiNO_2$ (d) NH_4I

Buffers (Sec. 14.14)

14.107 Predict whether each of the following pairs of substances could function as a buffer system in aqueous solution.

(a) HNO_3 and $NaNO_3$ (b) HF and NaF

(c) KCl and KCN   (d) H_2CO_3 and $NaHCO_3$

14.108 Predict whether each of the following pairs of substances could function as a buffer system in aqueous solution.

(a) HNO_3 and HCl   (b) HNO_2 and KNO_2

(c) $NaC_2H_3O_2$ and $KC_2H_3O_2$ (d) $HC_2O_3O_2$ and $NaNO_3$

14.109 Identify the two "active species" in each of the following buffer systems.

(a) HCN and KCN   (b) H_3PO_4 and NaH_2PO_4

(c) H_2CO_3 and $KHCO_3$ (d) $NaHCO_3$ and K_2CO_3

14.110 Identify the two "active species" in each of the following buffer systems.

(a) HF and LiF   (b) Na_2HPO_4 and KH_2PO_4

(c) K_2CO_3 and $KHCO_3$ (d) $NaNO_2$ and HNO_2

14.111 Write an equation for each of the following buffering actions.

(a) the response of a HF/F^- buffer to the addition of OH^- ions

(b) the response of a H_2CO_3/HCO_3^- buffer to the addition of H_3O^+ ions

(c) the response of a HCO_3^-/CO_3^{2-} buffer to the addition of acid

(d) the response of a H_3PO_4/$H_2PO_4^-$ buffer to the addition of base

14.112 Write an equation for each of the following buffering actions.

(a) the response of a HPO_4^{2-}/PO_4^{3-} buffer to the addition of OH^- ions

(b) the response of a HF/F^- buffer to the addition of H_3O^+ ions

(c) the response of a HCN/CN^- buffer to the addition of acid

(d) the response of a $H_2PO_4^-$/HPO_4^{2-} buffer to the addition of base

Acid–Base Titrations (Sec. 14.15)

14.113 What volume, in milliliters, of a 0.100-M NaOH solution would be needed to neutralize each of the following acid samples?

(a) 10.00 mL of 0.350 M H_2SO_4

(b) 50.00 mL of 1.500 M H_3PO_4

(c) 5.00 mL of 0.500 M HNO_3

(d) 75.00 mL of 0.00030 M HCl

14.114 What volume, in milliliters, of a 2.00-M KOH solution would be needed to neutralize each of the following acid samples?

(a) 50.00 mL of 6.50 M HCl

(b) 5.00 mL of 0.0500 M H_2SO_4

(c) 250.0 mL of 1.00 M H_3PO_4

(d) 75.00 mL of 2.50 M HNO_3

14.115 What volume, in milliliters, of 2.50-M aqueous ammonia [NH_3(aq)] would be needed to neutralize each of the following acid samples?

(a) 10.00 mL of 2.00 M HNO_3

(b) 20.00 mL of 6.00 M H_2SO_4

(c) 30.00 mL of 7.50 M H_3PO_4

(d) 100.0 mL of 0.100 M HCl

14.116 What volume, in milliliters, of a 1.30-M LiOH solution would be needed to neutralize each of the following acid samples?

(a) 7.50 mL of 0.100 M H_3PO_4

(b) 75.00 mL of 0.100 M H_2SO_4

(c) 150.0 mL of 7.50 M HNO_3

(d) 100.0 mL of 12.0 M HCl

14.117 It requires 34.5 mL of 0.102 M NaOH to neutralize each of the following acid solution samples. What is the molarity of each of the acid samples?

(a) 25.0 mL of H_2SO_4 **(b)** 20.0 mL of HClO

(c) 20.0 mL of H_3PO_4 **(d)** 10.0 mL of HNO_3

14.118 It requires 21.4 mL of 0.198 M NaOH to neutralize each of the following acid solution samples. What is the molarity of each of the acid samples?

(a) 30.0 mL of H_2CO_3 **(b)** 25.0 mL of $H_2C_2O_4$

(c) 25.0 mL of $HC_2H_3O_2$ **(d)** 20.0 mL of HCl

Acid and Base Stock Solutions (Sec. 14.16)

14.119 What is the molarity of each of the following stock solutions?

(a) dilute H_2SO_4 **(b)** dilute HCl

(c) concentrated HNO_3 **(d)** dilute NaOH

14.120 What is the molarity of each of the following stock solutions?

(a) dilute HNO_3 **(b)** concentrated NH_3

(c) concentrated H_2SO_4 **(d)** dilute $HC_2H_3O_2$

Additional Problems

14.121 Identify each of the following species as the conjugate base of a strong acid or the conjugate base of a weak acid.

(a) Br^- **(b)** CN^- **(c)** $H_2PO_4^-$ **(d)** NO_3^-

14.122 Identify each of the following species as the conjugate base of a strong acid or the conjugate base of a weak acid.

(a) $C_2H_3O_2^-$ **(b)** Cl^- **(c)** F^- **(d)** HPO_4^{2-}

14.123 A solution has a pH of 2.2. Another solution has a pH of 4.5. How many times greater is the $[H_3O^+]$ in the first solution than in the second one?

14.124 A solution has a pH of 3.4. Another solution has a pH of 6.7. How many times greater is the $[H_3O^+]$ in the first solution than in the second one?

14.125 What is the pH of an aqueous solution in which there are three times as many hydronium ions as hydroxide ions?

14.126 What is the pH of an aqueous solution in which there are three times as many hydroxide ions as hydronium ions?

14.127 Solution A has a pH of 3.20, solution B a pH of 4.20, solution C a pH of 11.20, and solution D a pH of 7.20. Arrange the four solutions in order of

(a) decreasing acidity   **(b)** increasing $[H_3O^+]$

(c) decreasing $[OH^-]$ **(d)** increasing basicity

14.128 Solution A has a pH of 1.25, solution B a pH of 12.50, solution C a pH of 7.00, and solution D a pH of 4.44. Arrange the four solutions in order of

(a) increasing acidity   **(b)** decreasing $[H_3O^+]$

(c) increasing $[OH^-]$ **(d)** decreasing basicity

14.129 In which of the following pairs of solutions does the first listed solution have a lower pH than the second listed solution?

(a) 0.1 M HCl and 0.2 M HCl

(b) 0.1 M HCl and 0.1 M H_2SO_4

(c) 0.20 M H_2SO_4 and 0.25 M HNO_3

(d) 0.2 M H_2SO_4 and 0.2 M H_2CO_3

14.130 In which of the following pairs of solutions does the first listed solution have a lower pH than the second listed solution?

(a) 0.2 M HNO_3 and 0.3 M HNO_3

(b) 0.1 M HNO_3 and 0.1 M H_2SO_4

(c) 0.1 M H_2SO_4 and 0.12 M HCl

(d) 0.2 M H_2SO_4 and 0.2 M $H_2C_2O_4$

14.131 What would be the pH of a solution that contains 0.1 mole of each of the solutes NaCl, HNO_3, HCl, and NaOH in enough water to give 3.00 L of solution?

14.132 What would be the pH of a solution that contains 0.1 mole of each of the solutes NaBr, HBr, KOH, and NaOH in enough water to give 5.00 L of solution?

14.133 Arrange the following 0.1-M aqueous solutions in order of decreasing pH: NH_4Br, $Ba(OH)_2$, $HClO_4$, K_2SO_4, and LiCN.

14.134 Arrange the following 0.1-M aqueous solutions in order of increasing pH: HBr, $HC_2H_3O_2$, $NaC_2H_3O_2$, KOH, and $Ca(NO_3)_2$.

14.135 Identify the buffer system(s) [conjugate acid–base pair(s)] present in solutions that contain equal molar amounts of the following.

(a) HCN, KCN, NaCN, and NaCl

(b) HF, HCl, $NaC_2H_3O_2$, and NaF

14.136 Identify the buffer system(s) [conjugate acid–base pair(s)] present in solutions that contain equal molar amounts of the following.

(a) HF, HCN, NaCN, and KF

(b) H_2CO_3, Na_2CO_3, KCN, HCN

14.137 It is possible to make two completely different buffers that involve the dihydrogen phosphate ion, $H_2PO_4^-$. Characterize each of the buffers by specifying the conjugate acid–base pair that is present.

14.138 It is possible to make two completely different buffers that involve the hydrogen carbonate ion, HCO_3^-. Characterize each of the buffers by specifying the conjugate acid–base pair that is present.

14.139 How many moles of $Ca(OH)_2$ would it take to completely neutralize 0.40 gram of H_3PO_4?

14.140 How many moles of $Sr(OH)_2$ would it take to completely neutralize 5.00 grams of HNO_3?

14.141 If 125 mL of 5.00 M HNO_3 are mixed with 125 mL of 6.00 M NaOH, what will be the

(a) $[H_3O^+]$ for the solution?
(b) $[OH^-]$ for the solution?
(c) pH of the solution?

14.142 If 125 mL of 6.00 M HNO_3 are mixed with 125 mL of 5.00 M NaOH, what will be the

(a) $[H_3O^+]$ for the solution?
(b) $[OH^-]$ for the solution?
(c) pH of the solution?

Cumulative Problems

14.143 Name and give the formula for each of the following species.

(a) conjugate base of hydroiodic acid
(b) conjugate base of the dihydrogen phosphate ion
(c) conjugate acid of the oxide ion
(d) conjugate acid of water

14.144 Name and give the formula for each of the following species.

(a) conjugate base of perchloric acid
(b) conjugate base of the bicarbonate ion
(c) conjugate acid of the hydroxide ion
(d) conjugate acid of ammonia

14.145 Write electron-dot structures for hydrocyanic acid and its conjugate base.

14.146 Write electron-dot structures for hydrochlorous acid and its conjugate base.

14.147 What is the pH of a solution obtained by each of the following operations?

(a) dissolving 4.8 g of HCl in enough water to obtain 0.40 L of solution
(b) dissolving 12.5 g of LiOH in enough water to obtain 255 mL of solution
(c) diluting 75 mL of 0.10 M HCl to a volume of 125 mL
(d) mixing equal volumes of 0.20 M HCl and 0.50 M HNO_3

14.148 What is the pH of a solution obtained by each of the following operations?

(a) dissolving 4.8 g of HBr in enough water to obtain 0.30 L of solution
(b) dissolving 3.50 g of NaOH in enough water to obtain 45 mL of solution

(c) diluting 25 mL of 0.10 HNO_3 to a volume of 375 mL
(d) mixing equal volumes of 0.20 M HCl and 0.20 M HNO_3

14.149 If 1.00 mL of 0.10 M HNO_3 is diluted to 100.0 mL, what is

(a) $[H_3O^+]$ before dilution?
(b) $[H_3O^+]$ after dilution?
(c) the pH before dilution?
(d) the pH after dilution?

14.150 If 100.0 mL of 0.10 M HCl is diluted to 500.0 mL, what is

(a) $[H_3O^+]$ before dilution?
(b) $[H_3O^+]$ after dilution?
(c) the pH before dilution?
(d) the pH after dilution?

14.151 The pH of a hydrochloric acid solution with a density of 1.10 g/mL is 3.40. Calculate the concentration of the acid in

(a) molarity units
(b) mass percent units

14.152 The pH of a nitric acid solution with a density of 1.20 g/mL is 2.70. Calculate the concentration of the acid in

(a) molarity units
(b) mass percent units

14.153 How many hydronium ions are present in a 10.0-mL sample of hydrochloric acid that has a pH of 5.42?

14.154 How many hydronium ions are present in a 50.0-mL sample of nitric acid that has a pH of 1.32?

14.155 How many ions are present in a 236-mL sample of a HNO_3 solution with a pH of 2.37 to which 0.100 mole of Na_2SO_4 has been added and dissolved?

14.156 How many ions are present in a 435-mL sample of a HCl solution with a pH of 1.54 to which 0.050 mole of $Mg(NO_3)_2$ has been added and dissolved?

14.157 How many liters of HCl gas, at STP, are dissolved in 7.50 L of aqueous HCl solution if the solution has a pH of 2.40?

14.158 How many liters of NH_3 gas, at STP, are dissolved in 7.50 L of aqueous NH_3 solution if the solution has a pH of 8.20?

14.159 An impure 1.00-g sample of the monoprotic acid potassium hydrogen phthalate $(KHC_8H_4O_4)$ is dissolved in water and titrated with 32.3 mL of 0.1000 M NaOH solution. Calculate the mass percent $KHC_8H_4O_4$ in the impure sample.

14.160 An impure 1.00-g sample of the diprotic oxalic acid $(H_2C_2O_4)$ is dissolved in water and titrated with 17.6 mL of 0.200 M NaOH solution. Calculate the mass percent $H_2C_2O_4$ in the impure sample.

Answers to Practice Exercises

14.1 (HCN, CN^-), (H_3O^+, H_2O)

14.2 (a) SO_4^{2-} (b) $H_2PO_4^-$

14.3 $Ag^+ + Cl^- \longrightarrow AgCl$

14.4 $H_2CO_3 + Mg^{2+} \longrightarrow MgCO_3 + 2\,H^+$

14.5 $H^+ + OH^- \longrightarrow H_2O$

14.6 (a) Molecular: $Fe(NO_3)_3 + Na_3PO_4 \longrightarrow FePO_4 +$

$$3\,NaNO_3$$

Ionic: $Fe^{3+} + 3\,NO_3^- + 3\,Na^+ + PO_4^{3-} \longrightarrow$

$$FePO_4 + 3\,Na^+ + 3\,NO_3^-$$

Net ionic: $Fe^{3+} + PO_4^{3-} \longrightarrow FePO_4$

(b) Molecular: $CaCl_2 + 2\,HNO_3 \longrightarrow Ca(NO_3)_2 + 2\,HCl$

Ionic: $Ca^{2+} + 2\,Cl^- + 2\,H^+ + 2\,NO_3^- \longrightarrow$

$$Ca^{2+} + 2\,NO_3^- + 2\,H^+ + 2\,Cl^-$$

Net ionic: all species cancel (no reaction occurs)

(c) Molecular: $2\,HCl + Na_2S \longrightarrow H_2S + 2\,NaCl$

Ionic: $2\,H^+ + 2\,Cl^- + 2\,Na^+ + S^{2-} \longrightarrow$

$$H_2S + 2\,Na^+ + 2\,Cl^-$$

Net ionic: $2\,H^+ + S^{2-} \longrightarrow H_2S$

14.7 2.22×10^{-13} M OH^-

14.8 (a) 6.0 (b) 12.0 (c) 11.0

14.9 10.10

14.10 4.0×10^{-9} M H_3O^+

14.11 (a) neutral (b) basic (c) acidic (d) neutral

14.12 (a) neither ion hydrolyzes

(b) NH_4^+; $NH_4^+ + H_2O \longrightarrow NH_3 + H_3O^+$

14.13 (a) no (b) yes (c) no (d) no

14.14 (a) $HCO_3^- + H_3O^+ \longrightarrow H_2CO_3 + H_2O$

(b) $H_2PO_4^- + OH^- \longrightarrow HPO_4^{2-} + H_2O$

14.15 0.194 M H_2SO_4

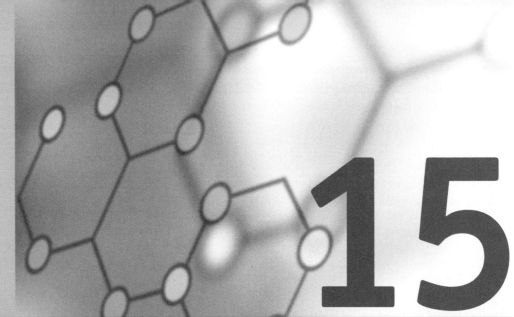

15

Oxidation and Reduction

15.1 Oxidation–Reduction Terminology

Oxidation–reduction reactions are a very important class of chemical reactions. They occur all around us and even within us. The bulk of the energy needed for the functioning of all living organisms, including humans, is obtained from food via oxidation–reduction processes. Such diverse phenomena as the electricity obtained from a battery to start a car, the use of natural gas to heat a home, iron rusting, and the functioning of antiseptic agents to kill or prevent the growth of bacteria all involve oxidation–reduction reactions. In short, knowledge of this type of reaction is fundamental to understanding many biological and technological processes.

The terms *oxidation* and *reduction*, like the terms acid and base (Sec. 14.1), have several definitions. Historically, the word *oxidation* was first used to describe the reaction of a substance with oxygen. According to this historical definition, each of the following reactions involves oxidation.

$$4\,Fe + 3\,O_2 \longrightarrow 2\,Fe_2O_3$$

$$S + O_2 \longrightarrow SO_2$$

$$CH_4 + 2\,O_2 \longrightarrow CO_2 + 2\,H_2O$$

The substance on the far left in each of these equations is said to have been *oxidized*.

Originally, the term *reduction* referred to processes where oxygen was removed from a compound. A particularly common type of reduction reaction, according to this original definition, is the removal of oxygen from a metal oxide to produce the free metal.

$$CuO + H_2 \longrightarrow Cu + H_2O$$

$$2\,Fe_2O_3 + 3\,C \longrightarrow 4\,Fe + 3\,CO_2$$

The word *reduction* comes from the reduction in mass of the metal-containing species; the metal has a mass less than that of the metal oxide.

Today the words *oxidation* and *reduction* are used in a much broader sense. Current definitions include the previous examples but also much more. It is now recognized that the same changes brought about in a substance from reaction with oxygen can be caused by reaction with numerous non-oxygen-containing substances. For example, consider the following reactions.

$$2\,Mg + O_2 \longrightarrow 2\,MgO$$

$$Mg + S \longrightarrow MgS$$

$$Mg + F_2 \longrightarrow MgF_2$$

$$3\,Mg + N_2 \longrightarrow Mg_3N_2$$

In each of these reactions magnesium metal is converted to a magnesium compound that contains Mg^{2+} ions. The process is the same—the changing of magnesium atoms, through the loss of two electrons, to magnesium ions; the only difference is the identity of the substance that causes magnesium to undergo the change. All these reactions are considered to involve *oxidation* by the current definition. **Oxidation** *is the process whereby a substance in a chemical reaction loses one or more electrons*. The current definition for *reduction* involves the use of similar terminology. **Reduction** *is the process whereby a substance in a chemical reaction gains one or more electrons*.

Oxidation and reduction are complementary processes rather than isolated phenomena. They *always* occur together; you cannot have one without the other. If electrons are lost by one species, they cannot just disappear; they must be gained by another species. Electron transfer, then, is the basis for oxidation and reduction. An **oxidation–reduction reaction** *is a chemical reaction in which transfer of electrons between reactants occurs*. The term *oxidation–reduction reaction* is often shortened to the term *redox reaction*.

There are two different ways of looking at the reactants in a redox reaction. First, the reactants can be viewed as being acted upon. From this perspective one reactant is *oxidized* (the one that loses electrons) and one is *reduced* (the one that gains electrons). Second, the reactants can be looked at as bringing about the reaction. In this approach the terms *oxidizing agent* and *reducing agent* are used. An **oxidizing agent** *is the reactant in a redox reaction that causes oxidation by accepting electrons from the other reactant*. Such acceptance, the gain of electrons, means that the oxidizing agent itself is reduced. Similarly, the **reducing agent** *is the reactant in a redox reaction that causes reduction by providing electrons for the other reactant to accept*. As a result of providing electrons, the reducing agent itself becomes oxidized. Note, then, that the reducing agent and substance oxidized are one and the same, as are the oxidizing agent and substance reduced.

Substance oxidized = reducing agent

Substance reduced = oxidizing agent

Table 15.1 summarizes the terms presented in this section.

Oxidation involves the *loss* of electrons, and reduction involves the *gain* of electrons. Students often have trouble remembering which is which. Two helpful mnemonic devices follow.

LEO the lion says *GER*.
Loss of Electrons: Oxidation.
Gain of Electrons: Reduction.
OIL RIG
Oxidation Is Loss (of electrons).
Reduction Is Gain (of electrons).

The terms *oxidizing agent* and *reducing agent* sometimes cause confusion because the oxidizing agent is not oxidized (it is reduced) and the reducing agent is not reduced (it is oxidized). By simple analogy, a travel agent is not the one who takes a trip—he or she is the one who causes the trip to be taken.

Table 15.1	**Oxidation–Reduction Terminology in Terms of Loss and Gain of Electrons**
Terms Associated with the Loss of Electrons	**Terms Associated with the Gain of Electrons**
Process of oxidation	Process of reduction
Substance oxidized	Substance reduced
Reducing agent	Oxidizing agent

15.2   Oxidation Numbers

Oxidation numbers are used to help determine whether oxidation or reduction has occurred in a reaction and, if such is the case, the identity of the oxidizing or reducing agents. An **oxidation number** *is the charge that an atom appears to have when the electrons in each bond it is participating in are assigned to the more electronegative of the two atoms involved in the bond.**

Consider an HCl molecule, a molecule in which there is one bond involving two shared electrons.

$$H\!:\!\ddot{C}l\!:$$

According to the definition for oxidation number, the electrons in this bond are assigned to the chlorine atom (the more electronegative atom; Sec. 7.18). This results in the chlorine atom having one more electron than a neutral Cl atom; hence, the oxidation number of chlorine is -1 (one extra electron). At the same time, the H atom in the HCl molecule has one fewer electron than a neutral H atom; its electron was given to the chlorine. This electron deficiency of one results in an oxidation number of $+1$ for hydrogen.

As a second example, consider the molecule CF_4.

1. **If the oxidation state of an atom in a molecule is $+n$, then the atom "owns" n fewer electrons in the molecule than it would as a free atom.**

2. **If the oxidation state of an atom in a molecule is $-n$, then the atom "owns" n more electrons in the molecule than it would as a free atom.**

Fluorine is more electronegative than carbon. Hence, the two shared electrons in each of the four carbon–fluorine bonds are assigned to the fluorine atom. Each F atom thus gains an extra electron, resulting in F having a -1 oxidation number. The carbon atom loses a total of four electrons, one to each F atom, as a result of the electron "assignments." Hence, its oxidation number is $+4$, indicating the loss of the four electrons.

As a third example, consider the N_2 molecule where like atoms are involved in a triple bond.

$$:N\!:::\!N\!:$$

Since the identical atoms are of equal electronegativity, the shared electrons are "divided" equally between the two atoms; each N receives three of the bonding electrons to count as its own. This results in each N atom having five valence electrons (three from the triple bond and two nonbonding electrons), the same number of valence electrons as in a neutral N atom. Hence, the oxidation number of N in N_2 is zero.

Before going any further in our discussion of oxidation numbers, it should be noted that *calculated* oxidation numbers are *not* actual charges on atoms. This is why the phrase

---

*In some textbooks the term *oxidation state* is used in place of oxidation number. In other textbooks the two terms are used interchangeably. We will use oxidation number.

"appears to have" is found in the definition of oxidation number given at the start of this section. In assigning oxidation numbers, we assume when we give the bonding electrons to the more electronegative element that each bond is ionic (complete transfer of electrons). We know that this is not always the case. Sometimes it is a good approximation, sometimes it is not. Why, then, do we do this when we know that it does not always correspond to reality? Oxidation numbers, as we shall see shortly, serve as a very convenient device for "keeping track" of electron transfer in redox reactions. Even though they do not always correspond to physical reality, they are very useful entities.

In principle, the procedures used to determine oxidation numbers for the atoms in the molecules HCl, CF_4, and N_2 can be used to determine oxidation numbers in all molecules. However, the procedures become very laborious in many cases, especially when complicated Lewis structures are involved. In practice, an alternative, much simpler procedure that does not require Lewis structures is used to obtain oxidation numbers. This alternative procedure is based on a set of operational rules that are consistent with and derivable from the general definition for oxidation numbers. The operational rules are

RULE 1 The oxidation number of an atom in its elemental state is zero.

For example, the oxidation number of Cu is zero, and the oxidation number of Cl in Cl_2 is zero.

RULE 2 The oxidation number of any monoatomic ion is equal to the charge on the ion.

For example, the Na^+ ion has an oxidation number of $+1$, and the S^{2-} ion has an oxidation number of -2.

RULE 3 The oxidation numbers of groups IA and IIA elements in compounds are always $+1$ and $+2$, respectively.

RULE 4 The oxidation number of fluorine in compounds is always -1 and that of the other group VIIA elements (Cl, Br, and I) is usually -1.

The exception for these latter elements is when they are bonded to more electronegative elements. In this case they are assigned positive oxidation numbers.

RULE 5 The usual oxidation number for oxygen in compounds is -2.

The exceptions occur when oxygen is bonded to the more electronegative fluorine (O then is assigned a positive oxidation number) or found in compounds containing oxygen–oxygen bonds (peroxides). In peroxides the oxidation number -1 is assigned to oxygen. Peroxides exist for hydrogen (H_2O_2), group IA elements (Na_2O_2, etc.), and group IIA elements (BaO_2, etc.).

RULE 6 The usual oxidation number for hydrogen in compounds is $+1$.

The exception occurs in hydrides, compounds where hydrogen is bonded to a metal of lower electronegativity. In such compounds hydrogen is assigned an oxidation number of -1. Examples of hydrides are NaH, CaH_2, and LiH.

RULE 7 In binary compounds, the element with the greater electronegativity is assigned a negative oxidation number equal to its charge as an anion in its ionic compounds.

For example, in the compound AlN, N (the more electronegative element) is assigned an oxidation number of -3, the charge on a nitride ion (N^{3-}).

RULE 8 The algebraic sum of the oxidation numbers of all atoms in a neutral molecule must be zero.

RULE 9 The algebraic sum of the oxidation numbers of all atoms in a polyatomic ion is equal to the charge on the ion.

The use of these rules is illustrated in Example 15.1.

EXAMPLE 15.1

Assigning Oxidation Numbers to Elements in a Compound or Polyatomic Ion

Assign oxidation numbers to each element in the following chemical species.

(a) SO_3 (b) N_2H_4 (c) $KMnO_4$ (d) ClO_4^-

SOLUTION

(a) Oxygen has an oxidation number of -2 (rule 5 or rule 7). The oxidation number of S can be calculated by rule 8. Letting x equal the oxidation number of S, we have

$$\begin{array}{ll} \text{S: 1 atom} \times (x) & = \quad x \\ \text{O: 3 atoms} \times (-2) & = \underline{-6} \\ \text{sum} & = \quad 0 \quad (\text{rule 8}) \end{array}$$

Solving for x algebraically, we get

$$x + (-6) = 0$$
$$x = +6$$

Consequently, the oxidation number of sulfur is $+6$ in the compound SO_3.

(b) Hydrogen has an oxidation number of $+1$ (rule 6). Rule 8 will allow us to calculate the oxidation number of N; the sum of the oxidation numbers must be zero. Letting x equal the oxidation number of N, we have

$$\begin{array}{ll} \text{H: 4 atoms} \times (+1) & = +4 \\ \text{N: 2 atoms} \times (x) & = \underline{2x} \\ \text{sum} & = \quad 0 \quad (\text{rule 8}) \end{array}$$

Solving for x algebraically, we get

$$2x + (+4) = 0$$
$$x = -2$$

Thus, the oxidation number of nitrogen in N_2H_4 is -2. Note that the oxidation number of N is not -4 (the calculated charge associated with two N atoms). Oxidation number is always specified on a *per atom* basis.

(c) Potassium has an oxidation number of $+1$ (rule 3), and oxygen has an oxidation number of -2 (rule 5 or rule 7). Letting x equal the oxidation number of manganese and using rule 8, we get

$$\begin{array}{ll} \text{K: 1 atom} \times (+1) & = +1 \\ \text{Mn: 1 atom} \times (x) & = \quad x \\ \text{O: 4 atoms} \times (-2) & = \underline{-8} \\ \text{sum} & = \quad 0 \quad (\text{rule 8}) \end{array}$$

Solving for x algebraically, we get

$$(+1) + x + (-8) = 0$$
$$x = +7$$

Thus, the oxidation number of manganese in $KMnO_4$ is $+7$.

(d) According to rule 9, the sum of the oxidation numbers must equal -1, the charge on this polyatomic ion. The oxidation number of oxygen is -2 (rule 5). Chlorine will have a positive oxidation number, since it is bonded to a more electronegative element (rule 4). Letting x equal the oxidation number of chlorine, we have

$$
\begin{aligned}
\text{Cl: 1 atom} \times (x) &= x \\
\text{O: 4 atoms} \times (-2) &= \underline{-8} \\
\text{sum} &= -1 \quad (\text{rule 8})
\end{aligned}
$$

Solving for x algebraically, we get

$$
x + (-8) = -1
$$

$$
x = +7
$$

Thus, chlorine has an oxidation number of $+7$ in this ion.

Practice Exercise 15.1

Assign oxidation numbers to each element in the following chemical species.

(a) SO_2 (b) NH_4^+ (c) $NaNO_3$ (d) N_2F_4

Answers to practice exercises are located at the end of the chapter.

Application of oxidation-number rules in the manner illustrated in Example 15.1 enables one to quickly assign oxidation numbers to the elements in a wide variety of compounds.

Many elements display a range of oxidation numbers in their various compounds. For example, nitrogen exhibits oxidation numbers ranging from -3 to $+5$ in various compounds. Selected examples are

NH_3	N_2H_4	N_2O	NO	N_2O_3	NO_2	HNO_3
-3	-2	$+1$	$+2$	$+3$	$+4$	$+5$

As shown in this listing of nitrogen-containing compounds, the oxidation number of an atom is written *underneath* the atom in the formula. This convention is used to avoid confusion with the charge on an ion.

Although not common, nonintegral oxidation numbers are possible. For example, the oxidation number of iron in the compound Fe_3O_4 is $+2.67$. The oxidation numbers of the oxygens in the compound add up to -8. Therefore, the iron atoms must have an oxidation number sum of $+8$. Dividing $+8$ by 3 (the number of iron atoms) gives $+2.67$.

Oxidizing and reducing agents were previously defined in terms of loss and gain of electrons (Table 15.1). They can also be defined in terms of changes in oxidation numbers. An **oxidizing agent** *is the reactant in a redox reaction that contains the element that shows a decrease in oxidation number.* Since the oxidizing agent is the substance reduced in a reaction, *reduction involves a decrease in oxidation number*; the oxidation number is reduced (decreased) in a reduction. A **reducing agent** *is the reactant in a redox reaction that contains the element that shows an increase in oxidation number.* Since the reducing agent is the substance oxidized in a reaction, *oxidation involves an increase in oxidation number.*

Table 15.2 summarizes the relationships between oxidation–reduction terminology and oxidation number changes. A comparison of Table 15.2 with Table 15.1 shows that the loss of electrons and oxidation number increases are synonymous as are the gain of electrons and oxidation number decreases. The fact that the oxidation number becomes more positive (increases) as electrons are lost is consistent with our understanding that electrons are negatively charged.

Table 15.2	Oxidation–Reduction Terminology in Terms of Oxidation Number Change	
Terms Associated with an Increase in Oxidation Number		**Terms Associated with a Decrease in Oxidation Number**
Process of oxidation		Process of reduction
Substance oxidized		Substance reduced
Reducing agent		Oxidizing agent

EXAMPLE 15.2

Determining Oxidation Numbers and Identifying Oxidizing Agents and Reducing Agents

Determine oxidation numbers for each atom in the following reactions, and identify the oxidizing and reducing agents.

(a) $2\,NO + O_2 \longrightarrow 2\,NO_2$ **(b)** $Zn + 2\,HCl \longrightarrow ZnCl_2 + H_2$
(c) $Cl_2 + 2\,I^- \longrightarrow I_2 + 2\,Cl^-$

SOLUTION

The oxidation numbers are calculated by the methods illustrated in Example 15.1.

(a) $2\,NO \;+\; O_2 \;\longrightarrow\; 2\,NO_2$
         $+2\;-2 \qquad 0 \qquad\quad +4\;-2$
      rules 5, 8 rule 1   rules 5, 8

The oxidation number of N has increased from $+2$ to $+4$. Therefore, the substance that contains N, NO, has been oxidized and is the reducing agent.
    The oxidation number of the O in O_2 has decreased from 0 to -2. Therefore, the O_2 has been reduced and is the oxidizing agent.

(b) $Zn \;+\; 2\,HCl \;\longrightarrow\; ZnCl_2 \;+\; H_2$
         $0 \qquad +1\;-1 \qquad\;\; +2\;-1 \qquad 0$
      rule 1 rules 6, 8   rules 4, 8 rule 1

The oxidation number of Zn has increased from 0 to $+2$. An increase in oxidation number is associated with oxidation. Therefore, the element Zn has been oxidized and is the reducing agent.
    The oxidation number of H has decreased from $+1$ to 0. A decrease in oxidation number is associated with reduction. Therefore, the HCl, the hydrogen-containing compound, is the oxidizing agent.

(c) $Cl_2 \;+\; 2\,I^- \;\longrightarrow\; I_2 \;+\; Cl^-$
         $0 \quad -1 \qquad\quad 0 \quad -1$
      rule 1 rule 2   rule 1 rule 2

The oxidation number of I has increased from -1 to 0. Thus, I^-, the iodine-containing reactant, has been oxidized and is the reducing agent.
    The oxidation number of Cl has decreased from 0 to -1. Thus, Cl_2, the chlorine-containing reactant, has been reduced and is the oxidizing agent.

Practice Exercise 15.2

Determine oxidation numbers for each atom in the following reactions, and identify the oxidizing and reducing agents.

(a) $2 SO_2 + O_2 \longrightarrow 2 SO_3$

(b) $2 Fe_2O_3 + 3 C \longrightarrow 4 Fe + 3 CO_2$

(c) $Pb + Cu^{2+} \longrightarrow Cu + Pb^{2+}$

15.3 Types of Chemical Reactions

Two classification systems for chemical reactions are in common use. We have now encountered both of them.

The first system, presented initially in Section 10.5, recognized five types of reactions.

1. Synthesis $(X + Y \longrightarrow XY)$
2. Decomposition $(XY \longrightarrow X + Y)$
3. Single-replacement $(X + YZ \longrightarrow Y + XZ)$
4. Double-replacement $(AX + BY \longrightarrow AY + BX)$
5. Combustion (reaction with O_2)

The second system involves two reaction types.

1. Oxidation–reduction (or redox)

2. Non-oxidation–reduction (or nonredox)

As we have just learned (Sec. 15.2), reactions in which oxidation numbers change are called oxidation–reduction reactions. A **non-oxidation-reduction reaction** *is a chemical reaction in which oxidation numbers do not change*.

These two classification systems are not mutually exclusive and are commonly used together. For example, a particular reaction may be characterized as a single-replacement redox reaction.

Synthesis reactions with only elements as reactants are always oxidation–reduction reactions. Oxidation number changes must occur because all elements (the reactants) have an oxidation number of zero and all of the constituent elements of a compound *cannot* have oxidation numbers of zero. Synthesis reactions in which compounds are the reactants may or may not be redox reactions.

$$S + O_2 \longrightarrow SO_2 \qquad \text{(redox synthesis)}$$

$$K_2O + H_2O \longrightarrow 2 KOH \qquad \text{(nonredox synthesis)}$$

$$2 NO + O_2 \longrightarrow 2 NO_2 \qquad \text{(redox synthesis)}$$

Both redox and nonredox decomposition reactions are common. At sufficiently high temperatures all compounds can be broken down (decomposed) into their constituent elements. Such reactions, where only elements are the products, are always redox reactions. Decomposition reactions where compounds are the products are most often nonredox reactions.

$$2 CuO \longrightarrow 2 Cu + O_2 \qquad \text{(redox decomposition)}$$

$$2 KClO_3 \longrightarrow 2 KCl + 3 O_2 \qquad \text{(redox decomposition)}$$

$$CaCO_3 \longrightarrow CaO + CO_2 \qquad \text{(nonredox decomposition)}$$

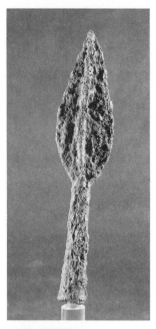

Figure 15.1

Spears like this one were made by prehistoric people by the reduction of iron ore with charcoal. Such a reaction is a single-replacement redox reaction:

$Fe_2O_3 + 3 C \longrightarrow 2 Fe + 3 CO$

(Erich Lessing/Art Resource, NY)

Single-replacement reactions are always redox reactions. By definition, an element and a compound are reactants and an element and a compound are products. The elements always undergo oxidation number change (see Fig. 15.1). Two of the reaction types studied in Chapter 14 are redox single-replacement reactions—the reaction between an acid and an active metal (Sec. 14.8) and the reaction between a metal and an aqueous salt solution (Sec. 14.10).

Double-replacement reactions generally involve acids, bases, and salts in aqueous solution. In such reactions ions, which maintain their identity, are generally trading places. Such reactions will always be nonredox reactions. All acid–base neutralization reactions (Sec. 14.8) are nonredox double-replacement reactions.

Combustion reactions (Sec. 10.5) are always redox reactions. However, as mentioned in Section 10.5, they do not fit any of the four general reaction patterns of synthesis, decomposition, single-replacement, and double-replacement. Features common to all combustion reactions are the necessity of oxygen (O_2) as a reactant and the presence of one or more oxides among the products.

EXAMPLE 15.3

Classifying Reactions as Redox or Nonredox

Classify the following reactions as redox or nonredox. Further classify them as synthesis, decomposition, single-replacement, double-replacement, or combustion.

(a) $Ni + F_2 \longrightarrow NiF_2$

(b) $Fe_2O_3 + 3 C \longrightarrow 2 Fe + 3 CO$

(c) $C_4H_8 + 6 O_2 \longrightarrow 4 CO_2 + 4 H_2O$

(d) $H_2SO_4 + 2 NaOH \longrightarrow Na_2SO_4 + 2 H_2O$

SOLUTION

The oxidation numbers are calculated by the method illustrated in Example 15.1.

(a)

$$Ni + F_2 \longrightarrow NiF_2$$

$$\begin{array}{cccc} 0 & 0 & & +2 \ -1 \\ \text{rule 1} & \text{rule 1} & & \text{rules 4, 8} \end{array}$$

This is a *redox* reaction; the oxidation numbers of both Ni and F change. Since one substance is produced from two substances, it is also a *synthesis* reaction. We thus have a redox synthesis reaction.

(b)

$$Fe_2O_3 + 3 C \longrightarrow 2 Fe + 3 CO$$

$$\begin{array}{cccc} +3 \ -2 & 0 & & 0 & +2 \ -2 \\ \text{rules 5, 8} & \text{rule 1} & & \text{rule 1} & \text{rules 5, 8} \end{array}$$

This is a *redox* reaction; carbon is oxidized, iron is reduced. Having an element and a compound as reactants and an element and compound as products is a characteristic of a *single-replacement* reaction. That is the type of reaction we have here: iron and carbon are exchanging places. We thus have a redox single-replacement reaction.

(c)

$$C_4H_8 + 6 O_2 \longrightarrow 4 CO_2 + 4 H_2O$$

$$\begin{array}{cccc} -2 \ +1 & 0 & & +4 \ -2 & +1 \ -2 \\ \text{rules 6, 8} & \text{rule 1} & & \text{rules 5, 7} & \text{rules 5, 8} \end{array}$$

This is a *redox* reaction; the oxidation numbers of both carbon and oxygen change. This reaction is also a *combustion* reaction. We thus have a redox combustion reaction.

(d) $H_2SO_4 \ + \ 2\,NaOH \ \longrightarrow \ Na_2SO_4 \ + \ 2\,H_2O$

$\quad \ \ +1\ +6\ -2 \quad +1\ -2\ +1 \qquad\ \ +1\ +6\ -2 \qquad +1\ -2$

rules 5, 6, 8   rules 3, 5, 6   rules 3, 5, 8   rules 5, 6

This is a *nonredox* reaction; there are no oxidation number changes. The reaction is also a *double-replacement* reaction; hydrogen and sodium are changing places, that is, "swapping partners." Thus we have a nonredox double-replacement reaction.

Practice Exercise 15.3

Classify the following reactions as redox or nonredox. Further classify them, as synthesis, decomposition, single-replacement, double-replacement, or combustion.

(a) $2\,KNO_3 \ \longrightarrow \ 2\,KNO_2 \ + \ O_2$

(b) $Zn \ + \ CuBr_2 \ \longrightarrow \ ZnBr_2 \ + \ Cu$

(c) $CH_4 \ + \ 2\,O_2 \ \longrightarrow \ CO_2 \ + \ 2\,H_2O$

(d) $NiCl_2 \ + \ 2\,NaOH \ \longrightarrow \ Ni(OH)_2 \ + \ 2\,NaCl$

15.4 Balancing Oxidation–Reduction Equations

Balancing an equation is not a new topic to us. In Section 10.3 we learned how to balance equations by the *inspection method*. With that method, we start with the most complicated compound within the equation and balance one of the elements in it. Then we balance the atoms of a second element, then a third, and so on until all elements are balanced. This inspection procedure is a useful method for balancing simple equations with small coefficients. However, it breaks down when applied to complicated equations.

Equations for redox reactions are often quite complicated and contain numerous reactants and products and large coefficients. Trying to balance redox equations such as

$$PH_3 \ + \ CrO_4{}^{2-} \ + \ H_2O \ \longrightarrow \ P_4 \ + \ Cr(OH)_4{}^- \ + \ OH^-$$

or

$$As_4O_6 \ + \ MnO_4{}^- \ + \ H_2O \ \longrightarrow \ AsO_4{}^{3-} \ + \ H^+ \ + \ Mn^{2+}$$

by inspection is a tedious, time-consuming, frustrating experience. Balancing such equations is, however, easily accomplished by systematic equation-balancing procedures that use oxidation numbers and focus on the fact that the numbers of electrons lost and gained in a redox reaction must be equal.

Two distinctly different approaches for systematically balancing redox equations are in common use; the *oxidation-number method* and the *half-reaction method*. Each method has advantages and disadvantages. We will consider both methods.

15.5 Oxidation-Number Method for Balancing Redox Equations

A useful feature of oxidation numbers is that they provide a rather easy method for recognizing and balancing redox equations. The steps involved in their use in this balancing process are as follows.

STEP 1   Assign oxidation numbers to all atoms in the equation and determine which atoms are undergoing a change in oxidation number.

STEP 2   Determine the magnitude of the change in oxidation number *per atom* for the elements undergoing a change in oxidation number.

Draw a large bracket from the element in the reactant to the element in the product, and write the increase or decrease in oxidation number at the middle of the bracket. (See the examples that follow.)

STEP 3   When more than one atom of an element that changes oxidation number is present in a formula unit (of either reactant or product), determine the change in oxidation number per *formula unit*.

Indicate this change per formula unit by multiplying the oxidation number change per atom, already written on the brackets, by an appropriate factor.

STEP 4   Determine multiplying factors that make the total increase in oxidation number equal to the total decrease in oxidation number.

Place them on the bracket also.

STEP 5   Place in front of the oxidizing and reducing agents and their products in the equation coefficients that are consistent with the total number of atoms of the elements undergoing oxidation-number change.

STEP 6   Balance all other atoms in the equation except those of hydrogen and oxygen.

In doing this, do not alter the coefficients determined previously.

STEP 7   Balance the charge (the sum of all the ionic charges) so that it is the same on both sides of the equation by adding H^+ or OH^- ions.

This step is necessary only when dealing with net ionic equations describing aqueous solution reactions. If the reaction takes place in acidic solution, add H^+ ion to the side deficient in positive charge. If the reaction takes place in basic solution, add OH^- ion to the side deficient in negative charge.

> For net ionic redox reactions, two "balances" must be made: atoms and charge. That is, there must be the same number of atoms of each element in the reactants and products, and the total charge on the reactants must equal the total charge on the products.

STEP 8   Balance the hydrogen atoms.

For net ionic equations, H_2O must usually be added to an appropriate side of the equation to achieve hydrogen balance. Water is, of course, present in all aqueous solutions and can be either a reactant or a product.

STEP 9   Balance the oxygen atoms.

The oxygens should automatically be balanced. If oxygens do not balance, there is a mistake in a previous step. Check your work.

Now let us consider some examples where these rules are applied. The first two examples will involve molecular equations. The third example involves a net ionic equation. In balancing net ionic equations, any H_2O, H^+, or OH^- present is usually left out of the unbalanced equation that we start with and then added as needed during the balancing process.

EXAMPLE 15.4

Balancing a Molecular Redox Equation Using the Oxidation-Number Method

Balance the following molecular redox equation by the oxidation-number method of balancing.

$$Cr + O_2 + HBr \longrightarrow CrBr_3 + H_2O$$

SOLUTION

STEP 1 We identify the elements being oxidized and reduced by assigning oxidation numbers.

$$Cr + O_2 + HBr \longrightarrow CrBr_3 + H_2O$$
$$0 \quad\quad 0 \quad\; +1-1 \quad\;\; +3-1 \quad +1-2$$

Chromium (Cr) and oxygen (O) are the elements that undergo oxidation-number change.

STEP 2 The change in oxidation number *per atom* is shown by drawing brackets connecting the oxidizing and reducing agents to their products and indicating the change at the middle of the bracket.

$$0 \quad\quad\;\; (+3) \quad\quad\;\; +3$$
$$Cr + O_2 + HBr \longrightarrow CrBr_3 + H_2O$$
$$0 \quad\quad\;\; (-2) \quad\quad -2$$

Change in oxidation number per atom

STEP 3 For Cr the change in oxidation number per formula unit is the same as the change per atom, since both Cr and CrBr$_3$, the two Cr-containing species, contain only one Cr atom. For O the change in oxidation number per formula unit will be double the change per atom since O$_2$ contains two atoms. The change per formula unit is indicated by multiplying the per-atom change by an appropriate numerical factor, which is 2 in this case.

$$(+3)$$
$$Cr + O_2 + HBr \longrightarrow CrBr_3 + H_2O$$
$$2(-2)$$

Change in oxidation number per formula unit

STEP 4 For Cr, the total increase in oxidation number per formula unit is +3. For oxygen, the total decrease in oxidation number per formula unit is −4. To make the increase equal to the decrease, we must multiply the oxidation-number change for the element oxidized (Cr) by 4 and the oxidation-number change for the element reduced (O) by 3. This will make the increase and decrease both numerically equal to 12.

$$4(+3)$$
$$Cr + O_2 + HBr \longrightarrow CrBr_3 + H_2O$$
$$3[2(-2)]$$

Oxidation-number increase equals oxidation-number decrease

STEP 5 We are now ready to place coefficients in the equation in front of the oxidizing and reducing agents and their products. The bracket notation indicates that four Cr atoms undergo an oxidation-number change. Place the coefficient 4 in front of both Cr and CrBr$_3$. The bracket notation also indicates that six O atoms (3 × 2) undergo an oxidation-number decrease of two units. Thus, we need six oxygen atoms on each side. Place the coefficient 3 in front of O$_2$ (6 atoms of O), and the coefficient 6 in front of H$_2$O (6 atoms of O).

$$4\,Cr + 3\,O_2 + HBr \longrightarrow 4\,CrBr_3 + 6\,H_2O$$

The equation is only partially balanced at this point; only Cr and O atoms are balanced.

STEP 6 We next balance the element Br (by inspection). There are twelve Br atoms on the right side. Thus, to obtain twelve Br atoms on the left side we place the co-efficient 12 in front of HBr.

$$4\,Cr + 3\,O_2 + 12\,HBr \longrightarrow 4\,CrBr_3 + 6\,H_2O$$

STEP 7 This step is not needed when the equation is a molecular equation.

STEP 8 In this particular equation the H atoms are already balanced. There are 12 hydrogen atoms on each side of the equation.

STEP 9 If all of the previous procedures (steps) have been carried out correctly, the O atoms should automatically balance. They do. There are six O atoms on each side of the equation. The balanced equation is thus

$$4\,Cr + 3\,O_2 + 12\,HBr \longrightarrow 4\,CrBr_3 + 6\,H_2O$$

Practice Exercise 15.4

Balance the following molecular redox equation by the oxidation-number method.

$$Zn + O_2 + HCl \longrightarrow ZnCl_2 + H_2O$$

EXAMPLE 15.5

Balancing a Molecular Redox Equation Using the Oxidation-Number Method

Balance the following molecular redox equation by the oxidation-number method of balancing.

$$HNO_2 + HI \longrightarrow NO + I_2 + H_2O$$

SOLUTION

STEP 1 $HNO_2 \ + \ HI \ \longrightarrow \ NO \ + I_2 + H_2O$
         $+1\,+3\,-2\ \ +1\,-1+2\,-2\ \ \ 0\ \ \ +1\,-2$

The two elements undergoing oxidation-number change are N and I.

STEP 2

$$\overset{+3 \qquad\quad (-1) \qquad\quad +2}{HNO_2 + HI \longrightarrow NO + I_2 + H_2O}$$
$$\underset{-1 \qquad (+1) \qquad 0}{}$$

Change in oxidation number per atom

STEP 3

$$\overset{(-1)}{HNO_2 + HI \longrightarrow NO + I_2 + H_2O}$$
$$\underset{2(+1)}{}$$

Change in oxidation number per formula unit

The iodine oxidation-number change per atom had to be multiplied by 2 since there are two iodine atoms per molecule in I_2. Thus, a minimum of two I atoms must undergo an oxidation-number increase. This illustrates that *both* reactant and product formulas must be considered when determining the change in oxidation number per formula unit.

STEP 4

$$\overset{2(-1)}{HNO_2 + HI \longrightarrow NO + I_2 + H_2O}$$
$$\underset{2(+1)}{}$$

Oxidation-number increase equals oxidation-number decrease

By multiplying the N per formula unit oxidation-number decrease of -1 by 2 we make the oxidation-number increase and decrease equal; both are at two units.

STEP 5 The coefficients for HNO_2, HI, NO, and I_2 in the equation are determined from the bracket information, which indicates that two N atoms undergo an oxidation number change for every two I atoms that change.

$$2\,HNO_2 + 2\,HI \longrightarrow 2\,NO + 1\,I_2 + H_2O$$

STEP 6 The only atoms left to balance are H and O.

STEP 7 This step is not needed for a molecular equation.

STEP 8 We balance the hydrogen, at four on each side, by placing the coefficient 2 in front of the H_2O on the right side.

$$2\,HNO_2 + 2\,HI \longrightarrow 2\,NO + 1\,I_2 + 2\,H_2O$$

STEP 9 If all of the procedures in previous steps have been carried out correctly, the O atoms should automatically balance. They do. There are four O atoms on each side of the equation.

$$2\,HNO_2 + 2\,HI \longrightarrow 2\,NO + I_2 + 2\,H_2O$$

Practice Exercise 15.5

Balance the following molecular redox equation by the oxidation-number method of balancing.

$$NF_3 + AlCl_3 \longrightarrow N_2 + Cl_2 + AlF_3$$

EXAMPLE 15.6

Balancing a Net Ionic Redox Equation Using the Oxidation-Number Method

Balance the following net ionic redox equation by the oxidation-number method of balancing.

$$Cu + NO_3^- \longrightarrow Cu^{2+} + NO_2$$

This reaction occurs in acidic solution.

SOLUTION

STEP 1 $Cu + NO_3^- \longrightarrow Cu^{2+} + NO_2$
　　　　$\ \ 0\ \ +5\,-2\ \ \ \ \ \ +2\ \ \ +4\,-2$

STEP 2 The two elements undergoing oxidation-number change are Cu and N.

$$
\begin{array}{c}
\ \ 0\ \ \ \ \ \ (+2)\ \ \ \ \ \ \ \ +2 \\
\overbrace{Cu\ +\ NO_3^-\ \longrightarrow\ Cu^{2+}}\ +\ NO_2 \qquad \text{Change in oxidation} \\
\underbrace{\phantom{Cu\ +\ NO_3^-\ \longrightarrow\ Cu^{2+}\ +\ NO_2}} \qquad \text{number per atom} \\
\ \ +5\ \ \ \ \ (-1)\ \ \ \ \ \ \ +4
\end{array}
$$

STEP 3 For both Cu and N the oxidation-number change per formula unit is the same as per atom. Both Cu and Cu^{2+} contain only one Cu atom; similarly, both NO_3^- and NO_2 contain only one N atom.

STEP 4 By multiplying the N oxidation-number decrease by 2 we make the oxidation-number increase and decrease per formula unit the same—two units.

$$\overset{(+2)}{\overbrace{Cu \; + \; \underset{\underset{2(-1)}{\underbrace{}}}{NO_3^-} \longrightarrow \; Cu^{2+}}} + \; NO_2$$

Oxidation-number increase equals oxidation-number decrease

STEP 5 The bracket notation indicates that two N atoms and one Cu atom undergo an oxidation-number change. Translating this information into coefficients, we get

$$1\,Cu + 2\,NO_3^- \longrightarrow 1\,Cu^{2+} + 2\,NO_2$$

STEP 6 The only atoms left to balance are H and O.

STEP 7 Since this is a net ionic equation, the charges must balance; that is, the sum of the ionic charges of all species on each side of the equation must be equal. (They do not have to add up to zero; they just have to be equal.) In acidic solution, which is the case in this example, charge balance is accomplished by adding H^+ ion.

As the equation now stands, we have a charge of -2 on the left side (two nitrate ions each with a -1 charge) and a charge of $+2$ on the right side (one copper ion). By adding four H^+ ions to the left side we balance the charge at $+2$.

$$-2 + (+4) = +2$$

The equation at this point becomes

$$1\,Cu + 2\,NO_3^- + 4\,H^+ \longrightarrow 1\,Cu^{2+} + 2\,NO_2$$

STEP 8 The hydrogen atoms are balanced through the addition of H_2O molecules. There are four H atoms on the left side ($4\,H^+$ ions) and none on the right side. Addition of two H_2O molecules to the right side will balance the H atoms at four per side.

$$1\,Cu + 2\,NO_3^- + 4\,H^+ \longrightarrow 1\,Cu^{2+} + 2\,NO_2 + 2\,H_2O$$

STEP 9 The O atoms automatically balance at six atoms on each side. This is our double check that previous steps have been correctly carried out. The balanced net ionic equation is thus

$$Cu + 2\,NO_3^- + 4\,H^+ \longrightarrow Cu^{2+} + 2\,NO_2 + 2\,H_2O$$

Practice Exercise 15.6

Balance the following net ionic redox equation by the oxidation-number method of balancing.

$$UO_2^+ + Cr_2O_7^{2-} \longrightarrow UO_2^{2+} + Cr^{3+}$$

This reaction occurs in acidic solution.

15.6 Half-Reaction Method for Balancing Redox Equations

The basis for the half-reaction method for balancing redox equations is the separation of the unbalanced redox equation into two *half-reactions*, one for oxidation and one for reduction. A **half-reaction** *is a chemical equation that describes one of the two parts of an overall oxidation–reduction reaction, either the oxidation part or the reduction part.* There are always two half-reactions associated with a given redox equation. These half-reactions, once obtained, are then balanced separately. The two balanced half-reactions are added together to generate the overall balanced equation.

The division of the original unbalanced redox equation into two parts (two half-reactions) is artificial. One half-reaction does not really take place independently of the other; we cannot have oxidation without reduction. Nevertheless, this method of balancing redox equations is preferred in certain areas of redox chemistry. In particular, it leads to an increased understanding of the reactions that take place in electrochemical cells such as batteries. Electrochemical cells are the topic of Section 15.8.

As we did with the oxidation-number method for balancing redox equations, we will break the half-reaction balancing process into a series of steps.

STEP 1 Using oxidation numbers, determine which atoms are oxidized and which are reduced. Based on this information, split the redox equation into two skeletal half-reaction equations:

(a) an *oxidation* half-reaction equation, which involves the formula of the substance containing the element oxidized along with other species associated with it.

(b) a *reduction* half-reaction equation, which involves the formula of the substance containing the element reduced along with other species associated with it.

STEP 2 Balance each of the half-reactions.

(a) First, balance the element oxidized or reduced and then balance any other elements present in the skeletal equation other than oxygen or hydrogen.

(b) Next, show the number of electrons lost or gained in the oxidation or reduction. Use the change in oxidation number and the number of atoms oxidized or reduced to determine the number of electrons lost or gained. Electrons lost (oxidation) are shown on the product side of the equation and electrons gained (reduction) on the reactant side of the equation.

(c) Balance the ionic charge by adding H^+ ions (acidic solution) or OH^- ions (basic solution) as reactant or product. (Remember that the electrons previously added must be considered in balancing charge.)

(d) Balance the hydrogen atoms by adding H_2O molecules as reactant or product.

(e) Verify that the oxygen atoms are balanced. (If they do not balance, a mistake has been made in a previous step.)

STEP 3 Multiply each balanced half-reaction by appropriate integers to make the total number of electrons lost equal the total number of electrons gained.

STEP 4 Add the two half-reactions together and cancel identical species, including electrons, on each side of the equation. See if the coefficients obtained can be simplified.

Examples 15.7 through 15.9 illustrate how the preceding guidelines for balancing redox equations are applied.

EXAMPLE 15.7

Balancing a Net Ionic Redox Equation That Involves Acidic Solution Using the Half-Reaction Method

Balance the following net ionic equation that occurs in acidic solution using the half-reaction method of balancing.

$$S^{2-} + NO_3^- \longrightarrow S + NO \quad \text{(acidic solution)}$$

SOLUTION

STEP 1 *Determine the oxidation and reduction skeletal half-reactions.* Assigning oxidation numbers, we get

$$S^{2-} + N\ O_3^{-} \longrightarrow S + N\ O$$
$$\phantom{S^{2}}{-2}\quad{+5}\ {-2}{0}\quad{+2}\ {-2}$$

Sulfur is oxidized, increasing in oxidation number from -2 to 0. Nitrogen is reduced, decreasing in oxidation number from $+5$ to $+2$.

The skeletal half-reactions for oxidation and reduction are

$$\text{oxidation:}\quad S^{2-} \longrightarrow S$$
$$\text{reduction:}\quad NO_3^{-} \longrightarrow NO$$

STEP 2 *Balance the individual half-reactions.*

(a) In both half-reactions, the element being oxidized or reduced is already balanced—one atom of S on both sides in the first half-reaction and one atom of N on both sides in the second half-reaction. There are no other elements present except oxygen.

$$\text{oxidation:}\quad S^{2-} \longrightarrow S$$
$$\text{reduction:}\quad NO_3^{-} \longrightarrow NO$$

(b) The oxidation number increase for S is $+2$. This is caused by the loss of two electrons, which are shown on the product side of the oxidation half-reaction.

$$\text{oxidation:}\quad S^{2-} \longrightarrow S + 2\,e^{-}$$

The oxidation number decrease for N is -3. This results from the gain of three electrons, which are shown on the reactant side of the reduction half-reaction.

$$\text{reduction:}\quad NO_3^{-} + 3\,e^{-} \longrightarrow NO$$

(c) Since this is an acidic solution reaction, charge balance is achieved by adding H^+ ions. In the oxidation half-reaction there is a charge of -2 on each side of the equation. No H^+ ions are needed, since the charge is already in balance.

$$\text{oxidation:}\quad S^{2-} \longrightarrow S + 2\,e^{-}$$

In the reduction half-reaction there is a charge of -4 on the left side of the equation (-1 from the NO_3^{-} ion and -3 from the three electrons). There is no charge on the right side of the equation. Charge balance is achieved by adding four H^+ ions to the left side of the equation. Each side of the equation will now have zero charge.

$$\text{reduction:}\quad NO_3^{-} + 3\,e^{-} + 4\,H^{+} \longrightarrow NO$$

(d) Water molecules are used to achieve hydrogen balance. Since no hydrogen is present in the oxidation half-reaction, no water molecules are needed.

$$\text{oxidation:}\quad S^{2-} \longrightarrow S + 2\,e^{-}$$

In the reduction half-reaction, two water molecules are added to the right side of the equation. We now have four hydrogen atoms on each side of the equation.

$$\text{reduction:}\quad NO_3^{-} + 3\,e^{-} + 4\,H^{+} \longrightarrow NO + 2\,H_2O$$

(e) There are no oxygen atoms present in the oxidation half-reaction. In the reduction half-reaction, the oxygen balances at three atoms on each side of the equation. The two balanced half-reactions are

oxidation: $S^{2-} \longrightarrow S + 2e^-$

reduction: $NO_3^- + 3\,e^- + 4\,H^+ \longrightarrow NO + 2\,H_2O$

STEP 3 *Equalize electron loss and electron gain.*

Two electrons are produced in the oxidation half-reaction and three electrons are gained in the reduction half-reaction. To equalize electron loss and electron gain, we multiply the oxidation half-reaction by three and the reduction half-reaction by two. We have then an electron loss of 6 and an electron gain of 6.

oxidation: $3(S^{2-} \longrightarrow S + 2\,e^-)$

reduction: $2(NO_3^- + 3\,e^- + 4\,H^+ \longrightarrow NO + 2\,H_2O)$

STEP 4 *Add the half-reactions and cancel identical species.*

Adding the two half-reactions together, we get

oxidation: $3\,S^{2-} \longrightarrow 3\,S + \cancel{6e^-}$

reduction: $2\,NO_3^- + \cancel{6e^-} + 8\,H^+ \longrightarrow 2\,NO + 4\,H_2O$

$$3\,S^{2-} + 2\,NO_3^- + 8\,H^+ \longrightarrow 3\,S + 2\,NO + 4\,H_2O$$

There are no species to cancel other than the electrons. The electrons must always cancel. If they do not, we have made a mistake in step 3.

Practice Exercise 15.7

Balance the following net ionic equation that occurs in acidic solution using the half-reaction method of balancing.

$$Fe^{2+} + Cr_2O_7^{2-} \longrightarrow Fe^{3+} + Cr^{3+} \quad \text{(acidic solution)}$$

EXAMPLE 15.8

Balancing a Net Ionic Redox Equation That Involves Basic Solution Using the Half-Reaction Method

Balance the following net ionic redox equation that occurs in basic solution using the half-reaction method of balancing.

$$S^{2-} + Cl_2 \longrightarrow SO_4^{2-} + Cl^- \quad \text{(basic solution)}$$

SOLUTION

STEP 1 *Determine the oxidation and reduction skeletal half-reactions.*

Assigning oxidation numbers, we get

$$\underset{-2}{S^{2-}} + \underset{0}{Cl_2} \longrightarrow \underset{+6\,-2}{SO_4^{2-}} + \underset{-1}{Cl^-}$$

Sulfur is oxidized, increasing in oxidation number from -2 to $+6$. Chlorine is reduced, decreasing in oxidation number from 0 to -1. The skeletal half-reactions for oxidation and reduction are

oxidation: $S^{2-} \longrightarrow SO_4^{2-}$

reduction: $Cl_2 \longrightarrow Cl^-$

STEP 2 *Balance the individual half-reactions.*

(a) In the oxidation half-reaction the S is already balanced.

$$\text{oxidation:} \quad S^{2-} \longrightarrow SO_4^{2-}$$

To balance the Cl in the reduction half-reaction the coefficient 2 must be added on the right side.

$$\text{reduction:} \quad Cl_2 \longrightarrow 2\,Cl^-$$

(b) The oxidation number increase for S is $+8$, which corresponds to the loss of 8 electrons.

$$\text{oxidation:} \quad S^{2-} \longrightarrow SO_4^{2-} + 8\,e^-$$

The oxidation number decrease for Cl is -1, which corresponds to a gain of one electron. Since there are two Cl atoms changing, the total electron gain is two electrons.

$$\text{reduction:} \quad Cl_2 + 2\,e^- \longrightarrow 2\,Cl^-$$

(c) Since this reaction occurs in basic solution, charge balance is achieved by adding OH^- ions. In the oxidation half-reaction there is a charge of -2 on the left side and a charge of -10 (one sulfate and eight electrons) on the right side. The charge is brought into balance, at a -10, by adding 8 OH^- ions to the left side of the equation.

$$\text{oxidation:} \quad S^{2-} + 8\,OH^- \longrightarrow SO_4^{2-} + 8\,e^-$$

In the reduction half-reaction, the charge is already balanced at a -2 on each side. No OH^- ions are needed.

$$\text{reduction:} \quad Cl_2 + 2\,e^- \longrightarrow 2\,Cl^-$$

(d) Hydrogen balance is achieved in the oxidation half-reaction by adding four H_2O molecules to the right side of the equation.

$$\text{oxidation:} \quad S^{2-} + 8\,OH^- \longrightarrow SO_4^{2-} + 8\,e^- + 4\,H_2O$$

Hydrogen balance is not needed in the reduction half-reaction since no hydrogen is present.

$$\text{reduction:} \quad Cl_2 + 2\,e^- \longrightarrow 2\,Cl^-$$

(e) Oxygen balances at 8 atoms on each side of the equation in the oxidation half-reaction. Oxygen is not present in the reduction half-reaction. The two balanced half-reactions are

$$\text{oxidation:} \quad S^{2-} + 8\,OH^- \longrightarrow SO_4^{2-} + 8\,e^- + 4\,H_2O$$
$$\text{reduction:} \quad Cl_2 + 2\,e^- \longrightarrow 2\,Cl^-$$

STEP 3 *Equalize electron loss and electron gain.*

Eight electrons are produced in the oxidation half-reaction and two electrons are gained in the reduction half-reaction. Multiplying the reduction half-reaction by 4 will cause electron loss and electron gain to be equal at eight electrons.

$$\text{oxidation:} \quad S^{2-} + 8\,OH^- \longrightarrow SO_4^{2-} + 8\,e^- + 4\,H_2O$$
$$\text{reduction:} \quad 4(Cl_2 + 2e^- \longrightarrow 2\,Cl^-)$$

STEP 4 *Add the half-reactions and cancel identical species.*

Adding the two half-reactions together, we get

oxidation: $S^{2-} + 8\,OH^- \longrightarrow SO_4{}^{2-} + \cancel{8\,e^-} + 4\,H_2O$

reduction: $4\,Cl_2 + \cancel{8\,e^-} \longrightarrow 8\,Cl^-$

$$S^{2-} + 8\,OH^- + 4\,Cl_2 \longrightarrow SO_4{}^{2-} + 4\,H_2O + 8\,Cl^-$$

There are no species to cancel other than the electrons.

Practice Exercise 15.8

Balance the following net ionic equation that occurs in basic solution using the half-reaction method of balancing.

$$Zn + MnO_4{}^- \longrightarrow Zn(OH)_2 + MnO_2 \quad \text{(basic solution)}$$

EXAMPLE 15.9

Balancing a Net Ionic Redox Equation That Involves Acidic Solution Using the Half-Reaction Method

Balance the following net ionic equation that occurs in acidic solution using the half-reaction method of balancing.

$$H_3AsO_3 + MnO_4{}^- \longrightarrow H_3AsO_4 + Mn^{2+} \quad \text{(acidic solution)}$$

SOLUTION

STEP 1 *Determine the oxidation and reduction skeletal half-reactions.*
Assigning oxidation numbers, we get

$$\underset{+1\ +3\ -2}{H_3AsO_3} + \underset{+7\ -2}{MnO_4{}^-} \longrightarrow \underset{+1\ +5\ -2}{H_3AsO_4} + \underset{+2}{Mn^{2+}}$$

Arsenic is oxidized, increasing in oxidation number from +3 to +5. Manganese is reduced, decreasing in oxidation number from +7 to +2. The skeletal half-reactions for oxidation and reduction are

oxidation: $H_3AsO_3 \longrightarrow H_3AsO_4$

reduction: $MnO_4{}^- \longrightarrow Mn^{2+}$

STEP 2 *Balance the individual half-reactions.*

(a) In both half-reactions, the element being oxidized or reduced is already balanced—one atom of As on both sides in the oxidation half-reaction and one atom of Mn on both sides in the reduction half-reaction.

oxidation: $H_3AsO_3 \longrightarrow H_3AsO_4$

reduction: $MnO_4{}^- \longrightarrow Mn^{2+}$

(b) The oxidation number increase for As is +2, which corresponds to the loss of two electrons.

oxidation: $H_3AsO_3 \longrightarrow H_3AsO_4 + 2\,e^-$

The oxidation number decrease for Mn is −5, which corresponds to the gain of five electrons.

$$\text{reduction:} \quad MnO_4^- + 5\,e^- \longrightarrow Mn^{2+}$$

(c) Since this reaction occurs in acidic solution, charge balance is achieved by adding H^+ ions. In the oxidation half-reaction there is a charge of zero on the left side and a charge of -2 (two electrons) on the right side. Charge balance, at zero, is achieved by adding two H^+ ions to the right side of the equation.

$$\text{oxidation:} \quad H_3AsO_3 \longrightarrow H_3AsO_4 + 2\,e^- + 2\,H^+$$

In the reduction half-reaction there is a charge of -6 on the left side (one MnO_4^- ion and 5 electrons) and a charge of $+2$ on the right side. Charge balance, at a $+2$, is achieved by adding 8 H^+ ions to the left side of the equation.

$$\text{reduction:} \quad MnO_4^- + 5\,e^- + 8\,H^+ \longrightarrow Mn^{2+}$$

(d) Hydrogen balance is obtained in the oxidation half-reaction by adding one H_2O molecule to the left side of the equation.

$$\text{oxidation:} \quad H_3AsO_3 + H_2O \longrightarrow H_3AsO_4 + 2\,e^- + 2\,H^+$$

Hydrogen balance is obtained in the reduction half-reaction by adding four H_2O molecules to the right side of the equation.

$$\text{reduction:} \quad MnO_4^- + 5\,e^- + 8\,H^+ \longrightarrow Mn^{2+} + 4\,H_2O$$

(e) Oxygen balances at 4 atoms on each side in both the oxidation and reduction half-reactions. The two balanced half-reactions are

$$\text{oxidation:} \quad H_3AsO_3 + H_2O \longrightarrow H_3AsO_4 + 2\,e^- + 2\,H^+$$

$$\text{reduction:} \quad MnO_4^- + 5\,e^- + 8\,H^+ \longrightarrow Mn^{2+} + 4\,H_2O$$

STEP 3 *Equalize the electron loss and electron gain.*
The lowest common multiple for an electron loss of 2 and an electron gain of 5 is 10 electrons. Thus, we multiply the oxidation half-reaction by 5 and the reduction half-reaction by two.

$$\text{oxidation:} \quad 5(H_3AsO_3 + H_2O \longrightarrow H_3AsO_4 + 2\,e^- + 2\,H^+)$$

$$\text{reduction:} \quad 2(MnO_4^- + 5\,e^- + 8\,H^+ \longrightarrow Mn^{2+} + 4\,H_2O)$$

STEP 1 *Add the half-reactions and cancel identical species.*
Adding the two half-reactions together, we get

$$
\begin{array}{ll}
\text{oxidation:} & 5\,H_3AsO_3 + 5\,H_2O \longrightarrow 5\,H_3AsO_4 + \cancel{10\,e^-} + 10\,H^+ \\
\text{reduction:} & 2\,MnO_4^- + \cancel{10\,e^-} + 16\,H^+ \longrightarrow 2\,Mn^{2+} + 8\,H_2O \\
\hline
& 5\,H_3AsO_3 + 5\,H_2O + 2\,MnO_4^- + 16\,H^+ \longrightarrow 5\,H_3AsO_4 + 10\,H^+ + 2\,Mn^{2+} + 8\,H_2O
\end{array}
$$

Both H^+ ion and H_2O are on both sides of the equation. We can cancel 5 H_2O molecules from each side and 10 H^+ ions from each side. The final balanced equation becomes

$$5\,H_3AsO_3 + 2\,MnO_4^- + 6\,H^+ \longrightarrow 5\,H_3AsO_4 + 2\,Mn^{2+} + 3\,H_2O$$

Practice Exercise 15.9

Balance the following net ionic equation that occurs in acidic solution using the half-reaction method of balancing

$$HNO_2 + Cr_2O_7^{2-} \longrightarrow Cr^{3+} + NO_3^- \quad \text{(acidic solution)}$$

In each of the three examples we have just considered, the oxidation and reduction half-reactions were simultaneously balanced. This approach was used in the examples to enable us to make comparisons. In practice, particularly when you are thoroughly familiar with the balancing procedure, one half-reaction is usually completely balanced before work begins on balancing the other half-reaction. Usually, it is better to work on just one reaction at a time.

A comparison of the two methods for balancing redox equations is in order. Basic to each method is being able to recognize the elements involved in the actual oxidation–reduction process. The oxidation-number method works on the principle that the increase in oxidation number must equal the decrease in oxidation number. The half-reaction method involves equalizing the number of electrons lost by the substance oxidized with the number of electrons gained by the substance reduced.

The oxidation-number method is usually faster, particularly for simple equations. This potential speed is considered the major advantage of the oxidation-number method. The half-reaction method's focus on electron transfer is its major advantage. The feature becomes particularly important in electrochemistry (Sec. 15.8). In this field it is most useful to discuss chemical reactions in terms of half-reactions occurring at different locations (electrodes) in an electrochemical cell.

15.7 Disproportionation Reactions

A *disproportionation reaction* is a special type of oxidation–reduction reaction. A **disproportionation reaction** *is a redox reaction in which some atoms of a single element in a reactant are oxidized and others are reduced.* For such reactant behavior to be possible, the reactant must contain an element that is capable of having at least three oxidation numbers: its original number plus one higher and one lower oxidation number. Note that any given atom is not both oxidized and reduced. Some atoms are oxidized, and other atoms of the same element are reduced.

An example of a disproportionation reaction is

$$3\,Br_2 + 3\,H_2O \longrightarrow HBrO_3 + 5\,HBr$$

Note that two bromine-containing products have been produced from one bromine-containing reactant. The reactant bromine atoms have an oxidation number of zero. Bromine in $HBrO_3$ has a +5 oxidation number (it has been oxidized), and bromine in HBr has a −1 oxidation number (it has been reduced).

$$3\,Br_2 + 3\,H_2O \longrightarrow HBrO_3 + 5\,HBr$$
$$0 \qquad\quad +5 \qquad -1$$

Thus, some of the reactant bromine atoms have been oxidized, while others have been reduced. A disproportionation reaction has taken place.

Example 15.10 shows how the procedures for balancing redox equations (Sec. 15.5 and 15.6) are slightly modified to balance disproportionation reaction equations.

EXAMPLE 15.10

Balancing a Disproportionation Redox Equation

Balance the following disproportionation redox reaction using (a) the oxidation-number method and (b) the half-reaction method.

$$NO_2 \longrightarrow NO_3^- + NO \quad \text{(acidic solution)}$$

SOLUTION

(a) *Oxidation-Number Method*

STEP 1 In assigning oxidation numbers we immediately become aware that this is a disproportionation reaction. Nitrogen is the only element for which an oxidation-number change occurs.

$$NO_2 \longrightarrow NO_3^- + NO$$
$$+4-2 \qquad +5-2 \quad +2-2$$

STEP 2 Since the species NO_2 is undergoing both oxidation and reduction, for balancing purposes we will write it twice on the reactant side of the equation. With the NO_2 in two places, brackets can then be drawn in the normal manner to connect the substances involved in oxidation and reduction.

$$\overset{+4}{NO_2} + \overset{(+1)}{NO_2} \longrightarrow \overset{+5}{NO_3^-} + NO$$
$$\underset{+4}{\qquad} \underset{(-2)}{\qquad} \underset{+2}{\qquad}$$

Change in oxidation number per atom

(The NO_2 molecules will be recombined later into one location.)

STEP 3 The change in oxidation number per formula unit in both cases is the same as per atom.

STEP 4 By multiplying the oxidation-number increase by 2 we equalize the oxidation-number increase and decrease per formula unit.

$$\overset{2(+1)}{NO_2} + NO_2 \longrightarrow NO_3^- + NO$$
$$\underset{(-2)}{\qquad}$$

Oxidation-number increase equals oxidation-number decrease

STEP 5 The bracket notation indicates that two N atoms undergo an increase in oxidation number for every one that undergoes a decrease in oxidation number. Translating this information into equation coefficients gives

$$2\,NO_2 + 1\,NO_2 \longrightarrow 2\,NO_3^- + 1\,NO$$

Now that the equation coefficients for the substance involved in oxidation and reduction, NO_2, have been determined, we can combine the NO_2 into one location, reversing the process carried out in step 2.

$$3\,NO_2 \longrightarrow 2\,NO_3^- + 1\,NO$$

STEP 6 The only atoms left to balance are oxygen atoms.

STEP 7 Since this is a net ionic equation, charge must be balanced. In acidic solution, which is the case here, we balance the charge by adding H^+ ion. As the equation now stands, we have a charge of -2 on the right side (two NO_3^- ions). By adding 2 H^+ ions to the right side of the equation we balance the charge at zero

$$3\,NO_2 \longrightarrow 2\,NO_3^- + 1\,NO + 2H^+$$

STEP 8 Hydrogen atom balance is achieved through the addition of H_2O molecules. There are no H atoms on the left side and 2 H atoms on the right side. Addition of one H_2O molecule to the left side will balance the H atoms at 2 on each side.

$$1\,H_2O + 3\,NO_2 \longrightarrow 2\,NO_3^- + 1\,NO + 2\,H^+$$

STEP 9 The oxygen atoms should automatically balance. They do, at 7 atoms on each side.

$$H_2O + 3\,NO_2 \longrightarrow 2\,NO_3^- + NO + 2\,H^+$$

(b) *Half-Reaction Method*

STEP 1 *Determine the oxidation and reduction skeletal half-reactions.*
Assignment of oxidation numbers is the same as in part (a).

$$\underset{+4\;-2}{NO_2} \longrightarrow \underset{+5\;-2}{NO_3^-} + \underset{+2\;-2}{NO}$$

Nitrogen is undergoing both oxidation and reduction. The skeleton half-reactions for oxidation and reduction are

$$\text{oxidation:}\quad NO_2 \longrightarrow NO_3^-$$
$$\text{reduction:}\quad NO_2 \longrightarrow NO$$

Note how disproportionation is handled at this point. The substance undergoing disproportionation appears as a reactant in both the oxidation and reduction half-reactions.

STEP 2 *Balance the individual half-reactions.*

(a) In both half-reactions, the element being oxidized or reduced is already balanced—one atom of N in both cases.

$$\text{oxidation:}\quad NO_2 \longrightarrow NO_3^-$$
$$\text{reduction:}\quad NO_2 \longrightarrow NO$$

(b) The oxidation number increase for the oxidized N is $+1$, which corresponds to the loss of one electron.

$$\text{oxidation:}\quad NO_2 \longrightarrow NO_3^- + 1\,e^-$$

The oxidation number decrease for the reduced N is -2, which corresponds to the gain of two electrons.

$$\text{reduction:}\quad NO_2 + 2\,e^- \longrightarrow NO$$

(c) Since this reaction occurs in acidic solution, charge balance is achieved by adding H^+ ions. In the oxidation half-reaction there is no charge on the left side of the equation and a charge of -2 on the right side. Adding two H^+ ions to the right side of the equation will balance the charge at zero on both sides.

$$\text{oxidation:}\quad NO_2 \longrightarrow NO_3^- + 1\,e^- + 2\,H^+$$

In the reduction half-reaction there is a charge of -2 on the left side of the equation and no charge on the right side. Charge balance is achieved by adding two H^+ ions to the left side of the equation.

$$\text{reduction:}\quad NO_2 + 2\,e^- + 2\,H^+ \longrightarrow NO$$

(d) Hydrogen balance is obtained in the oxidation half-reaction by adding one H_2O to the left side of the equation.

$$\text{oxidation:}\quad NO_2 + H_2O \longrightarrow NO_3^- + 1\,e^- + 2\,H^+$$

Hydrogen balance is obtained in the reduction half-reaction by adding one H_2O to the right side of the equation.

$$\text{reduction:}\quad NO_2 + 2\,e^- + 2\,H^+ \longrightarrow NO + H_2O$$

(e) Oxygen balances at 2 atoms on each side in both the oxidation and reduction half-reactions. The two balanced half-reactions are

oxidation: $NO_2 + H_2O \longrightarrow NO_3^- + 1\,e^- + 2\,H^+$

reduction: $NO_2 + 2\,e^- + 2\,H^+ \longrightarrow NO + H_2O$

STEP 3 *Equalize the electron loss and electron gain.*
The oxidation half-reaction involves the loss of one electron. The reduction half-reaction involves the gain of two electrons. Multiplying the oxidation half-reaction by a factor of 2 will cause electron loss and gain to be equal at two electrons.

oxidation: $2(NO_2 + H_2O \longrightarrow NO_3^- + 1\,e^- + 2\,H^+)$

reduction: $NO_2 + 2\,e^- + 2\,H^+ \longrightarrow NO + H_2O$

STEP 4 *Add the half-reactions and cancel identical species.*
Adding the two half-reactions together, we get

oxidation: $2\,NO_2 + 2\,H_2O \longrightarrow 2\,NO_3^- + 2\,e^- + 4\,H^+$

reduction: $NO_2 + 2\,e^- + 2\,H^+ \longrightarrow NO + H_2O$

$2\,NO_2 + NO_2 + 2\,H_2O + 2\,H^+ \longrightarrow 2\,NO_3^- + 4\,H^+ + NO + H_2O$

Both H_2O and H^+ are on both sides of the equation and some of each can be canceled. Also, NO_2 appears in two places on the left side of the equation and needs to be combined. The final balanced equation is

$3\,NO_2 + H_2O \longrightarrow 2\,NO_3^- + NO + 2\,H^+$

Practice Exercise 15.10

Balance the following disproportionation redox reaction using: (a) the oxidation-number method and (b) the half-reaction method.

$MnO_4^{2-} \longrightarrow MnO_2 + MnO_4^-$ (acidic solution)

15.8 Some Important Oxidation–Reduction Reactions

In this section we consider two important applications of redox reactions, both of which involve *electrochemical cells*. An **electrochemical cell** is a device in which chemical energy is converted into electrical energy or electrical energy is converted into chemical energy. All electrochemical cells involve redox reactions. The redox reactions involved may be *spontaneous* or *nonspontaneous*.

1. A *spontaneous* oxidation–reduction reaction can be used to convert chemical energy into electrical energy. For this to occur, a redox reaction must be carried out in a specially designed apparatus called a *galvanic cell*.

2. A *nonspontaneous* oxidation–reduction reaction can be caused to occur by using electrical energy to produce chemical energy. Such a process is called *electrolysis*, and the apparatus in which the reaction is carried out is called an *electrolytic cell*.

Galvanic Cells

When a strip of zinc metal is placed in a solution of copper(II) sulfate, a source of Cu^{2+} ion, a coating of copper metal forms on the zinc strip (see Fig. 15.2). At the same time this occurs, some of the zinc dissolves to give Zn^{2+} ions in solution. The reaction occurring is

$$Zn(s) + Cu^{2+}(aq) \longrightarrow Zn^{2+}(aq) + Cu(s)$$

The sulfate ions $(SO_4{}^{2-})$ present in the copper(II) sulfate solution remain unaffected by this change. This reaction occurs because Zn is more active than Cu and will replace it (activity series; Sec. 14.10).

This reaction is an example of a spontaneous oxidation–reduction reaction. When zinc metal and a solution of Cu^{2+} ions come into contact with each other, the Zn metal is spontaneously oxidized and the Cu^{2+} ions are spontaneously reduced; a direct transfer of electrons from the zinc atoms to the Cu^{2+} ions occurs. The products of this reaction are copper atoms (Cu^{2+} ions that have gained electrons) and Zn^{2+} ions (zinc atoms that have lost electrons). Heat is liberated as the reaction proceeds, as evidenced by a slight warming of the solution.

By separating the reactants for this spontaneous redox reaction, one can obtain energy in the form of electricity rather than heat. The desired arrangement, a simple galvanic cell, is shown in Figure 15.3. A **galvanic cell** *is an electrochemical cell in which a spontaneous redox reaction is used to convert chemical energy to electrical energy.*

The galvanic cell of Figure 15.3 has two compartments separated by a porous disk. One compartment contains a strip of zinc metal immersed in a solution of zinc sulfate $(ZnSO_4)$, and the other contains a strip of copper metal immersed in a solution of copper(II) sulfate $(CuSO_4)$. The porous disk prevents the solutions from mixing freely but it does allow for passage of ions from one compartment to the other, a necessity for proper operation of the cell. The two strips of metal, called *electrodes*, are connected by a wire. This wire allows electrons to be transferred from one electrode to the other. As the spontaneous reaction between Zn and Cu^{2+} ions occurs, the flow of electrons through the wire can be demonstrated by placing a light bulb in the external circuit (see Fig. 15.3). The light bulb glows.

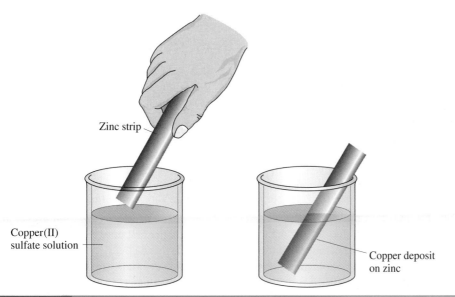

Zinc strip

Copper(II) sulfate solution

Copper deposit on zinc

Figure 15.2

A spontaneous redox reaction occurs when zinc metal is placed in a solution of Cu^{2+} ions. A coating of copper metal quickly deposits on the zinc.

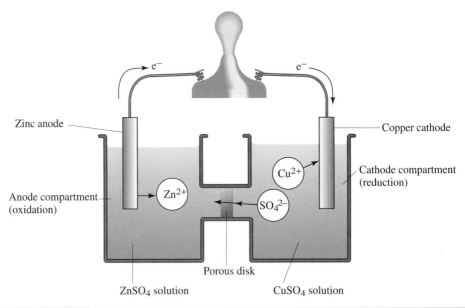

Anode compartment (oxidation)

Zinc anode

e^-

e^-

Copper cathode

Cathode compartment (reduction)

Zn^{2+}

Cu^{2+}

SO_4^{2-}

Porous disk

ZnSO₄ solution

CuSO₄ solution

Figure 15.3

A zinc-copper galvanic cell. The electrical energy produced by this cell is generated by the spontaneous redox reaction $Zn(s) + Cu^{2+}(aq) \longrightarrow Zn^{2+}(aq) + Cu(s)$.

What is actually happening in the cell to cause the light bulb to glow?

1. Electrons are produced at the zinc electrode through the process of oxidation.

$$Zn(s) \longrightarrow Zn^{2+}(aq) + 2\ e^-$$

2. The electrons pass from the zinc electrode to the copper electrode through the external circuit (the wire).
3. Electrons enter the copper electrode and are accepted by Cu^{2+} ions in solution adjacent to the electrode. This is a reduction reaction.

$$Cu^{2+}(aq) + 2\ e^- \longrightarrow Cu(s)$$

4. To complete the circuit, ions (both positive and negative) move through the solution, passing through the porous disk as needed.

> A mnemonic device can be helpful for remembering the relationship between cathode– anode and oxidation–reduction. The two words that begin with vowels (anode and oxidation) go together, and the two words beginning with consonants (cathode and reduction) go together.

Special names are given to the two electrodes in a galvanic cell; one is called the *cathode* and the other is the *anode*. The **cathode** *is the electrode in an electrochemical cell at which reduction takes place.* Here electrons enter the galvanic cell from the external circuit. The **anode** *is the electrode in an electrochemical cell at which oxidation takes place.* Here electrons leave the cell for the external circuit. In the cell now under discussion, the copper electrode is the cathode and the zinc electrode is the anode.

In principle, any spontaneous oxidation–reduction reaction can be used to build a galvanic cell, and many such cells have been studied in the laboratory. A selected few such galvanic cells are now used commercially. We will discuss two that are very common: (1) the dry cell and (2) the lead storage battery.

The Dry Cell

The *dry cell* is widely used in flashlights, portable radios and CD players, and battery-powered toys; it is often referred to as a flashlight battery. Two versions of the dry cell are marketed: an acidic version and an alkaline version.

The *acidic version* of the dry cell contains a zinc outer surface (covered with cardboard or paint for protection) that functions as the anode and a carbon (graphite) rod, in contact with a moist paste, that serves as the cathode (see Fig. 15.4). The paste (the cell is not truly dry) is a mixture of solid MnO_2, solid NH_4Cl, and graphite powder (C) moistened with water. In the operation of the cell, Zn is oxidized to Zn^{2+} ion, and MnO_2 (a paste component) is reduced to Mn_2O_3. The reduction takes place at the interface between the carbon cathode and the paste; the inert carbon electrode conducts the electrons to the external circuit. The electrode reactions are as follows.

Anode (oxidation):

$$Zn(s) \longrightarrow Zn^{2+}(aq) + 2\,e^-$$

Cathode (reduction):

$$2\,MnO_2(s) + 2\,NH_4^+(aq) + 2\,e^- \longrightarrow Mn_2O_3(s) + 2\,NH_3(aq) + H_2O$$

The useful life of acidic dry cells can be shortened if the slightly acidic paste corrodes the zinc. A protective paper is inserted between the paste and zinc to minimize this problem.

In the *alkaline version* of the dry cell, the solid NH_4Cl is replaced with KOH, and a steel rod rather than a graphite one is the cathode. The anode reaction still involves oxidation of zinc, but the zinc is present as a powder in a gel formulation. The cathode reaction also still involves the reduction of MnO_2. Equations for the electrode reactions are as follows.

Anode (oxidation):

$$Zn(s) + 2\,OH^-(aq) \longrightarrow ZnO(s) + H_2O(l) + 2\,e^-$$

Cathode (reduction):

$$2\,MnO_2(s) + H_2O(l) + 2\,e^- \longrightarrow Mn_2O_3(s) + 2\,OH^-(aq)$$

The alkaline dry cell costs roughly three times as much to produce as the acidic version. A major cost factor is the more elaborate internal construction needed to prevent leakage of the KOH solution. These cells provide up to 50% more total energy than the less expensive

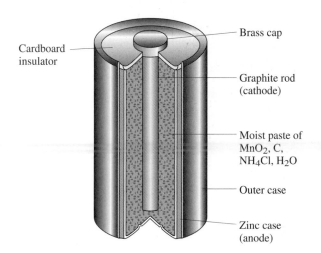

Cardboard insulator

Brass cap

Graphite rod (cathode)

Moist paste of MnO_2, C, NH_4Cl, H_2O

Outer case

Zinc case (anode)

Figure 15.4

The acidic Zn–MnO_2 dry cell is commonly known as a flashlight battery.

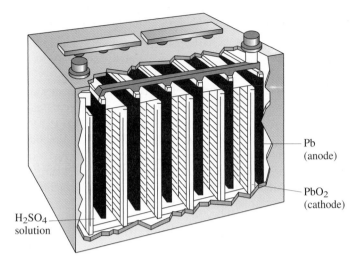

Figure 15.5

The six galvanic cells in a 12-volt automobile battery.

Pb (anode)

PbO₂ (cathode)

H₂SO₄ solution

acidic model because they maintain usable voltage over a larger fraction of the lifetime of the cathode and anode materials. Miniature alkaline cells find extensive use in calculators, watches, and camera exposure controls.

Lead Storage Battery

The lead storage battery provides the starting power for automobiles. A 12-volt lead storage battery, the standard size, consists of six galvanic cells connected together (see Fig. 15.5). Each cell generates 2 volts.

Both electrodes in a lead storage battery involve the element lead. Lead serves as the anode, and lead coated with lead (IV) oxide serves as the cathode. Glass fiber spacers separate electrodes to prevent them from touching each other. The electrodes are immersed in a 38% (m/m) sulfuric acid (H_2SO_4) solution. Details of one of the six cells of a lead storage battery are shown in Figure 15.6.

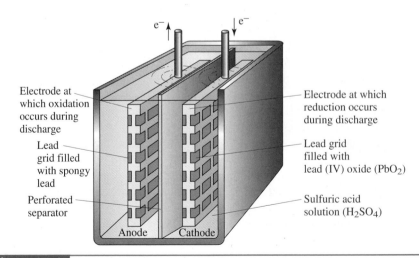

e⁻   e⁻

Electrode at which oxidation occurs during discharge

Lead grid filled with spongy lead

Perforated separator

Anode   Cathode

Electrode at which reduction occurs during discharge

Lead grid filled with lead (IV) oxide (PbO_2)

Sulfuric acid solution (H_2SO_4)

Figure 15.6

A single lead storage cell.

The electrode reactions when a lead battery is used to supply power (discharging) are as follows.

Anode (oxidation):

$$Pb(s) + SO_4^{2-}(aq) \longrightarrow PbSO_4(s) + 2 e^-$$

Cathode (reduction):

$$PbO_2(s) + 4 H^+(aq) + SO_4^{2-}(aq) + 2 e^- \longrightarrow PbSO_4(s) + 2 H_2O(l)$$

Battery discharge results in a buildup of $PbSO_4$ on the electrodes and a decrease in the density of the sulfuric acid. (The state of charge of a lead storage battery can thus be checked by a service station attendant by measuring the density [Sec. 3.8] of the sulfuric acid.)

Unlike dry cells, a lead storage battery can be recharged. This reverse process, which is nonspontaneous, uses an external source of electrical energy; in the automobile the external energy source is an alternator driven by the automobile engine.

In recharging, the $PbSO_4$ on the electrodes (formed during discharge) is converted back to Pb at one electrode and to PbO_2 at the other and H_2SO_4 is also produced. The electrode reactions are the reverse of what occurs during discharge.

Theoretically a lead storage battery should be rechargeable indefinitely. In practice, such batteries have a lifetime of 3–5 yr because small amounts of lead sulfate continually fall from the electrodes (to the bottom of the cell) as a result of "road shock" and chemical side reactions. Eventually the electrodes lose so much lead sulfate that the recharging process is no longer effective.

In "standard" lead storage batteries, water must be added to the individual cells on a regular basis. Recharging the battery, besides converting $PbSO_4$ back to Pb and PbO_2, also decomposes small amounts of water to give H_2 and O_2; hence, the H_2O must be replenished. Because of the possible presence of H_2 gas in a lead storage battery, a person should wear glasses (for eye protection) when releasing the cap of a battery since escaping gas can force sulfuric acid out. In addition, a person should not smoke while doing this, since hydrogen gas is flammable and forms explosive mixtures with oxygen. Newer automobile batteries have electrodes made of an alloy of calcium and lead. The presence of the calcium minimizes the decomposition of water during recharging. Thus, batteries with these alloy electrodes can be sealed; there is no need to add water.

Electrolytic Cells

The application of electrical energy from an external power source can be used to cause a nonspontaneous redox reaction to occur. The charging of the lead storage battery previously discussed is an example of this. The general term for such a process is *electrolysis*. **Electrolysis** *is the process in which electrical energy is used to cause a nonspontaneous redox reaction to occur.* During recharging, the lead storage battery is functioning as an *electrolytic cell.* An **electrolytic cell** *is an electrochemical cell in which chemical change is caused to occur through the application of electrical energy.*

Electrolytic cells have a number of important commercial applications, including (1) production of important industrial chemicals, (2) electrorefining and purification of metals, and (3) electroplating. Let us consider examples in all three of these areas.

Chlorine gas, hydrogen gas, and sodium hydroxide—three important industrial chemicals—can be produced simultaneously from the electrolysis of a concentrated aqueous sodium chloride solution (saltwater or brine solution). The type of electrolytic cell needed is shown in Figure 15.7.

In this cell the two electrodes are inert; that is, they do not participate in the redox reactions themselves but serve as surfaces on which the redox reactions occur. As with galvanic cells, reduction occurs at the cathode and oxidation at the anode.

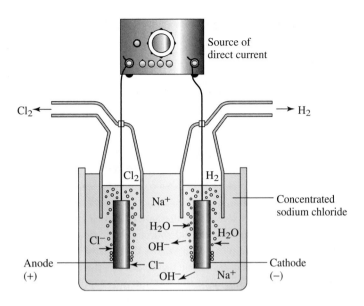

Figure 15.7

The electrolysis of aqueous sodium chloride solution (saltwater brine) produces hydrogen gas, chlorine gas, and sodium hydroxide solution.

The negative ions in the solution, the Cl^- ions, react at the anode, where they give up electrons and are oxidized to produce Cl_2 gas.

$$\text{anode (oxidation):} \quad 2\,Cl^-(aq) \longrightarrow Cl_2(g) + 2\,e^-$$

The positive ions in the solution, the Na^+ ions, do not react at the cathode to produce Na metal atoms. Instead, water, the solvent in the solution, reacts at the cathode; it is more easily reduced than Na^+ ion. The cathode reaction is

$$\text{cathode (reduction):} \quad 2\,H_2O(l) + 2\,e^- \longrightarrow H_2(g) + 2\,OH^-(aq)$$

This reduction yields H_2 gas as well as OH^- ions. The OH^- ions remain in solution, producing a solution that now contains Na^+ ions and OH^- ions; our solution of sodium chloride has been changed to one of sodium hydroxide. Thus, three substances—Cl_2 gas, H_2 gas, and aqueous NaOH solution—result from the electrolysis of a concentrated NaCl solution.

Metal purification frequently depends on electrolysis. All copper used in electrical wire is electrolytically purified; impurities decrease the electrical conductance of the wire. In an electrolytic cell used to purify copper, large slabs of impure copper (obtained from the reduction of copper ores) serve as anodes, and thin sheets of very pure copper serve as cathodes. The solution in which the electrodes are immersed is an acidic copper(II) sulfate solution. As the cell is operated, the anodes (the impure copper) decrease in size and the cathodes (the pure copper) increase in size (see Fig. 15.8). What is happening? At the

Figure 15.8

Cross section of an electrolytic cell for purifying copper.

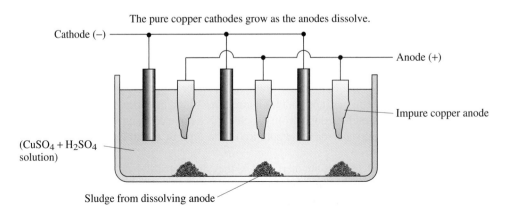

The pure copper cathodes grow as the anodes dissolve.

anodes, oxidation of Cu causes it to dissolve. The reaction is

$$\text{anode (oxidation):} \quad Cu(s) \longrightarrow Cu^{2+}(aq) + 2\,e^-$$

As the anodic copper dissolves, the impurities present also go into solution (some are also oxidized) or fall to the bottom of the cell. At the cathode, reduction causes copper to come out of solution.

$$\text{cathode (reduction):} \quad Cu^{2+}(aq) + 2\,e^- \longrightarrow 2\,Cu(s)$$

The net effect of this oxidation–reduction process is that copper is transferred via solution from one electrode (the impure one) to the other (the pure one). The electrical voltage supplied to the cell is set at a value that allows copper ions to be reduced but is not sufficient to reduce any dissolved impurities. Thus, only copper, and not impurities, deposits on the cathode.

Electroplating *is the deposition of a thin layer of a metal on an object through the process of electrolysis.* Electroplated objects are common in our society. Jewelry is plated with silver and gold. Tableware is often plated with silver. Gold-plated electrical contacts are used extensively. "Tin cans" are actually steel cans with a thin coating of tin. Chromium-plated steel automobile trim has been used for many years. The thin metallic layer deposited during electroplating is generally only 0.001–0.002 in. thick.

Figure 15.9 shows a simplified apparatus for electroplating items with silver. The object to be plated is made the cathode in a solution containing Ag^+ ions. The anode is a bar of the plating metal, silver in this case. At the cathode, Ag^+ ions from solution are deposited as metallic silver.

$$\text{cathode (reduction):} \quad Ag^+(aq) + e^- \longrightarrow Ag(s)$$

At the anode, silver from that electrode is oxidized to give Ag^+ ions in solution; this replenishes the supply of Ag^+ ions in solution needed for the plating process.

$$\text{anode (oxidation):} \quad Ag(s) \longrightarrow Ag^+(aq) + e^-$$

The electroplating bath, the solution around the electrodes, usually contains other chemicals besides the plating metal. For example, silver plating is usually done from a solution containing both AgCN and KCN.

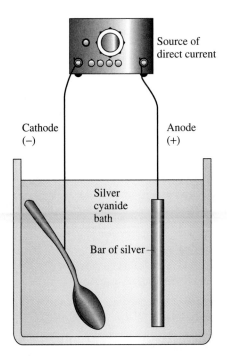

Cathode
(−)

Anode
(+)

Silver
cyanide
bath

Bar of silver

Figure 15.9

An apparatus for electroplating silver.

Source of
direct current

Summary

1. **Oxidation–Reduction Reaction Terminology** Oxidation is the loss of electrons by a reactant; reduction is the gain of electrons by a reactant. An oxidizing agent causes oxidation by accepting electrons from another reactant. A reducing agent causes reduction by providing electrons for another reactant to accept. An oxidation–reduction reaction is any reaction involving the transfer of electrons between reactants. Shortened terminology for an oxidation–reduction reaction is redox reaction.

2. **Oxidation Numbers** An oxidation number for an atom is a number that represents the charge that the atom appears to have when the electrons in each bond it is participating in are assigned to the more electronegative of the two atoms involved in the bond. Oxidation numbers are used to identify the electron transfer that occurs in a redox reaction. In terms of oxidation numbers, oxidation is associated with an increase in oxidation number and reduction involves a decrease in oxidation number.

3. **Redox and Nonredox Reactions** A redox reaction is a chemical reaction in which oxidation numbers change. A nonredox reaction is a chemical reaction in which there is no change in oxidation numbers.

4. **Methods for Balancing Redox Reactions** Two distinctly different approaches for systematically balancing redox equations are in common use: the oxidation-number method and the half-reaction method. Each method has advantages and disadvantages. The basis for the oxidation-number method is oxidation number increase must equal oxidation number decrease. In the half-reaction method for balancing redox equations the unbalanced redox equation is separated into two half-reactions, one for oxidation and one for reduction.

6. **Disproportionation Reactions** A disproportionation reaction is a redox reaction in which some atoms of a single element in a reactant are oxidized and others are reduced. For such reactant behavior to be possible, the reactant must contain an element that is capable of having at least three oxidation numbers: its original number plus one higher and one lower oxidation number.

7. **Spontaneous Redox Reactions** A spontaneous redox reaction can be used to convert chemical energy into electrical energy. For this to occur, a redox reaction must be carried out in a specially designed apparatus called a galvanic cell. Flashlight batteries and automobile batteries are examples of such cells.

8. **Nonspontaneous Redox Reactions** A nonspontaneous redox reaction can be caused to occur by using electrical energy to produce chemical energy. Such a process is called electrolysis, and the apparatus in which the reaction is carried out is called an electrolytic cell.

Key Terms

The new terms defined in this chapter are

anode   *Sec. 15.8*
cathode   *Sec. 15.8*
disproportionation reaction   *Sec. 15.7*
electrochemical cell   *Sec. 15.8*
electrolysis   *Sec. 15.8*
electrolytic cell   *Sec. 15.8*
electroplating   *Sec. 15.8*
galvanic cell   *Sec. 15.8*

half-reaction   *Sec. 15.6*
non-oxidation-reduction reaction   *Sec. 15.3*
oxidation   *Sec. 15.1*
oxidation number   *Sec. 15.2*
oxidation–reduction reaction   *Sec. 15.1*
oxidizing agent   *Secs. 15.1 and 15.2*
redox reaction   *Sec. 15.1*
reducing agent   *Secs. 15.1 and 15.2*
reduction   *Sec. 15.1*

Practice Problems

Oxidation–Reduction Terminology (Secs. 15.1 and 15.2)

15.1 Give definitions of *oxidation* in terms of

(a) loss or gain of electrons
(b) increase or decrease in oxidation number

15.2 Give definitions of *reduction* in terms of

(a) loss or gain of electrons
(b) increase or decrease in oxidation number

15.3 Give definitions of *oxidizing agent* in terms of

(a) loss or gain of electrons
(b) increase or decrease in oxidation number
(c) substance oxidized or substance reduced

15.4 Give definitions of *reducing agent* in terms of

(a) loss or gain of electrons
(b) increase or decrease in oxidation number
(c) substance oxidized or substance reduced

15.5 In each of the following statements, choose the word in parentheses that best completes the statement.

(a) An element that has lost electrons in a redox reaction is said to have been (oxidized, reduced).

(b) Reduction always results in an (increase, decrease) in the oxidation number.

(c) The substance oxidized in a redox reaction is the (oxidizing, reducing) agent.

(d) The reducing agent (gains, loses) electrons during a redox reaction.

15.6 In each of the following statements, choose the word in parentheses that best completes the statement.

(a) The reducing agent causes an (increase, decrease) in the oxidation number of the oxidizing agent in a redox reaction.

(b) The oxidizing agent (gains, loses) electrons during a redox reaction.

(c) Oxidation always results in an (increase, decrease) in the oxidation number.

(d) An element that has gained electrons in a redox reaction is said to have been (oxidized, reduced).

Assignment of Oxidation Numbers (Sec. 15.2)

15.7 Assign oxidation numbers to the atoms in each of the following compounds.

(a) NH_3 **(b)** H_2SO_3 **(c)** HNO_2 **(d)** Na_3PO_4

15.8 Assign oxidation numbers to the atoms in each of the following compounds.

(a) NO_2 **(b)** H_3PO_4 **(c)** AlF_3 **(d)** $KClO_4$

15.9 Assign oxidation numbers to the atoms in each of the following ions.

(a) P^{3-} **(b)** Mg^{2+} **(c)** NH_2^- **(d)** PO_4^{3-}

15.10 Assign oxidation numbers to the atoms in each of the following ions.

(a) Se^{2-} **(b)** Al^{3+} **(c)** SO_4^{2-} **(d)** ClF_4^-

15.11 What is the oxidation number of carbon in each of the following carbon-containing compounds?

(a) H_2CO_3 **(b)** $H_2C_2O_4$ **(c)** C_2H_4 **(d)** C_3H_8

15.12 What is the oxidation number of carbon in each of the following carbon-containing compounds?

(a) $HC_2H_3O_2$ **(b)** H_2CO **(c)** C_3H_6 **(d)** C_4H_{10}

15.13 What is the oxidation number of the metal present in each of the following compounds or complex ions?

(a) $Ni(NO_3)_2$ **(b)** $FeSO_4$
(c) $Al(OH)_4^-$ **(d)** $Zn(CN)_4^{2-}$

15.14 What is the oxidation number of the metal present in each of the following compounds or complex ions?

(a) $Fe(NO_3)_3$ **(b)** $CuSO_4$
(c) $Ag(CN)_2^-$ **(d)** $Zn(OH)_4^{2-}$

15.15 Determine the oxidation numbers of all of the elements present in each of the following polyatomic-ion-containing compounds.

(a) $Rh_2(CO_3)_3$ **(b)** $Ni_2(SO_4)_3$
(c) $Co_3(PO_4)_2$ **(d)** Cu_2SO_3

15.16 Determine the oxidation numbers of all of the elements present in each of the following polyatomic-ion-containing compounds.

(a) $Rh_3(PO_4)_2$ **(b)** $Cr_2(SO_4)_3$
(c) $Fe_2(CrO_4)_3$ **(d)** $K_2Cr_2O_7$

15.17 Indicate whether oxygen has a -2, -1, or positive oxidation number in each of the following oxygen-containing species.

(a) Na_2O **(b)** OF_2 **(c)** Na_2O_2 **(d)** BaO

15.18 Indicate whether oxygen has a -2, -1, or positive oxidation number in each of the following oxygen-containing species.

(a) O_2F_2 **(b)** SO_3 **(c)** BaO_2 **(d)** BeO

15.19 Indicate whether hydrogen has a $+1$ or -1 oxidation number in each of the following hydrogen-containing species.

(a) NaH **(b)** CH_4 **(c)** HCl **(d)** CaH_2

15.20 Indicate whether hydrogen has a $+1$ or -1 oxidation number in each of the following hydrogen-containing species.

(a) H_2Se **(b)** N_2H_2 **(c)** KH **(d)** MgH_2

Characteristics of Oxidation–Reduction Reactions (Sec. 15.2)

15.21 Identify which substance is oxidized and which substance is reduced in each of the following redox reactions.

(a) $N_2 + 3 H_2 \longrightarrow 2 NH_3$
(b) $Cl_2 + 2 KI \longrightarrow 2 KCl + I_2$
(c) $Sb_2O_3 + 3 Fe \longrightarrow 2 Sb + 3 FeO$
(d) $3 H_2SO_3 + 2 HNO_3 \longrightarrow 2 NO + H_2O + 3 H_2SO_4$

15.22 Identify which substance is oxidized and which substance is reduced in each of the following redox reactions.

(a) $2 Al + 3 Cl_2 \longrightarrow 2 AlCl_3$
(b) $Zn + CuCl_2 \longrightarrow ZnCl_2 + Cu$
(c) $2 NiS + 3 O_2 \longrightarrow 2 NiO + 2 SO_2$
(d) $3 H_2S + 2 HNO_3 \longrightarrow 3 S + 2 NO + 4 H_2O$

15.23 Identify which substance is the oxidizing agent and which substance is the reducing agent in each of the redox reactions in Problem 15.21.

15.24 Identify which substance is the oxidizing agent and which substance is the reducing agent in each of the redox reactions in Problem 15.22.

15.25 Identify the following species for the redox reaction

$$2 HNO_3 + SO_2 \longrightarrow H_2SO_4 + 2 NO_2$$

(a) substance that is oxidized
(b) oxidizing agent

(c) substance that contains the element that decreases in oxidation number

(d) substance that contains the element that loses electrons during the oxidation–reduction reaction

15.26 Identify the following species for the redox reaction

$$PH_3 + 2\,NO_2 \longrightarrow H_3PO_4 + N_2$$

(a) substance that is reduced

(b) reducing agent

(c) substance that contains the element that increases in oxidation number

(d) substance that contains the element that loses electrons during the oxidation–reduction reaction

15.27 For each of the following redox reactions that involve the reaction of a metallic element with a nonmetallic element, indicate whether the metal is oxidized or reduced and whether the nonmetal is oxidized or reduced.

(a) $3\,Zn + N_2 \longrightarrow Zn_3N_2$

(b) $2\,Ca + O_2 \longrightarrow CaO$

(c) $2\,Na + S \longrightarrow Na_2S$

(d) $Mg + Cl_2 \longrightarrow MgCl_2$

15.28 For each of the following redox reactions that involve the reaction of a metallic element with a nonmetallic element, indicate whether the metal is oxidized or reduced and whether the nonmetal is oxidized or reduced.

(a) $Cu + S \longrightarrow CuS$

(b) $4\,Ag + O_2 \longrightarrow 2\,Ag_2O$

(c) $Ni + F_2 \longrightarrow NiF_2$

(d) $2\,Al + N_2 \longrightarrow 2\,AlN$

Types of Chemical Reactions (Sec. 15.3)

15.29 Characterize each of the following reactions using one selection from the choices *redox* and *nonredox* combined with one selection from the choices *synthesis, decomposition, single-replacement,* and *double-replacement.*

(a) $H_2 + Cl_2 \longrightarrow 2\,HCl$

(b) $2\,HBr + Mg \longrightarrow MgBr_2 + H_2$

(c) $MgCO_3 \longrightarrow MgO + CO_2$

(d) $2\,KOH + H_2SO_4 \longrightarrow K_2SO_4 + 2\,H_2O$

15.30 Characterize each of the following reactions using one selection from the choices *redox* and *nonredox* combined with one selection from the choices *synthesis, decomposition, single-replacement,* and *double-replacement.*

(a) $Zn + Cu(NO_3)_2 \longrightarrow Zn(NO_3)_2 + Cu$

(b) $2\,SO_2 + O_2 \longrightarrow 2\,SO_3$

(c) $2\,CuO \longrightarrow 2\,Cu + O_2$

(d) $NaCl + AgNO_3 \longrightarrow AgCl + NaNO_3$

15.31 Characterize each of the following reactions as (1) a redox reaction, (2) a nonredox reaction, or (3) "can't classify" because of insufficient information.

(a) a synthesis reaction in which both reactants are elements

(b) a combustion reaction

(c) a decomposition reaction in which the products are all compounds

(d) a decomposition reaction in which an element and a compound are products

15.32 Characterize each of the following reactions as (1) a redox reaction, (2) a nonredox reaction, or (3) "can't classify" because of insufficient information.

(a) a synthesis reaction in which one reactant is an element and the other is a compound

(b) an acid–base neutralization reaction

(c) a decomposition reaction in which the products are all elements

(d) a single-replacement reaction involving an active metal and an acid

Balancing Redox Equations: Oxidation-Number Method (Sec. 15.5)

15.33 Balance the following equations by the oxidation-number method.

(a) $Cr + HCl \longrightarrow CrCl_3 + H_2$

(b) $Cr_2O_3 + C \longrightarrow Cr + CO_2$

(c) $SO_2 + NO_2 \longrightarrow SO_3 + NO$

(d) $BaSO_4 + C \longrightarrow BaS + CO$

15.34 Balance the following equations by the oxidation-number method.

(a) $Fe_2O_3 + CO \longrightarrow Fe + CO_2$

(b) $Al + MnO_2 \longrightarrow Al_2O_3 + Mn$

(c) $I_2O_5 + CO \longrightarrow I_2 + CO_2$

(d) $N_2H_4 + O_2 \longrightarrow N_2 + H_2O$

15.35 Balance the following equations by the oxidation-number method.

(a) $Br_2 + H_2O + SO_2 \longrightarrow HBr + H_2SO_4$

(b) $H_2S + HNO_3 \longrightarrow S + NO + H_2O$

(c) $SnSO_4 + FeSO_4 \longrightarrow Sn + Fe_2(SO_4)_3$

(d) $Na_2TeO_3 + NaI + HCl \longrightarrow NaCl + Te + H_2O + I_2$

15.36 Balance the following equations by the oxidation-number method.

(a) $HNO_3 + I_2 \longrightarrow NO_2 + H_2O + HIO_3$

(b) $As_4O_6 + Cl_2 + H_2O \longrightarrow H_3AsO_4 + HCl$

(c) $HI + HNO_3 \longrightarrow I_2 + NO + H_2O$

(d) $PbO_2 + Sb + NaOH \longrightarrow PbO + NaSbO_2 + H_2O$

15.37 Balance the following equations by the oxidation-number method. All reactions occur in acidic solution.

(a) $I_2 + Cl_2 \longrightarrow HIO_3 + Cl^-$

(b) $MnO_4^- + AsH_3 \longrightarrow H_3AsO_4 + Mn^{2+}$

(c) $Br^- + SO_4^{2-} \longrightarrow Br_2 + SO_2$

(d) $Au + Cl^- + NO_3^- \longrightarrow AuCl_4^- + NO_2$

15.38 Balance the following equations by the oxidation-number method. All reactions occur in acidic solution.

(a) $I^- + SO_4^{2-} \longrightarrow H_2S + I_2$

(b) $Mn^{2+} + BiO_3^- \longrightarrow MnO_4^- + Bi^{3+}$

(c) $Fe^{2+} + ClO_3^- \longrightarrow Fe^{3+} + Cl^-$

(d) $Pt + Cl^- + NO_3^- \longrightarrow PtCl_6^{2-} + NO_2$

15.39 Balance the following equations by the oxidation-number method. All reactions occur in basic solution.

(a) $S^{2-} + Cl_2 \longrightarrow SO_4^{2-} + Cl^-$

(b) $SO_3^{2-} + CrO_4^{2-} \longrightarrow Cr(OH)_4^- + SO_4^{2-}$

(c) $MnO_4^- + IO_3^- \longrightarrow MnO_2 + IO_4^-$

(d) $I_2 + Cl_2 \longrightarrow H_3IO_6^{2-} + Cl^-$

15.40 Balance the following equations by the oxidation-number method. All reactions occur in basic solution.

(a) $Zn + MnO_4^- \longrightarrow Zn(OH)_2 + MnO_2$

(b) $NO_2^- + Al \longrightarrow NH_3 + AlO_2^-$

(c) $NO_2^- + MnO_4^- \longrightarrow NO_3^- + MnO_2$

(d) $Al + NO_3^- \longrightarrow Al(OH)_4^- + NH_3$

Balancing Redox Equations: Half-Reaction Method (Sec. 15.6)

15.41 Balance the following half-reactions occurring in acidic solution.

(a) $MnO_2 \longrightarrow Mn^{3+}$

(b) $H_3MnO_4 \longrightarrow Mn$

(c) $MnO_4^- \longrightarrow Mn^{2+}$

(d) $MnO_4^- \longrightarrow MnO_2$

15.42 Balance the following half-reactions occurring in acidic solution.

(a) $V^{2+} \longrightarrow VO_2^+$

(b) $V^{3+} \longrightarrow VO^{2+}$

(c) $VO^{2+} \longrightarrow VO_2^+$

(d) $V \longrightarrow VO_2^+$

15.43 Balance the following half-reactions occurring in basic solution.

(a) $SeO_4^{2-} \longrightarrow Se$

(b) $Se^{2-} \longrightarrow SeO_3^{2-}$

(c) $SeO_4^{2-} \longrightarrow SeO_3^{2-}$

(d) $Se \longrightarrow SeO_3^{2-}$

15.44 Balance the following half-reactions occurring in basic solution.

(a) $H_3IO_6^{2-} \longrightarrow I_2$

(b) $IO_3^- \longrightarrow IO^-$

(c) $I^- \longrightarrow IO^-$

(d) $IO^- \longrightarrow H_3IO_6^{2-}$

15.45 Balance each of the following redox reactions by the half-reaction method. Each reaction occurs in acidic solution.

(a) $Zn + Cu^{2+} \longrightarrow Cu + Zn^{2+}$

(b) $Br_2 + I^- \longrightarrow Br^- + I_2$

(c) $S_2O_3^{2-} + Cl_2 \longrightarrow HSO_4^- + Cl^-$

(d) $Zn + As_2O_3 \longrightarrow AsH_3 + Zn^{2+}$

15.46 Balance each of the following redox reactions by the half-reaction method. Each reaction occurs in acidic solution.

(a) $Fe + Ag^+ \longrightarrow Fe^{3+} + Ag$

(b) $Cl_2 + Br^- \longrightarrow Cl^- + Br_2$

(c) $S_2O_3^{2-} + Cu^{2+} \longrightarrow S_4O_6^{2-} + Cu$

(d) $C_2O_4^{2-} + MnO_4^- \longrightarrow CO_2 + Mn^{2+}$

15.47 Balance each of the equations in Problem 15.37 using the half-reaction method for balancing.

15.48 Balance each of the equations in Problem 15.38 using the half-reaction method for balancing.

15.49 Balance each of the following redox reactions by the half-reaction method. Each reaction occurs in basic solution.

(a) $NH_3 + ClO^- \longrightarrow N_2H_4 + Cl^-$

(b) $Cr(OH)_2 + BrO^- \longrightarrow CrO_4^{2-} + Br^-$

(c) $CrO_2^- + H_2O_2 \longrightarrow CrO_4^{2-} + OH^-$

(d) $Bi(OH)_3 + Sn(OH)_3^- \longrightarrow Sn(OH)_6^{2-} + Bi$

15.50 Balance each of the following redox reactions by the half-reaction method. Each reaction occurs in basic solution.

(a) $Cr_2O_3 + ClO^- \longrightarrow CrO_4^{2-} + Cl^-$

(b) $NO + MnO_4^- \longrightarrow NO_3^- + MnO_2$

(c) $Al + PO_3^{3-} \longrightarrow PH_3 + AlO_2^-$

(d) $Bi(OH)_3 + SnO_2^{2-} \longrightarrow SnO_3^{2-} + Bi$

15.51 Balance each of the equations in Problem 15.39 using the half-reaction method for balancing.

15.52 Balance each of the equations in Problem 15.40 using the half-reaction method for balancing.

Balancing Redox Equations: Disproportionation Reactions (Sec. 15.7)

15.53 Balance each of the following redox reactions by the oxidation-number method.

(a) $HNO_2 \longrightarrow NO + NO_3^-$ (acidic solution)

(b) $ClO^- + Cl^- \longrightarrow Cl_2$ (acidic solution)

(c) $S \longrightarrow S^{2-} + SO_3^{2-}$ (basic solution)

(d) $Br_2 \longrightarrow BrO_3^- + Br^-$ (basic solution)

15.54 Balance each of the following redox reactions by the oxidation-number method.

(a) $NO + NO_3^- \longrightarrow N_2O_4$ (acidic solution)
(b) $H_5IO_6 + I^- \longrightarrow I_2$ (acidic solution)
(c) $P_4 \longrightarrow HPO_3^{2-} + PH_3$ (basic solution)
(d) $HClO_2 \longrightarrow ClO_2 + Cl^-$ (basic solution)

15.55 Balance each of the redox reactions in Problem 15.53 using the half-reaction method.

15.56 Balance each of the redox reactions in Problem 15.54 using the half-reaction method.

Important Oxidation–Reduction Processes (Sec. 15.8)

15.57 The spontaneous redox reaction between lead metal and copper(II) nitrate solution occurs according to the following net ionic equation.

$$Pb(s) + Cu^{2+}(aq) \longrightarrow Cu(s) + Pb^{2+}(aq)$$

Assume the reactants are separated into two compartments. A lead electrode is immersed in 1.00 M $Pb(NO_3)_2$ and a copper electrode in 1.00 M $Cu(NO_3)_2$. For this galvanic cell, indicate each of the following.

(a) equation for the oxidation half-reaction
(b) equation for the reduction half-reaction
(c) identity of anode and cathode
(d) direction of electron flow

15.58 The spontaneous redox reaction between nickel metal and cadmium nitrate solution occurs according to the following net ionic equation.

$$Ni(s) + Cd^{2+}(aq) \longrightarrow Cd(s) + Ni^{2+}(aq)$$

Assume the reactants are separated into two compartments. A nickel electrode is immersed in 1.00 M $Ni(NO_3)_2$ and a cadmium electrode in 1.00 M $Cd(NO_3)_2$. For this galvanic cell, indicate each of the following.

(a) equation for the oxidation half-reaction
(b) equation for the reduction half-reaction
(c) identity of anode and cathode
(d) direction of electron flow

15.59 What are the anode and cathode reactions during the operation of an acidic dry cell?

15.60 What are the anode and cathode reactions during the operation of an alkaline dry cell?

15.61 What are the anode and cathode reactions during the discharging of a lead storage battery?

15.62 What are the anode and cathode reactions during the charging of a lead storage battery?

15.63 Why does the density of the H_2SO_4 in a lead storage battery decrease as the cell discharges?

15.64 Why is it not possible to recharge a lead storage battery an infinite number of times?

15.65 Write an equation for what happens at the anode during the electrolysis of a concentrated aqueous NaCl solution.

15.66 Write an equation for what happens at the cathode during the electrolysis of a concentrated aqueous NaCl solution.

Additional Problems

15.67 Nitrogen forms a number of oxides including NO_2, N_2O_3, NO, N_2O, and N_2O_5. Arrange these oxides in order of increasing oxidation number of nitrogen.

15.68 Sulfur forms a number of oxides including S_2O, S_7O_2, SO_2, SO_3, and S_6O. Arrange these oxides in order of increasing oxidation number of sulfur.

15.69 Possible oxidation numbers for the element S range from +6 to –2. Based on this information, explain each of the following observations.

(a) The S^{2-} ion functions only as a reducing agent.
(b) The SO_4^{2-} ion functions only as an oxidizing agent.
(c) The SO_2 molecule can function as either a reducing agent or an oxidizing agent.
(d) The SO_3 molecule functions only as an oxidizing agent.

15.70 Possible oxidation numbers for the element N range from +5 to –3. Based on this information, explain each of the following observations.

(a) The N^{3-} ion functions only as a reducing agent.
(b) The NO_3^- ion functions only as an oxidizing agent.
(c) The NO_2^- molecule can function as either a reducing agent or an oxidizing agent.
(d) The NH_3 molecule can function as either a reducing agent or an oxidizing agent.

15.71 In which of the following pairs of ionic compounds is the oxidation number of the metal the same in both members of the pair?

(a) $CuSO_4$ and Cu_2SO_4 **(b)** $Fe(NO_3)_3$ and $FePO_4$
(c) $AuCl$ and $AgNO_3$ **(d)** $AlPO_4$ and GaN

15.72 In which of the following pairs of ionic compounds is the oxidation number of the metal the same in both members of the pair?

(a) CuO and $CuCl_2$ **(b)** Ni_2O_3 and NiN
(c) PbO_2 and $SnCl_4$ **(d)** Be_3N_2 and MgO

15.73 Classify each of the following pairs of balanced half-reactions as (1) *two reduction half-reactions,* (2) *two oxidation half-reactions,* or (3) *one reduction and one oxidation half-reaction.*

(a) $Fe^{3+} + e^- \longrightarrow Fe^{2+}$ and $Fe^{2+} + 2e^- \longrightarrow Fe$
(b) $Ni^{3+} + e^- \longrightarrow Ni^{2+}$ and $Ni \longrightarrow Ni^{2+} + 2e^-$
(c) $Cu \longrightarrow Cu^+ + e^-$ and $Cu \longrightarrow Cu^{2+} + 2e^-$
(d) $Au \longrightarrow Au^{3+} + 3e^-$ and $Au^{3+} + 3e^- \longrightarrow Au$

15.74 Classify each of the following pairs of balanced half-reactions as (1) *two reduction half-reactions,* (2) *two oxidation half-reactions,* or (3) *one reduction and one oxidation half-reaction.*

(a) $Sn^{2+} \longrightarrow Sn^{4+} + 2\,e^-$ and $Sn \longrightarrow Sn^{2+} + 2\,e^-$

(b) $Pb^{2+} \longrightarrow Pb^{4+} + 2\,e^-$ and $Pb^{2+} + 2\,e^- \longrightarrow Pb$

(c) $Co^{2+} + 2\,e^- \longrightarrow Co$ and $Co^{3+} + 3\,e^- \longrightarrow Co$

(d) $Ag \longrightarrow Ag^+ + e^-$ and $Ag^+ + e^- \longrightarrow Ag$

15.75 Write balanced equations for all possible redox reactions obtainable by combining the following balanced half-reactions in sets of two. The half-reactions must be used as written; they cannot be reversed in direction.

(1) $2\,H_2O + PH_3 \longrightarrow H_3PO_2 + 4\,H^+ + 4\,e^-$

(2) $3\,H_2O + As \longrightarrow H_3AsO_3 + 3\,H^+ + 3\,e^-$

(3) $MnO_4^- + 8\,H^+ + 5\,e^- \longrightarrow Mn^{2+} + 4\,H_2O$

(4) $SO_4^{2-} + 4\,H^+ + 2\,e^- \longrightarrow SO_2 + 2\,H_2O$

15.76 Write balanced equations for all possible redox reactions obtainable by combining the following balanced half-reactions in sets of two. The half-reactions must be used as written; they cannot be reversed in direction.

(1) $4\,OH^- + ClO_2^- \longrightarrow ClO_4^- + 2\,H_2O + 4\,e^-$

(2) $2\,H_2O + MnO_4^- + 3\,e^- \longrightarrow MnO_2 + 4\,OH^-$

(3) $6\,H_2O + NO_3^- + 8\,e^- \longrightarrow NH_3 + 9\,OH^-$

(4) $4\,OH^- + Al \longrightarrow AlO_2^- + 2\,H_2O + 3\,e^-$

15.77 Write the two balanced half-reactions associated with the following redox reaction.

$$4\,Zn + 10\,H^+ + NO_3^- \longrightarrow 4\,Zn^{2+} + NH_4^+ + 3\,H_2O$$

15.78 Write the two balanced half-reactions associated with the following redox reaction.

$$2\,NO_3^- + 3\,Cu_2O + 14\,H^+ \longrightarrow 6\,Cu^{2+} + 2\,NO + 7\,H_2O$$

Cumulative Problems

15.79 Classify each of the following water-producing reactions as an acid–base reaction or as an oxidation–reduction reaction. If it is an acid–base reaction, identify the acid; if it is an oxidation–reduction reaction, identify the oxidizing agent.

(a) $2\,HNO_3 + 3\,H_2S \longrightarrow 3\,S + 2\,NO + 4\,H_2O$

(b) $2\,KOH + H_2S \longrightarrow K_2S + 2\,H_2O$

(c) $2\,HI + H_2O_2 \longrightarrow I_2 + 2\,H_2O$

(d) $H_2SO_4 + 2\,NaOH \longrightarrow Na_2SO_4 + 2\,H_2O$

15.80 Classify each of the following water-producing reactions as an acid–base reaction or as an oxidation–reduction reaction. If it is an acid–base reaction, identify the acid; if it is an oxidation–reduction reaction, identify the oxidizing agent.

(a) $7\,HI + H_5IO_6 \longrightarrow 4\,I_2 + 6\,H_2O$

(b) $H_2CO_3 + 2\,KOH \longrightarrow K_2CO_3 + 2\,H_2O$

(c) $5\,HClO_2 + NaOH \longrightarrow 4\,ClO_2 + 3\,H_2O + NaCl$

(d) $14\,HNO_3 + 3\,Cu_2O \longrightarrow 6\,Cu(NO_3)_2 + 2\,NO + 7\,H_2O$

15.81 Convert each of the following balanced molecular redox equations to balanced net ionic redox equations.

(a) $SnSO_4(aq) + 2\,FeSO_4(aq) \longrightarrow Sn(s) + Fe_2(SO_4)_3(aq)$

(b) $PH_3(g) + 2\,NO_2(g) \longrightarrow H_3PO_4(aq) + N_2(g)$

(c) $S(s) + 3\,H_2O(l) + 2\,Pb(NO_3)_2 \longrightarrow$
$$2\,Pb(s) + H_2SO_3(aq) + 4\,HNO_3(aq)$$

(d) $4\,Zn(s) + 10\,HNO_3(aq) \longrightarrow$
$$4\,Zn(NO_3)_2(aq) + NH_4NO_3(aq) + 3\,H_2O(l)$$

15.82 Convert each of the following balanced molecular redox equations to balanced net ionic redox equations.

(a) $NaNO_3(aq) + Pb(s) \longrightarrow NaNO_2(aq) + PbO(s)$

(b) $3\,H_2S(aq) + 2\,HNO_3(aq) \longrightarrow$
$$3\,S(s) + 2\,NO(g) + 4\,H_2O(l)$$

(c) $2\,HNO_3(aq) + SO_2(g) \longrightarrow H_2SO_4(aq) + 2\,NO_2(g)$

(d) $10\,FeSO_4(aq) + 2\,KMnO_4(aq) + 8\,H_2SO_4(aq) \longrightarrow$
$$5\,Fe(SO_4)_3(aq) + 2\,MnSO_4(aq) + K_2SO_4(aq) + 8\,H_2O(l)$$

15.83 Balance each of the following net ionic redox reactions.

(a) hydrosulfuric acid plus dichromate ion produces chromium(III) ion plus sulfur (acidic solution)

(b) chlorate ion plus iodine produces iodate ion plus chloride ion (acidic solution)

(c) sulfide ion plus bromine produces sulfate ion plus bromide ion (basic solution)

(d) nitrogen dioxide disproportionates to produce nitrate ion plus nitrite ion (basic solution)
_____

15.84 Balance each of the following net ionic redox reactions.

(a) iron(II) ion plus permanganate ion produces iron(III) ion plus manganese(II) ion (acidic solution)

(b) iodine plus sulfur dioxide produces iodide ion plus sulfate ion (acidic solution)

(c) manganese(II) hydroxide plus nickel(IV) oxide produces manganese(III) oxide plus nickel(II) hydroxide (basic solution)

(d) chlorine disproportionates to produce chlorate ion and chloride ion (basic solution)
_____

15.85 The amount of $I_3^-(aq)$ in a solution can be determined by reacting it with a solution containing $S_2O_3^{2-}(aq)$.

$$I_3^-(aq) + 2\,S_2O_3^{2-}(aq) \longrightarrow 3\,I^-(aq) + S_4O_6^{2-}(aq)$$

Calculate the molarity of I_3^- in a solution given that 43.2 mL of 0.300-M $S_2O_3^{2-}$ solution reacts with a 20.0-mL sample of the I_3^- solution.

15.86 Oxalic acid, $H_2C_2O_4$, reacts with dichromate ion, $Cr_2O_7^{2-}$, in acidic solution as follows.

$$3\,H_2C_2O_4(aq) + Cr_2O_7^{2-}(aq) + 8\,H^+(aq) \longrightarrow$$
$$6\,CO_2(g) + 2\,Cr^{3+}(aq) + 7\,H_2O(l)$$

If 35.0 mL of an oxalic acid solution reacts completely with 25.0 mL of 0.0500-M $Cr_2O_7^{2-}$ solution, what is the molarity of the oxalic acid solution?

15.87 The amount of ozone, O_3, in polluted air can be determined in a two-step process. First the ozone is reacted with an acidic solution containing iodide ion.

$$O_3(g) + 2 I^-(aq) + 2 H^+(aq) \longrightarrow O_2(g) + I_2(s) + H_2O(l)$$

The iodine so produced is then reacted with thiosulfate, $S_2O_3^{2-}$, solution.

$$2 S_2O_3^{2-}(aq) + I_2(s) \longrightarrow S_4O_6^{2-}(aq) + 2 I^-(aq)$$

If 18.03 mL of a 0.00200-M $S_2O_3^{2-}$ solution completely reacts with the I_2 produced by a 28.09-g sample of polluted air, calculate the O_3 concentration, in ppm (m/m), in the sample of air.

15.88 The active ingredient in household bleach is the hypochlorite ion, ClO^-. The concentration of this ion in bleach can be determined using a two-step process. First, the bleach is reacted with an $I^-(aq)$ solution.

$$ClO^-(aq) + 2 I^-(aq) + 2 H^+(aq) \longrightarrow$$
$$I_2(s) + Cl^-(aq) + H_2O(l)$$

The iodine so produced is then reacted with thiosulfate, $S_2O_3^{2-}$, solution.

$$I_2(s) + 2 S_2O_3^{2-}(aq) \longrightarrow 2 I^-(aq) + S_4O_6^{2-}(aq)$$

A 50.00-g sample of a certain household bleach is found to react completely with 42.5 mL of a 0.0150-M $S_2O_3^{2-}$ solution. Calculate the concentration, in mass percent, of ClO^- ion in the bleach.

Answers to Practice Exercises

15.1 (a) +4 for S, −2 for O   **(b)** −3 for N, +1 for H
  (c) +1 for Na, +5 for N, −2 for O
  (d) +2 for N, −1 for F

15.2 (a) $2 SO_2 + O_2 \longrightarrow 2 SO_3$
      +4 −2   0   +6 −2
    reducing agent: SO_2; oxidizing agent: O_2

  (b) $2 Fe_2O_3 + 3 C \longrightarrow 4 Fe + 3 CO_2$
      +3 −2   0   0   +4 −2
    reducing agent: C; oxidizing agent: Fe_2O_3

  (c) $Pb + Cu^{2+} \longrightarrow Cu + Pb^{2+}$
      0   +2   0   +2
    reducing agent: Pb; oxidizing agent: Cu^{2+}

15.3 (a) redox decomposition
  (b) redox single-replacement

(c) redox combustion
(d) nonredox double-replacement

15.4 $2 Zn + O_2 + 4 HCl \longrightarrow 2 ZnCl_2 + 2 H_2O$

15.5 $2 NF_3 + 2 AlCl_3 \longrightarrow N_2 + 3 Cl_2 + 2 AlF_3$

15.6 $6 UO_2^+ + Cr_2O_7^{2-} + 14 H^+ \longrightarrow 6 UO_2^{2+} +$
$$2 Cr^{3+} + 7 H_2O$$

15.7 $6 Fe^{2+} + Cr_2O_7^{2-} + 14 H^+ \longrightarrow 6 Fe^{3+} +$
$$2 Cr^{3+} + 7 H_2O$$

15.8 $3 Zn + 2 MnO_4^- + 4 H_2O \longrightarrow$
$$3 Zn(OH)_2 + 2 MnO_2 + 2 OH^-$$

15.9 $3 HNO_2 + Cr_2O_7^{2-} + 5 H^+ \longrightarrow$
$$3 NO_3^- + 2 Cr^{3+} + 4 H_2O$$

15.10 $3 MnO_4^{2-} + 4 H^+ \longrightarrow$
$$2 MnO_4^- + MnO_2 + 2 H_2O$$

16 Reaction Rates and Chemical Equilibrium

16.1 Collision Theory

In Chapter 10 we learned how to write (and balance) chemical equations to represent chemical reactions and then use these equations to calculate amounts of products produced and reactants consumed in such reactions. We now concern ourselves with another important topic relative to chemical reactions: What causes a chemical reaction to occur?

A set of three generalizations, developed after study of thousands of different reactions, helps answer this question. Collectively, these generalizations are known as *collision theory*. **Collision theory** *is a set of statements that gives the conditions that must be met before a chemical reaction will take place.* Central to collision theory are the concepts of molecular collisions, activation energy, and collision orientation. The key concepts and statements of collision theory are

1. *Molecular collisions.* Reactant particles must interact (that is, collide) with one another before any reaction can occur.
2. *Activation energy.* Colliding particles must possess a certain minimum total amount of energy, called the activation energy, if the collision is to be effective (that is, result in reaction).
3. *Collision orientation.* Colliding particles must come together in the proper orientation unless the particles involved are single atoms or small, symmetrical molecules.

Let us look at these statements in the context of a reaction between two molecules or ions.

Molecular Collisions

When reactions involve two or more reactants, collision theory assumes that the reactant molecules, ions, or atoms must come into contact (collide) with each other in order for a reaction to occur (statement 1). The validity of this assumption is fairly obvious. Reactants cannot react with each other if they are miles apart.

Most reactions are carried out in liquid solution or in the gaseous phase. The reason for this is simple. In these situations reacting particles are more free to move about, and thus it is easier for the reactants to come into contact with each other. Reactions of solids usually take place only on the solid surface and therefore include only a small fraction of the total particles present in the solid. As the reaction proceeds and products dissolve, diffuse, or fall from the surface, fresh solid is exposed. Thus, the reaction eventually consumes all of the solid. The rusting of iron is an example of this type of process.

Activation Energy

Not all collisions between reactant particles result in the formation of reaction products. Sometimes reactant particles rebound from a collision unchanged. Statement 2 of collision theory indicates that for a reaction to occur, colliding particles must impact with a certain minimum energy; that is, the sum of the kinetic energies of the colliding particles must add to a certain minimum value. **Activation energy** *is the minimum combined kinetic energy reactant particles must possess in order for their collision to result in a reaction.* Each chemical reaction has a different activation energy.

In a slow reaction, the activation energy is far above the average energy content of the reacting particles. Only a few particles, those with well above average energy, will undergo collisions that result in a reaction—hence the slowness of the reaction.

It is sometimes possible to start a reaction by providing activation energy and then have it continue on its own. Once the reaction is started, enough energy is released to activate other molecules and keep the reaction going. The striking of a kitchen match is an example of such a situation. Activation energy is initially provided by rubbing the match head against a rough surface; heat is generated by friction. Once the reaction is started, the match continues to burn.

Collision Orientation

Even when activation energy requirements are met, some collisions between reactant particles still do not result in product formation. How can this be? Statement 3 of collision theory, which deals with the orientation of colliding particles at the moment of collision, relates to this situation. For nonspherical molecules and polyatomic ions, their orientation relative to each other at the moment of collision is a factor in determining whether a collision is effective.

As an illustration of the importance of proper collision orientation, consider the chemical reaction between NO_2 and CO to produce NO and CO_2.

$$NO_2(g) + CO(g) \longrightarrow NO(g) + CO_2(g)$$

In this reaction, an O atom is transferred from an NO_2 molecule to a CO molecule. The collision orientation most favorable for this to occur is one that puts an O atom from NO_2 near a C atom from CO at the moment of collision. Such an orientation is shown in Figure 16.1a. In Figure 16.1b–d, three undesirable NO_2–CO orientations are shown in which the likelihood of successful collision is very low. Thus, certain collision orientations are preferred over others. The undesirable collision orientations of Figure 16.1b–d, however, could still result in a reaction if the molecules collided with abnormally high energies.

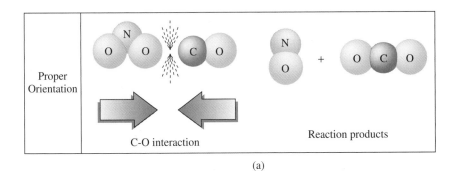

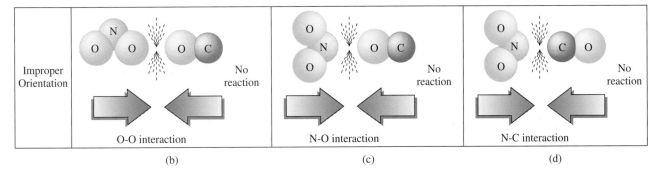

Figure 16.1

In the reaction between NO_2 and CO to produce NO and CO_2, the most favorable collision orientation is one that puts an O atom from NO_2 in close proximity to the C atom in CO.

16.2 Endothermic and Exothermic Reactions

In Section 11.8, we used the terms *endothermic* and *exothermic* to classify changes of state. Melting, sublimation, and evaporation are endothermic changes of state, while freezing, condensation, and deposition are exothermic changes of state. The terms *endothermic* and *exothermic* are also used to classify chemical reactions. An **endothermic chemical reaction** *is a chemical reaction that requires the continuous input of energy as the reaction occurs*. The photosynthesis process that occurs in plants is an example of an endothermic reaction. Light is the energy source for photosynthesis. Light energy must be continuously supplied in order for photosynthesis to occur; a green plant that is kept in the dark will die. An **exothermic chemical reaction** *is a chemical reaction in which energy is released as the reaction occurs*. The burning of a fuel (reaction of the fuel with oxygen) is an exothermic process.

What determines whether a chemical reaction is endothermic or exothermic? The answer to this question is related to the strength of chemical bonds—that is, the energy associated with chemical bonds. Different types of bonds, such as oxygen–hydrogen bonds and fluorine–nitrogen bonds, have different energies associated with them. In a chemical reaction, bonds are broken within reactant molecules (which requires energy), and new bonds are formed within product molecules (which releases energy). The energy balance between bond-breaking and bond-forming determines whether there is a net loss or a net gain of energy.

An endothermic reaction (absorption of energy) occurs when the energy required to break bonds in the reactants is greater than the energy released through bond formation in the products. The necessary additional energy must be supplied from external sources as the reaction proceeds. The opposite situation applies for an exothermic reaction (release of energy). Less energy is required to break reactant molecule bonds than is released through

> *Exothermic* means energy is released; energy is a "product" of the chemical reaction. *Endothermic* means energy is absorbed; energy is a "reactant" in the reaction.

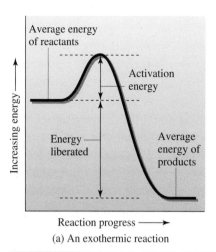

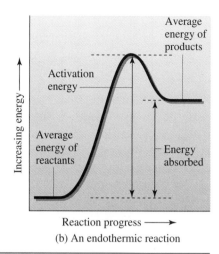

(a) An exothermic reaction   (b) An endothermic reaction

Figure 16.2

Energy diagram graphs showing the difference between an exothermic and an endothermic reaction. (a) In an exothermic reaction, the average energy of the reactants is higher than that of the products, indicating that energy has been released. (b) In an endothermic reaction, the average energy of the reactants is less than that of the products, indicating that energy has been absorbed in the reaction.

bond formation in product molecules. In this situation, the excess energy is released to the surroundings as the reaction proceeds. Figure 16.2 illustrates the energy relationships associated with endothermic and exothermic chemical reactions. Note that both of these diagrams contain a "hill" or "hump." The height of this "hill" corresponds to the activation energy needed for reaction between molecules to occur. This activation energy is independent of whether a given reaction is endothermic or exothermic.

16.3 Factors That Influence Reaction Rates

The **rate of a chemical reaction** *is the rate at which reactants are consumed or products produced in the chemical reaction in a given time period.* Natural processes have a wide range of reaction rates. A fire burns at a much faster reaction rate than the ripening of fruit, which is faster than the process of rusting, which is much faster than the process of aging in the human body. In this section we consider four different factors that affect reaction rates: (1) the physical nature of reactants, (2) reactant concentrations, (3) reaction temperature, and (4) the presence of catalysts.

Physical Nature of Reactants

The physical nature of reactants refers not only to the physical state of each reactant (solid, liquid, or gas) but also to the state of subdivision, that is, particle size. In reactions where the reactants are all in the same physical state, reaction rate is generally faster between liquid reactants than between solid reactants and faster still between gaseous reactants. Of the three states of matter the gaseous state is the one where there is the most freedom of movement; hence in this state there is a greater frequency of collision (reaction) between reactants.

In reactions involving solids and heterogeneous liquid mixtures, reaction occurs at the boundary surface between reactants. The greater the amount of boundary surface area, the greater the reaction rate. Subdividing a solid into smaller particles will increase surface area

Increasing surface area

Increasing contact area between reactants

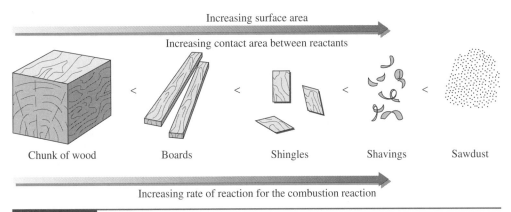

| Chunk of wood | | Boards | | Shingles | | Shavings | | Sawdust |

Increasing rate of reaction for the combustion reaction

Figure 16.3

Greater reactant surface area results in an increased reaction rate.

and thus increase reaction rate. For example, large pieces of wood are difficult to ignite, smaller pieces burn more rapidly, and wood shavings ignite instantaneously (see Fig. 16.3).

When particle size becomes extremely small, reaction rates can be so fast that an explosion results. A lump of coal is difficult to ignite; coal dust ignites explosively. The spontaneous ignition of coal dust is a real threat to underground coal-mining operations. Grain dust (very finely divided grain particles) is a problem in grain-storage elevators; explosive ignition of the dust from an accidental spark is always a possibility. Figure 16.4 shows the destruction that can result from the accidental ignition of grain dust in a storage elevator.

Reactant Concentration

An increase in the concentration of a reactant causes an increase in the rate of the reaction. Combustible substances burn much more rapidly in pure oxygen than they do in air (21% oxygen). Increasing the concentration of a reactant means that there are more molecules of that reactant present in the reaction mixture and therefore there is a greater possibility for

Figure 16.4

Extremely rapid combustion of grain dust produced the explosive effect that destroyed this grain elevator. (Science VU/Visuals Unlimited)

collisions between this reactant and other reactant particles. An analogy to the reaction-rate–reactant-concentration relationship can be drawn from the game of billiards. The more billiard balls there are on the table, the greater the probability of a moving cue ball striking one of them.

The actual quantitative change in reaction rate as the concentration of reactants is increased varies with the specific reaction. The rate usually increases, but not to the same extent in all cases. Simply looking at the balanced equation for a reaction will not enable you to determine how changes in concentration will affect the reaction rate. This must be determined by actual experimentation. In some reactions the rate doubles with a doubling of concentration; however, this is not always the case.

Reaction Temperature

The effect of temperature on reaction rates can also be explained by using the molecular-collision concept. An increase in the temperature of a system results in an increase in the average kinetic energy of the reacting molecules. The increased molecular speed causes more collisions to take place in a given time. Also, since the average kinetic energy of the colliding molecules is greater, a larger fraction of the collisions will have sufficient kinetic energy to equal or exceed the activation energy.

We make use of our knowledge of the effect of temperature on chemical reactions on a regular basis in our daily lives. The chemical reaction of cooking takes place faster in a pressure cooker because of a higher cooking temperature (Sec. 11.15). On the other hand, foods are cooled or frozen to slow down the chemical reactions that result in the spoiling of food, the souring of milk, and the ripening of fruit.

Presence of Catalysts

A **catalyst** *is a substance that, when added to a reaction mixture, increases the rate of the reaction but is itself unchanged after the reaction is completed.* Catalysts can be classified into two categories: (1) *homogeneous catalysts* and (2) *heterogeneous catalysts.* A **homogeneous catalyst** *is a catalyst that is present in the same phase as the reactants in a reaction mixture.* Homogeneous catalysts are usually dispersed uniformly throughout the reaction mixture. A **heterogenous catalyst** *is a catalyst that exists in a phase different from the reactants in a reaction mixture.* Heterogeneous catalysts are usually solids.

Catalysts increase reaction rates by providing alternative reaction pathways with lower activation energies than the original uncatalyzed pathway. This lowering of activation energy effect is illustrated in Figure 16.5.

> A rough generalization for many common reactions is that the rate of the chemical reaction doubles for every 10°C increase in temperature in the temperature range we normally encounter.

> You may have noticed that pictures from an "instant camera" develop more rapidly on warm days than on cold days. The chemical reactions involved in the development process occur faster at the higher temperatures.

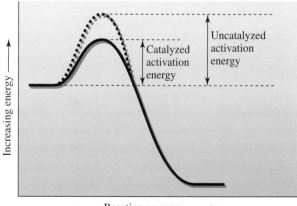

Figure 16.5

Catalysts lower the activation energy for chemical reactions. Reactions proceed more rapidly with the lowered activation energy.

In homogeneous catalysis the alternative pathway involves the formation of an intermediate "complex" that contains the catalyst. This catalyst-containing intermediate then breaks up to give the final products and regenerate the catalyst. The following equations, where C is the catalyst, illustrate this concept.

$$\text{Uncatalyzed reaction:} \quad X + Y \longrightarrow XY$$
$$\text{Catalyzed reaction:} \quad \textit{Step 1}: X + C \longrightarrow XC$$
$$\textit{Step 2}: XC + Y \longrightarrow XY + C$$

Catalysts that are solids are thought to provide a surface to which impacting reactant molecules are physically attracted and on which they are held with a particular orientation. Reactants so held are sufficiently close to, and thus favorably oriented toward, each other to allow the reaction to take place. The products of the reaction then leave the surface and make it available to catalyze other reactants.

Catalysts are used extensively in the chemical industry. Usually, very specific catalysts are used that accelerate one chemical reaction without influencing other possible competitive reactions. The small amounts of catalysts required, coupled with the fact that they are not used up, make the use of catalysts economically feasible in industrial processes. For example, a catalyst often makes it possible to avoid the high temperatures (costly) that would otherwise be necessary to cause a reaction with high activation energy to proceed.

Catalysts are a key element in the functioning of automobile emission control systems. In such systems, heterogeneous catalysts speed up reactions that convert air pollutants in the exhaust to less harmful products (see Fig. 16.6). For example, carbon monoxide is converted to carbon dioxide through reaction with the oxygen in air.

Catalysts are of extreme importance for the proper functioning of the human body and other biological systems. In the human body, catalysts called *enzymes*, which are proteins, cause many reactions to take place rapidly at body temperature and under mild conditions. These same reactions, uncatalyzed, proceed very slowly and then only under harsher conditions in a laboratory setting.

> Catalysts lower the activation energy for a reaction. Lowered activation energy increases the rate of the reaction as more reacting particles have the required activation energy.

> Biological catalysts, called *enzymes,* mediate nearly all the reactions that occur within a living organism.

16.4 Chemical Equilibrium

In our discussions of chemical reactions, up to this point, we have assumed that chemical reactions go to completion, that is, that reactions continue until one or more of the reactants is used up. This is usually not the case. Experiments show that in most chemical reactions the complete conversion of reactants to products does not occur regardless of the time allowed for the reactions to take place. The reason for this is that product molecules (provided they are not allowed to escape from the reaction mixture) begin to react with each other to again re-form the reactants. With time, a steady-state situation results where the rate of formation of products and the rate of re-formation of reactants are equal. At this point the concentrations of all reactants and all products remain constant; the reaction has reached a stage of *chemical equilibrium*. **Chemical equilibrium** *is the process wherein two opposing chemical reactions occur simultaneously at the same rate.* We have discussed equilibrium situations in previous chapters—see Sections 11.14 (vapor pressure), 13.2 (saturated solutions), and 14.3 (conjugate acids and bases). The first two of these previous equilibrium situations involved *physical* equilibrium (no chemical reaction) rather than *chemical* equilibrium. Conjugate acid–base relationships involve chemical equilibrium, a topic we now consider in detail.

The conditions that exist in a system in a state of chemical equilibrium can best be visualized by considering an actual chemical reaction. Suppose equal molar amounts of gaseous H_2 and I_2 are mixed together in a closed container and allowed to react.

$$H_2(g) + I_2(g) \longrightarrow 2\ HI(g)$$

Initially, no HI is present, so the only possible reaction that can occur is the one between H_2 and I_2. However, with time, as the HI concentration increases, some HI molecules collide with each other in a way that causes the reverse reaction to occur.

$$2\ HI(g) \longrightarrow H_2(g) + I_2(g)$$

The initial low concentration of HI makes this reverse action slow at first, but as the concentration of HI increases, so does the reaction rate. At the same time the reverse reaction rate is increasing, the forward reaction rate (production of HI) is decreasing as reactants are used up. Eventually, the concentrations of H_2, I_2, and HI in the reaction mixture reach a level at which the rates of the forward and reverse reactions become equal. At this point a state of chemical equilibrium has been reached.

Figure 16.7 shows graphically the behavior of reaction rates and reaction concentrations with time for both the forward and reverse reactions in the H_2–I_2–HI system. Figure 16.7a shows that the forward and reverse reaction rates become equal as a result of the forward reaction rate decreasing (as reactants are used up) and the reverse reaction rate increasing (as product concentration increases). Figure 16.7b shows the important point that reactant and product concentrations are usually not equal at the point at which equilibrium is reached. Rates are equal, but concentrations are not. For the H_2–I_2–HI system, much more product HI is present than reactants H_2 and I_2 at equilibrium. In

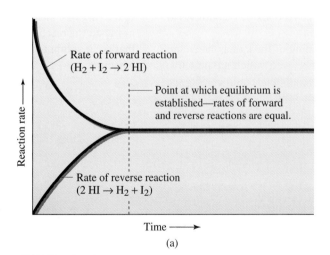

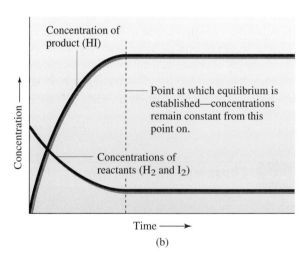

Figure 16.7

Graphs showing how reaction rates and reactant concentrations vary with time for the chemical system H_2–I_2–HI. (a) At equilibrium, rates of reaction are equal. (b) At equilibrium, concentrations of reactants and products remain constant but are not equal.

Figure 16.7b note that the point at which equilibrium is established is the point where the two curves become straight lines.

The equilibrium involving H_2, I_2, and HI could have been established just as easily by starting with pure HI and allowing it to change into H_2 and I_2 (the reverse reaction). The final position of equilibrium does not depend on the direction from which equilibrium is approached.

Instead of separate equations for the forward and reverse reactions for a system at equilibrium, it is normal procedure to represent the equilibrium by using a single equation and half-headed arrows pointing in both directions. Thus, the reactions between H_2 and I_2 and between 2 HI, at equilibrium, are written as

$$H_2(g) + I_2(g) \rightleftharpoons 2 HI(g)$$

The term *reversible* is often used to describe a reaction like the one we have just discussed. A **reversible reaction** *is a chemical reaction in which the conversion of reactants to products (the forward reaction) and the conversion of products to reactants (the reverse reaction) occur simultaneously.* When the half-headed double arrow notation is used in a chemical equation, it means that the chemical reaction is reversible.

> At chemical equilibrium, forward and reverse reaction rates are equal. Reactant and product concentrations, although constant, do not have to be equal.

16.5 Equilibrium Mixture Stoichiometry

Suppose that known amounts of reactants are placed in a reaction vessel and the system is allowed to reach chemical equilibrium (Sec. 16.4). To determine the composition of the resulting equilibrium mixture we need only experimentally determine the equilibrium concentration of one of the substances in the mixture. With this one value and the substance amounts originally present, the concentrations of all other substances present in the equilibrium mixture can be calculated. Example 16.1 shows how such a calculation is carried out.

EXAMPLE 16.1

Determining the Composition of an Equilibrium Mixture in Terms of Moles of Each Substance Present

0.0930 mole of NO and 0.0652 mole of Br_2 are placed in a container and allowed to react until equilibrium is established.

$$2 NO(g) + Br_2(g) \rightleftharpoons 2 NOBr(g)$$

At equilibrium 0.0612 mole of NOBr is present. What is the composition of the equilibrium mixture in terms of moles of each substance present?

SOLUTION

In solving this problem we will deal with three quantities for each of the substances involved in the equilibrium: (1) starting amount of each substance, (2) amount that changes (undergoes reaction), and (3) equilibrium amount of each substance. The following table, the starting point for our calculation, summarizes the known (given) quantities in terms of these three parameters.

	2 NO(g)	+	Br_2(g)	$\rightleftharpoons$	2 NOBr(g)
Start	0.0930 mole		0.0652 mole		0 mole
Change	—		—		—
Equilibrium	—		—		0.0612 mole

Four of the nine "blanks" in the table have numbers in them. The key observation is that two of the three "blanks" for NOBr are known. In a problem of this type, any time two of the three key items (start, change, and equilibrium) are known for a substance, the third can be calculated by addition or subtraction. For NOBr, we started with zero amount and ended up with 0.0612 mole at equilibrium. Obviously, the amount of change for NOBr is +0.0612 mole, the amount of NOBr formed.

	$2\,NO(g)$	$+$	$Br_2(g)$	$\rightleftharpoons$	$2\,NOBr(g)$
Start	0.0930 mole		0.0652 mole		0 mole
Change	—		—		+0.0612 mole
Equilibrium	—		—		0.0612 mole

Once one of the change values is known, all other change quantities can quickly be calculated. The molar-change values are related to each other in the same manner as the coefficients in the equation are related to each other. Thus, we know that

1. The molar amount of NO that reacts is the same as the molar amount of NOBr produced, since these two substances have the same coefficients in the equation, and

2. The molar amount of Br_2 that changes (reacts) is one-half the molar amount of NOBr produced since the Br_2/NOBr coefficient ratio is 1 to 2.

Placing this information into the "table" gives

	$2\,NO(g)$	$+$	$Br_2(g)$	$\rightleftharpoons$	$2\,NOBr(g)$
Start	0.0930 mole		0.0652 mole		0 mole
Change	−0.0612 mole		−0.0306 mole		+0.0612 mole
Equilibrium	—		—		0.0612 mole

Note the minus signs placed in front of the NO and Br_2 change amounts. This is because these amounts are consumed (used up in the reaction). The plus sign in front of the NOBr change amount indicates a gain in the amount of this substance.

The last two blanks in the table are now easily determined through subtraction. For NO, 0.0930 mole (start) − 0.0612 mole (change) = 0.0318 mole (equilibrium). Similarly, for Br_2 we have 0.0652 mole − 0.0306 mole = 0.0346 mole. Our completed table is

	$2\,NO(g)$	$+$	$Br_2(g)$	$\rightleftharpoons$	$2\,NOBr(g)$
Start	0.0930 mole		0.0652 mole		0 mole
Change	−0.0612 mole		−0.0306 mole		+0.0612 mole
Equilibrium	0.0318 mole		0.0346 mole		0.0612 mole

The equilibrium mixture composition, which is the bottom line of the table, is

0.0318 mole NO;   0.0346 mole Br_2;   0.0612 mole NOBr

Practice Exercise 16.1

Sulfur dioxide and oxygen react according to the following equation.

$$2\,SO_2(g) + O_2(g) \rightleftharpoons 2\,SO_3(g)$$

When 4.00 moles of SO_2 and 2.00 moles of O_2 are placed in an appropriate container and allowed to react until equilibrium is established, it is found that 2.96 moles of SO_3 have been formed. What is the composition of the equilibrium mixture in terms of moles of each substance present?

Answers to practice exercises are located at the end of the chapter.

16.6 Equilibrium Constants

The concentrations of reactants and products are constant (not changing) in a system at chemical equilibrium (Sec. 16.4). The numerical values of these equilibrium concentrations can be used to calculate an *equilibrium constant*, a single number that describes the extent to which the chemical reaction of concern has occurred. An **equilibrium constant** *is a numerical value that characterizes the relationship between the concentrations of reactants and concentrations of products in a system that is at chemical equilibrium.*

The equilibrium constant for a chemical reaction is obtained by writing an *equilibrium expression* and then evaluating it numerically. To illustrate the calculation of an equilibrium constant, let us consider a general gas phase reaction in which a moles of A and b moles of B react to produce c moles of C and d moles of D.

$$a A(g) + b B(g) \rightleftharpoons c C(g) + d D(g)$$

The equilibrium constant expression for this reaction is

$$K_{eq} = \frac{[C]^c [D]^d}{[A]^a [B]^b}$$

Note the following points about this general equilibrium constant expression:

1. The square brackets refer to molar (moles/liter) concentrations.
2. Product concentrations are always placed in the numerator of the equilibrium constant expression.
3. Reactant concentrations are always placed in the denominator of the equilibrium constant expression.
4. The coefficients in the balanced chemical equation for the equilibrium system determine the powers to which the concentrations are raised.
5. The abbreviation K_{eq} is used to denote an equilibrium constant.

An additional convention in writing equilibrium constant expressions, not apparent from the preceding equilibrium constant definition, is: *Only concentrations of gases and substances in solution are written in an equilibrium constant expression.* The reason for this convention is that other substances (pure solids and pure liquids) have constant concentrations. These constant concentrations are incorporated into the equilibrium constant itself. For example, pure water in the liquid state has a concentration of 55.5 moles/L. It does not matter whether we have 1.00, 50.0, or 750 mL of liquid water, the concentration will be the same. In the liquid state, pure water is pure water, and it has only one concentration. Similar reasoning applies to other pure liquids and pure solids. All such substances have constant concentrations.

The only information we need to write an equilibrium constant expression is a balanced chemical equation that includes information about physical state. Using the preceding generalizations about equilibrium expressions, for the reaction

$$4 NH_3(g) + 7 O_2(g) \rightleftharpoons 4 NO_2(g) + 6 H_2O(g)$$

we write the equilibrium expressed as

> The concentrations of *pure liquids* and *pure solids*, which are constants, are never included in an equilibrium constant expression.

Coefficient of NO_2 ↘   ↙ Coefficient of H_2O

$$K_{eq} = \frac{[NO_2]^4 [H_2O]^6}{[NH_3]^4 [O_2]^7}$$

Coefficient of NH_3 ⌐   ⌐ Coefficient of O_2

EXAMPLE 16.2

Using Balanced Chemical Equations to Determine Equilibrium Constant Expressions

Write the equilibrium constant expression for each of the following reactions.

(a) $4 NH_3(g) + 3 O_2(g) \rightleftharpoons 2 N_2(g) + 6 H_2O(g)$

(b) $6 Ca(s) + 2 NH_3(g) \rightleftharpoons 3 CaH_2(s) + Ca_3N_2(s)$

(c) $2 Ag_2CO_3(s) \rightleftharpoons 4 Ag(s) + 2 CO_2(g) + O_2(g)$

(d) $NaCl(aq) + AgNO_3(aq) \rightleftharpoons AgCl(s) + NaNO_3(aq)$

SOLUTION

(a) All of the substances involved in this reaction are gases. Therefore, each reactant and product will appear in the equilibrium constant expression.

   The product concentrations, each raised to the power of its coefficient in the balanced equation, are placed in the numerator.

$$K_{eq} = \frac{[N_2]^2[H_2O]^6}{-} \qquad \text{Equation coefficients}$$

The reactant concentrations, each raised to the power of its coefficient in the balanced equation, are placed in the denominator.

$$K_{eq} = \frac{[N_2]^2[H_2O]^6}{[NH_3]^4[O_2]^3}$$

Note that H_2O as a gas (water vapor or steam) is included in an equilibrium constant expression. The concentration of a gas can vary. Water, as a liquid, is never included in equilibrium constant expressions.

(b) Three of the four substances involved in this reaction are solids and thus will not appear in the equilibrium constant expression. Since both products are solids, the numerator of the equilibrium constant expression is 1. The concentration of NH_3 raised to the second power is the only factor in the denominator since the other reactant is a solid.

$$K_{eq} = \frac{1}{[NH_3]^2}$$

(c) The reactant Ag_2CO_3 is a solid and thus will not appear in the equilibrium constant expression. Since Ag_2CO_3 is the only reactant, this means there will be no denominator in the equilibrium expression. Two of the three products are gases and they appear in the numerator of the equilibrium constant expression.

$$K_{eq} = [CO_2]^2[O_2]$$

(d) All of the powers in this equilibrium constant expression are "1" because all of the coefficients in the balanced equation are ones.

$$K_{eq} = \frac{[NaNO_3]}{[NaCl][AgNO_3]}$$

AgCl is not included in the equilibrium expression because it is a solid.

Practice Exercise 16.2

Write the equilibrium constant expression for each of the following reactions.

(a) $2 NO_2(g) + 7 H_2(g) \rightleftharpoons 2 NH_3(g) + 4 H_2O(g)$

(b) $C(s) + H_2O(g) \rightleftharpoons CO(g) + H_2(g)$

(c) $NH_4Cl(s) \rightleftharpoons NH_3(g) + HCl(g)$

(d) $Fe_2(SO_4)_3(s) \rightleftharpoons 2 Fe^{3+}(aq) + 3 SO_4^{2-}(aq)$

At a given temperature, the numerical value of the equilibrium constant for a reaction is obtained by substituting the experimentally determined equilibrium concentrations at that temperature into the equilibrium constant expression for the reaction, as is shown in Example 16.3.

Equilibrium constant values vary with temperature changes. A change in temperature changes molecular energies, and molecular energies have a direct effect on the relative amounts of reactants and products present in an equilibrium mixture.

EXAMPLE 16.3

Using Equilibrium Concentrations to Calculate the Value of an Equilibrium Constant

At a temperature of 927°C, the equilibrium molar concentrations for the reaction

$$CO(g) + 3 H_2(g) \rightleftharpoons CH_4(g) + H_2O(g)$$

are

$$[CO] = 0.613, [H_2] = 1.839, [CH_4] = 0.387, \text{ and } [H_2O] = 0.387$$

Calculate the value of the equilibrium constant for this reaction at 927°C.

SOLUTION

The general expression for the equilibrium constant is

$$K_{eq} = \frac{[CH_4][H_2O]}{[CO][H_2]^3}$$

Substituting the known equilibrium concentrations into this expression gives

$$K_{eq} = \frac{[0.387][0.387]}{[0.613][1.839]^3}$$

$$= 0.039284051 \quad \text{(calculator answer)}$$

$$= 0.0393 \quad \quad \textbf{(correct answer)}$$

Practice Exercise 16.3

At a temperature of 350°C, the equilibrium concentrations for the reaction

$$N_2(g) + 3 H_2(g) \rightleftharpoons 2 NH_3(g)$$

are

$$[N_2] = 0.885, [H_2] = 0.665, \text{ and } [NH_3] = 1.230$$

Calculate the value of the equilibrium constant for this reaction at 350°C.

Equilibrium position *is a qualitative indication of the relative amounts of reactants and products present when a chemical reaction reaches equilibrium.* The terms *far to the right, to the right, neither to the right nor the left, to the left*, and *far to the left* are used in describing equilibrium position. In equilibrium situations where the concentrations of products are greater than those of reactants, the equilibrium position is said to lie to the *right* because products are always listed on the right side of a chemical equation. Conversely, when reactants dominate at equilibrium, the equilibrium position lies to the *left*. The terminology *neither to the right nor the left* indicates that significant amounts of both reactants and products are present in an equilibrium mixture.

Equilibrium position can also be indicated by varying the length of the arrows in the half-headed double-arrow notation for a reversible reaction. The longer arrow indicates the direction of the predominant reaction. For example, the arrow notation in the equation

$$CO_2 + H_2O \rightleftharpoons H_2CO_3$$

indicates that the equilibrium position lies to the right.

The magnitude of the equilibrium constant for a reaction gives information about how far a reaction proceeds toward completion, that is, about where the equilibrium position lies. A large value of K_{eq} (greater than 10^3) means that the numerical value of the numerator is significantly greater than that of the denominator. In terms of reactants and products, this means that the concentrations of the products are greater than those of the reactants. The equilibrium position lies to the right.

Conversely, if the equilibrium constant is small (less than 10^{-3}), we have a situation where reactants will predominate over products in the reaction mixture. The equilibrium position is said to lie to the left in this situation.

For equilibrium conditions where K_{eq} has a value between 10^3 and 10^{-3}, appreciable concentrations of both products and reactants are present. The reaction described in Example 16.3 falls into this category.

16.7 Le Châtelier's Principle

A chemical system in a state of equilibrium remains in that state until it is disturbed by some change of condition. Disturbing an equilibrium has one of two results: Either the forward reaction speeds up (to produce additional products) or the reverse reaction speeds up (to produce additional reactants). Then with time, the forward and reverse reactions again become equal and a new equilibrium, not identical to the previous one, is established. If more products have been produced as a result of the disruption, the equilibrium position is said to have *shifted to the right*. Similarly, when the disruption causes more reactants to form, the equilibrium position has *shifted to the left*.

Qualitative predictions about the direction in which chemical equilibria shift can be made using a guideline (principle) introduced in 1888 by the French chemist Henri-Louis Le Châtelier (1850–1936)—see "The Human Side of Chemistry 16." **Le Châtelier's principle** *states that if a stress (change of conditions) is applied to a chemical system in equilibrium, the system will readjust (change the position of the equilibrium) in the direction that best reduces the stress imposed upon the system.* We will use this principle in considering how four types of changes affect equilibrium position. The changes are: (1) concentration changes, (2) temperature changes, (3) pressure changes, and (4) addition of a catalyst.

Concentration Changes

Adding or removing a *gaseous* reactant or *gaseous* product from a reaction mixture at equilibrium will always upset the equilibrium. Le Châtelier's principle predicts that the reaction will shift in the direction that will minimize the change in concentration caused by the addition or removal. If an additional amount of any *gaseous* reactant or product has been *added* to the

The Human Side of Chemistry 16

Henri-Louis Le Châtelier (1850–1936)

Henri-Louis Le Châtelier (pronounced le-shot-lee-ay) was born in Paris in 1850. His father was Inspector General of Mines for France; his grandfather operated lime kilns. Henri would visit his grandfather's kilns during vacations. Contacts with his father, grandfather, and their associates were a major factor in shaping his career, a career most noteworthy for his constant application of chemical principles in industrial settings.

After obtaining a formal education, he registered as a mining engineer.

Later he obtained a degree in physical and chemical science. In 1887 he was appointed Professor of Industrial Chemistry and Metallurgy in the Écoles des Mines. Later came an appointment as Professor of Inorganic Chemistry and still later an appointment in General Chemistry.

Le Châtelier's interests were amazingly diverse. The chemistries of cement, of ceramics, and of glass occupied his attention for a time. Studies in combustion with the aim of preventing mine explosions were next. These led to studies of heat and its measurement. From his studies of heat came that for which he is best known, his principle of "stress and strain." This principle was

first proposed in 1884; a simplified version was presented in 1888. Metallurgy also occupied his attention. In this area he was most concerned with the chemistry and metallurgy of iron and steel.

In later life he was a great national hero in France. His linking of science with industry (especially during World War I) was very important to France. In 1916, the President of the United States (Woodrow Wilson) engaged him as a consultant when the United States' National Research Council was established. He held positions on many commissions and boards that advised the French government on scientific and technical questions.

system, the stress is relieved by shifting the equilibrium in the direction that *consumes* (uses up) some of the added reactant or product. Conversely, if a *gaseous* reactant or product is *removed* from an equilibrium system, the equilibrium will shift in a direction that *produces* more of the substance that was removed.

Let us consider the effect that selected concentration changes will have on the gaseous equilibrium.

$$N_2(g) + 3H_2(g) \rightleftharpoons 2NH_3(g)$$

Suppose some additional H_2 is added to the equilibrium mixture. The equilibrium will shift to the right; that is, the forward reaction rate will increase in order to use up additional H_2. Eventually a new equilibrium position will be reached. At this new position the H_2 concentration will still be higher than it was before the addition; that is, not all of the added H_2 is consumed. In addition, the N_2 concentration will have decreased (some N_2 had to react with the H_2) and the NH_3 concentration will have increased as the product of the H_2 and N_2 reacting.

Removal of some NH_3 from this newly established equilibrium position will cause an additional shift to the right. The concentration of H_2 and N_2 will decrease as the system attempts to replace the NH_3 that was removed by producing more of it. Again, not all of the removed NH_3 will be replaced. When the new equilibrium position is achieved, the NH_3 concentration will be less than it was before the NH_3 removal.

Figure 16.8 shows graphically the effects that the H_2 addition and NH_3 removal just discussed have on the concentration of all substances present in the N_2–H_2–NH_3 equilibrium mixture.

Throughout this discussion of the effect of concentration changes on an equilibrium system, we have referred to the effect of adding or removing a *gaseous* species. If we add a pure solid or liquid to a gas phase equilibrium system there will be no shift in the equilibrium position; nothing happens. This is because there will be no change in concentration. The solid or liquid was 100% pure before the addition and it is still "100%" after the addition.

More specifically, consider the equilibrium system

$$CaCO_3(s) \rightleftharpoons CaO(s) + CO_2(g)$$

Adding or removing $CaCO_3$ or CaO from this system will cause no change in the equilibrium position. In general, *adding or removing a species disturbs an equilibrium system only if the concentration of that species appears in the expression for the equilibrium constant* (Sec. 16.6).

> Thousands of chemical equilibria simultaneously exist in biological systems. Many of them are interrelated. When the concentration of a single substance changes, many equilibria are affected.

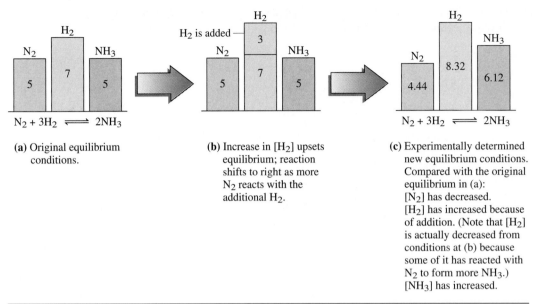

(a) Original equilibrium conditions.

(b) Increase in [H$_2$] upsets equilibrium; reaction shifts to right as more N$_2$ reacts with the additional H$_2$.

(c) Experimentally determined new equilibrium conditions. Compared with the original equilibrium in (a): [N$_2$] has decreased. [H$_2$] has increased because of addition. (Note that [H$_2$] is actually decreased from conditions at (b) because some of it has reacted with N$_2$ to form more NH$_3$.) [NH$_3$] has increased.

Figure 16.8

Concentration changes that result when H$_2$ is added to an equilibrium mixture involving the system N$_2$(g) + 3 H$_2$(g) $\rightleftharpoons$ 2 NH$_3$(g).

The concentrations of solids and liquids do not appear in equilibrium constant expressions. The concentrations of gases and species in *aqueous solution* do appear in equilibrium constant expressions. The changing of the concentration of one of the two ions in the equilibrium

$$Pb^{2+}(aq) + 2 Cl^-(aq) \rightleftharpoons PbCl_2(s)$$

would affect the equilibrium position.

Temperature Changes

Le Châtelier's principle can be used to predict the influence of temperature changes on an equilibrium provided it is known whether the reaction of concern is endothermic or exothermic.

Consider the *exothermic reaction*

$$H_2(g) + F_2(g) \rightleftharpoons 2 HF(g) + \text{heat}$$

Heat is produced when the reaction proceeds to the right. Thus, if we add heat to an exothermic system at equilibrium (by raising the temperature), the system will shift to the left in an attempt to use up the added heat. When equilibrium is reestablished, the concentrations of H$_2$ and F$_2$ will be higher and the concentration of HF will have decreased. Lowering the temperature of an exothermic system at equilibrium will cause the reaction to shift to the right as the system attempts to replace the lost heat.

The behavior, with temperature change, of an equilibrium reaction mixture involving an *endothermic* reaction such as

$$\text{Heat} + 2 CO_2(g) \rightleftharpoons 2 CO(g) + O_2(g)$$

> The effect of a temperature change on the position of an equilibrium depends on whether heat is a "reactant" or a "product" in the chemical reaction of concern.

is just the opposite of that of an exothermic reaction, since a shift to the left (rather than to the right) produces heat. Consequently, an increase in temperature will cause the equilibrium system to shift to the right (to use up the added heat), and a decrease in temperature will produce a shift to the left (to generate more heat).

Pressure Changes

Pressure changes affect systems at equilibrium only when gaseous substances are part of the equilibrium, and then only in cases where the chemical reaction is such that a change in the total number of moles of gaseous substances occurs. This latter point can be illustrated by considering the following two gas-phase reactions.

$$\underbrace{2\,H_2(g) + O_2(g)}_{3\text{ moles of gas}} \longrightarrow \underbrace{2\,H_2O(g)}_{2\text{ moles of gas}}$$

$$\underbrace{H_2(g) + Cl_2(g)}_{2\text{ moles of gas}} \longrightarrow \underbrace{2\,HCl(g)}_{2\text{ moles of gas}}$$

In the first reaction the total number of moles of gaseous reactants and products decreases as the reaction proceeds to the right, since 3 moles of reactants combine to give only 2 moles of products. In the second reaction there is no change in the total number of moles of gaseous substances present as the reaction proceeds, since 2 moles of reactants combine to give 2 moles of products. Thus, a pressure change will shift the position of equilibrium in the first reaction but not in the second reaction.

Pressure changes are usually brought about through volume changes. A pressure increase results from a volume decrease, and a pressure decrease from a volume increase (Sec. 12.3). The use of Le Châtelier's principle correctly predicts the direction of the equilibrium position shift resulting from a pressure change only when the pressure change is due to a volume change. It does not apply to pressure increases caused by the addition of a nonreactive (inert) gas to the reaction mixture. Such an addition has no effect on the equilibrium position. The partial pressure (Sec. 12.15) of each of the gases involved in the reaction remains the same.

> **Increasing the pressure associated with an equilibrium system by adding an inert gas (a gas that is not a reactant or a product in the reaction) does not affect the position of the equilibrium.**

According to Le Châtelier's principle, the stress of increased pressure is relieved by decreasing the number of moles of gaseous substances in the system. This is accomplished by the reaction shifting in the direction of the smaller number of moles, that is, to the side of the equation that contains the smaller number of moles of gaseous substances. For the reaction

$$2\,NO_2(g) + 7\,H_2(g) \rightleftharpoons 2\,NH_3(g) + 4\,H_2O(g)$$

an increase in pressure would shift the equilibrium position to the right, since there are 9 moles of gaseous reactants and only 6 moles of gaseous products. A stress of decreased pressure will result in an equilibrium system reacting in such a way as to produce more moles of gaseous substances.

Addition of a Catalyst

Catalysts do not change the position of equilibrium. This fact becomes clear when we remember that a catalyst functions by lowering the activation energy for a reaction (Sec. 16.3). The activation energy for the forward reaction is lowered but so is the activation energy for the reverse reaction. Hence, a catalyst speeds up both the forward and reverse reactions and has no effect on the position of equilibrium. However, the lowered activation energy allows equilibrium to be established more quickly than if the catalyst were not present (see Fig. 16.9).

EXAMPLE 16.4

Using Le Châtelier's Principle to Predict the Effects of Changes on the Equilibrium Position in an Equilibrium System

How will the gas-phase equilibrium

$$PCl_3(g) + Cl_2(g) \rightleftharpoons PCl_5(g) + \text{heat}$$

be affected by each of the following?

(a) removal of $PCl_5(g)$

(b) addition of $Cl_2(g)$

(c) temperature decrease

(d) an increase in the volume of the container (pressure decrease)

SOLUTION

(a) The equilibrium will shift to the right, according to Le Châtelier's principle, in an attempt to replenish the PCl_5 removed.

(b) The equilibrium will shift to the right in an attempt to use up the extra Cl_2 that has been placed in the system.

(c) Lowering the temperatures means that heat energy has been removed. The position of equilibrium will shift to the right in order to produce more heat to take the place of that removed.

(d) The system will shift to the left in an attempt to produce more moles of gaseous reactants; this will increase the pressure. In going to the left the reaction produces 2 moles of gaseous reactants for every 1 mole of gaseous product consumed.

Practice Exercise 16.4

How will the gas-phase equilibrium

$$CH_4(g) + 2\,H_2S(g) + \text{heat} \rightleftharpoons CS_2(g) + 4\,H_2(g)$$

be affected by each of the following?

(a) removal of $H_2(g)$

(b) addition of $CS_2(g)$

(c) temperature increase

(d) an increase in the volume of the container (pressure decrease)

16.8 Forcing Reactions to Completion

Reactions that would ordinarily reach a state of equilibrium can be forced to completion by using experimental conditions that place a "continual stress" on the potential equilibrium condition. Let us consider a few ways in which this "forcing" is done.

Continuous removal of one or more products of a reaction will force the reaction to completion. To reach equilibrium both reactants and products must be present. The removal of a product continually shifts the reaction to the right, that is, toward completion, according to Le Châtelier's principle. Eventually, one or more of the reactants is depleted, the sign of a completed reaction.

Product removal is very easy to arrange in situations that involve gaseous products. If the reaction is run in an open container, the gaseous products automatically escape to the atmosphere as fast as they are produced. Such a reaction will never reach equilibrium and will continue until the limiting reactant is used up.

Figure 16.9

A catalyst decreases the time required for equilibrium to be reached. It does not, however, change the position of equilibrium; the amount of product produced remains the same.

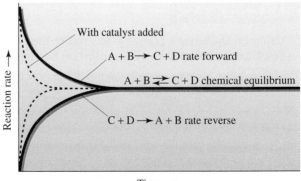

Sometimes another chemical reaction is used to remove a product. Consider the situation of a saturated solution of NaCl.

$$NaCl(s) \rightleftharpoons Na^+(aq) + Cl^-(aq)$$

In such a solution, as much NaCl is dissolved as is possible. Adding $AgNO_3$ to the saturated solution will cause more NaCl to dissolve. The Ag^+ ions from the $AgNO_3$ react with the Cl^- ions in the saturated solution to form insoluble AgCl.

$$Ag^+(aq) + Cl^-(aq) \rightleftharpoons AgCl(s)$$

This removes Cl^- (one of the products for the original equilibrium) from solution, thus upsetting the equilibrium. More NaCl will dissolve to compensate for the loss of the chloride ions (Le Châtelier's principle). Continued addition of $AgNO_3$ will eventually cause all of the NaCl to dissolve.

It is also possible to drive a reaction to completion by ensuring that an excess of one of the reactants is always present. The system will continually shift to the right (Le Châtelier's principle) to remove the stress caused by the excess reactant. Eventually other reactants will be depleted and the reaction will be completed. A procedure such as this is useful in a situation where one reactant is very expensive and others are much cheaper. To ensure that none of the expensive reactant goes unreacted, an excess of one of the less expensive reactants is used.

Summary

1. **Collision Theory** Collision theory is a set of statements that give the conditions that must be met before a chemical reaction will take place. The three basic tenets of collision theory are: (1) reactant molecules must collide with each other before any reaction can occur; (2) colliding particles must possess a certain minimum energy, called the activation energy, if the collision is to result in reaction; and (3) colliding particles must come together in the proper orientation if the reaction is to occur.

2. **Endothermic and Exothermic Reactions** An endothermic chemical reaction requires an input of energy as the reaction occurs. This is because the energy required to break bonds in the reactants is greater than the energy released through bond formation in the products. An exothermic chemical reaction releases energy as the reaction occurs. Less energy is required to break reactant molecule bonds than is released through bond formation in product molecules in an exothermic reaction.

3. **Factors That Affect Reaction Rate** The rate of a chemical reaction is the rate at which reactants are consumed or products produced in a given time period. Four factors that affect reaction rates are: (1) the physical nature of the reactants, (2) reactant concentrations, (3) reaction temperature, and (4) the presence of catalysts.

4. **Chemical Equilibrium** Chemical equilibrium is the condition in which two opposing chemical reactions (a forward reaction and a reverse reaction) occur simultaneously at the same rate. A state of chemical equilibrium is indicated in chemical equations by placing half-headed arrows pointing in both directions between reactants and products.

5. **Equilibrium Constant** An equilibrium constant is a numerical value that characterizes the relationship between the concentrations of reactants and products in a system at chemical equilibrium. The value of an equilibrium constant is obtained by writing an equilibrium expression and then numerically evaluating it. Equilibrium expressions can be obtained from the balanced chemical equation for a chemical reaction.

6. **Equilibrium Position** The relative amounts of reactants and products present in a system at equilibrium define the equilibrium position. The equilibrium position is toward the right when large amounts of products are present and is toward the left when large amounts of reactants are present.

7. **Le Châtelier's Principle** Le Châtelier's principle states that if a stress (change of conditions) is applied to a system in equilibrium, the system will readjust (change the position of the equilibrium) in the direction that best reduces the stress imposed on it. Stresses known to change equilibrium position include (1) concentration changes in reactants and/or products, (2) temperature changes, and (3) pressure changes. Catalysts do not change the position of an equilibrium.

Key Terms

The new terms defined in this chapter are

activation energy   *Sec. 16.1*
catalyst   *Sec. 16.3*
chemical equilibrium   *Sec. 16.4*
collision theory   *Sec. 16.1*
endothermic chemical reaction   *Sec. 16.2*
equilibrium constant   *Sec. 16.6*

equilibrium position   *Sec. 16.6*
exothermic chemical reaction   *Sec. 16.2*
heterogeneous catalyst   *Sec. 16.3*
homogeneous catalyst   *Sec. 16.3*
Le Châtelier's principle   *Sec. 16.7*
rate of a chemical reaction   *Sec. 16.3*
reversible reaction   *Sec. 16.4*

Practice Problems

Collision Theory (Sec. 16.1)

16.1 Why are reactions between substances in solution usually faster than reactions between solid-state reactants?

16.2 Why are gas-phase reactions usually faster than solid-phase reactions?

16.3 Under similar concentration and temperature conditions, would a reaction with an activation energy of 65 kJ/mole or one with an activation energy of 45 kJ/mole proceed at a faster rate? Explain your answer.

16.4 What is the relationship between activation energy and the minimum combined kinetic energy reactant particles must possess in order for their collision to result in a reaction?

16.5 What two factors determine whether a collision between two reactant molecules will result in a reaction?

16.6 What happens to the reactants in an ineffective molecular collision?

16.7 For the reaction

$$H_2 + Cl_2 \longrightarrow 2\,HCl$$

draw a sketch of a molecular orientation that is *highly favorable* for an effective collision.

16.8 For the reaction

$$H_2 + Cl_2 \longrightarrow 2\,HCl$$

draw a sketch of a molecular orientation that is *unfavorable* for an effective collision.

Endothermic and Exothermic Reactions (Sec. 16.2)

16.9 Classify each of the following reactions as *exothermic* or *endothermic*.

(a) $C_2H_4 + 3\,O_2 \longrightarrow 2\,CO_2 + 2\,H_2O + heat$
(b) $N_2 + 2\,O_2 + heat \longrightarrow 2\,NO_2$
(c) $2\,H_2O + heat \longrightarrow 2\,H_2 + O_2$
(d) $2\,KClO_3 + heat \longrightarrow 2\,KCl + 3\,O_2$

16.10 Classify each of the following reactions as *exothermic* or *endothermic*.

(a) $CaCO_3 + heat \longrightarrow CaO + CO_2$
(b) $N_2 + 3\,H_2 \longrightarrow 2\,NH_3 + heat$
(c) $CO + 3\,H_2 + heat \longrightarrow CH_4 + H_2O$
(d) $2\,N_2 + 6\,H_2O + heat \longrightarrow 4\,NH_3 + 3\,O_2$

16.11 Draw an energy diagram graph for a hypothetical chemical reaction that is *exothermic* by 35 kJ/mole and has an activation energy of 75 kJ/mole. Label the following on the diagram.

(a) average energy of the reactants
(b) average energy of the products
(c) activation energy
(d) amount of energy liberated during the reaction

16.12 Draw an energy diagram graph for a hypothetical chemical reaction that is *endothermic* by 35 kJ/mole and has an activation energy of 75 kJ/mole. Label the following on the diagram.

(a) average energy of the reactants
(b) average energy of the products
(c) activation energy
(d) amount of energy absorbed during the reaction

16.13 Reaction A occurs at room temperature and liberates 200 kJ of energy per mole of reactant. Reaction B does not occur until a temperature of 150°C is reached; it also liberates 200 kJ of energy per mole of reactant. Draw an energy diagram graph for each reaction, and indicate the similarities and differences between the two diagrams.

16.14 Reaction C occurs at room temperature and liberates 200 kJ of energy per mole of reactant. Reaction D also occurs at room temperature but absorbs 200 kJ of energy per mole of reactant. Draw an energy diagram graph for each reaction, and indicate the similarities and differences between the two diagrams.

Factors That Influence Reaction Rates (Sec. 16.3)

16.15 Using collision theory, indicate why each of the following factors influences the rate of a chemical reaction.

(a) temperature of reactants
(b) presence of a catalyst

16.16 Using collision theory, indicate why each of the following factors influences the rate of a chemical reaction.

(a) physical nature of reactants

(b) reactant concentrations

16.17 Why will a spark cause coal dust in a mine to explode and yet not cause an explosion with charcoal in a barbeque?

16.18 Milk will sour in a couple of days when left at room temperature yet can remain unspoiled for 2 weeks when refrigerated. Explain why.

16.19 The characteristics of four reactions, each of which involves only two reactants, are as follows.

Reaction	Activation Energy	Temperature	Concentration of Reactants
1	low	low	1 mole/L of each
2	high	low	1 mole/L of each
3	low	high	1 mole/L of each
4	low	low	1 mole/L of 1st reactant; 4 moles/L of 2nd reactant

For each of the following pairs of the preceding reactions, indicate which reaction is faster. The rates to be compared are the rates when the two reactants are first mixed. Explain each of your answers.

(a) 1 and 2 **(b)** 1 and 3

(c) 1 and 4 **(d)** 2 and 3

16.20 The characteristics of four reactions, each of which involves only two reactants, are as follows.

Reaction	Activation Energy	Temperature	Concentration of Reactants
1	high	low	1 mole/L of each
2	high	high	1 mole/L of each
3	low	low	1 mole/L of 1st reactant; 4 moles/L of 2nd reactant
4	low	low	4 moles/L of each

For each of the following pairs of the preceding reactions, indicate which reaction is faster. The rates to be compared are the rates when the two reactants are first mixed. Explain each of your answers.

(a) 1 and 2 **(b)** 1 and 3

(c) 1 and 4 **(d)** 3 and 4

16.21 Draw an energy graph for an *endothermic* reaction where no catalyst is present. Then draw an energy diagram for the same reaction when a catalyst is present. Indicate the similarities and differences between the two diagrams.

16.22 Draw an energy graph for an *exothermic* reaction where no catalyst is present. Then draw an energy diagram for the same reaction when a catalyst is present. Indicate the similarities and differences between the two diagrams.

Chemical Equilibrium (Sec. 16.4)

16.23 What condition must be met in order for a system to be in a state of chemical equilibrium?

16.24 What relationship exists between the rates of the forward and reverse reactions for a system in a state of chemical equilibrium?

16.25 What is the difference between a physical equilibrium and a chemical equilibrium?

16.26 What is the difference between *equal* product and reactant concentrations and *constant* product and reactant concentrations?

Equilibrium Mixture Stoichiometry (Sec. 16.5)

16.27 A 0.0200-mole sample of SO_3 is placed in a reaction container and allowed to decompose until equilibrium is established.

$$2 SO_3(g) \rightleftharpoons 2 SO_2(g) + O_2(g)$$

At equilibrium 0.0029 mole of O_2 is present. What is the composition of the equilibrium mixture in terms of moles of each substance present?

16.28 A mixture of 0.100 mole of SO_2 and 0.100 mole of O_2 is placed in a reaction container and allowed to react until equilibrium is established.

$$2 SO_2(g) + O_2(g) \rightleftharpoons 2 SO_3(g)$$

At equilibrium 0.0916 mole of SO_3 is present. What is the composition of the equilibrium mixture in terms of moles of each substance present?

16.29 A mixture of 0.296 mole of NH_3, 0.170 mole of N_2, and 0.095 mole of H_2 is allowed to reach equilibrium.

$$2 NH_3(g) \rightleftharpoons N_2(g) + 3 H_2(g)$$

At equilibrium, it is found that 0.268 mole of NH_3 is present. What is the composition of the equilibrium mixture in terms of moles of each substance present?

16.30 A mixture of 0.520 mole of NOCl, 0.010 mole of NO, and 0.053 mole of Cl_2 is allowed to reach equilibrium.

$$2 NOCl g \rightleftharpoons 2 NO(g) + Cl_2(g)$$

At equilibrium, it is found that 0.022 mole of NO is present. What is the composition of the equilibrium mixture in terms of moles of each substance present?

Equilibrium Constants (Sec. 16.6)

16.31 Write the expression for the equilibrium constant for each of the following reactions.

(a) $SO_2(g) + Cl_2(g) \rightleftharpoons SO_2Cl_2(g)$

(b) $2 NO_2(g) \rightleftharpoons N_2(g) + 2 O_2(g)$

(c) $2 SO_3(g) + CO_2(g) \rightleftharpoons CS_2(g) + 4 O_2(g)$

(d) $4 H_2(g) + CS_2(g) \rightleftharpoons CH_4(g) + 2 H_2S(g)$

16.32 Write the expression for the equilibrium constant for each of the following reactions.

(a) $PCl_5(g) \rightleftharpoons PCl_3(g) + Cl_2(g)$

(b) $2 NO(g) \rightleftharpoons N_2(g) + O_2(g)$

(c) $4 NH_3(g) + 7 O_2(g) \rightleftharpoons 4 NO_2(g) + 6 H_2O(g)$

(d) $CO(g) + 3 H_2(g) \rightleftharpoons CH_4(g) + H_2O(g)$

16.33 Write the expression for the equilibrium constant for each of the following reactions.

(a) $2 Pb(NO_3)_2(s) \rightleftharpoons 2 PbO(s) + 4 NO_2(g) + O_2(g)$
(b) $2 KClO_3(s) \rightleftharpoons 2 KCl(s) + 3 O_2(g)$
(c) $2 Ag(s) + Cl_2(g) \rightleftharpoons 2 AgCl(s)$
(d) $PCl_5(s) \rightleftharpoons PCl_3(l) + Cl_2(g)$

16.34 Write the expression for the equilibrium constant for each of the following reactions.

(a) $H_2SO_4(l) \rightleftharpoons SO_3(g) + H_2O(l)$
(b) $2 FeBr_3(s) \rightleftharpoons 2 FeBr_2(s) + Br_2(g)$
(c) $BaCl_2(aq) + Na_2SO_4(aq) \rightleftharpoons 2 NaCl(aq) + BaSO_4(s)$
(d) $2 Na_2O(s) \rightleftharpoons 4 Na(l) + O_2(g)$

16.35 At a particular temperature, a hypothetical chemical system has the following equilibrium molar concentrations: A = 3.00, B = 2.00, and C = 5.00. Calculate the value of the equilibrium constant for the system if the reaction occurring were each of the following.

(a) $A(g) \rightleftharpoons 2 B(g) + C(g)$
(b) $A(g) + 3 B(g) \rightleftharpoons 2 C(g)$
(c) $2 B(g) \rightleftharpoons A(g) + C(g)$
(d) $4 C(g) + B(g) \rightleftharpoons 3 A(g)$

16.36 At a particular temperature, a hypothetical chemical system has the following equilibrium molar concentrations: A = 2.00, B = 4.00, and C = 3.00. Calculate the value of the equilibrium constant for the system if the reaction occurring were each of the following.

(a) $A(g) + 2 B(g) \rightleftharpoons C(g)$
(b) $A(g) \rightleftharpoons B(g) + 3 C(g)$
(c) $2 C(g) + B(g) \rightleftharpoons 2 A(g)$
(d) $3 A(g) + 2 B(g) \rightleftharpoons 4 C(g)$

16.37 The equilibrium constant for the reaction

$$CS_2(g) + 4 H_2(g) \rightleftharpoons CH_4(g) + 2 H_2S(g)$$

is 0.0280 at a particular temperature. The system at equilibrium has $[H_2S] = 1.43$, $[H_2] = 1.00$, and $[CH_4] = 0.00100$.
    What is $[CS_2]$?

16.38 The equilibrium constant for the reaction

$$CH_4(g) + 2 H_2S(g) \rightleftharpoons CS_2(g) + 4 H_2(g)$$

is 3.30×10^4 at a particular temperature. The system at equilibrium has $[CH_4] = 0.709$, $[H_2S] = 0.0100$, and $[H_2] = 2.34$.
    What is $[CS_2]$?

16.39 A 6.00-L vessel contained 0.0222 mole of PCl_3, 0.0189 mole of PCl_5, and 0.1044 mole of Cl_2 at 230°C in an equilibrium mixture. Calculate the value of K_{eq} for the reaction

$$PCl_3(g) + Cl_2(g) \rightleftharpoons PCl_5(g)$$

16.40 An 8.00-L vessel contained 0.650 mole of H_2, 0.275 mole of I_2, and 2.86 moles of HI at 491°C in an equilibrium mixture.

Calculate the value of K_{eq} for the reaction

$$2 HI(g) \rightleftharpoons H_2(g) + I_2(g)$$

16.41 For reactions with each of the following equilibrium constants, describe the position of equilibrium as (1) mostly products, (2) mostly reactants, or (3) significant amounts of both reactants and products.

(a) 10^{-10} at 25°C   **(b)** 10^{30} at 25°C
(c) 10^9 at 127°C   **(d)** 10^2 at 327°C

16.42 For reactions with each of the following equilibrium constants, describe the position of equilibrium as (1) mostly products, (2) mostly reactants, or (3) significant amounts of both reactants and products.

(a) 10^{25} at 25°C   **(b)** 10^{-19} at 25°C
(c) 10^{-11} at 235°C   **(d)** 10^{-1} at 1235°C

Le Châtelier's Principle (Sec. 16.7)

16.43 For the reaction

$$CO(g) + 3 H_2(g) \rightleftharpoons CH_4(g) + H_2O(g)$$

determine the direction that the equilibrium will be shifted by each of the following changes.

(a) increase in CO concentration
(b) increase in CH_4 concentration
(c) decrease in H_2 concentration
(d) decrease in H_2O concentration

16.44 For the reaction

$$CH_4(g) + 2 O_2(g) \rightleftharpoons CO_2(g) + 2 H_2O(g)$$

determine the direction that the equilibrium will be shifted by each of the following changes.

(a) increase in O_2 concentration
(b) increase in CO_2 concentration
(c) decrease in CH_4 concentration
(d) decrease in H_2O concentration

16.45 For the reaction

$$2 C_2H_2(g) + 5 O_2(g) \rightleftharpoons 4 CO_2(g) + 2 H_2O(g) + heat$$

determine the direction that the equilibrium will be shifted by each of the following changes.

(a) increasing the concentration of C_2H_2
(b) decreasing the concentration of O_2
(c) increasing the temperature
(d) increasing the pressure by decreasing the volume of the container.

16.46 For the reaction

$$C_3H_8(g) + 5 O_2(g) \rightleftharpoons 3 CO_2(g) + 4 H_2O(g) + heat$$

determine the direction that the equilibrium will be shifted by each of the following changes.

(a) increasing the concentration of CO_2

(b) decreasing the concentration of C_3H_8

(c) decreasing the temperature

(d) increasing the pressure by decreasing the volume of the container

16.47 Consider the following chemical system at equilibrium.

$$2 H_2O(g) + 2 Cl_2(g) + heat \rightleftharpoons 4 HCl(g) + O_2(g)$$

For each of the following adjustments of conditions, indicate the effect on the position of equilibrium: shifts left, shifts right, no effect.

(a) heating the equilibrium mixture

(b) adding O_2 to the mixture

(c) increasing the pressure on the equilibrium mixture by adding an inert gas

(d) increasing the size of the reaction container

16.48 Consider the following chemical system at equilibrium.

$$2 N_2(g) + 6 H_2O(g) + heat \rightleftharpoons 4 NH_3(g) + O_2(g)$$

For each of the following adjustments of conditions, indicate the effect on the position of equilibrium: shifts left, shifts right, no effect.

(a) adding N_2 to the mixture

(b) decreasing the size of the container holding the mixture

(c) adding a catalyst to the mixture

(d) refrigerating the warm equilibrium mixture

16.49 For which of the following reactions is product formation favored by high temperature?

(a) $N_2(g) + O_2(g) \rightleftharpoons 2 NO(g) + heat$

(b) $N_2(g) + 3 H_2(g) \rightleftharpoons 2 NH_3(g) + heat$

(c) $CO(g) + 3 H_2(g) + heat \rightleftharpoons CH_4(g) + H_2O(g)$

(d) $2 H_2O(g) \rightleftharpoons 2 H_2(g) + O_2(g) + heat$

16.50 For which of the reactions in Problem 16.49 is product formation favored by high pressure?

Additional Problems

16.51 Write a balanced chemical equation for a totally gaseous equilibrium system that would lead to the following expressions for the equilibrium constant.

(a) $\dfrac{[NH_3]^2}{[N_2][H_2]^3}$

(b) $\dfrac{[N_2]^2[H_2O]^6}{[NH_3]^4[O_2]^3}$

(c) $\dfrac{[N_2][O_2]}{[NO]^2}$

(d) $\dfrac{[NO]^2}{[N_2][O_2]}$

16.52 Write a balanced chemical equation for a totally gaseous equilibrium system that would lead to the following expressions for the equilibrium constant.

(a) $\dfrac{[HCN]^2}{[H_2][C_2N_2]}$

(b) $\dfrac{[CH_4][H_2S]^2}{[CS_2][H_2]^4}$

(c) $\dfrac{[NOBr]^2}{[NO]^2[Br_2]}$

(d) $\dfrac{[NO]^2[Br_2]}{[NOBr]^2}$

16.53 The following reaction at a certain temperature has an equilibrium-constant value of 25.9.

$$2 CO_2(g) \rightleftharpoons 2 CO(g) + O_2(g)$$

For each of the following compositions, decide whether the reaction mixture is at equilibrium. If it is not, decide which direction the reaction shifts to reach equilibrium.

(a) $[CO_2] = 0.0300$, $[CO] = 0.350$, $[O_2] = 0.190$

(b) $[CO_2] = 0.0600$, $[CO] = 0.700$, $[O_2] = 0.380$

(c) $[CO_2] = 0.0280$, $[CO] = 0.356$, $[O_2] = 0.160$

(d) $[CO_2] = 0.0100$, $[CO] = 0.330$, $[O_2] = 0.180$

16.54 The following reaction at a certain temperature has an equilibrium-constant value of 0.016.

$$2 HI(g) \rightleftharpoons H_2(g) + I_2(g)$$

For each of the following compositions, decide whether the reaction mixture is at equilibrium. If it is not, decide which direction the reaction shifts to reach equilibrium.

(a) $[HI] = 0.080$, $[H_2] = 0.010$, $[I_2] = 0.010$

(b) $[HI] = 0.084$, $[H_2] = 0.012$, $[I_2] = 0.012$

(c) $[HI] = 0.076$, $[H_2] = 0.012$, $[I_2] = 0.012$

(d) $[HI] = 0.140$, $[H_2] = 0.031$, $[I_2] = 0.010$

16.55 At a given temperature, the equilibrium constant for the reaction

$$2 NO(g) + Br_2(g) \rightleftharpoons 2 NOBr(g)$$

is 2×10^3. What is the equilibrium constant, at the same temperature, for the following reaction?

$$2 NOBr(g) \rightleftharpoons 2 NO(g) + Br_2(g)$$

16.56 At a given temperature, the equilibrium constant for the reaction

$$CO(g) + H_2O(g) \rightleftharpoons CO_2(g) + H_2(g)$$

is 0.034. What is the equilibrium constant, at the same temperature, for the following reaction?

$$CO(g) + H_2(g) \rightleftharpoons CO_2(g) + H_2O(g)$$

16.57 Which of the following changes would change the *value* of a system's equilibrium constant?

(a) addition of a reactant or product

(b) increase in the total pressure

(c) decrease in the temperature

(d) addition of an inert gas

16.58 Which of the following changes would change the *value* of a system's equilibrium constant?

(a) removal of a reactant or product

(b) decrease in the total pressure

(c) increase in the temperature

(d) addition of a catalyst

16.59 Predict the direction in which each of the following equilibria will shift if the pressure on the system is decreased by expansion.

(a) $2 SO_3(g) \rightleftharpoons 2 SO_2(g) + O_2(g)$
(b) $2 HI(g) \rightleftharpoons H_2(g) + I_2(g)$
(c) $ClF_5(g) \rightleftharpoons ClF_3(g) + Cl_2(g)$
(d) $C(s) + CO_2(g) \rightleftharpoons 2 CO(g)$

16.60 Predict the direction in which each of the following equilibria will shift if the pressure on the system is increased by compression.

(a) $H_2(g) + C_2N_2(g) \rightleftharpoons 2 HCN(g)$
(b) $CO(g) + Br_2(g) \rightleftharpoons COBr_2(g)$
(c) $CS_2(g) + 4 H_2(g) \rightleftharpoons CH_4(g) + 2 H_2S(g)$
(d) $Ni(s) + 4 CO(g) \rightleftharpoons Ni(CO)_4(g)$

Cumulative Problems

16.61 Given the following descriptions of reversible reactions, write the equilibrium constant expression for each.

(a) In a decomposition reaction sulfur trioxide gas produces sulfur dioxide gas and oxygen gas.
(b) Hydrogen gas reduces nitrogen dioxide gas to produce ammonia gas and steam.
(c) Solid iron(II) oxide and carbon monoxide gas react to produce solid iron and carbon dioxide gas.
(d) Solid magnesium carbonate decomposes to produce solid magnesium oxide and carbon dioxide gas.

16.62 Given the following descriptions of reversible reactions, write the equilibrium constant expression for each.

(a) In a synthesis reaction gaseous hydrogen and gaseous bromine react to produce gaseous hydrogen bromide.
(b) In a redox reaction carbon disulfide gas reacts with hydrogen gas to produce methane gas and hydrogen sulfide gas.
(c) Solid sodium carbonate reacts with gaseous sulfur dioxide and oxygen gas to produce solid sodium sulfate and carbon dioxide gas.
(d) Chlorine gas reacts with liquid carbon disulfide to produce the liquids carbon tetrachloride and disulfur dichloride.

16.63 An equilibrium mixture, at 900°C in a 1725-mL container, involving the chemical system

$$CH_4(g) + 2 H_2S(g) \rightleftharpoons CS_2(g) + 4 H_2(g)$$

is found to contain 17.6 g CH_4, 50.8 g H_2S, 83.8 g CS_2, and 8.10 g H_2. Calculate the equilibrium constant for this reaction at the given temperature.

16.64 An equilibrium mixture, at 472°C in a 1325-mL container, involving the chemical system

$$N_2(g) + 3 H_2(g) \rightleftharpoons 2 NH_3(g)$$

is found to contain 4.23 g N_2, 0.915 g H_2, and 0.496 g NH_3. Calculate the equilibrium constant for this reaction at the given temperature.

16.65 For the chemical system

$$SbCl_5(g) \rightleftharpoons SbCl_3(g) + Cl_2(g)$$

it is found that an equilibrium mixture in a 1.00-L flask contains 2.48×10^{20} molecules of $SbCl_5$, 0.723 g of $SbCl_3$, and 0.00317 mole of Cl_2. Calculate the equilibrium constant for the reaction.

16.66 For the chemical system

$$PCl_5(g) \rightleftharpoons PCl_3(g) + Cl_2(g)$$

it is found that an equilibrium mixture in a 1.00-L flask contains 2.95×10^{20} molecules of PCl_5, 0.00451 mole of PCl_3, and 0.320 g of Cl_2. Calculate the equilibrium constant for the reaction.

16.67 Consider the following equilibrium situation, at constant temperature.

$$N_2O_4(g) \rightleftharpoons 2 NO_2(g)$$

An empty container is charged with pure $N_2O_4(g)$ until the pressure reaches 1.50 atm. The $N_2O_4(g)$ is then allowed to reach equilibrium with $NO_2(g)$, at which time the partial pressure of $N_2O_4(g)$ is 0.80 atm. What is the total pressure in atmospheres in the flask?

16.68 Consider the following equilibrium situation, at constant temperature.

$$2 O_3(g) \rightleftharpoons 3 O_2(g)$$

An empty container is charged with pure $O_3(g)$ until the pressure reaches 2.25 atm. The $O_3(g)$ is then allowed to reach equilibrium with $O_2(g)$, at which time the partial pressure of $O_3(g)$ is 0.11 atm. What is the total pressure, in atmospheres, in the flask?

16.69 At 750°C and 1.000 atm pressure, a gaseous mixture of carbon monoxide and carbon dioxide is in equilibrium with solid carbon. The gaseous mixture is 87.43% CO by mass.

$$C(s) + CO_2(g) \rightleftharpoons 2 CO(g)$$

Calculate the equilibrium constant for this reaction from the given information.

16.70 At 35°C and 1.00 atm pressure, a gaseous mixture of dinitrogen tetroxide and nitrogen dioxide at equilibrium is found to contain 32.7% by mass of N_2O_4.

$$N_2O_4(g) \rightleftharpoons 2 NO_2(g)$$

Calculate the equilibrium constant for the reaction from the given information.

Answers to Practice Exercises

16.1 1.04 moles SO_2; 0.52 mole O_2; 2.96 moles SO_3

16.2 **(a)** $K_{eq} = \dfrac{[NH_3]^2[H_2O]^4}{[NO_2]^2[H_2]^7}$ **(b)** $K_{eq} = \dfrac{[CO][H_2]}{[H_2O]}$
(c) $K_{eq} = [NH_3][HCl]$ **(d)** $K_{eq} = [Fe^{3+}]^2[SO_4^{2-}]^3$

16.3 5.81
16.4 **(a)** shift to the right **(b)** shift to the left
(c) shift to the right **(d)** shift to the right

17

Nuclear Chemistry

17.1 Unstable Nuclides and Radioactivity

Early in the text (Sec. 5.5) we considered the structure of the atom. An atom is composed of two regions: (1) a nuclear region where all protons and neutrons are found and (2) an extra-nuclear region where the electrons are found. Since that initial discussion of atomic structure, we have focused primarily on the behavior of electrons in atoms, ions, and molecules, because the arrangement of electrons determines the physical and chemical properties of substances. Little has been said about the nucleus because it remains unchanged in ordinary chemical reactions and is important only insofar as it influences the electrons.

In this chapter we examine a group of processes known as *nuclear reactions*. A **nuclear reaction** *is a reaction in which changes occur in the nucleus of an atom.* Nuclear reactions are not considered to be ordinary chemical reactions. It is in the field of nuclear reactions that we encounter the terms *radioactivity, nuclear power plant, A-bomb,* and *H-bomb.* All of these terms are now part of our everyday vocabulary.

A brief review of previously discussed concepts about atomic nuclei as well as some new material concerning them will serve as the starting point for the discussion of nuclear reactions.

Atomic nuclei are the very dense positively charged centers of atoms about which the electrons move. All nuclei of atoms of a given element contain the same number of protons. It is this characteristic number of protons that determines the identity of the element. The *atomic number* for an atom gives the number of protons in the nucleus. The number of neutrons associated with the nuclei of a given element may vary within a limited range. Atoms of a given element that differ in the number of neutrons in the nucleus are called *isotopes.* The *mass number* of an atom is equal to the total number of protons and neutrons present in the nucleus. Isotopes of an element have different mass numbers but the same atomic number.

In nuclear chemistry discussions, the term *nuclide* is used as an alternate designation for an atom. A **nuclide** *is an atom with a specific atomic number and a specific mass number*. The term *isotopes* (Sec. 5.7) refers to different forms of the same element; the term *nuclide* is used in describing atomic forms of different elements. The species $^{12}_{6}C$ and $^{13}_{6}C$ are isotopes of the element carbon. The species $^{12}_{6}C$, $^{15}_{7}N$, and $^{16}_{8}O$ are nuclides of different elements.

In order to uniquely identify a nucleus, or an atom for that matter, both the atomic number and mass number must be specified. Two notation systems exist for doing this. Consider a nuclide of nitrogen with seven protons and eight neutrons. This nuclide can be denoted as $^{15}_{7}N$ or nitrogen-15. In the first notation the superscript is the mass number and the subscript is the atomic number. In the second notation the mass number is appended to the name of the element with a hyphen. An advantage of the first notation is that the atomic number is shown; a disadvantage is the need for superscripts and subscripts. Both types of notation will be used in this chapter.

> In some textbooks the term *radioactive isotope* (or *radioisotope*) is used in place of *radioactive nuclide* (or *radionuclide*). We will use nuclide rather than isotope.

Studies concerning atomic nuclei show that nuclides may be divided into two categories on the basis of their stability. Some nuclides are stable and others are not. A **stable nuclide** *is a nuclide with a stable nucleus that does not readily undergo change*. Conversely, an **unstable nuclide** *is a nuclide with an unstable nucleus that spontaneously undergoes change*. The spontaneous change that occurs within an unstable nucleus involves radiation emission from the nucleus, a process by which the unstable nucleus can become more stable. The radiation emitted from unstable nuclides is called *radioactivity*. **Radioactivity** *is the radiation spontaneously emitted by an unstable nuclide*. Nuclides that possess unstable nuclei are said to be *radioactive*. A **radioactive nuclide** *is a nuclide with an unstable nucleus from which radiation is spontaneously emitted*. The term *radioactive nuclide* is often shortened to simply *radionuclide*.

Naturally occurring radionuclides exist for 29 of the 88 elements that are found in nature (Sec. 4.8). Radionuclides are known for all 113 elements, however, even though they occur naturally for only the above-mentioned 29 elements. This is because laboratory procedures have been developed by which scientists convert nonradioactive nuclides (stable nucleus) into radioactive nuclides (unstable nucleus). Such procedures are considered in Section 17.6.

17.2 Discovery of Radioactivity

> There are two key concepts for understanding the phenomenon of radioactivity: (1) certain nuclides possess unstable nuclei, and (2) nuclides with unstable nuclei spontaneously emit energy (radiation).

The fact that certain naturally occurring nuclides are radioactive was unexpectedly (accidentally) discovered by the French physicist and engineer Antoine-Henri Becquerel (1852–1908) while he was studying certain minerals called phosphors, which glow in the dark (phosphoresce) after exposure to radiation such as sunlight or ultraviolet light. The rays emitted by these phosphorescing minerals, like visible light, darken a photographic plate. One day in 1896, while working with a uranium ore sample that phosphoresced in the normal manner, he was interrupted. As he left, he inadvertently placed the uranium ore sample on top of an unexposed photographic plate packaged to protect it from light. Later it was determined that this photographic plate had been exposed by the uranium ore despite its being protectively wrapped. Becquerel correctly concluded that this plate exposure was due to radiation emitted by the uranium ore without external stimulus. As a result of this incident Becquerel is credited with having discovered the phenomenon we now call radioactivity. Further studies by Becquerel showed that this "radioactivity" was not unique to the one uranium ore he had used, but was a characteristic of all uranium-containing substances independently of whether they phosphoresced or not. (It seems strange now that this phenomenon was not detected earlier, since the element uranium had been isolated more than 100 years before Becquerel's discovery.)

The Human Side of Chemistry 17

Maria Sklodowska Curie (1867–1934)

Maria Sklodowska Curie, born in Warsaw, Poland, in 1867, was the daughter of parents formally involved with education. Her father was a physics teacher; her mother a principal of a girls' school. Times were hard for the Sklodowska family throughout Marie's younger years, mainly because of the Russian domination of Poland during this period.

Despite poverty, by 1891 Marie was able to go to France, where she remained the rest of her life. Living with the greatest frugality in Paris, she was able to obtain additional education, and graduate at the top of her class. (Once she fainted from hunger in the classroom.)

Her involvement with the field of radioactivity was the result of her being a student of Henri Becquerel's at the "right time." Her doctoral thesis dealt with radiations from uranium.

In 1894 Marie met Pierre Curie, who had just been appointed a professor of physics in Paris. They were married the next year. After their marriage the two scientists continued their research—at first, independently of each other. Marie's research proved the more productive. In 1898 Pierre abandoned his own research and started working on his wife's project as her assistant. Most of their work was carried out under miserable conditions—in a wooden shed with a leaky roof, no floor, and very inadequate heat. In 1902, after four years of work, the couple reported the isolation of 0.1 g of radium, a new highly radioactive element they had discovered. The 0.1 g of new element was the result of processing many tons of uranium ore.

Recognition came to the Curies in 1903 with their selection as corecipients (along with Henri Becquerel) of the Nobel Prize in Physics. After Pierre's accidental death in 1906 (he was run over by a heavy horse-drawn wagon), Marie was named to succeed him as a professor of physics, the first time a woman had occupied a professorial chair at the Sorbonne in Paris. She continued the work, often giving up recreational and social contacts for devotion to research.

In 1911 her further work was recognized with her selection to receive the Nobel Prize in Chemistry. She is the only person to receive two Nobel prizes for scientific research. In later years Irene, one of her two daughters, worked with her as an assistant. In 1934, Marie, now respectfully called Madame Curie, died of leukemia caused by overexposure to radiation.

In 1935, one year after her mother's death, Irene Joliot-Curie (1897–1956) and her husband were awarded the Nobel Prize in Chemistry, the third such prize for the Curie family. The prize was in recognition of further studies concerning radioactivity, this time in the field of artificial radioactivity.

Becquerel chose not to continue to work in this new field of study. Instead, he suggested to two of his colleagues, Pierre and Marie Curie (who had just recently been married), that they continue the project and find out what other substances besides uranium possessed "radioactive" properties.

Pierre Curie (1859–1906), a physicist, and Marie Sklodowska Curie (1867–1934), a chemist (see "The Human Side of Chemistry 17"), conducted a systematic search of the then-known elements to see how widespread this phenomenon was. The only other radioactive element they found was thorium. In further investigations, the Curies discovered a uranium ore sample that exhibited four times the radioactivity of a similar quantity of pure uranium or thorium, thus indicating the presence of a new substance more radioactive than either of these elements. From this ore the Curies were able to isolate, in 1898, two new radioactive elements: polonium, 400 times as radioactive as uranium, and radium, over a million times as radioactive as uranium. It was the Curies who coined the word *radioactive* to describe elements that spontaneously emit radiation.

17.3 Nature of Natural Radioactive Emissions

The first information concerning the nature of the radiation emanating from naturally radioactive materials was obtained by Ernest Rutherford in the years 1898–1899. Using an apparatus similar to that shown in Figure 17.1, he found that if a beam of radiation is passed between electric plates it is split into three components, indicating the presence of three different types of emissions from radioactive materials. A closer analysis of Rutherford's

Figure 17.1

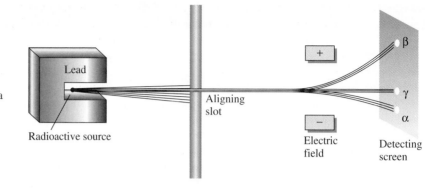

Figure 17.1

Effect of an electric field on radiation emanating from a naturally radioactive substance. Gamma rays are unaffected. The lighter beta particles are deflected considerably more than the heavier alpha particles.

experiment reveals that one radiation component is positively charged (it is attracted to the negative plate), a second component is negatively charged (it is attracted to the positive plate), and the third component carries no charge (it is unaffected by either charged plate). Rutherford chose to call the three radiation components alpha rays (α rays) (the positive component), beta rays (β rays) (the negative component), and gamma rays (γ rays) (the uncharged component). (Alpha, beta, and gamma are the first three letters of the Greek alphabet.) We mention Rutherford's nomenclature system because it "stuck"; we still use these Greek letter designations. Today, we speak of alpha particles, beta particles, and gamma rays. Further research has shown that both alpha and beta radiation involve particles with mass and that gamma radiation has no mass; that is, it is a form of energy.

The complete characterization of the three types of natural radioactive emissions required many years. Early work in the field was hampered by the fact that many of the details concerning atomic structure were not yet known. For example, recall (Sec. 5.5) that the neutron was not identified until 1932, 36 years after the discovery of radioactivity. In terms of modern-day scientific knowledge, Rutherford's three types of "radiation" are characterized as follows.

An **alpha particle** *is a particle in which two protons and two neutrons are present.* The notation used to represent an alpha particle is $^4_2\alpha$. The numerical subscript indicates that the charge on the particle is $+2$ (from the two protons). The numerical superscript indicates a mass of 4 amu. On the atomic mass scale (Sec. 5.8) protons and neutrons both have masses equal to 1.0 amu. Thus, the total mass of an alpha particle (two protons and two neutrons) is 4.0 amu. Alpha particles are identical with the nuclei of helium-4 ($^4_2$He) atoms.

A **beta particle** *is a particle whose charge and mass are identical to those of an electron.* However, beta particles are not extranuclear electrons; they are particles that have been produced inside the nucleus and then ejected. More concerning this process will be given in Section 17.4. The symbol used to represent a beta particle is $^0_{-1}\beta$. The numerical subscript indicates that the charge on the beta particle is -1, that of an electron. The use of the superscript zero for the mass of a beta particle is not to be interpreted as meaning that a beta particle has no mass, but rather that its mass number (protons + neutrons) is zero. The actual mass of a beta particle on the atomic mass scale is 0.00055 amu.

A **gamma ray** *is a form of high-energy radiation without mass or charge.* Gamma rays are similar to X rays, except they have higher energy. The symbol for gamma rays is $^0_0\gamma$.

17.4 Radioactive Decay

Alpha, beta, and gamma emissions come from the nucleus of an atom. These spontaneous emissions alter nuclei; obviously, if a nucleus loses an alpha particle (two protons and two neutrons), it will not be the same as it was before the departure of the particle. In the case of alpha and beta emissions, the nuclear alteration causes the identity of the atom to change;

that is, a new element is formed. Nuclear reactions thus differ dramatically from ordinary chemical reactions. In the latter, the identity of the elements is always maintained. This is not the case in nuclear reactions.

The term *radioactive decay* is used in describing a nuclear process where an element changes into another element as a result of radiation emission. **Radioactive decay** *is the process whereby a radionuclide is transformed into a nuclide of another element as a result of the emission of radiation from its nucleus.* The terms *parent nuclide* and *daughter nuclide* are often used in descriptions of radioactive decay processes. A **parent nuclide** *is the nuclide that undergoes decay in a radioactive decay process.* A **daughter nuclide** *is the nuclide that is produced as a result of a radioactive decay process.*

Radioactive decay occurs only in certain ways, called *modes of decay.* For naturally occurring radioactive substances the modes of decay are: (1) *alpha-particle decay* and (2) *beta-particle decay.* Gamma ray emission accompanies both of these decay modes. Separate consideration of these two modes of decay provides further insights into the nature of nuclear reactions and also illustrates how equations for nuclear reactions are written.

Alpha-Particle Decay

Alpha-particle decay *is the radioactive decay process in which an alpha particle is emitted from an unstable nucleus.* It always results in the formation of a nuclide of a different element. The daughter nuclide of such decay has an atomic number that is 2 less than that of the original nucleus and a mass number that is 4 less. We can represent alpha-particle decay in general terms by the equation

$$_Z^A X \longrightarrow {}_2^4\alpha + {}_{(Z-2)}^{(A-4)}Y$$

where X is the symbol for the nucleus of the original element undergoing decay and Y is the symbol for the nucleus of the element formed as a result of the decay.

As an introduction to nuclear equations, let us write equations for two actual alpha-decay processes. Both $_{83}^{211}$Bi and $_{92}^{238}$U are alpha emitters; that is, they are radionuclides that undergo alpha-particle decay. The nuclear equations for these two decay processes are

$$_{83}^{211}\text{Bi} \longrightarrow {}_2^4\alpha + {}_{81}^{207}\text{Tl}$$

$$_{92}^{238}\text{U} \longrightarrow {}_2^4\alpha + {}_{90}^{234}\text{Th}$$

In the first equation, $_{83}^{211}$Bi is the parent nuclide and $_{81}^{207}$Tl is the daughter nuclide; in the second equation, $_{92}^{238}$U is the parent nuclide and $_{90}^{234}$Th is the daughter nuclide.

The preceding two nuclear equations differ from ordinary chemical equations in three important ways:

1. The symbols in a nuclear equation represent nuclei rather than atoms. (We, thus, do not worry about electrons when writing a nuclear equation since there are no electrons in a nucleus.)

2. Atomic numbers (nuclear charge) and mass numbers are always specifically included in a nuclear equation.

3. The elemental symbols on the two sides of the equation are frequently not the same in a nuclear equation.

The procedures for balancing nuclear equations are different from those used for ordinary chemical equations. A **balanced nuclear equation** *is a nuclear equation in which the sums of the subscripts (atomic number or particle charge) on both sides of the equation are equal and the sums of the superscripts (mass number) on both sides of the equation are equal.* Both of our example equations are balanced. In the alpha decay of $_{83}^{211}$Bi, the subscripts on both sides total 83 and the superscripts total 211. For the decay of $_{92}^{238}$U, the subscripts total 92 on both sides and the superscripts total 238 on both sides.

> **Loss of an alpha particle from an unstable nuclide *always* results in (1) a decrease of 4 units in the mass number (A) and (2) a decrease of 2 units in the atomic number (Z).**

The rules for balancing nuclear equations are

1. **The sum of the subscripts must be the same on both sides of the equation.**
2. **The sum of the superscripts must be the same on both sides of the equation.**

The rule for balancing nuclear equations is based on both *charge* and *nucleon* conservation. The total charge is conserved (remains constant) during a nuclear reaction. This means that the sum of the subscripts (number of protons, or positive charges, in the nucleus) for the products must equal the sum of the subscripts for the reactants. Similarly, the total number of nucleons (protons and neutrons; Sec. 5.5) is conserved (remains constant) during a nuclear reaction. This means that the sum of the superscripts (the mass numbers) for the products equals the sum of the superscripts for the reactants.

Beta-Particle Decay

Beta-particle decay *is the radioactive decay process in which a beta particle is emitted from an unstable nucleus.* Beta-particle decay also always results in the formation of a nuclide of a different element. The mass number of the new nuclide is the same as that of the original atom. The atomic number, however, has increased by one unit. The general equation for beta decay is

$$\ _{Z}^{A}X \longrightarrow \ _{-1}^{0}\beta + \ _{(Z+1)}^{A}Y$$

Specific examples of beta particle decay are

$$\ _{4}^{10}Be \longrightarrow \ _{-1}^{0}\beta + \ _{5}^{10}B$$

$$\ _{90}^{234}Th \longrightarrow \ _{-1}^{0}\beta + \ _{91}^{234}Pa$$

Both of these nuclear equations are balanced; superscripts and subscripts add to the same sums on both sides of the equation.

It is not immediately apparent how a nucleus, composed only of neutrons and protons, ejects a negative particle (beta particle) when no such particle is present in the nucleus. The accepted explanation is that through a complex series of steps a neutron in the nucleus is transformed into a proton and a beta particle; that is,

$$\ _{0}^{1}n \longrightarrow \ _{1}^{1}p + \ _{-1}^{0}\beta$$

Loss of a beta particle from an unstable nucleus results in (1) no change in the mass number (*A*) and (2) an increase of 1 unit in the atomic number (*Z*).

Once formed within the nucleus, the beta particle is ejected with a high velocity. The net result of beta-particle formation is an increase by one in the number of protons present in the nucleus and a decrease by one in the number of neutrons present in the nucleus; the mass number is, however, constant as the total number of subatomic particles in the nucleus (protons and neutrons) has not changed. Note in our two examples of beta emission that the daughter nuclide has one more proton than the parent as evidenced by the atomic number of the daughter being greater than that of the parent by one unit. Subtraction of the atomic number of the daughter nuclide from its mass number in each case—to get the number of neutrons—will reveal that the daughter nuclide contains one fewer neutron than the parent. An explanation of why and when beta-particle formation occurs is considered in Section 17.8.

Gamma-Ray Emission

Among *synthetically* produced radionuclides (Section 17.6) pure "gamma emitters," radionuclides that give off gamma rays, but no alpha or beta particles, occur. These radionuclides are important in diagnostic nuclear medicine (Section 17.14). Pure "gamma emitters" are not found among naturally occurring radionuclides.

Gamma-ray emission *is the radioactive decay process in which gamma rays are emitted from an unstable nucleus.* For naturally occurring radionuclides, gamma-ray emission always occurs in conjunction with an alpha- or beta-decay process; it never occurs independently. Such gamma rays are most often not included in the nuclear equation, since they do not affect the balancing of the equation or the identity of the decay product. Gamma rays are to nuclear reactions what "heat" is to chemical reactions.

Just because gamma rays are usually left out of nuclear equations, it should not be assumed that they are not important. On the contrary, gamma rays are more important than alpha and beta particles when the effects of external radiation exposure on living organisms are considered (Sec. 17.11).

EXAMPLE 17.1

Writing Balanced Nuclear Equations Given the Parent Nuclide and Its Mode of Decay

Write a balanced nuclear equation for the decay of each of the following radioactive nuclides. The mode of decay is indicated in parentheses.

(a) $^{138}_{54}\text{Xe}$ (beta emission) (b) $^{142}_{58}\text{Ce}$ (alpha emission)

(c) $^{190}_{78}\text{Pt}$ (alpha emission) (d) $^{82}_{35}\text{Br}$ (beta emission)

SOLUTION

In each case the atomic and mass numbers of the daughter nuclide are obtained by first writing the symbols of the parent nuclide and the particle emitted by the nucleus (alpha or beta particle) and then balancing the equation.

(a) Let X represent the product of the radioactive decay, that is, the daughter nuclide. Then

$$^{138}_{54}\text{Xe} \longrightarrow \, ^{0}_{-1}\beta + \text{X}$$

Since the sums of the superscripts on both sides of the equation must be equal, the superscript for X must be 138. In order for the sums of the subscripts on both sides of the equation to be equal, the subscript for X must be 55. Then $54 = (-1) + (55)$. As soon as the subscript of X is determined, the identity of X can be determined from a periodic table. The element with an atomic number of 55 is cesium (Cs). Therefore,

$$^{138}_{54}\text{Xe} \longrightarrow \, ^{0}_{-1}\beta + \, ^{138}_{55}\text{Cs}$$

(b) Similarly, letting X represent the product of the radioactive decay, we have for the alpha decay of $^{142}_{58}\text{Ce}$

$$^{142}_{58}\text{Ce} \longrightarrow \, ^{4}_{2}\alpha + \text{X}$$

Balancing the equation, making the superscripts on the right side of the equation total 142 and the subscripts total 58, we get

$$^{142}_{58}\text{Ce} \longrightarrow \, ^{4}_{2}\alpha + \, ^{138}_{56}\text{Ba}$$

(c) Similarly, we write

$$^{190}_{78}\text{Pt} \longrightarrow \, ^{4}_{2}\alpha + \text{X}$$

Balancing superscripts and subscripts, we get

$$^{190}_{78}\text{Pt} \longrightarrow \, ^{4}_{2}\alpha + \, ^{186}_{76}\text{Os}$$

(d) Finally, we write

$$^{82}_{35}\text{Br} \longrightarrow \, ^{0}_{-1}\beta + \text{X}$$

In beta emission the atomic number of the daughter nuclide is always greater by one and the mass number does not change from that of the parent. The balancing procedure gives us this result.

$$^{82}_{35}\text{Br} \longrightarrow \, ^{0}_{-1}\beta + \, ^{82}_{36}\text{Kr}$$

Practice Exercise 17.1

Write a balanced nuclear equation for the decay of each of the following radioactive nuclides. The mode of decay is indicated in parentheses.

(a) $^{147}_{62}\text{Sm}$ (alpha emission) (b) $^{117}_{48}\text{Cd}$ (beta emission)

> Answers to practice exercises are located at the end of the chapter.

17.5 Rate of Radioactive Decay

Different kinds of radioactive nuclides do not decay at the same rate. Some decay very rapidly; others undergo disintegration at extremely slow rates. This indicates that radionuclides are not all equally unstable. The greater the decay rate, the lower the stability.

The concept of *half-life* is used to quantitatively express nuclear stability. A **half-life** *is the time required for one half of any given quantity of a radioactive substance to undergo decay*. For example, if a radionuclide's half-life is 12 days and you have a 4.00-g sample of it, then after 12 days (one half-life) only 2.00 g of the sample (half the original amount) will remain undecayed; the other half will have decayed into some other substance.

Half-lives of billions of years and as short as a fraction of a second have been determined. Table 17.1 contains examples of the wide range of half-life values.

Most naturally occurring radionuclides have long half-lives. Some radionuclides with short half-lives, however, are also found in nature. Such short-lived species, since they decay rapidly, must be continually produced in order to be present. Processes that result in their production are: (1) the decay of naturally occurring long-lived nuclides, (2) the decay of short-lived nuclides (daughter nuclides) that have been produced in the previous manner, and (3) bombardment reactions involving cosmic rays, which take place naturally in the upper atmosphere. Examples of the second method of producing short-lived nuclides are presented in Section 17.9.

The decay rate (half-life) of a radionuclide is constant. It is independent of outward conditions such as temperature, pressure, and state of chemical combination. It is dependent only on the identity of the radionuclide. For example, radioactive sodium-24, whether incorporated into $NaCl$, $NaBr$, Na_2SO_4, or $NaC_2H_3O_2$, decays at the same rate. If a nuclide is radioactive, nothing will stop it from decaying and nothing will increase or decrease its decay rate.

Figure 17.2 shows graphically the meaning of half-life. After one half-life has passed, one half of the original atoms have decayed, so half remain. During the next half-life, one-half of the remaining half will decay, so one fourth of the original atoms remain undecayed. After three half-lives, $\frac{1}{2} \times \frac{1}{2} \times \frac{1}{2} = \frac{1}{8}$ of the original atoms remain undecayed, and so on. Note from Figure 17.2 that only a very small amount of original material (less than 1%) remains after seven half-lives have elapsed.

Calculations involving amounts of radioactive material decayed, amounts remaining undecayed, and time elapsed can be carried out by using the following equation.

$$\left(\begin{array}{c} \text{Amount of radionuclide} \\ \text{undecayed after } n \text{ half-lives} \end{array} \right) = \left(\begin{array}{c} \text{original amount} \\ \text{of radionuclide} \end{array} \right) \times \frac{1}{2^n}$$

> **Half-life and rate of decay for a radionuclide are inversely related. The faster the rate of decay, the shorter the half-life.**

Table 17.1	Range of Half-Lives Found for Naturally Occurring Radionuclides
Element	**Half-life**
Vanadium-50	6×10^{15} yr
Platinum-190	6.9×10^{11} yr
Uranium-238	4.5×10^9 yr
Uranium-235	7.1×10^8 yr
Thorium-230	7.5×10^4 yr
Lead-210	22 yr
Bismuth-214	19.7 min
Polonium-212	3.0×10^{-7} sec

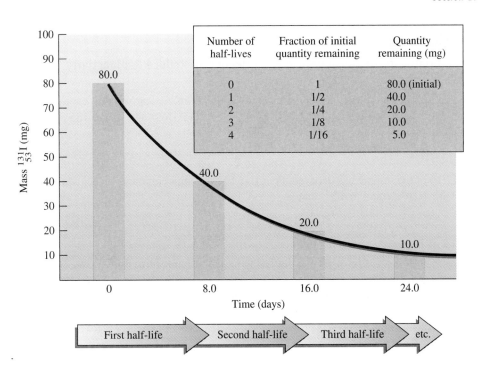

Figure 17.2

Decay of 80.0 mg of ^{131}I, which has a half-life of 8.0 days. After each half-life period, the quantity of original material present at the beginning of the period is reduced by half.

EXAMPLE 17.2

Using Half-Life to Calculate the Amount of Radionuclide That Remains Undecayed After a Certain Time

The half-life of cobalt-60 is 5.3 yr. If 2.0 g of cobalt-60 is allowed to decay for a period of 15.9 yr, how many grams of cobalt-60 remain?

SOLUTION

First, we must determine the number of half-lives that have elapsed.

$$15.9 \, \cancel{yr} \times \frac{1 \text{ half-life}}{5.3 \, \cancel{yr}} = 3.0 \text{ half-lives}$$

Knowing the number of elapsed half-lives and the original amount of radioactive cobalt present, we can use the equation

$$\binom{\text{Amount of radionuclide}}{\text{undecayed after } n \text{ half-lives}} = \binom{\text{original amount}}{\text{of radionuclide}} \times \frac{1}{2^n}$$

to get our answer.

$$\binom{\text{Amount of radionuclide}}{\text{undecayed after } n \text{ half-lives}} = 2.0 \text{ g} \times \frac{1}{2^3} \checkmark \text{ Three half-lives}$$

$$= 2.0 \text{ g} \times \frac{1}{8}$$

$$= 0.25 \text{ g (calculator and \textbf{correct answer})}$$

Practice Exercise 17.2

The half-life of iodine-131 is 8.0 days. If 12 g of iodine-131 is allowed to decay for a period of 32 days, how many grams of iodine-131 remain?

EXAMPLE 17.3

Using Half-Life to Calculate the Time Needed to Reduce Radioactivity to a Specific Level

Iodine-135 is a nuclide found in radioactive fallout from nuclear weapon explosions. Its half-life is 6.70 hr. How long will it take for 93.75% (15/16) of the iodine-135 atoms in a "fallout" sample to undergo decay?

SOLUTION

If 15/16 of the sample has decayed, then 1/16 of the sample remains undecayed. In terms of $1/2^n$, 1/16 is equal to $1/2^4$; that is,

$$\frac{1}{2} \times \frac{1}{2} \times \frac{1}{2} \times \frac{1}{2} = \frac{1}{2^4} = \frac{1}{16}$$

Thus four half-lives have elapsed in reducing the amount of iodine-135 to 1/16 of its original amount.

Since the half-life of iodine-135 is 6.70 hr, the total time elapsed will be

$$4 \text{ half-lives} \times \frac{6.70 \text{ hr}}{1 \text{ half-life}} = 26.8 \text{ hr} \quad \text{(calculator and \textbf{correct answer})}$$

Practice Exercise 17.3

Strontium-90 is a nuclide found in radioactive fallout from nuclear weapon explosions. Its half-life is 28.0 yr. How long will it take for 75.0% (3/4) of the strontium-90 atoms in a "fallout" sample to undergo decay?

In both Examples 17.2 and 17.3 the time elapsed was equivalent to a whole number of half-lives. In order to work problems involving a fractional number of half-lives, equations involving logarithms must be used. Such equations will not be presented in this text; hence only problems that involve a whole number of half-lives will be considered.

The half-life concept can be used to determine the age of objects that contain a radionuclide. The best-known radiochemical dating technique is radiocarbon dating. Radiocarbon dating involves measuring the amount of carbon-14 present in an object. Many archeological artifacts can be dated by carbon-14 techniques, since they were made from or contain once-living carbon-containing materials.

The isotope carbon-14, the only naturally occurring radionuclide of carbon, has a half-life of 5730 yr. It is continually produced in the upper atmosphere as a result of cosmic ray bombardment.

$$^{14}_{7}\text{N} + {}^{1}_{0}\text{n} \longrightarrow {}^{14}_{6}\text{C} + {}^{1}_{1}\text{p}$$

The steady-state (equilibrium) concentration of carbon-14, which reflects both its rate of formation and its rate of decay, is one carbon-14 atom for every 10^{12} nonradioactive carbon atoms. This trace amount of carbon-14 in the atmosphere reacts with oxygen to give carbon dioxide in the same manner that nonradioactive carbon does. Thus, approximately 1 out of every 10^{12} carbon dioxide molecules is radioactive. This radioactive carbon is incorporated into the structure of plants through photosynthesis and into animals and human beings through the food chain. A steady-state concentration of carbon-14, equal to that found in the atmosphere, is thus found in all living organisms. Upon the death of an organism, the intake of carbon-14 ceases and the natural level of radioactive carbon present within the structure begins to decrease as the result of carbon-14 decay.

$$^{14}_{6}\text{C} \longrightarrow {}^{0}_{-1}\beta + {}^{14}_{7}\text{N}$$

In carbon-14 dating the ratio of carbon-14 to total carbon in an object that contains once-living material (parchment, cloth, charcoal, etc.) is compared to the same ratio for living matter. A wooden object with a ratio of carbon-14 to total carbon one fourth that of a living tree would be approximately 11,400 years (two half-lives) old. An important assumption in the carbon-14 dating method is that the flow of carbon-14 into the biosphere is constant with time. There is some evidence, such as the carbon-14 content of the growth rings in old trees, to indicate that this is approximately true.

17.6 Transmutation and Bombardment Reactions

Radioactive decay, discussed in the previous sections, is an example of a *natural* transmutation reaction. A **transmutation reaction** *is a nuclear reaction in which a nuclide of one element is changed into a nuclide of another element.* It is also possible to cause transmutation to occur in a laboratory setting through use of *bombardment reactions*. A **bombardment reaction** *is a nuclear reaction in which small particles traveling at very high speeds are made to collide with target nuclei and cause them to undergo nuclear change.* Bombardment reactions involve *artificial* transmutation, artificial because the change does not occur naturally.

The first successful bombardment reaction was carried out in 1919, twenty-five years after the discovery of radioactive decay, by Ernest Rutherford, the same Rutherford who earlier had investigated the nature of alpha, beta, and gamma rays (Sec. 17.3). Rutherford's initial successful bombardment experiment consisted of letting alpha particles from a natural source (radium) bombard nitrogen gas. In this process he found that a new stable nuclide was formed: oxygen-17. The nuclear equation for this transmutation is

$$^{14}_{7}\text{N} + ^{4}_{2}\alpha \longrightarrow ^{17}_{8}\text{O} + ^{1}_{1}\text{p}$$

Further research carried out by many investigators has shown that numerous nuclei experience change under the stress of bombardment by small, high-energy particles. In most cases, the new nuclide that is produced is radioactive (unstable). Two examples of bombardment reactions now carried out in laboratories in which the product nuclide is radioactive are

$$^{44}_{20}\text{Ca} + ^{1}_{1}\text{p} \longrightarrow ^{44}_{21}\text{Sc} + ^{1}_{0}\text{n}$$

$$^{23}_{11}\text{Na} + ^{2}_{1}\text{H} \longrightarrow ^{21}_{10}\text{Ne} + ^{4}_{2}\alpha$$

Radioactive nuclides produced by bombardment reactions, like naturally occurring radionuclides, undergo radioactive decay. In many cases, the previously discussed alpha- and beta-particle modes of decay (Sec. 17.4) occur. Additional modes of decay, to be discussed in Section 17.7, are also encountered.

> In bombardment reactions, there are always two reactants (the target nuclide and the small, high-energy bombarding particle) and also two products (the daughter nuclide and another small particle such as a neutron or proton).

Synthetic Elements

Over 2000 bombardment-produced radionuclides that do not occur naturally are now known. This number is seven times greater than the number of naturally occurring nuclides (Sec. 5.7). In this total is at least one radionuclide of every naturally occurring element. In addition, nuclides of 25 elements that do not occur in nature have been produced in small quantities as the result of bombardment reactions. Four of these "synthetic" elements, produced between 1937 and 1941, filled gaps in the periodic table for which no naturally occurring element had been found. These four elements are technetium (Tc, element 43), an element with numerous uses in nuclear medicine (Sec. 17.14); promethium (Pm, element 61); astatine (At, element 85); and francium (Fr, element 87). The remainder of the synthetic elements, elements 93 to 112, and 114, are called the *transuranium elements* because of their occurrence immediately following uranium in the periodic table. (Uranium is the highest atomic-numbered, naturally occurring element.) All isotopes of all of the transuranium

> Production of the small, *high-energy* bombarding particles needed to effect a bombardment reaction requires use of a cyclotron or a linear accelerator (both very expensive, *large* pieces of equipment). Both use magnetic fields to accelerate charged particles to velocities at which the energy is sufficient to allow the particle to penetrate the nucleus and induce a nuclear change.

Table 17.2	Stability Characteristics of Transuranium Elements				
Name	Symbol	Atomic Number	Mass Number of Most Stable Nuclide	Half-life of Most Stable Nuclide	Discovery Year of First Isotope
neptunium	Np	93	237	2.14×10^6 yr	1940
plutonium	Pu	94	244	7.6×10^7 yr	1940
americium	Am	95	243	8.0×10^3 yr	1944
curium	Cm	96	247	1.6×10^7 yr	1944
berkelium	Bk	97	247	1400 yr	1950
californium	Cf	98	251	900 yr	1950
einsteinium	Es	99	252	472 days	1952
fermium	Fm	100	257	100 days	1953
mendelevium	Md	101	258	52 days	1955
nobelium	No	102	259	58 min	1958
lawrencium	Lr	103	262	3.6 hr	1961
rutherfordium	Rf	104	263	10 min	1969
dubnium	Db	105	262	34 sec	1970
seaborgium	Sg	106	266	21 sec	1974
bohrium	Bh	107	267	17 sec	1980
hassium	Hs	108	277	11 min	1984
meitnerium	Mt	109	268	0.042 sec	1982
darmstadtium	Ds	110	271	1.1 min	1994
element 111	—	111	272	0.002 sec	1994
element 112	—	112	285	10.7 min	1996
element 114	—	114	289	21 sec	1999

elements are radioactive. Table 17.2 gives information about the stability of the transuranium elements. Note the extremely short half-lives of the more recently produced elements.

Significant uses exist for some "synthetic" radionuclides, particularly in the field of medicine. For example, the synthetic radionuclides cobalt-60, yttrium-90, iodine-131, and gold-198 find use in radiotherapy treatment for cancer.

The element gold can be produced from platinum using bombardment. However, the process is astronomically expensive compared to the worth of the gold so produced. Platinum-196 is bombarded with deuterons (hydrogen-2 nuclei) to produce radioactive platinum-197.

$$^{196}_{78}\text{Pt} + ^{2}_{1}\text{H} \longrightarrow ^{197}_{78}\text{Pt} + ^{1}_{1}\text{p}$$

The platinum-197 decays through beta-particle emission to produce stable gold-197.

$$^{197}_{78}\text{Pt} \longrightarrow ^{0}_{-1}\beta + ^{197}_{79}\text{Au}$$

17.7 Positron Emission and Electron Capture

Laboratory-produced radionuclides undergo radioactive decay just as do radionuclides from nature. Four modes of decay are encountered. They are alpha particle emission and beta particle emission (the same as for naturally occurring radionuclides—Sec. 17.4) and two new modes not found for naturally radioactive substances: positron emission and electron capture.

Positron emission *is a radioactive decay process in which a positron is emitted from an unstable nucleus when a proton is converted to a neutron.* The particle involved in positron emission, the positron, is a particle we have not previously discussed. A **positron** *is a particle with the same mass as an electron or a beta particle, but with a positive charge.* The symbol for a positron is $^{0}_{1}\beta$. Its production in the nucleus is due to the conversion within the nucleus of a proton to a neutron.

$$^{1}_{1}p \longrightarrow\ ^{1}_{0}n\ +\ ^{0}_{1}\beta$$

This process is just the opposite of that occurring during beta-particle emission (Sec. 17.4). The net effect of positron emission is thus to decrease the atomic number (number of protons), while the mass number remains constant. An example of a radioactive decay process involving positron emission is

$$^{30}_{15}P \longrightarrow\ ^{0}_{1}\beta\ +\ ^{30}_{14}Si$$

Electron capture *is a radioactive decay process in which an electron in a low-energy orbital, such as the 1s orbital, is pulled into an unstable nucleus, converting a proton to a neutron.*

$$^{0}_{-1}e\ +\ ^{1}_{1}p \longrightarrow\ ^{1}_{0}n$$

An example of such a process is the reaction

$$^{87}_{37}Rb\ +\ ^{0}_{-1}\beta \longrightarrow\ ^{87}_{36}Kr$$

Why and when positron emission and electron capture occur is considered in Section 17.8.

EXAMPLE 17.4

Writing Balanced Nuclear Equations Given the Parent Nuclide and Its Mode of Decay

Write a balanced nuclear equation for the decay of each of the following radioactive nuclides. The mode of decay is indicated in parentheses.

(a) $^{62}_{29}Cu$ (positron emission)   **(b)** $^{118}_{52}Te$ (electron capture)

(c) $^{105}_{47}Ag$ (electron capture)   **(d)** $^{82}_{37}Rb$ (positron emission)

SOLUTION

In each case the atomic number and mass number of the daughter nuclide are obtained by first writing the symbols of the parent nuclide and the particle emitted (positron) or absorbed (electron) and then balancing the equation.

(a) Let X represent the product of the radioactive decay, that is, the daughter nuclide. Then

$$^{62}_{29}Cu \longrightarrow\ ^{0}_{1}\beta\ +\ X$$

Note that the Greek letter β is used to denote not only a beta particle but also a positron. The difference between the two particles is that the former is negatively charged ($^{0}_{-1}\beta$) and the latter is positively charged ($^{0}_{1}\beta$).

Since the sum of the superscripts on each side of the equation must be equal, the superscript for X must be 62. In order for the sums of the subscripts on both sides of the equation to be equal, at 29, the subscript for X must be 28. As soon as the subscript of X is determined, the identity of X is known. Looking at a periodic table we determine that the element with an atomic number of 28 is nickel (Ni). Therefore,

$$^{62}_{29}Cu \longrightarrow\ ^{0}_{1}\beta\ +\ ^{62}_{28}Ni$$

(b) Similarly, letting X represent the product daughter nuclide of the radioactive decay, we have for $^{118}_{52}Te$ decaying by the electron capture mechanism

$$^{118}_{52}Te\ +\ ^{0}_{-1}e \longrightarrow\ X$$

Note that in electron capture the electron appears on the reactant side of the equation. This makes equations for electron capture different from those for alpha, beta, and positron emissions where in each case the small particle involved is placed on the product side of the equation.

Balancing the above equation, making the superscripts on each side of the equation total 118 and the subscripts total 51, we get

$$^{118}_{52}\text{Te} + ^{0}_{-1}\text{e} \longrightarrow ^{118}_{51}\text{Sb}$$

(c) Similarly, for this electron capture we write

$$^{105}_{47}\text{Ag} + ^{0}_{-1}\text{e} \longrightarrow \text{X}$$

Balancing superscripts and subscripts, we get

$$^{105}_{47}\text{Ag} + ^{0}_{-1}\text{e} \longrightarrow ^{105}_{46}\text{Pd}$$

(d) Finally, we write

$$^{82}_{37}\text{Rb} \longrightarrow ^{0}_{1}\beta + \text{X}$$

In both positron emission and electron capture the atomic number of the daughter nuclide decreases by one and the mass number does not change from that of the parent. The balancing process gives results consistent with this generalization in this part as well as in the previous three parts of this problem.

$$^{82}_{37}\text{Rb} \longrightarrow ^{0}_{1}\beta + ^{82}_{36}\text{Kr}$$

Practice Exercise 17.4

Write a balanced nuclear equation for the decay of each of the following radioactive nuclides. The mode of decay is indicated in parentheses.

(a) $^{21}_{11}\text{Na}$ (positron emission)   **(b)** $^{7}_{4}\text{Be}$ (electron capture)

Figure 17.3 is a summary diagram contrasting the various modes of decay for radioactive nuclides that we have considered: alpha-particle emission and beta-particle emission (Sec. 17.3) and positron emission and electron capture (Sec. 17.7). It relates these various decay modes to periodic table position for daughter nuclides relative to a common parent nuclide. Note from Figure 17.3 that for a parent nuclide $\left(^{A}_{Z}\text{E}\right)$ the daughter nuclide has a lower atomic number for alpha-particle emission, positron emission, and electron capture and a higher atomic number only in the case of beta-particle emission.

Figure 17.3

Summary diagram for the various modes of radioactive decay. Alpha-particle emission, positron emission, and electron capture produce a daughter nuclide with an atomic number less than that of the parent nuclide. Beta-particle emission produces a daughter nuclide with an atomic number greater than that of the parent nuclide.

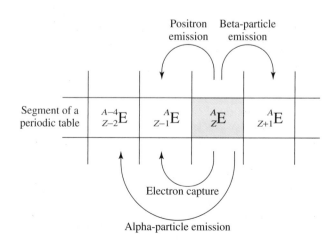

17.8 | Nuclear Stability

No simple rule exists that allows us to predict whether a particular nucleus is radioactive and how it might decay. However, a consideration of some observations about those nuclei that are stable is helpful in understanding why some nuclei are stable and others are not.

Two generalizations readily apparent from a study of the properties of the stable nuclei in nature are

1. *There is a correlation between nuclear stability and the total number of nucleons found in a nucleus.* All nuclei with 84 or more protons are unstable. The largest stable nucleus known is that of $^{209}_{83}$Bi, a nucleus that contains 209 nucleons. It thus appears that there is a limit to the number of nucleons that can be packed into a stable nucleus.

2. *There is a correlation between nuclear stability and neutron-to-proton ratio in a nucleus.* The number of neutrons necessary to create a stable nucleus increases as the number of protons increases. This pattern of increasing neutron-to-proton ratio with increasing atomic number is shown in Figure 17.4. For elements of low atomic number, neutron-to-proton

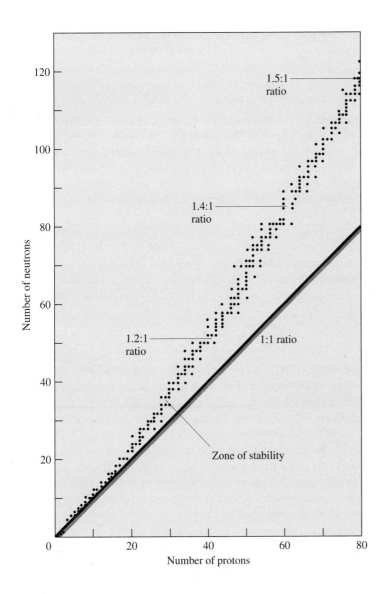

Figure 17.4

A graph of the number of neutrons versus the number of protons for stable nuclei.

ratios of stable nuclides are very close to 1. For heavier elements, stable nuclides have higher neutron-to-proton ratios, with the ratio reaching approximately 1.5 for the heaviest stable elements.

The fact that stable nuclei fall into a rather narrow zone of stability defined by neutron-to-proton ratios strongly suggests that neutrons are at least partially responsible for the stability of the nucleus. It should be remembered that like charges repel each other and that most nuclei contain many protons (with identical positive charges) squeezed together into a very small volume. As the number of protons increases, the electrostatic forces of repulsion between protons increase sharply. Therefore a greater number of neutrons is required to counteract the increased repulsive forces. Finally, at element 84, the repulsive forces become great enough that the nuclei are unstable regardless of the number of neutrons present.

The mode by which a radionuclide decays depends to a large extent upon how the neutron-to-proton ratio for the radionuclide compares with that of stable nuclei containing approximately the same number of nucleons. The driving force for radioactive decay is the tendency of unstable nuclei to adjust neutron-to-proton ratios in such a way that stability is achieved.

Unstable nuclei can be classed into three categories on the basis of their position on the graph in Figure 17.4 relative to where the stable nuclei are found (zone of stability). These categories are

1. *Unstable nuclei in which the neutron-to-proton ratio is too high.* Such nuclides, which can be considered proton-poor, lie to the left of the zone of stability in Figure 17.4.

2. *Unstable nuclei in which the neutron-to-proton ratio is too low.* Such nuclides, which can be considered proton-rich, lie to the right of the zone of stability in Figure 17.4.

3. *Unstable nuclei in which the total number of nucleons exceeds 209—the limit for a stable nucleus.* Such nuclei lie beyond the zone of stability in Figure 17.4.

Nuclei in each of these categories have a particular mode of decay that predominates over others.

Beta emission is the predominant decay mode for nuclides having neutron-to-proton ratios that are *too high* for stability. Such nuclides lie to the *left* of the zone of stability in Figure 17.4. In almost all cases, nuclei in this category have mass numbers greater than the atomic mass of the element. In the following example of beta emission, note that the mass number of the radioactive manganese nuclide (56) is greater than the atomic mass of manganese (54.94 amu).

$$^{56}_{25}Mn \longrightarrow \, ^{56}_{26}Fe + \, ^{0}_{-1}\beta$$

As discussed previously (Sec. 17.4), beta emission involves the transformation of a neutron into a proton. This increases the number of protons, decreases the number of neutrons, and causes a decrease in the neutron-to-proton ratio—the desired result. Note that in the above reaction for the beta decay of manganese-56, the neutron-to-proton ratio of the parent nuclide is 1.24 and that of the daughter iron-56 is 1.15.

Radionuclides lying to the *right* of the zone of stability have *too low* a neutron-to-proton ratio. They decay by converting a proton into a neutron—just the opposite of the process that occurs in beta-particle emission. Radionuclei in this category generally have mass numbers that are lower than the atomic mass of the element. The conversion of a proton into a neutron can be accomplished by either of two ways: (1) by positron emission or (2) by electron capture. These decay modes were described in Section 17.7. The process of electron capture seems to be preferred over positron emission for nuclides of high atomic number. For lighter nuclides numerous examples of both types of decay processes are known.

Alpha-particle emission is found primarily among elements in which the total number of nucleons exceeds 209. These are the nuclei that lie *beyond* the region of stability (Fig. 17.4). Alpha-particle emission is not, however, the only mode of decay for elements in this region; beta-particle emission is also common. Another characteristic of radionuclides of this type is that the decay process for reaching stability involves more than one step. This results in the formation of a *decay series,* the topic of Section 17.9.

Before we leave this section, we should note that the preceding guidelines work in most instances; however, there are exceptions. For example, both $^{146}_{60}$Nd and $^{148}_{60}$Nd are stable and lie in the region of stability in Figure 17.4, but $^{147}_{60}$Nd, which is radioactive, also lies in the region of stability (between the two stable nuclides).

17.9 Radioactive Decay Series

Radioactive nuclides with high atomic numbers cannot attain nuclear stability with a single emission; they are too far away from the zone of stability (Fig. 17.4) for this to occur. Instead, a series of decay steps is required for such nuclei to reach stability. A large radionuclide undergoes decay to produce a product nucleus that is also radioactive. This product in turn produces a third radionuclide; this in turn decays to produce a fourth nuclide, and so forth, until ultimately a stable nucleus is produced. Such a sequence of decay products is called a *radioactive decay series*. A **radioactive decay series** *is a sequence of nuclear reactions in which one radioactive nuclide decays to a second, which then decays to a third, and so forth, until a stable nuclide is finally produced.*

Three naturally occurring decay series are known. Each starts with a long-lived radionuclide and ends with a stable nuclide. A fourth decay series was discovered after the synthesis of certain elements not found in nature (transuranium elements—Sec. 17.6). Because the parent of this fourth series, plutonium-241, does not occur in nature to a measurable extent, the series is not classified as naturally occurring. General characteristics of the four decay series are given in Table 17.3.

Figure 17.5 shows all of the members of the uranium-238 decay series. It is representative of the other three series. Note, from Figure 17.5, that both alpha and beta particles are part of the decay sequence and that there is no pattern relative to which is emitted when.

One of the intermediate products in the uranium-238 decay series is the radionuclide radon-222 (see Fig. 17.5). This substance is a gas at normal temperatures, and is therefore a very mobile species. Its presence has been detected in both aqueous and atmospheric environments. For the typical American, radon-222 is a major radiation exposure source (see Sec. 17.13).

Table 17.3	The Four Known Radioactive Decay Series	
Parent	**Number of Decay Steps**	**Final Product of Series**
Uranium-238	14	lead-206
Thorium-232	10	lead-208
Uranium-235	11	lead-207
Plutonium-241	13	bismuth-209

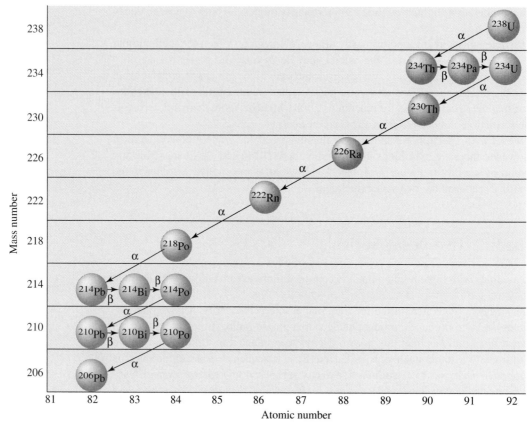

Figure 17.5

In this uranium-238 decay series, each nuclide, except lead-206 (the final product), is radioactive; the successive transformations continue until this stable product is obtained.

17.10 Chemical Effects of Radiation

The very energetic alpha, beta, and gamma radiations produced from radioactive decay travel outward from their nuclear sources into the material surrounding the radioactive substance. There, they interact with the atoms and molecules of the material, which dissipates their excess energy. Numerous interactions (collisions) between atoms or molecules and a "particle" or "ray" of radiation are required before the energy of the radiation is reduced to the level of surrounding materials. At this point the radiation is "harmless." Let us consider in closer detail the interactions that do occur between radiation and atoms or molecules.

It is the electrons of molecules that are most directly affected by radiation. In general, two things can happen to an electron subjected to radiation: excitation or ionization. *Excitation* occurs when radiation, through energy release, excites an electron from an occupied orbital into an empty, higher-energy orbital. *Ionization* occurs when the radiation carries enough energy to remove an electron from an atom or molecule.

Based on its effects on electrons, radiation of various types is classified into the categories *non-ionizing radiation* and *ionizing radiation*. **Non-ionizing radiation** *is radiation with insufficient energy to remove an electron from an atom or molecule*. Radio waves, microwaves, infrared light, and visible light are forms of non-ionizing radiation. The first three of these four types of non-ionizing radiation possess insufficient energy to excite electrons. Such radiation can, however, cause atoms to move, vibrate, and rotate more rapidly, causing

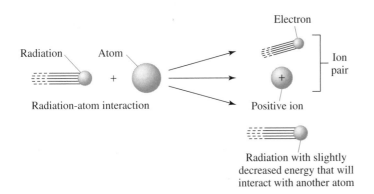

Figure 17.6

Ion-pair formation. When radiation interacts with an atom, an electron is often knocked away from the atom. The atom that loses the electron becomes an ion. An ion so produced and its "free electron" constitute an ion pair.

the material to increase in temperature. Visible light, the fourth type of non-ionizing radiation, possesses sufficient energy to excite electrons. Electrons that undergo excitation return, with time, to their normal states.

Ionizing radiation *is radiation with enough energy to remove an electron from an atom or molecule.* Radioactive emissions (alpha, beta, and gamma) as well as cosmic rays, X rays, and ultraviolet light are forms of ionizing radiation. (Cosmic "rays" are energetic particles coming from interstellar space; their makeup is primarily protons, alpha particles, and beta particles.)

The result of the interaction of ionizing radiation with matter is *ion-pair formation.* In ion-pair formation, the incoming radiation transfers sufficient energy into a molecule to knock an electron out of it, converting the molecule into a positive ion (see Fig. 17.6); that is, ionization occurs and an *ion pair* is formed. An **ion pair** *is the electron and positive ion that are produced during an interaction between a molecule and ionizing radiation.* This ionization process is not the normal voluntary transfer of electrons that occurs during ionic compound formation (Sec. 7.6) but rather the involuntary nonchemical removal of an electron from a molecule to form an ion. Many ion pairs are produced by a single "particle" of radiation because such a particle must undergo many collisions before its energy is reduced to the level of surrounding material. The electrons ejected from atoms or molecules during ion-pair formation frequently have enough energy to bombard neighboring molecules and cause additional ionization.

Free-radical formation, either directly or indirectly, usually accompanies ion-pair formation. A **free radical** *is a chemical species—either an atom or a molecule—that contains an unpaired electron.* The presence of the unpaired electron in a free radical usually causes it to be a very reactive species. (Recall from Sec. 7.9 that electrons normally occur in pairs in molecules.) Free radicals can rapidly react with other chemical species nearby, often precipitating a series of totally undesirable chemical reactions inside a living cell. Once formed, a free radical can either combine with another free radical to form a molecule in which the electrons are paired, or it can react with another molecule to produce a new free radical. The latter is a common occurrence. It is such production of new free radicals in a "chain-like" manner that causes major problems within a living cell. Such free-radical production is what makes the injury from radiation exposure far greater in magnitude than that expected merely on the basis of the energy of the incoming radiation.

Because water is the most abundant molecule in living organisms, the effects of ionizing radiation on water are of prime importance in assessing the effects of radiation exposure on health and life. The first step in the interaction of ionizing radiation with water is usually ion-pair formation.

$$H_2O + radiation \longrightarrow \underbrace{H_2O^+ + e^-}_{\text{ion pair}}$$

Just as the free radical H_2O^+ is not to be confused with the acidic species H_3O^+, the hydroxyl free radical (OH) is not to be confused with the hydroxide ion (OH⁻; the basic species in aqueous solution). The difference between these two species is seen by comparing their Lewis structures.

$$:\ddot{O}—H$$

hydroxyl free
radical

$$\left[:\ddot{O}—H\right]^-$$

hydroxide ion

The H_2O^+ ion formed in this interaction should not be confused with the H_3O^+ ion produced when acids dissolve in water (Sec. 14.2). The H_2O^+ ion is a free radical and is extremely reactive; the H_3O^+ ion is not a free radical. The Lewis structure for the H_2O^+ ion is

$$\left[\begin{array}{c} :\dot{O}\cdot \\ H \quad H \end{array}\right]^+ \quad \text{an unpaired electron}$$

The highly reactive H_2O^+ ion, a species not normally present in biological tissue, can then react with another water molecule causing further free-radical formation.

$$H_2O^+ + H_2O \longrightarrow H_3O^+ + OH$$

free
radical

new free
radical

The OH free radicals produced in this manner then interact with many different biomolecules to produce new free radicals, which in turn can react further. The result often devastatingly upsets cellular activity.

17.11 Biochemical Effects of Radiation

The three types of naturally occurring radioactive emissions—alpha particles, beta particles, and gamma rays—differ in their ability to penetrate matter and to cause ionization. Consequently, the extent of the biochemical effects of radiation depends on the type of radiation involved.

Alpha particles are the most massive and also the slowest moving particles involved in natural radioactive decay; they are emitted from nuclei at a velocity of about one tenth of the speed of light. They have low penetrating power and cannot penetrate the body's outer layers of skin. The major danger from alpha radiation occurs when alpha-emitting radionuclides are ingested, for example, in contaminated food, or inhaled in air. There are no protective layers of skin within the body.

Beta particles are emitted from nuclei at speeds of up to nine-tenths that of light. With their greater velocity, they can penetrate much deeper than alpha particles and can cause severe skin burns if their source remains in contact with the skin for an appreciable time. They do not ionize molecules as readily as alpha particles do because of their much smaller size. An alpha particle is approximately 8000 times heavier than a beta particle. It is estimated that a typical alpha particle travels about 6 cm in air and produces 40,000 ion pairs, and a typical beta particle travels 1000 cm in air and produces about 2000 ion pairs. Internal exposure to beta radiation is also serious.

Gamma radiation travels at the speed of light. Gamma rays readily penetrate deeply into organs, bones, and tissues.

Figure 17.7 contrasts the abilities of alpha, beta, and gamma radiations to penetrate paper, aluminum foil, and a thin layer of a lead-concrete mixture.

The minimum radiation dosage that causes human injury is unknown. However, the effects of larger doses have been studied and are listed in Table 17.4. As you can see, very serious injury or death can result from large doses of ionizing radiation. The doses causing the various effects listed in Table 17.4 are given in terms of the radiation unit called a *rem*, which is defined in the footnote to the table.

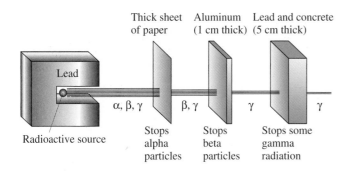

Thick sheet of paper   Aluminum (1 cm thick)   Lead and concrete (5 cm thick)

Lead

α, β, γ   β, γ   γ   γ

Radioactive source

Stops alpha particles

Stops beta particles

Stops some gamma radiation

Figure 17.7

Penetrating abilities of alpha, beta, and gamma radiation.

Table 17.4	The Effects of Short-Term Whole-Body Radiation Exposure on Humans
Dose (rems)[a]	**Effects**
0–25	no detectable clinical effects
25–100	slight, short-term reduction in number of some blood cells; disabling sickness not common
100–200	nausea and fatigue, vomiting if dose is greater than 125 rems; longer-term reduction in number of some blood cells
200–300	nausea and vomiting first day of exposure; up to 2-week latent period followed by appetite loss, general malaise, sore throat, pallor, diarrhea, and moderate emaciation; recovery in about 3 months unless complicated by infection or injury
300–600	nausea, vomiting, and diarrhea in first hours; up to a 1-week latent period followed by loss of appetite, fever, and general malaise in the second week, followed by hemorrhage, inflammation of mouth and throat, diarrhea, and emaciation; some deaths in 2–6 weeks, eventual death for 50% if exposure is above 450 rems; others recover in about 6 months
600 or more	nausea, vomiting, and diarrhea in first few hours; rapid emaciation and death as early as second week; eventual death of nearly 100%

[a]*A rem is the quantity of ionizing radiation that must be absorbed by a human to produce the same biological effect as 1 roentgen of high-penetration X rays. A roentgen is the quantity of high-penetration X rays that produces approximately 2×10^9 ion pairs per cubic centimeter in dry air at 0°C and 1 atm.*

17.12 Detection of Radiation

You cannot hear, feel, taste, see, or smell low levels of radiation. However, there are numerous methods for detecting the presence of radiation. Becquerel's initial discovery of radioactivity (Sec. 17.2) was the result of the effect of radiation on photographic plates. Radiation affects photographic film in the same way as ordinary light; the film is exposed. Technicians and others who work around radiation usually wear film badges (see Fig. 17.8) to record the extent of their exposure to radiation. When the film from the badge is developed, the degree of darkening of the film indicates the extent of radiation exposure. By using different filters, various parts of the film register exposures to the types of radiation (alpha, beta, gamma, and X rays).

Radiation can also be detected by making use of the fact that radiation ionizes atoms and molecules (Sec. 17.10). The Geiger counter operates on this principle. The basic components of a Geiger counter are shown in Figure 17.9. The detection part of such a counter is a metal tube filled with a gas (usually argon). The tube has a thin-walled window made of a material that can be penetrated by the radiation. In the center of the tube is a wire attached

Figure 17.8

Film badges are worn by technicians and others whose work requires them to be around radiation. The extent of exposure of all personnel is carefully logged to help prevent overexposure. (Yoav Levy/Phototake NYC)

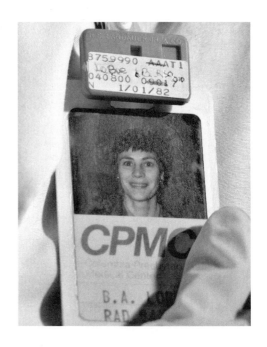

Figure 17.9

The principle of operation of a Geiger counter. When radiation enters through the window, it ionizes one or more gas atoms, producing ion pairs. The electrons from the ion pairs are attracted to the central wire, and the positive ions are drawn to the metal tube. This constitutes a pulse of electric current, which is amplified and displayed on the meter or other readout.

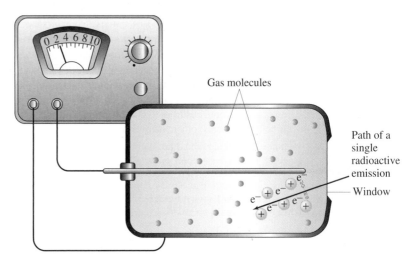

Gas molecules

Path of a single radioactive emission

Window

to the positive terminal of an electric power source. The shell of the metal tube is attached to the negative terminal of the same source. Radiation entering the tube ionizes the gas, which allows a pulse of electricity to flow. This pulse is amplified and displayed on a meter or some other type of readout display.

17.13 Sources of Radiation Exposure

Most of us will never come into contact with the radiation dosage necessary to cause the more serious effects listed in Table 17.4. Nevertheless, *low-level* exposure to ionizing radiation is something we constantly encounter. In fact, there is no way we can totally avoid this low-level exposure because much of it results from naturally occurring materials in the environment. Estimates of per capita radiation exposure from various sources are given in Figure 17.10.

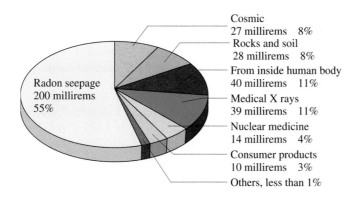

Figure 17.10

Estimated annual radiation exposure, in millirems, of an average American. Individual exposures vary widely.

Comparing the values of Figure 17.10, which are in *millirem* units, with those of Table 17.4, which are in *rem* units, shows that the current radiation exposure levels experienced by the general population are very small compared to those known to cause serious radiation sickness. Nevertheless, it is important to monitor carefully the amount of radioactive materials in the environment in order to ensure that radiation levels remain low.

With low-level radiation exposure, cell damage rather than cell death frequently occurs. If the damaged cells repair themselves improperly, which can occur, new, abnormal cells are produced when the cells replicate. Much still needs to be learned about the long-term effects of cell damage caused by low-level radiation exposure.

Cells that reproduce at a rapid rate, such as those in bone marrow, lymph nodes, and embryonic tissue, are the most sensitive to radiation damage. One of the first signs of overexposure to radiation is a drop in red blood cell count. This directly relates to the sensitivity of bone marrow, the site of red cell formation, to radiation. The sensitivity of embryonic tissue to radiation damage is the reason that pregnant women need to be protected from radiation exposure.

17.14 Nuclear Medicine

Radionuclides are used both diagnostically and therapeutically in medicine. In diagnostic applications, technicians use small amounts of radionuclides whose progress through the body or localization in specific organs can be followed. Larger quantities of radionuclides are used in therapeutic applications.

Diagnostic Uses for Radionuclides

The fundamental chemical principle behind the use of radionuclides in diagnostic medical work is the fact that a radioactive isotope of an element has the same chemical properties as a nonradioactive isotope of the element. (Radioactive and nonradioactive isotopes of an element differ in nuclear properties but not in chemical properties.) Thus, body chemistry is not upset by the presence of a small amount of a radioactive substance that is already present in the body in a nonradioactive form.

The criteria used in selecting radionuclides for diagnostic procedures include the following.

1. At low concentrations (to minimize radiation damage), the radionuclide must be detectable by instrumentation placed outside the body. Nearly all diagnostic radionuclides are gamma emitters because the penetrating power of alpha and beta particles is too low.

2. It must have a short half-life so that the intensity of the radiation is sufficiently great to be detected. A short half-life also limits the time period of radiation exposure.

3. It must have a known mechanism for elimination from the body so that the material does not remain in the body indefinitely.

4. Its chemical properties must be such that it is compatible with normal body chemistry. It must be able to be selectively transmitted to the part or system of the body that is under study.

The circulation of blood in the body can be followed by using radioactive sodium-24. A small amount of this isotope is injected into the bloodstream in the form of a sodium chloride solution. The movement of this radionuclide through the circulation system can be followed easily with a radiation detector. If the nuclide takes longer than normal to move through a particular part of the body, this is an indication that the circulation is impaired at that spot. Sodium-24 can also be used to locate blood clots because the amount of radiation will be high on one side and low on the opposite side of the clot.

Radiologists evaluate the functioning of the thyroid gland by administering iodine-131 to a patient, usually in the form of a sodium iodide (NaI) solution. The radioactive iodine behaves in the same manner as ordinary iodine and is absorbed by the thyroid at a rate related to the activity of the gland. If a hypothyroid condition exists, the amount accumulated is less than normal, and if a hyperthyroid condition exists, a greater than average amount accumulates.

The size and shape of organs, as well as the presence of tumors, can be determined in some situations by scanning the organ in which a radionuclide tends to concentrate. Iodine-131 and technetium-99, in the form of the polyatomic ion TcO_4^-, concentrate in brain tumors more than in normal brain tissue; this helps radiologists determine the presence, size, and location of brain tumors.

Table 17.5 lists a number of radionuclides that are used in diagnostic procedures. The half-life of the radionuclide, the parts of the body it concentrates in, and its diagnostic value are also given.

Table 17.5 Selected Radionuclides Used in Diagnostic Procedures

Isotope	Half-life	Part of Body	Use in Diagnosis
Barium-131	11.6 days	bone	detection of bone tumors
Chromium-51	27.8 days	blood	determination of blood volume and red blood cell lifetime
		kidney	assessment of kidney activity
Iodine-131	8.05 days	brain	detection of fluid buildup in the brain
		kidney	location of cysts
		lung	location of blood clots
		thyroid	assessment of iodine uptake by thyroid
Iron-59	45 days	blood	evaluation of iron metabolism in blood
Phosphorus-32	14.3 days	blood	blood studies
		breast	assessment of breast carcinoma
Potassium-42	12.4 hr	tissue	determination of intercellular spaces in fluids
Sodium-24	15.0 hr	blood	detection of circulatory problems; assessment of peripheral vascular disease
Technetium-99	6.0 hr	brain	detection of brain tumors, hemorrhages, or blood clots
		spleen	measurement of size and shape of spleen
		thyroid	measurement of size and shape of thyroid
		lung	location of blood clots

Table 17.6	Selected Radionuclides Used in Radiation Therapy		
Isotope	Type of Half-life	Emitter	Use in Therapy
Cobalt-60	5.3 yr	beta, gamma	external source of radiation for treatment of cancer
Iodine-131	8 days	beta, gamma	cancer of thyroid
Phosphorus-32	14.3 days	beta, gamma	treatment of some types of leukemia and widespread carcinomas
Radium-226	1620 yr	alpha, gamma	used in implantation cancer therapy
Radon-222	3.8 days	alpha, gamma	used in treatment of uterine, cervical, oral, and bladder cancers
Yttrium-90	64 hr	beta, gamma	implantation therapy

Therapeutic Uses for Radionuclides

When radionuclides are used for therapeutic purposes, the objectives are entirely different from those for diagnostic procedures. The main objective for radionuclides in therapeutic use is to *selectively* destroy abnormal (usually cancerous) cells. The radionuclide is often, but not always, placed within the body. There is no need to monitor the radiation produced with an external detector. Therapeutic radionuclides implanted in the body are usually alpha or beta emitters because an intense dose of radiation is needed in a small localized area.

A commonly used implantation radionuclide that is effective in the localized treatment of tumors is yttrium-90, a beta emitter with a half-life of 64 hr. Yttrium-90 salts are implanted by using small hollow needles that are inserted into the tumor.

External, high-energy beams of gamma radiation are also used extensively in the treatment of certain cancers. Cobalt-60 is frequently used for this purpose; a beam of radiation is focused on the small area of the body where the tumor is located. This therapy usually causes some radiation sickness because normal cells are also affected, although to a lesser extent. The operating principle here is that the more rapidly dividing abnormal cells are more susceptible to radiation damage than normal cells are (Sec. 17.13). Radiation sickness is the price paid for the destruction of abnormal cells. Table 17.6 lists selected radionuclides that are used in therapy.

17.15 Nuclear Fission and Nuclear Fusion

Our glimpse into the world of nuclear chemistry would not be complete without a brief mention of two additional types of nuclear reactions: *nuclear fission* and *nuclear fusion*. **Nuclear fission** *is a nuclear reaction in which the nucleus of a heavy element splits into two or more lighter nuclei as the result of nuclear bombardment.*

The first fissionable nucleus to be discovered, which remains the most important one, was uranium-235. Bombardment of this nucleus with neutrons causes it to split into two fragments. Characteristics of the uranium-235 fission reaction include the following.

1. There is no unique way in which the uranium-235 nucleus splits. The following are examples of ways in which this fission process may proceed.

$$^{235}_{92}U + ^{1}_{0}n \longrightarrow \begin{cases} ^{135}_{53}I + ^{97}_{39}Y + 4\,^{1}_{0}n \\ ^{139}_{56}Ba + ^{94}_{36}Kr + 3\,^{1}_{0}n \\ ^{131}_{50}Sn + ^{103}_{42}Mo + 2\,^{1}_{0}n \\ ^{139}_{54}Xe + ^{95}_{38}Sr + 2\,^{1}_{0}n \end{cases}$$

2. Very large amounts of energy, which are many times greater than that released by ordinary radioactive decay, are emitted during the fission process. It is this large release of energy that makes nuclear fission of uranium-235 the important process that it is. In general, the term *nuclear energy* is used to refer to the energy released during the nuclear fission process. An older term for this energy is *atomic energy*.

3. Neutrons, which are reactants in the fission process, are also produced as products. The number of neutrons produced per fission depends on the way in which the nucleus splits and ranges from two to four (as can be seen from the fission equations just given). On the average, 2.4 neutrons are produced per fission. The significance of the produced neutrons is that they can cause the fission process to continue by colliding with other uranium-235 nuclei. Figure 17.11 shows diagrammatically the chain reaction that can occur once the fission process is started.

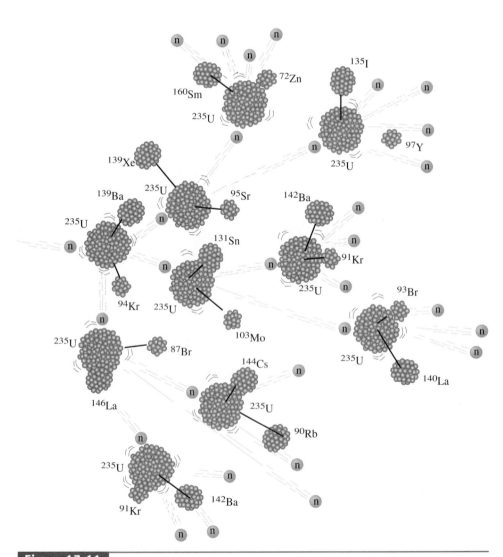

Figure 17.11

A fission chain reaction caused by further reaction of the neutrons produced during fission. (The "free" neutrons are shown here proportionally much larger than those contained within the nucleus.)

The process of nuclear fission, or "splitting the atom" as it is called in popularized science, can be carried out in both an uncontrolled manner and a controlled manner. The key question is "What happens to the neutrons produced during fission?" Do they react further, causing additional fission, or do they escape into the surroundings? If the majority of the neutrons produced react further (Fig. 17.11), an uncontrolled nuclear reaction (an atomic explosion) results. When only a few neutrons react further (on the average, one per fission), the fission reaction self-propagates in a controlled manner.

The process of nuclear fission is the basis for the operation of nuclear power plants to produce electricity. In this case, the fission process is carried out in a controlled manner. The reaction is controlled with rods that absorb excess neutrons (so that they cannot cause unwanted fissions) or with moderating substances that decrease the speed of the neutrons. The energy produced during the fission process, which appears as heat, is used to operate steam-powered, electricity-generating equipment.

There is another type of nuclear reaction that produces even more energy than nuclear fission. **Nuclear fusion** *is a nuclear reaction in which small nuclei are fused together to make a larger one*. This process is essentially the opposite of nuclear fission. Fusion requires a very high temperature—several million degrees.

One place that is hot enough for nuclear fusion to occur is the interior of the sun. Nuclear fusion is the process by which the sun generates its energy. Within the sun, in a three-step reaction, hydrogen-1 nuclei are converted to helium-4 nuclei with the release of extraordinarily large amounts of energy.

The use of nuclear fusion on Earth might seem impossible because of the high temperatures required. It has, however, been accomplished in a hydrogen bomb. In such a bomb, a *fission* device (an atomic bomb) is used to achieve the high temperatures needed to start the fusion reaction.

$$\text{}^3_1\text{H} + \text{}^2_1\text{H} \longrightarrow \text{}^4_2\text{He} + \text{}^1_0\text{n}$$

At the high temperature of fusion reactions, electrons completely separate from nuclei. Neutral atoms cannot exist. The high-temperature, gaslike mixture of nuclei and electrons that results is called a *plasma* and is considered by some scientists to represent a fourth state of matter (in addition to solids, liquids, and gases).

Table 17.7	**Differences Between Nuclear and Chemical Reactions**
Chemical Reaction	**Nuclear Reaction**
1. Different isotopes of an element have practically identical chemical properties.	1. Different isotopes of an element have different properties in nuclear processes.
2. The chemical reactivity of an element depends on the element's state of combination (free element, compound, etc.).	2. The nuclear reactivity of an element is independent of the state of chemical combination.
3. Elements retain their identity in chemical reactions.	3. Elements may be changed into other elements during nuclear reactions.
4. Energy changes that accompany chemical reactions are relatively small.	4. Nuclear reactions involve energy changes a number of orders of magnitude larger than those in chemical reactions.
5. Reaction rates are influenced by temperature, pressure, catalysts, and reactant concentrations.	5. Reaction rates are independent of temperature, pressure, catalysts, and reactant concentrations.

17.16 A Comparison of Nuclear and Chemical Reactions

As the discussions in this chapter have shown, nuclear chemistry is quite different from ordinary chemistry. Many of the laws of chemistry must be modified when we consider nuclear reactions. The major differences between nuclear reactions and ordinary chemical reactions are listed in Table 17.7. This table serves as a summary of many of the concepts presented in this chapter.

Summary

1. **Unstable Nuclides and Radioactivity** Some nuclides possess nuclei that are unstable. To achieve stability, these unstable nuclides spontaneously emit energy (radiation). Such nuclides are said to be radioactive.

2. **Radioactive Emissions** Three types of radiation are emitted by naturally occurring radioactive nuclides: alpha, beta, and gamma. These radiations are characterized by mass and charge values. Alpha radiation (alpha particle) has a mass of 4 amu and a +2 charge. Beta radiation (beta particle) has a very small mass (0.00055 amu) and a −1 charge. Gamma radiation, a form of energy, has no mass and no charge.

3. **Balancing of Nuclear Equations** The symbols in nuclear equations represent nuclei rather than atoms. Atomic numbers and mass numbers are always shown when representing the nuclei. The balancing of nuclear equations is based on both *charge* and *nucleon* conservation. Charge conservation means that the sum of the subscripts (number of protons, or positive charges, in the nucleus) for the products must equal the sum of the subscripts for the reactants. Similarly, the total number of nucleons (protons and neutrons) is conserved. This means that the sum of the superscripts (the mass numbers) for the products equals the sum of the superscripts for the reactants.

4. **Radioactive Decay Rate and Half-Life** Radioactive nuclides are not all equally unstable. Some have fast decay rates and others decay slowly. The rate of decay for a radionuclide is given by its half-life. A half-life is the time required for half of any given quantity of a radioactive substance to undergo decay.

5. **Transmutation and Bombardment Reactions** A transmutation reaction is a nuclear reaction in which a nuclide of one element is changed into a nuclide of another element. A bombardment reaction is a transmutation process in which small particles traveling at very high speeds are collided with stable nuclei; such collisions cause these nuclei to undergo nuclear change

(become unstable). Over 2000 synthetically produced radionuclides that do not occur naturally have been produced using bombardment reactions, including radionuclides of 25 elements not found in nature. Bombardment reactions involve *artificial* transmutation, artificial because the change does not occur naturally. Radioactive decay is a *naturally occurring* transmutation reaction.

6. **Positron Emission and Electron Capture** Laboratory-produced radionuclides undergo radioactive decay just as do naturally occurring radionuclides. Four modes of decay are encountered. They are alpha particle emission and beta particle emission and two new modes not found for naturally radioactive substances: positron emission and electron capture. Positron emission is a decay mode in which a positron is ejected from the nucleus when a proton is converted to a neutron. Electron capture is a decay mode in which an electron in a low-energy orbital is pulled into the nucleus, converting a proton to a neutron.

7. **Nuclear Stability** Two factors upon which nuclear stability depends are (1) the total number of nucleons present, and (2) the neutron-to-proton ratio in the nucleus. All nuclei with more than 209 nucleons are unstable. The neutron-to-proton ratio required for stability increases as the number of protons increases.

8. **Radioactive Decay Series** The product of the radioactive decay of an unstable nuclide is a nuclide of another element, which may or may not be stable. If it is not stable, it will decay and produce still another nuclide. Further decay will continue until a stable nuclide is formed. Such a sequence of reactions is called a radioactive decay series.

9. **Chemical Effects of Radiation** The radiation associated with radioactive decay is *ionizing* radiation, radiation with enough energy to remove an electron from an atom or molecule. The result of interaction of ionizing radiation with matter is ion-pair formation. Many ion pairs are produced by a single "particle" of radiation.

101. (a) Br, N, Cl, O, F (b) Na, Rb, Cs, Fr, Ba, Ra (c) F, O, N, Cl (d) 0.5 unit   **103.** (a) S (b) Br (c) N (d) N   **105.** Response (c) has the symbol backwards; the H is the + end.
107. (a) H—O has the greatest polarity (b) O—Al has the greatest polarity (c) B—N has the greatest polarity (d) Al—Cl has the greatest polarity   **109.** (a) nonpolar covalent (b) ionic (c) polar covalent (d) nonpolar covalent   **111.** (a) more covalent character (b) more ionic character (c) more covalent character (d) more ionic character   **113.** (a) nonpolar (b) polar (c) polar (d) polar   **115.** (a) nonpolar (b) polar (c) polar (d) nonpolar
117. (b and c). In (b) both are polar molecules. In (c) both are nonpolar molecules. In (a) and (d) one is polar and the other is nonpolar.   **119.** (a) Mg: $1s^22s^22p^63s^2$ Mg^{2+}: $1s^22s^22p^6$ (b) F: $1s^22s^22p^5$ F^-: $1s^22s^22p^6$ (c) N: $1s^22s^22p^3$ N^{3-}: $1s^22s^22p^6$ (d) Ca^{2+}: $1s^22s^22p^63s^23p^6$ S^{2-}: $1s^22s^22p^63s^23p^6$
121. (a) nonisoelectronic cations (b) nonisoelectronic anions (c) isoelectronic cations (d) nonisoelectronic anions   **123.** (a) ionic (b) molecular (c) ionic (d) molecular   **125.** (a) molecule (b) formula unit (c) formula unit (d) molecule
127. (a) monoatomic only (b) both monoatomic and polyatomic (c) polyatomic only (d) monoatomic only   **129.** (a) yes (b) yes (c) no (d) no   **131.** (a) O (b) N (c) O (d) F   **133.** (a) should be $BeCl_2$ (b) should be $CaCl_2$ (c) correct (d) should be Li_2S
135. (a) correct number of electron dots, but improperly placed (b) not enough electron dots (c) correct number of electron dots, but improperly placed (d) correct number of electron dots, but improperly placed   **137.** BA, CA, DB, and DA   **139.** (a) same, both single (b) different, single and triple (c) different, triple and double (d) different, triple and double   **141.** (a) x = Cl (b) x = Cl

143. (a)   (b)

Note that other resonance structures are possible for the $SO_4{}^{2-}$ and $NO_3{}^-$ ions.   **145.** (a) tetrahedral electron pair geometry; tetrahedral molecular geometry (b) tetrahedral electron pair geometry; trigonal pyramidal molecular geometry (c) linear electron pair geometry; linear molecular geometry (d) tetrahedral electron pair geometry; tetrahedral molecular geometry
147. (a) 109.5° because the electron group arrangement about the oxygen atom is tetrahedral (b) 120° because the electron group arrangement about the carbon atom is trigonal planar
149. (a) H—F bond is more polar than H—Cl, so HF molecule is more polar. (b) C—F bond is more polar than C—Cl, so H_3CF is the more polar molecule. (c) The symmetry of the linear CO_2 makes it nonpolar, so HCN molecule is more polar. (d) The symmetry of the trigonal planar SO_3 makes it nonpolar, so the angular SO_2 is more polar.   **151.** (a) HI, HBr, HCl (b) NH_3, H_2O, HF (c) SCl_2, PCl_3, $SiCl_4$ (d) CI_4, NI_3, HI   **153.** $CO_3{}^{2-}$, CO_2, CO   **155.** A = Al, D = N, formula = AlN
157. D = Ca, A = S, formula = CaS   **159.** A = O and D = F. The Lewis structure of the compound is :F:O:F:
161. A = H and D = O; x is 2 and y is 1; The formula is H_2O.

163. A = Be and D = F; The Lewis structure is

Chapter 8

1. (a) and (d)   **3.** (a) and (d)   **5.** (a) yes (b) yes (c) no, both molecular (d) no, both ionic   **7.** (a) fixed charge (b) variable charge (c) fixed charge (d) variable charge   **9.** (a) and (d)
11. (a) variable charge (b) variable charge (c) fixed charge (d) fixed charge   **13.** (a) potassium ion (b) magnesium ion (c) copper (I) ion (d) lead (IV) ion   **15.** (a) bromide ion (b) nitride ion (c) sulfide ion (d) selenide ion   **17.** (a) Zn^{2+} (b) Pb^{2+} (c) Ca^{2+} (d) N^{3-} **19.** (a) magnesium oxide (b) lithium sulfide (c) silver chloride (d) zinc bromide   **21.** (a) no (b) no (c) yes (d) yes   **23.** (a) +2 (b) +2 (c) +3 (d) +2   **25.** (a) iron(II) bromide, iron(III) bromide (b) copper(I) oxide, copper(II) oxide (c) tin(II) sulfide, tin(IV) sulfide (d) nickel(II) oxide, nickel(III) oxide   **27.** (a) aluminum chloride (b) nickel(III) chloride (c) zinc oxide (d) cobalt(II) oxide   **29.** (a) plumbic oxide (b) auric chloride (c) iron(III) iodide (d) tin(II) bromide
31. (a) FeS (b) SnS_2 (c) Li_2S (d) ZnS   **33.** In b and c both names denote the same compound   **35.** (a) $CaCl_2$ (b) Na_2O (c) $FeBr_3$ (d) Cu_2S **37.** (a) $PO_4{}^{3-}$ (b) $MnO_4{}^-$ (c) $NO_3{}^-$ (d) CN^- **39.** (a) peroxide (b) thiosulfate (c) oxalate (d) chlorate
41. (a) $SO_4{}^{2-}$, $SO_3{}^{2-}$ (b) $PO_4{}^{3-}$, $HPO_4{}^{2-}$ (c) OH^-, $O_2{}^{2-}$ (d) $CrO_4{}^{2-}$, $Cr_2O_7{}^{2-}$ **43.** In b and c both compounds contain polyatomic ions   **45.** (a) fixed charge (b) fixed charge (c) variable charge (d) variable charge   **47.** (a) +2 (b) +4 (c) +3 (d) +1   **49.** (a) zinc sulfate (b) barium hydroxide (c) iron(III) nitrate (d) copper(II) carbonate   **51.** (a) iron(III) carbonate, iron(II) carbonate (b) gold(I) sulfate, gold(III) sulfate (c) tin(II) hydroxide, tin(IV) hydroxide (d) chromium(III) acetate, chromium(II) acetate   **53.** (a) ammonium nitrate (b) ammonium chloride (c) sodium phosphate (d) copper(I) phosphate   **55.** (a) Ag_2CO_3 (b) $AuNO_3$ (c) $Cr_2(SO_4)_3$ (d) $NH_4C_2H_3O_2$ **57.** (a) $Fe_2(SO_4)_3$ (b) CuCN (c) $Sn(CO_3)_2$ (d) $Pb(OH)_2$ **59.** (a) 7 (b) 5 (c) 3 (d) 10   **61.** (a) tetraphosphorus decoxide (b) sulfur tetrafluoride (c) carbon tetrabromide (d) chlorine dioxide   **63.** (a) ICl (b) NCl_3 (c) SF_6 (d) OF_2 **65.** (a) hydrogen sulfide (b) hydrogen fluoride (c) ammonia (d) methane   **67.** (a) PH_3 (b) HBr (c) C_2H_6 (d) H_2Te **69.** The least electronegative nonmetal is listed first in the chemical formula and named first in the chemical name.   **71.** It is a polyatomic-ion-containing compound whose name is sodium nitrate.   **73.** (a) no (b) yes (c) yes (d) no
75. (a) cyanide comes from hydrocyanic acid (b) sulfate comes from sulfuric acid (c) nitrite comes from nitrous acid (d) borate comes from boric acid   **77.** (a) HCN (b) H_2SO_4 (c) HNO_2 (d) H_3BO_3 **79.** (a) nitric acid (b) hydroiodic acid (c) hypochlorous acid (d) acetic acid   **81.** (a) Cl^-, chloride ion (b) $ClO_2{}^-$, chlorite ion (c) $ClO_4{}^-$, perchlorate ion (d) $SO_4{}^{2-}$, sulfate ion   **83.** (a) arsenous acid (b) periodic acid (c) hypophosphorous acid (d) bromous acid   **85.** (a) hydrogen bromide (b) hydrocyanic acid (c) hydrogen sulfide (d) hydroiodic acid   **87.** (a) $HClO_3$ (b) HNO_2 (c) HF (d) $HC_2H_3O_2$ **89.** (a) 2 (b) 1 (c) 2 (d) 2
91. (a) no, 2+ and 1+ (b) both 3+ (c) both 2+ (d) no, 3+ and 1+   **93.** (a) no (b) no (c) yes (d) no   **95.** (a) yes (b) no (c) no (d) yes   **97.** (a) 4, $3K^+$ and N^{3-} (b) 2, K^+ and N^{3-} (c) 3, 2 Na^+ and O^{2-} (d) 3, 2 Na^+ and $O_2{}^{2-}$ **99.** (a) sulfate (b) perchlorate

(c) peroxide (d) dichromate **101.** (a) Ca_3N_2 (b) $Ca(NO_3)_2$ (c) $Ca(NO_2)_2$ (d) $Ca(CN)_2$ **103.** (a) K_3P (b) K_3PO_4 (c) K_2HPO_4 (d) KH_2PO_4 **105.** (a) N_2O, CO_2 (b) NO_2, SO_2 (c) SF_2, SCl_2 (d) N_2O_3 **107.** (a) $CaCO_3$, HNO_2 (b) $NaClO_4$, $NaClO_3$ (c) $HClO_2$, $HClO$ (d) Li_2CO_3, Li_3PO_4 **109.** (a) sodium nitrate (b) aluminum sulfide, magnesium nitride, beryllium phosphide (c) iron(III) oxide (d) gold(I) chlorate **111.** (a) $Ni_2(SO_4)_3$ (b) Ni_2O_3 (c) $Ni_2(C_2O_4)_3$ (d) $Ni(NO_3)_3$ **113.** The superoxide ion is O_2^-. The nitronium ion is NO_2^+. The formula of nitronium superoxide is NO_2O_2. **115.** (a) beryllium oxide (b) magnesium chloride (c) sodium carbonate (d) ammonium sulfate **117.** (a) SO_3, H_2O, H_2SO_4 (b) $HClO_4$, Cd, $Cd(ClO_4)_2$ (c) $ZnCO_3$, ZnO, CO_2 (d) Mg, $FeCl_2$, $MgCl_2$, Fe **119.** (a) ternary molecular (b) binary molecular (c) binary ionic (d) binary molecular **121.** (a) magnesium chloride, $MgCl_2$ (b) oxygen difluoride, OF_2 **123.** (a) $x = 4$; silicon tetrachloride (b) $x = 2$; magnesium chloride (c) $x = 3$; potassium nitride (d) $x = 3$; nitrogen trichloride **125.** beryllium bromate **127.** aluminum nitride **129.** carbon dioxide **131.** beryllium cyanide

Chapter 9

1. The same: 33.4% S and 66.6% O. Any size sample of SO_2 will have that composition. **3.** According to the law of definite proportions, the percentage A in each sample should be the same and the percentage D should also be the same. Sample I: % A = 57.81%, % D = 42.2% Sample II: % A = 57.81%, % D = 42.2% **5.** The mass ratio between X and Q present in the sample will be the same for samples of the same compound. Experiments 1 and 3 produced the same compound. **7.** The maximum possible is 100 g CO. **9.** (a) $Na_4SiO_4 = 184.05$ amu (b) $H_3BO_3 = 61.84$ amu (c) $C_{12}H_{11}NO_2 = 201.24$ amu (d) $C_{22}H_{30}ClNO_2 = 375.98$ amu **11.** (a) $C_2H_4(OH)_2 = 62.08$ amu (b) $(H_2N)_2CO = 60.07$ amu (c) $Mg_3(Si_2O_5)_2(OH)_2 = 379.28$ amu (d) $Al_2Si_2O_5(OH)_2 = 224.16$ amu **13.** $y = 4$ **15.** 62.1 amu **17.** (a) 75.751% Sn, 24.25% F (b) 36.76% Fe, 21.11% S, 42.13% O (c) 42.10% C, 6.49% H, 51.41% O (d) 47.43% C, 2.56% H, 50.00% Cl **19.** (a) 79.89% Cu, 20.1% O (b) 32.37% Na, 22.57% S, 45.06% O (c) 25.9% N, 74.08% O (d) 33.4% S, 66.6% O **21.** (a) yes (b) no (c) no (d) no **23.** $C_2H_2 = 92.24\%$ C, 7.76% H; $C_6H_6 = 92.24\%$ C, 7.76% H. Both compounds have the same ratio of carbon to hydrogen. Thus, the %C and %H will be the same in each. **25.** 363.1 g N **27.** (a) 6.02×10^{23} Ag atoms (b) 6.02×10^{23} H_2O molecules (c) 6.02×10^{23} $NaNO_3$ formula units (d) 6.02×10^{23} SO_4^{2-} ions **29.** (a) 1.51×10^{24} C atoms (b) 1.96×10^{24} C atoms (c) 1.4×10^{23} C atoms (d) 1.875×10^{23} C atoms **31.** (a) 9.03×10^{23} molecules CO_2 (b) 3.01×10^{23} molecules NH_3 (c) 1.40×10^{24} molecules PF_3 (d) 6.715×10^{23} molecules N_2H_4 **33.** (a) 63.6 g Cu (b) 137 g Ba (c) 28.1 g Si (d) 238 g U **35.** (a) 78.01 g/mole $Al(OH)_3$ (b) 100.92 g/mole Mg_3N_2 (c) 187.57 g/mole $Cu(NO_3)_2$ (d) 325.82 g/mole La_2O_3 **37.** (a) 79.30 g NaCl (b) 105.9 g Na_2S (c) 115.3 g $NaNO_3$ (d) 222.5 g Na_3PO_4 **39.** (a) 2.00 moles Cu greater (b) 1.00 mole Br greater (c) 1.50 moles N_2O greater (d) 4.87 moles B_2H_6 greater **41.** 48.0 g/mole **43.** 3.16×10^{-23} g **45.** The element with an atomic mass of 30.97 amu is phosphorus. **47.** 6.022×10^{23} atoms Al

49. (a) 3.0426:1 (b) 107.018:1 (c) 6.74202:1 (d) 3.482567:1 **51.** (a) 2.99390 (b) 5.98780 (c) 1.49645 (d) 11.9756 **53.** (a) Since the two ratios are the same, the samples contain an equal number of atoms. (b) Since the ratios differ, the samples do not contain an equal number of atoms. (c) Since the ratios are the same, the samples contain an equal number of molecules. (d) Since the ratios are the same, the samples contain an equal number of particles. **55.** (a) There are more Fe atoms than Cu atoms. (b) There are fewer Ni atoms than Cu atoms. (c) The ratios are equal; there are the same number of atoms present. (d) There are more Al atoms than Cu atoms. **57.** (a) 31.06 g S (b) 8.730 g Be (c) 230.6 g U (d) 27.21 g Si **59.** 3 moles Na/1 mole Na_3PO_4; 1 mole P/1 mole Na_3PO_4; 4 moles O/1 mole Na_3PO_4; 3 moles Na/1 mole P; 3 moles Na/4 moles O; 1 mole P/4 moles O **61.** (a) Both compounds contain the same moles of S. (b) They contain different moles of S. (c) Both compounds contain the same moles of S. (d) Both compounds contain the same moles of S. **63.** (a) $NaAuBr_4 = 12.0$ moles of atoms (b) $C_2H_2Cl_4 = 8$ moles of atoms (c) $Ba(NO_3)_2 = 27.0$ moles of atoms (d) $NH_4CN = 8.40$ moles of atoms **65.** (a) 1.408×10^{23} atoms S (b) 5.012×10^{23} atoms Be (c) 3.289×10^{22} atoms Ba (d) 2.293×10^{22} atoms Au **67.** (a) 7.52×10^{23} molecules HF (b) 4.70×10^{23} molecules N_2H_4 (c) 2.35×10^{23} molecules SO_2 (d) 1.53×10^{23} molecules H_2SO_4 **69.** (a) 107.9 g Ag (b) 4.071 g N (c) 4.8×10^8 g CO_2 (d) 1.131×10^{-19} g CO **71.** (a) 3.818×10^{-23} g Na (b) 4.035×10^{-23} g Mg (c) 9.655×10^{-23} g C_4H_{10} (d) 1.297×10^{-22} g C_6H_6 **73.** (a) 60.66 g Cl (b) 903.5 g Cl (c) 9.72 g Cl (d) 17.0 g Cl **75.** (a) 1.71×10^{23} atoms P (b) 3.38×10^{23} atoms P (c) 9.82×10^{22} atoms P (d) 1.23×10^{23} atoms P **77.** (a) 5.3×10^3 g O (b) 29.98 g O (c) 168 g O (d) 196 g O **79.** (a) 0.03119 mole H_2S (b) 2.348×10^{21} molecules S_8 (c) 3.042 g $S_3N_3O_3Cl_3$ (d) 1.878×10^{22} atoms S **81.** (a) 987 g CO_2 (b) 1970 g CO_2 (c) 658 g CO_2 (d) 1970 g CO_2 **83.** (a) 4.213 moles atoms (b) 1.005×10^{24} atoms C (c) 2.544 g O (d) 4.787×10^{22} molecules $C_{21}H_{30}O_2$ **85.** (a) 310. g Fe_2S_3 (b) 15 g impurities **87.** 15.8 g **89.** 99.19% **91.** 3.063×10^{23} atoms Cu **93.** 10.9 g Cr **95.** (a) HO (b) C_4H_7 (c) C_3H_8 (d) SN **97.** (a) Na_2S (b) $KMnO_4$ (c) H_2SO_4 (d) $C_2H_3O_5N$ **99.** (a) multiply by 3; 3 to 4 (b) multiply by 2; 3 to 4 (c) multiply by 4; 7 to 8 to 9 (d) multiply by 3; 4 to 7 to 6 **101.** (a) P_2O_5 (b) Mg_3N_2 (c) $Na_2S_2O_3$ (d) $Mg_2P_2O_7$ **103.** $NiCl_2$, $NiCl_3$ **105.** C_2H_6S **107.** BeO **109.** (a) CH_4 (b) CH (c) C_2H_5 (d) CH_3 **111.** CH_2 **113.** C_5H_6O **115.** (a) P_4O_{10} (b) S_4N_4 (c) $C_3H_6O_2$ (d) $B_3N_3H_6$ **117.** (a) $C_8H_8O_2$ (b) $C_8H_8O_2$ **119.** (a) $C_3H_6O_3$ (b) $C_3H_6O_3$ **121.** (a) gold (b) sulfur (c) Cl_2 molecules (d) Ne **123.** (a) P_4 (b) 1.00 mole Na (c) Cu (d) Be **125.** (a) false (b) true (c) false (d) false **127.** 2.837 g K and 1.163 g S **129.** 37.8 g B **131.** 4.48 g $C_6H_{12}O_6$ **133.** (a) CrO_3 has the largest mass. (b) CrO_3 sample contains the most oxygen atoms. **135.** 215 g **137.** (a) 107.9 amu = Ag (b) 10.81 amu = B **139.** (a) 8.37% H (b) 64.3% H (c) 64.3% H **141.** (a) Na_3AlF_6 (b) SO_2 (c) $BaCO_3$ (d) HClO **143.** 2.43×10^{23} atoms C **145.** (a) $C_4H_6O_2$ (b) $C_6H_9O_3$ (c) $C_{12}H_{18}O_6$ (d) $C_4H_6O_2$ **147.** 5.013 g sample burned **149.** (a) CH_4S (b) 7.221 g **151.** (a) 40.0% NaF, 40.0% $NaNO_3$, and 20.0% Na_2SO_4 **153.** $x = 3$ **155.** 40.08 g/mole

157. $C_3H_8O_2$ **159.** (a) 142 (b) 141.9 (c) 141.94 (d) 141.943
161. 11.3 g/cm^3 **163.** 69.4 mL water   **165.** 22.9 cm^3 Zn
167. (a) $Mg(NO_3)_2 = 148.32$ g (b) $NaN_3 = 65.02$ g
(c) $NiF_2 = 96.69$ g (d) $NH_4ClO_4 = 117.50$ g
169. 7×10^{17} atoms ^{40}K   **171.** (a) 1.40×10^{24} formula units
$Al_2(SO_4)_3$ (b) 2.81×10^{24} Al^{+3} ions (c) 4.21×10^{24} SO_4^{2-} ions
(d) 7.02×10^{24} ions   **173.** 32.7 g KCl   **175.** (a) 50.05% S
(b) 94.07% S (c) 32.69% S (d) 39.06% S   **177.** 4.52×10^{24} Cl$^-$
ions   **179.** 6.82×10^{26} atoms Ni   **181.** 81.0 mL solution
183. 1.39×10^{24} atoms total   **185.** 6.66×10^{22} atoms Cu
(The 1 cent is a counted number)

Chapter 10

1. $x = 14.33$ g   **3.** 2.2 g CO_2 **5.** (a) yes (b) no, Cl_2 (c) no, He
(d) yes   **7.** (a) (s) means solid, and (g) means gas (b) (g) means
gas, (l) means liquid, and (aq) means water (aqueous) solution
9. (a) balanced (b) balanced (c) unbalanced (d) balanced
11. (a) 4 N atoms and 6 O atoms on each side (b) 10 N atoms, 12
H atoms, and 6 O atoms on each side (c) 1 P atom, 3 Cl atoms,
and 6 H atoms on each side (d) 2 Al atoms, 3 O atoms, 6 H atoms,
and 6 Cl atoms on each side   **13.** (a) $2 Cu + O_2 \rightarrow 2 CuO$
(b) $2 H_2O \rightarrow 2 H_2 + O_2$ (c) $BaCl_2 + Na_2S \rightarrow BaS + 2 NaCl$
(d) $Mg + 2 HBr \rightarrow MgBr_2 + H_2$
15. (a) $3 PbO + 2 NH_3 \rightarrow 3 Pb + N_2 + 3 H_2O$
(b) $2 NaHCO_3 + H_2SO_4 \rightarrow Na_2SO_4 + 2 CO_2 + 2 H_2O$
(c) $TiO_2 + C + 2 Cl_2 \rightarrow TiCl_4 + CO_2$
(d) $2 NBr_3 + 3 NaOH \rightarrow N_2 + 3 NaBr + 3 HBrO$
17. (a) $CH_4 + 2 O_2 \rightarrow CO_2 + 2 H_2O$
(b) $2 C_6H_6 + 15 O_2 \rightarrow 12 CO_2 + 6 H_2O$
(c) $C_4H_8O_2 + 5 O_2 \rightarrow 4 CO_2 + 4 H_2O$
(d) $C_5H_{10}O + 7 O_2 \rightarrow 5 CO_2 + 5 H_2O$
19. (a) $Ca(OH)_2 + 2 HNO_3 \rightarrow Ca(NO_3)_2 + 2 H_2O$
(b) $BaCl_2 + (NH_4)_2SO_4 \rightarrow BaSO_4 + 2 NH_4Cl$
(c) $2 Fe(OH)_3 + 3 H_2SO_4 \rightarrow Fe_2(SO_4)_3 + 6 H_2O$
(d) $Na_3PO_4 + 3 AgNO_3 \rightarrow 3 NaNO_3 + Ag_3PO_4$
21. (a) $AgNO_3 + KCl \rightarrow AgCl + KNO_3$
(b) $CS_2 + 3 O_2 \rightarrow CO_2 + 2 SO_2$ (c) $2 H_2 + O_2 \rightarrow 2 H_2O$
(d) $2 Ag_2CO_3 \rightarrow 4 Ag + 2 CO_2 + O_2$ **23.** (a) synthesis
(b) synthesis (c) double-replacement (d) single-replacement
25. (a) $Zn + Cu(NO_3)_2 \rightarrow Zn(NO_3)_2 + Cu$
(b) $2 Ca + O_2 \rightarrow 2 CaO$
(c) $K_2SO_4 + Ba(NO_3)_2 \rightarrow BaSO_4 + 2 KNO_3$
(d) $2 Ag_2O \rightarrow 4 Ag + O_2$
27. (a) $K_2CO_3 \rightarrow K_2O + CO_2$ (b) $CaCO_3 \rightarrow CaO + CO_2$
(c) $NiCO_3 \rightarrow NiO + CO_2$ (d) $Fe_2(CO_3)_3 \rightarrow Fe_2O_3 + 3 CO_2$
29. (a) $CH_4 + 2 O_2 \rightarrow CO_2 + 2 H_2O$
(b) $2 C_6H_6 + 15 O_2 \rightarrow 12 CO_2 + 6 H_2O$
(c) $C_6H_{12} + 9 O_2 \rightarrow 6 CO_2 + 6 H_2O$
(d) $C_3H_4 + 4 O_2 \rightarrow 3 CO_2 + 2 H_2O$
31. (a) $CH_2O + O_2 \rightarrow CO_2 + H_2O$
(b) $C_3H_6O + 4 O_2 \rightarrow 3 CO_2 + 3 H_2O$
(c) $2 CH_2O_2 + O_2 \rightarrow 2 CO_2 + 2 H_2O$

(d) $C_4H_8O_2 + 5 O_2 \rightarrow 4 CO_2 + 4 H_2O$
33. (a) $4 C_2H_7N + 19 O_2 \rightarrow 8 CO_2 + 14 H_2O + 4 NO_2$
(b) $CH_4S + 3 O_2 \rightarrow CO_2 + 2 H_2O + SO_2$ **35.** (a) combination,
single-replacement, combustion (b) decomposition, single-
replacement (c) combination, decomposition, single-replacement,
double-replacement, combustion (d) combination, decomposition,
single-replacement, double-replacement, combustion
37. (a) 2 molecules B (b) 9 molecules A (c) 6 molecules D
(d) 2 moles B   **39.** 4 moles NH_3/3 moles O_2; 3 moles
O_2/4 moles NH_3; 4 moles NH_3/2 moles N_2; 2 moles
N_2/4 moles NH_3; 4 moles NH_3/6 moles H_2O; 6 moles
H_2O/4 moles NH_3; 3 moles O_2/2 moles N_2; 2 moles
N_2/3 moles O_2; 3 moles O_2/6 moles H_2O; 6 moles
H_2O/3 moles O_2; 2 moles N_2/6 moles H_2O; 6 moles
H_2O/2 moles N_2 **41.** (a) 2.00 moles NaN_3 (b) 9.00 moles CO
(c) 6.00 moles NH_2Cl (d) 2.00 moles $C_3H_5O_9N_3$ **43.** (a) 0.129
mole C_7H_{16} (b) 2.84 moles HCl (c) 0.710 mole Na_2SO_4 (d) 5.68
moles Na_2CO_3 **45.** (a) 6.12 moles products (b) 4.38 moles
products (c) 2.62 moles products (d) 4.81 moles products
47. (a) 19.0 moles Cl_2 (b) 0.33 mole HCl (c) 0.575 mole CH_4
(d) 0.308 mole CCl_4 **49.** (a) 8.20 g C (b) 4.80 g B
51. (a) 96.12 g reactants, 96.12 g products (b) 299.65 g reactants,
299.65 g products (c) 68.10 g reactants, 68.10 g products
(d) 97.22 g reactants, 97.22 g products   **53.** (a) 336 g O_2
(b) 112 g O_2 (c) 112 g O_2 (d) 56.0 g O_2 **55.** (a) 126 g HNO_3
(b) 126 g HNO_3 (c) 31.5 g HNO_3 (d) 210. g HNO_3
57. (a) 1.062 g C (b) 0.4885 g C_3H_8 (c) 31.31 g Cl_2
(d) 0.2313 g H_2O **59.** (a) 30.0 g SiO_2 (b) 76.7 g CO (c) 66.8 g
SiC (d) 20.7 g C   **61.** (a) 216 g LiOH (b) 239 g LiOH (c) 26.6 g
LiOH (d) 6.48 g LiOH   **63.** (a) 0.2977 mole Na_2SiO_3 (b) 79.95 g
HF (c) 9.867×10^{21} molecules H_2SiF_6 (d) 65.57 g HF
65. 23.7 g Na   **67.** 65.68 g Cr and 134.3 g Cl_2 **69.** 284 bolts
71. 213 kits   **73.** (a) 3.65 moles H_2 (b) 2.60 moles N_2
(c) 975 molecules H_2 (d) 3.00 moles H_2 **75.** (a) 13.8 g Mg_3N_2
(b) 27.7 g Mg_3N_2 (c) 36.0 g Mg_3N_2 (d) 36.0 g Mg_3N_2
77. 175 $CoCl_3$ formula units   **79.** 0 g Fe_3O_4, 9.6 g O_2
81. 2.57 g SF_4, 3.21 g S_2Cl_2, 5.56 g NaCl   **83.** 30.8%
85. (a) 209 g Al_2S_3 (b) 59.8%   **87.** 96.44%   **89.** 131 g
91. 31.8 g CO_2 **93.** 54.6 g SO_2 **95.** 293 g O_2
97. (a) 4.00 moles $NaClO_3$ (b) 22.2 g $NaClO_3$ **99.** (a) 4.00
moles HNO_3 (b) 9.00 g HNO_3 **101.** 6.69 g $AgNO_3$
103. (a) 6.00 moles O_2 (b) 9.00 moles H_2O (c) 1.00 mole H_3PO_3
(d) 1.50 moles Cl_2 **105.** 8.34 g N_2, 21.4 g H_2O, 45.2 g Cr_2O_3
107. 41.5 g H_2S **109.** (a) B is limiting (b) A is limiting (c) B is
limiting (d) A is limiting   **111.** (a) B (b) C (c) D
(d) B = 60.0 g, C = 135 g, D = 37.5 g   **113.** 80.1 g NO
115. 44.4%   **117.** 92.8%   **119.** 75.4%   **121.** 0.03677 ton ore
123. 38.3%   **125.** (a) $Zn + 2 AgNO_3 \rightarrow Zn(NO_3)_2 + 2 Ag$
(b) $HCl + NaOH \rightarrow NaCl + H_2O$ (c) $PCl_3 + Cl_2 \rightarrow PCl_5$
(d) $2 Cu + O_2 \rightarrow 2 CuO$ **127.** C_3H_6 **129.** The formula of the
copper oxide is Cu_2O; $2 Cu_2S + 3 O_2 \rightarrow 2 Cu_2O + 2 SO_2$
131. Molecular formula = C_6H_6; $2 C_6H_6 + 15 O_2 \rightarrow$
$12 CO_2 + 6 H_2O$ **133.** (a) 260.88 g BeF_2 (b) 0.26088 kg BeF_2
(c) 2.609×10^8 g BeF_2 (d) 0.5752 lb BeF_2 **135.** 2.040 moles
chlorine-containing products   **137.** 7.91 moles electrons
139. 8.139×10^{24} positive ions   **141.** 87.3 mL solution
143. 0.0788 g HNO_3 **145.** 0.16 ton $CaSO_3$

Chapter 11

1. (a) gaseous (b) liquid (c) gaseous (d) gaseous **3.** (a) both gases (b) liquid and solid (c) both gases (d) both solids **5.** (a) potential (b) kinetic (c) potential (d) potential **7.** (a) average velocity increases with increased temperature and vice versa (b) potential (attractive) (c) direct; higher temperature, higher disruptive forces (d) all three **9.** (a) solid (b) liquid (c) solid (d) solid or liquid **11.** (a) The predominant cohesive forces in the solid hold the particles in essentially fixed position. (b) The gas particles are widely separated (disruptive forces). The solid and liquid particles have very little space between them (cohesive forces). The space between the particles can be decreased greatly in gases, but not in solids or liquids. (c) The cohesive forces are dominant enough that changing the temperature has only a small effect on the space between particles. (d) The disruptive forces in a gas are so dominant that each particle can act independently of the others. **13.** (a) exothermic, endothermic (b) both endothermic (c) both exothermic (d) exothermic, endothermic **15.** (a) liquid, solid (b) both gases (c) liquid, solid (d) solid, gas **17.** (a) opposite changes (b) not opposite (c) not opposite (d) opposite changes **19.** (a) 2.0 calories (b) 1.0 kilocalorie (c) 100 Calories (d) 1000 kilocalories **21.** (a) 2.29×10^6 J (b) 547 kcal (c) 5.47×10^5 cal (d) 547 Cal **23.** 112 J **25.** 47.7°C **27.** 1.58 g **29.** 0.39 J/g°C **31.** 80.0 g copper **33.** 0.382 J/g°C **35.** 47 g **37.** 30.9 kJ/mole **39.** (a) heat of solidification (b) heat of condensation (c) heat of fusion (d) heat of vaporization **41.** (a) 1.96×10^4 J released (b) 1.13×10^5 J released (c) 1.02×10^4 J released (d) 1.13×10^5 J released **43.** 6680 J **45.** 5.125 times as much **47.** 3.83 moles CCl_4 **49.** A has the higher heat of fusion by 6 J/g **51.** 7.30×10^2 J

53.

55. (a) 2.4×10^3 J (b) 3.59×10^4 J (c) 2.30×10^5 J (d) 2.32×10^5 J **57.** (a) 8.32×10^3 J (b) 2.15×10^4 J (c) 8.03×10^3 J (d) 8.50×10^3 J **59.** 2.15×10^4 J **61.** (a) boiling point (b) vapor pressure (c) boiling (d) boiling point **63.** (a) Increasing the temperature increases the average kinetic energy of the particles, enabling more particles to evaporate. (b) The boiling point is lower; reactions occur more slowly at lower temperatures. (c) The boiling point is higher; reactions occur more rapidly at higher temperatures. (d) The particles leaving have higher than average kinetic energy reducing the average energy of particles of the liquid as they leave. **65.** (a) increase (b) no change (c) increase (d) no change **67.** (a) no change (b) decrease (c) no change (d) no change **69.** (a) increase (b) no change (c) no change (d) no change **71.** B must have lower cohesive forces between particles than A to make B evaporate faster. **73.** At the same temperature, the substance with the higher vapor pressure is more volatile. Thus,

CS_2 is more volatile. **75.** Polar molecules must be present. **77.** The stronger the intermolecular forces, the higher the boiling point. **79.** (a) London forces (b) hydrogen bonds (c) dipole–dipole interactions (d) London forces **81.** (a) no (b) yes (c) yes (d) no **83.** (a) Cl_2, larger mass (b) HF, hydrogen bonding (c) NO, dipole–dipole (d) C_2H_6, larger size **85.** (a) SiH_4 (b) $SiCl_4$ (c) $GeBr_4$ (d) C_2H_4 **87.** (a) true (b) false (c) true (d) false **89.** (a) NaCl, ionic solid vs. nonpolar molecule (b) SiO_2, macromolecular vs. nonpolar molecular (c) Cu, metallic vs. polar molecular and Cu is larger (d) MgO, higher ionic charges vs. Na^+ and F^- **91.** (a) $SnCl_4$ (b) SnI_4 (c) SnI_4 (d) SnI_4 **93.** Vaporizing 50.0 g water **95.** 9.52 g **97.** 38.3°C **99.** 0.421 J/g°C **101.** 1.4×10^3 J/g **103.** 6.5×10^2 J **105.** 25 kJ **107.** The unknown is likely a mixture of A and B. **109.** 0.775 g/mL **111.** 30.1 g/mole **113.** 9.45 kJ **115.** 2.38×10^3 J **117.** 3.1×10^7 gal/day

Chapter 12

1. (a) 4710 mm Hg (b) 186 in. Hg (c) 91.0 psi (d) 471 cm Hg **3.** (a) smaller (b) equal (c) larger (d) equal **5.** 999 mm Hg **7.** (a) decrease (b) increase (c) decrease (d) decrease **9.** (a) 1.28 L (b) 5.53 L (c) 4.00 L (d) 81.8 L **11.** (a) 2.6×10^2 mL (b) 16 mL (c) 6.8×10^2 mL (d) 2.9×10^3 mL **13.** 1.94×10^3 mm Hg **15.** (a) increase (b) decrease (c) increase (d) decrease **17.** (a) 6.46 L (b) 4.51 L (c) 4.35 L (d) 20.5 L **19.** (a) 19.3 mL (b) 1.11×10^3 mL (c) 1.52×10^3 mL (d) 1.36×10^4 mL **21.** −125°C **23.** (a) increase (b) decrease (c) increase (d) decrease **25.** (a) 1.34 atm (b) 1.68 atm (c) 2.36 atm (d) 3.37 atm **27.** (a) 3.70 atm (b) 1.23 atm (c) 2.26 atm (d) 0.608 atm **29.** 3.1 atm

31. (a) $T_1 \times \dfrac{P_2}{P_1} \times \dfrac{V_2}{V_1}$ (b) $\dfrac{V_2}{P_1} = \dfrac{V_1}{P_2} \times \dfrac{T_2}{T_1}$

33. (a) 1.78 mL (b) 0.900 mL (c) 0.348 mL (d) 0.153 mL **35.** (a) 5.90 L (b) 3.70×10^3 mL (c) 2.11 atm (d) −171°C **37.** (a) 235 mm Hg (b) 727°C **39.** (a) 2.48×10^3°C (b) −196.7°C (c) 186°C (d) 33°C **41.** (a) 3.15 L (b) 10.5 L (c) 3.64 L (d) 2.63 L **43.** (a) 0.953 L (b) 0.678 L (c) 1.89 L (d) 0.995 L **45.** $4\,NH_3(g) + 5\,O_2(g) \rightarrow 4\,NO(g) + 6\,H_2O(g)$ **47.** (a) 0.433 L C_3H_8 (b) 0.325 L C_3H_8 **49.** 0.11 L C_3H_8 **51.** (a) 1.50 L H_2 (b) 5.50 L H_2 (c) 1.00 L N_2 (d) 1.00 L H_2 **53.** (a) 22.5 L CO_2, 90.0 L H_2 (b) 33.0 L CO_2, 132 L H_2 (c) 64.0 L CO_2, 256 L H_2 (d) 16.0 L CO_2, 64.0 L H_2 **55.** 0.781 mole **57.** 2.80 L CO **59.** 1.5 g He **61.** 1.33 L **63.** (a) 0.450 mole O_2 (b) 2.32 moles CH_4 (c) 100.0 g N_2O (d) 100 g CO **65.** NF_3 **67.** 2.6 L **69.** (a) 28.0 L N_2 (b) 28.0 L NH_3 (c) 28.0 L CO_2 (d) 28.0 L Cl_2 **71.** (a) NH_3 (b) O_2 (c) NO_2 (d) NO **73.** (a) 42.2 g Ar (b) 46.6 g N_2O (c) 84.7 g SO_3 (d) 36.0 g PH_3 **75.** (a) 1.96 g/L (b) 4.11 g/L (c) 2.05 g/L (d) 1.34 g/L **77.** (a) O_3 (b) PH_3 (c) CO_2 (d) F_2 **79.** (a) 44.1 g/mole (b) 28.0 g/mole (c) 16.0 g/mole (d) 71.0 g/mole **81.** 2.50 L Xe **83.** (a) 26.9 L (b) 22.1 L (c) 5.24 L (d) 24.5 L **85.** 0.900 mole **87.** −2°C **89.** (a) 1.2 atm (b) 4.89 atm (c) 0.044 atm (d) 0.175 atm **91.** 0.0824 atm L/mole K **93.** 36.1 L **95.** (a) 29.5 g (b) 3.19 g (c) 15.7 g (d) 0.363 g **97.** 44 g Cl_2 added **99.** 28.0 g **101.** 103 g **103.** 90.0 g

105. CO gas   **107.** (a) 1.24 g/L (b) 1.68 g/L (c) 0.840 g/L
(d) 0.996 g/L   **109.** (a) 0.937 atm (b) 0.200 atm (c) 0.482 atm
(d) 0.597 atm   **111.** CO gas   **113.** 18.7 L NO_2 **115.** 61.1 g NO
117. 12.2 L   **119.** 259 L   **121.** 78.0 L   **123.** 7.39 L
125. (a) 5.51 L (b) 6.11 L (c) 76.8 L (d) 197 L   **127.** 8.01 atm
129. 9.0 atm for He, 5.0 atm for Ne, 15.0 atm for Ar
131. (a) 167 mm Hg (b) 354 mm Hg (c) 235 mm Hg (d) 86 mm
Hg   **133.** (a) 0.407 mole fraction CO, 0.259 mole fraction CO_2,
0.334 mole fraction H_2S (b) 0.700 atm P_{CO}, 0.445 atm P_{CO_2},
0.574 atm P_{H_2S} **135.** (a) 4.8 atm (b) 4.8 atm (c) 4.8 atm
(d) 0.015 mole O_2 **137.** (a) 0.60 atm (b) 0.40 atm (c) 0.400 atm
(d) 0.400 atm   **139.** 5.2 atm   **141.** (a) 5.10 atm (b) 10.3 atm
143. 0.455 atm for He, 0.227 atm for Ar, 0.318 atm for Xe
145. 1.28 atm   **147.** P_{N_2} = 213 mm Hg, P_{H_2} = 639 mm Hg
149. (a) 726 mm Hg (b) 617 mm Hg (c) 722 mm Hg (d) 690. mm
Hg   **151.** (a) 0.0909 (b) 72.7% Ar (c) 18.2% Ne (d) 9.09% He
153. (a) 25% O_2 (b) 37% Ar (c) 37% Ne (d) 0.25 atm
155. 2.69×10^{22} molecules CO_2 **157.** 3.3×10^{13} molecules O_2
159. 3.48 L NH_3 **161.** 0.14 g O_2 **163.** 43.8 mL   **165.** 14 mL
167. 25%   **169.** 50.0%   **171.** (a) 1.14:1 (b) 1.14:1
173. P_T = 1200.0 mm Hg   **175.** 1.3 atm   **177.** 2.5 atm
179. 1240 mm Hg   **181.** Steam   **183.** 75 amu   **185.** $x = 3$
187. (a) X_{HCl} = 0.216, X_{H_2} = 0.260, X_{He} = 0.524
(b) P_{HCl} = 0.259 atm, P_{H_2} = 0.312 atm, P_{He} = 0.629 atm
(c) 10.50 g/mole (d) 0.4685 g/L   **189.** C_4H_{10} **191.** $C_2H_4Cl_2$,
101 g/mole   **193.** 120.0 L products total   **195.** 1.06 g H_2O
197. 0.0868 m^3 HCl   **199.** 2.0 atm   **201.** 63.3%
203. 0.0147 cent/g

Chapter 13

1. (a) true (b) true (c) true (d) false   **3.** (a) solute (sodium
chloride), solvent (water)   (b) solute (sucrose), solvent (water)
(c) solute (water), solvent (ethyl alcohol)   (d) solute (ethyl
alcohol), solvent (methyl alcohol)   **5.** (a) unsaturated (b)
supersaturated (c) saturated (d) unsaturated   **7.** (a) soluble
(b) very soluble (c) very soluble (d) slightly soluble
9. (a) concentrated (b) concentrated (c) concentrated (d) dilute
11. (a) first solution (b) first solution (c) first solution (d) second
solution   **13.** (a) none (b) 44 g   **15.** (a) hydrated ion
(b) hydrated ion (c) oxygen atom (d) hydrogen atom
17. (a) decrease (b) increase (c) increase (d) increase
19. (a) very soluble (b) slightly soluble (c) slightly soluble
(d) very soluble   **21.** (a) generally soluble (b) generally soluble
(c) generally insoluble (d) generally soluble   **23.** (a) carbonate
rule (b) sulfide rule (c) ammonium ion rule (d) sulfate rule
25. (a) soluble with exceptions (b) soluble (c) insoluble with
exceptions (d) soluble   **27.** (a) both soluble (b) soluble,
insoluble (c) both soluble (d) both insoluble   **29.** (a) no,
$MgSO_4$ soluble (b) yes (c) no, both soluble (d) no, both soluble
31. (a) 7.03% NaI (b) 1.81% NaI (c) 19.4% NaI (d) 4.3% NaI
33. (a) 0.700 g NaCl (b) 4.38 g $AgNO_3$ (c) 135.5 g K_2SO_4
(d) 3.57 g HCl   **35.** (a) 5.18 g H_2O (b) 51.8 g H_2O
(c) 56.9 g H_2O (d) 2.3 g H_2O **37.** (a) 9.50×10^2 g H_2O
(b) 9.50×10^2 g H_2O (c) 9.50×10^2 g H_2O
(d) 9.50×10^2 g H_2O **39.** (a) 79.1% ethyl alcohol (b) 38.1%
ethyl alcohol (c) 20.9% ethyl alcohol (d) 5.14% ethyl alcohol

41. (a) 36.06% methyl alcohol (b) 66.72% H_2O **43.** 11.2 mL
H_2O_2 **45.** 1.40 gal H_2O **47.** (a) 2.67% (b) 1.20% (c) 3.32%
(d) 6.6%   **49.** (a) 750. mL solution (b) 33 mL   **51.** (a) 11.4 g
Na_3PO_4 (b) 3.75×10^3 g Na_3PO_4 **53.** 24%   **55.** (a) 1.79 ppm
(b) 170. ppm (c) 0.312 ppm (d) 2270 ppm   **57.** (a) 1790 ppb
(b) $17\overline{0},000$ ppb (c) 312 ppb (d) 2,270,000 ppb   **59.** Yes
61. 0.00044 mL   **63.** (a) 2.6×10^{-3} g NH_3
(b) 5.4×10^{-6} g NH_3 (c) 8.7 g NH_3 **65.** (a) 4.0 M (b) 3.81 M
(c) 1.06 M (d) 0.40 M   **67.** (a) 13.2 g HNO_3 (b) 0.378 g HNO_3
(c) 2.36×10^4 g HNO_3 (d) 150. g HNO_3 **69.** (a) 33.8 mL
(b) 226 mL   (c) 1.80×10^3 mL   (d) 1.52 mL   **71.** (a) 0.877 L
(b) 0.725 L (c) 0.385 L (d) 0.333 L   **73.** 22.72 M   **75.** 17.96%
77. 3.75 M   **79.** (a) 0.0357 m (b) 6.92 m (c) 0.00800 m
(d) 0.449 m **81.** (a) 10.7 g $Al(NO_3)_3$ (b) 4.76 g $MgCl_2$
(c) 8.20 g Na_3PO_4 (d) 8.71 g K_2SO_4 **83.** (a) 2.7×10^4 g H_2O
(b) 6.0×10^3 g H_2O (c) 1.02×10^3 g H_2O
(d) 4.9×10^2 g H_2O **85.** 0.584 m **87.** 0.762 M   **89.** 2.58 m
91. (a) 0.200 M   (b) 0.120 M   (c) 0.0267 M   (d) 0.00375 M
93. (a) 0.601 M (b) 0.638 M (c) 0.694 M (d) 2.27 M   **95.**
(a) 25 mL (b) 1.1×10^3 mL (c) 2.7×10^2 mL (d) 0.050 mL
97. (a) 380. mL H_2O (b) 30.0 mL H_2O (c) 3040 mL H_2O
(d) 1.4×10^5 mL H_2O **99.** (a) 3.3 M (b) 2.5 M (c) 8.86 M
(d) 0.00909 M   **101.** (a) 0.468 M (b) 6.00 M (c) 4.75 M
(d) 5.99 M   **103.** 1.00 L   **105.** 17.3 g S   **107.** 610 mL
109. 0.0632 M   **111.** 1.7 L NO   **113.** 0.0510 M
115. (a) $(NH_4)_3PO_4$ (b) $Ca(OH)_2$ (c) $AgNO_3$ (d) CaS,
$Ca(NO_3)_2$, $Ca(C_2H_3O_2)_2$ **117.** 80. g   **119.** (a) 17.8% (m/v)
(b) 1.05 M   **121.** 6.72 g Na_2SO_4 **123.** 1245 g H_2O
125. (a) 563 mL (b) 435 mL (c) 4540 mL (d) 896 mL   **127.**
(a) 0.245 M Al^{3+}, 0.735 M NO_3^- (b) 0.0735 M $Al(NO_3)_3$,
0.0735 M Al^{3+}, 0.221 M NO_3^- **129.** (a) 0.29 M K^+ (b) 0.027
M Cl^- (c) 0.0866 M PO_4^{3-} **131.** 3.74 mg solute/kg solution
133. 0.171 M   **135.** 8.06 mL   **137.** 340 g H_2O **139.** 298 g
solution, 266 mL solution   **141.** (a) 11.9% (m/v) (b) 12.2% (m/m)
(c) 15.1% (v/v)   **143.** 1.01 g/mL   **145.** (a) AgCl is insoluble
(b) $Ba_3(PO_4)_2$ is insoluble (c) $PbSO_4$ is insoluble (d) $BaSO_4$
and CuS are insoluble   **147.** 117 L   **149.** 2.32 g AgCl
151. 289 g $BaCrO_4$ **153.** 37.1% (m/m)   **155.** zinc
157. 0.0833 M

Chapter 14

1. (a) H^+ ion (b) dissociation   **3.** (a) Arrhenius acid
(b) Arrhenius acid   **5.** (a) $HBr \rightarrow H^+ + Br^-$
(b) $HClO_2 \rightarrow H^+ + ClO_2^-$ (c) $LiOH \rightarrow Li^+ + OH^-$
(d) $Ba(OH)_2 \rightarrow Ba^{2+} + 2\, OH^-$ **7.** (a) acid, NH_4^+ donates a
proton (b) base, HS^- accepts a proton (c) acid, $HClO_4$ donates a
proton (d) base, H_2O accepts a proton
9. (a) $HBr + H_2O \rightarrow H_3O^+ + Br^-$
(b) $H_2O + N_3^- \rightarrow HN_3 + OH^-$
(c) $H_2S + H_2O \rightarrow H_3O^+ + HS^-$
(d) $HClO_4 + NO_2^- \rightarrow HNO_2 + ClO_4^-$ **11.** (a) HSO_3^-
(b) HCN (c) HS^- (d) ClO^- **13.** (a) $H_2C_2O_4$ and $HC_2O_4^-$;
HClO and ClO^- (b) HSO_4^- and SO_4^{2-}; H_3O^+ and H_2O
(c) $H_2PO_4^-$ and HPO_4^{2-}; NH_4^+ and NH_3 (d) H_2CO_3 and
HCO_3^-; H_2O and OH^- **15.** (a) yes (b) no (c) no (d) yes
17. (a) (1) $HS^- + H_3O^+ \rightarrow H_2S + H_2O$

(2) $HS^- + OH^- \rightarrow H_2O + S^{2-}$

(b) (1) $HPO_4^{2-} + H_3O^+ \rightarrow H_2PO_4^- + H_2O$

(2) $HPO_4^{2-} + H_3O^+ \rightarrow H_2O + PO_4^{3-}$

(c) (1) $HCO_3^- + H_3O^+ \rightarrow H_2CO_3 + H_2O$

(2) $HCO_3^- + OH^- \rightarrow H_2O + CO_3^{2-}$

(d) (1) $H_2PO_3^- + H_3O^+ \rightarrow H_3PO_3 + H_2O$

(2) $H_2PO_3^- + OH^- \rightarrow H_2O + HPO_3^{2-}$

19. (a) monoprotic (b) diprotic (c) monoprotic (d) diprotic
21. (a) one acidic, zero nonacidic (b) two acidic, four nonacidic
(c) one acidic, seven nonacidic (d) zero acidic, four nonacidic
23. (a) $H_2C_4H_4O_4 + H_2O \rightarrow H_3O^+ + HC_4H_4O_4^-$
$HC_4H_4O_4^- + H_2O \rightarrow H_3O^+ + C_4H_4O_4^{2-}$
(b) $H_2SO_3 + H_2O \rightarrow H_3O^+ + HSO_3^-$
$HSO_3^- + H_2O \rightarrow H_3O^+ + SO_3^{2-}$ **25.** to emphasize the acidic
hydrogen as being different from the nonacidic ones
27. Monoprotic. Hydrogens attached to carbon atoms are not
acidic.   **29.** (a) $HClO_3$, strong (b) $H_2C_2O_4$, weak
(c) $HC_3H_3O_3$, weak (d) $H_2C_3H_2O_4$, weak   **31.** (a) strong, weak
(b) strong, weak (c) weak, strong (d) both weak   **33.** (a) strong,
strong (b) strong, strong (c) weak, strong (d) weak, strong
35. (a) HNO_3 (b) $HClO_4$ (c) H_3PO_4 (d) HF   **37.** (a) salt
(b) base (c) salt (d) acid   **39.** (a) Na^+, sodium; PO_4^{3-}, phosphate
(b) Li^+, lithium; NO_3^-, nitrate (c) NH_4^+, ammonium;
Cl^-, chloride (d) K^+, potassium; CN^-, cyanide   **41.** Each salt
is soluble in water.   **43.** (a) $NaI \rightarrow Na^+(aq) + I^-(aq)$
(b) $BaS \rightarrow Ba^{2+}(aq) + S^{2-}(aq)$
(c) $Li_2SO_4 \rightarrow 2\,Li^+(aq) + SO_4^{2-}(aq)$
(d) $Al(NO_3)_3 \rightarrow Al^{3+}(aq) + 3\,NO_3^-(aq)$
45. (a) molecular (b) net ionic (c) ionic (d) net ionic
47. (a) $Pb^{2+} + 2\,Br^- \rightarrow PbBr_2$ (b) $Fe^{3+} + 3\,OH^- \rightarrow Fe(OH)_3$
(c) $Zn + 2\,H^+ \rightarrow Zn^{2+} + H_2$
(d) $H_2S + 2\,OH^- \rightarrow S^{2-} + 2\,H_2O$
49. (a) $2\,Ag^+ + Pb \rightarrow 2\,Ag + Pb^{2+}$
(b) $Cl_2 + 2\,Br^- \rightarrow 2\,Cl^- + Br_2$ (c) $2\,Al^{3+} + 3\,S^{2-} \rightarrow Al_2S_3$
(d) (no net ionic reaction–dissociation only)   **51.** (a) yes
(b) yes (c) no (d) yes   **53.** (a) $Ni + 2\,HCl \rightarrow NiCl_2 + H_2$
(b) $Ca + 2\,H_2O \rightarrow Ca(OH)_2 + H_2$
(c) $Mg + 2\,HCl \rightarrow MgCl_2 + H_2$
(d) $Zn + 2\,H_2O \rightarrow Zn(OH)_2 + H_2$
55. (a) no (b) yes (c) yes (d) no   **57.** (a) 1 to 1 (b) 1 to 2
(c) 1 to 1 (d) 2 to 1   **59.** (a) $2\,HBr + Sr(OH)_2 \rightarrow SrBr_2 + 2\,H_2O$
(b) $HC_2H_3O_2 + LiOH \rightarrow LiC_2H_3O_2 + H_2O$
(c) $H_2SO_4 + Mg(OH)_2 \rightarrow MgSO_4 + 2\,H_2O$
(d) $H_3PO_4 + 3\,KOH \rightarrow K_3PO_4 + 3\,H_2O$
61. (a) $H^+ + OH^- \rightarrow H_2O$
(b) $HC_2H_3O_2 + OH^- \rightarrow C_2H_3O_2^- + H_2O$
(c) $H^+ + OH^- \rightarrow H_2O$
(d) $H_3PO_4 + 3\,OH^- \rightarrow PO_4^{3-} + 3\,H_2O$
63. (a) H_3PO_4, NaOH (b) HCN, KOH
(c) HCl, $Be(OH)_2$ (d) $HC_2H_3O_2$, $Ca(OH)_2$
65. (a) $Zn + 2\,HCl \rightarrow ZnCl_2 + H_2$

(b) $HCl + NaOH \rightarrow NaCl + H_2O$

(c) $2\,HCl + Na_2CO_3 \rightarrow 2\,NaCl + CO_2 + H_2O$

(d) $HCl + NaHCO_3 \rightarrow NaCl + CO_2 + H_2O$

67. (a) no (b) yes (c) no (d) yes
69. (a) $Fe + Cu^{2+} \rightarrow Fe^{2+} + Cu$

(b) $Sn + 2\,Ag^+ \rightarrow Sn^{2+} + 2\,Ag$

(c) $Zn + Ni^{2+} \rightarrow Zn^{2+} + Ni$

(d) $Cr + Pb^{2+} \rightarrow Cr^{2+} + Pb$

71. (a) an insoluble salt is formed (b) an insoluble salt is formed
(c) an insoluble salt is formed, weak acid is formed (d) a gas is
evolved   **73.** (a) $2\,Al^{3+} + 3\,S^{2-} \rightarrow Al_2S_3$
(b) $H^+ + OH^- \rightarrow H_2O$ (c) no reaction (d) no reaction
75. (a) 5.0×10^{-11} M   (b) 1.4×10^{-8} M (c) 3.3×10^{-5} M
(d) 4.0×10^{-7} M   **77.** (a) basic (b) basic (c) acidic (d) acidic
79. (a) 3.7×10^{-11} M (b) 1.3×10^{-6} M (c) 1.0×10^{-7} M
(d) 2.0×10^{-7} M   **81.** (a) acidic (b) basic (c) neutral (d) basic
83. (a) pH = 4.0 (b) pH = 9.0 (c) pH = 5.0 (d) pH = 9.0
85. (a) 1.4 (b) 3.2 (c) 9.1 (d) 6.3   **87.** (a) 2.5 (b) 2.52
(c) 2.523 (d) 2.5229 **89.** (a) both acidic (b) acidic, basic
(c) both basic   (d) acidic, neutral   **91.** (a) 1×10^{-3} M
(b) 1×10^{-5} M   (c) 2×10^{-6} M (d) 5×10^{-7} M
93. (a) 3.7×10^{-3} M (b) 3.7×10^{-4} M (c) 3.7×10^{-8} M
(d) 3.5×10^{-8} M   **95.** (a) Solution A (b) Solution B
97. (a) 4.199 (b) 3.900 (c) 3.500 (d) 1.500   **99.** (a) 2.20
(b) 0.70 (c) 4.38 (d) 10.37   **101.** (a) PO_4^{3-} (b) CN^-
(c) NH_4^+ (d) none   **103.** (a) neutral (b) basic (c) acidic
(d) neutral   **105.** (a) $NH_4^+ + H_2O \rightarrow H_3O^+ + NH_3$
(b) $C_2H_3O_2^- + H_2O \rightarrow OH^- + HC_2H_3O_2$
(c) $F^- + H_2O \rightarrow OH^- + HF$
(d) $CN^- + H_2O \rightarrow OH^- + HCN$ **107.** (a) no   (b) yes
(c) no   (d) yes   **109.** (a) HCN/CN^-
(b) $H_3PO_4/H_2PO_4^-$ (c) H_2CO_3/HCO_3^- (d) HCO_3^-/CO_3^{2-}
111. (a) $HF + OH^- \rightarrow F^- + H_2O$
(b) $HCO_3^- + H_3O^+ \rightarrow H_2CO_3 + H_2O$
(c) $CO_3^{2-} + H_3O^+ \rightarrow HCO_3^- + H_2O$
(d) $H_3PO_4 + OH^- \rightarrow H_2PO_4^- + H_2O$
113. (a) 70.0 mL (b) 2.25×10^3. mL (c) 25.0 mL (d) 0.22 mL
115. (a) 8.00 mL (b) 96.0 mL (c) 270. mL (d) 4.00 mL
117. (a) 0.0705 M (b) 0.176 M (c) 0.0587 M (d) 0.352 M
119. (a) 3 M   (b) 6 M (c) 16 M (d) 6 M **121.** (a) strong
(b) weak (c) weak (d) strong **123.** 200 **125.** 6.76
127. (a) A > B > D > C   (b) C < D < B < A
(c) C > D > B > A   (d) A < B < D < C   **129.**
(a) no, higher pH   (b) no, higher pH   (c) yes   (d) yes
131. pH = 1.5 **133.** $Ba(OH)_2$, LiCN, K_2SO_4, NH_4Br, $HClO_4$
135. (a) HCN/CN^- (b) HF/F^- **137.** $H_3PO_4/H_2PO_4^-$ and
$H_2PO_4^-/HPO_4^{2-}$ **139.** 0.0061 mole $Ca(OH)_2$
141. (a) 2.00×10^{-14} M   (b) 0.500 M   (c) 13.699
143. (a) the iodide, I^-, ion   (b) the hydrogen phosphate,
HPO_4^{2-}, ion   (c) the hydroxide, OH^-, ion   (d) the hydronium,
H_3O^+, ion   **145.** H:C:::N: and [:C:::N:]$^-$ **147.** (a) 0.48
(b) 14.312   (c) 1.22   (d) 0.46   **149.** (a) 0.10 M
(b) 0.0010 M   (c) 1.00   (d) 3.00   **151.** (a) 4.0×10^{-4} M HCl
(b) 0.0014% (m/m)   **153.** 2.3×10^{16} H_3O^+ ions
155. 1.82×10^{23} ions   **157.** 0.67 L HCl
159. 66.0% (m/m)

Chapter 15

1. (a) loss of electrons (b) increase in oxidation number
3. (a) an oxidizing agent gains electrons from another substance
(b) an oxidizing agent contains the atom that shows an oxidation number decrease (c) an oxidizing agent is itself reduced
5. (a) oxidized (b) decrease (c) reducing (d) loses
7. (a) $N = -3$, $H = +1$ (b) $H = +1$, $S = +4$, $O = -2$
(c) $H = +1$, $N = +3$, $O = -2$ (d) $Na = +1$, $P = +5$, $O = -2$
9. (a) $P = -3$ (b) $Mg = +2$ (c) $N = -3$, $H = +1$
(d) $P = +5$, $O = -2$ **11.** (a) $+4$ (b) $+3$ (c) -2 (d) $-8/3$
13. (a) $+2$ (b) $+2$ (c) $+3$ (d) $+2$ **15.** (a) $+3$ (b) $+3$ (c) $+2$
(d) $+1$ **17.** (a) -2 (b) $+2$ (c) -1 (d) -2 **19.** (a) -1 (b) $+1$
(c) $+1$ (d) -1 **21.** (a) H_2 oxidized, N_2 reduced (b) I^- oxidized, Cl_2 reduced (c) Fe oxidized, Sb reduced (d) S oxidized, N reduced
23. (a) N_2 is oxidizing agent, H_2 is reducing agent (b) Cl_2 is oxidizing agent, I^- is reducing agent (c) Sb_2O_3 is oxidizing agent, Fe is reducing agent (d) HNO_3 is oxidizing agent, H_2SO_3 is reducing agent **25.** (a) SO_2 (b) HNO_3 (c) HNO_3 (d) SO_2
27. (a) the metal is oxidized and the nonmetal is reduced (b) the metal is oxidized and the nonmetal is reduced (c) the metal is oxidized and the nonmetal is reduced (d) the metal is oxidized and the nonmetal is reduced **29.** (a) redox, synthesis (b) redox, single-replacement (c) nonredox, decomposition (d) nonredox, double-replacement **31.** (a) redox (b) redox (c) can't classify
(d) redox **33.** (a) $2\,Cr + 6\,HCl \rightarrow 2\,CrCl_3 + 3\,H_2$
(b) $2\,Cr_2O_3 + 3\,C \rightarrow 4\,Cr + 3\,CO_2$
(c) $SO_2 + NO_2 \rightarrow SO_3 + NO$
(d) $BaSO_4 + 4\,C \rightarrow BaS + 4\,CO$
35. (a) $Br_2 + 2\,H_2O + SO_2 \rightarrow 2\,HBr + H_2SO_4$
(b) $3\,H_2S + 2\,HNO_3 \rightarrow 3\,S + 2\,NO + 4\,H_2O$
(c) $SnSO_4 + 2\,FeSO_4 \rightarrow Sn + Fe_2(SO_4)_3$
(d) $Na_2TeO_3 + 4\,NaI + 6\,HCl \rightarrow 6\,NaCl + Te + 3\,H_2O + 2\,I_2$
37. (a) $I_2 + 5\,Cl_2 + 6\,H_2O \rightarrow 2\,HIO_3 + 10\,Cl^- + 10\,H^+$
(b) $8\,MnO_4^- + 5\,AsH_3 + 24\,H^+ \rightarrow 5\,H_3AsO_4 + 8\,Mn^{2+} + 12\,H_2O$
(c) $2\,Br^- + SO_4^{2-} + 4\,H^+ \rightarrow Br_2 + SO_2 + 2\,H_2O$
(d) $Au + 4\,Cl^- + 3\,NO_3^- + 6\,H^+ \rightarrow AuCl_4^- + 3\,NO_2 + 3\,H_2O$
39. (a) $8\,OH^- + S^{2-} + 4\,Cl_2 \rightarrow SO_4^{2-} + 8\,Cl^- + 4\,H_2O$
(b) $3\,SO_3^{2-} + 2\,CrO_4^{2-} + 5\,H_2O \rightarrow 2\,Cr(OH)_4^- + 3\,SO_4^{2-} + 2\,OH^-$
(c) $2\,MnO_4^- + 3\,IO_3^- + H_2O \rightarrow 2\,MnO_2 + 3\,IO_4^- + 2\,OH^-$
(d) $I_2 + 7\,Cl_2 + 18\,OH^- \rightarrow 2\,H_3IO_6^{2-} + 14\,Cl^- + 6\,H_2O$
41. (a) $MnO_2 + 4\,H^+ + e^- \rightarrow Mn^{3+} + 2\,H_2O$
(b) $H_3MnO_4 + 5\,H^+ + 5\,e^- \rightarrow Mn + 4\,H_2O$
(c) $MnO_4^- + 8\,H^+ + 5\,e^- \rightarrow Mn^{2+} + 4\,H_2O$
(d) $MnO_4^- + 4\,H^+ + 3\,e^- \rightarrow MnO_2 + 2\,H_2O$
43. (a) $SeO_4^{2-} + 4\,H_2O + 6\,e^- \rightarrow Se + 8\,OH^-$
(b) $Se^{2-} + 6\,OH^- \rightarrow SeO_3^{2-} + 3\,H_2O + 6\,e^-$
(c) $SeO_4^{2-} + H_2O + 2\,e^- \rightarrow SeO_3^{2-} + 2\,OH^-$
(d) $Se + 6\,OH^- \rightarrow SeO_3^{2-} + 3\,H_2O + 4\,e^-$
45. (a) $Zn + Cu^{2+} \rightarrow Cu + Zn^{2+}$
(b) $Br_2 + 2\,I^- \rightarrow 2\,Br^- + I_2$
(c) $S_2O_3^{2-} + 4\,Cl_2 + 5\,H_2O \rightarrow 2\,HSO_4^- + 8\,Cl^- + 8\,H^+$
(d) $6\,Zn + As_2O_3 + 12\,H^+ \rightarrow 2\,AsH_3 + 6\,Zn^{2+} + 3\,H_2O$
47. (a) $I_2 + 5\,Cl_2 + 6\,H_2O \rightarrow 2\,HIO_3 + 10\,Cl^- + 10\,H^+$
(b) $8\,MnO_4^- + 5\,AsH_3 + 24\,H^+ \rightarrow 5\,H_3AsO_4 + 8\,Mn^{2+} + 12\,H_2O$

(c) $2\,Br^- + SO_4^{2-} + 4\,H^+ \rightarrow Br_2 + SO_2 + 2\,H_2O$
(d) $Au + 4\,Cl^- + 3\,NO_3^- + 6\,H^+ \rightarrow AuCl_4^- + 3\,NO_2 + 3\,H_2O$
49. (a) $2\,NH_3 + ClO^- \rightarrow N_2H_4 + Cl^- + H_2O$
(b) $Cr(OH)_2 + 2\,BrO^- + 2\,OH^- \rightarrow CrO_4^{2-} + 2\,Br^- + 2\,H_2O$
(c) $2\,CrO_2^- + 3\,H_2O_2 + 2\,OH^- \rightarrow 2\,CrO_4^{2-} + 4\,H_2O$
(d) $2\,Bi(OH)_3 + 3\,Sn(OH)_3^- + 3\,OH^- \rightarrow 3\,Sn(OH)_6^{2-} + 2\,Bi$
51. (a) $S^{2-} + 4\,Cl_2 + 8\,OH^- \rightarrow SO_4^{2-} + 8\,Cl^- + 4\,H_2O$
(b) $3\,SO_3^{2-} + 2\,CrO_4^{2-} + 5\,H_2O \rightarrow 2\,Cr(OH)_4^- + 3\,SO_4^{2-} + 2\,OH^-$
(c) $2\,MnO_4^- + 3\,IO_3^- + H_2O \rightarrow 2\,MnO_2 + 3\,IO_4^- + 2\,OH^-$
(d) $I_2 + 7\,Cl_2 + 18\,OH^- \rightarrow 2\,H_3IO_6^{2-} + 14\,Cl^- + 6\,H_2O$
53. (a) $3\,HNO_2 \rightarrow 2\,NO + NO_3^- + H^+ + H_2O$
(b) $ClO^- + Cl^- + 2\,H^+ \rightarrow Cl_2 + H_2O$
(c) $3\,S + 6\,OH^- \rightarrow 2\,S^{2-} + SO_3^{2-} + 3\,H_2O$
(d) $3\,Br_2 + 6\,OH^- \rightarrow BrO_3^- + 5\,Br^- + 3\,H_2O$
55. (a) $3\,HNO_2 \rightarrow 2\,NO + NO_3^- + H^+ + H_2O$
(b) $ClO^- + Cl^- + 2\,H^+ \rightarrow Cl_2 + H_2O$
(c) $3\,S + 6\,OH^- \rightarrow 2\,S^{2-} + SO_3^{2-} + 3\,H_2O$
(d) $3\,Br_2 + 6\,OH^- \rightarrow BrO_3^- + 5\,Br^- + 3\,H_2O$
57. (a) $Pb \rightarrow Pb^{2+} + 2\,e^-$
(b) $Cu^{2+} + 2\,e^- \rightarrow Cu$ (c) anode is Pb, cathode is Cu (d) from anode to cathode **59.** anode: $Zn \rightarrow Zn^{2+} + 2\,e^-$; cathode: $2\,MnO_2 + 2\,NH_4^+ + 2\,e^- \rightarrow Mn_2O_3 + 2\,NH_3 + H_2O$
61. anode: $Pb + SO_4^{2-} \rightarrow PbSO_4 + 2\,e^-$; cathode: $PbO_2 + 4\,H^+ + SO_4^{2-} + 2\,e^- \rightarrow PbSO_4 + 2\,H_2O$
63. the H_2SO_4 in the solution leaves the solution, lowering the density of the solution **65.** $2\,Cl^- \rightarrow Cl_2 + 2\,e^-$ **67.** N_2O, NO, N_2O_3, NO_2, N_2O_5 **69.** (a) The oxidation number is already at its minimum value and cannot go any lower.
(b) The oxidation number is already at its maximum value and cannot go any higher.   (c) The oxidation number is at an intermediate value and can be either increased or decreased.
(d) The oxidation number is already at its maximum value and cannot go any higher.   **71.** (a) $+2$, $+1$ (b) $+3$ in both (c) $+1$ in both (d) $+3$ in both   **73.** (a) two reduction half-reactions (b) one reduction and one oxidation half-reaction (c) two oxidation half-reactions (d) one reduction and one oxidation half-reaction
75. $4\,MnO_4^- + 5\,PH_3 + 12\,H^+ \rightarrow 5\,H_3PO_2 + 4\,Mn^{2+} + 6\,H_2O$;
$2\,SO_4^{2-} + PH_3 + 4\,H^+ \rightarrow H_3PO_2 + 2\,SO_2 + 2\,H_2O$;
$3\,MnO_4^- + 5\,As + 3\,H_2O + 9\,H^+ \rightarrow 5\,H_3AsO_3 + 3\,Mn^{2+}$;
$3\,SO_4^{2-} + 2\,As + 6\,H^+ \rightarrow 2\,H_3AsO_3 + 3\,SO_2$
77. $Zn \rightarrow Zn^{2+} + 2\,e^-$; $NO_3^- + 10\,H^+ + 8\,e^- \rightarrow NH_4^+ + 3\,H_2O$
79. (a) redox, HNO_3 is the oxidizing agent (b) acid–base, H_2S is the acid (c) redox, H_2O_2 is the oxidizing agent (d) acid–base, H_2SO_4 is the acid **81.** (a) $Sn^{2+} + 2\,Fe^{2+} \rightarrow Sn + 2\,Fe^{3+}$
(b) $PH_3 + 2\,NO_2 \rightarrow H_3PO_4 + N_2$
(c) $S + 3\,H_2O + 2\,Pb^{2+} \rightarrow 2\,Pb + H_2SO_3 + 4\,H^+$
(d) $4\,Zn + 10\,H^+ + NO_3^- \rightarrow 4\,Zn^{2+} + NH_4^+ + 3\,H_2O$
83. (a) $3\,H_2S + Cr_2O_7^{2-} + 8\,H^+ \rightarrow 2\,Cr^{3+} + 3\,S + 7\,H_2O$
(b) $5\,ClO_3^- + 3\,I_2 + 3\,H_2O \rightarrow 6\,IO_3^- + 5\,Cl^- + 6\,H^+$
(c) $S^{2-} + 4\,Br_2 + 8\,OH^- \rightarrow SO_4^{2-} + 8\,Br^- + 4\,H_2O$
(d) $2\,NO_2 + 2\,OH^- \rightarrow NO_3^- + NO_2^- + H_2O$
85. $0.324\ M\ I_3^-$ **87.** 30.8 ppm (m/m) O_3

Chapter 16

1. Contact between molecules occurs with more ease in a solution. **3.** 45 kJ/mole; the lower activation energy is easier to overcome. **5.** molecular orientation at time of collision and molecular energy

7.

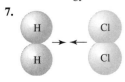

9. (a) exothermic (b) endothermic (c) endothermic (d) endothermic

11.

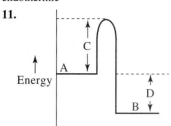

(a) The average energy of the reactants is shown as A; (b) the average energy of the products is shown as B; (c) the activation energy is shown as C; (d) the energy liberated in the reaction is shown as D, or A − B.

13.

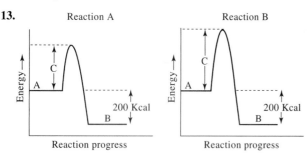

The reactions are exothermic to the same extent and the energy difference between reactants and products is the same. The reactions differ in activation energy (C). **15.** (a) With increased temperature molecules move faster and collide more often. (b) A catalyst allows for an alternate pathway, which requires less activation energy. **17.** Surface area is much greater for the coal dust. **19.** (a) 1 (b) 3 (c) 4 (d) 3

21.

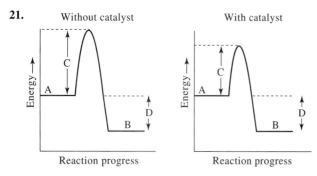

The energy difference between reactants and products (D) is the same. The activation energy (C) differs. **23.** The forward and reverse reaction rates must be equal. **25.** In a physical equilibrium there is no chemical reaction.

27. $SO_3 = 0.0142$ mole, $SO_2 = 0.0058$ mole, $O_2 = 0.0029$ mole

29. $NH_3 = 0.268$ mole, $N_2 = 0.184$ mole, $H_2 = 0.137$ mole

31. (a) $K = \dfrac{[SO_2Cl_2]}{[SO_2][Cl_2]}$ (b) $K = \dfrac{[N_2][O_2]^2}{[NO_2]^2}$

(c) $K = \dfrac{[CS_2][O_2]^4}{[SO_3]^2[CO_2]}$ (d) $K = \dfrac{[CH_4][H_2S]^2}{[H_2]^4[CS_2]}$

33. (a) $K = [NO_2]^4[O_2]$ (b) $K = [O_2]^3$ (c) $K = 1/[Cl_2]$
(d) $K = [Cl_2]$ **35.** (a) 6.67 (b) 1.04 (c) 3.75 (d) 0.0216
37. 0.0730 M **39.** 48.9 **41.** (a) mostly reactants (b) mostly products (c) mostly products (d) significant amounts of both reactants and products **43.** (a) to the right (b) to the left (c) to the left (d) to the right **45.** (a) to the right (b) to the left (c) to the left (d) to the right **47.** (a) shifts right (b) shifts left (c) no effect (d) shifts right **49.** (a) no (b) no (c) yes (d) no
51. (a) $N_2 + 3 H_2 \rightleftharpoons 2 NH_3$
(b) $4 NH_3 + 3 O_2 \rightleftharpoons 2 N_2 + 6 H_2O$
(c) $2 NO \rightleftharpoons N_2 + O_2$ (d) $N_2 + O_2 \rightleftharpoons 2 NO$
53. (a) at equilibrium (b) shifts to the left (c) at equilibrium
(d) shifts to the left **55.** 0.0005 **57.** (a) no (b) no (c) yes (d) no
59. (a) to the right (b) no shift (c) to the right (d) to the right

61. (a) $K = \dfrac{[SO_2]^2[O_2]}{[SO_3]^2}$ (b) $K = \dfrac{[NH_3]^2[H_2O]^4}{[H_2]^7[NO_2]^2}$

(c) $K = \dfrac{[CO_2]}{[CO]}$ (d) $K = [CO_2]$

63. $K = 38.9$ **65.** 0.0244 **67.** 2.20 atm **69.** 0.119

Chapter 17

1. A radioactive nuclide decays into other elements while a nonradioactive nuclide does not undergo decay. **3.** (a) ^9_5B or boron-9 (b) $^{44}_{19}K$ or potassium-44 (c) $^{96}_{45}Rh$ or rhodium-96 (d) $^{182}_{73}Ta$ or tantalum-182 **5.** (a) $^{14}_7N$ (b) $^{197}_{79}Au$ (c) rubidium-92 (d) tin-121 **7.** (a) $^4_2\alpha$ (b) $^0_{-1}\beta$ (c) $^0_0\gamma$ **9.** 2 protons and 2 neutrons **11.** (a) $^{192}_{78}Pt \rightarrow ^4_2\alpha + ^{188}_{76}Os$ (b) $^{217}_{86}Rn \rightarrow ^4_2\alpha + ^{213}_{84}Po$
(c) $^{212}_{85}At \rightarrow ^4_2\alpha + ^{208}_{83}Bi$ (d) $^{244}_{96}Cm \rightarrow ^4_2\alpha + ^{240}_{94}Pu$
13. (a) $^{48}_{21}Sc \rightarrow ^0_{-1}\beta + ^{48}_{22}Ti$ (b) $^{117}_{47}Ag \rightarrow ^0_{-1}\beta + ^{117}_{48}Cd$
(c) $^{92}_{36}Kr \rightarrow ^0_{-1}\beta + ^{92}_{37}Rb$ (d) $^{138}_{55}Cs \rightarrow ^0_{-1}\beta + ^{138}_{56}Ba$ **15.** mass number decreases by four; atomic number decreases by two
17. (a) $^0_{-1}\beta$ (b) $^{125}_{52}Te$ (c) $^4_2\alpha$ (d) $^{229}_{90}Th$
19. (a) $^{199}_{79}Au \rightarrow ^0_{-1}\beta + ^{199}_{80}Hg$ (b) $^{120}_{48}Cd \rightarrow ^0_{-1}\beta + ^{120}_{49}In$
(c) $^{152}_{67}Ho \rightarrow ^4_2\alpha + ^{148}_{65}Tb$ (d) $^{226}_{88}Ra \rightarrow ^4_2\alpha + ^{222}_{86}Rn$
21. (a) 1/32 (b) 1/16 (c) 1/8 (d) 1/256 **23.** (a) 6.0 yr (b) 2.4 yr (c) 1.5 yr (d) 1.2 yr **25.** (a) 2.0 g (b) 0.50 g (c) 0.062 g (d) 0.0078 g **27.** (a) 7.50 g (b) 9.84 g (c) 9.98 g (d) 10.0 g
29. (a) 16 hr (b) 32 hr (c) 4.0×10^1 hr (d) 56 hr **31.** (a) $^4_2\alpha$
(b) ^2_1H (c) $^{81}_{34}Se$ (d) ^9_4Be **33.** (a) $^9_4Be + ^4_2\alpha \rightarrow ^{12}_6C + ^1_0n$
(b) $^{58}_{28}Ni + ^1_1H \rightarrow ^{55}_{27}Co + ^4_2\alpha$ (c) $^{113}_{48}Cd + ^1_0n \rightarrow ^{114}_{48}Cd + ^0_0\gamma$
(d) $^{27}_{13}Al + ^4_2\alpha \rightarrow ^{30}_{15}P + ^1_0n$ **35.** (a) $^{242}_{96}Cm$ (b) $^{238}_{92}U$ (c) $^{252}_{98}Cf$
(d) $^{209}_{83}Bi$ **37.** 9 **39.** over 2000 **41.** mass number does not change; atomic number decreases by one **43.** $^0_{-1}\beta$ emission

45. (a) $^{29}_{15}P \rightarrow {^0_1\beta} + {^{29}_{14}Si}$ (b) $^{112}_{51}Sb \rightarrow {^0_1\beta} + {^{112}_{50}Sn}$
(c) $^{46}_{23}V \rightarrow {^0_1\beta} + {^{46}_{22}Ti}$ (d) $^{132}_{58}Ce \rightarrow {^0_1\beta} + {^{132}_{57}La}$
47. (a) $^{76}_{36}Kr + {^0_{-1}e} \rightarrow {^{76}_{35}Br}$ (b) $^{122}_{54}Xe + {^0_{-1}e} \rightarrow {^{122}_{53}I}$
(c) $^{100}_{46}Pd + {^0_{-1}e} \rightarrow {^{100}_{45}Rh}$ (d) $^{175}_{73}Ta + {^0_{-1}e} \rightarrow {^{175}_{72}Hf}$
49. (a) $^0_1\beta$ (b) $^0_{-1}e$ (c) $^{103}_{47}Ag$ (d) $^{133}_{55}Cs$ **51.** electron capture:
$^{63}_{30}Zn + {^0_{-1}e} \rightarrow {^{63}_{29}Cu}$; positron emission: $^{63}_{30}Zn \rightarrow {^0_1\beta} + {^{63}_{29}Cu}$
53. beta decay **55.** (a) 1.32 before, 1.24 after (b) 1.46 before,
1.47 after (c) 1.39 before, 1.43 after (d) 1.18 before, 1.23 after
57. (a) $^{87}_{36}Kr$, beta particle emission; $^{74}_{36}Kr$, positron emission; n/p
ratio is higher for $^{87}_{36}Kr$ (b) $^{84}_{34}Se$, beta particle emission; $^{63}_{33}As$,
positron emission; n/p ratio is higher for $^{84}_{34}Se$ (c) $^{74}_{31}Ga$, beta
particle emission; $^{64}_{31}Ga$, positron emission; n/p ratio is higher for
$^{74}_{31}Ga$ (d) $^{99}_{41}Nb$, beta particle emission; $^{99}_{46}Pd$, positron emission; n/p
ratio is higher for $^{99}_{41}Nb$ **59.** stable; a decay series always ends
with a stable nuclide **61.** $^{234}_{91}Pa$ **63.** (a) $^{220}_{86}Rn \rightarrow {^{216}_{84}Po} + {^4_2\alpha}$
(b) $^{216}_{84}Po \rightarrow {^{212}_{82}Pb} + {^4_2\alpha}$ (c) $^{212}_{82}Pb \rightarrow {^{212}_{83}Bi} + {^0_{-1}\beta}$ (d)
$^{212}_{83}Bi \rightarrow {^{212}_{84}Po} + {^0_{-1}\beta}$ **65.** an electron and the positive ion
produced during an ionization collision between radiation and an
atom **67.** (a) yes (b) no (c) yes (d) no **69.** The paper stops

alpha radiation but not beta or gamma radiation. **71.** alpha, 1/10
the speed of light; beta, up to 9/10 the speed of light; gamma, the
speed of light **73.** Only gamma radiation can be detected by
instrumentation placed outside the body. **75.** (a) implantation
cancer therapy (b) determination of intercellular spaces in fluids
(c) external radiation cancer therapy (d) determination of blood
volume and red blood cell lifetime **77.** (a) 4 (b) 4 (c) 2 (d) 3
79. $^{239}_{93}Pu$ **81.** (a) ^4_2He (b) ^2_1H (c) ^1_1H (d) ^3_2He **83.** (a) fusion
(b) both (c) both (d) fusion **85.** (a) fusion (b) neither (c) neither
(d) fission **87.** (a) stays the same (b) stays the same
(c) decreases by 4 (d) stays the same
89. (a) $^{228}_{88}Ra$, $^{228}_{89}Ac$, $^{228}_{90}Th$ (b) $^{228}_{90}Th$, $^{224}_{88}Ra$, $^{220}_{86}Rn$
91. six alpha particles and four beta particles **93.** phosphorus-28,
positron emission; phosphorus-34, beta particle emission
95. A, negligible amount (zero); B, approximately 0.250 mole; C,
negligible amount (zero); D, approximately 0.750 mole
97. 6.0 hr **99.** (a) 7.0 days (b) 3.80 g Q **101.** (a) isobars
(b) neither (c) isotopes (d) isotopes **103.** 19 days
105. (a) 50.0 g of UF_6 (b) 0.250 mole of UF_6 **107.** 0.364 L
109. 3.08×10^6 kg C

Names of Common Polyatomic Ions

Ion		Name	Ion		Name
N	NO_3^-	nitrate ion	**Cl**	ClO_4^-	perchlorate ion
	NO_2^-	nitrite ion		ClO_3^-	chlorate ion
	NH_4^+	ammonium ion		ClO_2^-	chlorite ion
	N_3^-	azide ion		ClO^-	hypochlorite ion
S	SO_4^{2-}	sulfate ion	**C**	CO_3^{2-}	carbonate ion
	HSO_4^-	hydrogen sulfate ion		HCO_3^-	hydrogen carbonate ion
	$S_2O_3^{2-}$	thiosulfate ion		$C_2O_4^{2-}$	oxalate ion
	SO_3^{2-}	sulfite ion		$C_2H_3O_2^-$	acetate ion
P	PO_4^{3-}	phosphate ion		CN^-	cyanide ion
	HPO_4^{2-}	hydrogen phosphate ion		OCN^-	cyanate ion
	$H_2PO_4^-$	dihydrogen phosphate ion		SCN^-	thiocyanate ion
	PO_3^{3-}	phosphite ion	**B**	BO_3^{3-}	borate ion
H	H_3O^+	hydronium ion	**Mn**	MnO_4^-	permanganate ion
	OH^-	hydroxide ion	**Cr**	CrO_4^{2-}	chromate ion
O	O_2^{2-}	peroxide ion		$Cr_2O_7^{2-}$	dichromate ion

Solubility Guidelines for Ionic Compounds in Water at 25°C

Soluble Compounds	Important Exceptions
Compounds containing the following ions are soluble with exceptions as noted.	
Group IA (Li^+, Na^+, K^+, etc.)	none
Ammonium (NH_4^+)	none
Acetate ($C_2H_3O_2^-$)	none
Nitrate (NO_3^-)	none
Chloride (Cl^-), bromide (Br^-), and iodide (I^-)	Ag^+, Pb^{2+}, Hg_2^{2+}
Sulfate (SO_4^{2-})	Ca^{2+}, Sr^{2+}, Ba^{2+}, Pb^{2+}

Insoluble Compounds	Important Exceptions
Compounds containing the following ions are insoluble with exceptions as noted.	
Carbonate (CO_3^{2-})	group IA and NH_4^+
Phosphate (PO_4^{3-})	group IA and NH_4^+
Sulfide (S^{2-})	groups IA and IIA and NH_4^+
Hydroxide (OH^-)	group IA, Ba^{2+}, Sr^{2+}, Ca^{2+}

Common Strong Acids

Formula	Name
HNO_3	nitric acid
H_2SO_4	sulfuric acid
$HClO_4$	perchloric acid
$HClO_3$	chloric acid
HCl	hydrochloric acid
HBr	hydrobromic acid
HI	hydroiodic acid

Common Strong Bases

Formula	Name
$LiOH$	lithium hydroxide
$NaOH$	sodium hydroxide
KOH	potassium hydroxide
$RbOH$	rubidium hydroxide
$CsOH$	cesium hydroxide
$Ca(OH)_2$	calcium hydroxide
$Sr(OH)_2$	strontium hydroxide
$Ba(OH)_2$	barium hydroxide

Mathematical Meanings of Metric System Prefixes

Prefix	Meaning	Prefix	Meaning
Tera (T)	10^{12}	Pico (p)	10^{-12}
Giga (G)	10^9	Nano (n)	10^{-9}
Mega (M)	10^6	Micro (μ)	10^{-6}
Kilo (k)	10^3	Milli (m)	10^{-3}
Hecto (h)	10^2	Centi (c)	10^{-2}
Deca (da)	10^1	Deci (d)	10^{-1}

Common Fixed-Charge Metallic Cations and Nonmetallic Anions

Cation	Name	Anion	Name
Li^+	lithium ion	F^-	fluoride ion
Na^+	sodium ion	Cl^-	chloride ion
K^+	potassium ion	Br^-	bromide ion
Rb^+	rubidium ion	I^-	iodide ion
Cs^+	cesium ion	O^{2-}	oxide ion
Be^{2+}	beryllium ion	S^{2-}	sulfide ion
Mg^{2+}	magnesium ion	N^{3-}	nitride ion
Ca^{2+}	calcium ion	P^{3-}	phosphide ion
Sr^{2+}	strontium ion	C^{4-}	carbide ion
Ba^{2+}	barium ion		
Ag^+	silver ion		
Zn^{2+}	zinc ion		
Cd^{2+}	cadmium ion		
Al^{3+}	aluminum ion		
Ga^{3+}	gallium ion		

Common Variable-Charge Metallic Cations

Cation	IUPAC Name	Older Name
Cu^+	copper (I) ion	cuprous ion
Cu^{2+}	copper (II) ion	cupric ion
Fe^{2+}	iron (II) ion	ferrous ion
Fe^{3+}	iron (III) ion	ferric ion
Sn^{2+}	tin (II) ion	stannous ion
Sn^{4+}	tin (IV) ion	stannic ion
Pb^{2+}	lead (II) ion	plumbous ion
Pb^{4+}	lead (IV) ion	plumbic ion
Au^+	gold (I) ion	aurous ion
Au^{3+}	gold (III) ion	auric ion